Proceedings of the International Asia Conference on Industrial Engineering and Management Innovation

For further volumes:
http://www.atlantis-press.com

About this Series

The proceedings of this series originate from the conference series 'International Asia Conference on Industrial Engineering and Management Innovation' (IEMI) whose aim is to share and disseminate information on the most recent and relevant researches, theories and practices in industrial and system engineering and to promote their development and application in university and enterprises. The IEMI-conferences are sponsored by Chinese Industrial Engineering Institution, CMES, which is the first and largest Chinese institution in the field of industrial engineering. CMES is also the sole national institution recognized by China Association of Science and Technology. The conferences are organized by CMES and the Tianjin University of Science and Technology.

Ershi Qi • Qin Su • Jiang Shen
Feng Wu • Runliang Dou
Editors

Proceedings of the 5th International Asia Conference on Industrial Engineering and Management Innovation (IEMI2014)

ATLANTIS
PRESS

Editors
Ershi Qi
Tianjin University
Tianjin
People's Republic of China

Feng Wu
Xi'an Jiaotong University
Xi'an
People's Republic of China

Qin Su
Xi'an Jiaotong University
Xi'an
People's Republic of China

Runliang Dou
Tianjin University
Tianjin
People's Republic of China

Jiang Shen
Industrial Engineering Institution of CM
Tianjin University
Tianjin
People's Republic of China

ISSN 2405-5891 ISSN 2405-5905 (electronic)
Proceedings of the International Asia Conference on Industrial Engineering and Management Innovation
ISBN 978-94-6239-099-7 ISBN 978-94-6239-100-0 (eBook)
DOI 10.2991/978-94-6239-100-0

Library of Congress Control Number: 2014958288

Printed on acid-free paper

Preface

At the time of July, Xi'an is brimming with energy and enthusiasm. In this vital season, I am honored to welcome all the colleagues from all over the word to attend the 5th International Asia Conference on Industrial Engineering and Management Innovation. It is your great efforts that make the proceedings of IEMI 2014 so fruitful, and the proceeding will serve as the guidebook for the potential development and application in IEMI, theoretically and technologically.

Established in June 1990, the Chinese Industrial Engineering Institution, CMES, is the first and largest institution with most members in the industrial engineering field. It is also the sole national institution recognized by the China Association of Science and Technology, and is called the Chinese Industrial Engineering Institution (CIEI). During the past 20 years, CIEI has held dozens of domestic and international academic conferences, including quite a number in Hong Kong, Macao, and Cross-Strait forums and seminars on industrial engineering. Sticking to the philosophy of "serving the country, serving the society, and serving the members", the CIEI has established extensive cooperation with institutes and organizations from the USA, UK, Russia, Japan, Korea, Hong Kong, Taiwan, and other countries or areas.

The 5th International Asia Conference on Industrial Engineering and Management Innovation is sponsored by the CIEI and organized by Xi'an Jiaotong University. The conference aims to share and disseminate information on the most recent and relevant researches, theories, and practices in industrial and systems engineering to promote their development and application in universities and enterprises. In the conference, some excellent scholars and experts both from home and abroad, as Keynote Speakers, share their glory of achievement on scientific research and application in Industrial Engineering and Management Innovation.

Here, I would like to extend my sincerest thanks to the Taiwan Institute of Industrial Engineers and IIE for their generous support; to the School of Management, and the Xi'an Jiaotong University for holding such an excellent event; and to Atlantis Press for the publication of the proceedings. I would also like to extend my sincerest thanks to all the delegates, keynote speakers, and staff of the organization committee for their contribution to the success of the conference in various ways.

Chinese Industrial Engineering Institution, CMES

Ershi Qi
Qin Su
Jiang Shen
Feng Wu
Runliang Dou

Organization

Honorary Chair

Yingluo Wang	Xi'an Jiaotong University

Conference General Chairs

Ershi Qi	Tianjin University
Qin Su	Xi'an Jiaotong University
Fugee Tsung	Hong Kong University of Science and Technology
Bopaya Bidanda	University of Pittsburgh

Program Chairs

Ching-Jong Liao	Taiwan University of Science and Technology
Jiang Shen	Tianjin University

Technical Committees

Benjamin Yen	University of Hong Kong
Bernad C. Jiang	Yuan Ze University
Ching-Jong Liao	Taiwan University of Science and Technology
Chung-Yee Lee	Hong Kong University of Science and Technology
Du-Ming Tsai	Yuan Ze University
Ershi Qi	Tianjin University
Gregory H. Watson	Business Excellence Solutions, Ltd.
James T. Lin	Hsinchu Tsing Hua University
Jeff Wu	Georgia Institute of Technology
Jiang Shen	Tianjin University
Jianjun Shi	Georgia Institute of Technology
Joon Young Park	KIIE
Kwang-Jae Kim	Pohang University of Science and Technology

Leyuan Shi	Peking University
Li Zheng	Tsinghua University
Loon Ching Tang	National University of Singapore
Pao-Long Chang	Feng Chia University
Peter W. Glynn	Stanford University
Shuo-Yan Chou	Taiwan University of Science and Technology
Takashi Irohara	Sophia University
Thong Ngee Goh	National University of Singapore
Way Kuo	City University of Hong Kong
Wei Huang (Wayne)	Ohio University and Xi'an Jiaotong University
Xiaofen Tang	Shanghai Association for Quality
Zhen He	Tianjin University
Zhibin Jiang	Shanghai Jiaotong University

Organizing Committee Chairs

Feng Wu	Xi'an Jiaotong University
Gengzhong Feng	Xi'an Jiaotong University
Runliang Dou	Tianjin University

Contents

Part II Industrial Engineering Practice

Part I

Industrial Engineering Theory

A Fuzzy Petri nets/ANP Evaluation Model on Business Process Reengineering Implementation

Qin Chen[1], Shi-long Liao[2,*], Shu-ping Yi[3]

[1]School of Materials Engineering, Lanzhou Institute of Technology, Lanzhou, China

[2]College of Mathematics and Computer Science, Northwest University for Nationalities, Lanzhou, China

[3]College of Mechanical Engineering, Chongqing University, Chongqing, China

(jslsl@xbmu.edu.cn)

Abstract – **Business Process Reengineering is a gradual process from its import to stability, and the relationship among different enterprise subsystems is interactive and correlative because of system complexity, a set of evaluation system were established, and an evaluation model based on fuzzy Petri nets and analytic network process was proposed. Finally, an example is presented to show that the model is effective for the BPR project evaluation.**

Keywords - **ANP, BPR, implementation, Petri net**

I. INTRODUCTION

Implementing Business Process Reengineering (BPR) for enterprises is a gradual and dynamic process toward the best performance; it is an essential task to evaluate the process of implementation BPR. Geetha proposed a generally object oriented, simulation-based decision support system, to simulate a variety of business processes and evaluate their dynamic behavior quantitatively in [1]. Oberweis proposed a higher order complex Petri nets model, evaluating the processes repeatedly in [2] as well as in [3]. Chen used the view based on strategy to select reengineering process with analytic hierarchy process model in [4]; Wu proposed using fuzzy comprehensive evaluation to assess business processes performance, which considered fuzzy Indexes and constructed fuzzy comprehensive judgment process evaluation system in [5]. An empirical analysis is presented for researching linkages between manufacturing strategy, performance measurement and business process reengineering. The results confirmed the need for a strategically-driven BPR approach and the positive impact of performance measurement on BPR performance in [6]. Wang and Yi used the Based DEA/AHP hybrid method to evaluate business process for machinery manufacturing enterprises in [7-10]. Altinkemer finds that enterprise-wide BPR projects are associated with more negative returns during project initiation than functionally focused project in [11]. Wang proved that the method of Six Sigma in business process reengineering has higher possibility, which make quality of products by the Six Sigma Level and improve the degree of risk acceptance of the enterprise in [12]. These evaluation methods in [1-13] provide the guidance for evaluation BPR implementation process but there are still problems: 1) the evaluation indexes of BPR implementation process focused on costs and time which are at tangible level, considering less for enterprise culture and organizational structure. 2) Enterprise is a complex system, the subsystems of enterprise are interaction and mutual influence, which are not a simple hierarchy system, and the existing evaluation methods of the system are simplified and did not evaluate BPR implementation comprehensive quantitatively. In order to acquire index data and select index comprehensively, it established the evaluation index system, which covers impact factors for BPR implementation comprehensively. This paper proposed fuzzy Petri net/ANP model to evaluate BPR project implementation quantitatively, and proved effectiveness and feasibility through the application.

II. BPR IMPLEMENTATION PROCESS EVALUATION INDEX SYSTEM

BPR implementation process is a complicated systematic project, according to our previous empirical analysis of research results, BPR and internal support system, ability support system and resources-input support system are closely related, each supporting system has certain promotion and support for the implementation of BPR in [7]. The supporting system is organic system that includes a number of key elements which are interrelated and interact with each other and mutual constraint which constitute a specific function and structure, which promote the success of BPR implementation. Therefore, BPR supporting system is operating carrier for the successful implementation of BPR which is diversity, hierarchy, dynamic and holistic

E. Qi et al. (eds.), *Proceedings of the 5th International Asia Conference on Industrial Engineering and Management Innovation (IEMI2014)*, Proceedings of the International Asia Conference on Industrial Engineering and Management Innovation 1, DOI 10.2991/978-94-6239-100-0_1,

system [7]. This BPR evaluation index system was built on top of BPR supporting system, in which are superior indexes, including corporate strategy, organizational structure, corporate culture, human resources, leadership ability, and the capacity of staff, organizational innovation, and organizational learning capability and so on. This BPR evaluation system reflects comprehensive, comparable, and levels system. As shown in Table I.

TABLE I
BPR IMPLEMENTATION PROCESS INDEXES SYSTEM

Support System	Superior indexes	Inferior indexes
Support system for internal environ-ment	Corporation Strategy P_1	Corporate long-term goal up to the extent P_{11} Knowledge of their own advantages and disadvantages of business P_{12} BPR implementation of the method mastery P_{13}
	Structure P_2	TABLE I Centralization P_{21} Standardization P_{22} Complexity P_{23}
	Corporation Culture P_3	Mission of P_{31} Adaptability P_{32} Consistency P_{33} Participatory P_{34}
	Human resources policy P_4	Reasonable remuneration policy P_{41} Incentives reasonable P_{42} Reasonable assessment policy P_{43} raining policy rationality P_{44}
support system for Resources input	Investment in human resources P_5	Senior leadership support and par-ticipation P_{51} Middle-level leadership support and participation P_{52} BPR team's support and participation BPR P_{53}
	time in-vestment P_6	Continuing education and training for staff P_{61} BPR project investment in time P_{62}
	funds in-vestment P_7	Fees for the process reengineering consulting P_{71} The purchase of computer and soft-ware costs P_{72}
Ability to support system	Innovation P_8	Ability of management innova-tion P_{81} Technological innovation ability P_{82}
	Organiza-tion-al learning capabilities P_9	Knowledge absorption capacity P_{91} Ability to apply knowledge P_{92} Ability to study knowledge P_{93}
	IT capability P_{10}	IT Environmental $P_{10,1}$ Strategic use of IT $P_{10,2}$ IT Planning and Control $P_{10,3}$ IT acquisition and implementation $P_{10,4}$
	Personal capacity P_{11}	Leadership ability $P_{11,1}$ The capacity of staff $P_{11,2}$ BPR team capacity $P_{11,3}$

Because of the systematically and complexity of the BPR evaluation, not only there is a hierarchical structure of the indexes system, but also there is inner loop and domination in index system for hierarchy structure in which exist dependence and feedback. For example, the "corporation strategy P_1" is dependent on "corporation culture P_3" in supporting system for internal environment, other indexes exist a strong relationship among "reasonable remuneration policy P_{41}", "Incentives reasonable P_{42}", "reasonable assessment policy P_{43}" in the "human resources policy P_4" index. At the same time there are interactions for two-level indexes under different superior indexes. For example, it is related with inferior index "BPR implementation of the method mastery P_{13}" under superior index "Enterprise Strategy P_1 "with inferior index "continuing education and training for staff P_{61} "under superior index of "company time input P_6 ". So how to determine the importance of each level indexes, whose weight becomes more difficult? It is not an internal independent hierarchical structure so that ANP method to solve complex weight problems is a good tool.

III. FUZZY PETRI NET/ANP MODEL AND ALGORITHMS

ANP network hierarchy structure, is not only consequences of a hierarchical structure, but also is within the cycle of mutual dominance hierarchy structure and there are dependency and feedback in the hierarchy structure. Control layer and network layer consist of typical ANP hierarchy structure [14].

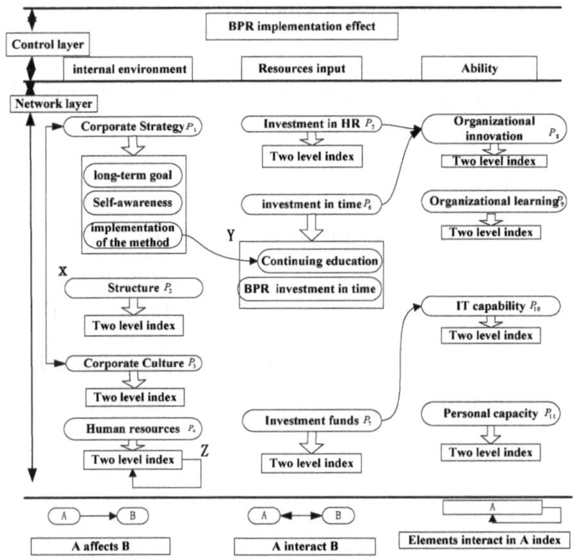

Fig.1. ANP construct model

Now we will give the evaluate model for BPR with the ANP method. As Shown in Fig.1, there is an interaction between some indexes with superior indexes, as shown in the X connection. There is an interaction for the inferior indexes relations in different superior indexes, as shown in the Y connection. The inferior indexes are reciprocal relationship under same superior index, as shown in the Z connection. Considering the complexity of the index system and limitations of the paper layout, the structure of ANP is not fully drawn, the logical relationship between the indexes is shown following Fig. 2. We use the ANP-specific software Super Decision to calculate related results.

Fuzzy Petri net (FPN) is much easier to evaluate some uncertain and fuzzy indexes. Compared to ordinary Petri net, the fuzzy Petri net has some special characters. FPN formal definition sees literature [15]. It is the mapping relation between FPN and corresponding evaluation model in Table II in [15].

The evaluation system of BPR implementation is transformed to FPN-based model by the mapping relation as shown in Fig.2.

TABLE II
THE MAPPING RELATION BETWEEN FPN AND EVALUATION SYSTEMS

FPN	Evaluation Systems
Place P_i	Indexes and the state remark level
Transition T	The transition of the evaluation status
Tokens value k	The member degree of evaluation index
Connection intensity a	The weight of evaluation index
Fuzzy indication M	The status distribution of evaluation systems

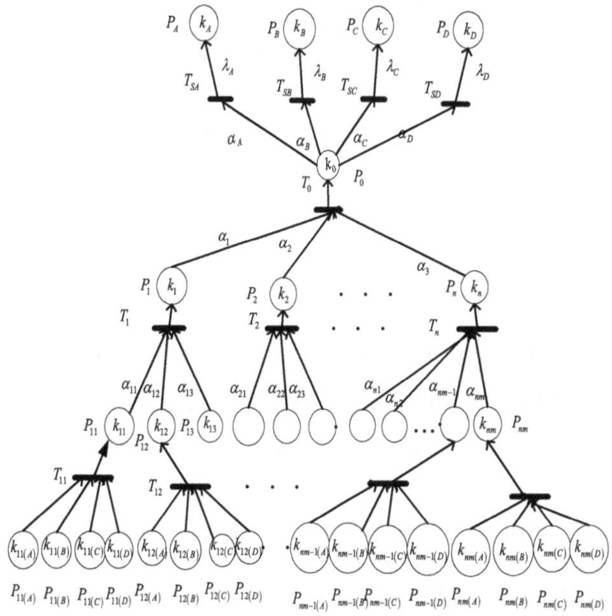

Fig.2. FPN evaluation model

The evaluation indexes in the FPN-based evaluation model are a hierarchy structure. In the Fig.2 p_0 called chief place corresponds to the chief evaluation index and p_i called superior place corresponds to the superior evaluation index;

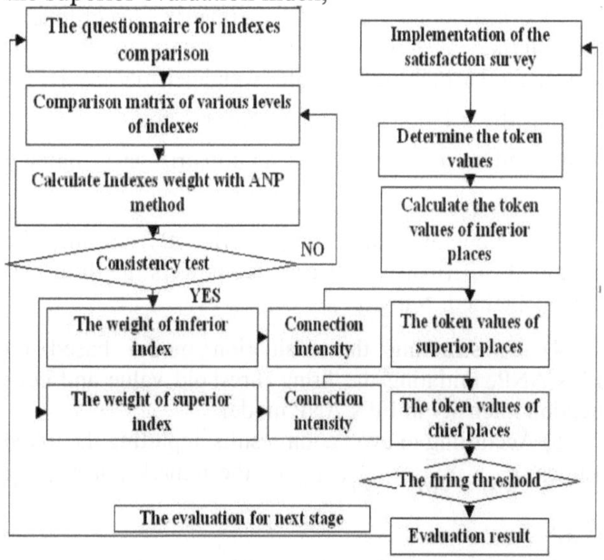

Fig.3. The FPN/ANP algorithm process

Likewise, P_{ij} called inferior place corresponds to the inferior evaluation index. Let V denotes remark set, and $V = \{A,B,C,D\} = \{Excellent, Good, Medium, Poor\}$. The Excellent, Good, Medium, Poor these four kinds of remarks correspond to the places P_A, P_B, P_C, P_D. Through fuzzy rules, $M(P_i)$ called fuzzy marking is possible token distribution, which reflects the final evaluation result, k is the token value of place, and $k_{ij(A)}$, $k_{ij(B)}$, $k_{ij(C)}$, $k_{ij(D)}$ are the token value of the remark places $P_{ij(A)}$, $P_{ij(B)}$, $P_{ij(C)}$, $P_{ij(D)}$, and $m_{ij(R)} = \frac{1}{N}\sum_{W=1}^{N} m_{ij(R)}^{W}$, $\sum_{R=1}^{4} m_{ij(R)} = 1$, and $m_{ij(R)}^{W}$ evaluated by the Wth appraiser denotes the member degree of the index p_{ij} to the remark level $L=(L = A,B,C,D)$, N denotes the number of appraisers in evaluation team. The a_{ij} connection intensity on the input arc (P_{ij},T_i) in the Fig.2 corresponds to the weight of the inferior index; the a_i connection intensity on the input arc (P_i,T_0) in the Fig.2 corresponds to the weight of the superior index. Usually, the unlabeled arcs denote that the intensities of input and output are 1. There is the ANP method to determine superior and inferior index weights.

The steps of FPN / ANP evaluation algorithm are as follows (Fig.3):

1) Constructing the network structure for decision Index, in the first step it consists of assigning priorities to related elements in order to build the unweighted supermatrix W_s.

2) Constructing weighted supermatrix. According to the criteria, it is done by means of the pairwise comparison method which obtained the weighted matrix A_s reflecting the relation for groups under some criteria. A_s Multiplied by W_s is weighted supermatrix $W_s^{(w)}$ namely: $W_s^{(w)} = A_s W_s$

3) Calculating the limit matrix. By raising the weighted supermatrix to successive powers the limit matrix is obtained. Namely $W_s^{(l)} = \lim_k W_s^{(k)}$

4) Constructing the evaluation model based on FPN/ANP, initiating the firing threshold value and connection intensity in FPN/ANP model.

5) According to evaluation results, inputting the token values $k_{ij(A)}$, $k_{ij(B)}$, $k_{ij(C)}$, $k_{ij(D)}$ of the remark place $P_{ij(A)}$, $P_{ij(B)}$, $P_{ij(C)}$, $P_{ij(D)}$.

6) Solving the token values of interior places: constitute vector k_{ij} with the token values $k_{ij(A)}$, $k_{ij(B)}$, $k_{ij(C)}$, $k_{ij(D)}$ of the remark places $P_{ij(A)}$, $P_{ij(B)}$, $P_{ij(C)}$, $P_{ij(D)}$, $k_{ij} = (k_{ij(A)}, k_{ij(B)}, k_{ij(C)}, k_{ij(D)})$;

7) Solving the superior places token values: $k_i = (a_{i1}\ a_{i2}\ \cdots\ a_{in}) \circ (k_{i1}\ k_{i2}\ \cdots\ k_{in})^T = (k_{i(A)}\ k_{i(B)}\ k_{i(C)}\ k_{i(D)})$

8) Solving the chief places token values: $k_0 = (a_1\ a_2\ \cdots\ a_n) \circ (k_1\ k_2\ \cdots\ k_n)^T$

9) Setting the firing threshold values, in transitions T_{SA}, T_{SB}, T_{SC}, T_{SD}. $l_A = l_B = l_C = l_D = l_{\max} = \max(k_{0(A)}, k_{0(B)}, k_{0(C)}, k_{0(D)})$, $a_A = (1\,0\,0\,0)^T$, $a_B = (0\,1\,0\,0)^T$, $a_C = (0\,0\,1\,0)^T$ $a_D = (0\,0\,0\,1)^T$. The input intensity function was defined as $ST_x = k_0 ?a_X\ (X\ \ A,B,C,D)$. The firing conditions $T_{SA}, T_{SB}, T_{SC}, T_{SD}$: If $ST_i = k_0$ 拙$_i$ $l_{\max}$ $(i = A,B,C,D)$, then firing transition T_{Si} $(i = A,B,C,D)$ and the token distribution of $P_A \sim P_D$ is $(ST_A\ 0\,0\,0)$, $(0\ ST_B\ 0\,0)$, $(0\,0\ ST_C\ 0)$, $(0\,0\,0\ ST_D)$.

IV. EXAMPLE OF EVALUATION

GH Company is a key enterprise in Guizhou Province in China, which is a mechanical, electronic, hydraulic and control for the integrated enterprise. GH company case is to illustrate the model evaluation process:

1) The appraisers were composed of the advisors management consultant, the member of BPR implementation team from enterprise.

2) Using the method of ANP, The appraisers acquired the weights of all evaluation indexes by "BPR questionnaire for indexes weight".

3) Through "questionnaire for BPR implementation", the member of the evaluation team evaluated the inferior indexes independently. Through the appraisers coordinated, each remark level for member degrees was acquired. For example, to evaluate the inferior index "Senior leadership support and participation" under the superior index "Investment in human resources", Four remark level "Excellent, Good, Medium and Poor" of the member degrees were severally $m_{51(A)} = 0.42$, $m_{51(B)} = 0.31$, $m_{51(C)} = 0.20$, $m_{51(D)} = 0.07$, then the token values of corresponding remark places $P_{51(A)}$, $P_{51(B)}$, $P_{51(C)}$, $P_{51(D)}$ were $k_{51(A)} = 0.42$, $k_{51(B)} = 0.31$, $k_{51(C)} = 0.20$, $k_{51(D)} = 0.07$, consequently, the four token value of the inferior place P_{51} constituted the vector, which was $k_{51} = (0.42\ \ 0.31\ \ 0.20\ \ 0.07)$. The token vector values of remainder inferior places P_{ij} were acquired in the same way.

4) According to the ANP algorithm, by calculation, it is determined by the weights of inferior index and Connection intensity. According to above step7, the token

vector value of the superior place was $k_5 = (a_{51}\ a_{52}\ \ a_{5n})(k_{51}\ k_{52}\ \ k_{5n})^T = (k_{5(A)}\ k_{5(B)}\ k_{5(C)}\ k_{5(D)})$.

5) On the basis of the FPN/ANP model and calculating, supposing that the chief place vector value was $k_0 = (a_1\ a_2\ \ a_{11})(k_1\ k_2\ \ k_{11})^T = (0.31\ \ 0.42\ \ 0.21\ \ 0.06)$.

6) On the basis of the firing condition for T_{SA}, T_{SB}, T_{SC}, T_{SD}, and the places of $P_A \sim P_D$, the firing threshold values of the transitions above are $l_A = l_B = l_C = l_D = l_{max} = \max(0.31\ 0.42\ 0.21\ \ 0.06) = 0.42$ $ST = k_0 ?a_B \quad 0.45 ?\ l_{max}$, so T_{SB} is fired, and the fuzzy marking is $(0\ 0.45\ \ 0\ \ 0)$. The place P_B token value was 0.45, the rest places token values were 0.Consequently, according to the final evaluation result, we knew that the process of BPR implementation was good. Then, to carry out evaluating for next stage BPR implementation.

V. CONCLUSION

The relationship among different enterprise subsystems is interactive and correlative with system complexity. A set of evaluation systems were established, and an evaluation model based on FPN/ANP was proposed. The evaluation model proposes a quantitative evaluation method for the implementation of one BPR project.

REFERENCES

[1] A. Geetha, K. Phalp, "Combining process modeling methods (Periodical style)," *Information and Software Technology,* vol.39, no.2, pp. 107-124, March 1997.

[2] A. Oberweis, "An integrated approach for the specification of processes and related complex structured objects in business applications (Periodical style)," *Decision Support Systems*, vol.17, no.1, pp. 31-53, June, 1996.

[3] Z. Y. Kang, "A Study of Business Process Reengineering Based on Petri Net (Published Conference Proceedings style)," in International Conference on Computing, Information and Control, ICCIC2011, Wuhan, PEOPLES R CHINA, pp.543-549, 2011.

[4] G. H. CHEN, Z. L. TAO, "Using the Analytic Hierarchy Process to Prioritize Business Processes for Reengineering Based on Business Strategy(Periodical style)," *Operations Research and Management Science*, vol.12, no.2, pp.101-105, October,2003.

[5] Z. Wu, L. Guo, S. D. Sun, Y. W. Peng,"Synthetical Fuzzy Evaluation of Business Process Performance(Periodical style)," *Journal of Northwestern Polytechnical University*. Vol.19, no.2, pp.186-189, February, 2001.

[6] N. V. Herzog, S. Tonchia, A. Polajnar, "Linkages between manufacturing strategy, benchmarking, performance measurement and business process reengineering (Periodical style)," Computers & Industrial Engineering, vol.57, no.3, pp.963-975. Oct 2009.

[7] J. Z. WANG, L. D. ZHAO, "Performance analysis method of business process reengineering based on queue theory (Periodical style)," *Journal of Southeast University Philosophy and Social Science Edition*, supplement, no.4, pp.119-122, 2002.

[8] T. WANG, S. P. YI, Y. Z. Yang, "Performance Evaluation Method for Business Process of Machinery Manufacture Based on DEA/AHP Hybrid Model (Periodical style)," *Chinese Journal of Mechanical Engineering*, 20, 3, April, 2007, pp.1128-1131.

[9] S. P. YI, T. WANG, Y. WANG, "Analysisand Empirical Study on the Impetusand Resistance of BPR based on a Mechanical Dynamics Model (Periodical style)," *Systems Engineering*, pp.89-93, vol.25, no.12, June, 2007.

[10] T. WANG, "Study on Supporting System and Performance Evaluation of Business Process Reengineering (Thesis or Dissertation style)," Ph.D. dissertation, Chongqing university, Chongqing, China, 2007.

[11] K. Altinkemer, Y. Ozcelik and ZD. Ozdemir, "Productivity and Performance Effects of Business Process Reengineering: A Firm-level Analysis (Periodical style)," Journal of Management Information System, vol.27, no4, pp.129-161, Spr. 2011.

[12] C. H. Wang, R. Y. Chen, "Enterprise Risk Research on the six sigma in the Business Process Reengineering(Published Conference Proceedings style)," in International Conference on Industrial Engineering and Management Science, ICIEMS, Shanghai, PEOPLES R CHINA, pp.889-897. 2013.

[13] W. Y. Wang, H. K. Chan, D. J. Pauleen, "Aligning business process reengineering in implementing global supply chain systems by the SCOR model (Periodical style)," International Journal of Production Research, vol.48, no.19, pp.5647-5669. 2010.

[14] G. J. ZHAO, X. Q. XING, "The Theoretical Analysis and Practical Application on ANP Method in Reginal Scientific Strength Evaluation (Periodical style)," Systems Engineering-theory Practice, vol.45, no.5, pp.41-45, 2004.

[15] S. J. Wang, G. Wang, G. A. Gao, "Fuzzy Petri Net-based Evaluation to the process of ERP implementation (Published Conference Proceedings style)," in *proceeding of the 2006 IEEE Asia-Pacific Conference on Service Computing,* Guangzhou, China, pp.11-17, *2006.*

Recent Advances in Concurrent Engineering Modeling

Yan-jun Qian[1,*], Thong Ngee Goh[2], Jun Lin[3]
[1]School of Management, Northwestern Polytechnical University, Xi'an, China
[2]Department of Industrial and Systems Engineering, National University of Singapore, Singapore
[3]School of Management, Xi'an Jiaotong University, Xi'an, China
([*]qian_yanjun@hotmail.com, ljun@mail.xjtu.edu.cn)

Abstract - **Over the last two decades, a number of studies have examined the trade-off involved in concurrent engineering (CE), time reduction versus additional effort for downstream rework. This study presents an overview of the recent CE modeling literature that examined this trade-off. We find that most CE models are built on the assumption that development stages are dependent where the principal information exchange between consecutive design stages is unidirectional, from upstream stage to downstream stage. According to literature review and field study, we believe such assumption is reasonable, because in many situations, current execution of design stages actually occurs within two sub-stages (Testing 1 and Development 2) which are sequentially dependent. In the future, we may also build analytical models based on interdependent stages so as to better understand the impact of project properties on best CE policies and product development performance.**

Keywords – **Concurrent engineering, overlapping, product development**

I. INTRODUCTION

Over the last two decades, concurrent engineering (CE), the practice of executing dependent product development stages simultaneously, has become the common mode of new product development (NPD) because of the increasing importance of time-to-market (Terwiesch et al., 2002 [1]; Lin et al., 2012 [2]). Although large reduction in project completion time may be achieved by applying CE approach, empirical studies also show that CE is not applicable to all NPD projects (Terwiesch and Loch, 1999 [3]; Mitchell and Nault, 2007 [4]). For example, based on the empirical study of 140 development projects in the electronics industries, Terwiesch and Loch (1999) concluded that CE was effective only if uncertainty resolution was fast. This is because CE requires that downstream stages start on preliminary information, and thus rework is often necessary to accommodate upstream design changes. Therefore, a key trade-off involved in CE is time reduction versus additional effort for downstream rework.

This study presents an overview of the recent CE modeling literature that examined the aforementioned trade-off. Our objective is to summarize the models' main assumptions and findings, and to identify possible future research directions. The rest of this study is organized as follows. Section 2 provides a taxonomy for grouping CE models. Section 3 reviews the relevant literature on CE. Finally, Section 4 summarizes the concluding comments.

II. A TAXONOMY FOR CE MODELS BASED ON INFORMATION DEPENDENCIES

An information-based view of product development has been widely adopted in the literature of CE (e.g. Clark and Fujimoto, 1991 [5]; Lin et al., 2008 [6]; Love et al., 2009 [7]). From this perspective, there are three general types of information dependencies, as shown in Fig.1 (Eppinger et al., 1994 [8]; Yassine et al., 1999 [9]; Bhuiyan et al., 2004 [10]). If stages A and B could be performed simultaneously with no interaction between them, then the two stages are said to be independent. On the other hand, the two development stages would be dependent (or sequentially dependent), if there is a unidirectional information flow between them. Finally, it is said to be interdependent if the two development stages are mutually dependent and the information flows in both ways.

According to this taxonomy, we discuss the recent CE models in the next section.

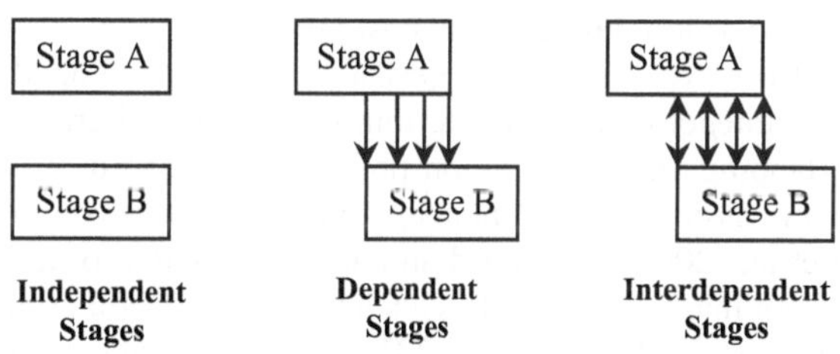

Fig.1. Independence, sequentia dependence, and interdependence

III. RECENT CE MODELS

Most CE models are built on the assumption that development stages are dependent or interdependent. Therefore, we categorize previous literature into two groups. Section *A* reviews the CE models for dependent development stages. Section *B* discusses the CE models for interdependent development stages.

A. CE Models based on Dependent Stages

Quite a few analytical models in CE are built on the assumption that development stages are dependent and the principal information exchange between consecutive design stages is unidirectional, i.e., from upstream stage to downstream stage.

For example, Krishnan et al. (1997) developed an integer program to determine the optimal number of information transfer between two consecutive development stages, as well as the start time of

E. Qi et al. (eds.), *Proceedings of the 5th International Asia Conference on Industrial Engineering and Management Innovation (IEMI2014)*, Proceedings of the International Asia Conference on Industrial Engineering and Management Innovation 1, DOI 10.2991/978-94-6239-100-0_2, © Atlantis Press and the authors 2015

downstream rework, such that project completion time would be minimized [11]. The authors proposed that the optimal degree of concurrency should be determined by two properties of the NPD process, "upstream evolution" and "downstream sensitivity", where "upstream evolution" denoted the speed at which upstream information narrows from an interval value to a final solution, and "downstream sensitivity" referred to the expected time needed for the downstream stage to incorporate upstream design changes. This principle was further developed by Loch and Terwiesch (1998), where they proposed that the optimal levels of concurrency and communication should be decided by the arrival rate of upstream design modifications, the impact of each modification (i.e. the percentage of downstream tasks would be affected by one upstream design change), and the downstream progress, i.e. the number of downstream tasks completed when the design change arrived [12].

Yassine et al. (1999) proposed a probabilistic model to determine the optimal degree of concurrency for a set of activities. Roemer et al. (2000) addressed the time-cost trade-off in CE and introduced an algorithm to determine the optimal concurrency level [13]. Chakravarty (2001) examined the optimal CE policies for three overlapping modes [14]. Roemer and Ahmadi (2004) examined the interactions between CE and crashing policies, and provided general guidelines for optimal CE and crashing policies [15]. Wang and Yan (2005) built an analytical model to determine the optimal concurrency level of two stages so as to minimize total cost [16].

Recently, Gerk and Qassim (2008) developed a mixed-integer model for determining the optimal crashing, overlapping, and substitution policies [17]. Based on the general assumption of nonnegative upstream evolution, Lin et al. (2009) developed an analytical model to derive the optimal degree of concurrency and functional interaction [18]. Lin et al. (2010) explicitly captures the interaction between project progress and CE and information exchange policies, and presented a model to derive optimal CE and communication policies [19]. Qian et al. (2010) proposed an analytical model to determine the optimal amount of time spent on upstream and downstream testing, as well as the optimal level of concurrency [20]. Lin et al. (2012) built an analytical model to determine the priority ordering of initial development and rework, and the optimal concurrency between development stages. They show that it is optimal to do the rework (resulting from upstream design changes) after the completion of initial development when learning effect exists.

The aforementioned studies are insightful in many ways, and are built on the assumption that development stages are dependent and the principal information exchange between design stages is unidirectional. However, many empirical studies argue that development stages are often interdependent for product development projects (e.g. Clark and Fujimoto, 1991; Bhuiyan et al., 2004; Lin et al., 2008). It is necessary to explain the contradiction of these two groups of studies. To do this, we reviewed the studies which include in-depth case studies and describe the information flows between development stages in detail.

Clark and Fujimoto (1991) were, perhaps, the first who systematically analyzed concurrent product development based on their field study in the world automotive industry. They investigated the information flows between two consecutive development stages, the design and development of dies for body panels. Wheelwright and Clark (1992) studied many other industries and generalized the process of new product development [21]. Swink et al. (1996) identified three levels of concurrency existed in the product development projects. The concurrency at the stage level was denoted as project phase concurrency which involves simultaneously developing market concepts, product designs, manufacturing processes, and product support structures [22]. Based on dozens of case studies in world class companies, Cooper (1994, 2007) concluded that flexibility is one of the key factors of the third generation stage-gate product development processes, i.e. projects can precede into the next stage even though the previous stage has not been totally completed [23, 24]. Bhuiyan et al. (2004) described the information flows and decision points in concurrent processes based on a study of six product development projects in a Canadian firm manufacturing printed circuit boards.

The above studies cover hundreds of product development projects in different industries. Although these studies may break new product project into different number of stages (typically four, five or six in number) and the content of each stage may be different, the information flows involved in two consecutive development stages are similar. Figure 2 shows a generic product development process adapted from Clark and Fujimoto (1991), Wheelwright and Clark (1992), Cooper (1994, 2007), Swink (1996), Bhuiyan et al. (2004), Mitchell and Nault (2007), and our experience in three consumer electronics companies. It is clear that the upstream stage and the downstream stage are interdependent. The downstream starts on the preliminary information of the upstream stages and the change of upstream will affects the progress of the downstream. On the other hand, the downstream supports feedback information to the upstream. However if we take an in-depth look at the development stages, we can see that concurrent execution of design stages actually occurs within two sub-stages (Testing 1 and Development 2) which are sequentially dependent. Therefore, if we can optimize the concurrency between Testing 1 and Development 2 and reduce the total development time of them, the project cycle time will be reduced accordingly.

The product development process shown in Figure 2 is generic and common to many organizations. Therefore, we conclude that concurrent execution often occurs between two dependent sub-stages and project cycle time can be reduced by optimizing the concurrency of these sub-stages. This conclusion supports previous CE studies based on dependent stages.

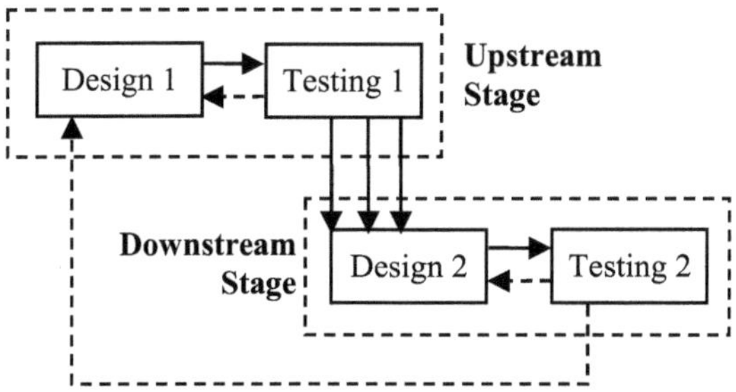

Fig.2. Information flows between two consecutive development stages

B. CE Models based on Interdependent Stages

Several CE models are built based on interdependent stages where the information flows in both ways. For example, Joglekar et al. (2001) proposed a performance generation model to determine the optimal degree of concurrency with the goal of maximizing project performance with deadline constraints [25]. Bhuiyan et al. (2004) proposed a discrete event simulation model to study the impact of CE and functional interaction on project performance. Lin et al. (2008) proposed a Dynamic Development Process Model for managing overlapped iterative product development, and validated the model with an in-depth case study at a handset design company.

IV. CONCLUSION

In summary, most CE models are built on the assumption that development stages are dependent where the principal information exchange between consecutive design stages is unidirectional, i.e., from upstream stage to downstream stage. According to literature review and field study, we believe such assumption is reasonable, because in many situations, current execution of design stages actually occurs within two sub-stages (Testing 1 and Development 2) which are sequentially dependent.

Several CE models are built based on interdependent stages where the information flows in both ways. Because of the complex information flows among interdependent stages, most of them use simulation to explore the linkage between CE policies and product development performance. In the future, we may build analytical models based on interdependent stages so as to better understand the impact of project properties on best CE policies and product development performance.

ACKNOWLEDGMENT

The work of Qian Y. was supported in part by National Natural Science Foundation of China (Grant 71101115), and by "the Humanities, Social Science and Management Research Funds of Northwestern Polytechnical University" (Grant 3102014RW0003). The work of Lin J. was supported in part by National Natural Science Foundation of China (Grants 71001084 and 71371149), and by "the Fundamental Research Funds for the Central Universities" (Grant sk2014043).

REFERENCES

[1] Terwiesch, C., Loch, C.H., De Meyer, A., 2002. Exchanging preliminary information in concurrent engineering: Alternative coordination strategies. Organization Science, 13 (4), 402-419.

[2] Lin, J., Qian Y., Cui, W., 2012. Managing the concurrent execution of dependent product development stages. IEEE Transactions on Engineering Management 59(1), 104-114.

[3] Terwiesch, C., Loch, C.H., 1999. Measuring the effectiveness of overlapping development activities. Management Science 45 (4), 455-465.

[4] Mitchell, V.L., Nault, B.R., 2007. Cooperative planning, uncertainty, and managerial control in concurrent design. Management Science 53 (3), 375-389.

[5] Clark, K.B., Fujimoto, T., 1991. Product development performance: Strategy, organization and management in the world auto industry. Boston: Harvard Business School Press.

[6] Lin, J., Chai, K.H., Wong, Y.S., Brombacher, A.C., 2008. A dynamic model for managing overlapped iterative product development. European Journal of Operational Research, 185 (1), 378-392.

[7] Love, P.E.D., Edwards, D.J., Irani, Z., Walker, D.H.T., 2009. Project pathogens: the anatomy of omission errors in construction and resource engineering project. IEEE Transactions on Engineering Management 56 (3), 425-435.

[8] Eppinger, S.D., Whitney, D.E., Smith, R.P., Gebala, D.A., 1994. A model-based method for organizing tasks in product development. Research in Engineering Design 6 (1), 1-13.

[9] Yassine, A.A., Chelst, K.R., Falkenburg, D.R., 1999. A decision analytic framework for evaluating concurrent engineering. IEEE Transactions on Engineering Management 46 (2), 144–157.

[10] Bhuiyan, N., Gerwin, D., Thomson, V., 2004. Simulation of the new product development process for performance improvement. Management Science 50 (12), 1690-1703.

[11] Krishnan, V., Eppinger, S.D., Whitney, D.E., 1997. A model-based framework to overlap product development activities. Management Science, 43 (4), 437-451.

[12] Loch, C.H., Terwiesch, C., 1998. Communication and uncertainty in concurrent engineering. Management Science, 44 (8), 1032-1048.

[13] Roemer, T.A., Ahmadi, R., Wang, R.H., 2000. Time-cost trade-offs in overlapped product development. Operations Research, 48 (6), 858-865.

[14] Chakravarty, A.K., 2001. Overlapping design and build cycles in product development. European Journal of Operational Research, 134 (2), 392-424.

[15] Roemer, T.A., Ahmadi, R., 2004. Concurrent crashing and overlapping in product development. Operations Research, 52 (4), 606-622.

[16] Wang, Z., Yan, H.S., 2005. Optimizing the concurrency for a group of design activities. IEEE Transactions on Engineering Management, 52 (1), 102-118.

[17] Gerk, J.E.V., Qassim, R.Y., 2008, Project acceleration via activity crashing, overlapping, and substitution. IEEE Transactions on Engineering Management 55(4), 590-601.

[18] Lin, J., Chai, K.H., Brombacher, A.C., Wong, Y.S., 2009. Optimal overlapping and functional interaction in product development. European Journal of Operational Research, 196 (3), 1158-1169.

[19] Lin, J., Qian, Y.J., Cui, W.T., Miao, Z.L., 2010. Overlapping and communication policies in product development. European Journal of Operational Research, 201 (3), 737-750.

[20] Qian, Y., Xie, M., Goh, T.N., Lin, J., 2010. Optimal Testing Strategies in Overlapped Design Process. European Journal of Operational Research 206 (1), 131-143.

[21] Wheelwright, S.C., Clark, K.B., 1992. Revolutionizing Product Development. The Free Press, New York.

[22] Swink, M.L., Sandvig, C., Mabert, V.A., 1996. Customizing concurrent engineering processes: five case studies. Journal of Production Innovation Management 13 (3), 229-244.

[23] Cooper, R.G., 1994. Third generation of new product processes. Journal of Production Innovation Management 11 (1), 3-14.

[24] Cooper, R.G., Kleinschmidt, E.J., 2007. Winning businesses in product development: The critical success factors. Research-Technology Management 50 (3), 52-66.

[25] Joglekar, N.R., Yassine, A.A., Eppinger, S.D., Whitney, D. E., 2001. Performance of coupled product development activities with a deadline. Management Science 47 (12), 1605-1620.

The Construction of the Streamline Curve and Surface

Xiu-juan Li
College of Science, Nanjing University of Aeronautics & Astronautics, Nanjing, China
(lixiujuan@nuaa.edu.cn)

Abstract - **The construction of the streamline curve and surface is a modeling technique of the free curve and surface. In this paper a streamline curve is regarded as an integral curve of the tangent vectors. Based on this method, the streamline curve and surface are constructed. At the same time, the revolution surface is constructed and deformed appropriately. And the streamline surface is used for surface splicing. Studies have shown that the constructed surface is easy to be deformed and convenient to be spliced.**

Keywords - **Quaternion, revolution surface, streamline curve and surface**

I. INTRODUCTION

The shape of a streamline curve is smooth, beautiful and pleasant, so it is favored by many shape designers. In the industrial design, the shape of a streamline curve is infiltrated into the appearance of many products. At present, due to the rapid development of computer and CFD, Streamline design attracts the attention of the shape designers and the researchers on CAGD once again. A streamline curve was represented as an integral curve of unit quaternion curve by Miura KT [1-5]. In references [2], a streamline curve is:

$$c(s) = c(s_0) + \int_{s_0}^{s} q(\mu) v q^{-1}(\mu) d\mu \tag{1}$$

In which, $q(\mu)$ in the form of B-spline is expressed as:

$$q(\mu) = \left(\prod_{i=n}^{1} \exp\left(\boldsymbol{\omega}_i \overline{B}_{i,k}(\mu) \right) \right) q_0 \tag{2}$$

In which,

$$\overline{B}_{i,k}(\mu) = \begin{cases} \sum_{j=i}^{i+k-1} B_{j,k}(\mu) & \mu_i < \mu < \mu_{i+k-1} \\ 1 & \mu \ge \mu_{i+k-1} \\ 0 & \mu \le \mu_i \end{cases} \tag{3}$$

$$\boldsymbol{\omega}_i = \frac{1}{2} \sin^{-1}\left(|\boldsymbol{t}_{i-1} \times \boldsymbol{t}_i| \right) \frac{\boldsymbol{t}_{i-1} \times \boldsymbol{t}_i}{|\boldsymbol{t}_{i-1} \times \boldsymbol{t}_i|} \tag{4}$$

Let the length of the streamline curve is L. That is $s_0 \le s \le s_0 + L$. Make $\mu = L\upsilon$. The expression (2) can be:

$$q(L\upsilon) == \left(\prod_{i=n}^{1} \exp\left(\boldsymbol{\omega}_i \overline{B}_{i,k}(L\upsilon) \right) \right) q_0 \tag{5}$$

So

$$c(s) = c(s_0) + \int_{s_0}^{s} q(\mu) v q^{-1}(\mu) d\mu = c(s_0) + L \int_{\frac{s_0}{L}}^{\frac{s}{L}} q(L\upsilon) v q^{-1}(L\upsilon) d\upsilon \tag{6}$$

Let $q(L\upsilon) = Q(\upsilon)$, $\frac{s}{L} = \kappa$. Since $s_0 \le s \le s_0 + L$, then $\frac{s_0}{L} \le \kappa \le \frac{s_0 + L}{L}$. So

$$c(L\kappa) = c(s_0) + L \int_{\frac{s_0}{L}}^{\kappa} Q(\upsilon) v Q^{-1}(\upsilon) d\upsilon \tag{7}$$

Let $c(L\kappa) = C(\kappa)$, then

$$c(s_0) = c\left(L \cdot \frac{s_0}{L} \right) = C\left(\frac{s_0}{L} \right) \tag{8}$$

And

$$C(\kappa) = C\left(\frac{s_0}{L} \right) + L \int_{\frac{s_0}{L}}^{\kappa} Q(\upsilon) v Q^{-1}(\upsilon) d\upsilon \tag{9}$$

Again let $u = \kappa$, then

$$C(u) = C\left(\frac{s_0}{L} \right) + L \int_{\frac{s_0}{L}}^{u} Q(\upsilon) v Q^{-1}(\upsilon) d\upsilon \tag{10}$$

In expression (10), $u = \frac{s}{L}$, that is

$$C\left(\frac{s}{L} \right) = C(u) = C\left(\frac{s_0}{L} \right) + L \int_{\frac{s_0}{L}}^{u} Q(\upsilon) v Q^{-1}(\upsilon) d\upsilon \tag{11}$$

Let the length of the integral interval of the above expression is δ, then $\delta = u - \frac{s_0}{L} = \frac{s}{L} - \frac{s_0}{L}$. So $0 \le \delta \le 1$.

Due to $u = \frac{s}{L}$, then $\frac{s_0}{L} \le u \le \frac{s_0 + L}{L}$. Suppose $s_0 = 0$, then $0 \le u \le 1$.

The expression (10) in comparison with (1) has one more parameter. The curve generated can have greater freedom by changing the parameters and result in a wide range of shapes.

The streamline surface is a set of the streamline curves along a parameter direction r. The streamline surface corresponds to expression (10) can be:

$$C(u,r) = C\left(\frac{s_0}{L}, r \right) + L \int_{\frac{s_0}{L}}^{u} Q(\upsilon, r) v Q^{-1}(\upsilon, r) d\upsilon \tag{12}$$

In which $u = \frac{s}{L}$. That is

$$C\left(\frac{s}{L}, r \right) = C(u,r) = C\left(\frac{s_0}{L}, r \right) + L \int_{\frac{s_0}{L}}^{u} Q(\upsilon, r) v Q^{-1}(\upsilon, r) d\upsilon \tag{13}$$

According to the expression (12), for any r, if $C\left(\frac{s_0}{L}, r \right)$ and $Q(\upsilon, r) v Q^{-1}(\upsilon, r)$ are known, and consider the integration of $Q(\upsilon, r) v Q^{-1}(\upsilon, r)$, the streamline surface can be constructed (Fig.1).

E. Qi et al. (eds.), *Proceedings of the 5th International Asia Conference on Industrial Engineering and Management Innovation (IEMI2014)*, Proceedings of the International Asia Conference on Industrial Engineering and Management Innovation 1, DOI 10.2991/978-94-6239-100-0_3, © Atlantis Press and the authors 2015

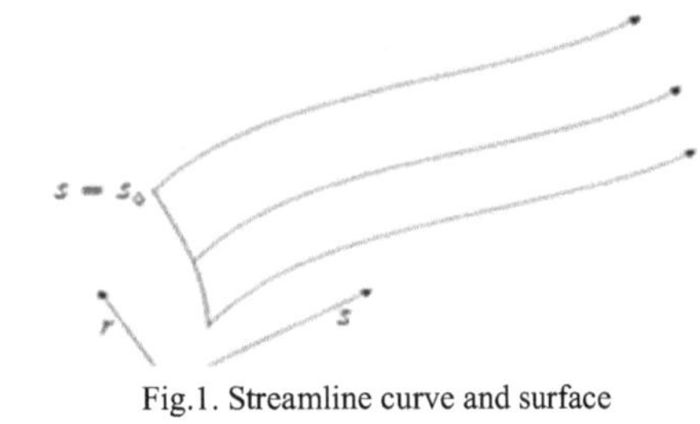

Fig.1. Streamline curve and surface

II. CONSTRUCTION OF STREAMLING SURFACE

In the above, the expression of the integral curve is analyzed. Through the variable transformation, the expression is reformatted, and an expression with a length parameter is obtained. For the expression, the literature [6] gives a subdivision method of the unit B spline curve interpolating the endpoints on a Gaussian sphere, a integral method of the streamline curve and surface, and so on. And it designs a construction method of the streamline curve and surface in the integral form.

Fig.2 is a section of a pipe. The dark curve is the parameter line, that is, $s = s_0$ of the streamline surfaces. The arrow is the direction of streamlined curve. Fig.3 is a petal. The construction is by selecting a parameter line r, that is, $s = s_0$ and giving some the tangent vectors of the streamline surface, the length parameter of the streamline curve and the integration interval, etc.

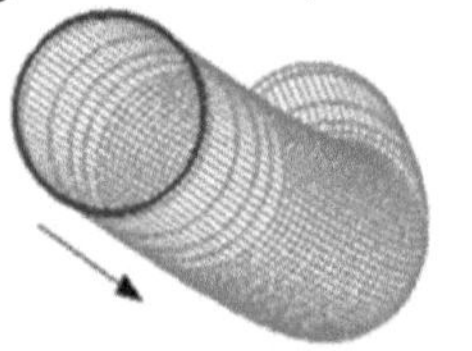

Fig.2. a pipe

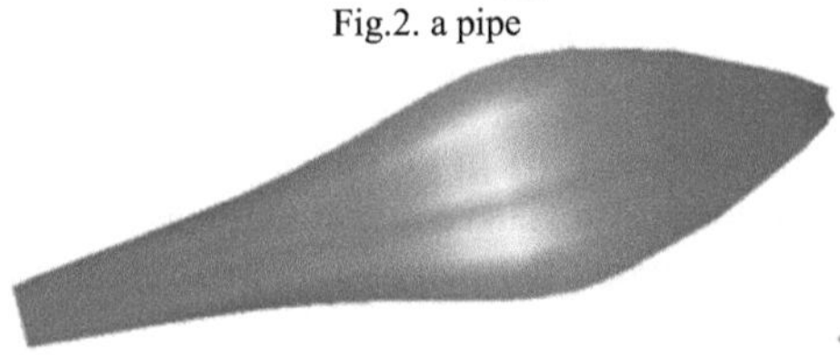

Fig.3. a petal

III. CONSTRRUCTION OF REVOLUTION SURFACE

The revolution surface is the most commonly used surfaces in engineering design and graphics. The existing representations of a revolution surface are the rational Bézier, NURBS, T-B spline, etc [6]. These methods are based on traditional methods of the curves and surfaces expressed by of the control points, and they cannot control the tangent vector and the curvature of a curve and surface directly. For the streamline surface, its control tangent vectors are mapped into Gaussian sphere and connected into a topology quadrangle grid. The corresponding subdivision surface is a unit quaternion surface on Gaussian sphere.

Two pictures in Fig.4 are constructed by a given set of tangent vectors, the equal length parameters of all the streamlined curves and the equal integral intervals. Among them, the streamline curves intersect the axis of rotation in (a), and they do not intersect in (b). In Fig.5, two revolution surfaces are constructed through a given a set of tangent vector, all equal length parameters of the streamlined curves, not all equal integration interval.

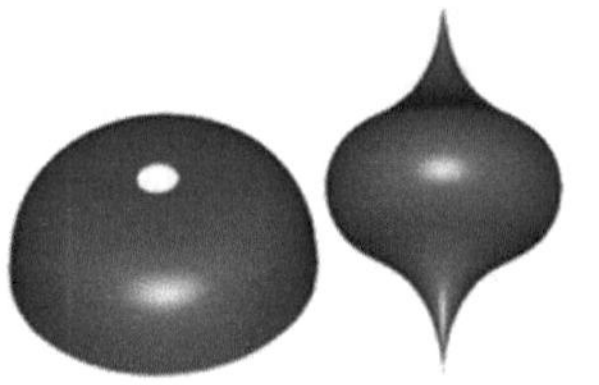

(a) (b)
Fig.4. Streamline surface

(a) (b)
Fig.5. Streamline surface

IV. SHAPE ADJUSTMENT

In this section, the directions of the tangent vectors are remained; the parameters including the integral interval and the length of the streamline curves are changed. Fig.6 only changes the integral interval, other conditions remain unchanged. A surface of revolution is generated. As the algorithm uses subdivision and numerical integration, the rotational surfaces are generated fast and easy. And the generated revolution surfaces keep the directions of the tangent vectors of the streamline curves at the two ends along the direction of the streamlined curve. In Fig.7, bee's abdomen is a streamline revolution surface. The conditions except the integral interval in the two pictures are the same. Fig.8 only changes the length of the streamline curves, the other conditions are the same.

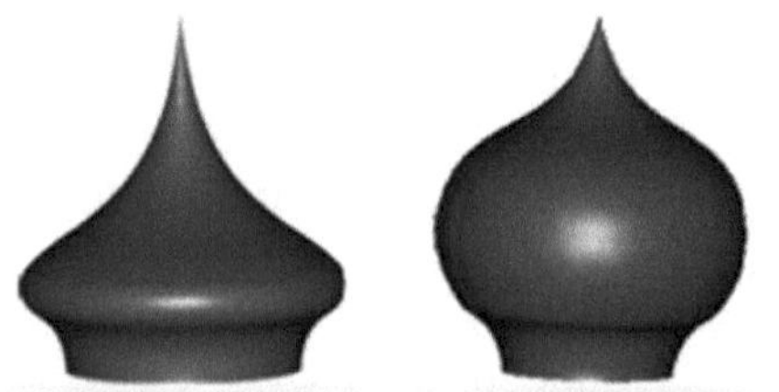

Fig.6. the streamline surface generated by changing the integration interval

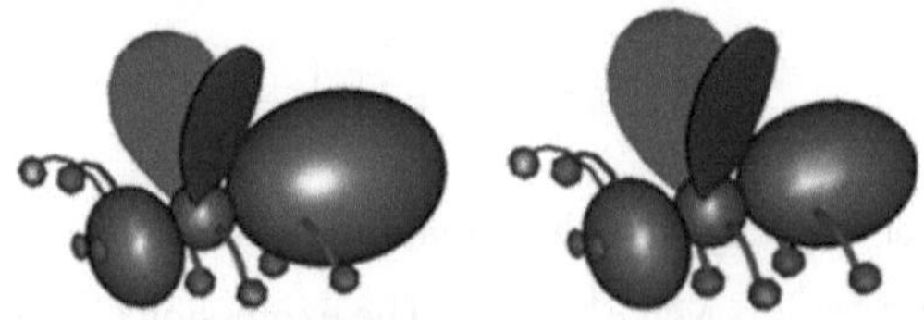

Fig.7. the streamline surface generated by changing the integration interval

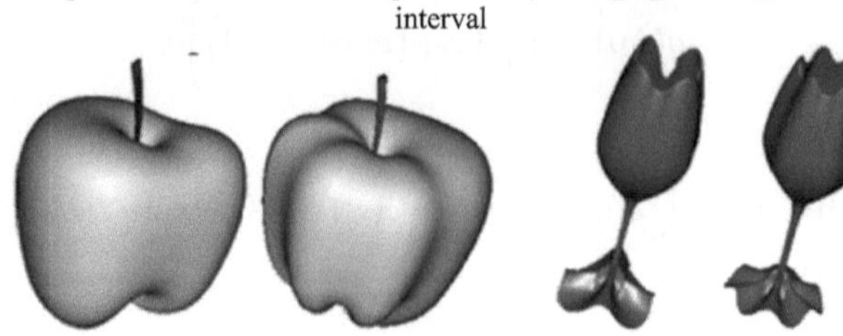

(a) (b)
Fig.8. the streamline surface generated by changing the length parameter

In above the shape adjustment to the integral surface is to generate the revolution surface firstly, and then to change a parameter. The above method of the shape adjustment to the revolution surface is simple, easy to operate. And it can quickly adjust the shape of the surface to some extent and enrich the surface modeling.

V. APPLICATION IN SURFACE SPLICING

Surface splicing plays an important role in geometric modeling, since 80's of the twentieth century; researchers have proposed a lot of surface splicing method [7, 8, 9]. Some of these methods only adapt to the surface in the implicit equation. Most cases can only get the numerical solution, and the range of application is restricted. Some have a complex form, not easy for practical application. At present, the development of the construction technology of the integral curve and surface is not mature. Because the curve and surface are expressed by the integral of the tangent vectors. The integrand, that is the linear interpolation of the tangent vectors, is calculated by the model of subdivision algorithm. In this way, it avoids the complex integral calculation, and improves the speed of generating streamline curve and surface.

Fig.9 (a) is a control grid of the non uniform C-C subdivision surfaces meeting the conditions. The two outer layers of the grid are regular. From outside to inside, the points in the third layer are not regular. (b) is a lighting effect after G^1 splicing of subdivision surface and streamline surface. (c) is a splice after changing the length parameter of the streamline surface arbitrarily.

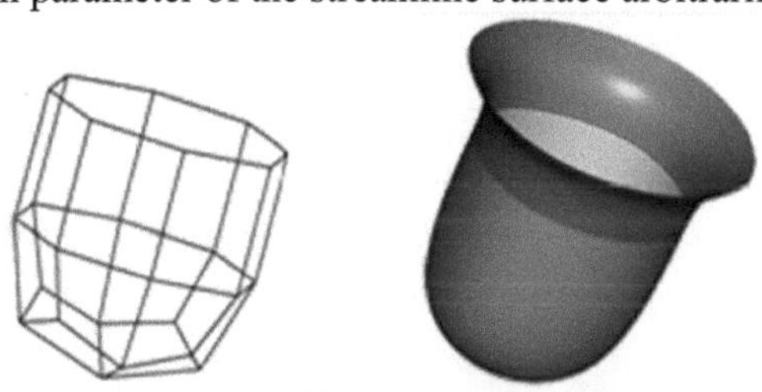

(a)Control mesh (b) a lighting effects after splicing

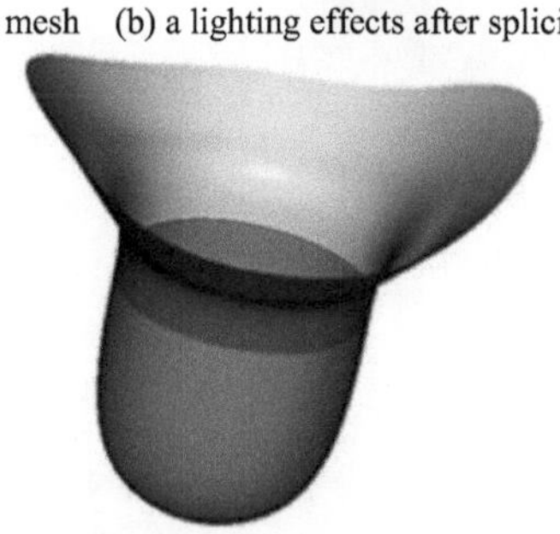

(c) a splicing after changing a parameter

Fig.9. a splicing of a non uniform C-C subdivision surface and a streamline surface

Fig.10 is a G^1 splicing of the elephant's nose. The root of the nose is a C-C subdivision surface, other parts is a streamline surface. In (a) and (b), the tangent vector directions of the streamline curve are the same, but the length parameters are not. (c) and (d) are so. The adjustment to the direction of tangent vector can change the direction of the nose, and the adjustment to the length of the parameters can easily change the length of the nose.

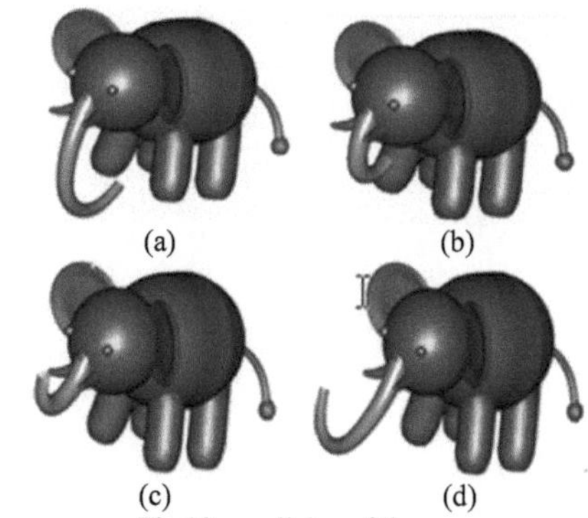

(a) (b) (c) (d)

Fig.10. a splicing of the nose

VI. CONCLUSION

In this paper, the streamline curve and surface in an integral form are reformatted. And then the streamline curve and surface controlled by parameters can be obtained. This method for constructing streamline curve and surface is rapid, simple and convenient. It can adjust the shape of the surface by changing the integration intervals and the length parameters. It is convenient to construct G^1 splicing surface of the streamline surface and other surfaces. The geometric shape of the splicing surface is also changed by altering the shape parameters. This reflects the advantage of the streamline surface.

REFERENCES

[1] K T Miura, N Sakiyama, T Kaneko. "Streamline modeling - designing free-form surfaces by blending tangent vectors," Sixth IFIP WG5.2 Workshops on Geometric Modeling, pp. 168~179, 1998.

[2] K T Miura. "Unit quaternion integral curve: a new type of fair free-form curves," Computer-Aided Geometric Design, vol.17, no. 1, pp.39~58, 2000.

[3] K T Miura, L Wang, F Cheng. "Streamline modeling with subdivision surfaces on the Gaussian sphere," Computer-Aided Design, vol.33, no.13, pp. 975~987, 2001.

[4] Y Suzuki, K T Miura. "Streamline modeling based on potential flow," Ninth Pacific Conference on Computer Graphics and Applications, pp. 16~18, 2001.

[5] K T Miura, F Cheng. "Fine tuning -curve and surface deformation by scaling derivations," Ninth Pacific Conference on Computer Graphics and Application, pp.150~159, 2001.

[6] X J Li. "Research on key technology of streamline curve and surface constructing," Ph. D. dissertation, Aeronautics and Astronautics Science and Technology. 2011.

[7] M Fang, G Wang. "Blending quadric surfaces with rational bicubic Bezier patches," Journal of Computer Aided Design & Computer Graphics, vol.19, no.9, pp.1148~1153, 2007.

[8] S H Cheng, Y L Zhang, Y W Zhou. "G1 continuity between two patches of adjacent cubic NURBS surface," in Proc. 10th IEEE Conf. Computer Aided Design and Computer Graphics. pp. 425~428.

[9] Q Song, J Wang. "Generation of parametric Gn blending surfaces based on relative direction of the base surfaces," Transactions of Beijing Institute of Technology, vol. 28, no.4, pp.334~337, 2008.

Research on Elements and Method of Evaluating B2C E-Commerce Service Quality Based on QFD

Zao-jie Kong*, Ming-liang Li
School of Economy Management, Hebei University of Technology, Tianjin 300401
(kongzj@sina.com)

Abstract - **This paper introduces quality function deployment into the research on the B2C e-commerce service quality evaluation for the corresponding problem that the determination of evaluation elements is not unified and heavily subjective. Above all, the paper defines 12 customer demands including "commodity searching convenience" after inquirying and summarizing customer demands information. Besides, the customers perception of demands' importance is investigated in the research. The weights of the customer demands are equal to the mean value of the demands' importance under the condition that the reliability of the data meets the standards. Secondly, the B2C e-commerce service quality evaluation elements are found according to literature sum-up and customers' suggestion. When the demands are determined, this research takes the service quality elements selecting probabilities as the degree of association between the demands and quality elements in the transition matrix. The paper largely overcomes the subjectivity in the traditional process of determining service quality evaluation elements, and obtains the importance of every element which paves the way for the evaluation.**

Keywords - **B2C e-commerce, customer demand, QFD, service quality elements**

I. INTRODUCTION

China's B2C e-commerce has developed rapidly, B2C e-commerce services companies such as Jingdong Mall, Taobao Mall, Suning Tesco and others rise rapidly. Due to the relatively good quality of goods, high degree of service standards, B2C e-commerce has become the most preferred shopping channel among consumers, especially the young groups. On the other hand, as a service industry, the service quality of B2C e-commerce is not only the primary consideration when consumers select online shopping merchants, but also the direction where the managers of e-commerce business company want to make great efforts to.

Recently, the study on the service quality of e-commerce has already conducted by a number of scholars, which mainly focus on the service quality evaluation of the e-commerce. Here are some introductions to these scholars: Hua Ying, one of the scholars, studied on the index of e-commerce service quality evaluation dimensions after analyzing the e-commerce service quality issues [1]; Other scholars like Zuo Wenming, Yang Wenfu, Huang Qiuping, Xie Zhenli and so on, took the B2C e-commerce industry's characteristics into consideration, referred to the relevant literatures and modeled on 36 preliminary evaluation indexes, finally finished a system of service quality evaluation for B2C e-commerce [2]; Improved the former scholar's model of e-commerce service quality evaluation, Yu Jiaze, rightly put forward a six-dimensional model of electronic service quality [3]. Generally, the elements of evaluating service quality of B2C e-commerce, which were defined in the studies mentioned above, were mainly based on the indicators about the service quality evaluation of e-commerce found in the literatures, and were often too subjective for different customer holds different knowledge level and preference. Besides, the credibility of the conclusions of the study is not ideal, because we are not sure whether these evaluating elements have reflected the customers' attention and whether they have highlighted the concerns of consumers. So, in order to address the heavily subjectivity on the determination of evaluation elements, this paper introduces quality function deployment into the research to obtain evaluation elements meets customer demands.

Quality Function Deployment was proposed by Dr. Yoji Akao (Professor of Japan University of Yamanashi,a quality management expert) in the sixty's of last century. To the up-stream steps of developing new product and assuring quality, the quality function deployment can

E. Qi et al. (eds.), *Proceedings of the 5th International Asia Conference on Industrial Engineering and Management Innovation (IEMI2014)*, Proceedings of the International Asia Conference on Industrial Engineering and Management Innovation 1, DOI 10.2991/978-94-6239-100-0_4, © Atlantis Press and the authors 2015

provide a good solution to the problem of customer satisfaction. In 1972, the phrase "quality deployment" was first used by Dr. Yoji Akao in his paper named "new product development and quality assurance--quality deployment system", whose essence thought lies in the use of a Cause & Effect diagram when considering how to plan, design, and manufacture to meet customer demands with quality in product development. Under the guidance of Prof. Shigeru Mizuno and Prof. Furukawa Kobe, the shipyard of Mitsubishi Heavy Industries designed the quality table which the western scholars called House of Quality and actually serves as a transition matrix from function to characteristics. This method solves the problem of complicated and huge tables caused by frequent use of Cause & Effect diagram. Later, combined with his own ideas and the House of Quality, Dr. Yoji Akao proposed the theory of quality function deployment (QFD) [4]. Then this kind of technology began spreading all over the world, and introduced into our country in the 1990s. Xiong Wei, who is the first one that studied the QDF in Japan and then introduced it to our country, has made many achievements in the researching of this field.

The QDF is a kind of quality analysis tool, which means analyzing customer's demands in the early period and then transforming the demands into the requirements of product design [5-6]. After the first idea that the quality gets from the examination, to the view that the quality comes from the manufacture of products, then the idea that get quality from design, people gradually realized that the design of the quality of product and service is the key to the customer perception of demands quality. In manufacturing industry, the quality function deployment has been widely and well applied. And, there are also a lot of the applications in such the service industry as health care, information services, government agencies etc. So the apply of quality function deployment in e-commerce service quality research is highly feasible. As the rapidly developed category of service industry, e-commerce service also needs reasonable design to be more in line with the customer's concerns.

II. THE ESSENCE OF THE QFD - HOUSE OF QUALITY

The Quality function deployment is completed through a series of graphs and matrices; the house of quality plays a crucial role in it. The House of Quality, which is also called "quality table", is described by Dr. Yoji Akao as a table where the quality that the customers really demand is showed in language and systematized and the relationship between the quality and the quality characteristics is also expressed for the purpose of changing the customers' demand into the alternative characteristics which can be used in the further quality design. The structure of house of quality is showed below:

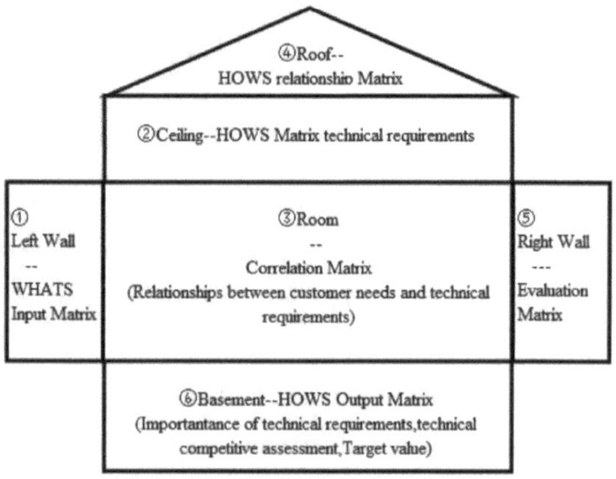

The structure of house of quality

(1) Left wall: Whats, on behalf of input items, namely, customer demands and its importance, is the "what" of House of quality. Customer demand: a customer determined characteristics of the products or services, is a technical description that transformed from non-technical language. At this point, suppose the importance of the customer i-th for degree Wi. Important value: quantitative evaluation for each customer's demands, which means that a certain degree of importance on customer demand.

(2) Ceiling: HOWS Matrix, representing what characteristics should have to meet each requirement of enterprise product or service, is the "how" of House of quality. Technical requirements (product/service features or engineering measures): transform the customer requirements into executable and measurable technical requirements or methods. Assume j-th service quality

elements for Fj.

(3) Room: Correlation Matrix, representing the strength of relationships between customer demands and technical characteristics. Assuming the relationship between the i-th customer demands Wi and the j-th quality elements Fj for Kij.

(4) Roof: HOWS relationship Matrix, representing the relationships between technology requirements.

(5) Right Wall: Evaluation Matrix, representing competing or competitive or comparative feasibility analysis, is the customer competitive assessment; evaluate the market competitiveness of products from the customer's point of view.

(6) Basement: HOWS output Matrix, representing the cost of technology assessment of the Hows items, etc. Including the importance of technology needs, target identification and technical competitive assessment. Used to determine the configuration of priority projects.

III. THE CONSTRUCTION METHODS OF B2C HOUSE OF QUALITY

In order to more objectively reflect the B2C e-commerce service quality, this paper introduces the theory of quality function deployment, and how to build the B2C House of Quality is the key to determine the evaluation elements of B2C e-commerce service quality. Therefore, it is necessary to consider the demands of customers, and also refer to the achievements of other researchers. There are five aspects to be done as follows.

A. Investigation and integration of customer demands

Customer demand is also called the voice of customer, which is the first input of house of quality. Thus whether it is accurate or not determines the effect of quality function deployment, which is one of the key tasks of this paper. There are a variety of methods to get the voice of customer, such as customer complaints, customer message, company service personnel interviews, customer interviews etc. In order to get a full range of customer demands categories, we need to follow certain investigative steps. Mainly are: Firstly, determine the appropriate respondents. This paper investigates the B2C e-commerce site's customers and we can take Jingdong Mall as an example; Secondly, choose reasonable investigative methods. We use 5W1H surveys to investigate customers; the last is the implementation of market surveys.

B. Access to importance degree of customer demands

There are two major methods to determine the weights of customer requirements. One is the Analytic Hierarchy Process, which refers to expert opinions to obtain the weights. Some researchers also combine rough set or fuzzy theory and AHP to investigate several key customers (generally 5 members) judgment matrix and determine the weights of customer demands, which makes the results more accurate. For example, Wang Xiaotun and Xiong Wei presented the rough AHP method to determine the weights of customer demands in the HOQ analysis, which combined the rough set theory with the analytic hierarchy process. It not only overcomes the ambiguity and subjectivity of AHP method, but also resolves the problems that multi-demand rough set theory cannot deal with [6].

However, both the expert opinions and the views of key customers, after all, are minority views, with very heavily subjectivity; and also, when there are more demand levels, or a certain level with huge demand species, the probability of passing consistency test sharply reduced. Thus this requires respondents to judge again, which changes the original judgment of the respondents and leads to inaccurate results. The other method is customer questionnaire, which requires more works, obtains more cumbersome, complex and irregular information, and needs researchers to make a conclusion. Nonetheless, the method of customer questionnaire refers to the original customer opinions, which is the most intuitive reflect of customer demands, and relatively more objective response to customer demands. Lin Liqiong used the direct survey method to determine the importance of customer demand when she studied the vocational education quality management system based on QFD [7]. Therefore, this paper adopts customer questionnaire to obtain the importance degree of customer demands.

C. The determination of B2C e-commence service quality elements

Customers' demand, the factor that in the point of customer's view and with heavy subjective feeling, is a kind of expectation that customers give to the service. Instead, the elements of service quality which are taken into consideration in the perspective of enterprises may do some influence on the customers' perceptions, and these factors are relatively objective, and can be determined by e-commence enterprise. At present, many scholars have conducted research on e-commerce service quality, since Parasuraman, the American service marketing master, published the first article on e-commerce service quality in 2005. But there are still few results of specific research on the B2C e-commerce service quality evaluation, and the established evaluation systems were mostly based on the research of others and did some modifications, which not only failed to reflect the characteristics of B2C e-commerce, but also could not reflect the voice of the customer. However, the research in this paper which is based on the customers' demands and having combined with the research in literatures to find the elements of service quality that may meet the need of customers, thus, make it be more flexible.

D. The determination of associated matrix of demand-quality elements

Uncertain relationship exists between customer demands and service quality elements, which is the correlation matrix of HOQ. Generally, the determination of correlation matrix between hardware products is easy, the simple products can be determined by Delphi method, while the complex products using the method of orthogonal experimental design. Some scholars also combine the rough set theory with experimental design methods, which can get better result. But in the process of constructing service quality house, it is difficult to determinate the correlation matrix, and the experimental design used in hardware products is not suitable, because the number of service quality elements is more than twenty, the number of trials will increase exponentially with the mounting of the quality elements, that will result in massive computation and heavy task. From the customer's point of view, different customer has a different view on the quality elements that can meet their demands. For example, there is a customer demand called "real and reliable product publicity," we can find a lot of quality elements meet this demand from the literature, among many quality elements, some customers think "commodity quality parameters authenticity" can meet this demand, and some customers think "product image authenticity" make them feel at ease when shopping. Of course, the customer options of quality elements to meet this demand may be one or several. So this selecting probability of demand - quality elements can well reflect the relationship between the demand and the quality elements, the greater the probability that the better can these quality elements able to meet certain demands of customers; conversely, it stating that the quality element is only a low degree of meeting the customer's demands.

E. The construction of B2C-HOQ

After the determination of customer demands and its importance, quality elements, correlation matrix, we can build the House of Quality for B2C e-commence. Customer demands and their importance for the left wall, quality elements for ceiling, correlation matrix for room computed quality elements importance for the floor. Because this paper focuses on exploring the quality elements, so the right wall (evaluation matrix) and roof (association matrix between quality elements) of the House of quality are not considered. Through the House we can distinguish the strength of relationship between customer demand and the quality elements, we can also get the important value of quality elements, so you can figure out in the customers ' minds, which quality elements are important when they evaluate an e-commence enterprises, which are not too concerned about.

IV. EMPIRICAL STUDY

A. The research of customer demand

The access of the customer demand is the most important step in the quality function deployment, is also a difficult step. For service products such as e-commerce service, need to survey customer demands, this paper used the method of 5W1H investigated on 80 Jingdong Mall customers, and received a variety of customer demands, and then use the Affinity Diagram to sort the customer demands. 5W1H question items have: WHO (Who? customer's personal information), WHEN (what time or in which situations he would choose the B2C shopping site), WHERE (The site can meet which of your needs), WHAT (experienced what kind of service issues, or what kind of service he expect?), WHY (Why do you have such a service problem or your service expectations?), HOW (how can the site do better to meet your service demand? That's the customer recommended). The statistics of the sample of customer demand information:

item		Number	rate
sex	male	37	46.25%
	female	43	53.75%
age	Bellow 20	8	10%
	20-30	64	80%
	30-40	6	7.5
	above40	2	2.5%
educational level	High school	2	2.5%
	Junior college	10	12.5%
	Bachelor	56	70%
	Master	12	15%
profession	student	60	75%
	staff	11	13.75%
	administrator	3	3.75%
	teacher	4	5%
	unemployed	0	0
	Soldier	2	2.5%
	others	0	0

The results of customer demand Affinity diagram analysis:

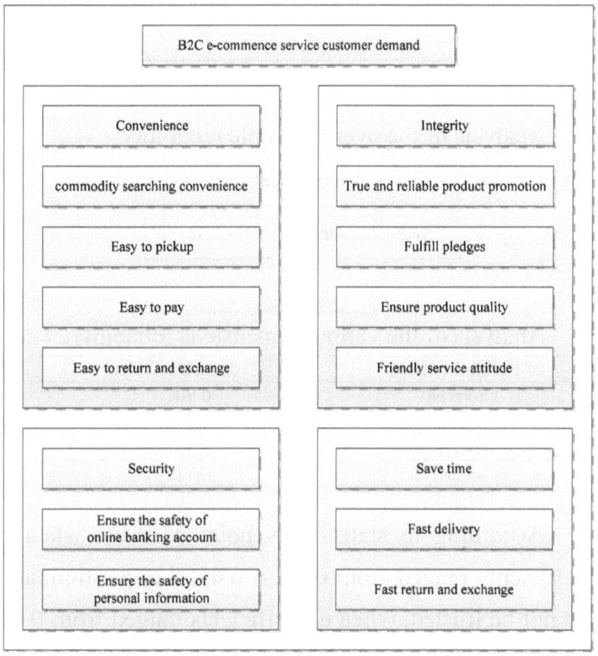

As can be seen from the affinity diagram analysis, the needs of convenience and integrity of B2C e-commerce are most. The first purpose customers choose online shopping is convenience, so convenience is customer demand for all forms of online shopping business, including typical B2C electronic business like Jingdong Mall. And because B2C e-commerce is of large scale, managements are more standardized, goods famous of better quality, therefore more customer demands in these two areas.

B. The survey of customer demands importance

In order to more accurately determine the importance of customer demands, the paper surveyed 80 Jingdong Mall customers, the questions is the important degree of each customer demand, by using five-point Richter Scale, customers think one demand is very important be rated as 5 points, important as 4 points, general assessment as 3 points, unimportant as 2 points, is not very important as1 point. Sample statistics:

item		num	rate
sex	male	35	43.75%
	female	45	56.25%
age	Bellow 20	11	13.75%
	20-30	61	76.25%
	30-40	5	6.25%
	above40	3	3.75%
Educational level	High school	4	5%
	Junior college	21	26.25%
	Bachelor	42	52.5%
	Master	13	16.25%
profession	student	67	83.75%
	staff	8	10%
	administrator	1	1.25%

teacher	2	2.5%
unemployed	0	0
Soldier	2	2.5%
others	0	0

Analysis of the overall sample reliability:

Reliability Statistics

Cronbach's Alpha	N of Items
.902	12

Analysis of the various dimensions reliability:

Reliability item	Cronbach's Alpha
overall	0.902
convenience	0.903
Save time	0.802
security	0.722
integrity	0.763

According to statisticians point of view, when α coefficients ranged from 0.60 to 0.65, the questionnaire can not be trusted, when α coefficients ranged from 0.65 to 0.75, the reliability of the questionnaire is the minimum acceptable value when α coefficients ranged from 0.70 to 0.80, the reliability of the questionnaire is quite good, when α coefficients ranged from 0.80 to 0.90, the reliability of the questionnaire is very good [8]. Statistical analysis results in this paper show that the reliability of this questionnaire is better and can be adopted.

Item Statistics

	Mean	Std. Deviation	N
True and reliable product promotion	4.7397	.55346	80
Fulfill pledges	4.7808	.50680	80
Ensure product quality	4.8493	.43048	80
Friendly service attitude	4.5068	.62635	80
Search products easy	4.3836	.71938	80
Easy to pickup	4.5068	.68967	80
Easy to pay	4.5068	.72884	80
Easy to return and exchange	4.6849	.52379	80
Fast delivery	4.3836	.73843	80
Fast return and exchange	4.4795	.62605	80
Ensure the safety of online banking account	4.9452	.28335	80
Ensure the safety of personal information	4.8493	.43048	80

As can be seen from the average score of the demand, customers seems to demand significant degree between 4.38 ~ 4.95, that is between "important" and "very important ", and therefore more emphasis on these needs. It also illustrates the pre-made customer demands investigation is valid; the customer demands summarized are consumer concerned. As can be seen from the statistics, the most important is network security, "to ensure the security of online banking accounts," the score reached 4.9452, "to ensure the security of personal information," the score reached 4.8493; Although the importance of integrity is not as much as security, its score is also very high, between 4.5~4.9, which have a direct relationship with the network transaction environment, because the e-commerce customer and merchant are not face-to-face, high degree of integrity can increase consumer's psychological trust; the important degree of convenience is between 4.3~4.7, below the security requirements and integrity demand; "fast delivery" is only 4.3836, "Fast return and exchange" is 4.4795, indicating that the customer requirements in time saving are not too high. Because of the fierce competition, B2C e-commerce businesses companies are generally faster in the delivery can be very good to meet the needs of customers.

C. Determination of B2C e-commerce service quality elements

Considering customer demands, evaluation indicators the relevant literature using, customer feedbacks, the paper sorting out the B2C e-commerce service quality elements to meet the customer demands.

Customer demands	Quality elements
commodity searching convenience	Completeness of the product quality parameters [3] [8], Reservation and open of customer evaluation information [5] [9], Webpage loading speed [1] [13], Site map definition [9][12], personalized recommendation service [1][8], diversity of merchandise [12]
Easy to pickup	Delivery on time, order tracking information provided timely [6], delivery point setting rationality
Easy to pay	Page loading speed [1] [4], online banking account protection measures [1] [9] [10], Payment diversity [7] [8], design of payment
Easy to return and exchange	customers waiting time when get help [1][4], convenience of contact online service [7], setting of delivery points, return policy [3], the design of return process
Fast delivery	Delivery on time, order tracking information provided [6], product ready for speed [1][4], delivery speed [1] [4] [8]
Fast return and exchange	return product prepare speed [1] [4], delivery on time, Customer waiting time to get help [1] [4], Online customer service convenience [7] [13], delivery speed [1] [4] [8]
Ensure the	Customer personal information protection measures, Online banking account protection

safety of online banking account	measures
Ensure the safety of personal information	Customer personal information protection measures, Online banking account protection measures
True and reliable product promotion	Completeness of product quality parameters [11], Product quality parameters authenticity [6] [11], Product Image authenticity [6], Reservation and open of customer evaluation information [3] [5]
Fulfill pledges	Customer waiting time to get help [1] [4], Online customer service convenience [7], Online customer service courtesy [5], Consistency of service before and after purchase, delivery ontime
Ensure product quality	Completeness of product quality parameters [11], Product quality parameters authenticity [6] [11], Product Image authenticity [6], return policy [3], Return Process Design, Online customer service courtesy [5], Consistency of service before and after purchase
Friendly service attitude	Reservation and open of customer evaluation information [3] [5], Online customer service courtesy [5], Consistency of service before and after purchase, Online customer service convenience [7]

D. Customer surveys to determine the correlation matrix, build house of quality

What quality elements can meet specific customer demands, there are different views for different customers. So this paper uses the customer survey method to determine the correlation matrix of house of quality, customer choice probability values to represent the demand quality factor correlation degree. According to a survey of 120 customers, obtained relevant data, statistic iced probability value. With the work finished can establish the B2C e-commerce service quality house (B2C-HOQ).

REFERENCES

[1] Hua Ying. Studies on e-commerce service quality evaluation [D]. Beijing: Beijing University of Posts and Telecommunications, 2011.

[2] Zuo Wenming, Yang Wenfu, Huang Qiuping, Xie Zhenli. System and model on B2C e-commerce website service quality evaluation [J]. Journal of Information, 2010

[3] Yu Jia. Study on the relationship between the electronic service quality dimensions of B2C and customer satisfaction [D]. Hefei: Anhui University, 2010

[4] Xiong Wei. Quality Function Deployment - From theory to practice [M]. Beijing: Science Press, 2009.3-4

[5] Xie M, Tan K C, Goh T N, Advanced QFD Applications. Milwankee: ASQ Quality Press, 2003.

[6] Moskowitz H, Kim K J. QFD optimizer: A novice friendly quality function deployment decision support system for optimizing product designs. Computers & Industrial Engineering, 1997, 32(3): 641-655.

[7] Wang Xiaotun, Xiong Wei. Rough AHP in determining customer needs importance of Quality Function Deployment [J]. Computer integrated manufacturing system, 2010

[8] Lin Liqiong. Research on quality management system of Higher Vocational Education Based on QFD [D]. Guangzhou: Guangdong University of Technology, 2011

[9] Chen Chao,Zou Ying. Common functions and application examples of SPSS15.0 (Chinese version) [M]. Beijing: Publishing House of electronics industry, 2008. 337

[10] Lan Qi. Model on e-commerce service quality factors [D]. Chengdu: University of Electronic Science and technology, 2010

[11] Lan qi, Yang Wenbo. Research on description and evaluation methods of e-commerce service quality [D]. Chengdu: University of Electronic Science and technology, 2010

[12] Li Wenwen. Construction of the quantitative model for service quality of online shopping store evaluation [J]. Science and Technology Management Research, 2011

[13] YangYang. The research of service quality evaluation in e-commerce environment [J]. Economic Tribune, 2010

integrity	Consistency of service before and after										0.68	0.36	0.78	8.51	
	Diversity of commo	0.43												2.04	
	Online customer service										0.3	0.23	0.53	4.94	
	Payment design			0.37										1.67	
	Diversity of paymen			0.68										3.07	

High efficiency				responsibility						empathy						security	
Product quality parameters	Product Image authentic	delivery ontim	Reservation and open of customer	Customer waiting time to	Online customer service	Order tracking information	webpage loading speed	delivery speed	return product prepare	Product prepare speed	Site navigation	Personalized recommendation service	Delivery point setting rationality	return policy	Return Process Design	Online banking account protection	Customer personal information protection
			0.47				0.38				0.53	0.33					
		0.53				0.69							0.56				
							0.27									0.61	
				0.47	0.53								0.31	0.58	0.62		
		0.58				0.53		0.73		0.23							
		0.23		0.41	0.35			0.31	0.64								
																0.94	0.78
																0.57	0.97
0.71	0.46		0.6														
		0.61		0.37	0.39												
0.75	0.54													0.52	0.25		
			0.57		0.63												
7	4.8	8.88	7.64	5.81	8.75	5.43	3.02	4.59	2.87	1.01	2.51	1.56	3.98	5.24	4.11	10.17	8.57

	Completeness of product quality	0.56								0.58		0.56		8.12	
Importance of customer demands		4.74	4..51	4.51	4.68	4.38	4.48	4.95	4.85	4.74	4.78	4.85	4.51		
Quality elements / Customer demand		search product easy	easy to pickup	easy too pay	easy to return and exchange	fast delivery	fast return and exchange	Ensure the safety of online banking account	Ensure the safety of personal information	True and reliable product promotion	Fulfill pledges	Ensure product quality	Friendly service attitude	importance of quality elements	weight of quality elements
		convience demand				time saving		security		integrity					

Histogram of quality elements importance:

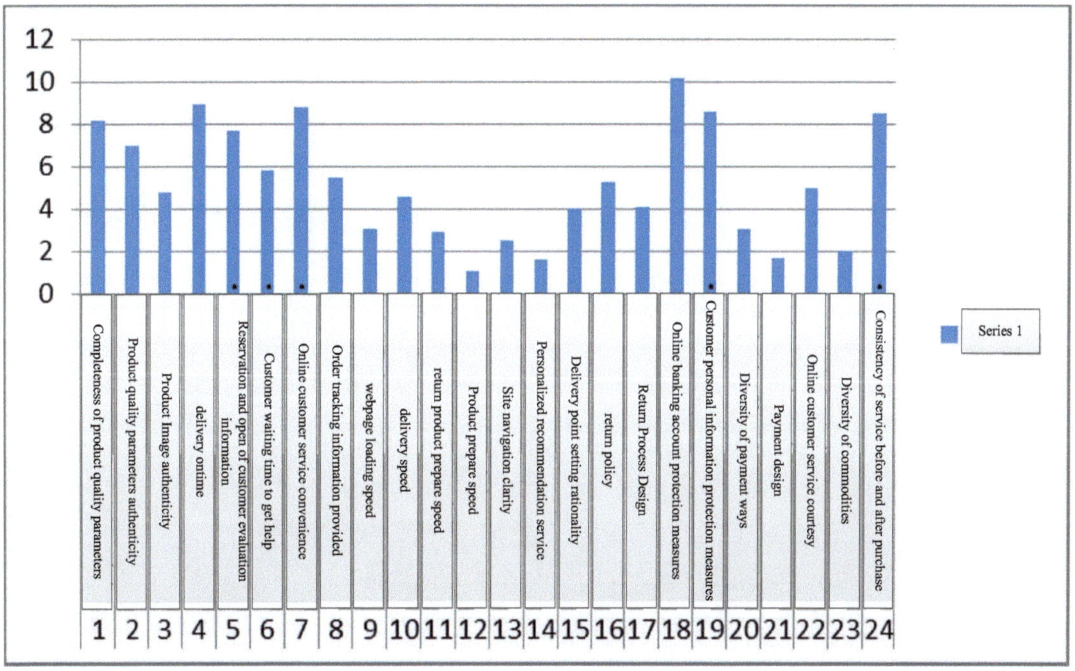

Applying Combined Hierarchical Analysis and Extension Methods for Siting Senior Day-Care Center

M.Y. Ku*, J. J. Jiang, H. P. Tsai, T. J. Ku, C. M. Kuo
Department of Industrial Engineering National Chin-Yi University of Technology Taichung Taiwan, 41170
(guumy@ncut.edu.tw)

Abstract - **A Senior Day-Care Center similar to daycare facilities that allows aging seniors to enjoy daytime living when their adult children are working and return home after office hours will enable a more harmonious family life between the aging parents and their adult children. In this research, the AHP (Analytic Hierarchy Process) theory is applied to calculate the weight of various relevant factors that affect the site selection of a proposed Senior Day-Care Center based on responses from experts to questionnaires so that important affecting factors can be identified. The Extension Excellence Evaluation Method (EM) was then used to select, compare and analyze several appropriate sites. The evaluation model comprises four facets including transportation, community environment, living functions and public facilities. The research results reveal that either the AHP or the EM method is capable of offering better answers.**

Keywords - **AHP, EM method, Senior Day-Care Center, Site selection**

I. INTRODUCTION

"Aging Population" is a world-wide problem to be faced by all developed nations; longer life-span leads to the problem of providing proper cares for seniors. Statistical data published by Directorate-General of Budget [1], Accounting and Statistics, Executive Yuan, R.O.C in 2011 show that the age of about 1.5 million citizens or 7.1% of total population in Taiwan exceeded 65 years in 1993. The senior population, reached 2.5 million in 2010 or 10.5% of the total population, has increased by 1 million in 17 years. Hence, Taiwan is officially considered as an "Ageing Society" so that proving the seniors with a safe life with dignity and their adult children with a worry-free environment is a dream to be pursued by every family.

The concept of nursing homes like those popular in developed nations such as the US that provide 24-hour assisting live-in services is not commonly accepted in the Chinese society. Hence, a senior center that functions like day-care nursery under-aged children to care for seniors during daytime allowing them to learn, rest and associate with one another is a most appropriate future tendency in Taiwan. It is similar to the "senior center" that popular in most US cities but provides more programmatic caring for seniors to alleviate caring burdens to their adult children during daytime. Seniors can be well cared for physically and mentally during daytime instead of being left alone at home while their adult children are working. The location for establishing such a senior day-care center directly affects the willingness of seniors and their family to use the facility and the sustainable management of the center as well. In this research, the factors that affect the siting of an appropriate center to care for ageing seniors are studied so that the center will be established at a location to entice seniors and ensure sustainable management for a win-win situation.

II. LITERATURE REVIEW

The available literature on the definition, investigation and siting of the center will be reviewed in order to provide more insight information on the siting of a proposed Senior Day-Care Center in Taiwan.

A. Investigation on the day-care Seni Day-Care Center:

In this section, the definition of senior day-care center, the importance of siting such a center and other relevant information are reviewed:

1) Definition of Senior Citizen

Article II of Senior Citizen Welfare Act promulgated by the Department of Social Affairs (The Ministry of Interior, Executive Yuan, ROC) [2] stipulates that seniors be defined as those with ages exceeding 65. Lin (1995) [3] pointed out that about 70%- 80% of seniors living in Taiwan are considered the "golden group" because they have good health with the ability to care for themselves.

In this research, the golden-group seniors that cannot be cared for by family members are targeted for conducting the studies.

2) The Senior Day-Care Center

Liao (2010) [4] has defined the senior day-care center as a facility that serves healthy seniors to take care of their daily living and provide various recreational activities such as gardening, exercise, education and social contact.

Ding (2012) [5] has also proposed the concept of "senior day-care center" based on day-care centers for children with the objective of providing day care for seniors as for children. Thus, seniors may congregate at a centralized location fully equipped for living, recreational and social activities. He advocates that the center will allow adult children to work without worries about their senior parents during daytime and also provide seniors with a dignified living during the temporary absence of their adult children.

In this research, the "senior day-care center" is defined as "a facility that targets healthy seniors with self-caring ability to provide daytime assistance, nursing and caring with emphases on sound mental health allowing them to enjoy companionship and recreational activities, leisure daytime living and education (e.g. senior class, gardening, and various recreational activities). After office hours, the seniors will return to their own homes to enjoy family

E. Qi et al. (eds.), *Proceedings of the 5th International Asia Conference on Industrial Engineering and Management Innovation (IEMI2014)*, Proceedings of the International Asia Conference on Industrial Engineering and Management Innovation 1, DOI 10.2991/978-94-6239-100-0_5, © Atlantis Press and the authors 2015

living so that they will have a safe living with dignity and superior quality."

3) Importance of the proposed senior day-care center

Ou (2009) mentioned that the closing of a senior day-care center in Chingshui township in Taichung county, Taiwan was caused by the failure of conducting market survey for satisfying the need of customers and that the concept of senior day-care center had not been accepted by the general public then.

Tseng (1997) stated that changes of family structure in Taiwan under the tendency of having fewer children in a family cause a heavy burden for the younger generation to care for the ageing population. On average, the number of working people to provide for every senior is dropping from 10 in 1992 to 7 in 2010, and is estimated to be 3 in 2030. Hence, establishing senior day-care centers is becoming an important future tendency.

B. Studies on site selection for locating the senior day-care center

As early as in 1909, Weber [6] developed the concept and theory to solve the problem for siting manufacturing plants.

1) Relevant theories for selecting location

According to Chapman and Walker (1991) [7], the theory for locating industries can be classified into the following four categories:

1. Normative industrial location theory: It is also known as Neoclassical Approach that emphasizes cost-minimization analysis, regional market analysis and profit-maximization analysis.
2. Behavioural Approach: Because many assumptions used in the normative industrial location theory do not conform to the real-world situation, a school of using questionnaire to carry out analyses merged in the 1960s by targeting industrial location for conducting empirical research.
3. Geography of Enterprise: Kong (1992) [8] proposed the theme concept for the school of industrial geology based on "how much industries are capable of improving and utilizing their surrounding environment." The conclusion is that the siting strategy is as important as the investing strategy; increasing products or making new products by industries is the cause, and selecting the location is the result.

2) Importance of site selection

Fu (2008) [9] stated that the siting strategy has a profound impact on the facility of an organization; a correct siting policy leads to sustainable prosperity for the organizational facility. Liu and Cheng (2000) [10] proposed that selecting the location will affect production and management cost, future expansion and development of plant, as well as environmental and ecological impacts. An enterprise survives by providing services to satisfy customers using the most economic method. A good location will bring prosperous business, sustainable development and profits to industries and investors whereas a poorly selected location¬ will cause excessively high costs. The business is thus difficult to manage for maintaining profits; it is eventually bankrupted.

3) Pitfall for selecting location

Inappropriate location will cause numerous unnecessary obstacles but mistakes in selecting location are frequent to cause more production cost. Fu and Meng (2004) [11] cited the common reasons for making the mistakes as:

1. Lacking thorough investigation and ignoring factors that need to be considered.
2. Making decisions subjectively based on personal feeling by managers or persons in charge instead of objectivey based on observations and facts.
3. Moving plants by unwilling top officials or managers from a traditional but well-constructed location into a relatively new and better location.
4. Plant being moved to an already crowded region or soon to be over-industrialized region.
5. Preferring structures that are readily available but their design does not comfort to the specifications and needs of the plant in question to cause difficulties for efficient and cost-effective operations in the future.
6. Selecting a location that is less expensive but ignoring the availability of qualified personnel available in the area because of lower cultural level and poorer education for the local residents.

C. The AHP (Analytical Hierarchy Process) method

AHP was proposed by Thomas L. Saaty in 1977 [12] in systemizing complicated problems so that weighing factors that are objective to individuals and cannot be quantified originally can be resolved at various levels, quantified and evaluated. Hsiao (2009) [13] stated that the AHP method uses hierarchical structures to organize various factors and adopts as much as possible all opposing concepts so that well-arranged hierarchical systems can be developed for a complicate project in order to compare factors with different scales.

1) Basic assumptions and procedures of AHP

Satty (1977) proposed the following basic procedures for using AHP to conduct decision analyses:

1. Decision making standards and preparing the hierarchical structure for a project,
2. Determinant matrix based on the weight ratio values of factors,
3. Analyses of the determinant matrix to obtain weight vectors with emphases on examining the determination of the matrix uniformity and solving the weight vectors,
4. Hierarchical integration of the weights for alternative plans, and
5. Final decision making to select the base decision plan or categorize various plans into different classes

The AHP proposed by Saaty (1977) is based on the following 4 basic assumptions:

1. Reciprocal comparison: AHP determinant matrices are positive reciprocal matrices.
2. Homogeneity: Disparities among factors in the same class of the hierarchical structure are not significant.
3. Independence: Evaluation of the relative importance among factors are influenced by the characteristics of selected plans.
4. Expectations: From the viewpoint of making decisions, the whole hierarchical structure is sufficiently complete; hence, major decision making norms that affect the decisions are included.

2) The AHP procedures

Simplified AHP procedures based on reviewing relevant literatures are listed as follows:

1. Establishing hierarchical structure: The second layer is formulated using the norms that need to be considered based on the decision-making problems.
2. Establishing pairs of comparative matrices: Each elements on the diagonal line of each matrix is the comparison of an element with itself so that their values are "1". The value of Aij denotes the relative importance of Ai with respect to Aj, and each layer much be compared by pairs.
3. Evaluating uniformity: Whether the results that are filled in the matrix conform to the uniformity requirement must be evaluated using CI (Consistency Index) and CR (Consistency Ratio). When CI and CR are less than 0.1, the following evaluation formulae are used within a certain acceptable scope:

$$C.I. = (\lambda max - n)/(n-1) \quad (1)$$

$$C.R. = C.I/R.I. \quad (2)$$

Where:

n = numerical criterion

λmax = maximum eigenvalues,

R.I. = Random Index for evaluating a matrix,

R.I. Values are influenced by the magnitude of "n" as shown in Table I.

TABLE I

VARIATION OF RI VALUES WITH RESPECT TO n VALUES

1	2	3	4	5	6	7	8	9	10
.00	.00	.58	.90	1.12	1.24	1.32	1.41	1.45	1.49

Source of information: Saaty (1980, 1986)

4. Determining weight: Numerical values of the established comparative matrix pairs are integrated to calculate eigenvalue vectors, weights in order to obtain the priority of attribute importance for norms and elements.
5. Selecting optimal plan: Weights of various factors for different stages are evaluated and then integrated to obtain the priority values between evaluating norms and projects. The most appropriate or the best project can then be selected based on the magnitude of priority values; greater priority value indicates better project.

D. Extension Method (EM)

EM is a new method proposed by Cai (1983) [14] for solving non-compatible and contradicting problems based on quantitative and qualitative viewpoints; it is capable of extending the scope of fuzzy scope from <0,1> to <-∞,+∞>. Yang and Cai (2000) [15] suggested that EM could be applied in the various fields such as industry, agriculture, military, economics, biology, medicine, design control search, marketing, planning, decision making and other related fields. In this research, the decision making model of EM is implemented to reveal factors that are relevant to evaluating the selection of the most appropriate location of a senior day-care center.

III. METHODOLOGY

In this research, the combined AHP method and the EM goodness evaluation is used for selecting the location to establish a senior day-care center. The AHP method is used to reveal the weights of various factors to be considered for the site selection; the resulting weight values are then included in the EM goodness calculations to select the best project for establishing the senior day-care center.

A. Structure of analyses

In this research, available relevant literature was reviewed first to establish the factors that need to be considered for selecting the location. Two type of questionnaires, one for the concerned general public and another for experienced scholars and experts, were then formulated and distributed. Responses from scholars and experts were evaluated using the AHP method to calculate the weights of various factors whereas responses from the general public were evaluated statistically. The weights for various factors were then used in the subsequent EM goodness calculations for obtaining the best locations. The schematic structural diagram for conducting the evaluation is shown in Fig.1:

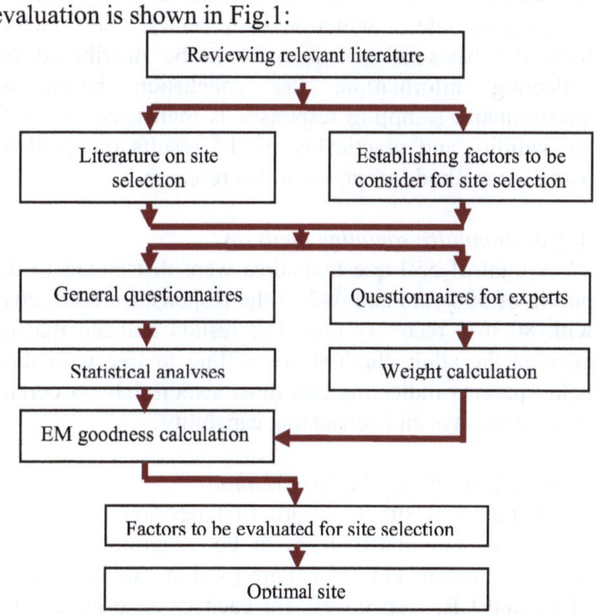

Fig.1. Schematic structural diagram for conducting the evaluation

B. Research Model

In this research, the factors that are considered for selecting the site of a senior day-care center are grouped in four concerned facets as shown in Fig. 2:

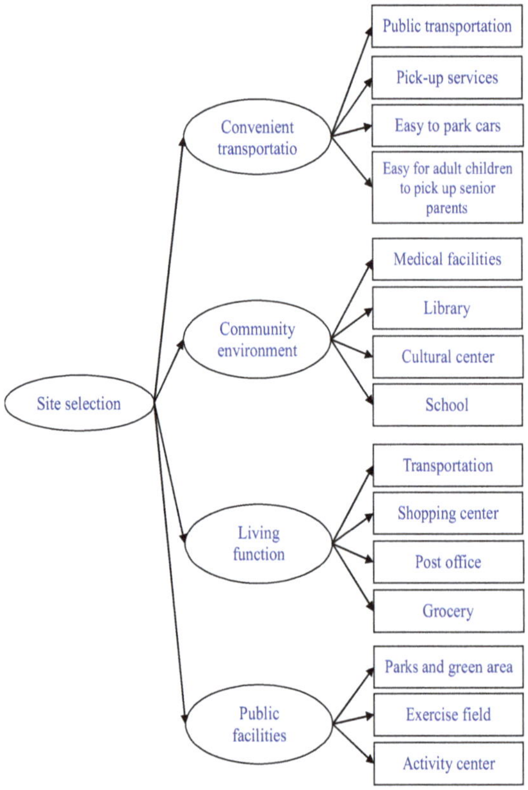

Fig.2. The four facets of factors to be considered for selecting the site of senior day-care

IV. VERIFYING RESULTS OF DATA ANALYSES

The relevant information in literature is reviewed and summed up to reveal the factors that are important to selecting the site of senior day-care center. These factors form the basis of questionnaire to be distributed for collecting information; the conclusion based on questionnaire sampling responses is then used to verify the validity and practicality of the results obtained by using the methods proposed in this research.

A. Questionnaire sampling analyses

A total of 550 questionnaires were distributed to the public of 50 years and 442 of the responses are effective with 80.36% recovery rate. The results indicate that on average the adult children are willing to live with their aging parents indicating that most seniors have a certain level of savings and consuming capability.

B. The Hierarchical Analysis Method

Hierarchical structures are first developed and then developed into matrix pairs to be integrated into two layers. Equations (1) and (2) are used to calculate values of CI and CR, respectively for each layer to assure that the results have consistency.

TABLE II
THE AHP-BASED ANALYSES ON THE OPTIMAL LOCATION

Project	A	B	C	D	E
Weighted Grades	0.1094	0.1641	0.2188	0.2735	0.0547
	0.2397	0.2397	0.2397	0.2397	0.2397
	0.1061	0.0796	0.0531	0.1326	0.0531
	1.0529	1.0529	1.0529	1.0529	1.0529
	1.1903	0.9522	0.2381	0.7142	0.4761
	0.0980	0.1307	0.1634	0.0653	0.0327
	0.1113	0.0742	0.0371	0.1485	0.1856
	0.0978	0.0326	0.0652	0.1303	0.1629
	0.0212	0.0170	0.0042	0.0127	0.0085
	0.0223	0.0179	0.0045	0.0089	0.0134
	0.1294	0.1294	0.0776	0.0259	0.0518
	0.3173	0.7933	0.7933	0.4760	0.1587
	0.0815	0.1087	0.1359	0.0544	0.0272
	0.1987	0.3975	0.3975	0.4968	0.1987
Total	3.7761	4.1899	3.4813	3.8319	2.7159
Ranking	3	1	4	2	5

C. The EM method

The EM method is applied to further process the AHP weighted into final results that are based on to identify the optimal location for establishing the proposed Senior Day-Care Center in Taichung using the 5 case-study projects selected.

The final evaluation results will lead to revealing the optimal projects and the poorest project that should never be adopted. This evaluation demonstrates the capability of the combined AHP and EM method to different the advantages and disadvantages of projects that are characterized by conflict and incompatible features.

D. Comparison between AHP method and the combination of AHP and EM method

TABLE III
COMPARISON BETWEEN THE AHP METHOD AND THE COMBINED AHP AND EM METHOD

Project	A	B	C	D	E
AHP	**3.7761**	**4.1899**	**3.4813**	**3.8319**	**2.7159**
	3	1	4	2	5
Combined AHP and EM	Fair	Suprtiot	Suprtiot	Poor	Very poor

The AHP method and the combined AHP and EM method may result in different priority for the various factors that influence the site selection for the proposed senior day-care center. The difference is caused by the fact that the AHP method is merely based on responses from experts whereas the AHP and EM method combines responsed from both experts and the general public.

V. CONCLUSION

The results obtained in this research show that both the AHP method alone and the combined AHP and EM are capable of revealing the optimal location. However, the combined method can be used to identify the best location so that it is obviously superior with more freedom for conducting the evaluation that keeps some degree of subjective selection with objective results. Thus, the combined AHP and EM method is a good evaluation method with the capability of differentiating good choice from bad choice for solving the contradictory and incompatible problem of selecting the location for the proposed Senior Day-Care Center.

Results of the questionnaire investigation reveal that most adult children in Taiwan are willing to live with their aging parents. Further, most aging parents own a certain level of cash savings and consuming ability so that establishing Senior Day-Care Center will be a future trend in Taiwan.

REFERENCES

[1] Directorate-General of Budget, Accounting and Statistics, Executive Yuan, R.O.C. (Taiwan), (2011), National Statistics, Republic of China, No.188, http://www.stat.gov.tw/public/Data/110418133071.pdf

[2] Department of Social Affairs, Welfare services (2007), Senior Citizens Welfare Act, http://sowf.moi.gov.tw/04/new04.asp

[3] Lin, Tsung-I, (1995), Elderly plight and wish in Taiwan: The main topics of the twenty-first century city, Taiwan of Mental Health Sciences, 8(2), 7-15

[4] Jhuo-Wei Liao, (2010), Apply Structural Equation Modeling and Fuzzy TOPSIS Method to Select Location Factor for Elderly Care Centers Facilities, Department of Industrial Engineering and Management National Chin-Yi University of Technology

[5] Pei-Yu Ding, (2012), Apply Structural Equation Modeling and Extension Method to Inspections into Location Selection Factors for Adult Day Care Centers, Department of Industrial Engineering and Management National Chin-Yi University of Technology

[6] Weber, A., (1909). "Uber den standort der industrien." Tubingen, Germany: J. C. B. Mohr. 1971. Theory of the location of industries.

[7] Chapman, K., & Walker, D. F. (1991). Industrial Location (2nd ed.). Cornwall: T. J. Press Ltd.

[8] Fan-chih Kung, (1992), The Study for Spatial Distribution Model of High-tech Industries in Taiwan, Department of Urban Planning, NCKU.

[9] H.Y. Fu, Production and Operations Management - building products and services benchmarking, Future Career Management Corporation, Taipei (2008)

[10]W J, Chen, S T, Liu, Facilities planning - plant layout and material handling, The ZTE management consultant company, Taipei (2000)

[11]H.Y. Fu, Chao-Yu Meng, (2004), Facilities Planning, Future Career Management Corporation, FCMC。

[12]Saaty, T.L., (1977), A scaling method for priorities in hierarchical structures, Journal of Mathematical Psychology 15 (3), 234–281.

[13]Hsiao, (2009). AHP method to investigate the industry's leading international express customers selection index weights decision analysis, Department of Bussiness Administration

[14]Cai, W., (1983). "Extension Set and Incompatible Problems," Science Exploration, 3(1), pp. 83-97

[15]Chun-Yan Yang, Wen Cai, (2000), Extension engineering Methods, China Engineering Science, 2 (12)

A Prospect Theory Based Approach to Heterogeneous Multiple Attribute Group Decision Making

Rong Liu[1], Yu Luo[2,*]
[1] Chongqing Industry Polytechnic College, Chongqing, China
[2]Chongqing University of Posts and Telecommunications, Chongqing, China
(luoyu@cqupt.edu.cn)

Abstract - **Heterogeneous multiple attribute group decision making (HMAGDM) problems are very complex and important in practical applications of decision making theory. In this paper, we proposed a prospect theory based method to solve HMAGDM problems. Specifically, the selection process under heterogeneous context is first proposed to obtain the individual ranking of alternatives. Then, the preference-approval structures are used to determine the reference points of the prospect theory. Next, the individual prospect values of the alternatives are calculated based on the prospect theory and the established reference points. Finally, the collective prospect values of the alternatives are derived to rank the alternatives.**

Keywords - **Heterogeneous context, multiple attribute group decision making, preference-approval structures, prospect theory**

I. INTRODUCTION

Multiple attribute group decision making (MAGDM) is an important research field of decision science, operational research and management science. Numerous approaches [1, 2] have been proposed to handle MAGDM problems. However, in some real decision situations, different decision makers may come from different areas with distinct knowledge and experience and use different judgments and evaluating methods. Evaluation information provided by different decision makers may be not only crisp, vague or uncertain but also various, including linguistic labels, interval numbers and real numbers. Such a type of MAGDM problems with various formats of information is called as heterogeneous MAGDM problems [3, 4], which is denote as HMAGDM. To solve HMAGDM problems are very complex and interesting in practical applications of decision making theory.

Traditionally, the existing approaches in MAGDM are focus on the mathematical optimal solutions to decision problems and don't consider the decision maker's psychological behavior. However, much empirical evidence [5, 6] has shown that the decision maker's psychological behavior would play an important role in decision analysis. Therefore, how to find the satisfactory solutions [7] involving the decision maker's psychological behavior has been an important problem in decision making. Particularly, in this paper, the prospect theory, which is one of the most influential psychological behavior theories, has been presented in [5, 8], is introduced into HMAGDM problems.

The purpose of this paper is to put forward a novel approach to solve the HMAGDM, in which the decision maker's psychological behaviors are considered. This approach seeks to maximize the satisfactory of all decision makers, based on the prospect theory. The rest of our paper is organized as follows. Section 2 introduces the preliminary knowledge regarding the interval numbers, linguistic variables and the prospect theory. HMAGDM problems are formulated in Section 3. Following this, a prospect theory based approach for solving HMAGDM problems is proposed in Section 4. Subsequently, Section 5 provides an illustrative example. Finally, concluding remarks are included in Section 6.

II. PRELIMINARIES

This section introduces some basic knowledge regarding the interval numbers, linguistic variables and the prospect theory, which will provide a basis for this study.

A. Interval numbers

An interval number is denoted by $I=[a, b]$, where a and b are the left and the right limits of interval I, respectively, and $a \le b$.

Let $I_1=[a_1, b_1]$ and $I_1=[a_2, b_2]$ be two arbitrary interval numbers, the distance between I_1 and I_2 is defined by [9]

$$d(I_1,I_2)=\sqrt{\frac{(a_1-a_2)^2+(b_1-b_2)^2}{2}}. \tag{1}$$

B. Linguistic variables

Let $S=\{s_i \mid i=0, 1, ..., g\}$ be a linguistic term set with odd cardinality, where the term s_i represents a possible value for a linguistic variable, and $s_i > s_j$ if and only if $i > j$. We denote $I(s)$ as the position index (or lower index) of s in S. For two linguistic variables s_i

E. Qi et al. (eds.), *Proceedings of the 5th International Asia Conference on Industrial Engineering and Management Innovation (IEMI2014)*, Proceedings of the International Asia Conference on Industrial Engineering and Management Innovation 1, DOI 10.2991/978-94-6239-100-0_6, © Atlantis Press and the authors 2015

and $s_j \in S$, Xu [10] defines the distance between s_i and s_j as follows:

$$d(s_i, s_j) = \frac{|i-j|}{g+1}. \tag{2}$$

C. Prospect theory

The prospect theory was initially proposed by Kahneman and Tversky [5] as a descriptive theory for forecasting individual's actual decision behavior under risk. Since then, a number of researchers have been investigated the prospect theory [8, 11, 12]. When using prospect theory in decision making process, two phases are composed. Namely, editing phase and evaluation phase. The outcomes of alternatives are coded as gains or losses relative to a reference point in editing phase. And in evaluation phase, the edited prospects are evaluated by a value function and a weighting function, and the alternative with highest prospect value is chosen [13]. The prospect theory consists of three crucial behavioral principles [5, 6], which are introduced below:

1) *Reference dependence.* The decision makers perceive the outcomes as gains or losses according to a reference point. Thus, the value function is divided into the gain domain and the loss domain relative to the reference point.
2) *Diminishing sensitivity.* The decision makers exhibit risk-averse tendency for gains and risk-seeking tendency for losses. According to the principle of diminishing sensitivity, the prospect value function is concave in the loss domain and convex in the gain domain, i.e., the marginal value of both gains and losses is decreasing with the size.
3) *Loss aversion.* The decision makers are more sensitive to losses than to totally identical gains [1]. In accordance with the principle of loss aversion, the prospect value function is steeper in the loss domain than in the gain domain.

In accordance with the above three principles, an S-shaped value function of the prospect theory is illustrated in Fig.1. In Fig.1, x denotes the gain ($x > 0$) or the loss ($x < 0$) of the outcome relative to the reference point. This form of function is given by Kahneman and Tversky [6]:

$$v(x) = \begin{cases} x^{\alpha}, & x \geq 0 \\ -\lambda(-x)^{\beta}, & x < 0 \end{cases} \tag{3}$$

In the function, α and β are the two parameters which determine the concavity and convexity of the function, respectively, and $0 \leq \alpha, \beta \leq 1$. λ is the coefficient of loss aversion, $\lambda > 1$.

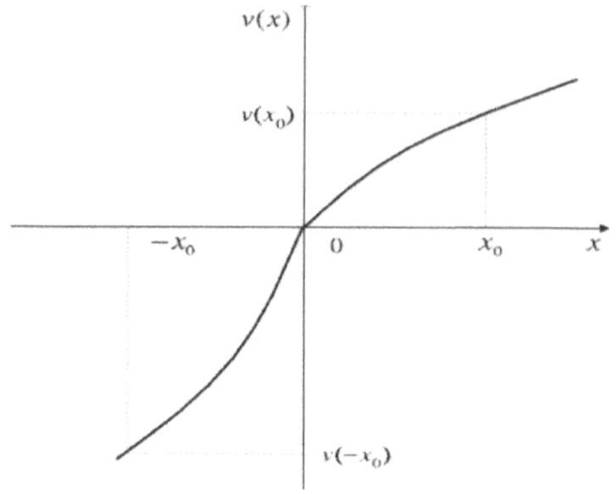

Fig.1. An S-shaped value function

III. HMAGDM PROBLEMS

Assume that there exists a group consisting of m decision makers d_k $(k = 1,...,m)$, denoted by $D = \{d_1,...,d_m\}$. Suppose that the weight vector over D is $\lambda = (\lambda_1, \lambda_2,...,\lambda_m)^T$, where $\lambda_t \in [0,1]$ is the weight of decision maker d_k, and $\sum_{k=1}^{m} \lambda_k = 1$. Let $X = \{x_1, x_2,..., x_n\}$ be a set of n alternatives. The group D has to choose one or rank n feasible alternatives x_i $(i = 1,2,...,n)$ based on m attributes a_j $(j = 1,2,...,K)$, both quantitatively and qualitatively. Denote the sets of attributes by $A = \{a_1, a_2,..., a_K\}$. Let $w = (w_1, w_2,..., w_K)^T$ be weight vector over A, where $w_j \in [0,1]$ is the weight of attribute a_j, and $\sum_{j=1}^{K} w_j = 1$.

In actual decision making situations, four types of attributes are often involved, i.e., real numbers, interval numbers, linguistic labels and fuzzy numbers. In this paper, we consider only three types of attributes, i.e., real numbers, interval numbers and linguistic variables. Let A^1, A^2 and A^3 be three subsets of A, representing the attributes whose values are denoted by real numbers, interval numbers and linguistic variables, respectively. Thus, $A^1 \cup A^2 \cup A^3 = A$ and $A^i \cap A^j = \phi$ $(i, j = 1,2,3.\ i \neq j)$, where ϕ is an empty set.

Let $V^{(k)} = [v_{ij}^{(k)}]_{n \times K}$ be a heterogeneous evaluation matrix, where $v_{ij}^{(k)}$ denotes the preference value assigned to the alternative x_i with respect to the attribute a_j provided by the decision maker d_k. For convenience, $v_{ij}^{(k)}$ can be represented as

$$v_{ij}^{(k)} = \begin{cases} r_{ij}^{(k)} \in R, & \text{if } a_j \in A^1 \\ [a_{ij}^{(k)},\ b_{ij}^{R(k)}], & \text{if } a_j \in A^2 \\ s_{t,ij}^{(k)} \in S, & \text{if } a_j \in A^3 \end{cases} \quad (4)$$

Since the physical dimensions and measurements of the K attributes are different, so the so the evaluation matrices $V^{(k)} = [v_{ij}^{(k)}]_{n\times K}$ $(t=1,2,...,m)$ need to be normalized. Let $\overline{V}^{(k)} = [\overline{v}_{ij}^{(k)}]_{n\times K}$ be the normalized evaluation matrix associated with the $V^{(k)} = [v_{ij}^{(k)}]_{n\times K}$ $(k=1,2,...,m)$.

$$\overline{v}_{ij}^{(k)} = \overline{r}_{ij}^{(k)} = \begin{cases} r_{ij}^{(k)} / r_{ij,\max}^{(k)}, & \text{if } a_j \in A^{1,b} \\ 1 - r_{ij}^{(k)} / r_{ij,\max}^{(k)}, & \text{if } a_j \in A^{1,c} \end{cases} \quad (5)$$

where, $r_{ij,\max}^{(k)} = \max\{r_{ij}^{(k)} \mid i=1,2,...,n\}$, $A^{1,b}$ and $A^{1,c}$ are subsets of benefit attributes and cost attributes belonging to A^1, respectively.

$$\overline{v}_{ij}^{(k)} = [\overline{a}_{ij}^{(k)}, \overline{b}_{ij}^{(k)}] = \begin{cases} [\frac{a_{ij}^{(k)}}{b_{ij,\max}^{(k)}}, \frac{b_{ij}^{(k)}}{b_{ij,\max}^{(k)}}], & \text{if } a_j \in A^{2,b} \\ [1-\frac{b_{ij}^{(k)}}{b_{ij,\max}^{(k)}}, 1-\frac{a_{ij}^{(k)}}{b_{ij,\max}^{(k)}}], & \text{if } a_j \in A^{2,c} \end{cases} \quad (6)$$

where, $b_{ij,\max}^{(k)} = \max\{b_{ij}^{(k)} \mid i=1,\ 2,\ ...,\ n\}$, $A^{2,b}$ and $A^{2,c}$ are subsets of benefit attributes and cost attributes belonging to A^2, respectively.

$$\overline{v}_{ij}^{(k)} = \overline{s}_{t,ij}^{(k)} = s_{t,ij}^{(k)},\ \text{if } a_j \in A^3 \quad (7)$$

IV. METHODS and PROCEDURES for HMAGDM

This section proposes a prospect theory based approach to solve HMAGDM problem. The basic principle of the method is that the best alternative should have the shortest distance from positive ideal solution (PIS) and farthest distance from negative ideal solution (NIS) [14, 15]. The main principles and procedure are summarized as follows:

Step 1: Normalize the evaluation matrices. The evaluation matrices $V^{(k)} = [v_{ij}^{(k)}]_{n\times K}$ $(k=1,2,...,m)$ can be transformed into the normalized evaluation matrices $\overline{V}^{(k)} = [\overline{v}_{ij}^{(k)}]_{n\times K}$ by using the Eqs. (5)-(7).

Step 2: Construct the positive ideal solution (PIS) and the negative ideal solution (NIS) for the decision makers. Let the PIS and NIS be $x^{(k+)} = \{v_1^{(k+)},..., v_K^{(k+)}\}$ and $x^{(k-)} = \{v_1^{(k-)},..., v_K^{(k-)}\}$ associated with the decision maker d_k, respectively.

If $a_j \in A^1$, then $v_j^{(k+)} = r_j^{(k+)} = \max_i\{\overline{r}_{ij}^{(k)}\}$ and $v_j^{(k-)} = r_j^{(k-)} = \min_i\{\overline{r}_{ij}^{(k)}\}$. If $a_j \in A^2$, then $v_j^{(k+)} = [a_j^{(k)},\ b_j^{(k)}] = [\max_i\{\overline{a}_j^{(k)}\}, \max_i\{\overline{b}_j^{(k)}\}]$ and $v_j^{(k-)} = [a_j^{(k)},\ b_j^{(k)}] = [\min_i\{\overline{a}_j^{(k)}\}, \min_i\{\overline{b}_j^{(k)}\}]$. If $a_j \in A^3$, then $v_j^{(k+)} = s_j^{(k+)} = \overline{s}^{(k)}_{\max_i I(\overline{s}_{t,ij})}$, and $v_j^{(k+)} = s_j^{(k+)} = \overline{s}^{(k)}_{\max_i I(\overline{s}_{t,ij})}$.

Step 3: Compute the distances between each alternative and the PIS as well as the NIS for the decision makers. Let $d(\overline{v}_{ij}^{(k)}, v_j^{(k+)})$ and $d(\overline{v}_{ij}^{(k)}, v_j^{(k-)})$ $(k=1,\ 2,\ ...,\ m)$ be the distance between the alternatives x_i and $x^{(k+)}$ and $x^{(k-)}$ with respect to the attribute $a_j \in A$, respectively.

If $a_j \in A^1$, then $d(\overline{v}_{ij}^{(k)}, v_j^{(k+)}) = \left|\overline{r}_{ij}^{(k)} - r_j^{(k+)}\right|$ and $d(\overline{v}_{ij}^{(k)}, v_j^{(k-)}) = \left|\overline{r}_{ij}^{(k)} - r_j^{(k-)}\right|$. If $a_j \in A^2$, then $d(\overline{v}_{ij}^{(k)},\ v_j^{(k+)})$ and $d(\overline{v}_{ij}^{(k)},\ v_j^{(k-)})$ are calculated based on Eq. (1). If $a_j \in A^3$, then $d(\overline{v}_{ij}^{(k)},\ v_j^{(k+)})$ and $d(\overline{v}_{ij}^{(k)},\ v_j^{(k-)})$ are calculated based on Eq. (2).

Let $D^{(k+)}(x_i,\ x^{(k+)})$, $D^{(k-)}(x_i,\ x^{(k-)})$ $(k=1,\ 2,\ ...,\ m)$ denote the distances between the alternatives x_i and $x^{(k+)}$, $x^{(k-)}$ respectively, where

$$D^{(k+)}(x_i, x^{(k+)}) = \sqrt{\sum_{j=1}^{K} (w_j d(\overline{v}_{ij}^{(k)}, v_j^{(k+)}))^2} \quad (8)$$

$$D^{(k-)}(x_i, x^{(k-)}) = \sqrt{\sum_{j=1}^{K} (w_j d(\overline{v}_{ij}^{(k)}, v_j^{(k-)}))^2} \quad (9)$$

Step 4: obtain the individual ranking of alternatives.

Let $c^{(k)} = (c_1^{(k)}, c_2^{(k)}, ..., c_n^{(k)})^T$ $(k=1,\ 2,\ ...,\ m)$, where $c_i^{(k)}$ denote the relative closeness degree of the alternative $x_i^{(k)}$ to the positive ideal solution, where

$$c_i^{(k)} = \frac{D^{(k-)}(x_i,\ x^{(k-)})}{D^{(k-)}(x_i,\ x^{(k-)}) + D^{(k+)}(x_i,\ x^{(k+)})} \quad (10)$$

Let $O^{(k)} = (o_1^{(k)}, o_2^{(k)}, ..., o_n^{(k)})^T$ be the individual ranking of the alternatives X associated with the decision maker d_k, where $o_i^{(k)}$ is the position of $c_i^{(k)}$ in $c^{(k)}$. Normalize the $c^{(k)} = (c_1^{(k)}, ..., c_n^{(k)})^T$ yields the $\overline{c}^{(k)} = (\overline{c}_1^{(k)}, ..., \overline{c}_n^{(k)})^T$, where

$$\overline{c}_i^{(k)} = \frac{c_i^{(k)}}{\sum_{i=1}^{n} c_i^{(k)}} \tag{11}$$

Step 5: Determine the reference points.

In this step, the decision makers need to provide their own preference-approval information according to the individual ranking of alternatives, then the preference-approval structures [16, 17] are expressed as:

$$PA^{(k)} = \frac{x_{\sigma(1)}^{(k)}, x_{\sigma(2)}^{(k)}, ..., x_{\sigma(l)}^{(k)}}{x_{\sigma(l+1)}^{(k)}, x_{\sigma(l+2)}^{(k)}, ..., x_{\sigma(n)}^{(k)}} \tag{12}$$

where $(\sigma(1), \sigma(2), ..., \sigma(n))$ is a permutation of $(1,2,...,n)$, and $x_{\sigma(i)}^{k} \in X$, $x_{\sigma(i+1)}^{k} \geq x_{\sigma(i)}^{k}$, i.e., if x_i is the j *th* largest based on the preference information provided by the decision maker e_k, then $x_{\sigma(j)}^{k} = x_i$.

The alternatives above the dash line are satisfactory (good), and the alternatives below the dash line are dissatisfactory (bad). Obviously, the reference point of the decision maker e^k is between $x_{\sigma(l)}^{k}$ and $x_{\sigma(l+1)}^{k}$. Suppose that $x_{\sigma(l)}^{(t)} = x_i$ and $x_{\sigma(l+1)}^{k} = x_j$. Let $c^{(k*)}$ be the reference point associated with the decision maker d_k, where

$$c^{(k*)} = \frac{\overline{c}_i^{(k)} + \overline{c}_j^{(k)}}{2} \tag{13}$$

Step 6: Calculate the individual prospect values of alternatives. Let $G^{(k)} = \{G_1^{(k)}, G_2^{(k)}, ..., G_n^{(k)}\}$ and $L^{(k)} = \{L_1^{(k)}, L_2^{(k)}, ..., L_n^{(k)}\}$, where, $G_i^{(k)}$ and $L_i^{(k)}$ be the gain and loss of the alternative x_i associated with the decision maker d_k, respectively.

$$G_i^{(k)} = \begin{cases} \overline{c}_i^{(k)} - c^{(k*)}, & \text{if } \overline{c}_i^{(k)} \geq c^{(k*)} \\ 0, & \textit{otherwise} \end{cases} . \tag{14}$$

$$L_i^{(k)} = \begin{cases} \overline{c}_i^{(k)} - c^{(k*)}, & \text{if } \overline{c}_i^{(k)} < c^{(k*)} \\ 0, & \textit{otherwise} \end{cases} . \tag{15}$$

Then, the prospect value of the alternative x_i associated with the decision maker d_k is given by:

$$PV_i^{(k)} = (G_i^{(k)})^{\alpha} + [-\lambda(-L_i^{(k)})^{\beta}], \quad \mathrm{i} = (1,2,...,\mathrm{n}) \tag{16}$$

where $0 \leq \alpha, \beta \leq 1$, and $\lambda > 1$. The smaller α is, the greater risk aversion in the gain domain is, and the smaller β is, the greater risk seeking in the loss domain is. About the values of α, β and λ, Tversky and Kahneman [6] have carried out a series of experiments to determine them. They discovered that the median values of α, β are both 0.88, and the value of λ is 2.25. Following that, some scholars obtained the same values for the parameters by experiments [11] [12]. Therefore, in this study, we take the values given by Tversky and Kahneman [6].

Step 7: Rank the alternatives.

Let OPV_i be the overall prospect value of the alternative x_i, which is computed by

$$OPV_i = \sum_{k=1}^{m} \lambda_k PV_i^{(k)} \tag{17}$$

Obviously, the lager the OPV_i indicates the more satisfactory the alternative x_i is. Thus, the collective ranking of the alternatives is obtain according to OPV_i $(i = 1, 2, ..., n)$.

V. ILLUSTRATIVE EXAMPLE

In this section, an example is provided to illustrate the use of the proposed approach. A company wants to choose a material supplier. Three decision makers, i.e., a technical consultant, a project manager and a purchasing manager, are invited to form a decision committee. The three decision makers are denoted as d_1, d_2 and d_3. After the investigation and survey, the four possible suppliers (x_1, x_2, x_3, x_4) are nominated as candidates. The committee finalizes the attributes after deep discussion. The attributes considered include:

a_1 : price, a_2 : order fulfillment rate, a_3 : quality assurance, a_4: after sales service.

where, the values of a_1 and a_2 are expressed by real numbers and interval numbers, respectively. Both the values of a_3 and a_4 are expressed by linguistic variables. Among the four attributes, attribute a_1 is of cost type, and the rest are of benefit type. Suppose that the linguistic term set used by the decision makers is as follows:

$S = \{s_0 = null, s_1 = very\ low, s_2 = low, s_3 = medium$
$s_4 = high, s_5 = very\ high, s_6 = perfect\}$.

Suppose that $\lambda = (1/3,\ 1/3,\ 1/3)^T$ and $w = (0.35, 0.2, 0.25, 0.2)^T$. The $V^{(1)}$, $V^{(2)}$ and $V^{(3)}$ are list as Tables I, II, III.

TABLE I
THE EVALUATION MATRIX $V^{(1)}$ PROVIDED BY d_1

	a_1	a_2	a_3	a_4
x_1	45	[0.95, 1]	s_4	s_3
x_2	48	[0.92,0.98]	s_5	s_4
x_3	41	[0.94, 0.96]	s_3	s_3
x_4	47	[0.96, 1]	s_6	s_5

TABLE II
THE EVALUATION MATRIX $V^{(2)}$ PROVIDED BY d_2

	a_1	a_2	a_3	a_4
x_1	45	[0.95, 1]	s_5	s_4
x_2	48	[0.92,0.98]	s_4	s_6
x_3	41	[0.94, 0.96]	s_5	s_5
x_4	47	[0.96, 1]	s_4	s_5

TABLE II
THE EVALUATION MATRIX $V^{(3)}$ PROVIDED BY d_3

	a_1	a_2	a_3	a_4
x_1	45	[0.95, 1]	s_5	s_4
x_2	48	[0.92,0.98]	s_5	s_5
x_3	41	[0.94, 0.96]	s_4	s_5
x_4	47	[0.96, 1]	s_6	s_4

Firstly, using steps 1-4, we can obtain that

$\overline{c}^{(1)} = (0.1643, 0.2829, 0.1586, 0.3943)^T$,

$\overline{c}^{(2)} = (0.2094, 0.2510, 0.3685, 0.1711)^T$,

$\overline{c}^{(3)} = (0.2321, 0.2234, 0.2377, 0.3068)^T$,

$O^{(1)} = (3,2,4,1)^T$,

$O^{(2)} = (3,2,1,4)^T$,

$O^{(3)} = (3,4,2,1)^T$,

According to the obtained individual ranking of the alternatives, assume that the decision makers give their preference-approval structures as follows:

$e_1 : PA^{(1)} = \dfrac{x_4, x_2, x_1}{x_3}$, $e_2 : PA^{(2)} = \dfrac{x_3, x_2}{x_1, x_4}$

$e_3 : PA^{(3)} = \dfrac{x_4, x_3, x_1}{x_2}$.

Then, compute $c^{(1*)}$, $c^{(2*)}$ and $c^{(3*)}$ according to Eq. (13), i.e.,

$c^{(1*)} = 0.1614$, $c^{(2*)} = 0.2302$, $c^{(3*)} = 0.2278$.

Next, based on Eqs. (14) - (15) obtains the gain and loss of the alternatives, i.e.,

$G^{(1)}=\{0.0029,\ 0.1214,\ 0,\ 0.2328\}$,

$L^{(1)}=\{0,\ 0,\ -0.0029,\ 0\}$,

$G^{(2)}=\{0,\ 0.0208,\ 0.1383,\ 0\}$,

$L^{(2)}=\{-0.0208,\ 0,\ 0,\ -0.0590\}$,

$G^{(3)}=\{0.0044,\ 0,\ 0.0099,\ 0.0790\}$,

$L^{(3)}=\{0,\ -0.0044,\ 0,\ 0\}$.

Use Eq. (16) to obtain the $PV_i^{(t)}$ $(t = 1,2,3.\ i = 1,2,3,4)$, i.e.,

$PV_1^{(1)} = 0.0058$, $PV_2^{(1)} = 0.1564$,

$PV_3^{(1)} = -0.0130$, $PV_4^{(1)} = 0.2773$,

$PV_1^{(2)} = -0.0745$, $PV_2^{(2)} = 0.0331$,

$PV_3^{(2)} = 0.1754$, $PV_4^{(2)} = -0.1865$,

$PV_1^{(3)} = 0.0084$, $PV_2^{(3)} = -0.0188$,

$PV_3^{(3)} = 0.0172$, $PV_4^{(3)} = 0.1072$.

Finally, using Eq. (17), the overall prospect values of the alternatives x_1, x_2, x_3, x_4 are obtained, respectively.

$OPV_1 = -0.0201$, $OPV_2 = 0.0569$, $OPV_3 = 0.0599$, $OPV_4 = 0.0660$.

Therefore, the most desirable supplier is x_4. The collective ranking of the alternatives is $x_4 \succ x_3 \succ x_2 \succ x_1$.

VI. CONCLUSION

In this paper, we proposed a novel approach based on prospect theory to solve HMAGDM problems. We first designed the method under heterogeneous context to obtain the individual ranking of alternatives. Based on this, the preference-approval structures are provided by the decision makers to determine the reference points. Then, the individual prospect values of alternatives are obtained. Finally, we derived the collective prospect values of alternatives by aggregating the individual prospect values of alternatives.

The main improvement of the proposed approach is that the decision makers' psychology behaviors are incorporated in decision process, and thus the alternative with highest prospect value is obtained. In terms of future research, the proposed approach can be extended to support MAGDM problems in which the attributes are in other forms, such as fuzzy numbers.

REFERENCES

[1] S.H. Kim, S.H. Choi, J.K. Kim, An interactive procedure for multiple attribute group decision making with incomplete information: Range-based approach, European Journal of Operational Research 118 (1999) 139-152.

[2] C. Macharis, L. Turcksin, K. urcksin, Multi actor multi criteria analysis (MAMCA) as a tool to support sustainable decisions: State of use, Decision Support Systems 54 (2012) 610–620.

[3] D.F. Li, Z.G. Huang, G.H. Chen, A systematic approach to heterogeneous multiattribute group decision making, Computers & Industrial Engineering 59 (4) (2010) 561-572.

[4] J.B. Yang, P. Sen, A general multi-level evaluation process for hybrid MADM with uncertainty, IEEE Transactions on Systems, Man, and Cybernetics 24 (10) (1994) 1458-1473.

[5] D. Kahneman, A. Tversky, Prospect theory: An analysis of decision under risk, Econometrica: Journal of the Econometric Society (1979) 263-291.

[6] A. Tversky, D. Kahneman, Advances in prospect theory: Cumulative representation of uncertainty.Journal of Risk and Uncertainty, 5(4) (1992) 297–323.

[7] H. A. Simon, A behavioral model of rational choice, The quarterly journal of economics 69(1) (1955) 99-118.

[8] M. Abdellaoui, H. Bleichrodt, C. Paraschiv, Loss aversion under prospect theory: A parameter-free measurement, Management Science 53(10) (2007) 1659-1674.

[9] H. Ishibuchi, H. Tanaka, Multiobjective programming in optimization of the interval objective function, European Journal of Operational Research 48 (2) (1990) 219-225.

[10] Z.S. Xu, Deviation measures of linguistic preference relations in group decision making, Omega 33 (2005) 249-254.

[11] X. D. He, X. Y. Zhou, Portfolio choice under cumulative prospect theory: An analytical treatment, Management Science 57(2) (2011) 315-331.

[12] P. P. Wakker, Prospect theory: For risk and ambiguity, Cambridge: Cambridge University Press (2010).

[13] Z. P. Fan, X Zhang, F. D. Chen, Y. Liu, Multiple attribute decision making considering aspiration-levels: A method based on prospect theory, Computers & Industrial Engineering (2013) 341-350.

[14] C.L. Hwang, K. Yoon, Fuzzy multiple attribute decision making: theory and applications, Springer, Berlin 1992.

[15] K. Yoon, C.L. Hwang, Multiple attribute decision making: methods and applications, Springer, Berlin, 1981.

[16] S. J. Brams, M. R. Sanver, Voting systems that combine approval and preference, The mathematics of preference, choice and order, Springer Berlin Heidelberg (2009) 215-237.

[17] B. Erdamar, J. L. García-Lapresta, D. Pérez-Román, M. Remzi Sanver, Measuring consensus in a preference-approval context, Information Fusion 17 (2014) 14–21.

ME-OWA based DEMATEL-ISM Accident Causation Analysis of Complex System

Song Wang
Equipment Engineering College, Engineering University of Armed Police Force of China, Xi'an Shaanxi, China
(onesoon@163.com)

Abstract - **Causation complexity of complex system safety accident impels human to deal with safety problem by the perspective of system science and complex science. The complex non-linear interactive action between causation factors is the main cause of the causation complexity of complex system safety accident. This paper proposes a novel technique, combining maximal entropy ordered weighted averaging (ME-OWA), decision making trial and evaluation laboratory (DEMATEL) approach and Interpretive Structural Modeling (ISM) method to analyze the accident causation of complex system. After reviewing the literature of the OWA operators and DEMATEL methodology application, offered the arithmetic steps of ME-OWA and DEMATEL. The proposed method can be used to find the accident key causation factors and analysis the hierarchical of causation network.**

Keywords - **Accident analysis, complex system, complex network, DEMATEL, ISM, ME-OWA**

I. INTRODUCTION

Major accidents keep occurring that seem preventable and that have similar systemic causes. Too often, we fail to learn from the past and make inadequate changes in response to accidents [1]. The complex non-linear interactive action between accident causation factors is the main cause of the causation complexity of complex system safety accident. The complexity of complex system accident causation not only related to subjective cognitive capacity defects, but also to objective complexity, both of the complexity comes from various complex nonlinear interactions among system components. Therefore, the study of complex systems safety can not only concern with whether the component is reliable, but apply the system science and complexity science to analysis and deal with safety problems of complex system, especially attention should be paied to association among safety causation factors. Causation complexity of complex system safety accident impels human to deal with safety problem by the perspective of system science and complex science.

Accident causation theory primary apply system theory and methods to study accident process, analysis accident causation and formation mechanism, study accident prevention and control strategies, as well as the rescue measures after accident.

The Heinrich domino theory [2] pointed out that a series of succession causation incident led to the last occurrence of the accident, there exist a chain of linked relations between the accident and the causation factors.

Reason model [3] thinks the occurrence of an accident follows the law of "decision-making errors, mismanagement, formation of a direct prerequisite for unsafe behavior, resulting in unsafe behavior, the defense system failure", the model pointed out that the occurrence of an accident not only exist a reaction chain, also exist a penetrating organizational deficiencies set, which means organization factors is the deeper reasons for accident.

Leveson [4] proposed a systems theory accident model, STAMP (Systems-Theoretic Accident Model and Processes), systems are viewed as interrelated components that are kept in a state of dynamic equilibrium by feedback loops of information and control. A system in this conceptualization is not a static design—it is a dynamic process that is continually adapting to achieve its ends and to react to changes in itself and its environment. The original design must not only enforce appropriate constraints on behavior to ensure safe operation, but the system must continue to operate safely as changes occur. The process leading up to an accident (loss event) can be described in terms of an adaptive feedback function that fails to maintain safety as performance changes over time to meet a complex set of goals and values.

The author put forward another systems theory accident model STCCM (System-theoretic Cognitive-constraint Model), in STCCM, the occurrence of accident due to the limited human cognitive level, resulting in can not enforce effective constraints to complex accident causation, and thus lead to safety accident.

These theories and models for the processing of a particular complex system is effective, but lack of consideration to micro-level accident causation factors, do not extract the key causing factors and offer effective method for analysis of accident causation network.

Decision-making trial and evaluation laboratory (DEMATEL) methodology was originally developed by the Science and Human Affairs Program of the Battelle Memorial Institute of Geneva between 1972 and 1976 to study and resolve the complicated and intertwined problem group [5]. The DEMATEL methodology has been developed initially to study the structural relations in a complex system. In 1988 Ronald R. Yager (1988) introduced a new aggregation technique based on the ordered weighted averaging (OWA) operators, which are important aggregation operators within the class of weighted aggregation methods. They have the ability to derive optimal weights of the attributes based on the rating of the weighting vectors after an aggregation process. In a recent study, OWA operators and DEMATEL methodology have been applied in many fields [6, 7, 8].

Although there are numerous excellent studies devoted to accident causation, few of these can explain

E. Qi et al. (eds.), *Proceedings of the 5th International Asia Conference on Industrial Engineering and Management Innovation (IEMI2014)*, Proceedings of the International Asia Conference on Industrial Engineering and Management Innovation 1, DOI 10.2991/978-94-6239-100-0_7, © Atlantis Press and the authors 2015

the interaction between causation factors in a systematic way. In addition, there is no application of ME-OWA based DEMATEL in analyzing complex system accident causation, especially analyzing the interaction between accident causation factors.

II. ME-OWA BASED DEMATEL-ISM METHODOLOGY

A. ME-OWA operators

The concept of OWA operators was first introduced by Yager in 1988 [9], which has the ability to get optimal weights of the attributes based on the rank of these weighting vectors after an aggregation process (see Definition 1).

Definition 1: An OWA operator of dimension n is a mapping $F: R_n \rightarrow R$, which has an associated n weighting vector $W=[w_1,w_2,\ldots,w_n]^T$ that has the properties

$$\sum_{i=1}^{n} w_i = 1,\ w_i \in [0,1],\ i = 1,2,\ldots,n,$$

such that

$$f(a_1,a_2,\ldots,a_n) = \sum_{i=1}^{n} w_i b_i \tag{1}$$

where b_i is the ith largest element in the vector $(a_1, a_2, \ldots, a_n)$, and $b_1 \geqq b_2 \geqq \ldots \geqq b_n$. The function value $f(a_1, a_2, \ldots, a_n)$ determines the aggregated value of arguments, $a_1, a_2, \ldots, a_n$. A fundamental aspect of OWA operator is the re-ordering step, in particular an aggregate a_i is not associated with a particular weight w_i but rather a weight is associated with a particular ordered position of aggregate.

Yager used *orness* and *dispersion* of the aggregation to measure weighting vector W of the OWA operator.

Definition 2. Assume F is an OWA aggregation operator with a weighting function $W= [w_1, w_2, \ldots , w_n]$. The degree of *orness* associated with this operator is defined as:

$$orness(w) = \frac{1}{n-1}\sum_{i=1}^{n}(n-i)w_i \tag{2}$$

where $orness(W) = \alpha$ is a situation parameter. It is easy to see that for any W the $orness(W)$ is always in the unit interval [0,1].

Definition 3. Assume W is a weighting vector with elements $w_1, w_2, \ldots, w_n$; then the measure of dispersion of W is defined as:

$$disp(w) = -\sum_{i=1}^{n} w_i \ln w_i \tag{3}$$

We can see when using the OWA operator as an averaging operator $Disp(W)$ measures the degree to which we use all the aggregates equally.

O'Hagan (1988) combined the principle of maximum entropy and OWA operators to propose a particular OWA weight that has maximum entropy with a given level of *orness*. This approach is based on the solution of the following mathematical programming problem:

$$Max.\quad -\sum_{i=1}^{n} w_i \ln w_i \tag{4}$$

$$s.t.\quad \frac{1}{n-1}\sum_{i=1}^{n}(n-i)w_i = \alpha, 0 \le \alpha \le 1 \tag{5}$$

$$\sum_{i=1}^{n} w_i = 1, \quad 0 \le w_i \le 1, \quad i = 1,2,\ldots,n \tag{6}$$

Fuller and Majlender [10] used the method of Lagrange multipliers on Yager's OWA equation to derive a polynomial equation, which can determine the optimal weighting vector under the maximal entropy. By their method, the associated weighting vector is easily obtained by Eqs. (7)–(9) [6]:

$$w_i = \sqrt[n-1]{w_1^{n-i} w_n^{i-1}} \tag{7}$$

and

$$w_n = \frac{((n-1)\alpha - n)w_1 + 1}{(n-1)\alpha + 1 - nw_1} \tag{8}$$

then

$$w_1[(n-1)\alpha + 1 - nw_1]^n = [(n-1)\alpha]^{n-1}[[(n-1)\alpha - n]w_1 + 1] \tag{9}$$

where w is the weight vector, n is the number of attributes, and a is the situation parameter.

B. DEMATEL-ISM Method

DEMATEL methodology applies matrices and digraphs for visualizing the structure of complicated causal relationships, constructs the direct impact on matrix by logical relationship between various factors in the system, calculates the direct influence of one factor to others and its indirect effects, then calculates the factor's centrality degree and reason degree. According to factor's centrality degree and reason degree, can gain the factor's type(whether is the reason factor or result factor).

Use the DEMATEL method to analyze the complex system safety accident causation factors can follow these steps:

Step 1: Determine complex system safety accident causation factors $\alpha_1, \alpha_2, \ldots, \alpha_n$, $\alpha_i \in A$ $(i=1,2,\ldots,n)$, where n is the causation factors number, A is the causation factors set;

Step 2: Setting evaluation scale of affect relationship(such as use Very High Influence, High Influence, Low Influence, Very Low Influence, No Influence five grades and assign 4,3,2,1,0), according to the experience of experts to gain the direct influence matrix $B(B=[\beta_{ij}]_{n\times n})$, β_{ij} said the directly affect degree of causation factor α_i to factor α_j, for the interaction between factors is not equivalent to each other, so $\alpha_{ij} \neq \alpha_{ji}$, if $i=j$, then $\alpha_{ij}=0$;

Step 3: Normalizing initial direct-relation matrix, gain normalized direct-relation matrix $C(C=[c_{ij}]_{n\times n})$

$$C = B \Big/ \max_{1 \le j \le n} \sum_{j=1}^{n} \beta_{ij} \tag{10}$$

Where, $\max_{1 \le i \le n} \sum_{j=1}^{n} \beta_{ij}$ is the maximum of row sum, normalization enable c_{ij} in the interval of [0,1].

Step 4: Computing total-relation influence matrix $T(T=[t_{ij}]_{n \times n})$. Total-relation influence matrix represents total cumulative of direct and indirect influence relationship between factors, which can be used to determine the final influence of every factor to the highest level factor, that is,

$$T = C + C^2 + ... + C^n = C\frac{I - C^{n-1}}{I - C} \tag{11}$$

Due to $0 \leqq c_{ij} \leqq 1$, so, if $n \to \infty$, $C^{n-1} \to 0$, then

$$T = C\frac{I}{I - C} = C(I - C)^{-1} \tag{12}$$

Step 5: Computing factor's degree of influences given f_i and degree of influences received e_i, sums of rows in total-relation influence matrix T gain f_i, sums of columns in total-relation influence matrix T gain e_i,

$$f_i = \sum_{j=1}^{n} t_{ij} (i = 1, 2, ..., n) \tag{13}$$

$$e_i = \sum_{j=1}^{n} t_{ji} (i = 1, 2, ..., n) \tag{14}$$

Step 6: Computing centrality degree M_i and reason degree N_i. The sum of degree of influences given and degree of influences received gain centrality degree M_i, and the subtraction of influences given and degree of influences received gain reason degree N_i.

$$M_i = f_i + e_i (i = 1, 2, ..., n) \tag{15}$$

$$N_i = f_i - e_i (i = 1, 2, ..., n) \tag{16}$$

Centrality degree M_i shows the total effects given and received by factor *i*, which also indicates the degree of importance that factor *i* plays in the entire system. On the contrary, reason degree N_i depicts the net effect that factor *i* contributes to the system. If N_i is positive, which means factor *i* is a causation cause, or else factor *i* is a result cause.

Step 7: Draw reason-result figure. Treat the centrality degree as abscissa, reason degree as ordinate, draw cartesian coordinate system, marked location of each cause factor in the coordinate system, analyzing the importance of each factor and obtain the key causation factors.

Thought the DEMATEL method can quarry the key causation factors, which cannot offer the relationship between them, so Interpretive Structural Modeling (ISM) method can be used to improve the solution results.

ISM is an effective methodology for dealing with complex issues. ISM first proposed by J. Warfield in 1973, ISM is a computer assisted learning process that enables individuals or groups to develop a map of the complex relationships between the many elements involved in a complex situation. ISM is often used to provide fundamental understanding of complex situations, as well as to put together a course of action for solving a problem. It has been used worldwide by many prestigious organizations including NASA.

Theoretical about integrated DEMATEL and ISM can refer to literature [11] The specific steps are conducted on the basis of above-mentioned seven steps:

Step 8: Computing integrated influence matrix $H(H=[h_{ij}]_{n \times n})$

$$H = I + T \tag{17}$$

Where, I is unit matrix。

Step 9: Given threshold λ, computing reachable matrix $K(K=[k_{ij}]_{n \times n})$

$$k_{ij} = 1, if \quad h_{ij} \ge \lambda (i, j = 1, 2, ..., n) \tag{18}$$

$$k_{ij} = 0, if \quad h_{ij} < \lambda (i, j = 1, 2, ..., n) \tag{19}$$

The value of λ directly influence the composition of reachable matrix and subsequent hierarchy division, in order to obtain satisfactory results, the specific value of λ can be gained after several times value analysis,

Step 10: Determine every causation factor reachable set and preceding set. Causation factor α_i's reachable set R_i and preceding set S_i can be calculated as:

$$R_i = \{\alpha_j | \alpha_j \in A, k_{ij} \ne 0\}, (i = 1, 2, ..., n) \tag{20}$$

$$S_i = \{\alpha_j | \alpha_j \in A, k_{ji} \ne 0\}, (i = 1, 2, ..., n) \tag{21}$$

Where A is the causation factors set.

Step 11: Verify whether that the formula $R_i = R_i \cap S_i$, $(i=1,2,\ldots,n)$ is set up. If the formula is set up, then induced causation factors α_i is an underlying factors, delete rows *i* and columns *i* in the matrix K;

Step 12: Repeat step 10 and step 11, until all factors are deleted;

Step 13: According to causation factors' deleted sequence, drawing causation factors' hierarchical structure graph.

III. NUMERICAL EXAMPLE

When investigating the causes of aviation accidents, the interaction between causation factors is always ignored by investigator. They more concern the direct accident cause chain, when someone or something should be condemned has been found, they will stop the investigation process and try to beautify the accident report.

A. Establishing causation factors system

Leveson (2011) pointed out complex system three types of accident causation, which is component failure (including physical component failure and human error), interactive disorders (dysfunctional interactions between components) and external environmental perturbations, when above three causation appear but lack of restraint or constraint become failure may lead to safety accident.

One of the most serious accidents occurred at Chernobyl in 1986 has been the subject of a great deal of investigation and has brought into being the concept of safety culture (M.A. Mariscal et al. 2012). Safety culture is reserved to the basic assumptions of the organisation, in other words to ''traits'' that are stable and deep-rooted.

Leveson's STAMP and author's STCCM points out that cognitive problem and constraints problem have close relationship with safety accident. The occurrence of accident due to the limited human cognitive level, resulting in can not enforce effective constraints to complex accident causation, and thus lead to safety accident.

Complex system is an open non-linear system, which exchange matter, energy and information with external world, must also have a certain degree of adaptability, such adaptations, including adaptation of system parameters and adaptation of system structure, so adaptability recession is the main causation of safety accident. Adaptability recession comes from the system brittle structure, and the lack of information in the cause of the accident also occupy an important position.

According to the characteristics of aviation complex system, we extracted eight main factors accident causation index system, which are component failure, human error, safety culture, constraints failure, communication obstacle, system brittle structure, cognitive deficits and uncertainties.

B. Data sources

Suppose there existed an aviation accident, six investigators were assigned to analyze the causation and provide the accident report. All of them thought that the eight factors, component failure α_1 , human error α_2, safety culture α_3, constraints failure α_4, communication obstacle α_5, system brittle structure α_6, cognitive deficits α_7 and uncertainties α_8, had curtain connection with the final aviation accident. In order to find the interaction between causation factors in a systematic way, the investigators were asked to offer the influence and direction between causation factors. After setting the evaluation scale of affect relationship (such as use Very High Influence, High Influence, Low Influence, Very Low Influence, No Influence five grades and assign 4,3,2,1,0), according to the experience of investigators to gain the six initial direct influence matrix.

$$X^1 = \begin{bmatrix} 0 & 3 & 0 & 4 & 3 & 2 & 1 & 2 \\ 3 & 0 & 2 & 3 & 3 & 2 & 3 & 2 \\ 1 & 2 & 0 & 1 & 2 & 3 & 3 & 1 \\ 3 & 2 & 1 & 0 & 2 & 1 & 1 & 1 \\ 1 & 3 & 2 & 1 & 0 & 3 & 3 & 3 \\ 3 & 2 & 2 & 2 & 2 & 0 & 2 & 2 \\ 3 & 3 & 2 & 3 & 2 & 2 & 0 & 4 \\ 2 & 4 & 1 & 2 & 1 & 1 & 3 & 0 \end{bmatrix}$$

$$X^2 = \begin{bmatrix} 0 & 4 & 1 & 3 & 2 & 4 & 2 & 1 \\ 2 & 0 & 1 & 2 & 4 & 1 & 2 & 3 \\ 2 & 3 & 0 & 2 & 3 & 2 & 4 & 1 \\ 4 & 1 & 2 & 0 & 2 & 1 & 2 & 3 \\ 2 & 4 & 1 & 3 & 0 & 4 & 3 & 2 \\ 3 & 1 & 2 & 3 & 1 & 0 & 3 & 1 \\ 4 & 4 & 1 & 4 & 1 & 1 & 0 & 3 \\ 3 & 3 & 2 & 2 & 2 & 1 & 4 & 0 \end{bmatrix}$$

$$X^3 = \begin{bmatrix} 0 & 2 & 1 & 2 & 1 & 1 & 2 & 3 \\ 2 & 0 & 1 & 2 & 1 & 1 & 2 & 3 \\ 2 & 1 & 0 & 2 & 3 & 2 & 2 & 2 \\ 4 & 1 & 2 & 0 & 3 & 2 & 1 & 2 \\ 2 & 4 & 1 & 3 & 0 & 3 & 4 & 3 \\ 4 & 1 & 2 & 3 & 2 & 0 & 1 & 2 \\ 2 & 4 & 2 & 2 & 2 & 2 & 0 & 3 \\ 3 & 3 & 2 & 2 & 2 & 1 & 4 & 0 \end{bmatrix}$$

$$X^4 = \begin{bmatrix} 0 & 2 & 0 & 3 & 4 & 1 & 2 & 1 \\ 4 & 0 & 1 & 2 & 3 & 2 & 4 & 3 \\ 1 & 1 & 0 & 1 & 1 & 2 & 2 & 1 \\ 2 & 1 & 1 & 0 & 3 & 1 & 2 & 2 \\ 2 & 4 & 2 & 2 & 0 & 3 & 4 & 4 \\ 2 & 2 & 4 & 2 & 3 & 0 & 2 & 4 \\ 2 & 4 & 1 & 2 & 2 & 1 & 0 & 4 \\ 1 & 3 & 1 & 2 & 1 & 2 & 4 & 0 \end{bmatrix}$$

$$X^5 = \begin{bmatrix} 0 & 2 & 2 & 3 & 2 & 1 & 2 & 3 \\ 4 & 0 & 1 & 3 & 4 & 2 & 2 & 3 \\ 2 & 2 & 0 & 3 & 2 & 4 & 3 & 2 \\ 2 & 2 & 2 & 0 & 3 & 1 & 2 & 1 \\ 2 & 3 & 3 & 2 & 0 & 3 & 2 & 4 \\ 4 & 2 & 1 & 2 & 3 & 0 & 3 & 2 \\ 2 & 3 & 2 & 4 & 2 & 1 & 0 & 3 \\ 3 & 4 & 2 & 2 & 4 & 1 & 4 & 0 \end{bmatrix}$$

$$X^6 = \begin{bmatrix} 0 & 4 & 0 & 3 & 2 & 1 & 3 & 2 \\ 2 & 0 & 1 & 4 & 3 & 3 & 3 & 1 \\ 2 & 3 & 0 & 1 & 2 & 3 & 4 & 2 \\ 3 & 3 & 2 & 0 & 2 & 3 & 1 & 2 \\ 2 & 4 & 2 & 2 & 0 & 3 & 4 & 2 \\ 4 & 2 & 3 & 2 & 1 & 0 & 3 & 2 \\ 3 & 4 & 1 & 3 & 1 & 3 & 0 & 3 \\ 1 & 2 & 1 & 3 & 2 & 1 & 4 & 0 \end{bmatrix}$$

C. Obtain the causation factors weight vector

This part, the ME-OWA operators will be used to obtain the causation factors weight vector. For there exists eight causation factors so n=8, choose situation parameter a=0.6. According to formula (7) to (9), the calculate result shows in Table I.

TABLE I
CAUSATION FACTORS WEIGHT VECTOR WHEN a =0.6

	w_1	w_2	w_3	w_4	w_5	w_6
a=0.6	0.15	0.156	0.161	0.167	0.186	0.18

D. Application of the decision-making method

From largest to smallest sort the value in X^k_{ij}, the j biggest element empowerment w_j (j=1,2,…,6), then gained the direct influence matrix B.

$$B=\begin{bmatrix} 0 & 2.773 & 0.617 & 2.970 & 2.276 & 1.606 & 1.970 & 1.940 \\ 2.773 & 0 & 1.150 & 2.617 & 2.946 & 1.784 & 2.617 & 2.454 \\ 1.634 & 1.940 & 0 & 1.617 & 2.126 & 2.617 & 2.940 & 1.467 \\ 2.940 & 1.617 & 1.634 & 0 & 2.467 & 1.456 & 1.467 & 1.784 \\ 1.820 & 3.634 & 1.784 & 2.126 & 0 & 3.150 & 3.287 & 2.940 \\ 3.287 & 1.634 & 2.276 & 2.306 & 1.940 & 0 & 2.287 & 2.120 \\ 2.617 & 3.634 & 1.467 & 2.940 & 1.634 & 1.617 & 0 & 3.306 \\ 2.101 & 3.126 & 1.467 & 2.150 & 1.934 & 1.150 & 3.820 & 0 \end{bmatrix}$$

According to formula (10) to (16) can gained the solution results, and the cause and effect diagram showed in Fig.1.

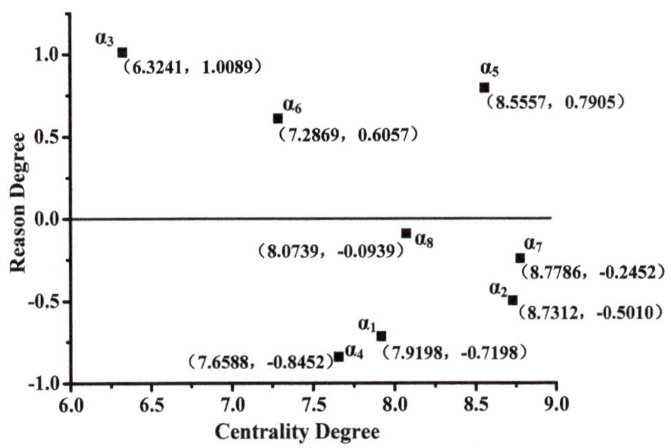

Fig.1. The cause-effect relationship diagram

According to formula $R_i= R_i \cap S_i,(i=1,2,\ldots,n)$, after calculation and analysis gained the first level causation factors sct $L_1=\{\alpha_1, \alpha_4\}$.Similarly available $L_2=\{\alpha_2, \alpha_5, \alpha_6\}$, $L_3=\{\alpha_3, \alpha_7, \alpha_8\}$. This is α_1 and α_4 are at the bottom, α_2, α_5 and α_6 at the second level, and α_3, α_7, α_8 at the top level.

Based on complex networks theory, treats the causation factor as the network node, and the influence relation as the edge, then from the reachable matrix K, sum of row should be the influences given f_i and of course it's the out-degree of the causation factor node, that is sum of column should be the influences received e_i and of course which is the in-degree of the causation factor node. Then the multi-level hierarchical structure model of causation factors can be gained, which shows in Fig.2.

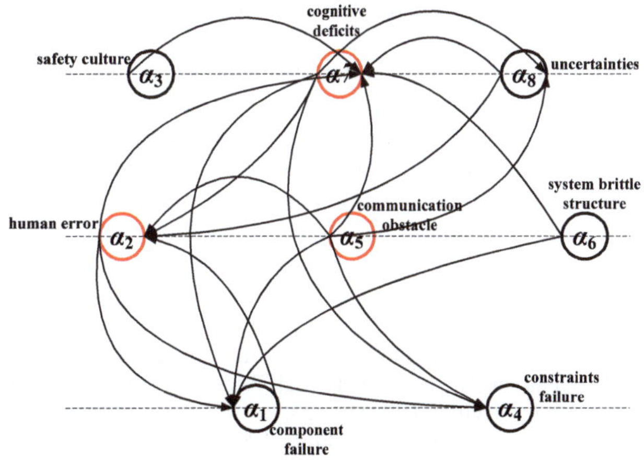

Fig.2. Multi-level hierarchical structure model of causation factors

In Fig.2, causation factors node 2, 5, 7 is the node with a bigger node degree, the greater number of node degree means the more connected edges, therefore the more the likelihood of complex non-linear interaction with other node, thereby increasing the possibility of major accidents. The figure illustrates not only the hierarchical relationship among the causation factors, at the same time identify the key causal factors and result factors.

If the sum of column and row should be the degree of causation factor node (removes the node itself), then from reachable matrix K (when λ=0.55), the degree of causation factor node can be gained, which show in Table II.

Table II
NODE DEGREE OF CAUSATION FACTOR

	α_1	α_2	α_3	α_4	α_5	α_6	α_7	α_8
Node degree	5	7	1	2	5	2	9	4

From table 2, node 7, 2, 1, 5 has bigger node degree, which further explanation that communication obstacle (α_5), cognitive deficits (α_7), human error (α_2) and component failure (α_1) are key causal factors, which would be the main reason for the said aviation accident.

IV. CONCLUSION

In order to analyze the accident causation network of complex system, this paper extract eight causation factors for aviation complex system safety accident analysis, which are component failure, human error, safety culture, constraints failure, communication obstacle, system brittle structure, cognitive deficits and uncertainties. Respectively used ME-OWA based DEMATEL method to gain potential key causation factors, used ISM method to construct the multi-level hierarchical structure model of complex system safety accident causation factors. Based on complex network theory, the bigger node degree factors are exactly the key causation factors in DEMATEL method. ME-OWA based DEMATEL-ISM method can be used in complex system safety accident later analysis and prevention analysis, but also in extracting key accident causation factors and analyzing the hierarchical structure among causation factors.

ACKNOWLEDGMENT

This study was partially supported by the National Natural Science Foundation of China with the Grant number of **71171199** and Fundamental Research Foundation of Engineering University of Armed Police Force of China with the Grant number of **WJY201410**.

REFERENCES

[1] Nancy G. Leveson, 2011. Applying systems thinking to analyze and learn from events. Safety Science 49, 55–64.

[2] Heinrich H, 1980. Industrial Accident Preventior. New York: McGraw-Hill, 112-123.

[3] Reason J, 1990. Human Error. UK, Cambridge: Cambridge University Press, 23-29.

[4] Leveson, N. G., 2004. A new accident model for engineering safer systems. Safety Science 42, 237-270.

[5] Fontela, E., A. Gabus, 1976. The DEMATEL observer, DEMATEL 1976 Report. Switzerland Geneva: Battelle Geneva Research Center, USA.

[6] Kuei-Hu Chang, Ching-Hsue Cheng, 2011. Evaluating the risk of failure using the fuzzy OWA and DEMATEL method. Journal of Intelligent Manufacturing 22,113–129.

[7] Ya-Ti Lin, Yeou-Herng Yang, Jin-Su Kang, et al., 2011. Using DEMATEL method to explore the core competences and causal effect of the IC design service company: An empirical case study. Expert Systems with Applications 38, 6262-6268.

[8] Lin, C. J., & Wu, W. W., 2008. A causal analytical method for group decision-making under fuzzy environment. Expert Systems with Application 34, 205–213.

[9] Yager, R. R., 1988. On ordered weighted averaging aggregation operators in multi-criteria decision making. IEEE Transactions on Systems, Man and Cybernetics18, 183–190.

[10] Fuller, R., & Majlender, P. 2001. An analytic approach for obtaining maximal entropy OWA operator weights. Fuzzy Sets and Systems 124, 53–57.

[11] Jianmin Hou, Dequn Zhou, 2011. Study on Influence Factors of Distributed Energy System Based on DEMATEL and ISM. International Journal of Nonlinear Science 12, 36-41.

The Optimal Consignment Policy for the Retailer Facing Multiple Manufacturers

Hui YANG
Department of Management Science and Technology, Nanjing University of Science and Technology, Nanjing, Jiangsu Province, China
(yanghui@njust.edu.cn)

Abstract - **Consignment is becoming an increasingly popular practice to supply chain management. We consider the optimal consignment policy for the retailer facing multiple manufacturers. We model the decision making of the retailer and the manufacturers as a Stackelberg game: The retailer, acting as the leader, offers the manufacturers a uniform consignment contract, which specifies the slotting fee and the percentage. The manufacturers, acting as the follower, choose whether to use the retailer. We formulate the retailer's decision process as a mixed integer programming (MIP) problem and solve it by graphical method. We propose the algorithm to determine the optimal consignment policy for monopoly retailer. This policy and each manufacturer's corresponsive behavior to it constitute the Stackelberg equilibrium. The numerical example shows that this optimal policy can improve the retailer's total charge.**

Keywords – **Consignment policy, game, optimization, supply chain management**

I. INTRODUCTION

Consignment policy is a new approach to supply chain management. Major retailers such as Wal-Mart, Carrefour, Amazon.com and Tmall.com are among the users of the consignment [1]. Taking the example of Amazon.com, it applies a special fee structure and provides a revenue sharing contract with consignment to its manufacturers. In such a business model, the manufacturer retains the ownership of the goods and provides a fixed fee (called slotting allowance) to the retailer. The retailer deducts a percentage of the selling price for each item sold and remits the balance to the manufacturer. This arrangement has many advantages [2-4]. It favors the retailer. Since no payment to the manufacture is made until the item is sold, the retailer has no money tied up in inventory and bears no risk associated with demand uncertainty [5]. Furthermore, it has been proved that the revenue sharing contract with consignment contract coupled with slotting fee will perfectly coordinate the channel and lead to Pareto improvements among channel participants [6]. Prior research highlighted the coordination mechanism in designing the consignment contracts. Cachon and Lariviere [7] study a VMI contract with revenue sharing and demonstrate that the decentralized system provides less capacity than the integrated system. Li and Hua [8] propose a cooperative game model to describe the payment bargaining process between the manufacturer and the retailer, and determine a new consignment contract with revenue sharing attached with the equilibrium payment scheme. However, no study explored the matching process between the consignment retailers and manufacturers before they become supply chain partners. In fact, the retailer can attract a quantity of manufacturers and select the right partners by designing appropriate consignment policy, while the manufacturers can compare different policies to determine which retailer to enter. This matching process is effective for resource allocation. This study focuses on the matching process between one retailer and multiple manufacturers, and develops a model for the retailer of how to design the optimal consignment policy. The paper proceeds as follows. Section II proposes a MIP model. Section III solves the model through graphical method and presents an algorithm to determine the optimal policy. A numerical example is placed in Section IV, followed by concluding remarks.

II. MODEL DESCRIPTION

Consider an environment comprised of N number of manufacturers, each producing a single product, and one retailer, which may carry one to N products. The manufacturers sell their products through consignment. The retailer charges a yearly slotting fee and a percentage of revenue whenever a product is sold. Assume the retailer use the same consignment policy for all of his manufacturers. We model the decision making of the retailer and the manufacturers as a Stackelberg game: The retailer, acting as the leader, offers the manufacturers a take-it-or-leave-it consignment contract, which specifies the slotting fee and the percentage. The manufacturers, acting as the follower, choose whether to use the retailer. In this game, the manufactures need to balance between the consignment charges and the benefit, while the retailer needs to balance the charges and the number of manufacturers to serve.

The following notations are used to formulate the mathematical model:

x Slotting allowance, a fixed fee paid yearly by the manufacturer to the retailer.

y The retailer's share of revenue generated from each unit; $0 \le y < 1$.

p_i Per-unit selling price for product i produced by manufacturer ; $i = 1,2...N$.

c_i Per-unit manufacturing cost for product i ; $i = 1,2...N$.

z_i Binary variables;

E. Qi et al. (eds.), *Proceedings of the 5th International Asia Conference on Industrial Engineering and Management Innovation (IEMI2014)*, Proceedings of the International Asia Conference on Industrial Engineering and Management Innovation 1, DOI 10.2991/978-94-6239-100-0_8, © Atlantis Press and the authors 2015

$$z_i = \begin{cases} 1, \text{if manufacturer } i \text{ uses the retailer} \\ 0, \text{otherwise} \end{cases} ; i = 1,2...N$$

q_i Product i's quantity of sales through consignment when manufacturer uses the retailer; $i = 1,2...N$.

π_i^m Manufacturer i's profit; $i = 1,2...N$.

π^r The retailer's total consignment charge.

In this dynamic game, the retailer offers all the manufacturers a revenue-sharing contract with a uniform consignment policy (x, y) . Given this policy, each manufacturer computes his profit and determines whether to take the contract (see Fig.1).

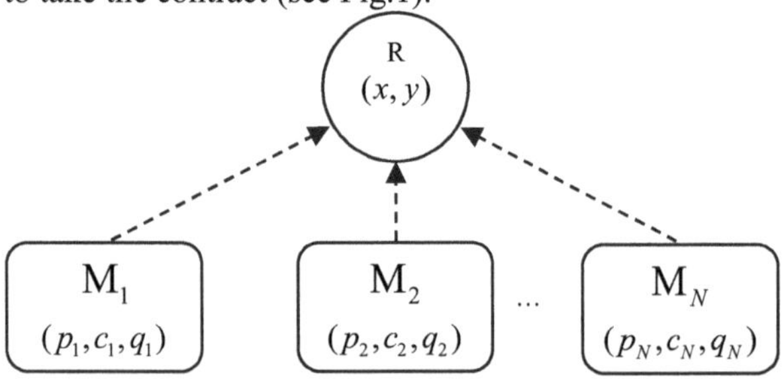

Fig.1. Channel structure

The profit of manufacturer i can be expressed as $\pi_i^m = q_i(p_i - c_i) - x - q_i p_i y$. If $\pi_i^m \geq 0$, manufacturer i chooses to take the contract then uses the retailer, and the binary variable $z_i = 1$; otherwise, he chooses not to use the retailer, and $z_i = 0$. The retailer's revenue is the sum of what she charges to those manufacturers who determines to take the contract. Her revenue function can be written as $\pi^r = \sum_{i=1}^{N} z_i(x + q_i p_i y)$.

To formulate the retailer's decision process as a mixed integer programming (MIP) problem, we identify the objective function and the constraints as follows:
MIP Model:

$$\text{maximize } \pi^r = \sum_{i=1}^{N} z_i(x + q_i p_i y)$$

$$\text{subject to: } M(1 - z_i) + q_i(p_i - c_i)z_i - (x + q_i p_i y) \geq 0$$
$$Mz_i - q_i(p_i - c_i)(1 - z_i) + (x + q_i p_i y) \geq 0$$
$$x \geq 0, y \geq 0, z_i = 0 \text{ or } 1 \tag{1}$$

In this model, the letter M represents a very large negative number. The first two constraints represent mathematical relationships between the value of z_i and the profit of manufacturer i . If manufacturer i's benefit $q_i(p_i - c_i)$ is greater than or equal to the consignment charge $x + q_i p_i y$, both constraints are satisfied when $z_i = 1$, while the equation $z_i = 0$ would violate the second constraint. If manufacturer i's benefit is less than the charge, both constraints are satisfied when $z_i = 0$, while the equation $z_i = 1$ would violate the first constraint. Using these constraints, we incorporate the manufacturer's decision problem into the retailer's decision model. The solution to this MIP Model would not only provide the retailer with an optimal consignment policy, but also provide the manufacturers with optimal responsive strategies. Therefore, it is the Stackelberg equilibrium of this game.

III. GRAPHICAL METHOD

There are some computer programs that are capable of solving this MIP problem, such as LINGO, MATLAB. We use the graphical method to solve it as there are two decision variables in a consignment policy. The graphical procedure is invaluable in providing us with insights into how the model works. For that reason alone, it is worthwhile to spend the rest of this section exploring graphical solutions as an intuitive basis for the analysis.

We plot the variable x as the horizontal axis of the graph and the variable y as the vertical axis. If we graph the linear equation $q_i(p_i - c_i) - x - q_i p_i y = 0$ in the first quadrant, we can get N *critical lines* where each manufacturer's profit is equal to zero. Without loss of generality, we draw two manufacturers' critical lines in the graph. There are two possibilities for the position of these two lines. First, the two critical lines have an intersection point in the first quadrant. As shown in Fig.2, manufacturer 1's horizontal intercept is smaller than that of manufacturer 2, but its vertical intercept is larger. This shows manufacturer 1's total profit is smaller than that of manufacturer 2, i.e. $q_1(p_1 - c_1) < q_2(p_2 - c_2)$, but its profit rate is larger than that of manufacturer 2, i.e. $\frac{p_1 - c_1}{p_1} > \frac{p_2 - c_2}{p_2}$. Second, the two lines have no intersection in the first quadrant (see Fig.3). This is because both the total profit and the profit rate of manufacturer 1 are larger than those of manufacturer 2. In such a case, we call manufacturer 1 the *strong* one and manufacturer 2 the *weak* one.

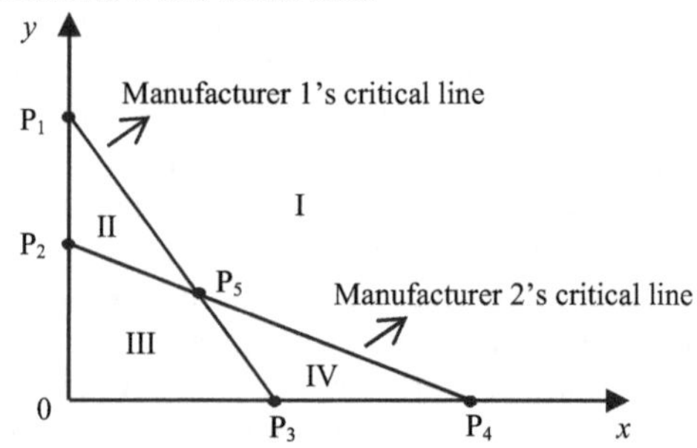

Fig.2. The two critical lines with an intersection point

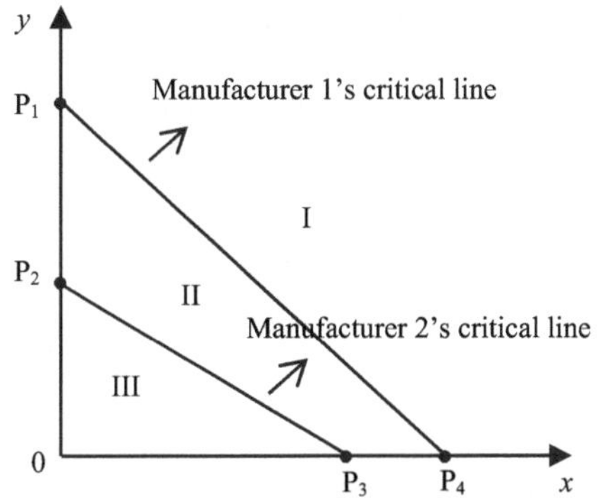

Fig.3. The two critical lines without an intersection

If the two lines have an intersection point, the first quadrant is divided into four regions. In region I, all the points satisfy $x+q_i p_i y > q_i(p_i - c_i), i=1,2$. This results in that neither of the manufacturers enters the retailer. Therefore, the retailer's charge is zero. In region II, Manufacturer 1 will decide to enter while manufacturer 2 will decide to leave. The retailer can get more charge when the solution point approaches to line P_1P_5 on the left side. Similarly, in region IV, manufacturer 2 will decide to enter while manufacturer 2 will decide to leave. The retailer can get more charge when the solution point approaches to line P_5P_4 on the left side. In region III, both manufacturers will decide to enter, and the retailer's charge approaches maximum when the solution point approaches point P_5 on the left side. Comparing the four regions, the retailer will choose the intersection point P_5 as her consignment policy. Subtracting a small number from the coordinate value of point P_5, she can get two manufacturers and approximately maximal charge.

If the two lines have no intersection, the first quadrant is divided into three regions. In region I, neither manufacturer enters the retailer. In region II, manufacturer 1 – the strong one will enter while manufacturer 2 will leave. The retailer gets more charge when the point approaches to line P_1P_4 on the left side. In region III, both manufacturers will choose to enter. It should be noted that different points of line P_2P_3 brings the retailer a fixed charge from manufacturer 2, that is, $q_2(p_2 - c_2)$, while different points means different things to manufacturer 1. The point which brings the retailer maximum charge from manufacturer 1 therefore represents the optimal policy for the retailer. This can be formulated as a simple linear programming model. That is

maximize $x + q_1 p_1 y$

subject to: $q_2(p_2 - c_2) - (x + q_2 p_2 y) = 0$

$0 \le x \le q_2(p_2 - c_2)$

$0 \le y \le 1 - c_2 / p_2$

It is easy to get the solution to this model. If $q_2 p_2 > q_1 p_1$, the optimal solution lies at the corner point P_2. If $q_2 p_2 < q_1 p_1$, the optimal solution lies at the corner point P_3. If $q_2 p_2 = q_1 p_1$, all points of line P_2P_3 are optimal solutions.

By comparing region II and region III, the retailer decides whether to get both manufacturers or to get one manufacturer. The optimal solution lies on segment P_1P_4, if the retailer charges more from one manufacturer by making policy on segment P_1P_4 than she charges from two manufacturers by making policy on segment P_2P_3. Otherwise, the optimal solution lies on segment P_2P_3. In either case, we can find the optimal solutions at the four corner points, i.e., point P_1 to P_4.

From above analysis, we can see that given two manufacturers' critical lines the optimal solutions can be found at the corner points. These points include two types: one is the intersection point of critical lines; the other is the intersection point of critical lines and coordinate axis. We can extend this result for two manufacturers to N manufacturers. Note that any two critical lines can be plotted as Fig.2. or Fig.3. Hence, we can get PROPOSITION 1.

PROPOSITION 1.The matching problem under monopoly-retailer-multiple-manufacturer settings can be formulated as the MIP Model. The optimal solution to this model can be found at the corner points, which are intersections of the manufacturers' critical lines or intersections of critical lines and coordinate axis in the first quadrant.

According to PROPOSITION1., the process for finding optimal solutions can be summarized as following four steps.

ALGORITHM1. Algorithm to Determine Optimal Consignment Policy for Monopoly Retailer:

Step1: Find the corner point set for N manufacturers using critical line method.

Step2: For each point, compute each manufacturer's profit $\pi_i^m = q_i(p_i - c_i) - x - q_i p_i y$. Judge whether he would enter or leave (z_i=0 or 1).

Step3: For each point, compute the retailer's total charge $\pi^r = \sum_{i=1}^{N} z_i(x + q_i p_i y)$.

Step4: Choose the maximal charge point.

This maximal charge point represents the optimal consignment policy for the retailer. This policy and each manufacturer's corresponsive behavior to it constitute the equilibrium of the Stackelberg game.

IV. NUMERICAL EXAMPLE

In this section, we provide an example to illustrate the matching between monopoly retailer and multiple manufacturers proposed in this study. Consider one retailer and four manufacturers. The selling price, the manufacturing cost and the forecasting quantity of sales for the four products are $p_1 = 8, c_1 = 3, q_1 = 10$; $p_2 = 9, c_2 = 5, q_2 = 20$; $p_3 = 6, c_3 = 1, q_3 = 15$; $p_4 = 5, c_4 = 2, q_4 = 20$.

To determine the optimal consignment policy, we apply Proposition 1. and Algorithm 1. described previously to the retailer. We find 12 corner points through computing. The matched manufacturers who choose to use the retailer and the retailer's charge at each point are shown in Table I.

Comparing these twelve points, the retailer's benefit varies greatly at different points. If she charges a slotting fee 26 and a percentage $\frac{3}{10}$, all the four manufacturers will accept the consignment contract, and the retailer gets the maximal charge 239.

TABLE I

MANUFACTURES' CHOICE AND RETAILERS' CHARGE AT CORNER POINT

Corner point	Matched manufacturers	Retailer's charge π^r
$(0, \frac{4}{9})$	1, 2, 3, 4	200
$(0, \frac{5}{6})$	3	75
$(0, \frac{5}{8})$	1, 3	106
$(0, \frac{3}{5})$	1, 3, 4	162.25
(50, 0)	1, 2, 3, 4	200
(80, 0)	2	80
(75, 0)	2, 3	150
(60, 0)	2, 3, 4	180
$(70, \frac{1}{18})$	2, 3	155
$(35, \frac{1}{4})$	2, 3, 4	197.50
$(26, \frac{3}{10})$	1, 2, 3, 4	239
$(10, \frac{1}{2})$	1, 3, 4	165

V. CONCLUSION

This study focuses on designing the optimal consignment policy for the retailer facing multiple manufacturers. We model the decision making of the retailer and the manufacturers as a Stackelberg game: The retailer, acting as the leader, offers all the manufacturers the same consignment contract, which specifies the slotting fee and the percentage. The manufacturers, acting as the follower, choose whether to use the retailer. We formulate the retailer's decision process as a mixed integer programming (MIP) problem and solve it by graphical method. We propose the algorithm to determine the optimal consignment policy for monopoly retailer. This policy and each manufacturer's corresponsive behavior to it constitute the Stackelberg equilibrium. Through a numerical example, we demonstrate that this optimal policy can improve the retailer's total charge. In future research, we will extend the model to a case where exist competitive retailers and find the equilibrium market segmentation: what types of manufacturers match what type of retailers.

ACKNOWLEDGMENT

This paper is substantially supported by research grants from the National Natural Science Foundation of China (Grant No. 71102073, 71172105).

REFERENCES

[1] S. L. Chen, C. L. Liu, "The optimal consignment policy for. the manufacturer under supply chain co-ordination," *International Journal of Production Research*, 46, 18, January 2008, pp. 5121-5143.

[2] H. Wang, "Slotting allowances and retailer market power," *Journal of Economic Studies*, 33, 1, January 2006, pp. 68-77.

[3] D. Kuksov, A. Pazgal, "The effects of costs and competition on slotting allowances," *Marketing Science*, 26, 2, March-April 2007, pp. 259-267.

[4] K. Pan, K. K. Lai, and S. C.Leung, "Revenue-sharing versus wholesale price mechanisms under different channel power structures," *European Journal of Operational Research*, 203, 2, March 2010, pp. 532-538.

[5] Y. Wang, J. Li, and Z. Shen, "Channel performance under consignment contract with revenue sharing," *Management Science*, 50, 1, January 2004, pp. 34-37.

[6] J. M. Chen, H. L. Cheng, and M.C. Chien, "Equilibrium analysis and Pareto improvement of a supply chain under revenue sharing contract with consignment and slotting allowances," *African Journal of Business Management*, 22, 5, September 2011, pp. 8763-8777.

[7] G. P.Cachon, M. A. Lariviere, "Supply chain coordination with revenue-sharing contracts: strengths and limitations," *Management science*, 51, 1, January 2005, pp. 30-44.

[8] S. Li, Z.Hua, "A note on channel performance under consignment contract with revenue sharing," *European Journal of Operational Research*, 184, 2, March 2008, pp. 793-796.

Emergency Decision-supporting System Based on Multi-Agents Negotiation

Qian-sheng Gong[1,2]
[1]Security Management Institute, Xi'an University of Science and Technology, Xi'an, China
[2]College of Economics and Business Management, Xi'an Shiyou University, Xi'an , China
(gongqiansheng1979@126.com)

Abstract - **In light of the high complexity of the Emergency Decision-making, this paper introduced the Agent technology in artificial intelligence theory on the basis of the traditional Emergency Decision-supporting System. Taking into full consideration of such aspects as the function structure and systematic structure, and the negotiation mechanism among Agents, an Emergency Decision-supporting System is built on the basis of Multi-Agent negotiation. The negotiation and calculation is achieved through the combination of the scenario analysis and case reasoning.**

Keywords - **Consultation, decision-supporting system, Multi-Agent, negotiation**

I. INTRODUCTION

The Emergency refers to the sudden incidents, which may threaten to cause serious harm to society, so as to taking the corresponding Emergency measures to cope with natural disasters, accidental disasters, public health Emergencies and social security incidents. After the occurring of the emergency, how to control its development in the shortest possible time and solve the resulting problems promptly and properly, as well as to minimize the loss of the event, are the hot issues in current academic research. However, due to the sudden occurrence, the incomplete information, the urgency of the time, the seriousness of the consequences and the limitation on the people's respond to Emergencies from the experience and rational aspects, making the Emergency Decision-making highly complicate. In addition, the Emergency Response is often related to a number of different departments, organizations and individuals, so how to coordinate their actions to improve the efficiency of Emergency Response is an urgent need to resolve the difficult problems. Therefore, the introduction of Multi-Agent technology in the artificial intelligence theory, an efficient computer Emergency Decision-supporting system is built for the decision-makers, and helping decision-makers make the right decisions in the short term have important theoretical and practical significance.

The concept of Agent first appeared in the artificial intelligence in the 1970s, and it has developed in the late 1980s, nowadays it has become the very active direction of the frontier research in the field of the computer science and technology, information engineering and network communication. As to the Agent technology application, wooldrige and Jelmings think it is mainly applied to the complex software system, which is complex structure, which can be decomposed into multiple subsystems, while the Emergency Decision-supporting System has the typical characteristics of complex software system, which can provide the new application fields for Multi-Agent technology [1]. In addition, in the Emergency Decision-supporting System based on Multi-Agent technology, the user only need to submit the task to the Agent on behalf of the user, the system will display the results to the user. Throughout the whole decision-making process, the mutual cooperation of the various types of Agents is achieved; and the users do not need to participate in the decision-making task decomposition and scheduling, as well as the information exchange between users and the collaboration of the knowledge-based system and the mission system, which can greatly improve the intellectualization of the decision-making system [2].

II. THE FUNCTION STRUCTURE OF THE EMERGENCY DECISION-SUPPORTING SYSTEM ON THE BASIS OF MULTI-AGENT NEGOTIATION

The Emergency Decision-support system should

E. Qi et al. (eds.), *Proceedings of the 5th International Asia Conference on Industrial Engineering and Management Innovation (IEMI2014)*, Proceedings of the International Asia Conference on Industrial Engineering and Management Innovation 1, DOI 10.2991/978-94-6239-100-0_9,

provide the technical support for Emergency Response and coordinated decision-making. Therefore, the characteristics and content of the Emergency Decisions directly determine the main function of the Emergency Decision-supporting System, whereby the design for the main functions of the Emergency Decision-supporting System based on Agent negotiation are as follows:

1) Information collection function

Through a variety of ways to collect down a large number of structured, semi-structured or unstructured event scenario information, and make it steamer, classified and formatted in a short period of time, and save it to the information database, so as to provide basic information for the scenario analysis and the Emergency decision

2) Communication and sharing information function

After the occurrence of the Emergency, information should be informed promptly to the relevant departments and individuals, achieving real-time inquiries of information, so that decision-making body and the relevant people can grasp the latest situation of the occurrence and the development of the Emergency at the first time so as to achieve the information sharing under the state of emergency.

3) Scenario Simulation and deduction function

The Emergency Simulation model is built on the basis of collected Scenario information, and predicting the future trends and deduction of the events by making use of the scientific methods.

4) Plan generation and evaluation function

The Emergency Response Plan is promptly generated by adopting the case reasoning method, and making a comprehensive evaluation of the rationality and effectiveness of the plan according to the results of scenario simulation and deduction, so as to correct the plan.

5) Decision-making body negotiation and the overall plan generation function

After the occurrence of the Emergency, coordinating the decision-making body on the basis of its simulation and deduction, and finishing the generation, submitting, modifying and integration process of sub-plans of the relevant departments, and making the overall assessment of the integration of the whole plan, which to form the final decision-making plan.

6) *Information maintenance and management function*

Efficient Emergency Decision-making Process is built on the basis of possession of sufficient information, therefore, the operation of the Emergency Decision -supporting system requires a large number of databases, and including information base, model base, and case base etc. also these databases as well as the safe operation of the entire system should be necessarily managed and maintained [3-5].

III. THE STRUCTURE OF THE SYSTEM

Because Agent has the ability in autonomy, adaptability and learning aspects, making agent as the basic integrated unit in designing Emergency Decision-supporting System. The following is the introduction of the whole Emergency decision-support system from two aspects in the constitution of the Agent entity and the establishment of the systematic structure.

A. The constitution of each Agent entity in the system

Agent is the individual, organization, or computer system which is able to adapt to the environment, and use flexible, independent action to achieve their designed goals, because Agent has a considerable degree of autonomy, initiative and adaptability, and to some extent it plays the role in human being to complete part of the task, the stability of the entire Emergency Decision-supporting System can be effectively guaranteed, so it will not lead to the collapse of the entire system just because of a subsystem error. In addition, a simple mechanical tasks such as information searching and selection can also be assigned to the relevant Agent in the problem-solution process, so the decision-making body can be free from the tedious day-to-day affairs to focus mostly on the decision-making issue itself.

Based on demands, the Intelligent Agent in the system mainly includes the following aspects:

1) Intelligent Interface Agent is the human-computer interaction platform between decision-making body and the system, which is used to understand the demands of the decision-making body, and accept the task given by decision-making body, and decompose the task. Decision-making body will visit the various components of the system through the intelligent interface Agent, and output the results of the visit. Intelligent Interface Agent is different from the traditional systematic interface, which is only emphasized on the human-computer interaction, whereas the intelligent interface agent focused on the Agent autonomy and learning, during the interactive process of the Auxiliary decision-making with the user, the agent can obtain the knowledge of certain characteristics of the decision-making body through continuous learning, which can independently make decisions consistent with the will of the user in the decision-making process; 2) Managing and mastering all the relevant information for the Agent, taking charge of the management, decision-making task decomposition, as well as coordination and cooperation among all the Agents in the system; 3) Scenario Analysis Agent, mainly to complete the emergency scenario content acquisition, scenario knowledge representation, the scenario network construction and scenario deduction and simulation function, so as to provide the necessary scenario input for the case reasoning. 4) Case Reasoning Agent is mainly based on the input scenario analysis results, in accordance with the scenario expression, case retrieval, protocol amendments, and case study to achieve the case reasoning tasks; 5) Knowledge Agent, its main function is based on the calculation request of the decision-making body or other Agents, obtaining the knowledge for emergency scenario analysis, Emergency Decision-making and management from the knowledge database; 6) Collaboration Agent, collaboration issue is one of the core issues of Multi-Agent research, to achieve coordinated operation in the Multi-Agent knowledge, intent, desire, action and planning in the center of the independent Agent. Establishing methods of collaboration in the process of the negotiation strategies, so as to reuse the corresponding algorithm, obtaining the knowledge from collaboration Agent to identify their own needs, and coming to a reasonable answer through their own reasoning; 7) Case Agent, case retrieval agent, retrieving the relevant case according to the demands of the decision-making body from the case repository, which can achieve real-time retrieval of cases, improve the efficiency, and make the simple processing for the retrieved cases and output information on the basis of the required format, which can be able to meet the specific demands [6].

B. The structure of Emergency Decision-supporting System based on Multi-Agent negotiation

The structure of Emergency Decision-supporting System based on Multi-Agent negotiation can be shown in Fig.1.

As shown in Fig.1, the negotiation body involved in the negotiation decision-making requires their proposal on the basis of their respective decision needs for negotiation, demanding for the coordination among different negotiation body through Collaboration Agent, and submitting the negotiation results to the Intelligent Interface Agent. Intelligent Interface Agent collect the emergency scenario data through the Scenario Information Agent in accordance with the negotiation requirements, Scenario Analysis Agent will collect data for analysis and tidy firstly, then grouping and clustering the useful scenario elements to express as scenario knowledge; On this basis, analyzing the cause-effect relationship among scenario elements, constructing scenario network diagram; accordingly, to predict the future development, evolution, structure, size and trends of the scenario information by using the relevant approaches, so as to forming the future development trend of the entire emergencies. Case Reasoning Agent, based on the results of scenario analysis, combined with the acquisition of the effective knowledge for the knowledge Agent, calculating the issues for retrieval and matching case from the Case Agent. If there are related cases to match in the previously typical case library, you can get the initial protocol; if the negotiation body is not satisfied with the initial program, you should identify the difference between current decision-making issue and similar cases, combined with information in the knowledge base to correct the program, until obtaining the satisfied protocol. If there is no matching case in the case base, regarding this case as the new case, and making the feature extraction appended to the case base. Improving the case base and putting the new knowledge into the Knowledge Base, to make preparation for the following issues [7-8].

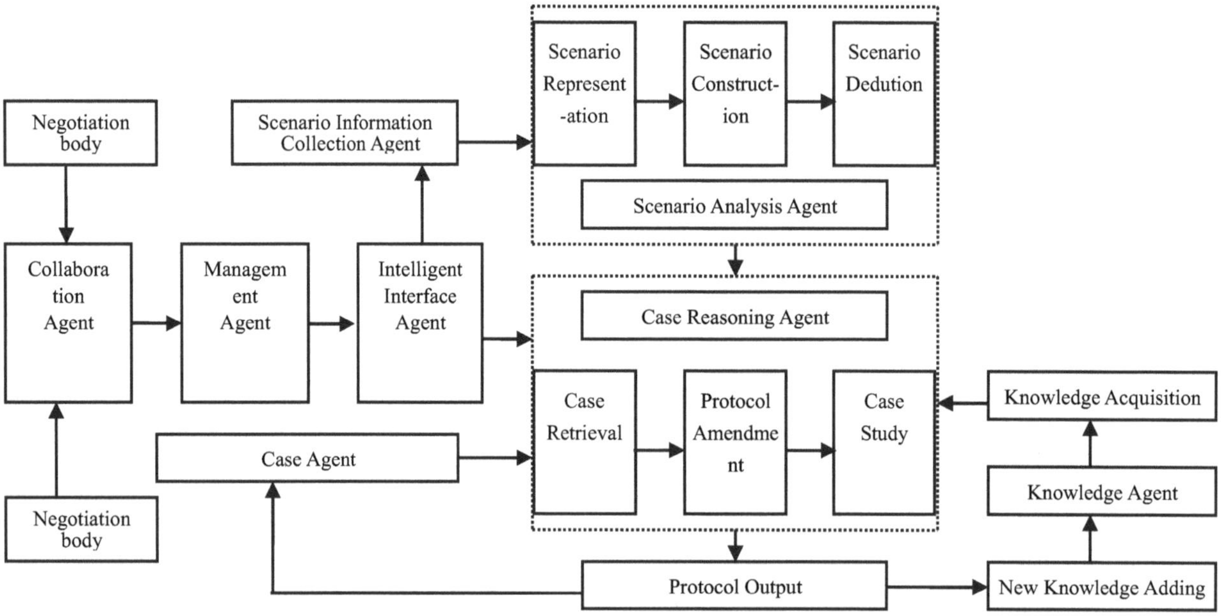

Fig.1. The structure of Emergency Decision-supporting System based on Multi-Agent negotiation

IV. NEGOTIATION AND COORDINATION MECHANISM AMONG AGENTS IN THE SYSTEM

Emergency Decision-making System is a complex issue, requiring the involvement of a number of organizations and individuals. The decision-making process involves the professional knowledge in many fields; therefore, how to coordinate the Agent's individual behavior in the Emergency Decision-supporting System is a key element in smooth operation for the whole system.

A. Negotiation model

The model can be expressed as a six-factors <A,T,C,E,P,K>, The specific meaning of each parameter is as follows: A is the collection of the involving Negotiation Agents. Each Agent has its unique ID number as their own identity in the system, and the Management Agent will establish the indexes for each type of Agent ,which is used for fast retrieval of the Agent; T is the collection of issues related to the negotiation; C is the collection of the negotiation outcomes; The so-called negotiation outcome is the contracts on every topic involved in the negotiation of each Agent; E is a non-empty collection of cases, The collection includes both the initial data in the case base, coupled with the case of the formation of new issues added to the case base for future problem-solving in accordance with the designed ideas in decision-making system. It also fully embodies the learning ability of the decision supporting system; P is the collection of negotiation protocol. The so-called negotiation protocol is the relevant rules required to participating in the negotiation among Agents. In this negotiation model, the negotiation protocol has such regulations: supposing a Negotiation Agent, it has the complete information about the Agent's preference involved in the negotiation, and resulting in the optimal contract by adopting it, and then submitting to each Agent, letting them decide whether to sign the contract, since each Agent has its own reservation utility, only when the negotiation results to its utility is greater than or equal to the reservation utility, it will sign the contract; K is a collection of knowledge, mainly related to the knowledge in relevant fields. Its origin may be aimed at the expert's negotiation results or the feature extraction and accumulation of the new problems [9].

B. Collaboration among Multi-Agents in the system

There are two main types of collaboration methods among Multi-Agents on the basis of the Emergency Decision-supporting System.

1) Based on collaboration for sharing task, namely, decision-making tasks can be decomposed into a series of relatively independent subtasks; each Agent can cooperate through calculating subtasks to achieve collaboration among Multi-Agents.

2) Based on sharing the results of the collaboration, which is similar to the principle of the expert scoring method, namely, when the decision-making task cannot be decomposed into a series of relatively independent subtasks, each Agent needs to share respectively different opinions about part issues in the whole task to achieve mutual coordination [10].

V. CONCLUSION

The paper proposes the Emergency Decision-supporting System on the basis of the Multi-Agent negotiation by adopting the Agent technology in artificial intelligence theory, and constructing the whole Emergency Decision-supporting System from the function structure and the systematic structure and Multi-Agent negotiation mechanisms and so on, and combining the methods of scenario analysis and the case reasoning to negotiate for calculation. This decision-supporting system highlights the characteristics of involving many departments in Emergency Decision-making, and demanding for the coordinating decision. It also makes up for the defects in the traditional decision-making system which only relies on the technology of the model and the data processing. It also provides the powerful technical support for making prompt and correct decisions in the emergencies, as well as providing the new research ideas and perspectives for the research of the decision-supporting system.

REFERENCES

[1] M. J .Wooldridge, N. R. Jennings, *"Intelligent agents: theory and practice"*. Knowledge Engineering Review, vol.10, no.2, pp.115-152, 1995.

[2] A. S .Haddadi, *Communication and cooperation in agent systems*. Berlin: Springer, 1996.

[3] P. J. Gmytrasiewicz, E. H. Durfee, D. K. Wehe, *A Decision-theoretic Approach to Coordinating Interactions*. Proc. 12th IJCAI, pp.62-68, 1991.

[4] M. J. Huber, E. H. Durfee, *Deciding When to Commit to Action During Observation-based Coordination*, ICMAS, 1995.

[5] E. Ephrati , *Deriving Multi-Agent Coordination Through Filtering Strategies*, IJCAI, 1995.

[6] Jing SHEN, Jun-cai YAO, *Research of Decision Support System Based on the Multi-Agent Negotiation*. Micro-electronics and Computer, pp.76-78, May.2009.

[7] M.Molina, G.Blasco, *A Multi-agent System for Emergency Decision Support*. LNCS, vol.2690, pp. 43-51,2003.

[8] O.B.Kwon, S.Norman, *Applying Case-based Reasoning and Multi-agent Intelligent System to Context-aware Comparative Shopping Original Research Article*. Decision Support Systems, vol.37, no 2, pp. 199-213, 2004.

[9] Yan CHEN, Xiao-chun HU, *Multi-Agent Automated Negotiation Model under Uncertainty*. Micro-electronics and Computer, pp.205-208, Jun.2008.

[10] Shan-lin YANG, Xiao-jian HU, *Establishing Models and Calculation Methods for the Complex Decision-making Tasks*. Beijing: Science publications, 2007.

Study on Consumers' Channel Choice Behavior Based on Fishbein's Model of Reasoned Action

Hui Zhang*, Chen Tian, Tiao-jun Xiao
School of Management and Engineering, Nanjing University, Nanjing, Jiangsu, China
(*quancheng2009@163.com, tracy_chen121@163.com, xiaotj@nju.edu.cn)

Abstract - **This paper studies the factors that may influence consumers' channel choice behavior. Taking smart phones as an example, we develop a structural equation model based on Fishbein's model and Technology Acceptance Model. We find that: (i) perceive risk of online shopping is negatively related to the attitude and subjective norms of online shopping; (ii) both the attitude and subjective norms of online shopping are positively related to the intention of online shopping and are negatively related to the intention of offline shopping, moreover, subjective norms of online shopping are positively related to the attitude toward online shopping; (iii) internet experience has no causal relationship with perceived risk of online shopping, attitude toward online shopping and subjective norms of online shopping.**

Keywords - **Perceive risk, structural equation model, TAM**

I. INTRODUCTION

With the rapid development of Internet, more and more internet users were appeared. According to the information released by China Internet Network Information Center, China's internet users have reached 618 million by the end of December 2013 [1]. Online shopping develops rapidly and many firms which open an online channel to sell products have got giant profits in recent years. Online shopping is different to the traditional (offline) shopping. People can easily search a lot of useful information from the websites and buy products very conveniently using clicking mouse. The greatest advantage of online shopping is that it is more efficient than offline shopping because it will not spend you a lot of time. More and more firms find this opportunity and try to open an online channel to expand market. Consumers have more choices of products under the dual-channel environment. When consumers decide to buy, different consumers may prefer to different types of channels. Some consumers want to measure the products' quality by experiencing and touching the product, this kind of consumers may like to buy products from offline channel. While some consumers pay much attention to the efficiency of shopping, this kind of consumers may like to choose online shopping. Consumers' behavior under the e-commerce environment shows the following characteristics: differentiation of consumer market, expansion of selection scope and sentimental of purchasing behavior [2]. Which factors can indeed affect consumers' channel choice under the dual-channel environment? We propose a structural equation model to address this question. Study consumers' channel choice has profound meanings. It can not only help us to find the factors that can indeed influence consumers' channel choice, but also provide supports for firms to carry out some efficient strategies to attract consumers. This paper constructs a structural model to study consumers' channel choice and also gets valuable conclusions.

II. LITERATURE REVIEW

A. Fishbein's Model

Fishbein's model was put forward by Ajzen and Fishbein. The behavioral intention, which refers to the will of performing a specific action under an established situation [3], is determined by the behavioral attitude and the subjective norms in this model. In Fishbein's model, the behavioral attitude and the subjective norms are endowed with weights, which can be expressed by the following formula:

$$BI = AW_1 + SW_2 . \qquad (1)$$

BI represents behavioral intention, A is behavioral attitude and S means subjective norms. W_1 and W_2 are the weights of behavioral attitude and subjective norms. The behavior attitude means that a person performs a specific behavior in a given environment, reflecting the degree of preference over the specific behavior [3]. The behavioral intention may be influenced by the social environment, thus the subjective norms means that persons who are important for him or her consider that he or she has high possibility to perform a specific behavior [3]. The influence degree to behavioral intention is different between the behavioral attitude and the subjective norms, which depends on the types of products and the environment and individual factors [4]. Fishbein's model is widely used in the field of sociology and psychology because it has a strong explanatory power to predict behaviors. Many scholars have used this model to study a lot of problems. While, this model is set in Western culture environment, can this model also apply to Oriental culture? There exist many differences between Western culture and Oriental culture: Western culture emphasizes individualism, freedom and equality, while Oriental culture attaches great importance to collectivism, authority and obedience [5]. Although Fishbein's model is set in the Western culture environment, lots of research results show that Fishbein's model can be also applied to Asian countries. Lee and Green [6] use Fishbein's model to study the purchasing behavior of consumers from South Korean, and find that Fishbein's model can be applied to South Korean. Chan and Lau [7] uses Fishbein's model to test local consumers' purchasing behavior of jewelry products, and also find that Fishbein's model has a general applicability. Zhang [8] studies consumers'

E. Qi et al. (eds.), *Proceedings of the 5th International Asia Conference on Industrial Engineering and Management Innovation (IEMI2014)*, Proceedings of the International Asia Conference on Industrial Engineering and Management Innovation 1, DOI 10.2991/978-94-6239-100-0_10,

purchase intention of foreign brand mobile phone in Chain to verify the applicability of Fishbein's model and also gets the same conclusion.

B. Technology Acceptance Model

Davis et al. [9] extends the relationship between behavioral attitude and behavioral intention, which puts forward the Technology Acceptance Model (TAM). Fig.1 shows the Technology Acceptance Model in detail.

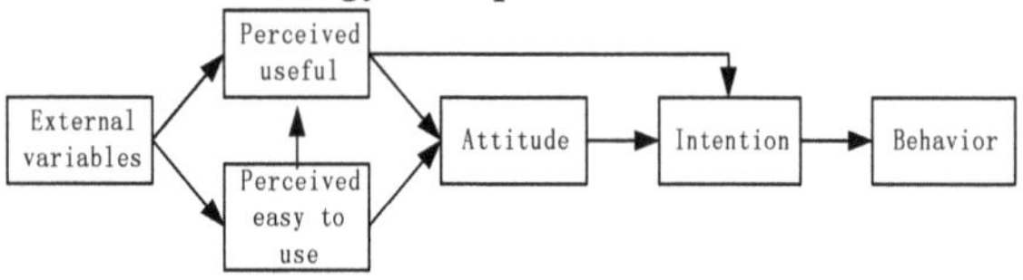

Fig.1. Technology Acceptance Model

In TAM, the behavior of an individual using system is decided by the intention which is decided by the attitude and perceived usefulness jointly. Furthermore, perceived usefulness of the system and perceived ease to use the system jointly decide the attitude. TAM is extensively used by scholars, not only because that it is simple and easy to understand, but also for its strong explanation of behavior and high reliability and validity [10]. In recent years, many scholars use TAM to analyze behaviors, and extend the Technology Acceptance Model [11-14]. Cheng and Bao [15] construct a structural model called consumers' online shopping model based on TAM, and find that perceived usefulness, perceived convenience and security have a significant effect on attitude and intention of online shopping. Jing and Zhou [16] also study consumers' behavior of online shopping based on TAM.

III. THE BASIC MODEL

Most quantitative researches study channel choice from the perspective of supply chain management by using game theory. A supplier not only can use traditional retail channels to sell products, but also can open an online channel. Can a firm get more profit if he opens an online channel? It has become an important problem that many scholars study from different perspectives [17-20]. In recent years, some scholars study channel choice from consumers' perspective. Gupta et al. [21] develop an economic model of consumers' channel choice and find that risk-averse consumers are more inclined to build channel loyalty than risk-neutral consumers. Empirical studies about channel choice are very few. Wang et al. [22] summarize the factors that can influence consumers' channel choice: channel factors, situational factors and consumers' own factors. Nicholson et al. [23] find that five kinds of situational factors can affect the customers' channel choice: social conditions, time conditions, physical condition, mental state and task. Yang et al. [24] develop a channel selection model of ULP (University-licensed-products) based on TAM, they divide the intention into the intention of online channel and the intention of offline channel.

This study develops a consumers' channel section model, combining with TAM and Fishbein's model and adding two external variables: perceived risk and internet experience, see Fig.2. Similar to Yang et al. [24], we also divide the intention into the intention of online shopping and the intention of offline shopping. This paper is different to Yang et al. [24] in the following aspects. Firstly, we study consumer's channel choice based on Fishbein's model and TAM, while Yang et al. [24] study consumers' channel choice only based on TAM. Secondly, we use smart phone to study consumers' channel choice, Yang et al. [24] use ULP to study consumers' behavior. Last but not least, we also consider internet experience in our model.

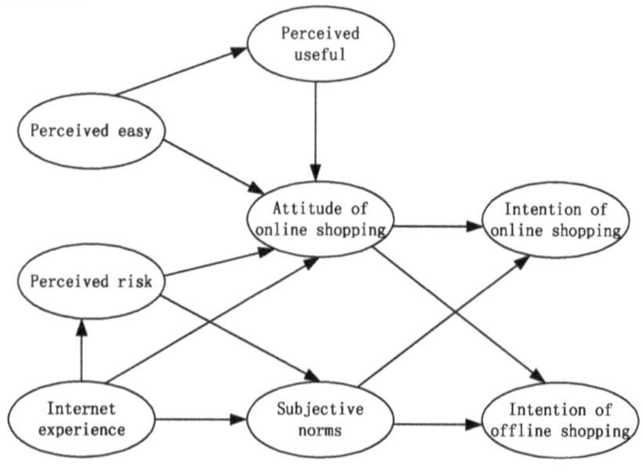

Fig.2. Consumers' Channel Choice Model

Based on Fishbein's model, TAM and Consumers' Channel Choice Model we construct, we give 12 hypotheses as follows.

H1: Perceived usefulness of online shopping is positively related to the attitude toward online shopping.

H2: Perceived ease to use of online shopping is positively related to the attitude toward online shopping.

H3: Perceived ease to use of online shopping is positively related to perceived usefulness of online shopping.

H4: The attitude toward online shopping is positively related to intention of online shopping.

H5: Subjective norms of online shopping are positively related to the intention of online shopping.

We give the above 5 hypotheses based on Fishbein's model and TAM. Based on experience of purchasing products through different channels and previous studies, we give the following hypotheses.

H6: The attitude toward online shopping is negatively related to the intention of offline shopping.

H7: Subjective norms of online shopping are negatively related to the intention of offline shopping.

The attitude toward online shopping and subjective norms generally have positive effects on the intention of online shopping, while the relationship between these two variables (the attitude toward online shopping and subjective norms) and the intention of offline shopping needs scholars' further studies. Yang et al [24] find that the attitude toward online shopping also can have positive effects on the intention of online shopping. Through a simple survey, we find that when consumers have a positive attitude toward online shopping and strong subjective norms, they generally have a high intention of online shopping and a low intention of offline shopping. Based on this, we give hypotheses H6 and H7.

H8: Perceived risk of online shopping is negatively related to the intention of online shopping.

H9: Perceived risk of online shopping is negatively related to subjective norms of online shopping.

H10: Internet experience is negatively related to perceived risk of online shopping.

H11: Internet experience is positively related to the attitude of online shopping.

H12: Internet experience is positively related to subjective norms of online shopping.

Stone et al [25] define perceived risk as the expectations of losses, the larger the expectations of losses are, the higher degree of risk consumers will perceive. Anne [26] considers that risk is a major obstacle to hinder the consumers from shopping online. Consumers face not only the uncertainty of products and service, but also privacy and security risks under the e-commerce environment. Based on this, we give hypotheses H8 and H9. Miyazaki and Fernandez [27] consider that most consumers' perceived risk of online shopping due to the unfamiliarity with this kind of shopping, and point out that internet experience can reduce the degree of perceived risk. Based on this, we give Hypotheses H10, H11 and H12.

IV. STUDY DESIGN AND EMIPRICAL ANALYSIS

A. Design of Questionnaire

There exist 8 variables in our model. Every variable needs items to measure. These items are mostly got with reference to previous studies. All items of variables except internet experience use 5-point Likert-type scales ranging from 1 to 5 where 1 means strongly disagree and 5 means strongly agree. With regard to internet experience, we use 4 items to measure and these 4 items are got with reference to experts' advice. Table I shows the items of 8 variables. We use the first letter to express the whole word, such as PU (perceived usefulness of online shopping), PE (perceived ease to use of online shopping). Items of perceived usefulness of online shopping, perceived ease to use of online shopping, internet experience, perceived risk of online shopping and subjective norms of online shopping are easy to understand. With regard to the intention of online shopping and the intention of offline shopping, we use 3 items to measure. Taking the intention of online shopping as an example, the first item is that I will choose online shopping when I want to buy a smart phone, the second item is that I will consider online shopping first when I want to buy a smart phone, the third item is that I will introduce my friends to buy from online channel if they want to buy a smart phone. With regard to the attitude toward online shopping, we use 2 items to measure, the first item is that I think it is right to choose online shopping and the second item is that I think it is smart to choose online shopping.

TABLE I
ITEMS OF RESEARCH VARIABLES OF CHANNEL SELECTION

PU	A1: save time	PE	B1: contact easy	IO	G1: will choose
	A2: high efficiency		B2: cancel order easy		G2: first consider
	A3:convenient		B3: order procedure easy		G3:introduce friends
			B4: study easy	IOF	H1: will choose
PR	C1: high risk	IE	D1: network age		H2: first consider
	C2: high uncertainty		D2: time duration		H3:introduce friends
	C3:insecurity		D3: time online per week	AO	K1: correct
	C4: low credibility		D4: age of shopping online		K2: smart
SN	F1: friends think that I should buy it from online channel				
	F2: relatives think that I should buy it from online channel				
	F3: persons who are important to me think that I should buy it from online channel				
	F4: as far as I know, persons around me have high probability to buy from online channel				

To study the hypotheses we give, we conducte a specific interview with students of Nanjing University and some social workers in Jiangsu. In this paper, we take smart phones as an example. As a kind of high-tech products, smart phones have the following characteristics: updating quickly, high uncertainty and high level of consumers' concern. In recent years, smart phones are widely popular with consumers. Famous brands, such as iphone, lenovo and samsung, occupy an important share of the smart phones' market. We use two channels (online and offline channel) to implement the survey of 450 questionnaires. At last, we recycled 380 questionnaires, and 359 valid questionnaires among them, thus the effective recovery rate reaches 79.7%. Table II shows the background of the respondents. E D in Table II means education degree.

TABLE II
BACKGROUND OF RESPONDENTS

Gender	Male 161(44.8%) Female 198(55.2%)			
Age	<20 age: 54(15%) 20-25 age: 259(72.1%) 25-30 age: 41(11.4%) >30 age 5(1.4%)			
E D	Other 13(3.65)	Undergraduate 153(42.6%)	Master 187(52.1%)	Doctor 6(1.7%)
Career	Students 316(88%) Social workers 43(12%)			

B. Reliablity analysis and Validity test

Reliability is an important index to evaluate the quality of the questionnaire. It can reflect the stability and consistency of the measure method. There are 4 kinds of reliability: split-half reliability, test-retest reliability, Cronbach's Alpha reliability and constructible reliability. Cronbach's Alpha reliability is widely used among the above 4 kinds of reliability. This research use Cronbach's Alpha reliability and this reliability needs Cronbach's Alpha coefficient to measure. Scholars generally believe that the Cronbach's Alpha reliability is very good when the Cronbach's Alpha coefficient is more than 0.7. Table III shows the Cronbach's Alpha coefficients of questionaries.

TABLE III
CRONBACH'S ALPHA COEFFICIENT OF QUESTIONNAIRE

Variables	Items	Cronbach's Alpha
Perceived Usefulness	A1, A2, A3	0.847
Perceived Easy to use	B1, B2, B3, B4	0.765
Perceived Risk	C1, C2, C3, C4	0.792
Internet Experience	D1, D2, D3, D4	0.623
Attitude of online shopping	K1, K2	0.875
Subjective Norms	F1, F2, F3, F4	0.819
Intention of online shopping	G1, G2, G3	0.902
Intention of offline shopping	H1, H2, H3	0.865

From Table III, we can get that all variables' Cronbach Alpha coefficients are higher than 0.7, it can indicate that all variables can be applied to analysis.

Validity refers to a degree of the psychological and behavioral characteristics which the measurement tools are able to test. There are many types of validity and 3 types of validity (content validity, criterion validity and structural validity) among them are widely used by scholars. In this study, most items of variables are got with reference to previous studies and experts' advice, thus this can guarantee the variables have a good content validity. In this research, we use the exploratory factor analysis to study the structural validity. Using Bartlett Test, we find KMO (Kaiser-Meyer-Olkin) is 0.826 which is closer to 1 and higher than 0.5 (the threshold is 0.5). It indicates that all variables can be applied to use the exploratory factor analysis. Using the exploratory analysis, we extracted seven factors and found that the accumulated explained variance is 66.698% which is higher than the threshold (50%). Moreover, every variables' factor load is higher than 0.5. Therefore, the questionnaire can guarantee a good structural validity.

C. Structral Equation Model Tseting

In this part, we will study the consumers' channel choice model combing with data from questionnaires. First, we use confirmatory factor method to analyze the fitness of the structural equation model we construct. There exist some indicators that can assess the fitness of the model: χ^2/df, GFI (goodness of fit index), AGFI (adjusted goodness of fit index), CFI (comparative fit index) and RMSEA (Root Mean Square Error of Approximation). A structural equation model can be very good when these indicators should meet the following conditions: $\chi^2/df < 3$, GFI>0.9, AGFI>0.8, CFI>0.9 and $0.05 \le \text{RMSEA} \le 0.08$. According to the results from AMOS, we add a hypothesis: H13 (Subjective norms of online shopping are positively related to attitude of online shopping). We analyze the model again by using the same method. Table IV shows the model's result of these indicators.

TABLE IV
GOODNESS OF FIT INDICES OF THE MODEL

χ^2	*df*	χ^2/df	*GFI*	*AGFI*	*CFI*	*RMSEA*
619.478	310	1.998	0.888	0.863	0.931	0.053

From Table IV, we can get that, $\chi^2/df = 1.998$ which meet the condition $\chi^2/df < 3$, AGFI, CFI and RMSEA meet the conditions, too. GFI is lower than 0.9, but very around to 0.9 and higher than the threshold 0.8. Therefore, the structural equation model we construct can be applied to analyze. Table V and Fig.3 show the result of empirical studies.

TABLE V
TEST RESULT OF HYPOTHESES

Hypotheses	relationship	S T C	*T*-value	*P*-value	result
H1	+	0.146	2.568	*	Support
H2	+	0.246	3.840	***	Support
H3	+	0.423	5.863	***	Support
H4	+	0.455	8.445	***	Support
H5	+	0.458	7.303	***	Support
H6	-	-0.328	-5.055	***	Support
H7	-	-0.261	-3.921	***	Support
H8	-	-0.180	-3.206	**	Support
H9	-	-0.177	-2.701	**	Support
H10	-	-0.094	-1.367	0.172	NO S
H11	+	0.090	1.598	0.110	NO S
H12	+	0.041	0.634	0.526	NO S
H13	+	0.469	7.247	***	Support

***means $P<0.001$, **means $P<0.01$, *means $P<0.05$

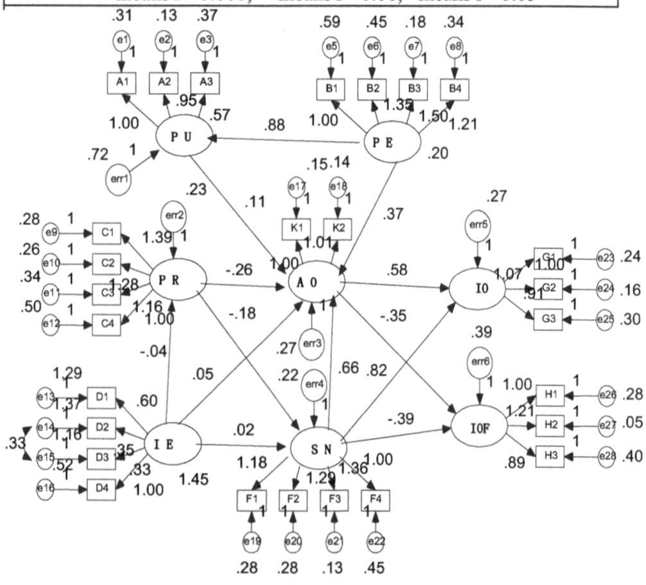

Fig.3. Estimates of Research Model of Chanel Selection

S T C in Table V means Standardized Path Coefficient. In Fig.3, we use the first letter to express the whole word, such as P U refers to perceived usefulness of online shopping, I O refers to the intention of online shopping and I O F means the intention of offline shopping. From Table V, we know that Hypotheses H1-H9 and H13 all got supported, while hypotheses H10, H11 and H13 are not supported. So, internet experience has no significant causal relationship with the attitude of online shopping, subjective norms of online shopping and perceived risk of online shopping.

V. DISCUSSIONS AND IMPLICATIONS

In this research, we get that the relationship between the attitude of online shopping and the intention of offline shopping is contrary to Yang et al. [24]. This may be caused by the products' type. As we known, smart phones are very expensive and consumers will pay much attention on it if they want to buy a smart phone. Thus, when they have a positive attitude of buying a smart phone through online channel, they may have a negative intention of offline shopping. Hypotheses we give based on TAM and Fishbein's model all get supported, which indicates that TAM and Fishbein's model have good applicability. Moreover, we also find that subjective norms of online shopping can be positively related to the attitude toward online shopping. To some extent, this expands Fishbein's model. With regard to the relationship between internet experience and the attitude toward online shopping, subjective norms of online shopping and perceive risk of online shopping, this needs more examinations. A person who has rich internet experience can have a short online shopping age at the same time, thus, he or she may not have a positive attitude toward online shopping, may not perceive low risk or obey subjective norms of online shopping.

Our study has some practical significance for enterprises who sell smart phones. Online retailers should do well with the work of website construction, and make consumers perceive the usefulness of online shopping and ease to use. For example, online retailers can design the website interface clearly, easy to understand and update information of products timely. It's necessary for online retailers to take some measures to reduce the consumers' perceived risk because perceived risk of online shopping is negatively related to the attitude toward online shopping. Online retailers can provide timely and reliable trading information and ensure the safety of trading capital and customer data privacy. Offline retailers should guarantee the high quality of services, and try their best to form a good reputation. Moreover, offline retailers can broaden the sales channels, for example, set up online sales channel. The attitude toward online shopping and subjective norms of online shopping are positively related to the intention of online shopping, and are negatively related to the intention of offline shopping, so the online and offline retailers should do good works of service, timely delivery and effective returns. Online and offline retailers should do something to make consumers satisfied. Online and offline retailers should understand how consumers evaluate the products and services after they receive the products. In this way, they can find the part in which they do not well and take some effective measures timely to improve.

VI. CONCLUSIONS

In this paper, we develop a structural equation model combing with TAM and Fishbein's model. We analyze this model by using AMOS and data from questionnaires and get a lot of conclusions. We get that: Perceived usefulness of online shopping is positively related to attitude toward online shopping; Perceived ease to use of online shopping is positively related to the attitude toward online shopping; Perceived ease to use of online shopping is positively related to perceived usefulness of online shopping; The attitude toward online shopping is positively related to the intention of online shopping; Subjective norms of online shopping are positively related to the intention of online shopping; The attitude toward online shopping is negatively related to the intention of offline shopping; Subjective norms of online shopping are negatively related to the intention of offline shopping; Perceived risk of online shopping is negatively related to the intention of online shopping; Perceived risk of online shopping is negatively related to subjective norms of online shopping; Subjective norms of online shopping are also positively related to the attitude toward online shopping. While, internet experience has no significant causal relationship with the attitude toward online shopping, subjective norms of online shopping and perceive risk of online shopping.

This study explores the factors that can influence consumers' channel choice, and this study is based on smart phones which are very popular with young consumers. Future research can be expanded to other products or directly regard products' type as a variable. The sample in this study is mainly made up of college students in Nanjing city. In future study, we can increase the ratio of social works. Moreover, items that measure internet experience needs to be improved, too.

ACKNOWLEDGMENT

This research was supported in part by the National Natural Science Foundation of China under Grants 70971060 and 71371093.

REFERENCES

[1] China Internet Network Information Center. The 33th statistical survey report on the internet development in China [R], Beijing, 2013, 12

[2] Z. C. Li, M L Liu. A study on the consumer behavior in electronic commerce. Chinese Journal of Management Science, 2002, 10(6): 88-91. (in Chinese)

[3] M. Fishbein, I. Ajzen. Belief, attitude, intention, and behavior: An introduction to theory and research mass: Addison-Wesley, 1975.

[4] Ajzen I, Fishbein M. Understanding attitudes and predicting social behavior [M]. Engwood Cliffs, NJ: Prentice Hall, 1980: 42-60.

[5] Q Y Hu. China's consumer market: Shanghai consumer's attitudes toward foreign-brand and Chinese-brand apparel . Unpublished Doctoral Dissertation, University of Missouri Columbia, 1994.

[6] C. Lee, R. T. Green. Cross-cultural examination of Fishbein behavior intention model. Journal of International Business Studies, 1991, 22(2):289-305.

[7] R. Y. Chan, L. Lau. A test of the Fishbein–Ajzen behavioral intention model under Chinese settings: Are there any differences between PRC and Hong Kong consumers?. Journal of Marketing Practice: Applied Marketing Science, 1998, 4(3): 85-101.

[8] L. Zhang. Applicability of Fishbein model and impacts of acculturation: A study of foreign-brand mobile phone

buying intentions. Journal of Management Sciences, 2007, 20(1): 30-37. (in Chinese)
[9] F. D. Davis, R. P. Bagozzi, P. R. Warshaw. User acceptance of computer technology: A comparison of two theoretical models. Management Science, 1989, 35(8): 982-1003.
[10] V. Venkatesh, F .D. Davis. A theoretical extension of the technology acceptance model: For longitudinal field studies. Management Science, 2006, 46(2): 186-204.
[11] W. Lewis, R. Agarwal, V. Sambamurthy. Sources of influence on beliefs about information technology use: An empirical study of knowledge workers. MIS Quarterly, 2003, 27 (4): 657 – 678.
[12] B. H. Wixom, P. A. Todd. A theoretical integration of user satisfaction and technology acceptance. Information Systems Research, 2005, 16 (1): 85 - 102.
[13] E. W. T. Ngai, J. K. L. Poon, Y. H. C. Chan. Empirical examination of the adoption of WebCT using TAM. Computers & Education, 2007, 48(2):250-267.
[14] I. F. Liu, M. C. Chen, Y.S Sun, D. Wible, C. H. Kuo. Extending the TAM model to explore the factors that affect intention to use online learning community. Computers & Education, 2010, 54(2):600-610.
[15] H. Cheng, G. M. Bao. The empirical study of determinants of online purchasing intention. Journal of Quantitative & Technical Economics, 2003, 11:150-155. (in Chinese)
[16] M. Jing, Y. Zhou. Study on online consumer behavior based on TAM and perceived risk. Shanghai Management Science, 2005, (5):5-7. (in Chinese)
[17] J. Huang, T J. Xiao, Z. H. Sheng. Study on the multi-channel supply chain management: An literature review. Science Research Management, 2009, 30(5):25-32. (in Chinese)
[18] H. T. K. Park, H. T. Keh. Modelling hybrid distribution channels: A game - theoretic analysis. Journal of Retailing and Consumer Services, 2003, 10(3):155- 167.
[19] W. K. Chiang, D. Chhajed, J. D. Hess. Direct marketing, indirect profits: A strategic analysis of dual- channel supply chain design. Management Science, 2003, 49(1): 1- 20.
[20] Y. C. Liu, J. Zhang. The benefits of personalized pricing in a channel. Marketing Science, 2006, 25(1): 97- 105.
[21] A. Gupta A, B-C. Su, Z. P Walter. Risk profile and consumer shopping behavior in electronic and traditional channels. Decision Support Systems, 2004 (38): 347-367.
[22] C. Wang, Y. J. Li. Research on internet consumer's behavior based on multiattribute utility theory. Journal of Systems & Management, 2010, 19(1): 62-67. (in Chinese)
[23] M. Nicholson, L. Clarke, M. Blakemore. One brand, three ways to shop: Situational variables and multichannel consumer behavior. International Review of Retail, Distribution and Consumer Research, 2002, 12(2): 131-148.
[24] S. J. Yang, J. K. Park, J. H. Park. Consumer's channel choice for university-licensed products: Exploring factors of consumer acceptance with social identification. Journal of Retailing and Consumer Services, 2007, 14(3): 165-174.
[25] N. G. K. Stone. Perceived rick: Further considerations for the marketing discipline. European Journal of Marketing, 1993, 27(3): 39-50.
[26] S. C. Anne. Perceived risk and risk reduction strategies in internet shopping. The International Review of Retail, Distribution and Consumer Research, 2002, 12(4): 375-394.
[27] A. D. Miyazaki, A. Fernandez. Consumer perception of privacy and security risks for online shopping. The Journal of Consumer Affairs, 2001, 35(1): 27-44.

A Study on the Spatial Disparity of the Regional Energy and Environment Efficiency

Jian Li*, Qin Su
School of Management, Xi'an Jiantong University, Xi'an, China
(ljlcxwxz@stu.xjtu.edu.cn)

Abstract - **Rapid economic growth caused lots of energy consumption and environment problems, growing researches began to pay attention to energy and environment efficiency. This study conducted a study on the spatial disparity of the regional energy and environment efficiency in China, using data envelopment analysis (DEA), we selected 30 provinces/cities in China and applied DEA model to analyze the regional energy/environment efficiency in the period of 2005 to 2010. Furthermore, the spatial analysis for the efficiency was conducted. Results highlighted that as a whole, the energy/environment efficiency in China was low. and the efficiency was decrease from the eastern to western areas. A high spatial clustering effect was identified between neighboring provinces, which presenting significant regional correlation. Finally, appropriate policy design and subsidies strategies are raised by considering the local realities.**

Keywords - **China, DEA, Energy/environment efficiency, spatial disparity**

I. INTRODUCTION

With the advancement of industrialization in China, our country's economy keeps grow continuously, and GDP has increased from 364.52 billion RMB in 1978 to 56.88 trillion RMB in 2013. Rapid economic growth has caused a lot of energy consumption and serious environmental problems. The Global Climate Change Conference was held in Copenhagen in 2009, Chinese government promised by 2020 to reduce carbon dioxide emissions of per unit GDP 40-45% compared with 2005 levels. The key factors are to improve energy efficiency and environment quality under the continuous growth of economy.

Energy efficiency is a relative concept, and at present, there are lots of different definitions of it, according to Ang (2006) [1], three indicators are used to calculate energy efficiency: thermodynamic indicators, physical-based indicators, and monetary-based indicators, both thermodynamic indicators and physical-based indicators are mainly taken consideration into project from micro perspective. And monetary-based indicators are referred to the energy consumption per unit current output from macro-economy level. Environmental efficiency is a measure of the actual pollutant emissions and potential pollution emissions; its economic implication is the potential for pollution emission reduction when the current levels of input and output keeps unchanged with the DMU of environmental production frontier as the benchmark. Farrell (1957) constructs a joint production framework incorporating both the "good" products and the "bad" ones, expanding the traditional efficiency evaluation system. Fare et al. (1989) [2] for the first time created a hyperbolic efficiency model, in which environmental efficiency is defined as to increase the "good" product yield while simultaneously reduce the "bad" (pollution) product, and he figured out the hyperbolic efficiency by linear programming. At macro-economy level, data envelopment analysis (DEA) (Charnes et al, 1978) [3] has widely taken consideration in studying energy and environment efficiency. Hu and Wang (2006) [4], within the total factor productivity framework, defined the total factor energy efficiency indicators (TFEE) by DEA method, obtained energy efficiency by calculating the ratio of "optimal energy inputs in the frontier" over "the actual energy invested" as a breakthrough to make up the deficiency in the traditional indicator of energy production that only considers energy as the only factor. However, Hu and Wang (2006) just take expected output in GDP as the sole output, not considering the CO_2, SO_2 and other unexpected outputs. In addition, Zhou et al. (2008) build environment and energy efficiency model, and contains energy inputs [5], and the expected and unexpected output model, but ignores the non-energy factor as inputs, such as labor and capital. Meanwhile, due to unbalanced regional economic level, climate conditions and resources endowment are different in China, thus form the regional diversity, and residents living consumption habits (embodying in energy consumption), and make the difference of energy and environmental efficiency among regions. At present, little literature take energy efficiency, environment efficiency and its spatial correlation consideration. With such background, in this research we systematically study energy efficiency and environmental efficiency of different provinces in China based on the review of earlier theories on economic growth, we further analyze spatial differences and correlation. Our research will produce practical significance for economic development and energy consumption in accordance with China's conditions.

E. Qi et al. (eds.), *Proceedings of the 5th International Asia Conference on Industrial Engineering and Management Innovation (IEMI2014)*, Proceedings of the International Asia Conference on Industrial Engineering and Management Innovation 1, DOI 10.2991/978-94-6239-100-0_11,

II. METHODOLOGY

A. Factors efficiency model

DEA model is a powerful tool in dealing with production with multi-input and output and widely applied to evaluate efficiency in energy and environmental fields. The basic idea is to conduct a comprehensive analysis of the input and output data of a number of decision making units (DMU), then compare each DMU with the efficient production frontier which is identified through linear programming, and finally figure out the efficiencies of each decision-making unit. The basic model is:

$$(I_{c^2R})\begin{cases}\underset{\theta,\lambda}{Min\,\theta} \\ -y_i + Y\lambda \geq 0, \\ \theta x_i - Y\lambda \geq 0 \\ \lambda \geq 0\end{cases}$$

Where in θ is scalar, λ is a constant vector, $N \times 1$, the value of θ is the efficiency of DMU_i, normally with $\theta \leq 1$. The scenario with $\theta = 1$ means that the DMU is technically efficient and right on the efficient frontier (Coelli, 1998 [6]). In this article we adopt the input DEA method, C^{2}R (Charnes et al, 1978)model under the CRS assumptions.

B. The model of energy and environment efficiency

According to the calculation method of energy efficiency and environment efficiency (Wei et al., 2007 [7]; Wang et al., 2010 [8]), We employ C-D production function $Y(t) = AK(t)^{\alpha} L(t)^{\beta}$, and establish the following functional form to calculate the total factor energy efficiency and environmental efficiency:

$$Y_{i,t} = A_{i,t} K_{i,t}{}^{\alpha} L_{i,t}{}^{\beta} E_{i,t}{}^{1-\alpha-\beta}$$

Where, i and t stands for decision making units(DMU) and time, Y, A, K, L and E stands for output (including both the "good" and the "bad" outputs), technological progress, capital stock, labor, and energy consumption input respectively.

C. Spatial correlation model

Global spatial autocorrelation is the whole-spaced spatial distribution description of regional economic development. In actual spatial correlation analysis application research, $Moran'sI$ index is mostly used. $Moran'sI$ is defined below:

$$Moran'sI = [\sum_{i=1}^{n}\sum_{j=1}^{n} W_{ij}(Y_i - \bar{Y})(Y_j - \bar{Y})] / [S^2 \sum_{i=1}^{n}\sum_{j=1}^{n} W_{ij}]$$

Wherein, $S^2 = 1/n\sum_{i=1}^{n}(Y_i - \bar{Y})$, $\bar{Y} = 1/n\sum_{i=1}^{n} Y_i$,

Y_i is proxy for the observation of the i th region, n is the total number of regions as this study is the provincial spatial correlation nation-wide. W_{ij} is spatial weight matrix, $Moran'sI$ can be considered as the sum of the products of each observation, with a range of $-1 \leq Moran'sI \leq 1$. If the observations are spatial positively related, the index would be larger, and smaller when they are negatively related.

III. EMPIRICAL ANALYSIS

Data Sources

We select 30 provinces (municipalities, excluding Tibet, Hong Kong, Macao and Taiwan) and 2005-2010 as the sample period, the data sources are "*China Statistical Yearbook*" (2006-2011), "*China Energy Statistical Yearbook*" (2006-2011). Data were processed with ArcGIS, GEODA and Deap 2.1.

(1) Input Indicators:

1. Capital Stock(K): relevant data is from *China Statistical Yearbook* (2006-2011); "perpetual inventory mechanism" is employed to estimate the actual capital stock, with $K_{i,t} = I_{i,t} + (1-\delta_i)K_{i,t-1}$, therein we extend the capital stock series to 2010 in strict accordance with their methods and convert the results into the capital stock in constant prices of year 2000.

2. Labor (L): data source is "China Statistical Yearbook"(2006-2011), employment of a year is the value of the average of employment at the end of that year and the employment at the end of the previous year. Since per capita years of education or other data is inaccessible, provincial differences of labor quality is not included.

3. Energy (E): relevant data is from "China Statistical Yearbook"(2006-2011), using energy consumption at the end of year as energy inputs for all provinces (municipalities), converted into standard coal.

(2) Output Indicators:

1. The "Good" output (Y_G): provincial GDP is selected, from "China Statistical Yearbook" 2011, and with GDP deflator derived from constant prices of year 2000 we make the adjustments.

2. The "bad" output (Y_B): the total CO_2 and SO_2 emissions present this indicator, and the data is from "China Statistical Yearbook" (2006-2011).

Notes: when calculating energy efficiency, we use provincial capital stock, labor, and energy consumption as input variables, provincial GDPs as output variables. For environmental efficiency, we use provincial capital stock, labor, and energy consumption as input variables, and provincial GDPs and their total CO_2 and SO_2emissions as the output variable.

IV. RESULTS AND DISCUSSION

A. Analysis on energy and environment efficiency

TABLE I lists the provincial or municipal energy efficiency values from 2005 to 2010. As can be found that Beijing, Shanghai and Guangdong province had the highest energy efficiencies from2005 to 2010, with their efficiency values all being 1duringthe six years and located at the frontier, and these three provinces or cities are all in the eastern or eastern coast area; Tianjin, Fujian and Chongqing also have a number of values at the frontier; the two provinces performed worst in terms of energy efficiency are Qinghai and Ningxia, with their values both lower than 0.6, and the two are in the northwest region.

TABLE II shows the environmental efficiency values. Beijing, Shanxi, Inner Mongolia, Shanghai, Guangdong, Guangxi, Chongqing, Guizhou, Ningxia are among the top in the list and their environmental efficiency values are all at the environmental frontier from 2005 to2010, and can be considered as the benchmark in evaluating environmental efficiency levels of other provinces. The geoFig.ical distribution is three in North China, one in the East, two in the South, two in the Southwest, and one in the Northwest. Among them Beijing, Shanghai, and Guangdong are also the three with the highest energy efficiency values and they are the best practitioners in energy-saving and being environment-friendly. Tianjin, Fujian provinces are alternately on the environmental frontier and Qinghai has the lowest environmental efficiency value, the six-year average of which is only0.696, indicating that the back ward provinces are faced with high pressure in emission reduction.

TABLE I
PROVINCIAL AND MUNICIPAL ENERGY EFFICIENCY (2005-2010)

Province	2005	2006	2007	2008	2009	2010	average
Beijing	1.000	1.000	1.000	1.000	1.000	1.000	1.000
Tianjin	0.980	1.000	0.975	1.000	0.921	0.926	0.967
Hebei	0.800	0.736	0.788	0.725	0.619	0.642	0.718
Shanxi	0.846	0.797	0.931	0.923	0.674	0.665	0.806
Inner mongolia	0.679	0.700	0.979	0.987	0.670	0.710	0.788
Liaoning	0.778	0.717	0.679	0.648	0.600	0.564	0.664
Jilin	0.759	0.677	0.661	0.639	0.639	0.692	0.678
Heilongjiang	0.889	0.850	0.806	0.757	0.692	0.688	0.780
Shanghai	1.000	1.000	1.000	1.000	1.000	1.000	1.000
Jiangsu	0.862	0.856	0.885	0.922	0.823	0.822	0.862
Zhejiang	0.891	0.866	0.846	0.832	0.835	0.842	0.852
Anhui	0.776	0.688	0.793	0.790	0.701	0.793	0.757
Fujian	0.988	0.983	1.000	1.000	0.873	0.835	0.947
Jiangxi	0.821	0.804	0.911	0.911	0.862	0.902	0.869
Shandong	0.765	0.734	0.723	0.722	0.700	0.708	0.725
Henan	0.885	0.802	0.913	0.865	0.718	0.727	0.818
Hubei	0.800	0.762	0.818	0.808	0.799	0.796	0.797
Hunan	0.837	0.813	0.910	0.910	0.863	0.892	0.871
Guangdong	1.000	1.000	1.000	1.000	1.000	1.000	1.000
Guangxi	0.989	1.000	1.000	1.000	0.851	0.771	0.935
Hainan	1.000	0.841	0.789	0.800	0.761	0.750	0.824
Chongqing	0.754	0.880	1.000	1.000	1.000	1.000	0.939
Sichuan	0.848	0.877	0.967	0.945	0.846	0.886	0.895
Guizhou	0.608	0.606	0.789	0.814	0.603	0.629	0.675
Yunnan	0.721	0.657	0.935	0.891	0.652	0.669	0.754
Shaanxi	0.643	0.670	0.615	0.599	0.597	0.604	0.621
Gansu	0.726	0.720	0.687	0.643	0.591	0.569	0.656
Qinghai	0.508	0.512	0.519	0.548	0.539	0.541	0.528
Ningxia	0.464	0.456	0.479	0.517	0.513	0.513	0.490
Xinjiang	0.621	0.622	0.609	0.612	0.581	0.613	0.610
Nationwide	0.808	0.788	0.834	0.827	0.751	0.758	----

TABLE II
PROVINCIAL AND MUNICIPAL ENVIRONMENTAL EFFICIENCY (2005-2010)

Province	2005	2006	2007	2008	2009	2010	average
Beijing	1.000	1.000	1.000	1.000	1.000	1.000	1.000
Tianjin	1.000	1.000	0.988	1.000	0.970	0.974	0.989
Hebei	0.853	0.827	0.808	0.754	0.753	0.754	0.792
Shanxi	1.000	1.000	1.000	1.000	1.000	1.000	1.000
Inner mongolia	1.000	1.000	1.000	1.000	1.000	1.000	1.000
Liaoning	0.869	0.810	0.803	0.774	0.758	0.769	0.797
Jilin	0.781	0.736	0.740	0.721	0.762	0.784	0.754
Heilongjiang	0.906	0.879	0.848	0.805	0.766	0.767	0.829
Shanghai	1.000	1.000	1.000	1.000	1.000	1.000	1.000
Jiangsu	0.930	0.927	0.941	0.956	0.931	0.951	0.939
Zhejiang	0.921	0.925	0.913	0.907	0.919	0.914	0.917
Anhui	0.784	0.759	0.820	0.817	0.776	0.825	0.797
Fujian	1.000	0.983	1.000	1.000	0.904	0.874	0.960
Jiangxi	0.948	0.999	1.000	1.000	1.000	1.000	0.991
Shandong	0.817	0.797	0.796	0.800	0.784	0.789	0.797
Henan	0.944	0.878	0.914	0.865	0.791	0.832	0.871
Hubei	0.827	0.805	0.821	0.812	0.848	0.858	0.829
Hunan	0.886	0.884	0.911	0.910	0.937	0.996	0.921
Guangdong	1.000	1.000	1.000	1.000	1.000	1.000	1.000
Guangxi	1.000	1.000	1.000	1.000	1.000	1.000	1.000
Hainan	1.000	0.841	0.789	0.800	0.761	0.765	0.826
Chongqing	1.000	1.000	1.000	1.000	1.000	1.000	1.000
Sichuan	0.854	0.877	0.967	0.945	0.905	0.938	0.914
Guizhou	1.000	1.000	1.000	1.000	1.000	1.000	1.000
Yunnan	0.734	0.706	0.935	0.891	0.731	0.779	0.796
Shaanxi	0.930	0.983	0.975	0.996	0.978	0.994	0.976
Gansu	0.840	0.867	0.844	0.801	0.780	0.836	0.828
Qinghai	0.616	0.711	0.662	0.693	0.728	0.763	0.696
Ningxia	0.914	1.000	1.000	1.000	1.000	1.000	0.986
Xinjiang	0.752	0.762	0.811	0.829	0.845	0.894	0.816
Nationwide	0.904	0.899	0.910	0.903	0.888	0.902	----

B. Analysis on energy and environment efficiency

From above analysis about energy and environment efficiency, it is showed that there are a huge differences among different provinces, energy and environmental efficiencies vary substantially and to account for that, it is necessary to discuss the spatial correlation between energy efficiency and environmental efficiency.

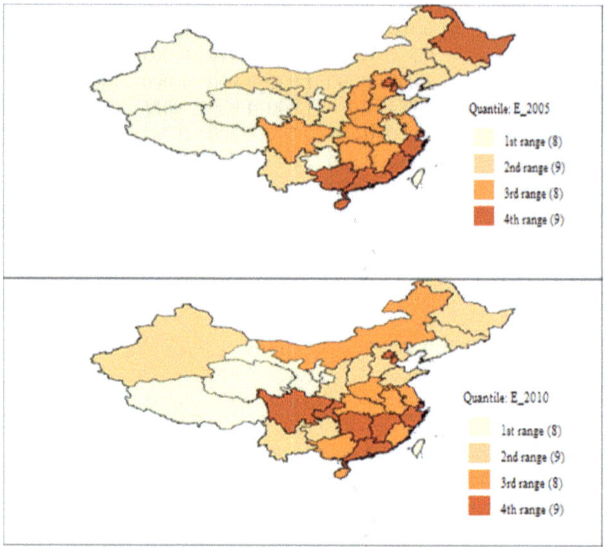

Fig.1. Energy Efficiency Spatial Distribution (Up: 2005, down: 2010)

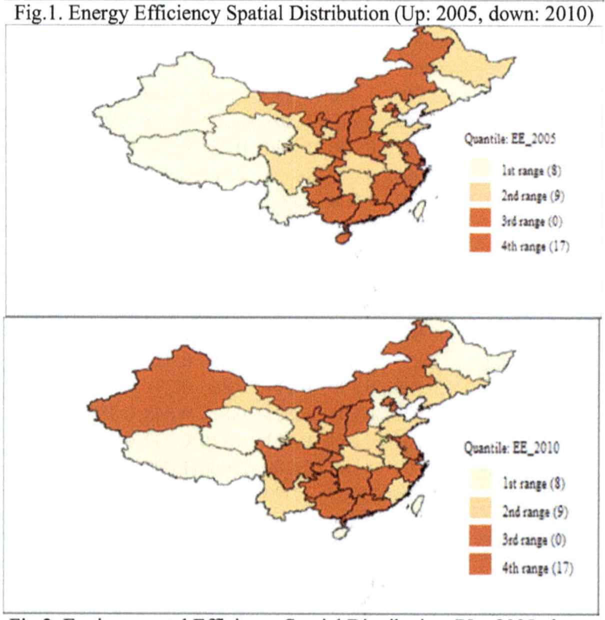

Fig.2. Environmental Efficiency Spatial Distribution (Up: 2005, down: 2010)

Fig.1 and 2 respectively are the comparison diagrams of the national energy and environmental efficiency spatial distribution as of 2005 and 2010, with a four-different-level division for better illustration. As can be seen from Fig.1(up) in 2005 there was a highly strong regional correlation in energy efficiency distribution and a progressively decreasing characteristic from the east to the west, which is slightly weakened in 2010; Likewise, Fig.2 (down) presents a ever a stronger regional correlation in environmental efficiency distribution than energy efficiency, as pollution emissions are highly regional. However in 2010 the spatial distribution of environmental efficiency was nationwide and the decreasing trend from the east to the west disappeared.

Based on above we explore deeper into the regional correlation. TABLE III shows that all statistics of $Moran'sI$ are positive values and pass the 5% significance level test, indicating that the energy efficiency and environmental efficiency are regionally positively related in spatial distribution, also there is space gathering phenomenon in regional energy and environmental efficiency distribution. Provinces (municipalities) with high/low efficiency normally have a surrounding group with high/low efficiency, showing a convergence trend in neighboring areas. In other words, provinces (municipalities) share similar energy and environmental efficiency characteristics with their neighbors, showing a significant spatial correlation. According to the efficiency data of the national 30 provinces (municipalities), spatial correlation index formula and the $Moran'sI$ of 2005-2010 by spatial (four) adjacency matrix, $Moran'sI$ for energy efficiency during the 6-year period basically stay constant, while for environmental efficiency $Moran'sI$ gradually shrink over the 6 years, meaning that environmental pollution has been transforming from a convergence to a divergence gradually.

TABLE III
ENERGY EFFICIENCY AND ENVIRONMENTAL EFFICIENCY $Moran'sI$ INDICES (2005~2010)

Spatial Correlation	Year	Moran'I	E(moran'I)	Mean	Sd	P-value
Spatial Correlation Of energy efficiency	2005	0.3714	-0.0303	-0.0321	0.1157	0.01
	2006	0.3571	-0.0303	-0.026	0.1038	0.01
	2007	0.3535	-0.0303	-0.0606	0.089	0.01
	2008	0.3521	-0.0303	-0.025	0.1124	0.01
	2009	0.367	-0.0303	-0.0253	0.1227	0.02
	2010	0.3644	-0.0303	-0.0344	0.1015	0.01
Spatial Correlation Of environmental efficiency	2005	0.412	-0.0303	-0.0338	0.1082	0.01
	2006	0.4032	-0.0303	-0.027	0.1131	0.01
	2007	0.3787	-0.0303	-0.0401	0.0951	0.01
	2008	0.3695	-0.0303	-0.0071	0.1217	0.02
	2009	0.3739	-0.0303	-0.0393	0.0952	0.01
	2010	0.3669	-0.0303	-0.0361	0.1036	0.02

Note: The K-Nearest Neighbor spatial weights matrix is utilized, k=4

Fig.3 shows a scatter plot of the energy efficiency. Shanghai, Beijing, Jilin, Tianjin, Anhui, Shandong, Shanxi, Guangdong, Guangxi, Jiangsu, Jiangxi, Hebei, Henan, Zhejiang, Hainan, Hubei, Hunan, Fujian, Liaoning, Chongqing, Heilongjiang, all together 21 provinces are in quadrant I, the high-high positive autocorrelation cluster (HH). They were amid a neighboring group with relative high efficiency in year 2005. Inner Mongolia, Guizhou, Shaanxi are in quadrant II, the low-high negative space autocorrelation cluster (LH); they were surrounded by areas with higher efficiency though they were low themselves. Yunnan, Ningxia, Shanxi, Xinjiang, Gansu, Qinghai are located in quadrant III, the low-low spatial correlation cluster (LL); these provinces, together with their surrounding areas were low in energy efficiency as of 2005. Yunnan, Inner Mongolia, Jilin, Sichuan, Shanxi, Hainan are in quadrant IV, the high-low spatial autocorrelation relationship (HL), with higher efficiencies within themselves but lower neighboring regional efficiencies. Jilin, Shanxi and Hainan are in both quadrants I and IV, Inner Mongolia, at the same time is in both quadrant II ad IV, Yunnan is in both quadrant III and IV. In 2010 there was not an obvious sign of change. It can be seen that the East and Central China possess higher energy efficiency and the West are low in that.

Fig.4 shows the space scatter plot of environmental efficiency. In year 2005, Shanghai, Inner Mongolia, Beijing, Tianjin, Ningxia, Shandong, Shanxi, Guangxi, Jiangsu, Jiangxi, Hebei, Henan, Zhejiang, Hainan, Hubei, Hunan, Gansu, Fujian, Guizhou, Liaoning, Chongqing, Shaanxi, Heilongjiang were in quadrant I, and Jilin and Anhui were in quadrant II; Yunnan, Xinjiang, and Qinghai were in quadrant III; Sichuan, Guangdong, and Hainan were in quadrant IV. In 2010, Heilongjiang and Liaoning shifted from high – high to low – high type, in accordance with above analysis that environmental efficiency in the old northeastern industrial area has been low. Hainan turns from the high – low type to low – high type.

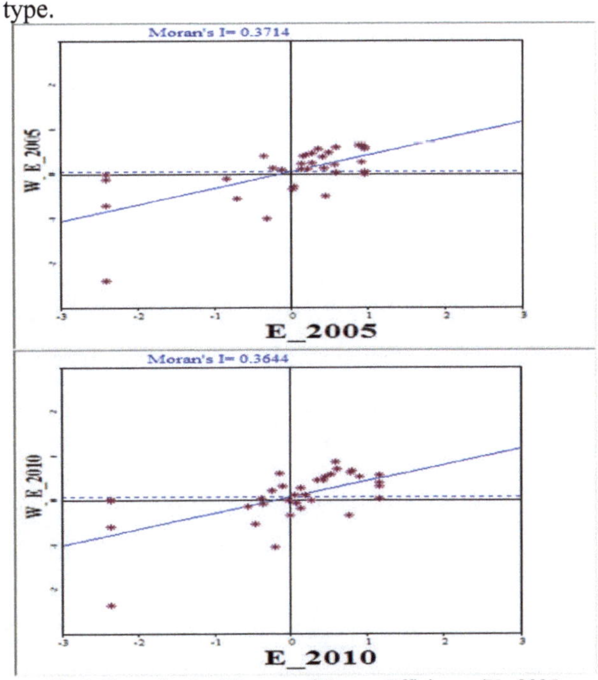

Fig.3. Space Scatter Diagram of Energy Efficiency (Up:2005, down:2010)

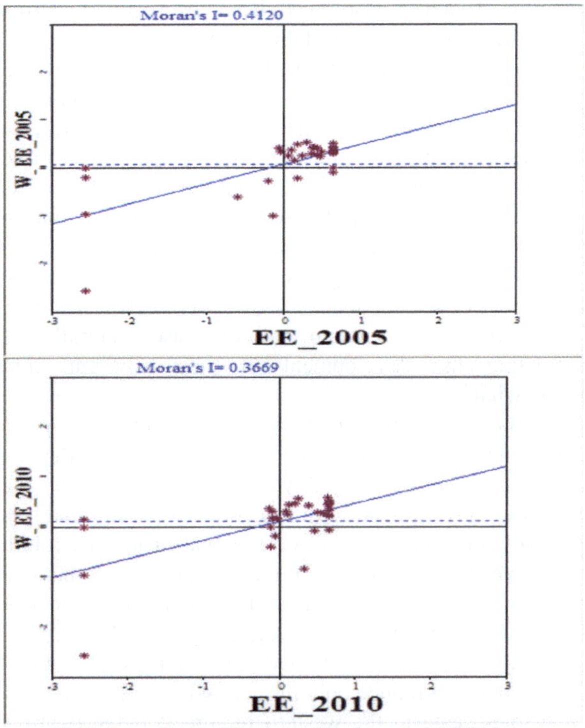

Fig.4. Space Scatter Diagram of Environmental Efficiency (Up: 2005, down: 2010)

To summarize from all the spatial correlation analysis on efficiency, we can see, from the east to the west, a progressive ladder-like energy efficiency distribution, with little changes from 2005 to 2010, indicating that China has been in a sTABLE situation in terms of energy efficiency. Environmental efficiency holds the same distribution characteristics with energy cfficiency, but the ladder-like distribution was weakened during the period of 2005 to 2010 and some provinces (such as the three in the Northeast) declined. The fact that environmental efficiency distribution has gradually become well-distributed despite of the distinct national energy efficiency ladder-like situation suggests that environmental pollution is more regional.

V. CONCLUSION

Through this study we find that: 1) the overall national level of energy and environment efficiency is not high, but highly regional. The average energy efficiency level over the six years is 0.7952 and that for environmental efficiency is relatively higher as of 0.8952. In addition, under the circumstances of limited energy efficiency level, in recent years some changes have taken place for environmental protection and emission reduction does work to a certain degree, and 2) the regional distribution of energy efficiency presents a progressively decreasing trend from the east to the west, while this type of trend for environmental efficiency is relatively weak. The spatial distribution disparity of energy efficiency is derived from the relatively strong spatial distributional correlation in energy efficiency, that is, strong agglomeration effects in both the areas of high efficiency and low efficiency rates. The weakening of the spatial distributional differences in environmental efficiency reveals that the regional spatial spillover effect is obvious, in other words the improved efficiency of the eastern region will rapidly improve those of the adjacent areas. The low efficiency values of some provinces in central and western regions are closely associated with their backward economies. Their lower energy efficiency results in a large number of pollution gases and pollutants, affecting the development in their adjacent areas somewhat.

Based on the above analysis, to further improve efficiency, and promote economy to develop harmoniously and steadily, there is necessity that improve government subsidies for energy, stimulate the energy consumption, balance regional differences: 1) to improve energy and environmental efficiency: national energy efficiency level stays at a relatively lower level and shows a gradually decreasing trend, therefore subsidy policy must be carried out to improve the energy efficiency and reduce pollution emissions. 2) To increase subsidies scale in the western region: it is backward in terms of economic development in the western region, which is still in the initial stage of economic development and requires tremendous energy consumption as the driving force. For the region with current extremely low energy and environmental efficiency, it requires high volume of subsidies to promote the economic development and thus to balance the economic gap between the East and the West in the economic ladder. 3) To unify regional subsidy policy: via analysis we find spatial clustering effects in China's energy and environmental efficiency, that is, agglomeration effect is bred among provinces with similar economic structures. The unification of subsidy policy within a region, breaking the provincial boundaries, with full consideration of the spillover effects between neighboring provinces, would jointly improve the efficiency level.

ACKNOWLEDGMENT

This study was supported by the Major Program of the National Natural Science Foundation of China (Grant No. 71390333) and the National Natural Science Foundation of China(Grant No. 71371151). We are so thankful for co-workers in our research group with who we make some constructive comments and helpful discussions and we would like to thank the anonymous reviewers' work and those of editors relevant to this paper.

REFERENCES

[1] B.W. Ang, Monitoring changes in economy-wide energy efficiency: from energy-GDP ratio to composite efficiency index, Energy Policy, 34, 2006, pp: 574–582.

[2] R.Fare, S.Grosskopf, E.Kokkelenberg, Measuring Capacity, Utilzation and Technical Change: a Nonparametric Approach, August 1989, pp:655-666.

[3] A. Charnes, W.W Cooper, E.Rhodes. "Measuring the efficiency of decision making units", European Journal of Operation Research, November, 1978, pp: 429-444.

[4] Hu, J. L., Wang, S. C.. Total-factor energy efficiency of regions in China.Energy Policy, 34, 2006, pp: 3206–3217

[5] P.Zhou, B.W.Ang, K.L.Poh, "Measuring environmental performance under different environmental DEA technologies", Energy Economics, 30, 2008, pp: 1-14.

[6] T.Coelli., A multi-stage methodology for the solution of orientated DEA models, October, 1998, pp: 143-149.

[7] C.Wei, M.H.Shen, "Energy efficiency and its influencing factors: the empirical analysis based on DEA", Management World, August, 2007, pp: 66-76.

[8] B.Wang, Y.R.Wu, P.F.Yan. "Environmental Efficiency and Environmental Total Factor Productivity Growth in China's Regional Economies" Economic Research Journal, May, 2010, pp: 95-109.

Research on Credibility Support Mechanism of Manufacturing Cloud Service Based on Classified QoS

Qiu-yun ZHAO*, Le WEI, Hong-ping SHU
Department of Computer, Chengdu University of Information Technology, Chengdu, China
(zqyuen@cuit.edu.cn)

Abstract - **In order to make the manufacturing progress go on smoothly for cloud manufacturing environment, a credibility support mechanism of the manufacturing cloud service is raised based on classified QoS. With the mechanism, a manufacturing cloud service model describes the environment information and the QoS level explicitly, and builds the adaptive control logic in itself. According to environment changes and rules in the rule library, the manufacturing cloud service uses an adaptive algorithm and adjusts the QoS level. There are three dynamic adjustion types which drive services to transform between different states. Besides to guarantee the service performance, a credibility evaluation system of services is introduced in to quantize the QoS.**

Keywords - **Classified QoS, cloud manufacturing, credibility support, manufacturing cloud service**

I. INTRODUCTION

Cloud manufacturing is a new manufacturing model [1], in which web and the cloud manufacturing platform are applied. Through this model, manufacturing resources are organized according to users' requirement, and various manufacturing services are provided on-demand. Cloud manufacturing is service – oriented. In cloud manufacturing, all manufacturing resources and capacities are virtualized and applied as services to manage productions centrally and intelligently [2]. Users can choose suitable cloud services on demand to carry manufacturing activities going on. However, manufacturing cloud services perform well or bad from different providers. Besides, it's inevitable to face unpredictable issues such as the quit or fault for cloud services, which would cause the relevant manufacturing actions infected. Therefore, it has important realistic significance to research the credibility support mechanism of manufacturing cloud services for cloud manufacturing environment.

As a newly arisen conception, researches about cloud manufacturing mostly focus on its conception and architecture, discussion of it is far from enough, and studies on the credibility support mechanism of it have been done few. Existing researches [3-6] on cloud manufacturing are still lack of pertinence. Affecting factors on the QoS of the cloud manufacturing haven't been analyzed thoroughly, and the credibility support system of manufacturing cloud services is not proposed either. Based on technical literatures and with the combination of the classified QoS thought and adaptive technology, this paper gives a manufacturing cloud service model, which has built-in sensors and decision-making units and is able to describe the environment information explicitly. In the manufacturing cloud service model, running states and transformations between different states are defined. The credibility evaluation method and QoS adaptive algorithm are also given.

II. MANUFACTURING CLOUD SERVICE MODEL

For cloud manufacturing environment, there are enormous manufacturing cloud services which correspond varies resources such as manufacturing resources and capacities [7]. The environment is open, dynamic, isomorous and distributed, and the traditional service model can't effectively support manufacturing activities going on. Hence, the classified QoS and adaptive technology are introduced to the cloud manufacturing definition.

Definition 1 manufacturing cloud service. Manufacturing cloud service means the manufacturing resources and capacities [8], and is simply called MCS below. MCS can be described as a 6-tuple. That is S=(ID, Attribute, Event, Environment, Control, Level).

1) ID is the service identification.

2) Attribute is the service attribute set which mainly includes the basic and functional attributes. The basic attributes have the name, usability, provider, and contact information. The functional ones mean functional parameters of manufacturing services, such as the processing shape, accuracy, roughness and flatness.

3) Event means the set of received and sent events. The exact definition is Event=(ES, ER).

4) As the environment in which MCS works, Environment is concretely revealed by the usage of all resources at a moment while the service running. Then we have Environment =(Name, Sum, Ability, t), where Name refers to the resource name, Sum is the resource total number and Ability is the usable resource set at t moment.

5) We have Control=(Sensor, DecisionMaker, Executor, RuleBase), where Control is the adaptive control logic, Sensor is used to sense the external changes, DecisionMaker is the decision-making unit which can receive environment information and make decisions according to the preset rule, Executor is the executing unit to execute decisions from DecisionMaker, and RuleBase means rule libraries. Rules are basis to make decisions and are also the presentation that MCS has adaptive ability.

6) Level is the set of QoS parameters. That is, Level=(ID, $Contract_p$, $Contract_r$), where ID is the level identification, $Contract_p$ means quality parameters of a MCS which provides to other services, meanwhile $Contract_r$ refers to quality parameters of a MCS which requires from others.

E. Qi et al. (eds.), *Proceedings of the 5th International Asia Conference on Industrial Engineering and Management Innovation (IEMI2014)*, Proceedings of the International Asia Conference on Industrial Engineering and Management Innovation 1, DOI 10.2991/978-94-6239-100-0_12, © Atlantis Press and the authors 2015

MCS's are not isolated but able to call each other. Sometimes a MCS needs other services while providing its own services. This situation is called service dependence. To ensure the QoS, the interdependent QoS levels should be the same. Supposing that MCS_1, MCS_2 both are MCS's and MCS_1 uses services from MCS_2, the QoS of MCS_1 depends on the quality of necessary services. That is,

$MCS_1.Level.Contract_r = MCS_2.Level.Contract_p$.

About the MCS sensory perception of the environment, a function named sensor is defined as,

sensor: Event→ Environment.

Here sensor senses the environment through Event. Sensor receives the environment information and DecisionMaker makes decisions autonomously. Then the decision-making function can be defined as,

Decision: Environment ×Rule→Action.

Here Environment×Rule refers to the map from the environment to rules, Action represents suitable actions according to rules. The effect of actions on the MCS running state is described too, that is,

Eexecute: S_i×Action→S_j.

III. MCS CREDIBILITY EVALUATION AND ADAPTIVE ADJUSTION BASED ON CLASSIFIED QOS

A. MCS Credibility evaluation

In the cloud manufacturing activities, a QoS level needs to be built by users according to the application requirement. Consequently, the MCS QoS in the application should meet the level too. A running MCS doesn't always provide services at the preset level but adjusts services dynamically with the environment changing. Theoretically, the QoS level is adjusted to better the QoS. However with some reasons, the MCS QoS level may be reduced, although the decreased degree is modest and the application function is still usable. Now the MCS can be thought in a risk state. Furthmore, if the QoS level gets too low to ensure the application function, the MCS could be considered in a fault state. Finally if the MCS QoS is at the preset or a higher level, the MCS would be in a normal state. The different MCS states are defined as following.

Definition 2 The normal state of MCS is,

$$MCS^{(t1,t2)}_{normal} \xrightarrow{define} MCS \cdot Level^{(t1,t2)} \geq \xi .$$

That is to say from time t1 to t2, MCS is in the normal state if the QoS level of MCS is higher than ξ or equal to ξ. Here is the preset QoS level.

Definition 3 The risk state of MCS is,

$$MCS^{(t1,t2)}_{risk} \xrightarrow{define} (MCS \cdot Level^{(t1,t2)} < \xi) \wedge (\forall x \in P, \forall y \in R \mid I(x) \Rightarrow O(x,y))$$

This means that from time t1 to t2, MCS is in the risk state if the QoS level is lower than ξ and the expected value of y is outputted with the valid input parameter x. ξ is the same as the one defined in Definition 2, P is the precondition which consists of the input parameter of x, and R is the post condition which contains output parameter y.

Definition 4 The fault state of MCS is,

$$MCS^{(t1,t2)}_{risk} \xrightarrow{define} (MCS \cdot Level^{(t1,t2)} < \xi) \wedge (\forall x \in P, \forall y \in R \mid I(x) \Rightarrow O(x,y))$$

, From time t1 to t2, MCS would be in the fault state if the QoS level is lower than ξ and the expected value of y can't be obtained with the valid input parameter of x. Variables of ξ, P and R are same as those in Definition 3.

The MCS states are not isolated. States can transform to each other as shown in Fig.1. In Fig.1, every route is a state transformation. The trigger conditions for transformation are the environment changes and the QoS evaluation. Conditions are mapped as rules in the RuleBase. The transformation is the rule execution, i.e. the dynamic adjustion of the MCS QoS level. The MCS would be in one of the three states after a transformation.

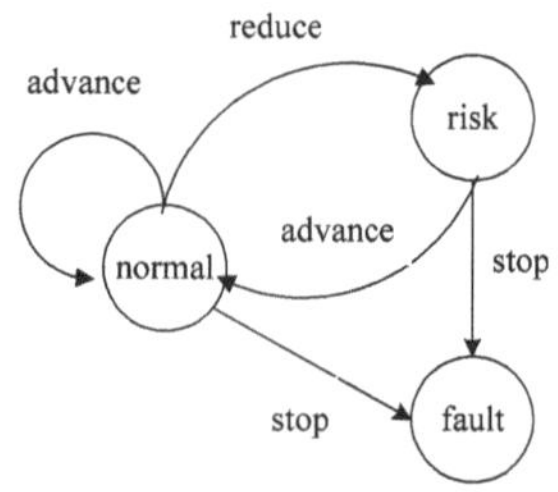

Fig.1. The MCS state transformations

To quantize the QoS, the MCS credibility can be evaluated by the formula as,

$$CDW(\text{MCS}, t) = \sum_{d \in DS} W_d \times DW(d,t) \qquad (1)$$

where *CDW*(MCS, *t*) is the comprehensive credibility of the MCS at *t* time, *DS* is the trusted characteristic set, W_d is the weighted coefficient of the trusted characteristic of *d* in the comprehensive value and W_d meets $\sum_{d \in DS} W_d = 1$, and *DW*(*d*,*t*) means the credibility value of *d* at *t* time.

As a general concept such as the usability, reliability, testability, confidentiality, security, timeliness and maintainability, these all can be thought as the trusted characteristics to compute the credibility in (1), i.e. one of *DS*. For simplicity, the reliability, timeliness and price are chosen as the evaluation indicators of MCS's.

1) reliability. The service reliability is the MCS ability to offer services normally within a period. This conception represents the frequency of the MCS normal running in the period. Supposing that a MCS is requested *N* times and responses normally for *Nr* times, the reliability can be quantized as,

$$Reliability(\text{MCS}) = Nr/N \qquad (2)$$

Corresponding to the Fig. 1, the MCS reliability is the frequency of keeping the normal and risk states during the period of (0,t). That is, *Reliability*(MCS)=P(q(τ)≠fault, τ∈(0,t)).

2) timeliness. It is the MCS timeliness that whether the service requests are responded timely. On the premise that a MCS is executed for a long time of *t*, the service timeliness is quantized as,

$$Time(\text{MCS}) = 1/t \qquad (3)$$

In the MCS, the timeliness is guaranteed by priority-based scheduling. If a service is demanded with higher timeliness, priorities of the corresponding methods could be improved in advance to make the service use resources with priority. Because the QoS is ensured not only when the service serves at the preset QoS level but also when the QoS level is improved, as reflected in the Fig.1, the MCS timeliness is the frequency of keeping the normal state. That is, $Time(\text{MCS})=\text{P}(\text{q}(t)=\text{normal})$.

3) cost. The cost is the total expenses needed from the service submission to the return of the service executing result. *Cost* consists of the service executing cost and the logistics one. Let *Ce* be the service executing cost and *Ct* be another. Then cost can be quantized as,

$$Cost(\text{MCS})=1/(Ce+Ct) \quad (4)$$

B. Adaptive Adjustion Based on Classified QoS

To make the MCS work well in the cloud manufacturing environment and provide high QoS, the adaptive control logic is introduced in except for the QoS level. The adaptive control logic senses the environment changes and the QoS evaluation results, maps the changes and results as rules in the ruleBase. Then the control logic takes the nonlinear decision to bind a rule, executes the responding adaptive actions and finishes the adaptive adjustion. The adaptive actions are defined as following.

Definition 5 Suppose expressions of $X\in\text{MCS}$, $\text{Level}_i<\text{Level}_j<\text{Level}_k$ and $\text{X.Level}=\text{Level}_j$, and Level_{min} is the lowest QoS level.

Improve the QoS level. If the environment of X satisfies the condition of $\text{X.Environmment.Ability} \geq \text{X.Level}_k\text{.Contractr}$, the QoS level of X could be improved from Level_j to Level_k, which is donated as $\text{advance}(\text{Level}_j\rightarrow\text{Level}_k)$.

Reduce the QoS level. If the environment of X meets conditions of $\text{X.Environmment.Ability}<\text{X.Level}_k\text{.Contractr}$ and $CDW(\text{X},t)\geq\xi$, the QoS level of X could be adjusted from Level_j down to Level_i, which is donated as $\text{reduce}(\text{Level}_j\rightarrow\text{Level}_i)$ and ξ is the lowest credibility allowed by the QoS.

Stop the service. If the QoS level is up to relations of $\text{X.Level}=\text{Level}_{min}$ and $CDW(\text{X},t)<\xi$, the service of X should be stopped, which is donated as stop(X).

Three adaptive actions can make the MCS state change as illustrated in Fig. 1. The advance action drives the state to be normal from the risk or normal state; The reduce action makes the state get risk from normal and the stop action could transform the state to be fault from the normal or risk state. Effects of adaptive actions on the MCS QoS is described as,

Proposition 1 Suppose $X\in\text{MCS}$, t_1 is a time before an adaptive action happening and t_2 is a time after the action.

If the advance action is executed, then the service of X would meet the expression of $CDW(\text{X},t_1)>CDW(\text{X},t_2)$.

If the reduce action is executed, then X would meet the expression of $CDW(\text{X},t_1)<CDW(\text{X},t_2)$.

If the stop action is executed, then X would meet expressions of $CDW(\text{X},t_1)>CDW(\text{X},t_2)$ and $CDW(\text{X},t_2)=0$.

To continue serving while the MCS executes adaptive actions, the MCS should be brought into a special state in which the MCS doesn't accept new service requests. So states named as Active and Inactive are defined. In the Active state, the MCS serves normally, and in the Inactive state MCS carries adaptive adjustions. Accordingly, the state driving functions of activate and deactivate are defined. We have,

$$\text{active:MCS}\rightarrow\text{Active},$$
$$\text{deactive: MCS}\rightarrow\text{Inactive}.$$

After a reduce action, the MCS QoS needs to be revaluated to decide whether to keep the MCS in the risk state or transform it into the fault state through the stop action. Therefore the QoS evaluation can be calculated with the function of evaluate. That is,

$$\text{Evaluate: S}_i\times\text{CDW}\rightarrow\text{S}_j.$$

The MCS makes decisions adaptively according to rules in RuleBase. The rule syntax is established as

When event If environment expressions Then *op**, where the *op* is one of transformation actions. The formal description of *op* is,

$$op^*\mapsto\{\text{advance},\text{reduce},\text{stop}\}^*$$

Based on the above statements and analysis, the adaptive adjustion algorithm of the MCS is given. We have,

```
1 Function Adaptive
2 Begin
3  receive(event) ;
4  MCS.Environment :=sensor(environment ) ;
5  rule:=select(MCS.RuleBase);
6  action:=decision(rule);
7  deactivate(MCS);
8  if action=advance then
9       excuate(advance);
10      active(MCS);
11 elseif action=reduce then
12      excuate(reduce);
13      if evaluate(CWD(MCS,t))< ξthen
14           excuate(stop);
15      else
16           active(MCS);
17      end if
18 elseif action=stop then
19      excuate(stop);
20 end if
21 end Function Adaptive
```

The MCS receives event in line 3. The MCS senses the current external environment, and sets the Environment of the service according to the perceived information in line 4. Decision-maker queries the rule library with the received events and perceived environment, and consequentially gets the rule in line 5. Control carries the adaptive decision out according to the rule and receives the action to be performed in line 6. Control drives the MCS into the Inactive state and prepares for the next adaptive action in line 7.

In line 8-20, the concrete action is executed. For the advance or setPriority action, Control would execute the action and drive the MCS into the active state. For the reduce action, Control would execute the action, and estimate the QoS, and then decides that whether to stop the service or drive it into the *active* state. If the evaluation result of the QoS is lower then the standard value, the MCS would be stopped.

C. Instance

There is a service provided by an enterprise to process gaskets. The service is used by the enterprise itself and also is published into the cloud manufacturing platform as a MCS. The MCS would stop serving outwards when its capacity is taken more then 80%, and it would serve outwards mainly when capacity is occupied less then 20%, and then it would serve selectively when capacity is between the above numbers. This MCS has three QoS levels. Then,

Levels=($Level_1$, stop, capacity<=20%);

Levelc=($Level_2$, common, 20%<capacity<80%);

Levelh=($Level_3$, high, capacity>=80%);

Here the Levels， Levelc and Levelh identify different QoS levels, i.e. stoppting service, common speed process and fast process. The capacity is the capacity remainder. About the three levels, four rules are derived in ruleBase. That is,

When $event_1$ If capacity <=0.2 Then stop

When $event_2$ If MCS.Level.ID=$Level_2$ and 0.2<capacity <0.8 Then advance

When $event_3$ If MCS.Level.ID=$Level_3$ and 0.2<capacity <0.8 Then reduce

When $event_4$ If capacity≥0.8 Then advance

Here the event of $event_1$, $event_2$, $event_3$ and $event_4$ are produced by the environment. Any environment change could cause some event. The MCS receives an event and responds whether resources are up to critical values. The value of the capacity parameter is derived from Environment. That is capacity =MCS.$Environment_{capacity}$.Ability. In this way by the event mechanism and adaptive actions, the QoS level can be adjusted easily according to the adaptive adjustion algorithm described in section III.B while the MCS capacity changes.

IV. CONCLUSION

By using the thought of the classified QoS in the MCS model, it becomes practicable to control the QoS at the service level, and then make the application operation intuitionally and rationally. The explicit environment description mechanism and the adaptive control logic support the MCS with abilities to sense dynamic environment changes, make decisions according to changes and rules in ruleBase autonomously, and perform adaptive actions. As a result, the MCS adapts to dynamic and open environments. Besides, the MCS QoS evaluation ensures the dependability of the MCS running. Therefore, it's the next work to research how to combine different services, what influences of combined services would be exerted on the credibility and how to realize basic development tools and prepare for the implementation and application.

ACKNOWLEDGMENT

The paper is under the auspices of the Sichuan soft science research project No.2012ZR0116, two projects No.12ZB199 and No.13ZB0081 supported by Youth Foundation of Sichuan education department, and the application achievement transformation project No.12DXYB100JH-002 of Chengdu universities, colleges and the research institutes. Here, we acknowledge great thanks to them.

REFERENCES

[1] LI Bohu, ZHANG Lin, WANG Shilong, TAO Fei, CAO Junwei, JIANG Xiaodan, SONG Xiao, et al. "Cloud manufacturing: a new service-oriented networked manufacturing model," *Computer Integrated Manufacturing Systems*, vol. 16, no. 1, pp. 1-7, Jan. 2010. (In Chinese)

[2] ZHANG Lin, LUO Yongliang, TAO Fei, REN Lei, GUO Hua. "Key technologies for the construction of manufacturing cloud," *Computer Integrated Manufacturing Systems*, vol. 16, no.11, pp. 2510-1520, Nov. 2010. (In Chinese)

[3] GAN Jia, DUAN Guijiang. "Method of cloud manufacturing sercice trust evaluation," *Computer Integrated Manufacturing Systems*, vol. 18, no. 7, pp. 1527-1535, Jul. 2012. (In Chinese)

[4] LIU Zhizhong, XUE Xiao,AN Jiyu, LU Baoyun. "Research on manufacturing cloud service optimal composition based on CL-MMAS algorithm," *Computer Engineering and Applications*, vol. 48, pp. 239-242, 2012. (In Chinese)

[5] LIU Weining, LI Yiming, LIU Bo. "Service composition in cloud manufacturing based on adaptive mutation particle swarm optimization," *Journal of Computer Applications*, vol. 32, no. 8, pp. 2869-2874, 2878, Oct. 2012. (In Chinese)

[6] LV Qingcong. "Research on Semantic Service Discovery and Service Composition for Pervasive Computing," .Shanghai: Dong hua University, 2009.

[7] HU Anrui, ZHANG Lin, TAO fei, LUO Yongliang. "Resource service management of cloud manufacturing based on knowledge," *Journal of Tongji University (Natural Science)*, vol. 40, no. 7, pp.1094-1101, Jul.2012. (In Chinese)

[8] LI Huifang, DONG Xun, SONG Changgang. "Intelligent searching and matching approach for cloud manufacturing service," *Computer Integrated Manufacturing Systems*, vol. 18, no. 7, pp. 1485-1493, Jul. 2010. (In Chinese)

Mean-Variance Newsvendor Model with a Background Risk

Jiang-feng LI*, Qiong WU
Department of Sciences, Jiujiang University, Jiujiang, China
(jiangfengli2000@163.com)

Abstract - **This paper examines the effects of an additive background risk on the optimal order quantity of a risk-averse newsvendor with Mean-Variance utility. We derive several unambiguous comparative statics results with the additive background risk with the use of the concept of the Mean-Variance vulnerability.**

Keywords - **Background risk, mean-variance newsvendor, risk averse, risk vulnerability**

I. INTRODUCTION

The original newsvendor model is based on the assumption that a risk-neutral newsvendor determines his optimal order quantity by minimizing the expected cost or maximizing the expected profit. However, the assumption is inconsistent with many empirical and experimental observations. Motivated by these inconsistency, an alternative approach incorporates newsvendor's risk preferences into the model. There are two popular approaches to capture newsvendor's risk preference so far: the general utility function based on expected utility theory [1-6] and the Mean-Variance analysis [7-10].

The expected utility (EU) is the most predominant decision model under risk, however, its application is always limited because of the difficulty in choosing specific utility function. On the other hand, the Mean-Variance (MV) analysis as an alternative to EU has been highly popular in the financial economics, since it only needs the first two moments of a distribution and is simplicity and tractability. Recently, the model aroused the interest of researchers again due to the contributions made by Lajeri-Chaherli and Nielsen [11], Wagener [12], Ormiston and Schlee [13], Lajeri-Chaherli [14], Eichner and Wagener [15-16]. Therefore, this paper uses MV tool to model newsvendor's risk preference.

However, most of extant risk-averse models under Mean-Variance preference have neglected an important fact that newsvendor usual face several risk sources which cannot been hedged and traded in the market, such as income risk, human capital risk. This type of unhedgable risk usual is labeled background risk in economic literature. The presence of background risk complicates the decision problem and yields significant impact on people's decision behavior. However, fewer works address this issue in the context of newsvendor problem with two exceptions, paper [2] and [5]. But they dealt with this problem with a particular type of background risk separately under EU framework.

The present paper aims to analyze the effects of an additional background risk on the order quantity of a newsvendor with MV preference, which is not yet known in existent inventory management literature.

The rest of this article is organized as follows. Section 2 presents a general MV newsvendor model with a background risks, and conducts a simple comparison between the risk-neutral and the risk aversion model. Section 3 investigates the effects of the background risk on the newsvendor's decision. Our conclusion is drawn in the final section.

II. BASIC MODEL WITH A BACKGROUND RISK

In the classical single-period newsvendor setting, a newsboy has to determine the number of newspapers to order ex ante to maximize his objective function over final profit. The newsvendor buys newspapers at an unit price c and resells them at a price $p > c$. The unsold newspapers are disposed with a price $s < c$ which is called as the salvage value. A natural assumption is that $0 < s < c < p$ without loss the generality. Now, we consider a new circumstance in which the newsvendor not only faces an uncertain demand $\tilde{x}$, but also confronts an uninsured risk $\tilde{y}$ called background risk in relevant literatures. The random variables $\tilde{x}$ and $\tilde{y}$ are fully characterized by their cumulative distribution functions (c.d.f) $F_{\tilde{x}}(x)$ with support [0,T] and $F_{\tilde{y}}(x)$. We further suppose that the inverse function of $F_{\tilde{x}}(x)$ exists and denote it by $F_{\tilde{x}}^{-1}(x)$. Moreover, the independence between $\tilde{x}$ and $\tilde{y}$ is assumed in the remainder of the paper. At first, considering a benchmark situation without any background risk, the newsvendor then is endowed with the following profit function:

$$\tilde{\pi}(q) = (p-c)q - (p-s)\max(0, q-\tilde{x}). \qquad (1)$$

When the background risk $\tilde{y}$ is interpreted as uncertainty incomes or health status, et al. Then, his profit function can be presented as follows:

$$\tilde{z}(q, \lambda, \tilde{y}) = \tilde{\pi}(q) + \lambda\tilde{y} \qquad (2)$$

With (1), the expectation $\mu[\tilde{\pi}(q)]$ and the variance $\sigma^2[\tilde{\pi}(q)]$ of the profit are written as follows:

$$\begin{aligned}\mu[\tilde{\pi}(q)] &= (p-c)q - (p-s)\mu[\max(0, q-\tilde{x})] \\ &= (p-c)q - (p-s)\int_0^q F_{\tilde{x}}(x)dx,\end{aligned} \qquad (3)$$

$$\begin{aligned}\sigma^2[\tilde{\pi}(q)] &= (p-s)^2\sigma^2[\max(0, q-\tilde{x})] \\ &= (p-s)^2\left[2\left(\int_0^q (q-x)F_{\tilde{x}}(x)dx\right) - \left(\int_0^q F_{\tilde{x}}(x)dx\right)^2\right]\end{aligned} \qquad (4)$$

With the additive background risk $\tilde{y}$, the expected profit and the variance are respectively,

$$\mu[\tilde{z}(q, \lambda, \tilde{y})] = \mu[\tilde{\pi}(q)] + \lambda\mu[\tilde{y}] \qquad (5)$$

$$\sigma^2\left[\tilde{z}(q, \lambda, \tilde{y})\right] = \sigma^2\left[\tilde{\pi}(q)\right] + \lambda^2\sigma^2\left[\tilde{y}\right] \qquad (6)$$

E. Qi et al. (eds.), *Proceedings of the 5th International Asia Conference on Industrial Engineering and Management Innovation (IEMI2014)*, Proceedings of the International Asia Conference on Industrial Engineering and Management Innovation 1, DOI 10.2991/978-94-6239-100-0_13, © Atlantis Press and the authors 2015

In terms of (3) and (5),the first-order derivatives of $\mu[\tilde{\pi}(q)]$ equals to that of $\mu[\tilde{z}(q)]$. Similarly, combining (4) and (6), we know that the first-order condition of $\sigma^2[\tilde{\pi}(q)]$ also equals to that of $\sigma^2[\tilde{z}(q)]$. That is,

$$\frac{\partial\mu[\tilde{\pi}(q)]}{\partial q}=\frac{\partial\mu[\tilde{z}_a(q)]}{\partial q}=(p-c)-(p-s)F_{\tilde{x}}(q) \quad (7)$$

$$\frac{\partial\sigma^2[\tilde{\pi}(q)]}{\partial q}=2(p-s)^2(1-F_{\tilde{x}}(q))\int_0^q F_{\tilde{x}}(x)dx. \quad (8)$$

The newsvendor has MV preference over the final random profit. In the two different decision settings, without background risk, with an additive background risk, his decision objective can be expressed as respectively,

$$\text{Model 1} \quad \underset{q}{Max}\, U[\mu_{\tilde{\pi}}(q), \sigma^2_{\tilde{\pi}}(q)] \quad (9)$$

$$\text{Model 2} \quad \underset{q}{Max}\, U[\mu_{\tilde{z}}(q), \sigma^2_{\tilde{z}}(q)] \quad (10)$$

where the function $U[\mu,\sigma^2]$ is a general MV utility function and assumed to be four times differentiable with the following properties for all $(\mu,\sigma^2)\in R\times R_+$:

$$(a)U_{\sigma^2}[\mu,\sigma^2], \quad (b)U_{\mu}[\mu,\sigma^2]$$

The conditions (a) and (b) reflect the newsvendor's attitude toward risk, in fact, which imply the slope of the indifference curve are upward sloping in the (μ,σ^2) plane. Paper [16] presents the following function:

$$A(\mu,\sigma^2)=-\frac{U_{\sigma^2}[\mu,\sigma^2]}{U_{\mu}[\mu,\sigma^2]}, \quad (11)$$

which is just slope of the indifference curve and an analog of Arrow-Pratt risk-averse coefficient under MV framework.

Theorem 1: Let q^*_{mv} denote the MV newsvendor's optimal order quantity without background risk, symbols q^*_{abr} represent the optimal quantities with the background risk, then the q^*_{mv} and q^*_{abr} are uniquely contained in the interval $[0,q^*_e]$, where $q^*_e=F^{-1}_{\tilde{x}}(\frac{p-c}{p-s})$ denotes the solution of maximizing expectation profits.

Proof: We only demonstrate $q^*_{abr}\in[0,q^*_e]$, another is similar.

$$\frac{dU\,[\mu_{\tilde{z}}(q),\sigma^2_{\tilde{z}}(q)]}{dq}$$

$$=U_{\mu}[\mu_{\tilde{z}}(q),\sigma^2_{\tilde{z}}(q)]\frac{\partial\mu_{\tilde{z}}(q)}{\partial q}+U_{\sigma^2}[\mu_{\tilde{z}}(q),\sigma^2_{\tilde{z}}(q)]\frac{\partial\sigma^2_{\tilde{z}}(q)}{\partial q}$$

$$\lim_{q\to0^+}\frac{\partial\mu_{\tilde{z}}(q)}{\partial q}=(p-c)>0\,,\ \lim_{q\to0^+}\frac{\partial\sigma^2_{\tilde{z}}(q)}{\partial q}=0$$

Therefore, using these information to evaluate the limits yields

$$\lim_{q\to0^+}\frac{dU[\mu_{\tilde{z}}(q),\sigma^2_{\tilde{z}}(q)]}{dq}=U_{\mu}[\mu_{\tilde{z}}(0),\sigma^2_{\tilde{z}}(0)](p-c)>0$$

By (7) and (8) again,

$$\lim_{q\to q^*_e}\frac{\partial\mu_{\tilde{z}}(q)}{\partial q}=\frac{\partial\mu_{\tilde{\pi}}(q^*_e)}{\partial q}=0\,,$$

$$\lim_{q\to q^*_e}\frac{\partial\sigma^2_{\tilde{z}}(q)}{\partial q}=2(p-s)^2(1-F_{\tilde{x}}(q^*_e))\int_0^{q^*_e}F_{\tilde{x}}(x)dx>0$$

This implies in turn,

$$\lim_{q\to q^*_e}\frac{dU[\mu_{\tilde{z}}(q),\sigma^2_{\tilde{z}}(q)]}{dq}=U_{\sigma^2}[\mu_{\tilde{z}}(q^*_e),\sigma^2_{\tilde{z}}(q^*_e)]\frac{\partial\sigma^2_{\tilde{z}}(q^*_e)}{\partial q}<0$$

Because of the concavity of $U[\mu,\sigma^2]$ about (μ,σ^2) and increasing of $\mu_{\tilde{z}}(q)$ and $\sigma_{\tilde{z}}(q)$ in q over $[0,q^*_e]$, there exists an unique solution $q^*_{abr}\in[0,q^*_e]$.

Q.E.D.

III. COMPARATIVE STATICS ANALYSIS WITH BACKGROUND RISK

Assume the newsvendor faces an unfair additive background risk $\tilde{y}$, which means that $\mu_{\tilde{y}}\le0$, we will examine the effects of the newsvendor' attitude toward risk and an increasing in background risk on his optimal ordering. An increasing in background risk, here, has dual implications that an new background risk is incorporated into the model without background risk and a stochastically dominating shift in the background risk. Thus, the newsvendor's decision problem can be denoted as Model 2 in (10).

Differentiating $U\left[\mu_{\tilde{z}}(q),\sigma^2_{\tilde{z}}(q)\right]$ about q , we obtain the first-order condition of the Model 2 as following,

$$\begin{aligned}&\frac{dU\,[\mu_{\tilde{z}}(q^*_{abr}),\sigma^2_{\tilde{z}}(q^*_{abr})]}{dq}\\&=U_{\mu}[\mu_{\tilde{z}}(q^*_{abr}),\sigma^2_{\tilde{z}}(q^*_{abr})]\frac{\partial\mu_{\tilde{z}}(q^*_{abr})}{\partial q}\\&+U_{\sigma^2}[\mu_{\tilde{z}}(q^*_{abr}),\sigma^2_{\tilde{z}}(q^*_{abr})]\frac{\partial\sigma^2_{\tilde{z}}(q^*_{abr})}{\partial q} \quad (12)\\&=U_{\mu}[\mu_{\tilde{z}}(q^*_{abr}),\sigma^2_{\tilde{z}}(q^*_{abr})]\frac{\partial\mu_{\tilde{\pi}}(q^*_{abr})}{\partial q}\\&+U_{\sigma^2}[\mu_{\tilde{z}}(q^*_{abr}),\sigma^2_{\tilde{z}}(q^*_{abr})]\frac{\partial\sigma^2_{\tilde{\pi}}(q^*_{abr})}{\partial q}=0\end{aligned}$$

where q^*_{abr} indicates the optimal solution of the Model 2. Manipulating (12) simply, we define a new function:

$$\begin{aligned} & \Upsilon\left(q,\lambda,\mu_{\tilde{y}},\sigma^2_{\tilde{y}}\right) \\ & = \frac{\partial \mu_{\tilde{\pi}}(q)/\partial q}{\partial \sigma^2_{\tilde{\pi}}(q)/\partial q} - \left(-\frac{U_{\sigma^2}[\mu_{\tilde{z}}(q),\sigma^2_{\tilde{z}}(q)]}{U_{\mu}[\mu_{\tilde{z}}(q),\sigma^2_{\tilde{z}}(q)]} \right) \\ & = \frac{\partial \mu_{\tilde{\pi}}(q)/\partial q}{\partial \sigma^2_{\tilde{\pi}}(q)/\partial q} - A\left(\mu_{\tilde{z}}(q),\sigma^2_{\tilde{z}}(q)\right). \end{aligned} \tag{13}$$

It is obvious that $\Upsilon\left(q^*_{abr},\lambda,\mu_{\tilde{y}},\sigma^2_{\tilde{y}}\right)=0$. Using symbols $A_{\mu}(\mu,\sigma^2)$ and $A_{\sigma^2}(\mu,\sigma^2)$ to denote the two partial derives of $A(\mu,\sigma^2)$ in the equation (10), we have following result.

Theorem 2: Let $q^*_{abr}(\lambda)$ denote the optimal order of $U\left[\mu_{\tilde{z}},\sigma^2_{\tilde{z}}\right]$ with a parameter λ ($\lambda..0$),

(1) then $\frac{d\,q^*_{abr}(\lambda)}{d\lambda}>0$, if and only if $A_{\mu}(\mu,\sigma^2)„\ 0$ and $A_{\sigma^2}(\mu,\sigma^2)..0$ for all $(\mu,\sigma^2)\in R\times R_+$;

(2) let q^*_{mv} be the optimal order of $U\left[\mu_{\tilde{\pi}},\sigma^2_{\tilde{\pi}}\right]$ then $q^*_{abr}(\lambda)„\ q^*_{mv}$ for all $\lambda..0$ if $A_{\mu}(\mu,\sigma^2)„\ 0$ and $A_{\sigma^2}(\mu,\sigma^2)..0$ for all $(\mu,\sigma^2)\in R\times R_+$.

Proof: (1) Because of

$$\begin{aligned} & 0=\Upsilon\left(q^*_{abr}(\lambda),\lambda,\mu_{\tilde{y}},\sigma^2_{\tilde{y}}\right) \\ & = \frac{\partial \mu_{\tilde{\pi}}(q^*_{abr}(\lambda))/\partial q}{\partial \sigma^2_{\tilde{\pi}}(q^*_{abr}(\lambda))/\partial q} - A\left(\mu_{\tilde{z}}(q^*_{abr}(\lambda)),\sigma^2_{\tilde{z}}(q^*_{abr}(\lambda))\right) \end{aligned} \tag{14}$$

thus, by (5) and (6),

$$\begin{aligned} & \frac{\partial \Upsilon^a\left(q^*_{abr}(\lambda),\lambda,\mu_{\tilde{y}},\sigma^2_{\tilde{y}}\right)}{\partial \lambda} \\ & = -\mu_{\tilde{y}}A_{\mu}\left(\mu_{\tilde{z}}(q^*_{abr}(\lambda)),\sigma^2_{\tilde{z}}(q^*_{abr}(\lambda))\right) \\ & \quad -2\lambda\sigma^2_{\tilde{y}}A_{\sigma^2}\left(\mu_{\tilde{z}}(q^*_{abr}(\lambda)),\sigma^2_{\tilde{z}}(q^*_{abr}(\lambda))\right) \end{aligned} \tag{15}$$

In terms of the implicative function theorem,

$$\frac{\partial q^*_{abr}(\lambda)}{\partial \lambda} = -\frac{\partial \Upsilon^a\left(q^*_{abr}(\lambda),\lambda,\mu_{\tilde{y}},\sigma^2_{\tilde{y}}\right)/\partial\lambda}{\partial \Upsilon^a\left(q^*_{abr}(\lambda),\lambda,\mu_{\tilde{y}},\sigma^2_{\tilde{y}}\right)/\partial q} \tag{16}$$

Furthermore, according to local second-order condition

$$\frac{\partial \Upsilon\left(q^*_{abr}(\lambda),\lambda,\mu_{\tilde{y}},\sigma^2_{\tilde{y}}\right)}{\partial q}<0. \tag{17}$$

From (16), we have

$$sign\left(\frac{\partial q^*_{abr}(\lambda)}{\partial \lambda}\right) = sign\left(\frac{\partial \Upsilon^a\left(q^*_{abr}(\lambda),\lambda,\mu_{\tilde{y}},\sigma^2_{\tilde{y}}\right)}{\partial \lambda}\right)$$

Therefore, for the unfair background risk $\tilde{y}$ ($\mu_{\tilde{y}}„\ 0$), it is obvious that

$$\frac{\partial \Upsilon\left(q^*_{abr},\lambda,\mu_{\tilde{y}},\sigma^2_{\tilde{y}}\right)}{\partial \lambda}„\ 0 \Leftrightarrow \begin{cases} A_{\mu}(\mu,\sigma^2)\le 0 \\ A_{\sigma^2}(\mu,\sigma^2)\ge 0 \end{cases} \tag{18}$$

(3) if $A_{\mu}(\mu,\sigma^2)„\ 0$ and $A_{\sigma^2}(\mu,\sigma^2)..0$ for all $(\mu,\sigma^2)\in R\times R_+$, then $q^*_{abr}(\lambda)„\ q^*_{abr}(0)=q^*_{mv}$ from assertion (1), which implies in turn that introducing an "unfair" additive background risk will induce the newsvendor to reduce his order quantity.

Q.E.D.

Theorem 3: $\frac{d\,q^*_{abr}(\mu_{\tilde{y}})}{d\mu_{\tilde{y}}}..0$, if and only if $A_{\mu}(\mu,\sigma^2)„\ 0$.

Proof: To examine effects of changing in the mean of the background risk $\tilde{y}$ on q^*_{abr}, differentiating the function $\Upsilon\left(q^*_{abr},\lambda,\mu_{\tilde{y}},\sigma^2_{\tilde{y}}\right)$ about $\mu_{\tilde{y}}$ will yields

$$\frac{\partial \Upsilon\left(q^*_{abr},\lambda,\mu_{\tilde{y}},\sigma^2_{\tilde{y}}\right)}{\partial \mu_{\tilde{y}}} = -\lambda A_{\mu}\left(\mu_{\tilde{z}}(q^*_{abr}),\sigma^2_{\tilde{z}}(q^*_{abr})\right) \tag{19}$$

By the implicit function theorem,

$$\frac{\partial q^*_{abr}(\mu_{\tilde{y}})}{\partial \mu_{\tilde{y}}} = -\frac{\partial \Upsilon\left(q^*_{abr},\lambda,\mu_{\tilde{y}},\sigma^2_{\tilde{y}}\right)/\partial\mu_{\tilde{y}}}{\partial \Upsilon\left(q^*_{abr},\lambda,\mu_{\tilde{y}},\sigma^2_{\tilde{y}}\right)/\partial q} \tag{20}$$

Then, the denominator is negative from the second-order condition. As a result,

$$\begin{aligned} & sign\left(\frac{\partial q^*_{abr}(\mu_{\tilde{y}})}{\partial \mu_{\tilde{y}}}\right) = sign\left(\frac{\partial \Upsilon\left(q^*_{abr},\lambda,\mu_{\tilde{y}},\sigma^2_{\tilde{y}}\right)}{\partial \mu_{\tilde{y}}}\right) \\ & = sign\left(-A_{\mu}\left(\mu_{\tilde{z}}(q^*_{abr}),\sigma^2_{\tilde{z}}(q^*_{abr})\right)\right) \end{aligned} \tag{21}$$

Therefore,

$$\frac{\partial q^*_{abr}(\mu_{\tilde{y}})}{\partial \mu_{\tilde{y}}}\ge 0 \Leftrightarrow A_{\mu}\left(\mu,\sigma^2\right)\le 0 \tag{22}$$

Q.E.D.

Theorem 4: $\frac{d\,q^*_{abr}(\sigma^2_{\tilde{y}})}{d\sigma^2_{\tilde{y}}}\le 0$, if and only if $A_{\sigma^2}(\mu,\sigma^2)..0$.

Proof: To differentiate $\Upsilon\left(q^*_{abr},\lambda,\mu_{\tilde{y}},\sigma^2_{\tilde{y}}\right)$ about $\sigma^2_{\tilde{y}}$, there has

$$\frac{\partial \Upsilon\left(q^*_{abr},\lambda,\mu_{\tilde{y}},\sigma^2_{\tilde{y}}\right)}{\partial \sigma^2_{\tilde{y}}} = -\lambda^2 A_{\sigma^2}\left(\mu_{\tilde{z}}(q^*_{abr}),\sigma^2_{\tilde{z}}(q^*_{abr})\right) \tag{23}$$

Using the implicit function theorem and the second-order condition again, we have

$$sign\left(\frac{\partial q^*_{abr}(\sigma^2_{\tilde{y}})}{\partial \sigma^2_{\tilde{y}}}\right) = sign\left(-A_{\sigma^2}\left(\mu_{\tilde{z}}(q^*_{abr}),\sigma^2_{\tilde{z}}(q^*_{abr})\right)\right) \tag{24}$$

Thus, it is easy to see

$$\frac{\partial q^*_{abr}(\mu_{\tilde{y}})}{\partial \sigma^2_{\tilde{y}}} \leq 0 \Leftrightarrow A_{\sigma^2}\left(\mu,\sigma^2\right) \geq 0 \qquad (25)$$

Q.E.D.

Remark: In financial literature, it has been shown that the index $A(\mu,\sigma^2)$ defined in equation (10) corresponds to the Arrow-Pratt risk-averse coefficient under Expected Utility framework, and $A_\mu(\mu,\sigma^2) \leq 0$ denote the decreasing absolute risk-averse under Mean-Variance.

IV. CONCLUSION

The preference of a background risk complicates the decision-making problem of the newsvendor, which can be seen from our main results. In this paper, we examine the effects of a background risk on newsvendor's ordering decision, we present a set of conditions which are corresponding to the concept of Arrow-Pratt risk-averse coefficient and DARA. The newsvendor will behave in more cautious way if his Mean-Variance preference satisfies those conditions when there is an additional background risk. Furthermore, we also examine the effects of changes in the distribution of the background risk in the sense of first-order and second-order stochastic dominance.

ACKNOWLEDGMENT

The authors are grateful for financial supports from Foundation of Educational Committee of Jiangxi Province (Grant: GJJ14726), and Science Foundation of Jiujiang University (Grant: 8899209).

REFERENCES

[1] T. Eichner, "Mean variance vulnerability", Management Science, vol.54, no.3, 2008, pp.586-593.
[2] L.Eeckhoudt, C.Gollier, and H. Schlesinger, "The risk-averse (and prudent) newsboy", Management Science, vol.41, no.5, 1995, pp. 786-794.
[3] V. Agrawal and S. Sridhar, "Impact of uncertainty and risk aversion on price and order quantity in the newsvendor problem", Manufacturing and Service Operations Management, vol.2, no.4, 2000, pp.410-424.
[4] B.Keren and J.S. Pliskin, "A benchmark solution for the risk-averse newsvendor problem", European Journal of Operational Research, vol.174, no.3,2006,pp.1643-1650.
[5] S. Benoit, "The newsvendor problem under multiplicative background risk", European Journal of Operational Research, vol.200, no.3, 2010, pp.918-923.
[6] H.S. Lau, "The newsboy problem under alternative optimization objectives", Journal of the Operational Research Society, vol.31, no.6, 1980, pp. 525-535.
[7] F. Chen and A.Federgruen, "Mean-variance analysis of basic inventory models". Working Paper, Columbia Business School, New York, USA, 2000.
[8] T.M. Choi, D. Li and H. Yan, "Mean-variance analysis for the newsvendor problem". IEEE Transactions on Systems, Man, and Cybernetics-Part A: Systems and Humans, vol.38, no.5, 2008, pp.1169-1180.
[9] J. Wu, J. Li, S. Wang and T.C.E. Cheng, " Mean-variance analysis of the newsvendor model with stockout cost", Omega, vol.37, no.3, 2009,pp.724-730.
[10] T.M. Choi and C.H. Chiu, "Mean-downside-risk and mean-variance newsvendor model: implications for sustainable fashion retailing", International Journal Production Economics, vol.135, no.2, 2012, pp.552-560.
[11] F. Lajeri-Chaherli, F and L.T. Nielsen," Parametric characterizations of risk aversion and prudence, Economic Theory, vol.15, no.2, 2000, pp.469-476.
[12] A. Wagener, "Prudence and risk vulnerability in two-moment decision models", Economics Letters, vol.74, no.2, 2002, pp.229-235.
[13] M. B. Ormiston, and E. E. Schlee, "Mean-variance preferences and investor behaviour", The Economic Journal, vol.111, no.474, 2001, pp.849-861.
[14] F. Lajeri-Chaherli, "More on properness: the case of mean-variance preferences", The Geneva Papers on Risk and Insurance Theory, vol.27, no.1, pp.49-60, 2002.
[15] T. Eichner and A.Wagener, "Multiple risks and mean-variance preferences", Operations Research, vol.57, no.3 2009, pp.1142-1154.
[16] T. Eichner and A.Wagener, "Variance vulnerability, background risks, and mean-vaiance preferences", The Geneva Papers on Risk and Insureance Theory, vol.28, no.2, 2003, pp.173-184.

Optimization of Multiple Responses Considering both Model Uncertainty and Regret Effect

Lin-han Ouyang, Yi-zhong Ma*, Ting-ting Huang, Jian-jun Wang
Department of Management Science and Engineering, Nanjing University of Science and Technology, Nanjing, China
(yzma-2004@163.com)

Abstract - **Loss function approach is effective for multi-response optimization. However, previous loss function approach ignore model uncertainty and decision-makers' regret effect when determining the optimal input settings. In this paper, an overall loss function is proposed to optimize correlated multiple responses via confidence intervals, and we refer to the idea of worst case strategy used in decision theory to incorporate model uncertainty and decision-makers' regret effect into the loss function. The new loss function is composed of three sub-functions: expect performance, robustness performance, and regret performance, through which model uncertainty and regret effect are quantified. It is found that the traditional loss function will be a special case of the proposed approach if engineers could collect sufficient experimental data to estimate model parameters. An example is employed to illustrate the effectiveness of the proposed approach. The results show that the proposed approach can achieve reliable optimal operating conditions in comparison with those of traditional loss function approach.**

Keywords – **Loss function, Model uncertainty, Regret effect**

I. INTRODUCTION

In multi-response optimization problem, which has received considerable attention from academia and industry, one determines the optimal operating condition for the process variables in order to deal with robustness and optimization simultaneously for multiple responses (Apley and Kim [1]). Here, robustness means that the responses are not sensitive to variability which derives from model parameters. Hence, minimizing response variability amounts to selecting the inputs so that the output responses are robust or insensitive to variations in the model parameters. The main approach to this problem is loss function (Pignatiello [2], and Vining [3]), which mainly convert a multi-response problem into one with a single objective and then optimize the objective function. Main critiques for these techniques are that they cannot cope with model uncertainty, decision makers' regret effect, or correlations among responses.

At the stage of product design, the loss function is an effective approach to help quality engineers to design products or processes. The merit of the loss function approach is that it takes into account not only the correlation among responses but also the process economics.

Pignatiello [2] proposes a general loss function by extending Taguchi's [4] univariate loss function to resolve the correlation among responses. Elsayed and Chen [5] present a multi-characteristic model based on the loss function, which is similar to Pignatiello's approach. Jayaram and Ibrahim [6] extend Taguchi's loss function based on the desirability function to optimize the multi-response optimization problem. Their approach can easily be understood and applied by practitioners. Vining [3] constructs an improved quadric loss function to resolve simultaneously the correlation among responses, the process economics and the quality of predictions. Wu and Chyu [7] present an asymmetric loss function based on a mathematical programming model. The goal of their approach is to minimize the average quality loss of total experiments. Ko et al. [8] consider bias, the quality of predictions and robustness through integrating predicted future responses into the loss function. The main advantage of the loss function approach is that it incorporates the variance-covariance structure of responses as well as the quality of predictions. The merit of their loss function is that it integrates the features of robustness and reliability into a single model. Köksoy and Fan [9] develop an upside-down normal loss function that presents a reasonable risk measurement for the losses of being off-target in a product or process design.

Model uncertainty is an important issue in a product or process design. Failure to account for model uncertainty may lead to low customer satisfaction and operational performance. During the multi-response optimization process, model uncertainty problems have been studied from different aspects. We distinguish two general categories: (a) single model method: in this category, researchers attempt to construct the confidence intervals of predictive values or regard model parameters as random variables. Then they will obtain some information from the confidence intervals or the posterior distributions of model parameters to optimize the multi-response optimization problem. (b) multi-model ensemble method: in this category, researchers attempt to explore the possibility of using a weighted average surrogate model instead of a single surrogate model. When deciding which method is more appropriate, one needs to consider whether he can accurately select out the true significant variables.

Peterson et al. [10] present an alternative confidence region that distinguishes saddle points from minima/maxima and that can handle constraints on the inputs. Chiao and Hamada [11] propose a procedure to maximize the probability that multivariate normal responses will satisfy user-specified conditions. Peterson [12] suggests a Bayesian posterior predictive approach to cope with model uncertainty for the multi-response optimization problem. Zhang et al. [13] derive the mean and variance of response by discussing the compound effect of both the parameter uncertainty in design variables and meta-modeling uncertainty obtained from a kriging model. Erfani and Utyuzhnikov [14] implement

E. Qi et al. (eds.), *Proceedings of the 5th International Asia Conference on Industrial Engineering and Management Innovation (IEMI2014)*, Proceedings of the International Asia Conference on Industrial Engineering and Management Innovation 1, DOI 10.2991/978-94-6239-100-0_14,

the fuzzy value of parameters to handle model uncertainty. The major advantage of their method is that it provides an opportunity to control the extent of robustness according to designers' psychological preference. He et al. [15] define the robustness measure for the traditional desirability function to deal with model uncertainty. The proposed method takes account of all values in the confidence interval rather than a single predicted value for each response. Rajagopal et al. [16] and Ng [17] incorporate Bayesian model averaging method into multi-response optimization in order to consider model uncertainty. Zhou et al. [18] construct ensemble of surrogates aiming to reduce the impact of model uncertainty on product design.

Fig.1 illustrates the potential danger of traditional approaches ignoring model uncertainty. The dashed curve represents the true, but unknown model, and the solid curve is the fitted model. If the goal is to maximize response performance, the optimal value of the design variable is point A. However, traditional approaches based on the fitted model select point B for the design variable, which results in the response performance being well below the true optimal. Thus, even a slight deviation of the fitted model from the true model might result in unacceptable performance. Furthermore, it can be seen from Fig.1 that the difference between the predicted value and the true value of point B is larger than that of point A.

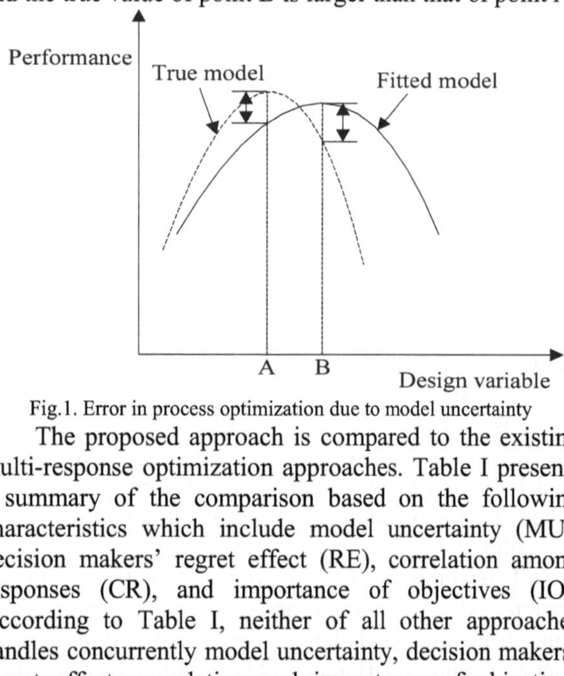

Fig.1. Error in process optimization due to model uncertainty

The proposed approach is compared to the existing multi-response optimization approaches. Table I presents a summary of the comparison based on the following characteristics which include model uncertainty (MU), decision makers' regret effect (RE), correlation among responses (CR), and importance of objectives (IO). According to Table I, neither of all other approaches handles concurrently model uncertainty, decision makers' regret effect, correlation and importance of objectives except the proposed approach. Therefore, based on the existing approaches, we propose a new loss function method incorporating the above all characteristics into multi-response optimization process.

This paper is organized as follows. In Section 2, we review the modeling process multi-response optimization problem. The proposed method for multi-response optimization problem is presented in Section 3, and an illustrated example is employed to verify the effectiveness of the proposed approach in Section 4. Finally, conclusion and some discussions are summarized in Section 5.

TABLE I
COMPARISON OF THE OPTIMIZATION APPROACHES

Method	MU[1]	RE[2]	CR[3]	IO[4]
He et al.	√		√	
Peterson			√	
Zhan et al.			√	
Erfani and Utyuzhnikov	√	√		
Goethals and Cho	√			√
Pinatiello	√	√		
Vining	√	√		
Ko et al.	√	√		
Ng		√	√	√
Chiao		√		
Rajagopal			√	√
Proposed method	√	√	√	√

(1: model uncertainty, 2: decision-makers' regret effect, 3: correlation among responses, 4: importance of objectives)

II. MODELING PROCESS IN RESPONSE OPTIMIZATION

Consider the following regression model, which is widely used in response surface methodology. The parametric model can be represented as

$$\mathbf{y} = \mathbf{X}\boldsymbol{\beta} + \boldsymbol{\varepsilon} \tag{1}$$

In general, $\mathbf{y}$ is an $(n\times 1)$ vector of observations, $\mathbf{X}$ is an $(n\times p)$ model matrix of $(p\text{-}1)$ variables augmented with a column of ones, $\boldsymbol{\beta}$ is a $(p\times 1)$ vector of the regression coefficients, and $\boldsymbol{\varepsilon}$ is an $(n\times 1)$ vector of random errors. The regression coefficients estimated by ordinary least squares (OLS) can be expressed as

$$\hat{\boldsymbol{\beta}} = (\mathbf{X}'\mathbf{X})^{-1}\mathbf{X}'\mathbf{y} \tag{2}$$

To construct confidence interval for model parameters, we assume that the errors are independently and normally distributed with a mean 0 and a variance σ^2. Therefore, the observations are also independently and normally distributed. Since the least-squares estimator $\hat{\boldsymbol{\beta}}$ is a linear combination of the observations, it follows that $\hat{\boldsymbol{\beta}}$ is normally distributed with a mean $\boldsymbol{\beta}$ and a covariance matrix $\sigma^2(\mathbf{X}'\mathbf{X})^{-1}$. It is obvious that the marginal distribution of each model parameter β_j is normal with a mean $\hat{\beta}_j$ and a variance $\sigma^2 C_{jj}$, where C_{jj} is the jth diagonal element of the $(\mathbf{X}'\mathbf{X})^{-1}$ matrix. Consequently, each of the statistics

$$\frac{\hat{\beta}_j - \beta_j}{\sqrt{\hat{\sigma}^2 C_{jj}}}, \qquad j = 1,2,\cdots p \tag{3}$$

is distributed as t with $(n\text{-}p)$ degree of freedom, where σ^2 is the estimate of the error variance.

Based on the result given in (3), the $(1-\alpha)$ confidence interval for the model parameter β_j $(j=1,2,\ldots p)$ is obtained as follows

$$\hat{\beta}_j - t_{\alpha/2,n-p}\sqrt{\hat{\sigma}^2 C_{jj}} \le \beta_j \le \hat{\beta}_j + t_{\alpha/2,n-p}\sqrt{\hat{\sigma}^2 C_{jj}} \tag{4}$$

where $t_{\alpha/2,n-p}$ is the $(1-\alpha/2)$ quantile for the t distribution with (n-p) degrees of freedom.

III. PROPOSED APPROACH

An optimization strategy is proposed for multi-response problem, in which the effect of model uncertainty and decision-makers' regret effect on product design are considered as different objective functions. With respect to the multi-objective optimization problem, linear weighted method is adopted to transform it to be a single objective optimization problem.

A. Traditional Loss Function

Loss function is an effective approach to seek the optimal solution for multi-response optimization problem. The solution represents some explicit compromise among conflicting conditions. Consider the multivariate analog for the squared error loss

$$L(\mathbf{y},\boldsymbol{\theta}) = (\mathbf{y}-\boldsymbol{\theta})'\mathbf{C}(\mathbf{y}-\boldsymbol{\theta}) \qquad (5)$$

where $\mathbf{C}$ is a $p\times p$ positive definite matrix of costs, $\boldsymbol{\theta}$ is a $p\times 1$ vector of target values and $\mathbf{y}$ is a vector of predicted values.

Generally, an appropriate objective is to select the optimal parameter setting $\mathbf{x}$, which minimizes the expected loss $E(L(\mathbf{y},\boldsymbol{\theta}))$.

$$E(L(\mathbf{y},\boldsymbol{\theta})) = (E[\mathbf{y}(\mathbf{x})]-\boldsymbol{\theta})'\mathbf{C}(E[\mathbf{y}(\mathbf{x})]-\boldsymbol{\theta}) + trace[\mathbf{C}\Sigma_{\hat{\mathbf{y}}(\mathbf{x})}] + trace[\mathbf{C}\Sigma_{\mathbf{y}(\mathbf{x})}] \qquad (6)$$

where $\boldsymbol{\Sigma}_{\hat{\mathbf{y}}(\mathbf{x})}$ and $\Sigma_{\mathbf{y}(\mathbf{x})}$ are the variance-covariance matrix of the estimated mean responses and the true responses, respectively. Suppose $\mathbf{X}_i$ be the model matrix for the ith response, and suppose $\mathbf{x}_i$ be the optimal setting expanded corresponding the model matrix for the ith response. Let $\boldsymbol{\Sigma} = [\sigma_{ii^*}]$ be the variance-covariance matrix for the responses.

Then, the prediction variance at the optimal setting with respect to the ith response is

$$h_{ii} = \text{var}[\hat{y}_i] = \sigma_{ii}\mathbf{x}_i'(\mathbf{X}_i'\mathbf{X}_i)^{-1}\mathbf{x}_i \qquad (7)$$

The covariance of two responses, $\hat{y}_i$ and $\hat{y}_{i^*}$ will be given by

$$h_{ii^*} = \sigma_{ii^*}\mathbf{x}_i'(\mathbf{X}_i'\mathbf{X}_i)^{-1}\mathbf{X}_i'\mathbf{X}_{i^*}(\mathbf{X}_{i^*}'\mathbf{X}_{i^*})^{-1}\mathbf{x}_{i^*} \qquad (8)$$

Thus the $\boldsymbol{\Sigma}_{\hat{\mathbf{y}}(\mathbf{x})}$ *and* $\boldsymbol{\Sigma}_{\mathbf{y}(\mathbf{x})}$ can be written by $[h_{ii^*}]$ and $\boldsymbol{\Sigma}+\boldsymbol{\Sigma}_{\hat{\mathbf{y}}(\mathbf{x})}$, respectively.

B. The New Single Objective Optimization Function

In this section, three single objective functions will be constructed based on the confidence intervals of model parameters and the traditional loss function approach. The proposed loss function combines the strengths of Vining's and Ko et al's methods. As such, it allows engineers to consider the expect performance and uncertainty performance as well as the regret performance of loss function.

1) Expect performance of loss function

Suppose $E[\mathbf{y}(\mathbf{x})] = f(\mathbf{x},\boldsymbol{\beta})$, (6) can be represented as

$$E(\mathbf{x},\boldsymbol{\beta},\boldsymbol{\theta}) = (f(\mathbf{x},\boldsymbol{\beta})-\boldsymbol{\theta})'\mathbf{C}(f(\mathbf{x},\boldsymbol{\beta})-\boldsymbol{\theta}) + trace[\mathbf{C}\Sigma_{\hat{y}(\mathbf{x})}] + trace[\mathbf{C}\Sigma_{y(\mathbf{x})}] \qquad (9)$$

If true model parameters $\boldsymbol{\beta}$ can be estimated, a solution will be obtained via minimizing (9). However, model parameters are rarely estimated precisely owing to the process complexity and limited experimental data. Therefore, we construct a new expect function, averaging the largest loss value and the smallest loss value under model parameters' confidence intervals. Then the worst case strategy is used to construct corresponding optimization strategy. The new expect performance of loss function is presented as follows:

$$Ex(\mathbf{x}) = \frac{1}{2}(\max_{\boldsymbol{\beta}\in\boldsymbol{\beta}^I}[E(\mathbf{x},\boldsymbol{\beta},\boldsymbol{\theta})] + \min_{\boldsymbol{\beta}\in\boldsymbol{\beta}^I}[E(\mathbf{x},\boldsymbol{\beta},\boldsymbol{\theta})]) \qquad (10)$$

where $\boldsymbol{\beta}^I$ are confidence intervals of model parameters. $E(\mathbf{x})$ is presented by averaging the worst case scenario and the best case scenario under the assumption that true model parameters will be located in the corresponding confidence intervals.

2) Robustness performance of loss function

In this subsection, the effect of robustness denotes the biggest difference between the two model parameter vectors used in (9). Contrary to the current methods, the proposed function tends to underestimate the robustness of a nominal solution in order to guarantee its reliability. The effect of robustness can be expressed by

$$Ro(\mathbf{x}) = \max_{\boldsymbol{\beta},\boldsymbol{\beta}'\in\boldsymbol{\beta}^I}\|E(\mathbf{x},\boldsymbol{\beta},\boldsymbol{\theta}) - E(\mathbf{x},\boldsymbol{\beta}',\boldsymbol{\theta})\| \qquad (11)$$

where $\boldsymbol{\beta}$ or $\boldsymbol{\beta}'$ are model parameters, and $\boldsymbol{\beta}^I$ are confidence intervals of model parameters.

3) Regret performance of loss function

As mentioned before, when uncertainty on model parameters is a crucial process, engineers may not feel confident using the optimal solution derived from inaccurate parameters. Generally, compared with the solution itself, the engineers are more concerned with how the output of an optimal solution is going to make them feel. Minimizing regret criterion is used for the determination of the final solution when a reference solution is given. Therefore, in this subsection, we take decision makers' regret effect into account via the regret theory to determine the optimal solution in the multi-response optimization problem. A new regret performance is proposed by introducing the minimax regret criterion used in decision theory. The regret performance can be expressed by

$$Re(\mathbf{x}) = \max_{\substack{\boldsymbol{\beta}\in\boldsymbol{\beta}^I \\ \mathbf{z}\in\Omega}}\{E(\mathbf{z},\boldsymbol{\beta},\boldsymbol{\theta}) - E(\mathbf{x},\boldsymbol{\beta},\boldsymbol{\theta})\} \qquad (12)$$

C. Construction of a New Optimization Strategy

As mentioned above, the aforesaid three objectives will be aggregated by the linear weighting method. In this subsection, the proposed optimization strategy is presented by:

$$\min_{\mathbf{x}} \ \omega_1 Ex(\mathbf{x}) + \omega_2 Ro(\mathbf{x}) + \omega_3 Re(\mathbf{x}) \quad s.t.\ \mathbf{x} \in \Omega \tag{13}$$

where ω_1, ω_2 and ω_3 are weights of expect performance, robustness performance, and regret performance, respectively. Ω is the acceptable region of design variables. The optimal setting can be obtained by minimizing (13). In the optimization of product or process design, we use the standard optimization function *fmincon* in Matlab. Since the optimization result is easily influenced by the initial point, 100 initial points are sampled from the acceptable region of input variables and then obtain a best recommended solution.

IV. AN ILLUSTRATED EXAMPLE

A. Problem Descriptions

In this section, the effectiveness of the proposed method is illustrated by an example presented in [17]. Engineers aim to select the optimal setting for reaction time (x_1), reaction temperature (x_2), and the amount of catalyst (x_3) which maximizes the conversion (y_1) of a polymer and achieves a target value of 57.5 for the thermal activity (y_2). The experimental data is presented in [17]. The target values are set by T_1=100 and T_2=57.5, and the cost matrix **C** is

$$\mathbf{C} = \begin{bmatrix} 0.100 & 0.025 \\ 0.025 & 0.500 \end{bmatrix}.$$

To solve the multi-response optimization problem, the statistical models are built for two responses, respectively, as follows:

$$\hat{y}_1 = 81.091 + 4.040x_2 + 6.204x_3 + 2.125x_1x_2 + 11.375x_1x_3 - 3.875x_2x_3 - 1.834x_1^2 + 2.938x_2^2 - 5.191x_3^2$$

$$\hat{y}_2 = 59.948 + 3.583x_1 + 2.230x_3 + 0.823x_1^2$$

In this example, since the variance-covariance matrix $\mathbf{\Sigma}$ is unknown, it is estimated by the maximum likelihood method and is given as follows:

$$\hat{\mathbf{\Sigma}} = \begin{bmatrix} 11.846 & -0.545 \\ -0.545 & 1.704 \end{bmatrix}.$$

B. Process Optimization

Given the fitted models and the value of the significance level, we can construct the confidence intervals of model parameters and regard the three functions are the equal importance. It is worth noting that if the confidence level is zero, indicating that the effect of model uncertainty is neglected, the confidence intervals of model parameters would be degenerated into real numbers. Three scenarios can be expressed as

Considering neither the model uncertainty nor the regret effect, the traditional loss function (TLF) is:

$$\min_{\mathbf{x}} \ J_{TLF} = E(\mathbf{x}) \quad s.t.\ \mathbf{x} \in \Omega \tag{14}$$

Considering both the model uncertainty and the regret effect, the overall loss function (OLF) is:

$$\min_{\mathbf{x}} \ J_{OLF} = Ex(\mathbf{x}) + Ro(\mathbf{x}) + Re(\mathbf{x}) \quad s.t.\ \mathbf{x} \in \Omega \tag{15}$$

The first optimization strategy in (17) is the traditional loss function proposed by Ko et al. Meanwhile, it can be regarded as special cases of the proposed approach when the confidence level is zero. Neglecting model uncertainty and regret effect, the optimal parameter settings are (-0.354, 1.682, -0.408).

Without loss of generality, we firstly construct the 90% confidence intervals for model parameters and the results are summarized in Table II.

Based on the confidence intervals and the proposed optimization strategy, the optimal parameter settings of the overall loss function are (0.976, 0.075, 0.345). The comparing results are presented in Table 3 with the 90% confidence intervals for model parameters.

TABLE II
CONFIDENCE INTERVALS FOR MODEL PARAMETERS

Responses	Parameters	Lower	Estimated	Upper
	β_0	77.692	81.091	84.490
	β_1	1.785	4.041	6.296
	β_2	3.949	6.204	8.459
	β_3	-0.822	2.125	5.072
y_1	β_4	8.428	11.375	14.322
	β_5	-6.822	-3.875	-0.928
	β_6	-4.029	-1.834	0.361
	β_7	0.744	2.939	5.134
	β_8	-7.388	-5.193	-2.997
	β_0	59.219	59.948	60.677
	β_1	2.894	3.583	4.273
y_2	β_2	1.541	2.230	2.919
	β_3	0.158	0.823	1.488

TABLE III
OPTIMAL SOLUTIONS FOR TWO APPRAOCHES

Scenarios	Expect	Robustness	Regret
Traditional approach	3981.5	1151.10	13, 646
Overall loss function	61.153	104.500	18, 116

From Table III, we know that the optimal solution performs better than that of the traditional approach in terms of expect performance and uncertainty performance. Hence, it can be seen that a modest 90% improvement in the value of overall loss value is achieved by considering model uncertainty and regret effect when selecting the optimal settings. As to this case, the traditional loss function method gives an optimal solution at which regret performance is better while expect and robustness performances are worse comparing to the proposed approach.

In all fairness, we should admit that we are not strictly comparing "apples to apples" in this case. Each optimal solution is decided by the corresponding objective function. The settings are "optimal" for different optimal strategies because they analyze the problem from different aspects and points. To be specific, for some problems, minimizing the traditional loss function is more meaningful than minimizing the overall loss function.

V. CONCLUSION AND DISCUSSION

Most existing research on this topic adopts loss function to deal with the correlation among responses, however, they ignore model uncertainty or regret effect. In this paper, the multi-response optimization problem has been resolved by proposing an overall loss function approach in which model uncertainty and regret effect are considered. As a remedy, the proposed overall loss function has the following methodological advantages:

(1) Via integrating the confidence intervals of model parameters into the construction of optimization strategy, the proposed approach takes into account the effect of model uncertainty on the optimal solution.

(2) The proposed approach considers decision-makers' regret effect by introducing the regret concept used in multi-objective decision literature into the multi-response optimization problem.

For each multi-response optimization problem, there exist three stages: collecting data, building model, optimizing the objective function. It should be noted that the stage of model building has great influence on the optimal solution obtained in the third stage. Therefore, the proposed method introduces confidence intervals of model parameters into the construction of optimization strategy. The optimization strategy incorporates expect, robustness, and regret performances together rather than just expect performance in the traditional loss function. It needs to be pointed out that the proposed approach has limitations as well. The loss function optimization requires that the cost matrix should be determined prior to the optimization of the objective function. Though some researchers have developed several methods to resolve it, it is not easy for decision makers to specify the value of cost matrix in some case.

ACKNOWLEDGMENT

This work was partially supported by the National Natural Science Foundation of China under grant Nos. 71371099 and 71211140350, and colleges and universities in Jiangsu Province plans to graduate research and innovation projects (CXZZ13_0226), which are gratefully acknowledged by the authors.

REFERENCES

[1] D.W. Apley, J. Kim, "A cautious approach to robust design with model parameter uncertainty", *IIE Transactions*, vol. 43, no. 7, pp. 471-482, 2011.

[2] J. J. Pignatiello, "Strategies for robust multiresponse quality engineering", *IIE Transactions*, vol. 25, no.3, pp. 5-15, 1993.

[3] G. G. Vining, "A compromise approach to multiresponse optimization", *Journal of Quality Technology*, 30, 4, pp. 309-313, 1998.

[4] G.Taguchi, E. A.Elsayed, C.H. Thomas, "*Quality engineering in production systems*", McGraw-Hill College, 1989.

[5] E. A.Elsayed, A. Chen, "Optimal levels of process parameters for products with multiple characteristics", *International Journal of Production Research*, vol. 31, no. 5, pp. 1117-1132, 1993.

[6] J. S. R. Jayaram, Y. Ibrahim, "Multiple response robust design and yield maximization", *International Journal of Quality & Reliability Management* , vol. 16, no 9, pp. 826-837, 1999.

[7] F. C. Wu, C. C. Chyu, "Optimization of robust design for multiple quality characteristics", *International Journal of Production Research*, vol. 42, no. 2, pp. 337-354, 2004.

[8] Y. H. Ko, K. J. Kim, C. H. Jun, "A new loss function-based method for multiresponse optimization", *Journal of Quality Technology*, vol. 37, no. 1, pp. 50-59, 2005.

[9] O. Köksoy, S. K. S. Fan, "An upside-down normal loss function-based method for quality improvement", *Engineering Optimization*, vol. 44, no. 8, pp. 935-945, 2012.

[10] J. J. Peterson, C. Suntara, E. Del Castillo, "A general approach to confidence regions for optimal factor levels of response surfaces", *Biometrics*, vol. 58, no. 2, pp. 422-431, 2002.

[11] C. H. Chiao, M. Hamada, "Analyzing experiments with correlated multiple responses", *Journal of Quality Technology*, vol. 33, no. 4, pp. 451-465, 2001.

[12] J. J. Peterson, "A posterior predictive approach to multiple response surface optimization", *Journal of Quality Technology*, vol. 36, no. 2, pp. 139-153, 2004.

[13] S. Zhang, P. Zhu, W. Chen, P. Arendt, "Concurrent treatment of parametric uncertainty and metamodeling uncertainty in robust design", *Structural and Multidisciplinary Optimization*, vol. 47, no. 1, pp. 63-76, 2013.

[14] T. Erfani, S. V.Utyuzhnikov, "Control of robust design in multiobjective optimization under uncertainties", *Structural and Multidisciplinary Optimization*, vol. 45, no. 2, pp. 247-256, 2012.

[15] Z. He, P. F. Zhu, S. H. Park, "A robust desirability function method for multi-response surface optimization considering model uncertainty", *European Journal of Operational Research*, vol. 221, no. 1, pp. 241-247, 2012.

[16] R. Rajagopal, E. Del Castillo, "Model-robust process optimization using Bayesian model averaging", *Technometrics*, vol. 47, no. 2, pp. 152-163, 2005.

[17] S. H. Ng, "A Bayesian-model averaging approach for multiple-response problem", *Journal of Quality Technology*, vol. 42, no. 1, pp. 52-68, 2010.

[18] X. J. Zhou, Y. Z Ma, Y. L.Tu, Y. Feng, "Ensemble of surrogates for dual response surface modeling in robust parameter design", *Quality and Reliability Engineering International*, vol. 29, no. 2, pp. 173-197, 2013.

Determinants of Productivity in the Ready-made Garments SMEs of Lahore, Pakistan

Amir Ikram*, Qin Su
[1]School of Management, Xi'an Jiaotong University, Shaanxi, Xi'an, China
(amirikram@stu.xjtu.edu.cn)

Abstract - **The study has been undertaken to investigate the determinants of productivity in the readymade garments SMEs of Lahore. It's a survey whereby 55 SMEs of Lahore were contacted by employing Stratified Random Sampling technique. For data analysis, various quantile regression techniques along with its diagnostic analyses have been used to identify the association and its significance between the organizational variables and firm productivity. Firms' productivity has been measured through output per labor of each SME. The significant determinants of productivity are found out to be ambition and motivation, age of the enterprise, installation of generator facility, bank facility, and working capital management. Implications of law of diminishing marginal returns were also examined; increase in number of machines leads to augment productivity up to a certain level and then starts a diminishing return. The findings of this study have good implications for the policy makers and the entrepreneurs alike.**

Keywords - **Entrepreneurs, Law of Diminishing Returns, Productivity, Readymade garments, SMEs**

I. INTRODUCTION

This study is undertaken with respect to the Small & medium garments manufacturing firms working in the city of Lahore, Pakistan to examine the determinants of productivity of SMEs. The goal of this research is to scrutinize the determinants of productivity in the SMEs of ready-made garments (RMG) in Pakistan. The main research question is 'What are the determinants of productivity in the export-oriented garments' firms in Lahore, Pakistan'? While a paper on 'Organizational determinants of firm performance; a case of Garments Cluster Lahore' has already been prepared, therefore 'determinants of productivity of SMEs in RMGs' is another major aspect of Pakistan's textile industry to be researched in. Therefore, there was a need to determine the extent of emphasis on cost effectiveness, efficiency and quality management systems in effect in Pakistan's readymade garments' industry.

Pakistan is primarily an agricultural country; this agricultural base provides conducive environment to its textile sector. The high labor intensity of this sector makes it a great source of strong comparative advantage for South Asian countries. Thus, it is an important source of economic well-being by generating employment and foreign exchange earnings [1]. The textile sector is a mainstay in the economic development of Pakistan; export of textile to different countries is the backbone of its economy. Textile industry contributes about 8.5% in the GDP, 54% of total exports, and 38% of total labor force are from this sector (Pakistan Economic Survey, 2010). Knitwear, ready-made garments and cotton yarn also have important shares in total exports. Overall, the US and the EU are Pakistan's largest trading partners accounting for 25% and 20% shares of Pakistani exports respectively [2]. Pakistan's textile sector continues its healthy growth in the fiscal year of 2010-2011, as its export went up by over 23 per cent in first four months as against the corresponding period of the preceding year [3]. Despite the gloomy outlook of the overall industry, this turnaround is indeed comforting news. There is satisfactory growth both in terms of quantity as well as value: readymade garments exports in 2011 raised by 25% in terms of quantity and 38% in terms of value; knitwear's exports improved by 28% in quantity and 32% in value [4].

The readymade garment industry in Pakistan commonly runs on small, medium and large units; majority are having 50 machines and below. Now large units are forthcoming in the organized sector of the industry [4]. The RMG industry has faced several international and domestic challenges over the last decade. In the international arena, implementation of the rules and regulations of the World Trade Organization (WTO) and preferential trade arrangements among different groups of countries are of special concern [5]. As per the Global Competitiveness Index of World Economic Forum, international economic competitiveness of Pakistan has dropped to 118th in 2010-11 from 83rd position in 2006-07 [2]. Though some literature is available on the quality management issues and determinants of the SMEs performance in the readymade garments sector of Pakistan, determinants of firm's productivity and analysis of cost-effectiveness is the area to be researched in.

A. Objectives

1. Identify the determinants of productivity in RMG SMEs of Lahore.
2. Identify the factors affecting the profitability of garments' firms, and particularly the implications of working capital management.
3. Ascertain specific issues relating to the entrepreneur and productivity of the enterprise.
4. Identify problems arising out of energy crisis and inflation, especially for the entrepreneurs who work on orders.
5. Make recommendations to establish & thereby promote the competitiveness of SMEs in RMG sector of Lahore.
6. Make recommendations for future actions, addressed to the federal and provincial institutions.

E. Qi et al. (eds.), *Proceedings of the 5th International Asia Conference on Industrial Engineering and Management Innovation (IEMI2014)*, Proceedings of the International Asia Conference on Industrial Engineering and Management Innovation 1, DOI 10.2991/978-94-6239-100-0_15,

B. *Hypotheses*

1. There is positive association between age of SMEs and its productivity.
2. The relationship between bank loan facility and productivity is positive.
3. There is positive relationship between installation of generator unit and productivity of the enterprise.
4. Motivation of the entrepreneurs (by-choice entrepreneurs) 'by-choice entrepreneurs (ambition-oriented)' is greater than that of 'forced entrepreneurs'.
5. Law of marginal returns to scale prevails in the SMEs of RMG sector (the increase in the number of employees will first enhance the productivity at increasing rate and then increase the productivity at decreasing rate depicting diminishing returns to scale)
6. Working capital management (A/R turnover, A/P turnover, and inventory management) enhances productivity.
7. The 'family-entrepreneurship' improves the productivity of the enterprise.

II. LITERATURE REVIEW

In this informative and competitive age, efforts are being made to take care of the cost, quality and thereby the productivity. Most of the theoretical models on industry dynamics assume that firms are born with an intrinsic ability, their productivity [6, 7]. A competitive market condition, innovativeness, R&D and export activities enhances the productivity growth [8]. The vast degree of variation in productivity is being experienced, even within the same manufacturing industries, both in developing and the developed countries [9]. Relevant study with respect to Pakistan shows that the majority of the SMEs lack formal training and performance appraisal practices [10]. It was noticed that organizations are better off if they increase labor size in proportion to the area; overcrowding of labor results in negative impact on motivation [11].

In the garment industry, the investment in net fixed assets, raw material, wages and salaries, and power and fuel consumption contributes 75 per cent of total inputs [12]. Various studies [13, 14, 15] have depicted that SME's profitability can be improved by effective working capital management; it has been observed that profitability can be increased by reducing the accounts receivable turnover, inventory turnover, and cash conversion cycle. It has been seen that the firms who have taken long term financing, i.e. bank loans, exhibits less productivity; while those who rely on the short-term financing or line of credit demonstrate improvement in the productivity [16]. The literature addressing the relationship between the area of SMEs and the productivity is quite limited; this makes the study of this independent variable of area of the SME even more interesting. Bardhan [17] found inverse relationship between area and productivity but it was primarily attributed to negative relationship between size and other inputs than of scale diseconomies; thus his findings supported the law of diminishing returns to some extent and imply that the higher the land, higher the productivity. Nonetheless, it was analyzed that if economies of scale are in place, then the proposition of direct relationship between area and productivity is effective. It would be fair to say that the studies, though limited in this regard, are tilted towards exhibiting negative correlation between the size with respect to the area and the productivity. The reason might be that the small land owners out of compulsion want to maximize their land efficiency [18]. With respect to Pakistan, the entrepreneurs are wary because of the frequent electric shutdowns, energy crisis, custom delays, ever rising inflation, and lack of government support [19].

III. SIGNIFICANCE

SMEs, especially in the readymade garments sector, are playing a vital role in the development of developing Asian countries. The recent crisis of power outages and inflationary impact on inputs has increased the cost of doing business, thus compounding the complications for exporters. This sector of readymade garments mainly relies upon export of their output, but there are hindering factors which are influencing the productivity of export-oriented SMEs. All these predicaments lead to low productivity; thus there arouse a need to analyze dynamics of productivity and the factors of input, thereby suggesting measures to enhance productivity.

IV. DESIGN & SOURCE OF THE DATA

It is cross-sectional study as the data has been collected from different firms in the sector in a given time period. Due to the descriptive nature of the research, we primarily used quantitative techniques. The survey was conducted via mailed, telephone and self-administered questionnaire. The final sample was 44 proprietors of SMEs of ready-made garments (RMG) sector in the city of Lahore, Pakistan. Firms' productivity has been measured through output (quantity per SME); as the concerned entrepreneurs were either unaware or hesitant in disclosing their output in monetary terms. Since population size is, to some extent, identifiable, we went for probability sampling, and sampling technique being employed in this study is stratified random sampling, with small and medium enterprises being two 'strata'. Moreover, we performed 'multicollinearity test' to make sure that our final sample size of 44 SMEs gives robust findings and output.

Sample is based on the list of Punjab Government Industrial Directory 2007 and Pakistan Readymade Garments Manufacturers and Exporters Association (PRGMEA). Sample for this study is almost 26 % of the population. After gathering the contact information of the readymade garments sector, 55 entrepreneurs were contacted. Out of the 55 contacted, 4 units found to have become sick or closed down, while 7 did not agree to respond, thereby 44 SMEs agreed and answered the survey questionnaire; thus the healthy response rate of almost 86% has been experienced.

Data analysis plan

Data was analyzed by using the statistical soft-wares, EViews and in some instances SPSS. We first resorted to traditional analysis approach of correlation matrix, primarily to test multicollinearity among the variables and thereby validating our sample size. Then a rather technical method of 'quantile regression estimates' has been employed and compared with OLS estimates; the quantile regression analysis is supplemented by the diagnostic analysis and examination of the process coefficients.

V. THEORETICAL FRAMEWORK

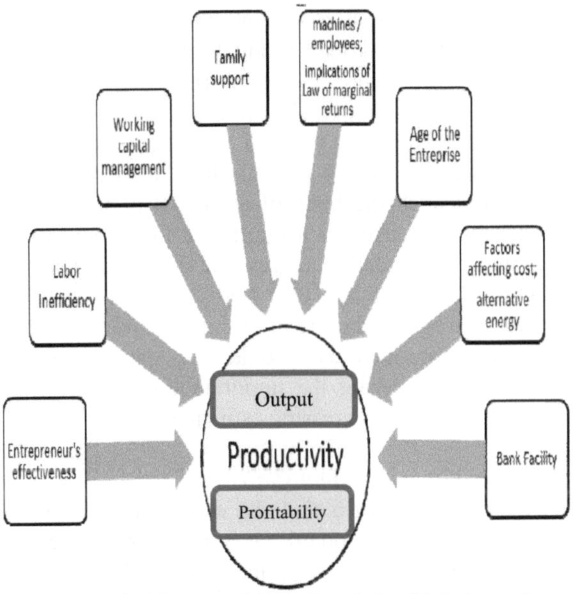

Fig.1. Theoretical Framework depicting relationship between dependent and independent variables

A schematic Fig. 1 summarized the relation of all the factors (the independent variables) discussed above with the productivity of the firm (the dependent variable). And labor efficiency comprises variables like number of workers, relative number of per-piece employees and salaried employees. In working capital management, we took variables accounts receivable turnover, accounts payable turnover, and inventory management. While other variables mentioned in Fig.1 does not contain sub categories.

VI. DESCRIPTIVE STATISTICS

Table I describes the characteristics of the data; the SMEs on average produce a monthly output of 4373 units, which is decent considering the grave situation the textile sector is in. The average age of the SMEs is found to be 15; this is understandable as hardly any new venture has been established in this sector in the period 2006-2011 primarily because of worsening energy crisis and ever increasing cost of doing business.

TABLE I
DESCRIPTIVE STATISTICS

Variables	N	Min	Max	Mean	Std. Deviation
Productivity (Units per month)	44	1700	12000	4373.3	2547.4
No. of workforce	44	12	240	70.4	55.1
Stitching Machines	44	8	217	61.0	50.3
Age of the enterprise	44	6	44	15.7	8.4

Source: Author's own calculations

VII. TECHNICAL ANALYSIS

A. Correlation Matrix

TABLE II
CORRELATION MATRIX [ASSOCIATION AMONG THE VARIABLES]

Variables	*Machines*	*Generator*	*Stock*	*Bank loan*	*A/R Turn*	*A/P Turn*	*Age*
Machines	1.00						
Generator	-0.19	1.00					
Stock	0.20	0.06	1.00				
Bank facility	-0.03	-0.09	0.01	1.00			
A/R Turnover	0.16	-0.23	0.30	-0.21	1.00		
A/P Turnover	-0.16	-0.34	-0.11	-0.04	0.09	1.00	
Age	0.18	-0.23	0.07	-0.05	-0.04	0.10	1.00
Family support	-0.42	0.15	0.02	-0.17	-0.10	-0.13	-0.19

Source: Author's own calculations

Table II exhibits the association among all the variables, excluding the dependent variable (productivity). There is weak pair-wise association among the variables (i.e. $r \leq 0.5$ for all the variables), which indicate that there is very low association between regressors, hence no multicollinearity among the variables in Table II. Hence the analysis based on these variables will give us robust results. It is also an indication that our sample size is appropriate and giving reliable findings.

B. Linear Regression in Contrast with Quantile Regression

The quantile regression coefficients allows comparing how some percentiles of the productivity may be more affected by certain determinants/ independent variables than other percentiles. This is reflected in the change in the size of the regression coefficient. The Table III gives OLS estimate and thereby compares it with the coefficients for the 5th, 10th, 50th, 90th, 95th quantile regression for the productivity.

The OLS estimate for 'Age' states that one year increase in the age of the enterprise will bring around 58 units of output increase in productivity; the relationship is positive and in line with our hypothesis. Notably the quantile regression analysis shows that coefficient value for 'age' is diminishing and eventually becomes negative in the 90th quantile regression; consequently the positive relationship between age and productivity ceases to exist for the SMEs in highest growth bracket which indicates that these particular SMEs have become stagnant over time and are not that much growth-oriented as they used to be. The relationship is significant at 10 % of significance for both the OLS and quantile regression estimates.

TABLE III
LINEAR REGRESSION & QUANTILE REGRESSION TABULATION

Variables	Linear Regression (OLS Estimates)	Quantile Regression				
		5th	10th	50th	90th	95th
Intercept	1811	622	622	1645	3869	4793
	(0.08)	(0.00)	(0.00)	(0.00)	(0.00)	(0.00)
Age	58	24	24	16	-10	-19
	(0.03)	(0.00)	(0.00)	(0.09)	(0.00)	(0.00)
Ambition	211	113	113	429	456	393
	(0.15)	(0.00)	(0.00)	(0.00)	(0.00)	(0.00)
A/P Turnover	-1246	-239	-239	-362	-590	-534
	(0.00)	(0.00)	(0.00)	(0.01)	(0.00)	(0.00)
A/R Turnover	299	-177	-177	-36	256	265
	(0.09)	(0.00)	(0.00)	(0.77)	(0.00)	(0.00)
Bank Facility	1416	152	152	401	842	680
	(0.06)	(0.00)	(0.00)	(0.03)	(0.00)	(0.00)
Generator	-684	-116	-116	-441	-1032	-1075
	(0.00)	(0.00)	(0.00)	(0.00)	(0.00)	(0.00)
Machines	46	34	34	46	40	38
	(0.00)	(0.00)	(0.00)	(0.00)	(0.00)	(0.00)
Family Support	250	33	331	139	-284	-347
	(0.24)	(0.00)	(0.00)	(0.29)	(0.00)	(0.00)

Source: Author's own calculations
Note: the figures in parentheses are the probability values

As we can see in Table III, the ambition-oriented entrepreneurs experience increase in the productivity; though there is a decrease in the coefficient value of 'ambition' from 90th to 95th quantile, which more probably suggests that the entrepreneurs who experience highest level of productivity are somewhat in a state of indifference. The A/P turnover value(s) suggest the negative relationship at all conventional level of significance ($p < 0.01$). The less the A/P turnover, the higher the productivity; thus the SMEs which extend their account payable experience increase in productivity. Whereas the values for A/R turnover suggests that the SMEs which receive their payments quickly shows positive effect on productivity, and this scenario holds true for OLS estimate and SMEs in upper quantile range. There exists a contradiction for the SMEs in lower & median quanitles; this inconsistency can be attributed as indication that though the low to average productive SMEs are receiving payments on their accounts receivable on regular basis, but they are somehow unable to make productive use of the resultant boost in the cash. Since for A/R turn, the median regression line (298.8206) differs from the mean (OLS) regression line (-36.27) significantly, this distribution is asymmetric. The values against Bank facility in Table III indicate significant positive relationship with productivity ($p<0.05$). It suggests that SMEs take bank loans primarily to enhance growth and productivity, and it is fair enough to say that they are succeeding in this endeavor. There is significant negative correlation between electric 'generator' and productivity; this is pretty understandable because such alternative energy sources increase the cost of doing business; but SMEs have to swallow the bitter pill of installing generator facility as they are left with no other choice. The values for machines in the Table III portray significant positive relationship ($p<0.01$) between productivity and number of machines and there is a diminishing return from 90th quantile onwards. This is consistent with the principle of law of marginal returns, whereby increases in variable factor (machines in this study) leads to augment productivity up to a certain level and then starts a diminishing return. The last variable in Table 3 depicts family support or family entrepreneurship for that matter enhances productivity. This is consistent with the viewpoint that two minds are better than one; and this holds true for the SMEs having low to average productivity. However for SMEs in the upper quantiles of productivity, the situation is otherwise; family support does not positively affect productivity anymore. This could be the case that the enterprises having higher productivity need singular direction to keep abreast with the progress; as there is a proverb 'too many crooks spoil the broth'; same scenario prevails here with more than one entrepreneur at the helm of affairs. It is pertinent to mention that relationship is insignificant for OLS estimate (mean) and 50th quantile regression (median); whereas it's significant for 5th, 10th, 90th and 95th quantile estimates. Overall, the relationship between family support and productivity would be categorized as insignificant considering the importance of implications for mean and median regression estimates.

C. *Specification and Diagnostic Tests*

In order to confirm that the estimated models are consistent with the standard OLS assumption, several diagnostic tests have been applied. As per the results in Table 4, we fail to reject null hypothesis in both cases Normality test and Ramsey's RESET test; this indicates that residuals are normally distributed and model is correctly specified. In the case of Wald test, we reject null hypothesis at all conventional levels of significance and conclude that conditional quantiles are not identical and that quantiles differ across quantile values. In nutshell, we can say those diagnostic and specification tests show that results of this study are vigorous.

TABLE IV
SPECIFICATION AND DIAGNOSTIC TESTS

Tests	**Equation**
Normality Test	
Jarque-Bera	2.2655
	(0.322)
Ramsey's RESET Test (Specification Test)	
F-statistic	1.119907
	(0.2899)
Log Likelihood Ratio	1.103359
	(0.2935)
Slope Equality Test (Wald Test)	
Chi-square statistic	225.0966
	(0.0000)

VIII. FINDINGS

This study strengthens the notion that proper working capital management is of equal importance to SMEs as well. The faster the payments are being received on account receivables, the higher the productivity. The cash received from fast recovery of accounts receivables could thus be used in managing the affairs effectively and efficiently. Likewise the SMEs which extends their bills on accounts payable experience higher productivity. Working capital management also encompasses proper management of inventory, with respect to reordering and visit to the suppliers; the lesser the visit(s) or reordering, the higher the efficiency. This is quite understandable as visits to the market and reordering the suppliers incurs cost and time, thus affects the efficiency.

In SMEs where generator facility is being installed, decline in productivity is experienced. Installation and functioning of alternative energy source results in constant monetary cost, which increases the cost of doing business manifold. Taking bank facility is seen to be widespread practice by SMEs in this sector; the SMEs which have taken the bank loan experience a growth in their productivity. This implies that loans taken are being utilized efficiently for the growth of the SMEs. Entrepreneurs own characteristics are of paramount importance because in such SMEs, entrepreneurs are seen to be at the center-stage. The productivity of the 'by-choice entrepreneurs' is greater than that of 'forced entrepreneurs'. This comes as no surprise because by-choice entrepreneurs are supposed to be ambitious and motivated, and relationship between motivation and productivity is well established [20]. Furthermore, productivity is found to be positively associated with the age of the SME. As we, the human beings get mature with time and learn from our mistakes; similar is the case with SMEs. Thus as a result SMEs experience continual growth as they mature but after accomplishing a certain high level of productivity, the positive relationship between age and productivity ceases to exist which is an indication that these particular SMEs have become stagnant over time and not that much growth-oriented as they used to be.

IX. CONCLUSION

The findings of this study have implications for the policy makers and both the prospective and current entrepreneurs. The manufacturing sector is faced with many challenges because of growing energy crisis, ever increasing cost of doing business, institutional and governance weaknesses, global economic crisis and unfavorable security situation. In spite of these problems, there are a lot of firm-level success stories. Meanwhile many SME units have ceased to operate. It has also been seen that many SMEs, especially in the readymade garments sector, have shifted and are shifting their units to low cost destinations like Bangladesh, primarily because of energy crisis & deteriorating law and order situation; the SMEs in this sector are operating at just 50% to 60% capacity level[1]. Though entrepreneurs are eager to pursue continual growth, yet it has been observed that they are content with relying on customer's/customer-given brand names, instead of their own. Among the internal factors affecting profitability are the lack of market research, promotion and branding.

Some organizational and entrepreneurial variables are found to be of paramount importance and thus needed to properly taken care of. The proper management of working capital was found to be effective in enhancing productivity. However, it was observed that working capital management is taken for granted by many SMEs, which needs to be addressed and pondered upon. The entrepreneurs should know that their small size, both with respect to area and number of employees is not a drawback; in fact capitalizing on these factors would lead to enhancing productivity. Majority of SMEs responded to load shedding by running their own generators, but for them the cost of maintaining a power generator is quite high and ultimately results in lower productivity. Thus government should make pragmatic measures to ensure continuous supply of energy to this export-oriented sector. The readymade garments firms of Lahore exists far between and follows more of a pattern of urbanization economy. Government should take measures to narrow the cluster by dedicating some areas to this sector. Considering the internal and external dynamics, 'localization economy' would be more facilitative for the entrepreneurs of this sector.

Government needs to take drastic steps to make doing business a relaxed and hassle-free job. Considering the regional scenario, doing business in Pakistan is difficult, inflated and inefficient. As per the World Bank Doing Business Report 2012, Pakistan's ranking has gone from bad to worse; it moved from 96th position in 2011 to 105th position in 2012 [2]. Thus the government should come up with a policy that ensures interruption-free energy supplies to the areas where readymade garments' firms are denser. Measures should be taken to remove or at least reduce the bureaucratic hurdles, decreasing the time that it takes to clear exports & imports through customs, and provision of 'one-window operation' for exporters. Most of the raw-material used in readymade garments SMEs is being imported from China; free trade agreements with China in this regard is recommended. In order to upgrade the labor's skill and efficiency, the private-public partnerships can be of great advantage.

Limitations & Future Research

In the developing or under-developed countries, some SMEs more often than not don't get themselves registered and more often they are also not legally responsible to do so. Moreover, these very SMEs are not used to keep financial records and maintain any specific accounts. Thus in our study, we followed the subjective

[1] Information deduced by the author himself during survey

productivity approaches in order to somehow minimize the effects of non-registration by SMEs. The productivity of the enterprise significantly depends on the business-life phase the country is in. This study is conducted in the period of 2010-2012, a roller-coaster period in Pakistan. Thus the findings might not be relevant to some other time period. Nevertheless, the future researchers are advised to compare the results of this study with any such relevant study of some stable period to know the discrepancies.

REFERENCES

[1] P. Argawal, "Improving India's Exports of Textiles and Garments," *Economic and Political Weekly*, pp. 3886-3888, 2001.

[2] A. Bashir, "State of Manufacturing in Pakistan," *Lahore Chamber News*, pp. 2-5, Dec. 2011.

[3] M.Z. Khan. (2011, August 18). "Manufacturing manages paltry growth in FY11," [Online]. Available: Dawn.com, http:// www.dawn.com/ 2011/08/18/manufacturing-manages-paltry-growth-in-fy11.html [August 18, 2011]

[4] Dr. A.M. Noor, (2011, January). "Readymade garment manufacturers face high cost of production," *Pakistan Trade Journal.* [Online]. Available: http://www.ptj.com.pk/Web-2011/01-2011/Dr.Noor-2.htm. [December 06, 2011]

[5] N. Ahmed, "Sustaining Ready-made Garment Exports from Bangladesh," *Journal of Contemporary Asia*, vol. 39, no. 4, pp. 597-618, 2009.

[6] H. Hopenhayn, "Entry, exit and firm dynamics in long run equilibrium," *Econometrics,* vol. 60, no. 5, pp. 1127-1150, 1992.

[7] L. Sels, S. Winne, J. Delmotte, J. Maes, D. Faems, and A. Forrier, "Linking HRM and Small Business Performance: An Examination of the Impact of HRM," *Small Business Economics*, pp. 83-101, 2006.

[8] Oh. Dong-hyun, "Productivity growth, efficiency change and technical progress of the Korean manufacturing industry," *Journal of the Asia Pacific Economy*, vol 16, no. 1, pp. 50-70, 2011.

[9] E. Bartelsman, and M. Doms, "Understanding Productivity: Lessons from Longitudinal Micro data," *Journal of Economic Literature,* vol. 38, pp. 569-594, 2000.

[10] R. Arif, K. Jamshed, A. Usman and A.M. Sarfraz, "An exploratory study of the performance of SMEs in Pakistan and the characteristics of successful firms" *Unpublished empirical study by LUMS, Lahore, Pakistan*, 2005.

[11] R.H. Vadhani, "Need to equip garment industry for facing challenges of post-MFA era," *Textile Magazine*, pp. 57-62, August, 2002.

[12] R. Joshi, and S. Singh, "Estimation of TFP in the Indian garment industry," *Journal of Fashion Marketing and Management,* pp. 145-160, 2010.

[13] P.J.G. Teruel, and P. M. Solano, "Effects of working capital management on SME profitability," *International Journal of Managerial Finance*, vol.3, no. 2, pp. 164-177, 2007.

[14] K. Padachi, "Trend in working capital management and its impact on firms' performance: An analysis of Mauritian small manufacturing firms," *International Review of Business Research Papers*, Vol.2, No. 2, pp. 45-58, 2006.

[15] T. Afza, and M. S. Nazir, "Working Capital Approaches and Firm's Returns in Pakistan," *Pakistan Journal of Commerce and Social Sciences*, pp. 25-36, 2008.

[16] A.M. Fernandes, "Firm Productivity in Bangladesh Manufacturing Industries," *Washington: The World Bank*, pp. 128-132, 2011.

[17] P.K. Bardhan, "Size, Productivity, and Returns to Scale: An Analysis of Farm-Level Data in Indian Agriculture," *Journal of Political Economy*, pp. 1370-1386, 1973.

[18] A.K. Kiani, "Farm Size and Productivity in Pakistan," *European Journal of Social Sciences,* vol. 7, no. 2, pp. 42-52, 2008.

[19] M.A. Rauf, M.J.V Riemsdijk, and J.K. Looise, "Investigating Human Resource Management formalization in SMEs; The case of the ready-made garments manufacturers and exporters in Lahore, Pakistan," *Netherlands: University of Twente, School of Management and Governance,* 2009.

[20] J. Michie, C. Oughton, and Y. Benn, "Employee ownership, motivation," *Berkbeck, University of London,* 2002.

Identification and Prevention of Safety Hazards on Construction Based on Fuzzy Recognition

Yan-guo ZAN*, Yue LI, Xue-liang HOU
Research Institute of Engineering Technology and Management, North China Electric Power University, Beijing, China
(yanguozan@126.com)

Abstract - **For many safety hazards, difficult to identify and resulting in a large number of engineering problems of the accident, mathematical model is established by using fuzzy recognition, which can recognition construction safety hazards. Using a lot of architectural engineering data, collecting real-time continuous information and extracting the corresponding characteristic set, identify security level, and thus realize the engineering safety hazard identification and prevention. Finally, fuzzy recognition is proved to be feasible on the engineering safety hazard identification problem by analyzing an example.**

Keywords - **Characteristic set, construction project, dangerous source, fuzzy recognition**

I. INTRODUCTION

In recent years, with the construction industry develops rapidly, construction techniques and the external environment of modern construction projects are becoming more and more complex, which resulting in engineering safety problems increasingly prominent. Engineering accidents have occurred frequently, which gives life and property of the people brought serious damage [1,2]. There are two main reasons, firstly, most construction workers is not enough emphasis on engineering safety. Secondly, more importantly, much research on engineering accidents is mainly pure theory, so the quantitative analysis study of engineering safety is not enough [3]. Due to construction tends to experiential operations, most did not realize the security risks before accident. The project lacks safety hazard identification and prevention. Engineering construction is a complex process that involves many aspects, it is difficult to practice engineering safety issues for effective management and control alone qualitative analyze or subjective experience. Thus, to solve this problem, this article will use modern science and technology. Using scientific fuzzy identification methods by collecting a large number of data, scientifically and quantitatively, identify a project safety hazard, and then identify the risks. Many appropriate measures should be taken in order to achieve the effect of the reduction and prevention of engineering accidents [4, 5].

II. SIGNIFICANCE OF THE RESEARCH

Fuzzy recognition mainly represents various types of patterns by fuzzy sets, and then the theory and methods of identification. As too many hazards of the influence construction accident, it not only look at one or two factors, but consider multiple factors to identify hazards of engineering project. Therefore, it is necessary to introduce fuzzy recognition method, which the quantitative characteristics factors of a project compared with the quantitative features of standards mode, and then draw the risk level of the project [6-8]. Finally, the risk of the project can be recognized according to the risk level.

In order to make fuzzy problems clearly shown, introducing the membership function, that is, a degree of object belongs to a set. To the uncertainty problems in the past mainly depend on experience, but so many prove experience is not always reliable. Through the establishment of the membership degree of fuzzy function, to make a more accurate quantitative judgment from the numerical, it has higher accuracy, which is a process of fuzzy recognition. Fuzzy recognition makes the more obscure and difficult to identify problems achieve a more accurate identification. Engineering safety hazards identification is a complicated problem, and the introduction of fuzzy recognition method will play a better effect of identification and prevention [9,10].

III. METHODS AND FUZZY RECOGNITION

Construction engineering safety is related to many factors, the identification of engineering brought certain difficulty whether safety accidents occur, together with the impact of many factors, which the role played by the various factors on the engineering safety are also different.

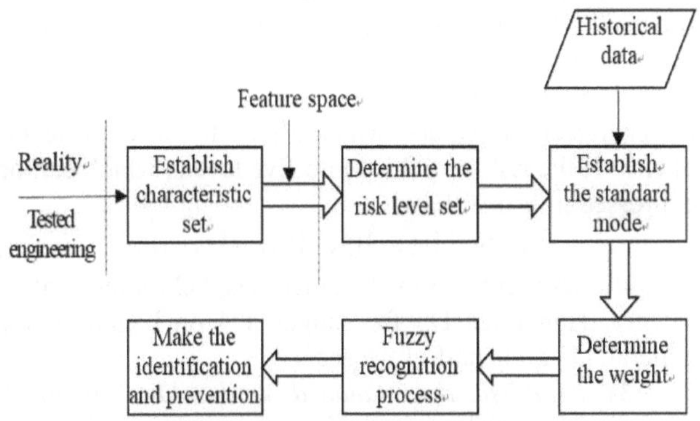

Fig.1. The process of fuzzy recognition method for construction safety hazard identification and prevention

It increases the difficulty to identify engineering hazards. In order to reduce or avoid safety accidents occur, it need a specific method for engineering hazards identification, and the risk level of engineering division, determine their level of security, and then look for hazards and take corresponding measures. If the risks can be detection early, it can be early prevention. It can reduce or avoid the occurrence of engineering accidents.

The process of fuzzy recognition method for construction safety hazard identification and prevention as shown in Fig.1:

1) *Establish Characteristic Set*: Feature extraction is particularly important, the good characteristics can better performance factors affecting the safety of engineering characteristics, and according to the characteristics of the appropriate classification, it can get the lower error rate and a finer resolution level. In the practical problems, the people generally want to collect characteristics as much as possible, because too many characteristics can be better to distinguish the various patterns, but too many characteristics will be brought difficulty to identifying. The more characteristics computation, the greater workload, even more can't calculate, neither economic nor the necessary.

From the general situation, the project has many safety hazards. There are more dangerous levels reflected by the characteristics of a project, which can be represented in a multi-dimensional vector. By methods of fuzzy membership degree or expert scoring, these characteristics were quantified, and then the qualitative attribute values are translated into quantitative numerical values. Get numerical feature set as follows:

$$X = (x1, x2, x3, x4, \cdots) \quad (1)$$

2) *Determine the Risk Level Set*: In order to ensure the accuracy of fuzzy recognition, it should not put risk level too fine division. The division of risk level is based on the degree of the accident. In construction, engineering accident are divided into different levels according to different risk levels, extract characteristic values of various factors about a certain level of construction projects. The establishment of eigenvalues database, statistical analysis these data can draw the level dangerous set. Similarly, draw other level dangerous sets, then come to a set of dangerous levels. Each element in the set is a special set of feature, these elements of the feature sets are the characteristic values, which affect the different levels of risk. If the risk is divided into five levels, which can be expressed in Y as follow:

$$Y = \{D_1, D_2, D_3, D_4, D_5\} \quad (2)$$

D1 represent extremely dangerous, D2 representative is very dangerous, D3 for danger, D4 on behalf of the general, D5 on behalf of the security.

3) *Establish the standard mode*: Due to many characteristics factors affecting a certain engineering safety, the relationship among characteristics is very complex, so determining a membership to engineering safety characteristics also has different methods, commonly used the expert evaluation, AHP, expert scoring method, etc. Different methods are using for the specific problems. For the problem of a single factor, based on experience the dangerous degree of certain specific factor can be reasonably determined, such as the greater supervision, the more conducive to construction safe. Generally, regular operation in engineering construction, based on experience, the experts can give some simple factors of fuzzy membership by scoring, and to determine the set D_j (j=1, 2, 3,…, k).

4) *Determining Weights*: Many characteristics factors affect the safety of a project, but the different characteristic factors have different effects, in order to more accurately express the different characteristics factors varying degrees of engineering safety, to take the weights of each characteristic factors, getting weight is a long-term process. Through a myriad of engineering practice, a large number of engineering accident hazards are analyzed, it can draw the evaluation index weight set. At the same time make the characteristic values quantification, also draw a specific the weight of each characteristic factors in the engineering, its representation is as follows:

$$W = (W_1, W_2, \cdots, W_i) \quad i = (1, 2, \cdots, n) \quad (3)$$

$$\sum_{i=1}^{n} W_i = 1, \quad W_1, W_2, \cdots, W_n \in [0\ 1],$$

5) *Identification Process*: Identification process is to choose a kind of fuzzy recognition algorithm, and according to the multiple characteristics nearly principles of pattern recognition. First, determine the kind of near-degree, near-degree is used to measure close degree between the fuzzy sets. It is necessary to consider the weight when the near-degree is determined.

If there are n characteristics, each membership of every feature factors can be concluded through the membership function, also can use other methods to make various characteristics of hazard factor quantification. Each project has a set to belong to their own characteristics. That is the difference between the value of affecting this project safety and the value of the standard model of the characteristic set about the same characteristic factors [11]. Take the absolute value of the difference, if the absolute value is larger, it means that the greater gap between the feature set with standard model, then their near-degree is smaller. Since different degree of the various influence engineering safety characteristics influence engineering safety, it is necessary to consider weight, then gaining the sum of absolute difference value, which the engineering characteristics of safety factors set and the standard model. The process is expressed as follows:

$$q(D, B) = W_1 \times |d_1 - b_1| + W_2 \times |d_2 - b_2| + \cdots + W_n \times |d_n - b_n| = \sum_{i=1}^{n} W_i |d_i - b_i| \quad i = (1, 2, \cdots, n). \quad (4)$$

If sum of absolute difference is much greater, it described a large degree of difference between them, the near-degree much smaller, so the near-degree can be expressed as follows:

$$N(D, B) = 1 - q(D, B)$$

$$= 1 - \sum_{i=1}^{n} W_i \mid d_i - b_i \mid \quad (5)$$

6) *Identification and Prevention*: According to the near-degree ,it is known the current status of the measured project to determine the risk of the project, identify the project in which the level of risk, evaluate the current state and further analysis the impact of the project safety hazards. And then many appropriate measures should be taken for specific problems to control the risk and to prevent the effect [12].

IV. RESULTS AND DISCUSSION

A. Engineering Background

Within a district in Handan, there is an 11-story ordinary residential building to be built, which has a layer underground and 11 floors above ground, the main structural height is 43.4m. Before the excavation of foundation pit, it is a necessary to measure the centerline, direction and elevation of foundation pit. Through the geotechnical investigation, the soil condition in which the foundation located is silty clay, belonging to the second category soil. The supporting condition will concrete filling pile, near the parts of status quo floor for steel pipe piles, the groundwater depth is -4.5 m. Then according to the geological and hydrological data and combined with the field situation, determine the scope of the slope excavation and support scheme, excavation scope, prevention and drainage measures. Finally the decision will plan the excavation of foundation pit, its elevation changes is more, which is the range from 3.3 meters to 6 meters (sump pit), the plat size of foundation pit is 53700mm* 80100mm, foundation pit adopt the method of no support excavation, not to take other measures for reinforcement of the foundation soil, and did not put slope. In order to speed up the construction speed, the use of mechanical construction, the top of the pit has a dynamic load, so the berm should be left greater than lm between the top edge of the pit and the dynamic load [13, 14].

B. Fuzzy Recognition

TABLE I
FIVE EXPERTS ON INFLUENCING FACTERS OF FOUNDATION PIT ENGINEERING SAGETY RATINGS

Characteristic factors	A	B	C	D	E
Geological conditions X_1	80	70	75	50	75
Foundation pit depth X_2	50	40	45	20	45
Foundation pit slope X_3	65	60	40	70	65
Groundwater conditions X_4	55	50	30	55	60
Support measures X_5	70	60	65	40	65
Surroundings X_6	90	80	85	60	85
Construction supervision X_7	60	50	55	30	55

In order to identify the safety of the construction and foundation pit engineering, there are seven characteristic factors selected, which including soil conditions, pit depth, pit slope, groundwater conditions, supporting measures, the environment and construction supervision. The seven characteristic factors affecting foundation pit engineering safety are rated by five experts who invited, score results in Table I.

By the weighted average of the five expert scoring above it is concluded that the characteristics factors affect the situation of engineering safety, then put the various factors and subordinate function correspond each other. So that make each characteristic factor fuzzification, the process is as follows:

$$\overline{x_1} = 80+70+75+50+75 = 70,$$

$$\overline{x_1}' = \overline{x_1}/100 = 0.7,$$

Similarly, $\overline{x_2}' = \overline{x_2}/100 = 0.4$, $\overline{x_3}' = 0.6$, $\overline{x_4}' = 0.5$, $\overline{x_5}' = 0.6$, $\overline{x_6}' = 0.8$, $\overline{x_7}' = 0.5$.

So it can get fuzzy feature set of the project hazards:

$$B = (0.7, 0.4, 0.6, 0.5, 0.6, 0.8, 0.5).$$

The foundation pit safety level is divided into many grades in different parts, sometimes even in the same area, the standards are not the same division, division level will be different, in order to illustrate the practicability of fuzzy recognition, also to show the fuzzy identification can be used for other general construction engineering safety identification and prevention. According to the damage of engineering accident, the security level is divided into the following: extremely dangerous (D_1), very dangerous (D_2), dangerous (D_3), general (D_4), security (D_5).

The statistical foundation data of ordinary residential building in the region for analysis, determine the fuzzy set to which the level of security standards mode set is corresponding. First, collect relevant data, classify the various features of the factors, the analysis of the various characteristics of factors, identify the factors corresponding to the characteristics of the membership, which can come in standard mode that are fuzzy sets corresponding to security level [15, 16].

For supervision, it is easy to know the supervision stricter, the better to ensure the safety of the project. Although the different supervision for affecting safety is different, their relationship is probably a positive correlation. It was found that the presence the following function between supervision and safety, which is shown in Fig.2:

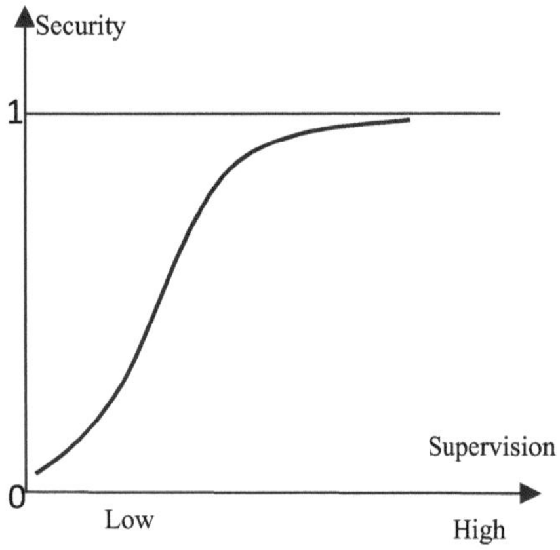

Fig.2. The relationship between supervision and security

Through collect and statistical analysis large amount of data of the foundation pit, it can draw a specific function between supervision and security. Then find the different levels of security corresponding protection and security. Similarly, the value of other features factors set can be obtained by this method, Expert scoring method can also be used to get the membership of each feature derived factors. Assumes identified five levels of engineering safety standards mode fuzzy feature set as follows:

$$D1 = (0.2, 0.3, 0.3, 0.3, 0.2, 0.4, 0.2)$$
$$D2 = (0.4, 0.4, 0.5, 0.5, 0.5, 0.6, 0.4)$$
$$D3 = (0.6, 0.5, 0.7, 0.6, 0.6, 0.7, 0.5)$$
$$D4 = (0.9, 0.7, 0.8, 0.8, 0.8, 0.9, 0.7)$$
$$D5 = (1.0, 0.9, 0.9, 0.9, 1.0, 1.0, 0.9).$$

Determine the characteristic factors of the Foundation Pit Engineering weight, mainly by experts about similar projects in the past, the dependence of the individual characteristics determined, but also to get through statistical experiments. For the safety of the excavation project, take the weight vector is:

$$W = (0.20, 0.09, 0.08, 0.21, 0.22, 0.12, 0.08).$$

Then near-degree of Foundation Pit Engineering is calculated for each of the excavation and the standard model:

$$N(D1, B) = 1 - q(D1, B)$$
$$= 1 - \sum_{i=1}^{n} Wi \mid di - bi \mid = 0.665$$

Similarly, $N(D2, B) = 0.878,$

$N(D3, B) = 0.930,$

$N(D4, B) = 0.782,$

$N(D5, B) = 0.643.$

The near-degree shows that safety level of the foundation pit engineering is dangerous. It can be further analyzed to determine hazards, and targeted to take measures. The fuzzy recognition can identify the risk of the foundation pit engineering in advance. Better to the engineering accident prevention.

V. CONCLUSIONS

Through calculation and analysis above, we can get the following conclusion:

1) The rate of foundation pit accidents is higher in architectural engineering, but for the modern science and the construction technology, it is not preventable and inevitable. Thus, selecting seven characteristics factors of affecting foundation pit engineering, identifies security level of projects. Once achieving information, the project can use a large amount of data and appropriately consider more characteristic factors to improve the accuracy of the fuzzy recognition on construction safety and provide a strong basis to ensure the safety of engineering.

2) The fuzzy recognition is not only apply to the excavation engineering hazards identification and prevention, but also to the other construction engineering. The construction engineering accidents are caused by many factors. Through the statistical analysis of a large number of engineering safety accidents, one can find out the causes of accidents, make the classification of the data, set up a specific engineering standard mode, and divide the appropriate level of security, so as to identify the construction hazard and take correspond measures for the specific problems.

3) Fuzzy recognition make the project achieve the safety level identification from the quantitative analysis to qualitative analysis of the level of security project under the multiple characteristics factors. Firstly, quantitative analysis is better to scientifically identify the engineering level of security, which is good for dealing with the possible accidents in construction. Secondly, quantitative analysis emphasize on the numerical analysis and minimize the interference of the experience, which makes the identification process is more scientific and accurate reducing the interference of experience. And the identification process is more scientific and accurate, which provides a strong favorable guarantee to reduce the occurrence of engineering accidents.

4) Modern construction project is relatively easy to collect a large amount of data. Only using the scientific method, it can take full advantage of data to form a more effective decision-making. Fuzzy Mathematics provides a channel for fuzzy intelligent development, the fuzzy recognition used to identify safety hazards of the project, which is conducive to realize intelligent computing and will also promote informationization of project security management and the development of intelligent.

ACKNOWLEDGMENT

This work was financially supported by the National Natural Science Foundation of China (71171081),

Program for New Century Excellent Talents in University (NCET-11-0633).

REFERENCES

[1] Ji-jian XIE, Cheng-ping LIU. Fuzzy mathematics and its application (Fourth edition) [M]. WuHan: Huazhong University of Science and Technology Press (Chinese), 2012.
[2] Xue-ling HOU. Theory and application implementation status diagnosis based on evidence-based science of construction projects [M]. Beijing: Electronic Industry Press(Chinese), 2011.
[3] Zhan-peng XU, Qi-fei YIN. Fuzzy identification method and its application [J]. Artificial Intelligence and Recognition Technology, 2008, pp. 75-76.
[4] Zhi-yuan GOU, Shao-hua WANG, Jun-yi XIAO. Explore the Residential Function to Determine the Degree of depreciation based on triangular fuzzy number [J]. Engineering Management Journal, 2013, 27(3), pp. 023-026.
[5] Zeng-fu WANG. Pattern recognition [M]. Anhui: University of Science and Technology of China Press (Chinese), 2010.
[6] Peng WU. Fuzzy recognition intelligent system based on cash character [J]. Advanced Materials Research, 2013(748), pp.675-678.
[7] Lin-gang TIAN, Cong-cong JIN, Chao BA. Application of improved fuzzy AHP in seawall safety assessment [J]. Journal of Wuhan University (Chinese), 2013, 46(3), pp.317-320.
[8] Wen-fang ZHAO, Gu-jun WANG, Guang-wen ZHANG, Yi ZHANG. Quantitative risk assessment of major hazard to resolve instances [J]. Journal of Safety and Environment, 2013, 13(5), pp.239-243.
[9] Xue DENG, Jia-ming LI. Analytic hierarchy weight calculation method analyzes rights law and its application [J]. Mathematics in Practice and Theory, 2012, 42(7), pp.93-99.
[10] Zhen YANG, Dong-tao HU, Ting-ting TAO, Ying Sun. Discussion and recommendations of mine safety management are based on hazard identification [J]. Industrial Minerals and Processing, 2013(12), pp.49-51.
[11] Yao-na GONG. Hazard identification and control measures [J]. Engineering Quality (Chinese), 2012, 30(6), pp.70-73.
[12] Ni WU, Qi-jie LIANG, Zhi-hua CHENG. Water conservancy and hydropower project construction in hazard identification problems and countermeasures [J]. Construction Technology (Chinese), 2013, 42(12), pp.101-103.
[13] Dong ZHENG, Ci ZHOU. Construction site hazard identification, risk assessment and control [J]. Construction Safety (Chinese), 2012(4), pp. 45-48.
[14] Kai TANG. Research identification problem building construction site dangerous source [J]. Construction and Engineering (Chinese), 2012 (10), pp. 150-151.
[15] Yan-fang WEN, Hui-min QIN. Building construction hazard identification assessment and control [J]. Construction Safety (Chinese), 2012(4), pp.39-42.
[16] Jian CHEN. Construction safety issues with the pre-construction control measures [J]. Architectural Design Management, 2009(9), pp.61-63.

Analysis of the Ecological Migration's Negative Benefit of China's Large-Scale Water Conservancy and Hydropower Projects under the Perspective of Low Carbon

Wei-jun HE[1,*], Min AN[1], Liang YUAN[2]
[1]Department of Economics and Management, Three Gorges University, Yichang, China
[2]Business School, Hohai University, Nanjing, China
(*616614429@qq.com, 398827404@qq.com)

Abstract - **Immigration is one of the main negative ecological impact of large-scale water conservancy and hydropower project construction. This paper aims at finding the method to quantify the negative benefit of ecological immigration. Firstly, using the ecological footprint model to calculate the biological productive land area of negative effect on Ecological generated by each of immigration; Secondly , uniting the biological productive land to forest land area under the entry of ecological footprint method by using simple mathematical conversion formula; Finally, on the basis of a large study of Chinese scholars on forest carbon sinks , calculating the transformed land area of carbon sink under the forest entry of each immigrant, and using carbon trading price of the international carbon market to monetize it.**

Keywords - **Carbon sinks, negative effects of ecological migration, the method of ecological footprint, water conservancy and hydropower engineering**

I. INTRODUCTION

The ecological benefit of large-scale water conservancy and hydropower projects in China has always been the hot topic in the government and the academic circles, especially the negative effects of ecological benefit is needed precautionary. Immigration as one of the most important negative ecological factors in large-scale water conservancy and hydropower project construction, aroused the attention from all sectors of society. In the background of the global market economy, how to quantify the negative benefit of ecological immigration, is a problem worthy of attention and thinking.

In recent years, the international community is strongly advocating low carbon economy, the universe is widespreading implementation of low carbon concept and the international carbon market is increasingly mature, especially the establishment of China's carbon trading platform, to provide external conditions in monetization of ecological benefits from the perspective of low carbon. In 2011, the National Development and Reform Commission identified seven carbon emissions pilot provinces and cities, including Beijing, Shanghai, Tianjin, Guangdong, Shenzhen, Chongqing, and Hubei [1]. June 18, November 26, November 28, December 26 in 2013, China's Shenzhen, Shanghai, Beijing, Tianjin carbon trading market has started carbon trading. That makes it possible to monetizate ecological benefits through carbon sequestration. So can we use of carbon sinks to quantitative out the ecological benefit of large-scale water conservancy and hydropower project?

By reading lots of related literature, learned that most scholars focus on the large-scale water conservancy and hydropower project resettlement, compensation measures and other related economic benefits, while on the research of the quantitative negative ecological immigration is relatively weak. Chinese scholars, Professor Guoqing Shi through using the theory of environmental capacity, puts forward the concept of environmental capacity, which has become an important theory basis for resettlement. However, due to lack of practical applications, and is used in combination of qualitative and quantitative methods, this theory's quantitative results are not obvious [2]. It just considered the resettlement from the perspective of environmental problems, and cannot be said to consider immigration from the overall ecosystem perspective. So how to quantitative or monetize the ecological migration's ecological migration of large-scale water conservancy and hydropower engineering in the whole ecosystem? We introduced the concept of ecological footprint and carbon sinks, and through the method of ecological footprint to monetize the negative benefit of ecological immigration in China's large-scale water conservancy and Hydropower Engineering from the perspective of low carbon.

II. RELEVANT THEORETICAL BASIS

A. The summary of ecological migration in China's large-scale water conservancy and hydropower project

Since entering the new century, with the implementation of the western development, the south to North Water Diversion Project, power transmission from west to East and other major construction projects, China's water conservancy is facing unprecedented development , at the same time, it also brings the great migration of local area population. And as people paid unprecedented attention to the ecological environment, when treated the development plan of water conservancy and hydropower project, people are more cautious. In the next period, ecological issues will be an important constraint of water conservancy and hydropower construction and even the further development of whole water conservancy, and migration is one of the main factors affecting the quality of the environment of water conservancy and Hydropower Engineering, it is essential to make our country's large-scale water conservancy and hydropower project of ecological migration in-depth analysis [3].

The so-called ecological migration refers to the population migration occurred form the deterioration of the ecological environment or in order to protect and

E. Qi et al. (eds.), *Proceedings of the 5th International Asia Conference on Industrial Engineering and Management Innovation (IEMI2014)*, Proceedings of the International Asia Conference on Industrial Engineering and Management Innovation 1, DOI 10.2991/978-94-6239-100-0_17,

restore the ecological environment [4]. Ecological migration caused by the large water conservancy and hydropower project construction is an important part of the ecological migrants, the study of ecological migration is beneficial to alleviate environment pressure, and it helps the immigration building a social safety net of their own in the new social environment, with these promoting the stability and development of the reservoir area. The research on ecological immigration is migration on the ecological environment pressure and realize the effective attempt of interest mechanism.

How to quantify the negative benefit of ecological migration in large-scale water conservancy and hydropower project? We introduced the ecological footprint method to quantify ecological migrants of large-scale water conservancy and hydropower engineering.

B. Introduction of ecological footprint method

Ecological footprint method was developed by the Canadian ecological economist William in the 1990s, and proposed by Wackernage [5]. It is an approach to measure the extent of human use of natural resources. It is based on the land area of the ecological footprint to measure the ecological and natural impact of human activity in a certain area, and is an important topic in the research field of sustainable development and ecological economy. The use of the ecological footprint method is to convert food consumption to arable land required for the production of the food, while absorbing carbon dioxide required forest, cropland or grassland area may represent the total amount of carbon dioxide emissions. A common understanding of the ecological footprint, is refers to human activities (cities, factories, farms, railway, etc.) leave footprints on the earth, which is the higher the value of ecological footprint, the more serious human damage to the ecological environment [6-7].

The method of ecological footprint are based on six basic assumptions: 1. most of the natural resources and waste's quantitation can be determined for human consumption in social production and consumption; 2. these natural resources and waste can be converted to the corresponding productive land area; 3. all types of productive land area can be converted into the equivalent of the average global land biomass production which the standard unit is "global hectares" (ghm2); 4. various types of land space is mutually exclusive, we can obtain human needs productive land area after weighted sum; 5. the production area can be represented by a global hectare which can be said the natural supply of ecosystem services; 6. ecological deficit can be produced when the ecological footprint is greater than the ecological carrying capacity [8].

The use of the ecological footprint is through the accounting, in the accounting accounts, all sorts of natural resources and energy consumption is converted into the six kinds of basic biological productive land area, which are forest, grassland, arable land, construction land, the fossil energy land and waters. However, since the average biological productivity of different types of land vary widely, in order to make these kinds of land area and the calculation results can be compared and accumulation, we use global average productivity of land and water area as a benchmark and before these kinds of land area calculation results give out, they were multiplied by corresponding equilibrium factor (unified equilibrium factor announced by FAO and WWF) respectively when calculating the ecological footprint, to be converted into comparable biological productive and equilibrium land area [9].

C. The theory of forest carbon sinks

In this paper, using the method of ecological footprint to quantify the negative benefit of ecological migration in large-scale water conservancy and hydroelectric engineering, we began to think, after the ecological benefits of immigration was expressed in biological productive land area, whether it could be in a more intuitive form of currency to expresse in the discussion of large-scale water hydropower project's ecological benefits of migration?

International community's vigorously promote low carbon attracted the attention of scholars ,with the establishment of international and China's carbon trading market and gradually mature, coupled with scholars do a lot of research in the carbon things that could trade, all these opens up a new train of thought for us in this paper. Among them, due to the forest's good ecological benefit and the forest carbon sink raises the earliest studies .Now the current research on forest carbon sinks has been relatively mature , and parts of forest carbon sinks can be traded on the international market .In the above, there is a forest entry when using the ecological footprint method to calculate the immigration's various types of productive land area, while using the equilibrium factor convert them into the same land area, we can unify all of them to the forest land area. In this perspective, we can monetize the negative benefit of ecological emigration.

For forest carbon sinks, Cao Jixin and his parner make an objective evaluation on three kinds of forest carbon sink estimation method, and according to the actual situation put forward to choose a single method or a variety of methods for estimating forest carbon sinks [10]; Li Shunlong carried out the detailed research on forest carbon sinks in his doctoral thesis, and estimate the annual one hectare of forest carbon sinks are 27 tons [11]; Zhang Ying and other scholars use the forest carbon accounting through the regression model, and roughly estimate the optimal accounting prices of forest carbon sinks in our country are slightly higher than the international forest carbon sink [12].

III. THE NEGATIVE ECOLOGICAL IMMIGRATION MONETIZATION PROCESS BASED ON THE PERSPECTIVE OF LOW

A. The introduction of the Ecological Footprint Model

Form the earlier summary of the ecological footprint in this approach, we introduce the ecological footprint model, and the calculation formula of the ecological footprint model is expressed as:

$$EF = N \times \text{ef} = N \times \sum (aa_i)$$
$$= N \times \sum (C_i / P_i)$$

Where: i is the type of consumer goods and inputs; EF is the total ecological footprint; ef is the per capita ecological footprint; N is the total population; aa_i refer to the converted biological production area of the ith commodity; c_i is the total consumption of the i-th consumer goods; p_i is the global average consumer production of i-th consumer goods.

B. Analysis of ecological migration's negative benefit by the method of ecological footprint

According to the theory and method of ecological footprint, the immigration account of large-scale water conservancy and hydropower projects in China are divided into biological resource consumption and energy consumption two parts, we can use the following TABLE I and TABLE II to calculate the negative benefits of ecological immigrant in detail. In the case of known a total population of immigrants, we can also obtain per capita ecological footprint by divide the corresponding total number of immigrants, this paper does not detail.

TABLE I
THE ECOLOGICAL FOOTPRINT OF CONSUME BIOLOGICAL RESOURCES

Product Name	Global average yield/(kg·hm^{-2})	Consumption/(t)	The total ecological footprint/(hm^2)	Land Type
Paddy	X1	Y1	Y1/X1	Arable land
Wheat	X2	Y2	Y2/X2	
Tubers	X3	Y3	Y3/X3	
Corn	X4	Y4	Y4/X4	
Soybean	X5	Y5	Y5/X5	
Cotton	X6	Y6	Y6/X6	
Oil	X7	Y7	Y7/X7	
Hemp	X8	Y8	Y8/X8	
Sugar	X9	Y9	Y9/X9	
tobacco	X10	Y10	Y10/X10	
Vegetables	X11	Y11	Y11/X11	
Melon class	X12	Y12	Y12/X12	
Pork	X13	Y13	Y13/X13	Grassland
Beef and mutton	X14	Y14	Y14/X14	
Poultry	X15	Y15	Y15/X15	
Sheep wool	X16	Y16	Y16/X16	
Goat wool	X17	Y17	Y17/X17	
milk	X18	Y18	Y18/X18	
honey	X19	Y19	Y19/X19	
Eggs	X20	Y20	Y20/X20	
Tung seed	X21	Y21	Y21/X21	Woodland
Camellia seed	X22	Y22	Y22/X22	
Dried bamboo shoots	X23	Y23	Y23/X23	
Chinese chestnut	X24	Y24	Y24/X24	
Tea	X25	Y25	Y25/X25	
Fruit	X26	Y26	Y26/X26	
cocoon	X27	Y27	Y27/X27	
Log	X28	Y28	Y28/X28	
Freshwater products / sea products	X29	Y29	Y29/X29	Waters

TABLE II
THE ECOLOGICAL FOOTPRINT OF CONSUME ENERGY RESOURCES

Consumption items	Consumption /(t)	Conversion factor /(GJ·t^{-1})	Global average energy footprint /(GJ·hm^{-2})	The total ecological footprint /(hm^2)	Land Type
Coal	Y30	R1	X30	(Y30*R1)/X30	Fossil Energy
Coke	Y31	R2	X31	(Y31*R2)/X31	
Crude	Y32	R3	X32	(Y32*R3)/X32	

Gasoline	Y33	R4	X33	(Y33*R4)/X33	
Kerosene	Y34	R5	X34	(Y34*R5)/X34	
Diesel	Y35	R6	X35	(Y35*R6)/X35	
Fuel Oil	Y36	R7	X36	(Y36*R7)/X36	
LPG	Y37	R8	X37	(Y37*R8)/X37	
Electric power	Y38	R9	X38	(Y38*R9)/X38	**building**

LPG: Liquefied petroleum gas

Through the above analysis and calculation, we can obtain the ecological footprint of immigration in large-scale water conservancy and hydropower project. Finally use the ecological footprint model to get the ecological footprint of land area with summarize and addition, before this, according to the relationship between the equilibrium factor which mentioned above, converted the various types of biological productive land area to the item of forest land area, specific methods are as follows:

The total ecological footprint represented by the biologically productive arable land area is as follow:

$$EF_{arable\ land}=\sum_{i=1}^{12} Y_i \div \sum_{i=1}^{12} X_i$$

The unit is hm^2, the balance of arable land factor is 2.8, while the balance of woodland factor is 1.1, so such a biological productive farmland conversion into forest land area is:

$$\left(\sum_{i=1}^{12} Y_i \div \sum_{i=1}^{12} X_i\right) \times (28 \div 11) \tag{1}$$

The total ecological footprint represented by the biologically productive grassland land area is as follow:

$$EF_{grassland}=\sum_{i=13}^{20} Y_i \div \sum_{i=13}^{20} X_i$$

The unit is hm^2, the balance of grassland factor is 0.5, while the balance of woodland factor is 1.1, so such a biological productive grassland conversion into forest land area is:

$$\left(\sum_{i=13}^{20} Y_i \div \sum_{i=13}^{20} X_i\right) \times (5 \div 11) \tag{2}$$

The total ecological footprint represented by the biologically productive water area is as follow:

$$EF_{water}=Y_{29} \div X_{29}$$

The unit is hm^2, the balance of water factor is 0.5, while the balance of woodland factor is 1.1, so such a biological productive water area conversion into forest land area is:

$$(Y_{29} \div X_{29}) \times (2 \div 11) \tag{3}$$

The total ecological footprint represented by the biologically productive fossil energy area is as follow:

$$EF_{fossil\ energy}=\sum_{i=30\,j=1}^{i=37\,j=8} Y_i R_j \div \sum_{i=30}^{37} X_i$$

The unit is hm^2, the balance of fossil energy factor is 1.1, while the balance of woodland factor is 1.1, so such a biological productive fossil energy area conversion into forest land area is:

$$\sum_{i=30\,j=1}^{i=37\,j=8} Y_i R_j \div \sum_{i=30}^{37} X_i \tag{4}$$

The total ecological footprint represented by the biologically productive construction land area is as follow:

$$EF_{construct}=(Y_{38} \times R_9) \div X_{38}$$

The unit is hm^2, the balance of construction land factor is 1.1, while the balance of woodland factor is 1.1, so such a biological productive construction land area conversion into forest land area is:

$$[(Y_{38} \times R_9) \div X_{38}] \times (28 \div 11) \tag{5}$$

The total ecological footprint represented by the biologically productive woodland area is as follow:

$$\sum_{i=21}^{28} Y_i \div \sum_{i=21}^{28} X_i \tag{6}$$

Above all, by using the method of ecological footprint, the negative ecological benefits of immigration in large-scale water conservancy and hydropower engineering is represented as the biological productive woodland area, and the negative ecological benefit is the sum of (1) ~ (6), the unit is hm^2:

$$\begin{aligned} EF_{total} = \{ & \left(\sum_{i=1}^{12} Y_i \div \sum_{i=1}^{12} X_i\right) \times (28 \div 11) \\ & + \left(\sum_{i=13}^{20} Y_i \div \sum_{i=13}^{20} X_i\right) \times (5 \div 11) \\ & + (Y_{29} \div X_{29}) \times (2 \div 11) \\ & + \sum_{i=30\,j=1}^{i=37\,j=8} Y_i R_j \div \sum_{i=30}^{37} X_i + \\ & [(Y_{38} \times R_9) \div X_{38}] \times (28 \div 11) \\ & + \sum_{i=21}^{28} Y_i \div \sum_{i=21}^{28} X_i \} \end{aligned}$$

C. Analysis the benefit of ecological migration in large-scale water conservancy and hydropower engineering based on the perspective of low carbon

In the theoretical basis of this paper has introduced the research for forest carbon sink of our scholars', using previous studies that one hectare of forest output of 27 tons of carbon sinks, the negative ecological benefits of immigration in the large-scale water conservancy and hydropower project's carbon sink generated as follows:

$$C = 27 * EF\ (t).$$

Although the current international carbon price is 10~15 dollars, analysis the situation of our country, obtain the optimal price of forest carbon sink in China is \$10.11~15.17, here we take \$12(can be selected by the market based on the specific situation around), the total price of carbon sink M=12*C (\$).

In summary, the monetary value of negative ecological benefits of immigrants in China's large-scale water conservancy and hydropower project is expressed as:

$$M = 12 \times 27 \times EF_{total}$$
$$=\{(\sum_{i=1}^{12} Y_i \div \sum_{i=1}^{12} X_i)\times(28 \div 11) + (\sum_{i=13}^{20} Y_i \div \sum_{i=13}^{20} X_i)\times(5 \div 11)$$
$$+(Y_{29} \div X_{29})\times(2 \div 11) + \sum_{i=30 j=1}^{i=37 j=8} Y_i R_j \div \sum_{i=30}^{37} X_i$$
$$+[(Y_{38} \times R_9) \div X_{38}]\times(28 \div 11) + \sum_{i=21}^{28} Y_i \div \sum_{i=21}^{28} X_i\}$$

(the values of X, Y were as shown in the table) , the result which is calculated is the monetization of the negative ecological benefit of immigrants in large-scale water conservancy and hydropower project.

IV. CONCLUSION

Because of the large-scale water conservancy and hydropower engineering has an important strategic position in the process of China's socialist construction, and immigrants is undoubtedly one of the most important negative ecological effects of China's large-scale water conservancy and hydropower engineering construction, then the importance of study negative ecological migration in China's large-scale water conservancy and hydropower project is self-evident. This paper attempts to quantify the negative ecological immigration of large-scale water conservancy and Hydropower Engineering in our country from the angle of carbon sink, and monetize it with the international market transaction price on carbon sink to get the value of negative ecological immigration. This process provides a way for the currency of negative ecological benefits of immigrants in China's large-scale water conservancy and hydropower project. It helps us to consider the ecological loss of immigration more fully and objectively, what's more it provides a reference for government to formulate more reasonable and more comprehensive ecological compensation policy.

This paper provides ideas and methods to quantify and monetize the negative ecological benefits for immigrants. We can do much more scientific and persuasive research on the basis of this study later.

ACKNOWLEDGMENT

This work was supported by the longitudinal project of National Social Science Fund Project, namely "The mechanism research of ecological environment protection in Wuling Mountain Area," project number (13BMZ057).

REFERENCES

[1] Zhou Hongchun. Development and Enlightenment of the world's carbon trading market [J]. Chinese soft science, 2009, 12: 39-48

[2] Zhu Dongkai, Shi Guoqing, Zhang Bin. The present situation and Prospect of immigration problem in economics research of Water Resources and Hydropower Engineering [J]. China Soft Science, 2005, 03: 55-60.

[3] Yang Wenjian, Liu Yaoxiang. The benefit sharing and resettlement mode research in Reservoir immigration and hydropower engineering [J].Population and economic, 2002, 04: 9 -14.

[4] Li Dong. The research of the ecological migration in China - a literature review [J] Northwest population, 2009, 01: 32-35.

[5] Wei Xiaoyan, Xia Jianxin, Wu Yanhong. The study on ecological compensation standard in water diversion project based on the ecological footprint theory [J]. Journal of soil and water conservation research, 2012, 19 (5): 214-222.

[6] Wackernagel M, Rees W E. Our ecological footprint: reducing human impact on the earth [M]. New Society Publishers, 2013.

[7] Wackernagel M, Rees W E. Perceptual and structural barriers to investing in natural capital: Economics from an ecological footprint perspective [J]. Ecological economics, 1997, 20(1): 3-24.

[8] Wackernagel M, Yount J D. The ecological footprint: an indicator of progress toward regional sustainability [J]. Environmental Monitoring and Assessment, 1998, 51(1-2): 511-529.

[9] Ma Qiong, Su Meiling.The population and water and soil resources coordinated development research in Tarim Basin based on the ecological footprint [J]. Journal of arid land resources and environment, 2014, 3: 006.

[10]Cao Jixin, Tian Zan, Wang Xiaoping. Estimation methods and their development trend of forest carbon sinks [J]. Journal of ecological environment, 2009, 18 (5): 2001-2005.

[11]Li Shunlong. The economics research of forest carbon sinks [D]. PhD thesis of northeast forestry university, 2005.

[12]Zhang Ying, Wu Lili, Su Fan, et al. Research on the measurement model of China's forest carbon accounting [J]. Journal of Beijing Forestry University, 2010(2):194-200.

Research on Networked Collaborative Technical Service of Manufacturing Enterprise in Cloud Computing Environment

Dong-dong Cai[1], Tian-rui Zhang[2,*], Zhen-yu Wang[3], Zi-xuan Wang[3], Tian-biao Yu[3]
[1]School of Economics, Shenyang University of Technology, Shenyang, China
[2]School of Mechanical Engineering, Shenyang University, Shenyang, China
[3]School of Mechanical Engineering and Automation, Northeastern University, Shenyang, China
(tianjiangruixue@126.com)

Abstract - **On the basis of the modern manufacturing mode and the technical service under the cloud computing conditions, the networked cooperative technical service mode based on cloud computing was put forward for manufacturing enterprise to quickly response to the market demand. Firstly, ontology-based semantic identification technology and the basic concept of ontology were studied by expounding the principles and method of ontology and describing the algorithm of semantic identification for technical service information in production process. Secondly, the queue and assignment of the technical services task in production process were researched respectively by using the fuzzy analytic hierarchy process and the Hungarian algorithm. Finally, the prototype system of networked technical service oriented product process was carried out to verify the correctness and feasibility of the researched techniques.**

Keywords - **Cloud computing, networked collaborative, semantic identification, task assigned, task marched, technical service**

I. TECHNOLOGY SERVICE MODE BASED ON CLOUD COMPUTING

A. Technical Service

Technical service means one technical party uses their technical knowledge for another party to solve a particular technical problem on the services they provide. It usually covers how to improve the product structure, process, and product quality, reduce product cost, save resources, protect the energy consumption of resources and environment, realize safe operation, and increase the economic benefit and social benefit, as well as other professional technical work. With the science and technology developing to the social production and social life in all areas, technical service itself has rich contents and various forms, and the general social services contain technical content.

Along with the technical service development in production in the recent years, the technology services plays a great role to the great extent, and its main functions include technology teaching, skills, communication, technology planning, technology assessment, technical services and technical training and other activities[1-2].

B. Technology service mode based on cloud computing

Product life cycle of the network technology services is also a kind of distributed parallel technology service mode, therefore, technology service based on cloud computing network platform aims to improve the efficiency and save resources. The platform level model is shown as shown in Fig.1. Due to the limitation of length, the process of production of technical services is introduced for example [3-6].

II. TECHNOLOGY OF SEMANTIC IDENTIFICATION IN TECHNICAL SERVICE

The semantic Identification in technical service is a kind of analytical identification based on knowledge. On the basis of the natural language understanding, the semantic identification is completed through statistical models and computational linguistics application of knowledge.

A. Method to establish ontology

Ontology is a philosophical concept, and with the application of artificial intelligence, it has been given a new definition which considers that ontology is field of a semantic basis for the communication between the different entities, which can obtain a defined consensus.

As the areas and needs for the use of ontology are different, the process of ontology construction is different. The general five Ontology criteria presented by Gruber: (1) Clarity and objectivity, (2) Consistency, (3) The maximum monotone scalability, (4) The body agreed minimum principle and (5) The minimum dependence on codes.

There are accordinly five general ontology construction methods: Assessment method, Bernaras method, Methonotology method, Senses method and framework law method.

B. Knowledge representation of technology services based on ontology

Technical services knowledge in the production process is the basis of technical services. In order to reuse the model of Technical services knowledge and enhance the flexibility of knowledge representation, this paper presents a three-layer framework model of technical service-oriented knowledge representation for the expressed types of know-ledge of technical services.

The three-layer structure is used to ensure standardization, reuse and share of the knowledge of technical services, which also reduces the difficulty of modeling. Meanwhile, the body element does not depend on the establishment of a specific language, so this choice of representation makes technology services model flexible [7].

E. Qi et al. (eds.), *Proceedings of the 5th International Asia Conference on Industrial Engineering and Management Innovation (IEMI2014)*, Proceedings of the International Asia Conference on Industrial Engineering and Management Innovation 1, DOI 10.2991/978-94-6239-100-0_18, © Atlantis Press and the authors 2015

C. Algorithm of semantics identification in technical service

In practical engineering, the algorithm of semantics identification in technical service can generally be divided into four parts: similarity calculation of text, similarity calculation of sentence, similarity calculation of value and similarity calculation of ontology. In this paper, the last one is adopted.

Ontology similarity refers to the degree of similarity between two cases. If the similarity reaches to pre-set threshold, the two cases is similar, otherwise is not. Setting A and B are two cases of technical service to be compared with each other. A_0 and B_0 respectively describe the synonyms and attributes of A and B. The similarity between A and B can be calculated using the following formula:

$$EetSim(A,B) = \frac{|A \cap B|}{|A \cap B| + \alpha(A_0,B_0)|A/B| + (1-\alpha(A_0,B_0))|B/A|} \quad (1)$$

$$\alpha(A,B) = \begin{cases} \dfrac{depth(A_0)}{depth(A_0)+depth(B_0)} & (depth(A_0) \le depth(B_0)) \\ 1-\dfrac{depth(A_0)}{depth(A_0)+depth(B_0)} & (depth(A_0) \succ depth(B_0)) \end{cases} \quad (2)$$

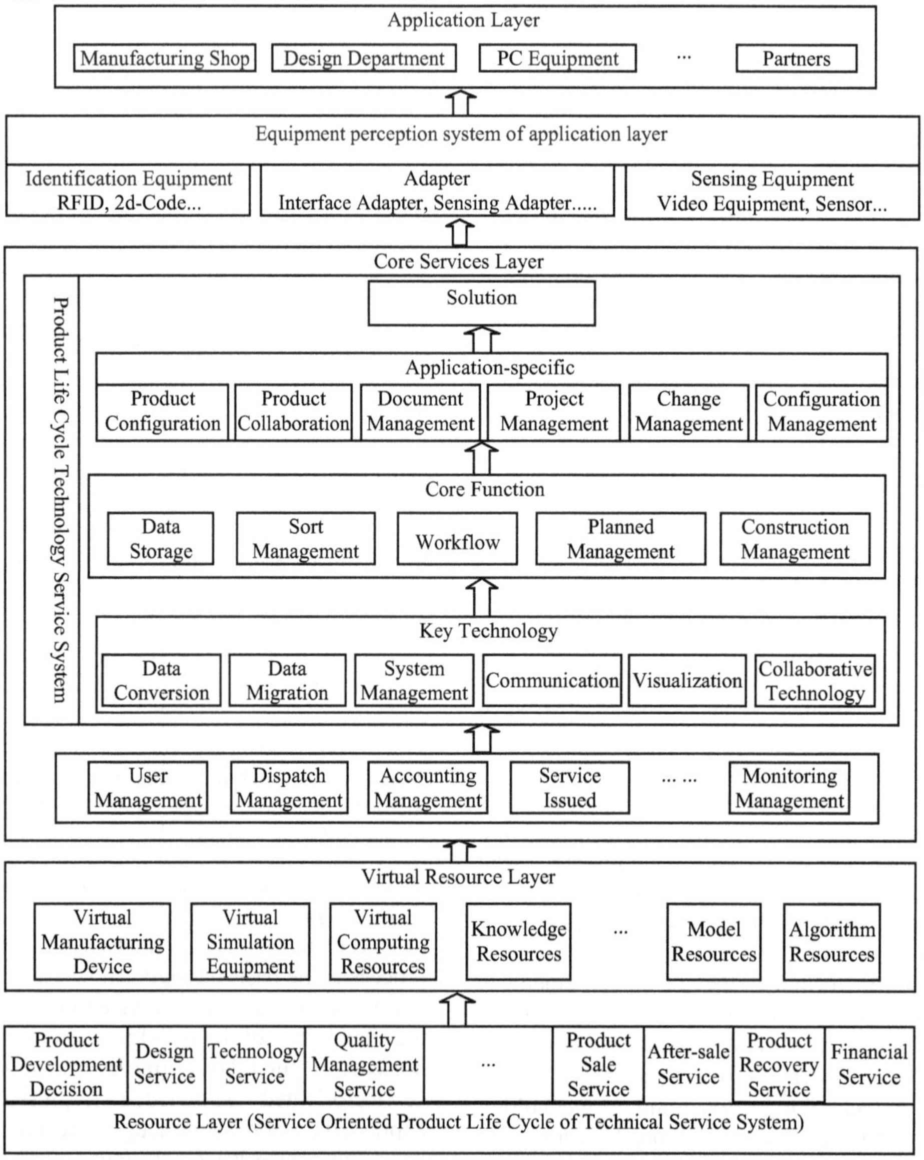

Fig.1. Level model of based on cloud computing network collaborative technology service platform

III. TASK MARCHED TECHNICAL SERVICE BASED ON AHP

The task marched technical service refers to the sorting of every subsystem in accordance with the time of service request, the importance of the service request, and the urgent degree of the service request.

A. Principle of Fuzzy Analytic Hierarchy Process

Assume matrix $R = (r_{ij})_{m\times m}$, if it satisfies the following condition:

$$0 \le r_{ij} \le 1, (i=1, 2, \dots, m;\ j=1, 2, \dots, m) \quad (3)$$

$$r_{ij} = r_{ik} - r_{jk} + 0.5 \quad (4)$$

Then, it can be called fuzzy matrix.

In common sense, the relative matrix is fuzzy complementary matrix in decision process. But, the fuzzy complementary matrix is not sure to be consistent, which makes the matrix not satisfy the center division transitivity. So, it should transfer fuzzy complementary matrix to be the fuzzy consistent matrix. Sum the every line of the matrix, and mark it $r_i=\sum_{j=1}^{m} r_{ij}$, $i=1, 2, \ldots, m$. Carry out the following Mathematical transformation: $a_{ij}=\frac{r_i-r_j}{2m}+0.5$. Then, the built matrix $A=(a_{ij})_{m\times m}$ is consistent [8].

B. Construction of Task Marched Model

(1) Establish the feature set matrix

After the feature set $Q=\{O_1, O_2, \ldots, O_k\}$ ($k=1, 2, \ldots, n$) is established, the feature set matrix $A=(a_{ij})_{n\times n}$ is established by using the fuzzy value express the relation of every target feature.

(2) Establish the precedence relation matrix

Under the feature O_k ($k=1, 2, \ldots, n$), task T_i ($i=1, 2, \ldots, m$) should be carried out multi-objective decision. Then, the fuzzy precedence relation matrix $B_k=(b_{ij}^k)_{m\times m}$ ($k=1, 2, \ldots, n$) can be obtained.

(3) The fuzzy consistent matrix

The matrix $A=(a_{ij})_{n\times n}$ and matrix $B_k=(b_{ij}^k)_{n\times n}$ ($k=1, 2, \ldots, n$) are transferred to its respective fuzzy consistent matrix $R^0=(r_{ij}^0)_{n\times n}$ (where $r_{ij}^0=\frac{r_i-r_j}{2n}+0.5$, $r_i=\sum_{j=1}^{n} a_{ij}$) and $R^k=(r_{ij}^k)_{m\times m}$ (where $r_{ij}^k=\frac{r_i-r_j}{2m}+0.5$, $r_i=\sum_{j=1}^{m} b_{ij}^k$).

(4) The sorting of single feature

Root method is used to calculate the prior value of task T_i under the feature O_k: S_i^k, $S_i^k=\overline{S_i}/\sum_{j=1}^{m}\overline{S_j}$ (where $\overline{S_i}=\left[\prod_{j=1}^{m} r_{ij}^k\right]^{1/m}$). The bigger the value S_i^k is, the higher the priority is.

The same method can be used to calculate the value of weight.

(5) The sorting of multi feature

Calculate the total prior value V_t, $V_t=\sum_{k=1}^{n} w_k s_i^k$, and make the sorting.

C. Sorting calculation of marched task

Nine tasks can be sorted, and the result is shown in Table I.

(1) Establish the precedence relation matrix.

$$A=\begin{bmatrix}0.5 & 1.0 & 1.0 & 1.0\\ 0.0 & 0.5 & 0.0 & 1.0\\ 0.0 & 1.0 & 0.5 & 1.0\\ 0.0 & 0.0 & 0.0 & 0.5\end{bmatrix}$$

(2) Establish the precedence relation matrix of the common feature for every task.

$$B_1=\begin{bmatrix}0.5 & 1.0 & 1.0 & 1.0 & 1.0 & 1.0 & 1.0 & 1.0 & 1.0\\ 0.0 & 0.5 & 1.0 & 1.0 & 1.0 & 1.0 & 1.0 & 1.0 & 1.0\\ 0.0 & 0.0 & 0.5 & 1.0 & 1.0 & 1.0 & 1.0 & 1.0 & 1.0\\ 0.0 & 0.0 & 0.0 & 0.5 & 1.0 & 1.0 & 1.0 & 1.0 & 1.0\\ 0.0 & 0.0 & 0.0 & 0.0 & 0.5 & 1.0 & 1.0 & 1.0 & 1.0\\ 0.0 & 0.0 & 0.0 & 0.0 & 0.0 & 0.5 & 1.0 & 1.0 & 1.0\\ 0.0 & 0.0 & 0.0 & 0.0 & 0.0 & 0.0 & 0.5 & 1.0 & 1.0\\ 0.0 & 0.0 & 0.0 & 0.0 & 0.0 & 0.0 & 0.0 & 0.5 & 1.0\\ 0.0 & 0.0 & 0.0 & 0.0 & 0.0 & 0.0 & 0.0 & 0.0 & 0.5\end{bmatrix}$$

(3) The fuzzy consistent matrix R^0.

$$R^0=\begin{bmatrix}0.5000 & 0.7500 & 0.6250 & 0.8750\\ 0.2500 & 0.5000 & 0.3570 & 0.6250\\ 0.3750 & 0.6250 & 0.5000 & 0.7500\\ 0.1250 & 0.3750 & 0.2500 & 0.5000\end{bmatrix}$$

TABLE I
HIERARCHICAL ORGANIZATION FORM OF TASK SCHEDULING MODEL

Feature Layer (O)		Request Time A	Production Type B	Product Type C	Production Quantity D
Task Layer (T)	Task 1	①	*a*	*a*	*b*
	Task 2	②	*a*	*d*	*a*
	Task 3	③	*c*	*c*	*a*
	Task 4	④	*b*	*b*	*b*
	Task 5	⑤	*c*	*a*	*b*
	Task 6	⑥	*a*	*c*	*a*
	Task 7	⑦	*b*	*d*	*b*
	Task 8	⑧	*c*	*c*	*a*
	Task 9	⑨	*b*	*b*	*b*

(4) Single feature sorting and calculation of each task priority value are as shown in Table II.

$$\omega=(0.3527, 0.2169, 0.2854, 0.1450)$$

TABLE II
EACH TASK IN A SINGLE FEATURE OF PRIORITY VALUES

S_i^k	1	2	3	4
1	0.1942	0.1515	0.1585	0.1375
2	0.0057	0.1515	0.0625	0.0781
3	0.1629	0.0703	0.0981	0.0781
4	0.1472	0.1116	0.1319	0.1375
5	0.1312	0.0703	0.1585	0.1375
6	0.1151	0.1515	0.0981	0.0781
7	0.0987	0.1116	0.0625	0.1375
8	0.8167	0.0703	0.0981	0.0781
9	0.0632	0.1116	0.1319	0.1375

(5) Many feature sorting and calculation of task's overall priority value:

$$V=(0.1665, 0.0640, 0.1120, 0.1338, 0.1267, 0.1180, 0.0968, 0.0834, 0.1041);$$

So: $V_1 > V_4 > V_5 > V_6 > V_3 > V_9 > V_7 > V_8 > V_2$

IV. TECHNICAL SERVICE TASK ASSIGNMENTS BASED ON HUNGARY ALGORITHM

According to the parade priority principle of technical service task, technical service task is assigned.

A. Technical service task assignment problem about mathematical model

According to the results of task sorting, technical service task is assigned to the most appropriate technical service personnel. The principle formula of the assignment problem is as follows [9-12]:

$$\text{Set, } x_{ij} = \begin{cases} 1 & \text{Assigning } i \text{ people to do } j \text{ things} \\ 0 & \text{Assigning } i \text{ people to do } j \text{ things} \end{cases},$$

$(i, j = 1, 2, \ldots, n)$.

The mathematical model of standard assignment problem could be written for:

$$\min z = \sum_{i=1}^{n}\sum_{j=1}^{n} c_{ij}x_{ij} \tag{5}$$

$$s.t.\begin{cases} \sum_{i=1}^{n} x_{ij} = 1 & j = 1, 2, \cdots, n \\ \sum_{j=1}^{n} x_{ij} = 1 & i = 1, 2, \cdots, n \qquad C_{ij} = \begin{bmatrix} c_{11} & \cdots & c_{1n} \\ \cdots & \cdots & \cdots \\ c_{n1} & \cdots & c_{nn} \end{bmatrix} \\ x_{ij} = 0 \text{ or } 1 & i, j = 1, 2, \cdots, n \end{cases}$$

In the model, constraints (1) shows that everything must be done by a person and only a person, and constraints (2) shows that everyone will do and do only one thing. Formula (5) is the coefficient matrix of assignment problem.

B. Principal and application of Hungarian Algorithm

The Hungarian Algorithm of the assignment problem was proposed in 1995 by Kuhn.

(1) Simplification - Make new coefficient matrix (C_{ij}) in the entire column all appear 0 elements.

(2) To undertake trying assignment.

(3) To cover all elements with the least amount of linear.

(4) To simplify again.

C. Assigned calculation of technical service task

With technical service mission number n less than technical service personnel number m for example, supposing there is 3 tasks T_1, T_2, T_3 and four people of technical service M_1, M_2, M_3, M_4,their costing time C_{ij} to finish each work was shown in Table III, assigned mission results are calculated.

TABLE III
CALCULATED MISSION ASSIGNED RESULTS

Staff	T_1	T_2	T_3
M_1	2	15	13
M_2	10	4	14
M_3	9	14	16
M_4	7	8	11

(1) Adding m - n virtual task and give costing time 0 to finish the virtual task. At present, the problem was translated into assignment problem with the same personnel number and mission number.

(2) Using Hungarian method to solve the problem.

1) In the transformation matrix, all the columns are about the entire current minimum element.

2) Conducting task assignment

$$C_2 = \begin{bmatrix} 0 & 13 & 7 & 7 \\ 6 & 0 & 5 & 5 \\ 0 & 5 & 3 & 0 \\ 0 & 1 & 0 & 2 \end{bmatrix} \longrightarrow \begin{bmatrix} \textcircled{0} & 13 & 7 & 7 \\ 6 & \textcircled{0} & 6 & 5 \\ \not{0} & 5 & 3 & \textcircled{0} \\ \not{0} & 1 & \textcircled{0} & 2 \end{bmatrix} = C_3$$

Now, there were four independent elements of 0 in the coefficient matrix C_3 . Then the result of matrix was $x_{11} = x_{22} = x_{34} = x_{43} = 1$, $x_{ij} = 0$, which were M_1 to fulfill task T_1, M_2 to fulfill task T_2, and M_4 to fulfill task T_3; however that of M_3 was not allocated to a task.

V. CONCLUSION

In the paper, technology service mode based on cloud computing is constructed; an ontology-based semantic identification technology is illustrated, and the production process for the network technology service system of intelligent search function is introduced through the study of technical service task rank and appointment which allows the user to submit technical service task accurately to the corresponding technical personnel; When multiple tasks arrive at the same time, the system will assign the tasks in the accordance with the weight, lining up time and efficiency to solve the pressing technical problem, improve the production process of the intelligent, and improve the technical service efficiency and quality. In order to realize the production process for the network, the collaborative change technical service should be applied to solve the problems between the heterogeneous systems of data exchange in the future.

ACKNOWLEDGMENT

The paper was supported by the Shenyang Science and Technology Research Projects (Grant No. Shen D&R [2012] 212) and Shenyang Science and Technology Plan Projects (Grant No. F12-277-1-58).

REFERENCES

[1] J. Zhou, "Research on Network Technology Service Oriented Product Life Cycle", Northeastern University Ph.D. Thesis, 2009.

[2] B.H. Li, L. Zhang, S.L. Wang, "Cloud manufacturing: a new service-oriented networked manufacturing model" , Computer Integrated Manufacturing System, Vol.16, No. 1, 2010, pp.1-8.
[3] L.Y. Zuo, L.H. Wu, "Database Query Scheduling Algorithm Based on the Cloud Computing", Journal of Zhengzhou University(Engineering Science), Vol. 31, No. 4, 2010, pp. 65-68.
[4] Mont, "Clarifying the concept of product-service system", Journal of Cleaner Production, Vol. 10, No. 3, 2002, pp. 237-245.
[5] L.C. Arbósyi, "Design of a rapid response and high efficiency service by lean production principles: Methodology and evaluation of variability of performance", Int. J. Production Economics, No.80, 2002, pp.169–183.
[6] J.C. Aurich, C. Fuchs, "An approach to life cycle oriented technical service design", Annals of the CIRP, Vol. 53, No. 1, 2004, pp. 151-154.
[7] R. Neches, R.E Fikes, T.R Gruber, "Enabling Technology for Knowledge Sharing", AI Magazine, Vol. 12, No. 3, 1991, pp. 36-56.
[8] R. Studer, V.R. Benjamins, D, "Fensel. Knowledge Engineering Principles and Methods", Data and Knowledge Engineering, Vol. 25, No. 122, 1998, pp. 161-197.
[9] M. Uschold, "Building Ontologoes: Towards a Unified Methodology", 16th Annual of the British Computer Society Specialist Group on Expert Systems, 1996, pp. 2-14.
[10] X.F. Xu, J.Z. Wang, T. Mo, "Design of service engineering", Computer Integrated Making System, Vol.13, No. 8, 2007, pp. 1454-1464.
[11] W.J. Zhang, W. Liu, Z.J. Zhang, "Research of task assignment problem", Computer Integrated Making System, Vol.15, No. 6, 2009, pp. 1231-1236.
[12] B. Shen, D. Gong, H. Zhao, "Research on networked manufacturing platform under the support of product life cycle", Manufacturing automation, Vol. 28, No. 2, 2006, pp. 17-20.

Advancement Degree of Difficulty Assessment Method for Complex Products

Dong Zhao*, Fa-jie Wei
School of Economics and Management, University of Beihang, Beijing, China
(luckystone2005@163.com)

[1] *Abstract* - **In order to supervise and control the technical risk of the complex products, the paper proposed an assessment method of the advancement degree of difficulty for complex products based on the Fuzzy Integrated Evaluation. On the basis of the assessment process of the technology readiness, a three-tiered fuzzy integrated evaluation model was established according to the advancement degree of difficulty level standard. Two methods, the Analytic Hierarchy Process and the confidence criterion, were used to ensure the effectiveness and objectivity of the assessment result. Finally, the applicability of the evaluation method was verified by an actual application sample.**

Keywords - **Advancement degree of difficulty, complex products, technology readiness level**

I. INTRODUCTION

The complex products, which are widely distributed in national defense, electricity, transportation, and many other domains, have a direct effect on the comprehensive national strength and the core competitiveness of the country. As the difficulty and the complexity of complex products in customer demand, product technology, manufacturing process, and work environment, many kinds of risk, such as performance degradation, Schedule delays, and cost increase, occur more easily during the developing process [1] [2]. According to the analysis results of the research and development programs about complex products released by U.S. Government Accountability Office, it is the immature critical technology element moved into the next development stage in advance which can causes the occurrence of significant risk frequently. According to the scientific laws of technological development, the technology readiness assessment based on TRL appraises the technology readiness extent using standardized measurement level. It can offer the effective reference to manager and technician as a risk identification tool [3-5].

AD^2 (Advancement Degree of Difficulty) which can reflect the difficulty level of technology development to the target state objectively gets a close attention by forefront research institutions. NASA (National Aeronautics and Space Administration) has already analyzed the necessity and the significance of the AD^2 in their white paper of technology readiness assessment, and a combination of the TRL and the AD^2 method had been adopted in both "Mars Exploration Program" and "Sunjammer Mission". The development trend and efficiency of the projects can be ensured effectively with this combination mode [6] [7].

The work was supported by the National Natural Science Foundation (Grant NO.71332003) and the Basic Defense Technology Research Projects (Z132012A002).

The objective assessment of the AD^2 during the development process of complex products not only can make further perfection on the technology readiness assessment system, but also can provide more accurate support of details for management organization and researcher to reduce technology research and development risk [8] [9]. However, the assessment method of the AD^2 is very limited at the present stage, and the existing ones are built on account of form evaluation. The result of the AD^2 has more subjective and one-sidedness, and it is difficult to realize the comprehensive and objective assessment. For that reason, synthesizing the technology readiness assessment and the advancement degree of difficulty level standard, the paper proposes an assessment method of advancement degree of difficulty for complex products based on the Fuzzy Integrated Evaluation.

II. METHODOLOGY

A. Technology Readiness Level

The TRL (Technology Readiness Level) adopted by NASA has the vastest application scale among a variety of TRL standard. Using the track of the seed of technology developing to the final validation as a thread, the TRL of NASA totally includes 9 levels. Level 1 means the lowest, and Level 9 means the highest [10].

TRL 1. Basic principles observed and reported. This is the lowest "level" of technology maturation. At this level, scientific research begins to be translated into applied research and development.

TRL 2. Technology concept and/or application formulated. Once basic physical principles are observed, then at the next level of maturation, practical applications of those characteristics can be 'invented' or identified. At this level, the application is still speculative: there is not experimental proof or detailed analysis to support the conjecture.

TRL 3. Analytical and experimental critical function and/or characteristic proof-of-concept. At this step in the maturation process, active research and development (R&D) is initiated. This must include both analytical studies to set the technology into an appropriate context and laboratory-based studies to physically validate that the analytical predictions are correct. These studies and experiments should constitute "proof-of-concept" validation of the applications/concepts formulated at TRL 2.

TRL 4. Component and/or breadboard validation in laboratory environment. Following successful "proof-of-concept" work, basic technological elements must be integrated to establish that the "pieces" will work together to achieve concept-enabling levels of performance for a

E. Qi et al. (eds.), *Proceedings of the 5th International Asia Conference on Industrial Engineering and Management Innovation (IEMI2014)*, Proceedings of the International Asia Conference on Industrial Engineering and Management Innovation 1, DOI 10.2991/978-94-6239-100-0_19,

component and/or breadboard. This validation must be devised to support the concept that was formulated earlier, and should also be consistent with the requirements of potential system applications. The validation is "low-fidelity" compared to the eventual system: it could be composed of ad hoc discrete components in a laboratory.

TRL 5. Component and/or breadboard validation in relevant environment. At this level, the fidelity of the component and/or breadboard being tested has to increase significantly. The basic technological elements must be integrated with reasonably realistic supporting elements so that the total applications (component-level, sub-system level, or system-level) can be tested in a "simulated" or somewhat realistic environment.

TRL 6. System/subsystem model or prototype demonstration in a relevant environment (Ground or Space). A major step in the level of fidelity of the technology demonstration follows the completion of TRL 5. At TRL 6, a representative model or prototype system or system - which would go well beyond ad hoc, "patch-cord" or discrete component level bread boarding - would be tested in a relevant environment. At this level, if the only "relevant environment" is the environment of space, then the model/prototype must be demonstrated in space.

TRL 7. System prototype demonstration in a space environment. TRL 7 is a significant step beyond TRL 6, requiring an actual system prototype demonstration in a space environment. The prototype should be near or at the scale of the planned operational system and the demonstration must take place in space.

TRL 8. Actual system completed and "flight qualified" through test and demonstration (Ground or Flight). In almost all cases, this level is the end of true "system development" for most technology elements. This might include integration of new technology into an existing system.

TRL 9. Actual system "flight proved" through successful mission operations. In almost all cases, the end of last "bug fixing" aspects of true "system development". This might include integration of new technology into an existing system. This TRL does not include planned product improvement of ongoing or reusable systems.

B. Advancement Degree of Difficulty

As an effective measuring means of technology readiness, TRL has been mainly used to assess the current readiness state. It is so hard for TRL to appear the degree of difficulty and the required work when the technology expands to the target that the technology readiness assessment cannot present the technology state comprehensively. The definition of Advancement Degree of Difficulty is the difficulty level when technology spans to the target TRL from the current TRL. AD^2 based on full life-cycle of technology development reflect the uncertainty during technology R&D process. It enriches the technology readiness assessment system [11]. At the same time, it is conducive to carry out the quantification and the management of the technology risk.

Some natures of the AD^2 can be summarized based on its connotation [12]. Aiming at the same technology with the same present TRL, the higher the value of the target TRL is, the higher the value of the AD^2 is. Aiming at the same technology with the same target TRL, the higher the value of the present TRL is, the lower the value of the AD^2 is. Aiming at the different technology, even though both the present TRL and target TRL are at the same levels, the value of the AD^2 may be different in consideration of the complexity and diversity of technology domains.

In order to show the advancement degree of difficulty intuitively and clearly, the AD2 levels were also divided in 9 levels, same as TRL. Level 9 means the biggest, and Level 1 means the smallest. The 9 levels of the AD^2 as shown in follows:

AD^2 9. Requires new development outside of any existing experience base. No viable approaches exist that can be pursued with any degree of confidence. Basic research in key areas needed before feasible approaches can be defined.

AD^2 8. Requires new development where similarity to existing experience base can be defined only in the broadest sense. Multiple development routes must be pursued.

AD^2 7. Requires new development but similarity to existing experience is sufficient to warrant comparison in only a subset of critical areas. Multiple development routes must be pursued.

AD^2 6. Requires new development but similarity to existing experience is sufficient to warrant comparison on only a subset of critical areas. Dual development approaches should be pursued in order to achieve a moderate degree of confidence for success. (Desired performance can be achieved in subsequent block upgrades with high degree of confidence.

AD^2 5. Requires new development but similarity to existing experience is sufficient to warrant comparison in all critical areas. Dual development approaches should be pursued to provide a high degree of confidence for success.

AD^2 4. Requires new development but similarity to existing experience is sufficient to warrant comparison across the board. A single development approach can be taken with a high degree of confidence for success.

AD^2 3. Requires new development well within the experience base. A single development approach is adequate.

AD^2 2. Exists but requires major modifications. A single development approach is adequate.

AD^2 1. Exists with no or only minor modifications being required. A single development approach is adequate.

C. AD^2 Assessment Based on Fuzzy Integrated Evaluation

(1) Assessment dimension

Once a TRL has been established for the various elements of the technology under development, it becomes necessary to assess what will be required to advance the technology to the level required, that is the value of the AD[2] However, the difficulty of this step cannot be overemphasized because the difference of the technologies is large and it requires the art of "prediction" and "prospect". So it is important to assess the AD[2] effectively and smoothly that the assessment dimension has been formulated reasonably. Combining Integration Readiness Levels, Design Readiness Levels, and Materials Readiness Levels, etc. into one overarching process, the focus is on the design & analysis, developing & manufacturing, operations & integration, test & evaluation, and validation in determining the AD[2] based on the systems engineering approach, as shown in Table I.

TABLE I
DESCRIPTION OF ASSESSMENT DIMENSION

Assessment dimension	Description
Design &Analysis	The influences on hardware and software are mainly from design methods, models, design tools, analysis tools, and data bases, etc. The possibility of achieving the experimental results also needs to be considered.
Developing & Manufacturing	Two aspects: 1) the necessary of the manufacturing for technology unit. 2) the manufacturability of the final product supporting technology development.
Operation & Integration	The operations during design, manufacturing, and evaluation. Maintainability, supportability, reliability, testability, etc. will be contained. Life cycle cost also will be considerate.
Test &Evaluation	Two kinds of questions: DT&E focus on the applicability questions that whether the product can meet the performance requirements enacted in development plan. OT&E focus on the questions that whether the product can meet the job requirements in various conditions. Facilities, infrastructure, and available personnel should be confirmed.
Validation	A conventional method to validate the research and development plan. Laboratory environment, operating environment, real environment etc.

(2) Fuzzy Integrated Evaluation Model

Establish the factors set of the AD[2] evaluation object based on the actual development process of technology, $X=\{X_1,X_2,X_3,X_4,X_5\}$. X_2 is the factors set of developing & manufacturing. X_3 is the factors set of operations & integration. X_4 is the factors set of test & evaluation. X_5 is the factors set of validation. A series of evaluation indicators are contained in every dimension, and the quantity and content is depending on the real condition of the technology. Supposing each assessment dimension has five indicators considering the computational simplicity, $X_i=\{x_{i1},x_{i2},x_{i3},x_{i4},x_{i5}\}$.

Above all, set up the Fuzzy Integrated Evaluation model, as shown in Fig.1.

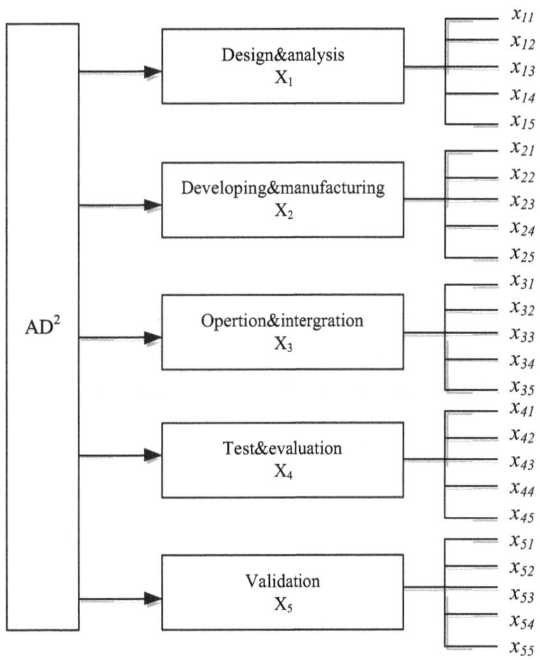

Fig.1. The model of the Fuzzy Integrated Evaluation for the AD[2]

Set the assessment level of all evaluation indicators as 9 levels on the basis of AD[2] levels. Evaluation language set shown as follows.

$$G_1=\{VVS(G_{ij}^{(1)}),VS(G_{ij}^{(2)}),S(G_{ij}^{(3)}),MS(G_{ij}^{(4)}),M(G_{ij}^{(5)}),$$
$$MB(G_{ij}^{(6)}),B(G_{ij}^{(7)}),VB(G_{ij}^{(8)}),VVB(G_{ij}^{(9)})\}$$
$$i=1,2,3,4,5;j=1,2,3,4,5$$

The probability of the indicator $x_{ij}\left(i,j=1,2,3,4,5\right)$ assessed to be $k\left(k=1,2,3,4,5,6,7,8,9\right)$ level is $r_{ij}^k\left(i,j=1,2,3,4,5\right)$, the fuzzy matrix of each assessment dimension can be shown as follows.

$$R_i=\begin{bmatrix} r_{11} & r_{12} & \cdots & r_{18} & r_{19} \\ r_{21} & r_{22} & \cdots & r_{28} & r_{29} \\ \vdots & \vdots & \ddots & \vdots & \vdots \\ r_{41} & r_{42} & \cdots & r_{48} & r_{49} \\ r_{51} & r_{52} & \cdots & r_{58} & r_{59} \end{bmatrix} \quad (1)$$

Calculate the weight vector W_i of every evaluation indicators x_{ij} , $W_i=\left(\omega_1,\omega_2,\omega_3,\omega_4,\omega_5\right)$, the integrated evaluation fuzzy subsectors of assessment dimension X_i can be computed synthesizing W_i and R_i .

$$E_i = W_i \circ R_i$$

$$= (\omega_1, \omega_2, \omega_3, \omega_4, \omega_5) \circ \begin{bmatrix} r_{11} & r_{12} & \cdots & r_{18} & r_{19} \\ r_{21} & r_{22} & \cdots & r_{28} & r_{29} \\ \vdots & \vdots & \ddots & \vdots & \vdots \\ r_{41} & r_{42} & \cdots & r_{48} & r_{49} \\ r_{51} & r_{52} & \cdots & r_{58} & r_{59} \end{bmatrix} \quad (2)$$

$$= [e_{i1}, e_{i2}, e_{i3}, e_{i4}, e_{i5}, e_{i6}, e_{i7}, e_{i8}, e_{i9}]$$

Establish fuzzy evaluation matrix of evaluation object $R = (E_1, E_2, E_3, E_4, E_5)^T$, and Calculate the weight vector $W = (\omega_1', \omega_2', \omega_3', \omega_4', \omega_5')$ of each assessment dimension X_i. The integrated evaluation fuzzy vector of evaluation object X can be computed synthesizing R and W.

$$E = W \circ R$$

$$= (\omega_1', \omega_2', \omega_3', \omega_4', \omega_5') \circ \begin{bmatrix} e_{11} & e_{12} & \cdots & e_{18} & e_{19} \\ e_{21} & e_{22} & \cdots & e_{28} & e_{29} \\ e_{31} & e_{32} & \cdots & e_{38} & e_{39} \\ e_{41} & e_{42} & \cdots & e_{48} & e_{49} \\ e_{51} & e_{52} & \cdots & e_{58} & e_{59} \end{bmatrix} \quad (3)$$

$$= [e_1, e_2, e_3, e_4, e_5, e_6, e_7, e_8, e_9]$$

Unitize the value of the integrated evaluation fuzzy vector to be 1-9. The integrated fuzzy evaluation results of the AD^2 is shown as follows.

$$\tilde{E} = [\tilde{e}_1, \tilde{e}_2, \tilde{e}_3, \tilde{e}_4, \tilde{e}_5, \tilde{e}_6, \tilde{e}_7, \tilde{e}_8, \tilde{e}_9] \\ (\tilde{e}_i \subseteq N,\ N = 1, 2, \cdots 9) \quad (4)$$

(3) Weight Coefficient Calculation

The factors sets of the AD^2 evaluation object are $X = \{X_1, X_2, X_3, X_4, X_5\}$, and evaluation indicators are $X_i = \{x_{i1}, x_{i2}, x_{i3}, x_{i4}, x_{i5}\}$. Set up judgment matrix of each hierarchy [13].

$$C_i = \begin{bmatrix} 1 & a_{12}^{(i)} & a_{13}^{(i)} & a_{14}^{(i)} & a_{15}^{(i)} \\ a_{21}^{(i)} & 1 & a_{23}^{(i)} & a_{24}^{(i)} & a_{25}^{(i)} \\ a_{31}^{(i)} & a_{32}^{(i)} & 1 & a_{34}^{(i)} & a_{35}^{(i)} \\ a_{41}^{(i)} & a_{42}^{(i)} & a_{43}^{(i)} & 1 & a_{45}^{(i)} \\ a_{51}^{(i)} & a_{52}^{(i)} & a_{53}^{(i)} & a_{54}^{(i)} & 1 \end{bmatrix} \quad (5)$$

Thereinto, $a_{ij}^{(i)} \rangle 0, a_{ij}^{(i)} = \dfrac{1}{a_{ij}^{(i)}}$.

Calculate weight vector of each evaluation indicator using characteristic root method on account of Formula 6.

$$\tilde{\omega}_i = \sum_{j=1}^{n} \frac{a_{ij}^{(i)}}{\sum_{k=1}^{n} a_{kj}^{(i)}} \qquad i = 1, 2, \cdots n. \quad (6)$$

Assessment dimension can be calculated at the same time.

$$W_i = \frac{\tilde{\omega}_i}{\sum_{i=1}^{n} \tilde{\omega}_i} \quad (7)$$

Calculate value of the maximum eigenvalue based on Formula 8.

$$\lambda_{max} = \frac{1}{n} \sum_{j=1}^{n} \frac{(AW_i)_j}{W_j} \quad (8)$$

Calculate the consistency ratio of each judgment matrix using Formula 9.

$$CR_i = \frac{CI_i}{RI} \quad (9)$$

Thereinto, RI means Random Index, $CI_i = \dfrac{\lambda_{max} - n}{n - 1}$.

A computational process should be carried out to test the consistency of the weight coefficient based Formula 6, 7, 8, 9. If the validation passes, the results can be used as the weight coefficient of the AD^2 assessment, and if not, the results will be analyzed again until that validation passes.

(4) Confidence Criterion

The Fuzzy Integrated Evaluation method above is established according to the principle of maximum membership degree. The principle will be limited when the membership degree of the evaluation results is equal in value. So the method above should be amended with the confidence criterion [14].

Define language set $G = \{G_1, G_2, \cdots G_k, \cdots G_s\}$ as an ordered evaluation one. α is Confidence, and $\mu_x(G_i)$ is membership degree, $\sum_{i=1}^{s} \mu_x(G_i) = 1$.

When $G_1 \rangle G_2 \rangle \cdots \rangle G_s$,

$t_0 = \min\left\{t : \sum_{i=1}^{t} \mu_x(G_i) \geq \alpha, 1 \leq t \leq n\right\}$ is contented.

When $G_1 \langle G_2 \langle \cdots \langle G_s$,

$t_0 = \max\left\{t : \sum_{i=1}^{t} \mu_x(G_i) \geq \alpha, 1 \leq t \leq n\right\}$ can be met.

Then, the evaluation object is be part of Class A.

The confidence criterion focuses on "strong" more. The kinds of "strong" account for more percent of the total. The value of the confidence is 0.6-0.8 normally [15].

III. RESULTS

Comparing with the form evaluation method, the actual application simple is carried out choosing critical technology elements of airplane control system as assessment objects to verify the assessment method of the AD^2.

There were eleven critical technology elements under assessment whose TRL are all below Level 7. Twenty experts are involved in the whole assessment process. The value of the AD^2 is assessed with two methods considering the process of the technology being developed to the TRL 7 from the present TRL. Part of the result of comparison of the two methods is shown in Table II.

TABLE II
COMPARISON OF THE AD^2

NO.	AD^2 Level								
	Result based on form evaluation								Fuzzy Integrated Evaluation
	1	2	3	4	5	6	7	8	
1	3	5	3	3	2	2	3	4	3
2	2	2	1	2	3	2	2	1	2
3	3	2	2	3	2	2	1	2	2
4	5	5	4	4	4	3	4	3	4
5	5	7	6	5	5	3	5	5	5
6	1	3	1	1	1	1	2	1	1

According the result of the application simple, the data with the form evaluation method present a larger dispersion. The differences among expert subjective assessment for the same critical technology element are larger due to the effect of experience and comprehension. The results with the form evaluation method are more ex parte. However, it can be seen that a mainstream value exist in the result according to the analysis of the overall data, around which the expert assessment fluctuate. The result with the method established in the paper is close to the mainstream value, which indicates that the assessment method based Fuzzy Integrated Evaluation can assess the AD^2 of the complex products objectively.

IV. CONCLUSION

The paper proposes an assessment method of the advancement degree of difficulty for complex products based on the Fuzzy Integrated Evaluation. According to the advancement degree of difficulty level standard, a three-tiered fuzzy integrated evaluation model was established using two methods, the Analytic Hierarchy Process and the confidence criterion, to improve the objectivity of the method. There still need some improvement in this article due to technologies' variety and complexity. A further study about these will be done in further research.

REFERENCES

[1] Hansen. K, Rush. H, "Hotspots in Complex Product Systems: Emerging Issues in Innovation Management", *Technovation*. vol. 9, no. 18, pp. 555-561, 1998.

[2] Xin Li, "Research on System Model in CoPS", *Co-Operative Economy & Science*, vol. 11, pp. 103-104, 2009.

[3] Mankins J C, "Technology Readiness Assessments: A Retrospective", *Acta Astronautica*, vol. 9, no. 65, pp. 1216-1223, 2009.

[4] Britt B L, Berry M W, "Document Classification Techniques for Automated Technology Readiness Level Analysis", *Journal of the American Society for Information Science and Technology*, vol. 59, pp. 675-680, 2008.

[5] Maochun An, Jianguo Wang, "The Foreign Evaluation Method of Technology Maturity and Its Application", *Evaluation & Management*, vol. 2, no. 6, pp. 1-3, 2008.

[6] Mankins J C, *Technology Readiness Levels: A White Paper*, Washington, DC: NASA, 1995.

[7] Yansheng Wu, *Technology Readiness and Its Assessment Method*, Beijing: National Defense Industry Press, 2012, pp. 158-163.

[8] Bilbro. J. W, "Using the Advancement Degree of Difficulty ($AD^{2,}$) as an Input to Risk Management", presented at the *Proceedings of Multi-Dimensional Assessment of Technology Maturity Conference*, Piscataway, USA. IEEE, 2008.

[9] Bilbro. J. W, "A suite of tools for technology assessment, Technology Maturity Conference: Multi-Dimensional Assessment of Technology Maturity", Virginia Beach, VA: Air Force Research Laboratory, 2007.

[10] "Best Practices—Better Management of Technology Development Can Improve Weapon System Outcomes", Washington, DC: GAO, July, 1999.

[11] Da Li, Kunsheng Wang, Kuan Ma, "Review on the Evaluation Method of Technology Maturity", *Scientific Decision Making*, vol. 85, no. 10, pp. 85-94, 2012.

[12] Smith, J. E., "Decision analysis in management science", *Management Science*, vol. 5, no. 50, pp. 561-574, 2004.

[13] Hexiang Liu. et, *The Theory and Application of Fuzzy Mathematics*, Beijing, Science Press. 2012, pp. 253-255.

[14] Qiansheng Cheng, "Attribute Recognition Theoretical Model With Application", *Acta Scientiarum Naturalium Universitatis Pekinensis*, vol. 1, no. 33, pp. 12-20, 1997.

[15] Guangzhi Pu, Yuwen Zhang, "Grey Fuzzy Comprehensive Evaluation Method Based on Interval Numbers of Three Parameters", *Systems Engineering and Electronics*, vol. 23, no. 9, pp. 43-45, 2001.

Analysis of Financing Pledged Accounts Receivable in Supply Chain by Duplex Stackelberg Game Models

Li-qun Ma, Yu-xiao Li*
School of Management, Tianjin Polytechnic University, Tianjin, China
(lyx8831@163.com)

Abstract - **Supply chain finance is a financing mode which involves core enterprise, the member enterprises, banks and other subjects of the supply chain. On the basis of symbiotic relationship, each subject is based on their own situation to make financing decision. To analyze the finance decision making of member enterprises based on the Duplex Stackelberg Models with assuming that member enterprises only would be default. In the game of core enterprise and member enterprise, the results shows that the factor which influence the guarantee probability of the core enterprise is the member enterprise's external lost, and the factor which influence the following treaty probability of member enterprise is caution money.**

Keywords - **Accounts Receivable Financing, Supply Chain Finance, Stackelberg Game**

I. INTRODUCTION

Supply chain finance is a new financing mode to solve the funds demand of enterprises in supply chain, its four financing patterns are accounts receivable pledge financing, stock secured financing and advance payment financing. Haul Lee pointed that the competition of the market is not only the competition between enterprises, but the competition between supply chains. William Atkinson (2008) [1] proposed four keys of Supply Chain Finance, namely obtaining the necessary supply chain financial knowledge, integrating payable accounts receivable and inventory, technology as a catalyst, as well as using of financial options for global operations. Chang Hwan Lee and Byong-Duk Rhee (2010) [2] reviewed the inventory financing business of financing costs and profit maximization problem. Hu Yuefei and Huang Shaoqing (2009) [3] started from the real economy background of supply chain finance business development, giving an academic definition for connotation and extension of the supply chain finances. Yan Guangle (2011) [4] analyzed credit financing mode and property financing mode with game model, and studied the supply chain finance ecosystem relationships among members; Lu Qihui, Zeng Lifei, and Zhou Weihua (2012) [5] showed that Supply Chain Financing with Accounts Receivables can make suppliers using financing funds for continuous production, so that it can achieve the optimal production in a short period when facing rapid growth of the market, and when the supplier initial cash scale is larger, the supplier meets the market demand more quickly. Cao Wenbin and Ma Cuixiang (2013) [6] introduced the default penalty by using the method of game theory to solve the traditional accounts receivable financing and Nash equilibrium of the supply chain financing with accounts receivable. Through comparing the supply chain financing with accounts receivable, it can achieve the Pareto equilibrium, and also can reduce financing costs and improve lending rate [7, 8].

Research on Supply chain finance has made fruitful results in the areas of concept connotation, financing instruments and their practical application, but on financing decision of the member enterprise in supply chain need for further research [9-11]. This paper will build Duplex Stackelberg Models to analysis the financing decision-making of the member enterprise with accounts receivable pledge in the supply chain.

II. THE ACCOUNTS RECEIVABLE PLEDGE AND THE GAME SUBJECTS

The accounts receivable financing is one of three main financing modes in supply chain finance. In order to obtain operating funds, as the basis of genuine trade

E. Qi et al. (eds.), *Proceedings of the 5th International Asia Conference on Industrial Engineering and Management Innovation (IEMI2014)*, Proceedings of the International Asia Conference on Industrial Engineering and Management Innovation 1, DOI 10.2991/978-94-6239-100-0_20, © Atlantis Press and the authors 2015

between seller and buyer, accounts receivable financing are provided by the seller, and contract's accounts receivable as the first source of repayment.

Based on this definition, the main subjects of accounts receivable financing include core enterprise member enterprises, supply chain enterprise and commercial banks serving as the representatives of financial institutions.

Game theory is the study of rational decision-making of each individual in their behavior when interacting directly in decision-making and decision equilibrium problems. Therefore, we make the following research hypothesis in the model: (1) The hypothesis of "rational man". (2)Game participants are aware of other parties of strategy space, the payoff function and characteristics etc., it is a perfect information game. (3) The game order hypothesis. In the process of the game, it is assumed to be member enterprises applicant loan first, then the core enterprise choose whether guarantee, last bank choose whether to provide a loan, it is a dynamic game. Based on these, we construct the game model, Game: core enterprise and member enterprises, led by core enterprise, member enterprises as followers.

III. THE GAME OF CORE ENTERPRISE AND MEMBER ENTERPRISE

A. Variable Definition

Assuming I represents loan scale, r_0 represents loan rate, r represents using funds rate of return by financing enterprises, r_b represents risk-free return rate, B represents accounts receivable, D represents margin which is given by core enterprise to bank, Y represents reward which is given to Supply Chain enterprises who don't get loan but still non-default by bank, such as member enterprises can directly obtain loans, and are simplified lending procedures, etc. p represents providing loan probability of bank, $1-p$ represents no providing loan probability, q represents non-default probability of member enterprise, $1-q$ represents default probability.

When bank loans and member enterprise default, member enterprise will suffer external loss, Assuming L represents this external loss, it has two sides, the one is affecting the future of the bank's loan amount and lending rates to member enterprise, the other one is this member enterprise's loss after losing the core enterprise business and cooperation brought by core enterprise. y represents reward which is given to member enterprises who don't get loan but always non-default by core enterprise, such as core enterprise can expand the opportunities for cooperation and increase the amount of raw materials ordering, S represents that the stability of supply chain operation of the core enterprise reduce transaction costs and obtain production stable income, ω represents guarantee probability of core enterprise, $1-\omega$ represents not guarantee probability.

B. Model Construction

The dynamic game is constructed in core enterprise and member enterprise, the payoff matrix of game model is shown in Table I:

TABLE I
PAYOFF MATRIX OF PERFECT INFORMATION GAME OF CORE ENTERPRISE AND MEMBER ENTERPRISE

	Non-default	Default
Guarantee	$S, I(r-r_0)$	$-D, I(1+r_0+r)-L$
Not guarantee	$0, y$	$0,0$

C. Model Analysis

1) *Core Enterprise's Expected Return:* Obtained from Figure 2, core enterprise's expected return is:

$$V_{(G)} = qS-(1-q)D \tag{1}$$

$$V_{(N)} = 0 \tag{2}$$

Let $V_{(G)} = V_{(N)}$, we can get $q = \frac{D}{S+D}$

When $q > \frac{D}{S+D}$, $V_{(G)} > V_{(N)}$, that is non-default probability of member enterprise is larger than $\frac{D}{S+D}$, the expected return which core enterprise guarantee is larger than the expected return which core enterprise doesn't, at this point core enterprise will guarantee for member enterprise, otherwise core enterprise will not guarantee.

Further, from derivation to q, we can get $\frac{\partial q}{\partial D}=\frac{S}{(S+D)^2}$, obviously $\frac{\partial q}{\partial D}>0$, that is non-default probability of member enterprise increase with the increase of margin.

2) *Member Enterprise's Expected Return Analysis:* Obtained from Fig.2, member enterprise's expected return is:

$$W_{(N)}=\omega I(r-r_0)+(1-\omega)y \tag{3}$$

$$W_{(D)}=\omega[I(1+r_0+r)-L] \tag{4}$$

Let $W_{(N)}=W_{(D)}$, we can get $\omega=\frac{y}{I(1+2r_0)-L+y}$

Considering $0<\omega\le 1$, so there is $I(1+2r_0)-L\ge 0$.

When $\omega>\frac{y}{I(1+2r_0)-L+y}$, $W_{(N)}>W_{(D)}$, that is guarantee probability of core enterprise is larger than $\frac{y}{I(1+2r_0)-L+y}$, the expected return which member enterprise non-default is larger than the expected return which member enterprise doesn't, at this point member enterprise will non-default, otherwise member enterprise will not.

Further, from derivation to ω, we can get $\frac{\partial\omega}{\partial L}=\frac{y}{[I(1+2r_0)-L+y]^2}$, obviously $\frac{\partial\omega}{\partial L}>0$. The larger the member enterprise's external loss due to default, the larger guarantee probability of core enterprise is, increase of external loss will limit member enterprise's non-default, and thus core enterprise will increase the likelihood of guarantee.

IV. THE STRATEGY OF GAME ON ACCOUNTS RECEIVABLE PLEDGE

In the two games, subjects of the supply chain check and balance, connect with each other. First member enterprise applies for funding, followed by the core enterprise decides whether to guarantee, in the end bank decides whether to provide loan, according to the bank's decision member enterprise decides whether to obtain loan. Throughout the supply chain, the subjects make maximizing expected return as principle to choose action. Core enterprise will not guarantee for all member enterprises, in the process of financing the core enterprise's credit and its guarantee for member enterprises are the key of bank loans, bank makes a decision in defining the core enterprise strategy. According to the backward induction of Stackelberg game. First of all is the optimal decision of bank and supply chain enterprise. Then the core enterprise and member enterprise bond together to game with bank, which ensures the stability and profitability of the whole supply chain. When the bank chooses to loan, from the perspective of long-term development supply chain enterprise will choose to non-default strategy. It does not only enhance the competitiveness of the chain, but also ensure effective and stable operation of the entire supply chain. The occurrence requirement of strategy (loan, non-default) is $q>\frac{I(1+r_0+r_b)-B-D}{I(1+2r_0)-B-D}$, and $p>\frac{Y}{I(1+2r_0)-B-D+Y}$, that is non-default probability of member enterprise is larger than $\frac{I(1+r_0+r_b)-B-D}{I(1+2r_0)-B-D}$ and provide-loan probability of bank is larger than $\frac{Y}{I(1+2r_0)-B-D+Y}$, at this time the two sides income is $(Ir_0, I(r-r_0))$. In the game of core enterprise and member enterprise, based on the game 1, core enterprise choose to guarantee, in order to maintain trade relations with core enterprise trade relations and ensure financing to continue. Member enterprise choose non-default, the occurrence requirement of strategy (guarantee, non-default) is $q > \frac{D}{S+D}$ and $\omega > \frac{y}{I(1+2r_0)-L+y}$, that is non-default probability of member enterprise is larger than $\frac{D}{S+D}$ and guarantee probability of core enterprise is larger than $\frac{y}{I(1+2r_0)-L+y}$, at this time the two sides income is $(S, I(r-r_0))$. Through the game it achieves better economic status, it helps member enterprise improve its credit level, access to bank financial support, revitalize the sedimentary money effectively, and improve the transfer efficiency of funds.

V. CONCLUSION

In this paper, through normative analysis, under the premise of Member enterprise default, we construct Duplex Stackelberg Game Models, core enterprise and member enterprise with pledge of accounts receivable financing mode. Based on game analysis of core enterprise and member enterprise, we can get that factor which influences the guarantee probability of core enterprise is member enterprise's external lost, and the factor which influences the non-default probability of member enterprise is margin. The Duplex Stackelberg game analysis provides the conditions of subjects' selections, and forms a basis for the analysis of member enterprise financing decision process. In addition, we assume that information is completely symmetrical, but in practice, incomplete information is common. Therefore, further study is considering the situation of asymmetric information.

REFERENCES

[1] William Atkinson. Supply Chain Finance: The Next Big Opportunity [J]. Supply Chain Management Review, 2008, (04):57-60.

[2] Chang Hwan Lee and Byong-Duk Rhee. Coordination contracts in the presence of positive inventory financing costs [J]. Production economics, 2010, (02):124-138.

[3] Hu Yuefei and Huang Shaoqing. Supply Chain Finance: Background, Innovation and Concept [J]. Research on Financial and Economic Issues, in Chinese, 2009, (08):76-82.

[4] Yan Guangle. The Game Analysis of Supply Chain Financing Modes [J]. Enterprise Economy, in Chinese, 2011, (04):7-11.

[5] Lu Qi-hui, Zeng Li-fei and Zhou Wei-hua. Research on Decision-making and Value of Supply Chain Financing with Accounts Receivables [J]. Journal of Management Sciences in China, in Chinese, 2012, (05): 14-22.

[6] Cao Wen-bin and Ma Cui-xiang. A Game Analysis of Account Receivable Financing based on Supply Chain Finance [J]. Commercial Research, in Chinese, 2013, (03): 174-179.

[7] Liu Linyan and Song Hua. The Framework and Development of Supply Chain Financing [J]. Research of Finance and Education, in Chinese, 2011, (02):15-22.

[8] Allen N.Berger, Gregory F.Udell. A More Complete Conceptual Framework for SME Finance [C]. World Bank Conference on Small and Medium Enterprise: Overcoming Growth Constraints, 2004, (10):14-15.

[9] Li Zhan-lei, Hou Cuicui and Sun Hongzhe. A Cusp Catastrophe Model of Supply Chain Finance Ecosystem [J]. Commercial Times, in Chinese, 2013, (03):54-56.

[10] Wang Yan. The Game Analysis of Supply Chain Financing Subjects [J]. Communication of Finance and Accounting, in Chinese, 2012, (03):29-30.

[11] Shenzhen development bank-China Europe International Business School Supply Chain Finance team. Supply Chain Finance: New finance under the new economy [M]. Shanghai: Shanghai Far East Publishers, 2009.

Retailer's Optimal Ordering Decision with Trade Credit Financing

Xiao-li Wu*, Jia-cheng Zhou
Department of Industrial Engineering, School of Business Administration, South China University of Science & Technology, Guangzhou, China
(*pearvoo@hotmail.com, franky.zhou@foxmail.com)

Abstract - **As competition becomes more intensive and the capital issue attracting more attention in the operation, trade credit has become important to achieve competition advantage. The trade credit issue, the discount on price of items, the present level of holding capital and some other key factors on business decision are firstly specialized on in this paper, and then, on the basis of the classical EOQ model, a simple useful mathematic mode is established with the retailer's capital consideration. The retailer's optimal ordering quantity can be characterized by three situations. At last, the influence of the payment delay, as well as the present capital on hand, on the final decision is discussed by analyzing the several solutions.**

Keywords - **Discount on price, EOQ, supply chain finance, trade credit**

I. INTRODUCTION

With many researchers' discussion and exploration, economic order quantity models have been developed into several expansion models, including the purchasing decisions of the durable goods and consumable goods, the decentralized decision-making models among multiple independent and deterministic demands, the centralized decision-making models among multiple uncertain demands, multiple stochastic inventory decision-making models, and so on. No matter how complex the model of the above problem is, the core idea evolved is to develop based on the economic order quantity, EOQ, model.

However, the above models involve only start-up costs, holding costs and the purchase costs are left alone , without considering the capital level in the ordering process as well as the effective utilization of cash, which are directly related to the corporation's cash flow level. Therefore, a more reasonable model that can reflect a corporate cash position is needed for further research, which considers the inventory related decisions of the companies, such as the optimal order cycle decisions.

Given the interest in inventory financing, trade credit has brought into the front. Followed by Wikipedia, trade credit is the largest use of capital for a majority of business to business (B2B) sellers in the United States and is a critical source of capital for a majority of all businesses. For example, Wal-Mart, the largest retailer in the world, has used trade credit as a larger source of capital than bank borrowings. Trade credit is common in current trading markets since buyers may prefer to trade credit in terms of guaranteeing product qualities, after sale services and so on. Consequently, it is worth analyzing the optimal decisions under trade credit.

Our problem is most related to the EOQ model as well as the trade credit concern. Starting from the simple EOQ model, the earliest research on trade credit by Goyal [1] builds a single product inventory model, while Khouja et al. [2] prove the retailer's optimal inventory policy given the credit period is or is not related to the order quantity. Hwanglt et al. [3] analyze the retailer's optimal sales price and order quantity problem under constant demand elasticity function and constant deterioration rate of commodities. Jamal et al. [4] and other studies the retailer's optimal order quantities and payment time decisions for perishable product under no trade credit punishment.

Huang and Chung [5] further extend EOQ model to gain the retailer's optimal replenishment and payment policies by allowing trade credit and prepayment discount to the retailer. Chung et al. [6] analyze the retailer's optimal ordering when credit period is related to the ordering quantity under certain deterioration rate. They also design the solving process to achieve optimal order decisions effectively.

Some work focus on the trade credit period issue, such as Y-H Huang [7], Jie Yan et al. [8], Tao Zhang et al. [9] and Cheng Hong [10]. Other research papers assume there are linkage between trade credit opportunity and purchasing price discount, which is closely related to our work. This kind of discussion is more practical and implementable. The recent research papers include Youjun Zhou et al. [11], Yinchun Cha [12], Zhen Yang [13], Mengzhang Liu et al. [14], Bin Luo et al. [15] and Yunmiao Gui et. al. [16]. Most of these researches ignore the on-hand capital level; the loan and risk free interest rates, if not all, but only focus on the tradition costs in EOQ models. No paper calculates the party's interests due to all the cash flow on hand.

The remainder of this paper is organized as follows. Section 2 we provide a model description. Based on Section 2, Section 3 derives that the optimal ordering policy can be characterized by three scenarios in terms of the retailer's initial asset level. Finally, Section 4 provides the numerical study as well as conclusions.

II. MODEL DESCRIPTION

The paper discusses the retailer's ordering decision problem at a two-level supply chain system that consists of a retailer and a supplier. At the beginning, the retailer M, who has initial cash on hand, will make order from his supplier N and sale to his downstream consumer. The supplier N provides trade credit to her downstream party M with certain cost. In other words, the retailer M needs to pay more for the unit purchasing cost in terms of his payment delay. The later deferred time the retailer M pays, the higher unit purchasing cost the retailer M needs to pay. The supplier N will replenish the retailer's ordered products right after M's ordering. Moreover, under the

E. Qi et al. (eds.), *Proceedings of the 5th International Asia Conference on Industrial Engineering and Management Innovation (IEMI2014)*, Proceedings of the International Asia Conference on Industrial Engineering and Management Innovation 1, DOI 10.2991/978-94-6239-100-0_21, © Atlantis Press and the authors 2015

trade credit opportunity provided by N, the retailer will decide when to pay to maximize his profit. This is different from the traditional product ordering. In this paper, we assume that the market demand follows a constant rate. We build the mathematical model based on traditional EOQ model but with consideration of both interest rate and loan interest.

In the following, we list symbols that will be immediately needed in our formulation of the problem.

A: initial capital the retailer has on hand
P: retail price of the product
Q: the retailer's order quantity
R: retailer's revenue from sales
β: proportional payment time by retailer, $0 \le \beta \le 1$
C (Q): unit wholesale price of the product
C_0: unit wholesale price of the product at time 0, i.e., $\beta = 0$
k: coefficient of wholesale price in time
t: time point on payment
T: time interval between ordering
D: demand rate
h: unit holding cost
H: inventory holding cost
B: purchasing cost
K: ($K \le A$) fixed ordering cost
I: total deposit or loan interest during the whole cycle
In specific, r_c is deposit interest rate, r_d is loan interest
Y: final net income of the retailer
G: average income during the whole cycle

From the above notations, the inventory holding cost during time interval *T* is

$$H = \frac{1}{2}Q$$

In terms of the unit purchasing cost, the retailer can defer the ordering payment but at the same time with a higher ordering cost. Hence, the cost is proportional to the payment time. That is,

$$C(Q) = C_0 + k \times t = C_0 + \frac{k\beta}{D}Q;$$

Therefore, the retailer's purchasing cost is $B = C(Q) \times Q$; in this paper, we assume that the small and medium sized retailer has good credit in the bank and can borrow from bank whenever necessary. Before the ordering payment time point βT, retailer M would not pay the ordering since he has the trade credit advantage and his cash on hand is $A - K + DtP$. After the time point βT, *M* has already deducted ordering payment from his account and hence $A - K - Q\,C(Q) + DtP$ at any time *t* is his on hand cash. Therefore, the total interest rate is

$$I(Q) = \int_0^{\beta T} (A - K + DtP) r_c dt +$$
$$\int_{\beta T}^{T} \left[(A - K - QC(Q) + DtP)^+ r_c - (A - K - QC(Q) + DtP)^- r_d \right] dt.$$

Knowing from the above expressions, the final net income of the retailer as well as the average one are the following:

$$Y = f(Q) = R - K - H - B + I$$
$$G(Q) = f(Q)/T$$

The model is to maximize G(Q) by choosing a proper Q for any given β,

$$\max_{Q \ge 0} G(Q)$$

Suppose $\bar{Q} = \arg\max_{Q \ge 0} G(Q)$.

III. RETAILER'S ORDERING DECISION

After building the modified EOQ model for the retailer, we try to characterize the retailer's ordering decision under different situations. Before presenting the situations, let's define two notations

$$Q_1 = \frac{(P\beta - C_0) + \sqrt{(P\beta - C_0)^2 + 4\frac{k\beta}{D}(A - K)}}{2\frac{k\beta}{D}}$$

$$Q_2 = \frac{(P - C_0) + \sqrt{(P - C_0)^2 + 4\frac{k\beta}{D}(A - K)}}{2\frac{k\beta}{D}}$$

Obliviously, $Q_2 \ge Q_1$ as $0 \le \beta \le 1$.

Since *I(Q)* is complicated in terms of the sign of $A - K - QC(Q) + DtP$, we provide three scenarios to simplify the function *I(Q)* so as to write out the expression of *G(Q)*. By taking the first and second order derivatives, we can gain the results one by one. In specific, let Q*={Q|G'(Q)=0}.

Theorem 1 If $0 < Q \le Q_1$, $\bar{Q} = \min\{Q^*, Q_1\}$.

Proof. If we can show that $A - K + DP\beta T - QC(Q) \ge 0$, we can write out G(Q) and show the optimal Q easily. Let $g_1(Q) = A - K + DP\beta T - QC(Q)$, then we have

$$g_1(Q) = -\frac{k\beta}{D}Q^2 + Q(P\beta - C_0) + A - K,$$

which holds due to *Q=DT.*

As $\Delta = (P\beta - C_0)^2 + 4 \times \frac{k\beta}{D}(A - K) \ge 0;$

there are two roots for g_1(Q), that is, Q_0 and Q_1. We can get that $Q_0 \le Q \le Q_1$ due to $g_1(Q) \ge 0$. And $0 \le Q \le Q_1$ with $g_1(0) \ge 0$. Under this constraint, we get Hence, the average income G (Q) is

$$G(Q) = PD - \frac{KD}{Q} - \frac{1}{2}Qh - C(Q)D + r_c(A-K)$$
$$-r_c QC(Q)(1-\beta) + \frac{1}{2}r_c PQ$$

By taking the first and second order derivatives, we can show the result accordingly.

Under situation $0 < Q \le Q_1$, the inequality $A - K + DP\beta T - QC(Q) \ge 0$ and Q>0 are guaranteed, which implies that the retailer has nonnegative cash flow on hand after paying his ordering. As a result, the retailer is self-financed and does not need to borrow from the bank.

Similar to situation $0 < Q \le Q_1$, we have

Theorem 2 If $Q \ge Q_2$, $\bar{Q} = \max\{Q^*, Q_2\}$.

Under situation $Q > Q_2$, the inequality $A - K + DPT - QC(Q) \le 0$ and Q>0 are guaranteed, which implies that the retailer has negative cash flow on hand after paying his ordering. Different from previous scenario, the retailer does need to borrow from the bank. For the rest situation, we have the following theorem.

Theorem 3 If $Q_1 < Q < Q_2$, then $\bar{Q} = Q_2$ when $Q^* \ge Q_2$; $\bar{Q} = Q^*$ when $Q_1 \le Q^* \le Q_2$; and $\bar{Q} = Q_1$ when $Q^* \le Q_1$.

From the situation $Q_1 < Q < Q_2$, we can derive that $A - K + P\beta Q_1 - QC(Q_1) \le 0$ and $A - K + DPT - QC(Q) \ge 0$. This relation indicates that though retailer M has negative cash flow after paying back the ordering, he can achieve nonnegative cash flow in the end. So in this situation, retailer M will borrow money under negative cash flow and then switch to deposit money when he has positive cash flow.

Overall, although we cannot derive the single closed form order quantity, the optimal decision belongs to one of the three situations. We should choose the one with the largest average income $G(\bar{Q})$.

IV. NUMERICAL STUDY AND CONCLUSION

To get a better understanding of our theoretical results, we will test two groups of numerical studies. Since we set the proportional payment time β fixed, we use the following example to derive the impact of βon the optimal order quantity.

Example 1: Suppose P = \$10/unit, C_0 = \$5/unit, k = \$2/week, K = \$50, h = \$2/unit/week, r_d =0.35%, r_c=0.038%, A =\$50.

TABLE I
IMPACT ON β

β	0.05	0.10	0.15	0.20	0.25
Q	30	29	28	27	26
G(Q)	33.67	30.72	27.89	25.17	22.51

From Table I, we test the optimal order quantity Q as well as the average income value G(Q) for different β. The following graphs provide more clear trends of the data in Table I.

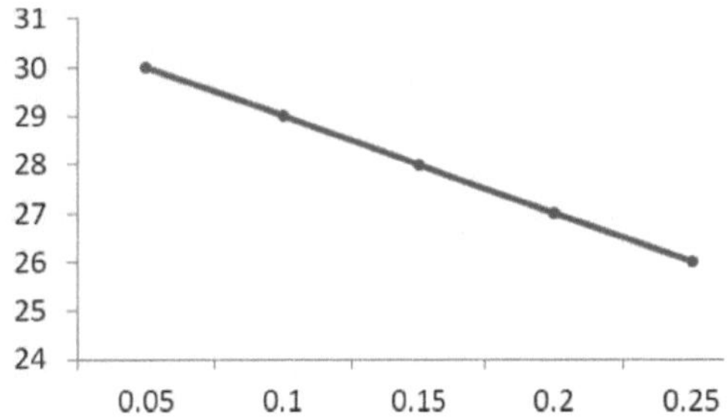

Fig.1. The optimal order quantity in β

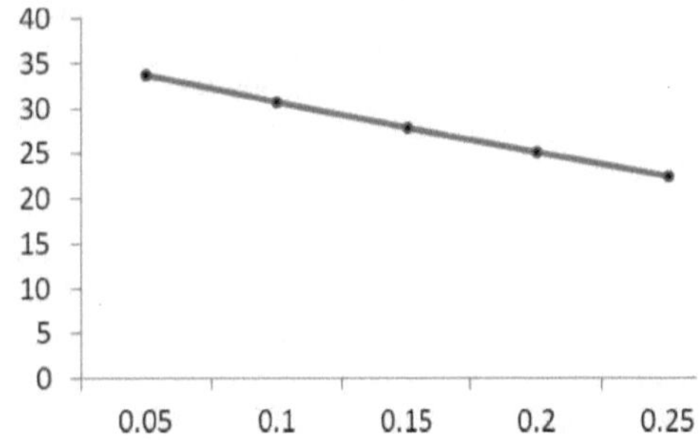

Fig.2. The optimal average income in β

From Fig.1 and 2, when β increases from 0.05 to 0.25, the optimal Q drops from 30 to 26, while G (Q) drops from 33.67 to 22.51. The figures indicate the same trend that the optimal Q and G (Q) decrease with respect to β.

In summary, our research gives the optimal ordering policy under the consideration of on hand cash flow and trade credit issue, which provides a new insight to such EOQ models. But the research is limited for given proportion of payment. The further research can be extending to multiple decisions. Besides the ordering quantity decision, the proportion of payment is also a decision variable. Meanwhile, we may consider the supplier's decision for providing trade credit penalty cost. In this paper we suppose the retailer has very good credit in the bank and ignore the difficulties that a retailer need consider when borrowing from the bank. Considering more complex or realistic facts in the model is worth to be further studied.

ACKNOWLEDGMENT

This research was partially supported by National Science Foundation of China under Grant No.71131003, 71201061, Seedling project (humanity) of Guangdong Province, China (2012WYM0012), the Fundamental Research Funds for the Central Universities (2014ZZ0072).

REFERENCES

[1] Goyal S K, "Economic order quantity under conditions of permissible delay in payments", *Journal of the Operational Research Society*, 1985: 335-338.

[2] Khouja M, Mehrez A, "Optimal inventory policy under different supplier credit policies", *Journal of manufacturing Systems*, 1996, 15(5): 334-339.

[3] Hwang H, Shinn S W, “Retailer's pricing and lot sizing policy for exponentially deteriorating products under the condition of permissible delay in payments”, *Computers & Operations Research*, 1997, 24(6): 539-547.
[4] Jamal A M M, Sarker B R, Wang S, "Optimal payment time for a retailer under permitted delay of payment by the wholesaler", *International Journal of Production Economics*, 2000, 66(1): 59-66.
[5] Huang Y F, Chung K J. “Optimal replenishment and payment policies in the EOQ model under cash discount and trade credit”, *Asia Pacific Journal of Operational Research*, 2003, 20(2): 177-190.
[6] Chung K J, Liao J J. “Lot-sizing decisions under trade credit depending on the ordering quantity”, *Computers & Operations Research*, 2004, 31(6): 909-928.
[7] Huang Y F. "Optimal retailer's ordering policies in the EOQ model under trade credit financing", *Journal of the Operational Research Society*, 2003, 54(9): 1011-1015.
[8] Min J, Zhou Y WQ, Liu Y X, "Supply chain inventory model with two levels of trade credit and time varying demand",*Journal of Systems Science and Information*, 2011, 31(2): 262-269.
[9] Zhang T, Dai G X, Jiang X L, "A Partial Trade Credit Model with Cash Discount Under Supply Chain Management", *Journal of Qingdao University (Natural Science Edition)*, 2011, 1: 018.
[10] Cheng H, Wang X Y, Su Y S, "Optimal Trade Credit Policy for Supplier under Asymmetric Information in the Supply Chain", *Journal of Donghua University*, 2011, 28(4).
[11] Zhou Y J, Meng C, Cao L, "The Inventory Model for Deteriorating Items with Trade Credit and Cash Discounts Policy", *Logistics Sci-tech*, 2010 (004): 15-19.
[12] Za Y C, Hua Z S, Wu J, "Analysis of Order Policy under Delay in Payment and Temporary Price Discount", *Chinese Journal of Management*, 2008, 5(2): 188-192.
[13] Yang Z, Luo B, "Optimal Payment Strategy for Retailer under Postponement of Disbursement", *Commercial Research*, 2009 (012): 182-185.
[14] Liu M Z, Shan M, "Choices of Retailer Ordering Policy Under Advance Purchasing Price Discount and Trade Credit", *Guide To Business*, 2013 (2).
[15] Luo B, Cheng J H, Wan S Y, "Research on the Inventory Model of Deteriorating Items under Delay Payments and Partial Price Discount", *Technology Economics*, 2010, 29(007): 117-119.
[16] Gui Y M, Qin L P, Zhang T L,"A Research into the Supply Chain Coordination with Supplier' s Delay in Payments and Retailer's Sales Effort", *Economic Survey*, 2011, 4: 022.

Kinemics Sensors Application in Urban Train Control

Xiao-feng Zhao
PRD, Shanghai Fuxin Intelligent Transportation Solutions Co., Ltd, Shanghai, China
(simonzhaoxf@126.com)

Abstract - **Urban train control system usually adopts all types of kinemics sensors to acquire train speed, acceleration, position, distance and travel direction. This paper summarizes mainstream signalling supplier's speed measurement configuration, suggests sensors installation layout, and illustrates speed acquisition redundant architecture. From the view of software processing, discuss raw kinemics data calculation, vehicle zero velocity detection and command, as well as wheel slip determination and compensation.**

Keywords - **Kinemics sensors, redundant architecture, raw speed measurement, speed acquisition, sensors installation layout, wheel slip compensation, zero velocity control**

I. INTRODUCTION

Urban rail train speed measurement is vital function, exactness and accuracy of speed measurement will affect the safety of automatic train protection, especially in moving block CBTC signalling system.

As the important part of on-board subsystem in signalling system, train speed measurement system comprise kinemics sensors, interface unit and main process unit. Kinemics sensors sense train speed, acceleration, position, distance and travel direction; Interface unit receives and deal with raw data from these sensors; Main process unit is responsible for whole speed measurement algorithms, generates actual train speed, detect train wheel slip, and makes the compensation under specific conditions.

II. KINEMICS SENSORS

Common train control kinemics sensors have WIG (Wheel Impulse Generator), OPG (Optical Pulse Generator), HES (Hall-Effect Sensor), DRS (Doppler Radar Sensor), ACC (Accelerometer) and GNSS (Global Navigation Satellites System).

Wheel Impulse Generator, also known as Speed Measuring Motor, belongs to passive non-contact sensor, is earliest involved speed acquisition device in urban rail, which is used on the vehicle in audio frequency track circuit based ATC system. WIG is installed on vehicle axle head without power supply. WIG is only used in the lines going into revenue service before 2000, and new built lines will not use it[1].

Optical Pulse Generator belongs to active contact sensor, is the most widely used speed acquisition device now, and comprise photovoltaic module, optical grating plate, transmission shaft, etc. Common type is infrared reflection type. OPG is also installed on vehicle axle head. When train wheel moves, transmission shaft makes internal optical grating plate rotate. There are two circles holes at the outer edge of optical grating plate, each circle have 200 holes, and these holes have different size which means different codes, so some signalling suppliers call OPG as coding odometer. While working, OPG needs external direct current power supply to drive internal infrared diode, which generates reflection light. The light goes through the holes of optical grating plate, and is reflected on the bottom, then accepted by photosensitive tube, at last transited to square wave voltage pulse which outputs to signalling system[2]. OPG outputs signal have 6 channels. Because mechanical rotation wear and blocked optical grating leading to pulse lost, OPG only can be used on the vehicle whose highest speed is lower than 140km/h[3]. The suppliers using OPG have Alstom / CASCO, Siemens, Ansaldo and Beijing Traffic Control Technology.

Hall-Effect Sensor belongs to active non-contact sensor. Compares with optical pulse generator, HES is smaller and lighter, but more complex installation which needs customized matching speed measuring gear and axle head cover. Speed measuring gear is mechanical component, installed on the head of axle, and mantled by the cover which is fastened on the bogie. The sensor is installed on one side of the cover, and keeps about 1mm distance with the gear. HES works with external direct current power supply. While train is moving, the gear rotates with axle of wheel, and the sensor detects the teeth of gear by hall-effect, then outputs current signal to signalling system. HES outputs signal only have 2 channels. Because good dustproof, anti-oil and pulse lost prevention, HES is not only suitable for urban rail, but also fits intercity train particularly whose highest speed more than 180km/h[4]. The supplier using HES has Thales.

Doppler Radar Sensor is considered as auxiliary wheel slip detection device, and comprises DSP processing system, detection dresser, mixer, high frequency generator, electromagnetic collector, antenna, etc. DRS is installed on the bottom under the train, and antenna is towards to train positive operation direction. DRS works with external direct current power supply. While train is moving, high frequency generator creates 24GHz micro-wave signal, most of signal emits through the antenna, a little signal couples into mixer as local signal; Emitted signal is reflected by rail surface, and accepted by mixer through the antenna. After mixing these two signals, difference value is got, i.e. Doppler frequency shift, which is proportional to train speed[5]. At last DSP processing system sends the data

E. Qi et al. (eds.), *Proceedings of the 5th International Asia Conference on Industrial Engineering and Management Innovation (IEMI2014)*, Proceedings of the International Asia Conference on Industrial Engineering and Management Innovation 1, DOI 10.2991/978-94-6239-100-0_22, © Atlantis Press and the authors 2015

to signalling system by serial communication. DRS has high accuracy, wide range, extents to 0.2~600km/h. The suppliers using DRS have Siemens and Beijing Traffic Control Technology.

Accelerometer is also auxiliary wheel slip detection device, and comprises cantilever block, torque motor, non-contact displacement sensor. Common type is force balance servo angle type[6]. ACC is installed on the bottom of signalling rack in train body. While working ACC needs external direct current power supply, cantilever block is linked with armature of torque motor, and keeps balance initially. When train is moving, cantilever block deviates from original position, then non-contact displacement sensor detects this change and activates coil of torque motor which generates counterforce making new balance between cantilever block and torque motor, and outputs voltage signal to signalling system. ACC only has one channel output signal. The suppliers using ACC have Thales and Ansaldo.

Global Navigation Satellites System is satellites based radio navigation system. Through receiving unique coding sequence from satellites, train calculates current three-dimensional position (longitude, latitude and height), speed, direction and time[7]. GNSS is mainly used as train localization in modern tram signalling system, and real-time measured speed is not involved automatic train protection function, but considered as reference data for operation dispatch. GNSS have American GPS and China Beidou.

III. TRAIN SPEED MEASUREMENT SYSTEM

In most cases, train speed measurement system's main process unit and interface unit, i.e. signalling system on-board mainframe, respectively deploys one set in A1 and A2 car of one train. Each on-board signalling mainframe adapts 2oo3 architecture. Two mainframes at two ends are connected with network link, and become hot standby mutually. Few signalling supplier use 2oo2 architecture, and connected two ends mainframes form 2×2oo2 architecture.

Train speed measurement system usually adapts multi-sensors fusion technology; while protecting the safety and availability of train operation, deploy different types and quantities of kinemics sensors.

TABLE I
TRAIN SPEED MEASUREMENT SYSTEM CONFIGURATION [8][9]

A1-car	A2-car	TYPE	SIG
OPG×1	OPG×1	Dual	2oo3
OPG×1+DRS×1	OPG×1+DRS×1	Dual	2oo2
OPG×2+DRS×1	OPG×2+DRS×1	Single	2oo3
OPG×2+ACC×4	OPG×2+ACC×4	Single	2oo3
HES×2+ACC×3	None	Single	2oo3

Table I summarizes urban rail mainstream signalling suppliers' train speed measurement system configuration solution. Divided by implementation model, train speed measurement system includes single-end and dual-end. Single-end system means on-board signalling mainframe only uses current end's kinemics sensors to implement speed measurement function; Dual-end system means on-board signalling mainframe has to use current end and the other end's kinemics sensors to implement complete speed measurement function. In dual-end system, current end signalling mainframe receives the other end's speed data by on-board network transmission.

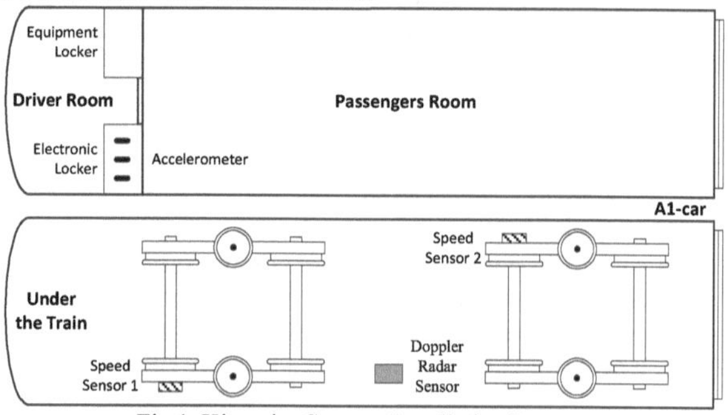

Fig.1. Kinemics Sensors Installation Layout

Kinemics sensors installation position is seen in Fig.1 (Look down at the top of train). Speed sensor in the figure could be OPG or HES. Single-end system's two speed sensors are installed on braking axles of one end, and lay at two sides of vehicle; Dual-end system's two speed sensors are installed on braking axles of two ends, and lay at two sides of vehicle. Other devices have the same layout in A2 car as above figure.

Take 2oo3 on-board signalling mainframe connection for example. Speed sensor in the figure could be OPG or HES. Each speed sensor connects to interface unit, and is divided to 3 channels signal involving into interface processor 1, 2, 3. DRS's RS-485 interface supports one leader with multi-followers mode, so it also can be divided to 3 channels signal involving into interface processor. ACC usually adapts one-to-one connection[10][11].

IV. SPEED MEASUREMENT ALGORITHMS

Speed measurement algorithms mainly provide credible speed and distance for automatic train protection function based on the inputs of kinemics sensors, velocity zero detection and protection, as well as wheel slip determination and compensation.

A. Raw Speed Measurement

OPG optical grating plate has two circles holes: clock pulse hole and coding hole. The number of clock pulse hole is 100 or 200, and the number of coding hole is 100. OPG has 5 photovoltaic sensors: C1~C5. C1, C2 and C3 scan clock pulse holes, calculate actual wheel speed, and determine actual wheel direction as per the pulse's phases; C4 scans coding holes, which have different sizes and constitute fake random sequence, i.e. code, when wheel rotates. Different wheel rotation direction is related to different code.

OPG calculates actual wheel speed with the formation:

$$V_{measure} = \frac{\pi \cdot d}{N \cdot P_{width}} \quad (1)$$

Whereby:

- $V_{measure}$: actual wheel line speed;
- d: wheel diameter;
- N: the number of pulses generated when train wheel rotates for one circle;
- P_{width}: the width of pulse, unit: mm.

OPG determines actual wheel direction as below:

- ✓ Forward: when C1 signal changes from low to high, C2 signal is low, and C3 signal is high;
- ✓ Reverse: when C1 signal changes from high to low, C2 signal is high, and C3 signal is low.
- ✓ Only when C4 code verification passes, OPG raw speed measurement would be considered valid by on-board signalling mainframe.

The teeth number of speed measuring gear matching with HES has 100, 110 or 120. HES has 2 sensors: S1 and S2, which detect the number of teeth and determine actual wheel direction.

The formation of actual wheel speed calculation for HES is similar to OPG.

HES determines actual wheel direction as below:

- ✓ Forward: S1 signal phase is 90° faster than S2 signal;
- ✓ Reverse: S2 signal phase is 90° faster than S1 signal.

DRS completes raw speed measurement by itself, and communicate with on-board signalling mainframe by RS-485, whose transmission rate is 19200 baud. DRS provides speed, distance, direction and acceleration as per customer's requirements.

ACC provides acceleration's polarity and value to on-board signalling mainframe. ACC usually adapts scale factor with ±5V/g, and measurement range with ±1g. It's worth noting that measured acceleration includes gravity component along the grade that train is operating.

B. Velocity Zero Detection and Protection

When train speed is less than 0.5km/h and doesn't detect any pulses from kinemics sensors, on-board signalling mainframe considers train is velocity zero.

Besides, on-board signalling mainframe also acquire velocity zero status (VZS) from train lines of vehicle. On-board signalling mainframe provides velocity zero command (VZC) to vehicle, which is used to control train door circuit and traction enable circuit.

C. Wheel Slip Determination and Compensation

Wheel slip has two scenarios: spin and slide. Spin means wheel speed is bigger than train speed, and usually happens during train traction; Slide means wheel speed is smaller than train speed, and generally appears during train braking. Either OPG or HES is installed on braking axle of trailer, so train slides more frequently[12][13].

There are two types of wheel slip determination algorithms: Vehicle Parameter method and Real-time Detection method. The former determines wheel slip with experimental parameter converted from adhesion - slip rate curve, which is suitable for the solution only with OPG or HES; the latter determines wheel slip as per real-time detected acceleration and speed, which is used by the solution with DRS or ACC[14][15].

Train speed measurement system using real-time detection method usually equips two OPGs or HESs, so it has two scenarios: single wheel slip and dual wheel slip. When single wheel slip happens, the method compensate slipped wheel with measured distance from OPG or HES which doesn't slip. When dual wheel slip happens, each wheel calculates compensated speed with current period measured acceleration and last period measured speed, then gets compensated distance.

V. EPILOGUE

Kinemics sensors have many types with different performance. Optical Pulse Generator is easy installation; more sensors with coding check function, and particularly suitable for urban vehicle with the highest speed no more than 120km/h. Hall-Effect Sensor is simple, reliable, installed with speed measuring gear, and fits urban rail, intercity rail as well as high speed rail. Doppler Radar Sensor is high accuracy, large scope of speed detection, powerful functions, advanced interface, and becomes ideal speed acquisition auxiliary device. Accelerometer is simple structure, easy installation, steady performance, single interface, and shall be first choice of acceleration detection.

Train speed measurement system usually adapt multi-sensors information fusion technology, the architecture of single-end solution is more clear, but speed reference data must not be transmitted between main processors of on-board controller, which is more beneficial to speed measurement implementation. Using OPG or HES with ACC is better configuration of the system.

It's hard to choose between two speed measurement algorithms. Vehicle Parameter method is based on the research of vehicle anti-skid control, refine experimental parameters used to determine wheel slip for signalling system, which reveals the integration trend of signalling and vehicle. Real-time Detection method uses signalling system's kinemics sensors, more accurate and closer to site working conditions, but complexity of the algorithm is higher accordingly.

REFERENCES

[1] Xiao-yue Li, "Shanghai line1 on-board signaling system", *Railway Operation Technology*, 4, 16, December 1998, pp. 172-177.
[2] Yan-ming Li, Pei-sun Ma, Jun Xu, and Xue-guan Gao, "A DC motor speed detecting and controlling system based on reflecting optoelectronical sensor", *Machinery & Electronics*, March 2002, pp. 24-27.
[3] Wei-min Zhen, "Mechanism and applications of novel speed sensor at the shaft ends", *Technological Development of Enterprise*, 26, 9, September 2007, pp. 70-72.
[4] TB/T 2760, *Vehicle Axle Head Photovoltaic Tachometer*, China Ministry of Railways, 1998.
[5] Lu Lu, Zheng-guo Xu, Wen-hai, Wang, and You-xian Sun, "Performance degradation monitoring and speed reading compensation method for on-board radar speed sensors of trains", *China Railway Science*, 32, 4, July 2011, pp. 93-98.
[6] Xiao-feng Su, "Xi an metro line 2 on-board signaling system", *Science & Technology Information*, May 2012, pp. 396-397.
[7] Jian Wang, Hui Zhang, Bo-gen Cai, and Wei Shang-guan, "A survey of track occupancy recognition technology based on GNSS", *Journal of the China Railway Society*, 32, 2, April 2010, pp. 94-97.
[8] Yang Lao, "Shen yang metro line 1 signalling system on-board controller", *Technological Development of Enterprise*, 29, 4, February 2010, pp. 116-117.
[9] Lian-zi Zhang, "Siemens signalling system on-board redundancy function", *Railway Signalling & Communication Engineering*, 9, 4, August 2012, pp. 44-46.
[10] Zhang Yang, Zhou Da-Tian, Liu Hong-Jie and Liu Bo, "Research on multi-sensor information fusion based CBTC train speed and position measurement", *Urban Rapid Transit*, 24, 4, August 2011, pp. 30-32.
[11] Da-tian Zhou, *Research on Multi-sensor Information Fusion Based Train Location Method*, Beijing Jiaotong University, Beijing, China, Doctoral Thesis, 2007.
[12] Wang Xiu-Hong, *Research on the Key Technology of Anti-skid Control for Locomotive*, Beijing Jiaotong University, Beijing, China, Master Thesis, 2010.
[13] Ying Lin, and Chang-lin Wang, "Research of slip and slide detection and adjust method in on-board ATP", *Urban Mass Transit*, March 2011, pp. 28-31.
[14] UIC 544-1, *Brakes - Braking Power*, International Union of Railways, 2004.
[15] TB/T 1407, *Train Traction Calculation Regulation*, China Ministry of Railways, 1998.

Study of Fraud Behaviors in Online Auction Based on Game Theory

Qi Tang*, Xu-hua Lv, Li-Li Chen
College of Science, Wuhan University of Science and Technology, Wuhan 430065, PR China
(498007504@qq.com)

Abstract - **In this paper, a shill bidding regulatory game model is designed, which considers the cost factor and detection accuracy at the same time, and quantifies these two factors base on literature [1]. It analyzed the possible decision behaviors of regulator and seller's decision under the interdependence case and the mixed Nash equilibrium problem. It obtained the optimal penalty factor analytic by building the game and researching the strategies between regulatory authorities and sellers according to which the regulatory authorities punish the sellers using the shill bidding.**

Keywords - **Nash equilibrium, online auction, regulatory game, shill bidding**

I. INTRODUCTION

Shill bidding in online auction has increased and drawn the attention of many researchers. The online auction is to use the Internet to sell some goods or services information, intends to sell to the highest bidder by way of competitive bidding, its essence is to increase price. Auctioneer and bidders in online auctions in order to maximize themselves utility, may cause serious fraud that interfere network transaction integrity. Shill bidding is a common fraud practice of seller in online auction. The so-called shilling bidding refers to the seller disguised himself as buyers bid, or employ other buyers bid, to mislead the actual bidders into giving a higher price.

How to effectively prevent fraud in online auctions, purify network online auction environment is a very important issue that need to be carefully addressed in the auction theoretical studies. About discussion and application method of effectively suppress shill bidding, the existing research methods are mainly redesign fee structure and the identification of shill bidding [1-5]. Shah [6] use association rules to verify the strong correlation between participation in the shill bidding of sellers and bidders; Luo Zhangzhou [7] use the data sequence of bidders to detect relationship between shill bidding sequence , and gives a data mining method of identifying shill bidding behaviors ; Li Xuefeng [8] using association rules to analysis asking price of sellers and bidders in online auction , and find a balanced strategy to identify the shill bidding; Ji Shuxian [9] get conclusions through research that the reserve price of shill bidding about setting the changing expectations of sellers; Above studies concern the behavior of bidders and sellers of individual decisions based on cost, rather less concerned the influence to the decision-making behavior caused by that regulatory authorities regulate shill bidding fraud. In addition to the above mentioned study about identifying the shill bidding, the existence of interdependent game relationship between regulator and seller's decision behavior make it necessary to present shill bidding regulatory game model to find the optimal strategy of the auction site. The literature [1] have a try in this respect, but the literature [1] did not introduce check accuracy and cost factor simultaneously into the shill bidding regulatory game model , so there a little distance between the literature [1] and the reality. Based on the above considerations, the paper has been wrote on the basis of the literature [1], taking the cost of factor and detection accuracy into account about the impact on the game players, in more general terms and conditions established a new model of shill bidding regulatory game, get its fine mixed Nash equilibrium and optimal coefficients, in order to provide a theoretical basis for the management of the online auction site regulators [10, 11].

II. DESCRIPTION OF A REGULATORYGAME MODEL

A. Model Assume

(1) There are two rational participants A and B, where A is the site regulators, B for the seller;

(2) It needs costs when the regulatory institutions check and seller shill bids. Set seller publication fee is C1 , seller commission is C2 , seller shill bids cause the damages fee on their own integrity is C3, the seller shill bids cause the damages fee on the integrity of the site is C4 , the fee that site regulatory institutions to check whether the seller shill bids is C5 .

(3) Although the shill bidding can be identified, because it needs costs when the regulatory institutions check, it is impossible to complete examination of all transactions. Set checking probability t, unchecked probability 1-t, where $0 \leq t \leq 1$ and it represents the optimal inspection probability of regulatory institutions.

(4) Due to the seller shill bids not only obtain illegal income, but also need costs, it is impossible to always shill bids. Set sellers Choose shill bidding probability is s, the probability of not is 1-s, where $0 \leq r \leq 1$. r^* represent the optimal probability which the seller shill bids.

(5) When regulatory institutions unchecked and seller is shill bidding, the seller will receive the illegal income M;

(6) When regulatory institutions check whether the seller shill bids, the accuracy of be found is k.

(7) When regulatory institutions check out the

E. Qi et al. (eds.), *Proceedings of the 5th International Asia Conference on Industrial Engineering and Management Innovation (IEMI2014)*, Proceedings of the International Asia Conference on Industrial Engineering and Management Innovation 1, DOI 10.2991/978-94-6239-100-0_23, © Atlantis Press and the authors 2015

seller is shill bidding, the penalty coefficient is q.

B. Establish regulatory game model

This model assume the independent private values and risk neutral, consider the cost of online auction regulatory and seller shill bids and the accuracy to detect, when seller be found for the use of shill bidding, he will be Confiscated the illegal income and fined at a certain ratio of illegal gains fines .Under these conditions, we establish the regulatory game model.

According to the above assumptions, when the seller shill bids, the loss of regulatory institutions take inspection strategy is $M\text{-}kM - kqM + C_4 + C_5$: the loss of not is their inspection costs C5; when the seller don't shill bids, the loss of regulatory institutions take inspection strategy is $M{+}C_4$ which equal to the fee seller shill bids cause the damages on the integrity of the site, the loss of not is zero.

Then the payments of participants A, B at different strategies are shown in Table I. As can be seen from Table I, the payment of regulatory institutions is the negative losses on regulators caused by the seller shill bids; the payment of seller is the illegal gains by shilling bids.

TABLE I
THE PAYMENT OF THE SHILL BIDDING UNDER THE DIFFERENT STRATEGIES

regulatory institutions	Sellers	
	Shill bids	not
check	$kM + kqM - M - C_4 - C_5$, $M - kM - kqM - C_1 - C_2 - C_3$	$-C_5$, 0
not	$-M - C_4$, $M - C_1 - C_2 - C_3$	0 , 0

III. SOLUTION OF THE SURERVISION ON GAME MODEL

Theorem 1 Shill bidding regulation game model exists mixed strategy Nash equilibrium. it is $\left[(t^*,1-t^*),(r^*,1-r^*)\right]$, and $t^* = \dfrac{M - c_1 - c_2 - c_3}{(q+1)kM}$, $r^* = \dfrac{c_5}{(q+1)kM}$.

Proof: the Shill bidding payments under the different strategies is available:

(1) Participants A's expected revenue is given by

$$\Pi_A(t,r) = r \cdot t[(q+1)kM - M - C_4 - C_5] + t(1-r)(-C_5) + r(1-t)\cdot(-M - C_4)$$
$$= [(q+1)kM + C_4]rt - C_5 t - (M - C_4)r \quad (1)$$

(2) Participants B's expected revenue is given

$$\Pi_B(t,r) = -rt[(q+1)M - M + C_1 + C_2 + C_3] + r(1-t)[M - C_1 - C_2 - C_3]$$
$$= -[(q+1)kM + C_3]rt + [M - C_1 - C_2 - C_3]r \quad (2)$$

Let the mixed strategy Nash equilibrium of regulatory game is (t^*, r^*), you need to solve the optimization problem:

$$\max_t \Pi_A(t, r^*) = \max_t\{[(q+1)kM + C_4]r^*t - C_5 t - (M - C_4)r^*\}$$
$$\max_r \Pi_B(t^*, r) = \max_s\{-[(q+1)kM + C_3]\cdot rt^* + [M - C_1 - C_2 - C_3]r\} \quad (3)$$

Obtained from the optimization of first-order conditions

$$\frac{\partial \pi_A}{\partial t} = [(q+1)kM + c_4]\cdot r^* - c_5 = 0,$$
$$r^* = \frac{c_5}{(q+1)kM} \quad (4)$$
$$\frac{\partial \pi_B}{\partial r} = -(q+1)kMt^* + M - c_1 - c_2 - c_3 = 0$$
$$t^* = \frac{M - c_1 - c_2 - c_3}{(q+1)kM} \quad (5)$$

Can get mixed strategy Nash equilibrium points:

$$\begin{cases} r^* = \dfrac{c_5}{(q+1)kM} \\ t^* = \dfrac{M - c_1 - c_2 - c_3}{(q+1)kM} \end{cases} \quad (6)$$

That mixed strategy Nash equilibrium is: $\left[(t^*,1-t^*),(r^*,1-r^*)\right]$

In this case the expected revenue in the Nash equilibrium point

$$\Pi_A(t^*, r^*) = [(kq + k)M][\frac{1}{(q+1)kM} - 1] \times C_5 \cdot \frac{M - C_1 - C_2 - C_3}{(q+1)kM} - (M + C_4)\cdot\frac{C_5}{(q+1)kM} = \frac{C_5(M + C_4)}{(q+1)kM} \quad (7)$$

$$\Pi_B(t^*, r^*) = -[(q+1)kM]\cdot\frac{C_5}{(q+1)M} \times \frac{M - C_1 - C_2 - C_3}{(q+1)kM} + [M - C_0 - C_2 - C_3]\cdot\frac{C_5}{(q+1)kM} = 0 \quad (8)$$

Nash equilibrium showed that: considered the inspection cost of regulatory institutions and detection accuracy at the same time, regulatory institutions accord the probability $(t^*,1-t^*)$ randomly select checks or not; seller accord the probability $(r^*,1-r^*)$ randomly select Shill Bidding or not. At this point both players are unable to improve their expect earnings by changing their separate randomly selected probability distribution.

The economic meaning of the above theorem is: under the condition that website regulatory institutions inspect and sellers Shill Bids need costs, furthermore the check accuracy rate is k, regulatory institutions check Shill Bidding according to the optimal probability $t^* = \frac{M - c_1 - c_2 - c_3}{(q+1)kM}$, seller Shill Bids according to the optimal probability $r^* = \frac{c_5}{(q+1)kM}$. At this time website regulator's expected revenue is $\frac{c_5(M + c_4)}{(q+1)kM}$, and the seller's expected revenue is 0. Thus, the website regulatory institutions intend to improve expects earnings, should not only reduce their inspection costs, but also use the fine coefficient of seller Shill Bidding and improve inspection accuracy.

Corollary 1: When website regulatory institutions inspection costs C5 = 0, the mixed strategy Nash equilibrium of game is $[(t^*,1-t^*),(r^*,1-r^*)]$ and

$$R^* = (r^*, 1 - r^*) = (0,1) \tag{9}$$

The economic meaning of this Corollary is: even the check cost is zero, regulators also need to check Shill Bidding according to the probability $t^* = M - c_1 - c_2 - c_3/(q+1)kM$, then the probability of seller Shill Bids strategy is r^*=0, Thus achieve the purpose of suppressing Shill bidding fraud.

Corollary 2: website regulatory institution and seller's best response function to each other possible selection strategy probability are as follows:

(1) The regulatory institution's reaction function to the seller is

$$t = \begin{cases} 1, & r < \frac{c_5}{(q+1)kM} \\ [0,1], & r = \frac{c_5}{(q+1)kM} \\ 0, & r > \frac{c_5}{(q+1)kM} \end{cases} \tag{10}$$

(2) The seller's reaction function to the regulatory institution is

$$r = \begin{cases} 1, & t < \frac{M - c_1 - c_2 - c_3}{(q+1)kM} \\ [0,1], & t = \frac{M - c_1 - c_2 - c_3}{(q+1)kM} \\ 0, & t > \frac{M - c_1 - c_2 - c_3}{(q+1)kM} \end{cases} \tag{11}$$

The regulatory institution response to the seller function (11) showed that:

When $r < \frac{c_5}{(q+1)kM}$, the optimal strategy of the regulatory institution is to check;

When $r = \frac{c_5}{(q+1)kM}$, regulatory institution randomly select check or not;

When $r > \frac{c_5}{(q+1)kM}$, the optimal strategy of the regulatory institution is not to check;

The seller response to the regulatory institution function (12) showed that:

When $t < \frac{M - c_1 - c_2 - c_3}{(q+1)kM}$, the optimal strategy of the seller is fraudulent behavior which equal to Shill bids;

When $t = \frac{M - c_1 - c_2 - c_3}{(q+1)kM}$, the optimal strategy of the seller is randomly select Shill bids or not.

When $t > \frac{M - c_1 - c_2 - c_3}{(q+1)kM}$, the optimal strategy of the seller is honest bidding which equal to don't Shill bids.

Corollary2 shows: websites posting fee, inspection costs, commissions, fines coefficient, illegal income of Shill bidding and so on have an impact on balanced type of participants' behavior. So regulators can reduce costs by changing the parameter values of influenced factors to curb online fraud and achieve win-win situation.

Theorem2 the optimal fines coefficient that regulators treat to seller shill bidding is $q_2^* = \frac{M - c_1 - c_2 - c_3}{t^* kM} - 1$

Proof: For regulators, the use of optimal inspection probability $t^* = \frac{M - c_1 - c_2 - c_3}{(q+1)kM}$, make the seller's expected revenue is zero, then the seller's expected revenue is zero whether shill bids or not, rational person would not choose to shill bids, thus achieve the purpose of suppressing shill bidding fraud effectively. The optimal fine coefficient is $q_2^* = \frac{M - c_1 - c_2 - c_3}{t^* kM} - 1$ which can be obtained by the optimal inspection probability equation.

QED.

Theorem 2 shows that: To curb online auction fraud, for shill bidding seller the fine penalty coefficient has nothing to do with the seller's bidding probability, but related to regulators' optimal check probability, the check accuracy, and the corresponding cost of shill bidding.

IV. ANALYSIS AND DISCUSSION OF THE MODEL

A. Analysis of affecting factors about seller's Decision Behavior

Formula (5) obtain the optimal solution r^* which refer seller shill bids probability, indicate that the optimal bid probability is affected by punishment parameter and the check accuracy k and inspection costs c_5 and other factors in the whole game model.

(1) $\frac{\partial r^*}{\partial q} < 0$ express seller shill bids be found ,the higher of the penalty coefficient is, the lower of the probability seller shill bids. Penalty coefficient increases, the punishment increased after be found shill bids, seller's illegal income reduce. The initiation of selecting shill bidding reduced, the probability reduce.

(2) $\frac{\partial r^*}{\partial k} < 0$ express the higher check accuracy of the regulatory, the smaller probability seller shill bids.

(3) $\frac{\partial r^*}{\partial c_5} > 0$ express with the increase costs about regulators to check on shill bidding, the probability of the seller shill bids have being corresponding increased.

B. Analysis of affecting factors about Auction site supervisor Decision Behavior

Formula (7) obtain the optimal solution t^* which refersite regulator's check probability, indicate that the optimal check probability is affected by publish feeC1, commissions C2, the seller's integrity lossC3 and the penalty coefficient q and the check accuracy etc.

(1) $\frac{\partial t^*}{\partial c_1} < 0$ express the higher publication fee charged by the auction site , the lower probability of the optimal inspection of regulator . This is because the higher publication fee of seller , the higher the cost of shill bidding , is not conducive for sellers to obtain illegal profits, and therefore may corresponding reduce shill bidding, the optimal inspection probability for regulators also reduced accordingly.

(2) $\frac{\partial t^*}{\partial c_3} < 0$ express that the higher integrity loss for seller, the lower check probability for regulator. When the seller integrity loss increased, the loss increases after be found shill bidding, thus the check probability for regulator reduce.

(3) $\frac{\partial t^*}{\partial q} < 0$ express that the greater punishment coefficient for seller, the check probability reduced accordingly. Punishment multiple increases, namely, strengthen penalties of seller's shill bidding, regulator can reduce the check probability accordingly.

(4) $\frac{\partial t^*}{\partial k} < 0$ express that the higher check accuracy of the regulator, the smaller probability t seller shill bidding.

V. CONCLUSION

The basic theory and methods of game theory was used in this paper, according to the identification of shill bidding, in consideration of the regulatory costs and shill bidding costs, and the check accuracy, establish a regulatory shill bidding game model, and do a quantitative analysis, we have done a more detailed analysis of the website regulatory decision-making behavior and sellers , which obtained its mixed strategy Nash equilibrium and a fine corresponding coefficient foe shill bidding and also obtained factors affect both the decision-making behavior. The model is applied to the actual online auction which has certain guiding significance for the policy decisions of suppressing fraud:

(1) the optimal probability analytic expression of shill bidding and the optimal check probability analytic expression that website regulatory check whether seller shill bids pointed out a number of factors that influence the regulatory authorities to check shill bidding, so provide a theoretical basis for online auction regulatory depending on the operation of the management department .

(2) the regulatory shill bidding model established in this paper about how to effectively curb other similar fraud problem which share the decision interdependence has a certain significance, thus we have expanding the scope of application of the non-cooperative game.

REFERENCES

[1] Lv Xuhua, Cheng Lili. Study on supervising game of shill bidding in auctions [J], Journal of Wuhan University of Science and Technology, 2013, 06

[2] Kosmopoulou G, Dakshina G, Slive D. The effect of shill bidding upon price: experimental evidence [J]. Industrial Journal Industrial Organization, 2007, 25(2):291-313.

[3] Vickrey W. Counter speculation, auctions and competitive sealed tenders [J]. Journal of Finance, 1961, 16(1): 235-256.

[4] Ba S, Pavou P. A evidence of the effect of trust building technology in electronic markets [J]. Price Premiums and Buyer Behavior 2002, (3):289-301.

[5] Ba S, Whinston A B, Zhang H. Building trust in online auction markets through an economic incentive Mechanism [J]. American Business Law Journal, 2003, 3:121-155.

[6] Shah H S, Joshi N R, Wurman P R. Mining for bidding st rategies on eBay [C] //SIGKDD'2002 Workshop on Web Mining for Usage Patterns and User Profiles. USA: Springer, 2002.

[7] Zhang Hua Luo, Lu Liu. The "shill" bidding in the online auction bid research (Chinese) [J], Journal of management engineering, 2006, 20(1): 7-10.

[8] Li Xuefeng, Liu Lu, Zhang Zhao. Shill bidding in online auctions recognition model [J], Journal of Systems Management, 2007, 16 (4):360-364.

[9] Recognition bid [J], Business Studies, 2009, 04:170 -171. Jishu Xian, Dong Wei. Online auction marketplace care.

[10]Kauffman R J, Wood C A. The effects of shilling on final bid prices in online auctions [J]. Electronic Commerce Research and Applications, 2005, 4(1): 21-34.

[11]Weiying Zhang, Game theory and information economics (Chinese) [M]. Shanghai people's publishing house, 1996: 167-185

Pricing Strategy for Remanufactured Products Based on Consumer Preference

Xue-mei Du*, Zheng-pin Xie, Wen Hu
School of Economics & Management, Tongji University, Shanghai, 200092, China
(*xuemei.du@163.com, xzpin90@163.com, waitingforfox@163.com)

Abstract - **With continuous deepening of China's sustainable development strategy, the recycling of waste materials gradually becomes an effective way to protect the environment, promote resource regeneration and save costs. It is vital for the firms to decide their pricing strategy of remanufactured products, which impacts the sales performance of new products and remanufactured products. This paper considers remanufactured product heterogeneity and consumer preference by introducing the advertising effectiveness and consumer quality expectations coefficient and builds a Static Game Model and a Two-stage Dynamic Model. Then numerical simulation analysis on coefficient of consumer quality expectations is conducted for further study. Research findings show that the consumer quality expectations coefficient has various impacts on manufacturer's price decisions and profits.**

Keywords - **Advertising quality, consumer preference, price strategy, remanufactured products**

I. INTRODUCTION

At present, in developed countries, recycling of the used materials has become convention in manufacturing industry and retail industry. UK remanufacturer Recoup and Eprop published a report that the EU's energy recovery rate is 34.5% in 2012 which is 33.2% in 2011; However, the recovery rate of waste products in China remained low.

In this context, an important problem for modern manufacturers to address is how to determine the price of remanufactured products. Researches in the field of pricing strategy making for remanufactured products emerged only several years age. Robotis focused on the problem of the recycled product with different quality and introduced the integral equation to apply quantitative analysis [1], Gu Qiaolun studied the price decisions of recycled products based on the reverse supply chain between the manufacturer and the retailer by game theory and obtained Two non-cooperative game equilibrium and a cooperative game equilibrium [2] .

Gong Yande considered the strategy of pricing and coordination of supply chain under the joint action of outsourcing logistics and waste materials recycling [3]. Chen Qiushuang studied the pricing decisions of recycled products by game theory with the limit of minimal recycled quantity [4]. Li Xiang researched the recovery pricing decisions of remanufacturing reverse logistics system under the random quantity and demand [5]. The sales price strategy research on remanufactured products also attracts lots of academic attention. In general, research in this area can be divided into two parts: one type of the studies mainly focused on the various factors that exert impact on the sales price of remanufactured products departed from the perspective of manufacturer maximize profits while took learning curve, cost saving and other internal controllable factors into consideration. Those researches assumed that the demands obey a certain distribution. For example, Ferrer and Swaminathan analyzed the profit activities, from the perspective of monopoly environment, by an independent operator or OEM manufacturer responsible for remanufacturing without considering the differentiation of new product and remanufactured products [6]. Ferrer made a further study on the differentiation and analyzed the monopoly environment in two-period, multi-period (three, four and five) and infinite planning horizons, and characterize the optimal remanufacturing and pricing strategy for the firm [7]. Bao Xiaoying gained maximum expected profit, the pricing model of remanufactured products was proposed from the perspective of remanufacturers. The result showed the optical price decreased with disposal cost of worn-out products [8]. Ding Xuefeng took the capacity factors into account and analyzed the capacity's impact on the optimal prices and profits of remanufacturing. The other type of researches considered the customer demands and other market factors when enterprises determine their price strategy of remanufactured products [9]. Xu Feng established a t w o - period model of a manufacturer who makes and sells a new product and a remanufactured products simultaneously [10]. Saadany suggested that the flow of returned items is variable and is controlled by two decision variables which are the purchasing price for returned items corresponding to an acceptance quality level and presented deterministic mathematical models for multiple remanufacturing and production cycles [11]. Cao Jun mainly studied the remanufactured products pricing discrimination strategy in closed-loop supply chain under the consideration of the impact of advertising on consumer expectations of product quality [12]. Dan Bin and Ding Xuefeng studied the static optimal pricing and two-period dynamic pricing of new products and remanufactured products for a monopolist and they recommend that company can arouse environmental awareness of consumers, as well as improve the return rate to increase the profits of company [13].

Wu Cheng-Han formulated a two-period supply chain model consisted of two chain members, an OEM and a remanufacturer in order to investigate the product design decision of the OEM and both two chain members' competitive pricing strategies [14].

This context will research the remanufactured products' pricing strategy which is effected by consumer preference under the consideration of products remanufactured product heterogeneity and consumer heterogeneity.

E. Qi et al. (eds.), *Proceedings of the 5th International Asia Conference on Industrial Engineering and Management Innovation (IEMI2014)*, Proceedings of the International Asia Conference on Industrial Engineering and Management Innovation 1, DOI 10.2991/978-94-6239-100-0_24, © Atlantis Press and the authors 2015

II. MODEL FOUNDATION

A. Assumptions

This paper assumes: (1) the market includes new products and remanufactured products whose quality and price are different. In addition, customers are free to make choices within the market; (2) there is only one manufacturer in the market and no upper limit to the manufacturer's production capacity; (3) the potential market demand (Q) in each period is a constant and follows a uniform distribution in the interval $[0, Q]$, we use a linear demand function: $q_i = Q - p_i$ to describe it; (4) the new product can be remanufactured only once.

B. Consumer Behavior Analysis

In the aspect of consumer behavior, we mainly consider the impact of advertisement on consumers' quality expectation of remanufactured products. The variables and parameters are listed as follows:

- I Advertised quality of the manufactured product
- R Real quality of the manufactured product on average.
- Q The expectation of consumer i on remanufactured product quality.
- a_i The quality expectations coefficient of Consumer i, which is determined by the advertising, represents the difference of consumer expectation on the products quality. The subscript is i = 1, 2, ... ($0 \le a_i \le 1$).
- V_i The value evaluation of the remanufactured products by consumer i.
- b_i The value evaluation of the unit product quality by consumer i ($b_i > 0$).
- b_0 Intrinsic utility of the product.
- b_1 Price coefficient ($b_1 > 0$).
- U_i The utility of consumer i.

According to Boulding William [15], Consumer i' s expectation of remanufacturing quality (Q_i) is determined by both advertised product quality and real product quality. The quality expectation function is: $Q_i = a_i I + (1 - a_i) R$. Among them, advertised qualities of the manufactured product (I) are the same to every consumer, as well as real qualities of the manufactured product on average (R). The value evaluation function is: $V_i = b_0 + b_i Q_i$. The consumer's utility function is: $U_i = V_i - b_1 p = b_0 + b_i Q_i - b_1 p$. Here we assume $b_0 = 0, b_1 = 1$, therefore the consumer's utility function is:

$$U_i = b_i Q_i - p = b_i [a_i I + (1 - a_i) \mathrm{R}] - \mathrm{p} \quad (1)$$

C. Manufacturer Behavior Analysis

The manufacturer divides the market into new manufactured product market and remanufactured product market with correspondingly demand q_n and q_r. During the manufacturer's quality and price decision-making process, we take the advertising impact into consideration. The variables and parameters are listed as follows:

This paper views the decision-making process between manufacturer and consumers as Static Game Model and Two-stage Dynamic Model respectively. Manufacturer decides the price and quality of the product both in the reality and advertisement. Consumers decide whether to buy or not according to the quality expectation and the price. The difference between advertising quality and real quality is decided by the manufacturer.

- $q_{i,j}$ Demand of product j in period of i, $i = 1, 2 \ldots.$, j=n (new product)or r (remanufactured product). In period 1, there is only new product, so $q_1 = q_{1n}$.
- $p_{i,j}$ Price of product j in period i, $i = 1, 2 \ldots.$, j=n (new product) or r (remanufactured product) and $p_1 = p_{1n}$
- C_n New product cost (C_n) and remanufactured products cost (C_r).
- β Discount factor per period (β≤1). The Discount rate increase as β decreases. E.g, if the discount rate = 5%, then β=0.95.
- γ Core collection yield γ≤1, defined as the fraction of new products made in period i that is available for remanufacturing in period i +1. Hence, $q_{ir} \le \gamma q_{i-1n}$.

D. Static Game Model

Firstly, the demand of new products and remanufactured products should be identified. We determine the demands of two markets on the basis of the constraints of the consumer behavior. We assume that: (1) the adverting qualities of new products and remanufactured products are uniform; (2) the new products' advertising quality and real quality are equal to I; (3) We define the population as consumer quality expectation coefficient which is impacted by the remanufactured products' quality in advertisement: $a_n = \sum_{i=1}^{k} a_{i,n} / k$, $a_r = \sum_{i=1}^{h} a_{i,r} / \mathrm{h}$. (4) $b_{i,n} \ne b_{j,n}$, $b_{i,r} \ne b_{j,r}$, $i \ne j$. So the consumers' value evaluation to the remanufactured products is: $\begin{cases} V_{i,n} = b_{i,n}[a_n I + (1 - a_n) R] \\ V_{i,r} = b_{i,r}[a_r I + (1 - a_r) R] \end{cases}$.

(5) consumers' utility(U_i) in the two market should be greater than zero, based on formula(1):

$$\begin{cases} U_{i,n} = b_{i,n} I - p_n \ge 0 \\ U_{i,r} = b_{i,r}[a_r I + (1 - a_r) R] - p_r \ge 0 \end{cases}$$

(6) in order to distinguish the new manufacturing market from remanufacturing market, the new manufacturing market should meet the incentive compatibility constraints:

$$\begin{cases} U_{i,n} = b_{i,n} I - p_n \ge b_{i,n}[a_n I + (1 - a_n) \mathrm{R}] - \mathrm{p}_r \\ U_{i,r} = b_{i,r}[a_r I + (1 - a_r) R] - p_r \ge b_{i,r} I - p_n \end{cases}$$

(7) the demands of the two market are:

$$q_n = q_n = Q - \frac{I(\mathrm{p}_n - \mathrm{p}_r)}{(1 - a_n)(1 - \mathrm{R})} \quad (2)$$

$$q_r = \frac{I(p_n - p_r)}{(1-a_n)(1-R)} - p_r \quad (3)$$

(8) the manufacturer's profit function in a period (excluding the first period) is:

$$\max_{p_n, p_r} ((p_n - C_n)q_n + (p_r - C_r)q_r)$$

$$s.t \begin{cases} q_n \geq 0 \\ q_r \geq 0 \end{cases}$$

(9) we can obtain the optional price and quantity demands of the products under profit maximization, according to the equations (2) and (3). For convenience, we define $\frac{I-R}{I} = 1 - \frac{R}{I} = d$, d represents the diversity of advertised products quality and real products quality, the results are as follows:

Interval 1. When $d \leq \frac{C_n - C_r}{Q(1-a_n)}$, manufacturer only produces remanufactured products:

$$q_n = 0;\ q_r = \frac{Q - C_r}{2},\ p_r = \frac{Q + C_r}{2}$$

Interval 2. When $\frac{C_n - C_r}{Q(1-a_n)} < d < \frac{C_n - C_r}{C_r(1-a_n)}$, manufacturer produces both new products and remanufactured products:

$$q_n = \frac{Q}{2} - \frac{C_n - C_r}{2(1-a_n)d},\ p_n = \frac{Qd(1-a_n + C_n + Q)}{2};$$

$$q_r = \frac{C_n - C_r}{2(1-a_n)d} - \frac{C_r}{2},\ p_r = \frac{Q + C_r}{2}$$

Interval 3.When $d \geq \frac{C_n - C_r}{C_r(1-a_n)}$, manufacturer only produces new products:

$$q_n = \frac{Q}{2} - \frac{C_n}{2 + 2(1-a_n)d},$$

$$p_n = \frac{Qd(1-a_n) + C_n + Q}{2};\ q_r = 0$$

E. Two-stage Dynamic Model

In this model we divide the manufacturer's decision-making process into two periods. In period 1, manufacturer only produces new products, from the period 2, manufacturer produces both new products and remanufactured products. Here we consider the time value of money (β) and the yield constrain of remanufacturing (γ).

(1) The profit function based on the manufacture profit maximization in the two periods is:

$$\max_{p_1, p_{2n}, p_{2r}} (p_1 - C_n)q_1 + \beta((p_{2n} - C_n)q_{2n} + (p_{2r} - C_r)q_{2r})$$

$$s.t. \begin{cases} q_1 = Q - p_1 \\ \gamma q_1 \geq q_{2r} \end{cases}$$

(2) In the profit function, the variables q_{2n} and q_{2r} satisfy formula (1) and (2). So the manufacturer decision interval can be demarcated as follows:

Interval 1. When $d \leq \frac{(Q - C_n)(1-\gamma) + \beta\gamma^2(C_n - C_r)}{(1-a_n)(Q + 2Q\beta\gamma^2 - \beta\gamma^2)}$, in period 2, manufacturer only produces the remanufactured products while the remanufacturing process uses all cores collected every period:

$$q_1 = \frac{q_{2r}}{\gamma};\ p_1 = Q = \frac{q_{2r}}{\gamma};\ q_{2n} = 0;$$

$$q_{2r} = Q - \frac{\beta\gamma^2(Q + C_r - Q) + 2Q}{2\beta\gamma^2 + 2},$$

$$p_{2r} = \frac{\beta\gamma^2(Q + C_r) + \gamma(C_r - Q) + 2Q}{2\beta\gamma^2 + 2}$$

Interval 2. When $\frac{(Q - C_n)(1-\gamma) + \beta\gamma^2(C_n - C_r)}{(1-a_n)(Q + 2Q\beta\gamma^2 - \beta\gamma^2)} < d < \frac{C_n - C_r}{(1-a_n)(\gamma Q - \gamma C_n + C_r)}$, in period 2, manufacturer produces both new products and remanufactured products, the remanufacturing process uses all cores collected every period:

$$q_1 = \frac{q_{2r}}{\gamma},\ p_1 = Q - \frac{q_{2r}}{\gamma};$$

$$q_{2n} = Q - \frac{p_{2n} - p_{2r}}{(1-a_n d)},\ p_{2n} = \frac{Qd(1-a_n) + C_n + Q}{2}$$

$$q_{2r} = \frac{p_{2n} - v_{2r}}{(1-a_n)d} - p_{2r},$$

$$p_{2r} = \frac{(2 + 2\beta\gamma^2(1-a_n)d)p_{2n} - \gamma(1-a_n)d[Q - C_n + \beta\gamma(C_n - C_r) + \beta\gamma(1-a_n)d]}{2 + 2(1-a_n)d + 2\beta\gamma^2(1-a_n)d}$$

Interval 3. When $\frac{C_n - C_r}{(1-a_n)(\gamma Q - \gamma C_n + C_r)} < d < \frac{C_n - C_r}{C_r(1-a_n)}$, manufacturer produces both new products and remanufactured products. Meanwhile, in this period, the remanufacturing process uses just a part of the cores in period 1:

$$q_1 = \frac{Q - C_n}{2},\ p_1 = \frac{Q + C_n}{2};$$

$$q_{2n} = \frac{Q}{2} - \frac{C_n - C_r}{2(1-a_n)d},\ p_{2n} = \frac{Qd(1-a_n) + C_n + Q}{2};$$

$$q_{2r} = \frac{C_n - C_r}{2(1-a_n)d} - \frac{C_r}{2},\ p_{2r} = \frac{Q + C_r}{2}.$$

Interval 4. When $d \geq \frac{C_n - C_r}{C_r(1-a_n)}$, in period 2, manufacturer only produces remanufactured products:

$$q_1 = \frac{Q - C_n}{2},\ p_1 = \frac{Q + C_n}{2},\ q_{2r} = 0;$$

$$q_{2n} = \frac{Q}{2} - \frac{C_n}{2 + 2(1-a_n)d},\ p_{2n} = \frac{Qd(1-a_n) + C_n + Q}{2};$$

Form the results we can see that the interval 3 and 4 in Two-stage Dynamic Model are uniform to the interval 2 and 3 in Static Game Model respectively. Because the remanufactured products' yield in second period is constrained by the new products in first period, the interval 2 in Two-stage Dynamic Model is relatively different to the Static Game Model.

III. NUMERICAL SIMULATION ANALYSIS

In this part, further study about the consumer expectation coefficient's impact on manufacturer's price decision based on the two models is discussed. The analysis above shows that the result of Two-stage Dynamic Model is similar to Static Game Model. Therefore, we divide our study into two parts: firstly, analyze the Static Game Model; secondly, analyze the interval 2 of the Two-stage Dynamic Model.

We set a_n =0.6 and a_n =0.2 for comparison, and define Q=1000, C_n =600, $C_r = 500$. Fig.1 shows that under different a_n , there emerged two trends: (1) the interval for manufacturer to produce remanufactured products decreases as a_n decreases. In other words, d decreases in the first and second interval. If the difference between advertising quality and real quality of remanufactured products is too large, consumers will not choose to buy. When a_n decreases, consumers have a higher requirement to real product quality of remanufacturing; (2) the new products' price and yield increases as a_n decreases. Fig.1 shows that, when a_n =0.2; p_n, q_n are bigger than a_n =0.6 at the same level of d. It reflects that the new products' demand and price become greater as consumers tend to be more rational.

Fig.2 shows the manufacturer's short-term profits based on the manufacturer's optimal yield and price under different a_n . It shows that, when a_n =0.2 and d is fixed, the profit is higher.

Afterwards, further study on the interval 2 of the Two- stage Dynamic Model is carried out. Based on the deduced function from the interval 2 of the Two-stage Dynamic Model, we define d =0.2 and the other variables are constant (Q=1000, C_n =600, $C_r = 500$), a_n is an in dependent variable, the results are shown in Fig.3.

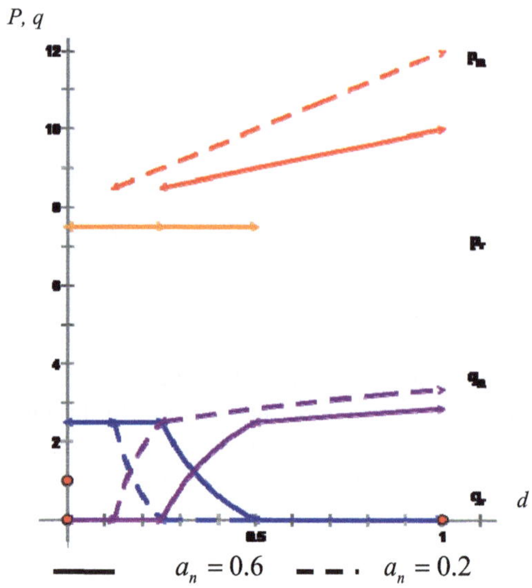

Fig.1. Manufacturer's decision under different a_n based on the Static Game Model

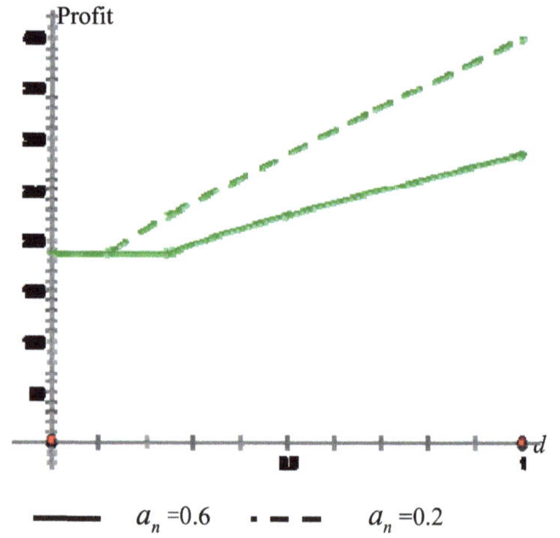

Fig.2. Manufacturer's profit under different a_n based on the Static Game Model

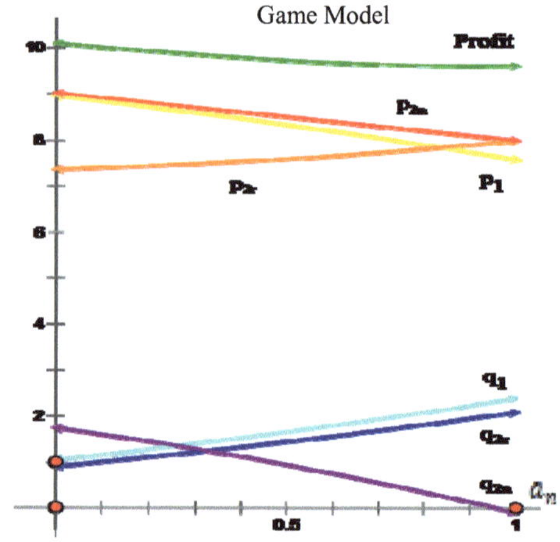

Fig.3. When d=0.2, a_n 's impact on the several variables in the interval 2 of Two-stage Dynamic Model

Fig.3 shows that all the functions in the value interval of a_n are monotone functions. Based on the analysis of the functions' trend, main conclusions are drawn as follows: (1) if d is a constant, new products' price (p_{2n}) and yield (q_{2n}) increase respectively as a_n decreases. New products' demand and price become greater as consumers are more rational; (2) if d is a constant, remanufactured products' price (p_{2r}) and quantity demand (q_{2r}) decrease respectively.

IV. ANALYSIS AND CONCLUSION

Considering the analysis results of static Game model and Two-Stage Dynamic Model, the coefficient of consumer quality expectations (a_n) can affect manufacturer's decision-making through the following ways:

(1) Coefficient of consumer quality expectations (a_n) has influence on the production decision of remanufactured products. It is the primary problem for manufacturers to consider whether to produce remanufactured products or not when making production,

management decisions and arranging production plan. Through the analysis above, we find that manufacturer's selective areas of producing remanufactured products decreases as a_n decreases. This shows that consumers have a higher requirement to the real product quality of remanufacturing as the consumers tend to be rational. When deciding whether to produce remanufactured products, manufacturers should conduct an investigation about consumer preferences. If consumers appear more willing to believe the advertising quality of remanufactured products, even there's certain difference between the actual quality and advertising quality, market demand will also be generated, which plays a positive role in promoting manufacturers to produce remanufactured products.

(2) Coefficient of consumer quality expectations (a_n) has influence on the pricing strategy of new products. According to the analysis results of Static Game Model and Two-Stage Model, for any interval of any stage, the optimal price of new products will increase with the decrease of a_n, which suggests that manufacturer should raise the price of new products in order to obtain higher profits as the consumers tend to be rational.

(3) Coefficient of consumer quality expectations (a_n) has influence on the pricing strategy of remanufactured products. Since yield constraints of remanufactured production are taken into consideration in the two-stage model, a_n 's influence on pricing strategy of remanufactured products becomes relatively complicated. Without yield constraints, the optimal price of remanufactured products is a constant and not affected by a_n; with yield constraints, the optimal price decreases as a_n decreases. Therefore manufacturers should reduce the price of remanufactured products in order to achieve greater profits when consumers tend to be rational.

It is obvious that the coefficient of consumer quality expectations (a_n) do have certain influences on all aspects of remanufactures' decision-making process. It plays an important role in the pricing strategy of remanufactured products, and provides valuable references for manufacturers to price subsequent remanufactured products.

REFERENCES

[1] Andreas Robotis, Shantanu Bhattacharya, Luk N. Van Wassenhove. The effect of remanufacturing on procurement decisions for resellers in secondary markets [J]. *European Journal of Operational Research*, 2005, 163(3):688-705.

[2] Gu Qiaolun, Gao Tegang, Shi lianshuang. Price decision analysis for reverse supply chain based on game theory [J]. *Systems Engineering – Theory & Practice*, 2005, 25(3): 20-25

[3] .Gong Yande, Li Bangyi, Liu Tao. Pricing and coordination strategy of TPL- CLS under double effects of logistics qutsourcing and products recovery [J]. *Chinese Journal of Management Science*, 2008, 16(6): 46-53().

[4] Chen Qiushuang,Gu Qiaolun,Sun Guohua. Pricing decisions of reverse supply chains with minimal recycled quantity Limit [J]. *Mathematics in Practice and Theory*, 2009, 39(3): 35-44.

[5] Li Xiang, Li Yongjian, Cai Xiaoqiang. Collection pricing decision in a remanufacturing system considering random yield and random demand [J]. *Journal of Systems Science and Information,* 2009, 29(8): 19-26.

[6] Geraldo Ferrer, Jayashankar M. Swaminathan. Managing new and remanufactured product [J]. *Management Science*, 2006, 56(1): 15-26.

[7] Geraldo Ferrer, Jayashankar M. Swaminathan. Managing new and differentiated remanufactured product [J]. *European Journal of Operational Research*, 2010, 203: 370-379.

[8] Bao Xiaoying, Yang Chenghu, Pu Yun. Optimal pricing policy of the remanufactured products [J]. *Computer Integrated Manufacturing Systems*, 2008, 14(12): 2436-2440.

[9] Ding Xuefeng, Xie Wuzhou, Guo Chenghen. Discussions on urban grass-roots management system reform [J]. *Journal of Chongqing University of Technology (Social Science)*, 2010, 24(5):44-47.

[10] Xu Feng,Sheng Shaohan,Chen guohua. The remanufactured products pricing strategy in a heterogeneous market [J]. Chinese Journal of Management Science, 2008, 16(6): 130-136.

[11] Ahmed M.A. El Saadany, Mohamad Y. Jaber. A production/remanufacturing inventory model with price and quality dependant return rate [J]. *Computers & Industrial Engineering*, 2010(58): 352-362.

[12] Cao Jun, Xiong Zhongkai, Tao Xuefei. The remanufactured products prices discrimination strategy in closed-loop supply chain [J]. Science & Technology Progress and Policy, 2009, 26(9): 30-33

[13] Dan Bin, Ding Xuefeng. Optimal pricing for the remanufactured products and analyzing the effect of cannibalization and market growth [J]. Systems Engineering - Theory & Practice, 2010, 30(8): 1371-1379

[14] Cheng-Han Wu. OEM product design in a price competition with remanufactured products[J].*Omega*, 2013, 41: 287-298

[15] Boulding William, Ajay Kalra, Richard Staelin. A dynamic process model of service quality: From expectations to behavioral intentions [J]. *Marketing Research*, 1993, 32(2):7-27.

Different Performance on Auditory Discrimination between Sighted and Blind Individuals

Min-Sheng Chen*, Jian-Shan Liu, Wei-Ru Chen
Department of Industrial Engineering & Management, Taiwan Yunlin University of Science and Technology, Douliou, Taiwan
(chens@yuntech.edu.tw)

Abstract - **The purpose of this study is to examine differences in auditory discrimination ability between sighted and blind individuals under different surrounding sound and noise conditions. In this experiment, an auditory-words-perception task was manipulated to investigate performance of visually impaired and sighted individuals on the discrimination of aurally presented words within different conditions. The results showed that sighted individuals have a 0.8 of accuracy on their performance with a 25% increment necessary in sound level. On the other hand, blind individuals have a 0.7 of accuracy with a 15% increment in sound level. These results also can provide industrial designer with guidance for interface designs for products intended for the visually impaired.**

Keywords - **Auditory discrimination, environmental sound, noise, visually impaired**

I. INTRODUCTION

The majority of the literature on hearing ability has indicated that blind individuals are more sensitive to sound compared to the sighted individuals [1, 2, 3]. Additionally, other research also has showed that blind individuals also have better performance on spatial sound source discrimination tasks compared to the sighted individuals [4, 5, 6]. Fewer studies took background noise into account to investigate hearing discrimination. To understand the difference of hearing perception ability between blind and sighted individuals is the focus of this study.

To explore the performance of word discrimination, different background noises were manipulated [7, 8, 9, 10, 11]. Furthermore, Patterson (1982) [12] suggested that the volume of signal at least should be over 15dB of masking sound, but can't over the 25dB, would be detected by participants in the background noise. If the volume of signal was over the 25dB, the signal would distract the participants.

Wan et al. (2010) [3] expected to understand the difference of hearing ability between those with different statuses of vision. In their experiment of tone discrimination, participants would hear two pure sounds in each trial. The first sound which had four levels (500, 750, 1000 and 1500Hz) was the baseline to the second sound. After thirty minutes of the first sound, it was followed by the second sound. Second sound also had four levels (2%, 1%, 0.5% and 0.25%) which were over the baseline sound. Participants were asked to answer whether the second sound was a high tone or low tone compared to the baseline sound. The results showed that blind individuals had better performance compared to the sighted individuals overall. Especially, the congenitally blind were more sensitive to tone discrimination. But there was no difference between the late-blind and sighted individuals, indicating that their ability of tone discrimination were similar. Bonnel et al. (2003) [13] also manipulated the different increasing levels of tone (3%, 2% and1%). They found that the performance would be better with each increment change. This means that the diversity of sounds should over a threshold for people to be able to clearly discriminate sounds clearly.

Rokem & Ahissar (2009) [14] used the hearing stimuli of words which appeared continuously in a quiet environment or 60dB background noise. The number of the words was from one to five. Participants answered the words which they heard. The results showed that the performance of memory was poorer with the increase in the number of words. Furthermore, the threshold of the blind individuals was better than the sighted individuals in both quiet and background noise conditions.

To investigate the perception ability and attention of semantic sound in the blind group, the hearing stimuli included six consonant-vowel sounds (ba, da, ga, pa, ta, and ka) [15]. However, the consonant-vowel sounds would be broadcast on both the right and left sides simultaneously. Participants had to achieve three tasks. First was non-forced (NF) condition where the participants wouldn't be told in advance to notice which side of speaker. Next was the forced-right (FR) condition where participants would be informed to pay attention to the right side in advance. Finally, forced-left (FL) condition where participants would be informed to pay attention to the left side in advance. Participants should repeat the words which they heard. The results indicated that participants performed better on forced-right condition than on the other conditions. Additionally, blind individuals had higher accuracy of perceiving the sound than did sighted individuals.

Because of the developmental and learning backgrounds were different, the cognitive ability of blind individuals were different individually. Röder & Rösler (2003) [16] proposed that the factor which affected the performance of hearing memory was age and not the duration of blindness. But most of the literature has shown that the ability of sound discrimination of congenitally blind individuals was better than for those late-blind individuals. Similarly, congenitally blind individuals were more sensitive to perceive the sound compared to the late-blind individuals.

How to assist blind individuals to better comprehend auditory information clearly is the goal of the present study. To achieve this, our study explored which degree

E. Qi et al. (eds.), *Proceedings of the 5th International Asia Conference on Industrial Engineering and Management Innovation (IEMI2014)*, Proceedings of the International Asia Conference on Industrial Engineering and Management Innovation 1, DOI 10.2991/978-94-6239-100-0_25, © Atlantis Press and the authors 2015

of volume increment could best be discriminated by sighted and blind individuals for the sound of words at different volumes of background noise. Further, we attempted to understand whether the different volume of background noise would affect the ability of hearing discrimination or not.

II. METHODOLOGY

To explore the differences in auditory-word perception ability between sighted and visually impaired individuals under different surrounding sound and noise conditions, different volumes of diverse background noises were manipulated in the experiment. In the experiments, participants repeated the word they heard in different conditions.

A. Participants

Twenty-four individuals participated in the experiment. They were divided into two groups: sighted (n = 12; mean age =24.92 years, SD = 4.17; 4 women), and blind (n = 12; mean age = 24.83 years, SD = 6.78; 3 women). The twelve blind individuals included six late-blind individuals (mean age 24.17 years, SD=7.46; mean blind: 12.5 years) and six congenitally blind individuals (mean age 25.50 years, SD = 6.65; mean blind: 25.17 years). Congenitally blind individuals included those who had lost their visual perception before two years old and they did not have any visual memory. All participants had normal hearing and at least twelve years of education. None of the participants had perceptual disabilities or music training.

B. Materials

Fixity target stimuli of the experiment were selected from the "Speech Discrimination Form" and all the target stimuli were presented one second intervals from notebook computer placed 30 centimeters in front of participants. Additionally, there were four speakers set far from the participants: to the left, right, front and rear. Each speaker would broadcast the background noise which is street noise at the following volume levels: 60dB, 70dB, and 80dB.

C. Experimental Design

The present study was a 2×3×3 factorial design experiment.

(1) Independent variables

The between factor was vision status. Participants were divided into two groups, one was the blind group and the other was the sighted group. The within factors were the volume level of background noise and increasing level of target stimuli volume. In other words, there were three different background noises and their volume levels were 60dB, 70dB and 80dB. However, according to Bonnel et al. (2003) [13] and Wan et al. (2010) [3] research, there should be three target stimuli volume levels: 15%, 20% and 25%. It means the volume of the target stimuli would increase with the increasing of background noise. For example, in the 15% level the volume level of the background noise was at the 60 dB condition; then, the volume of the target stimuli was 69dB (60(1+15%) = 69 dB).

(2) Dependent variables

The dependent variable was the grade participants got from each trial. If participant repeated the target word accurately, they would get one point for each trial. They completed five trials of each experimental condition (total grade was 45 points).

D. Procedure

The experiment was started by explaining to the participants the purpose and procedure of the experiment and seeking their written consent. The participants were given five practice trials and the experimenter would confirm they could repeat the target accurately before the formal experiment. There were four speakers set at distance of three meters from the participants to the right, left, front, and rear. These four speakers broadcast the background noise continuously (Fig.1). The target stimuli would broadcast randomly from the notebook in front of the participants.

The volume of the target stimuli which depend on the volume of background noise would be increased by three levels 15%, 20% and 25%. Each target stimuli was presented one time for one second. Participants then needed to repeat the target word which they heard from the notebook as soon as possible after the target stimuli. There was no limitation of reaction time. There were five words for each condition. They would take two minutes break when the background noise was changed. It should take about thirty minutes for them to finish the experiment.

To reduce the learning and fatigue effect, the present study used counterbalanced design to allot the participants. In background noise factor, four participants were in the 60 dB condition, four participants are in the 70dB condition and four participants in 80 dB condition for the first part of the experiment. This allotment pattern continued with each increase in volume level.

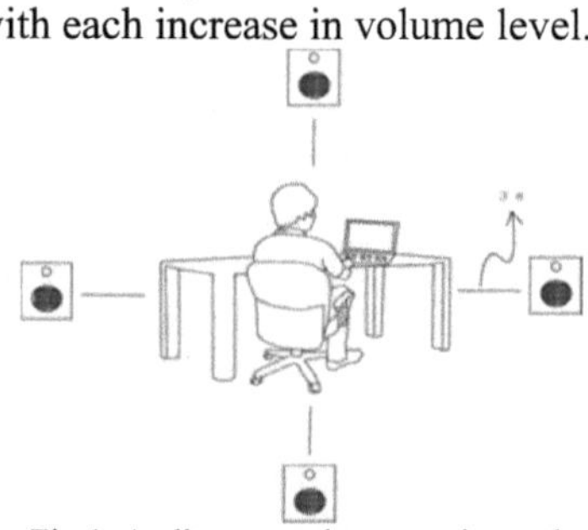

Fig.1. Auditory-words-perception task

III. RESULTS

A. Sighted individuals and blind individual

To measure accuracy, we analyzed the total grades of correct answers. We conducted an analysis of variance (ANOVA) on the accuracy rates of participants. The results revealed the main effect of vision status was significant (p=0.000), indicating that there was a statistically significant difference between blind and sighted individuals in auditory-words perception ability. The accuracy rate of blind individuals (mean=4.222, SD=0.135) was better than that of sighted individuals (mean=2.481, SD=0.135).

However, the interaction between vision status and background noise level showed no significance (p=0.195). Background noise had no significant impact on the perception performance of auditory-words between blind and sighted individuals. But, the interaction between vision status and the increase volume level was significant (p=0.000). A LSD post-hoc test revealed that each increase in the volume level resulted in a significant difference (p<0.05). The best performance on the increase in the volume levels was for the 25% level (0.833), followed by the 20% level (0.411) and the 15% level (0.244). There was a significant difference, even over 50%, between the performances of the 20% and 25% level. Further, sighted individual scored 80% correct hearing perception when the volume was increased to at least 25% relative to the background noise. Furthermore, there were also significant differences between the increased volume levels in the blind individuals (p<0.05). Blind individuals had better performance than sighted individuals. They also had higher correct accuracy (0.722) than sighted individuals (0.244) when the volume level increased to 25%. Thus, blind individuals can clearly hear the words (70%) when the volume was increased only 25% related to the background noise.

TABLE I
ANALYSIS OF VARIANCE RESULTS REGARDING THE ACCURACY FOR BLIND AND SIGHTED PARTICIPANTS ON AUDITORY-WORDS-PERCEPTION TASK

Source	Type III SS	df	MS	F	p
Between					
SV	163.63	1	163.63	83.358	0.000*
Error	43.185	22	1.963		
Within					
BN	0.176	2	0.088	0.141	0.869
BN × SV	2.120	2	1.060	1.697	0.195
Error	27.481	44	0.625		
IL	169.565	2	84.782	126.29	0.000*
IL × SV	27.343	2	13.671	20.366	0.000*
Error	29.537	44	0.671		
BN × IL	21.185	4	5.296	10.891	0.000*
BL × IL× SV	8.241	4	2.060	4.236	0.003*
Error	42.769	88	0.486		
Total	535.259	215			

Notes: SV = status of vision; BN=background noise; IL=increment in the volume level; *p < .05

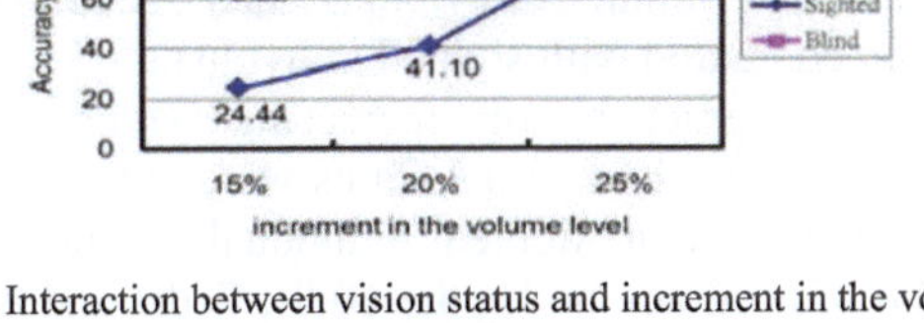

Fig.2. Interaction between vision status and increment in the volume level

The main effect of volume level increases was significant (p=0.000), indicating that there was a difference in performance between the different volume levels in auditory-words perception task. Participants performed better with each increase in volume level (Accuracy, 15%: 0.242, 20%: 0.310 and 25%: 0.454).

Additionally, the interaction of the increase in volume levels, background noise and visual-status were significant (p=0.003). We analyzed the accuracy of sighted and blind individuals separately. Firstly, we analyzed the performances of sighted individuals. Fig.3. shows that the interaction effect appeared at the 15% and 20% level of increases in volume when the background noise was 60 dB. But the post-hoc test revealed these two conditions were not significantly different (p=0.389). However, we found that the accuracy decreased with the volume of background noise increment when increase in volume was 15% level. There were significant differences between 70dB and 80dB (p=0.006) and 60dB and 80dB (p=0.000). When the increased volume was at the 20% condition, sighted individuals performed differently, but only on the 60dB and 80dB of background noises (p=0.019). Nevertheless, there were no differences noted in performance when the volume level was increased to the 25% level.

Then, we analyzed the accuracy of different increasing levels of volume with the same volume of background noise. Their performances were significantly different between 15% and 25% levels (p=0.000) and 20% and 25% levels (p=0.000) when the background noise was 60dB (Accuracy, 15%: 0.3833, 20%: 0.3067 and 25%: 0.85). With 70dB of background noise, there was a significant difference between 15% and 25% levels (p=0.000) and 20% and 25% levels (p=0.000) (Accuracy, 15%: 0.283, 20%: 0.417 and 25%: 0.767). All conditions had significantly different performances (p=0.000) when the background noise was 80dB (Accuracy, 15%:0.667, 20%:0.5 and 25%:0.883).

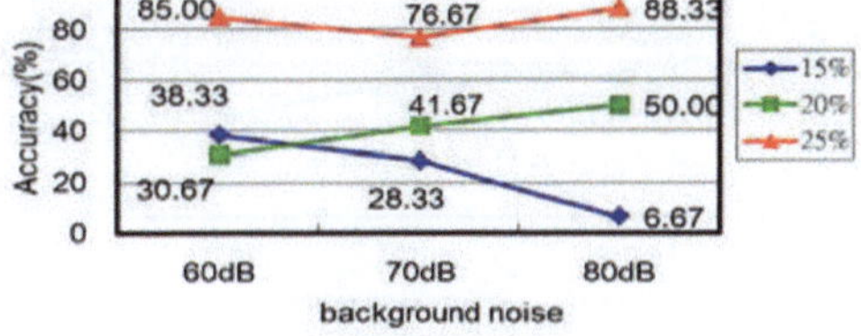

Fig.3. Interaction between background noise and increment in the volume level (sighted)

In addition, we also analyzed the performance of blind individuals. Fig.4 shows that an overlap appeared at the 15% and 20% levels of increase volume when the background noise was 70 dB. And the post-hoc test revealed these two conditions were not significantly different (p=1.000). In increasing volume levels to the 15% and 20% conditions, performance of blind individuals exhibited no difference for each background noise. However, there was a difference between 70dB and 80dB (p=0.007) and 60dB and 80dB (p=0.001) when the volume level was increased to the 20% level.

Similarly, we analyzed the accuracy of different increasing volume levels in the same volume of background noise. The blind group's performances were significantly different between the 15% and 25% levels (p=0.000) and 20% and 25% levels (p=0.000) when the background noise was 60dB (Accuracy, 15%: 0.6933, 20%: 0.75 and 25%: 1.00). With 70dB of background noise, there was also had significant difference between 15% and 25% levels (p=0.001) and 20% and 25% levels (p=0.001) (Accuracy, 15%: 0.7833, 20%: 0.7833 and 25%: 0.9833). The performance of auditory-words perception also had a significant difference between 15% and 25% levels (p=0.000) and 20% and 25% levels (p=0.000) when the background noise was 80dB (Accuracy, 15%: 0.70, 20%: 0.95 and 25%: 0.967).

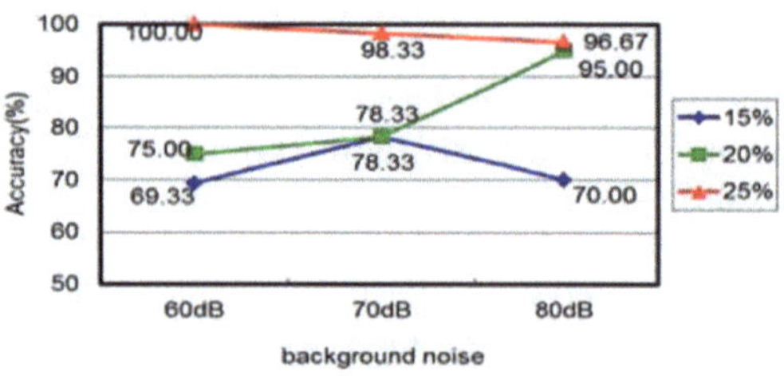

Fig.4. Interaction between background noise and increment in the volume level (blind)

Overall, there were interaction between the background noise level ×increasing volume level (p=0.000). Fig.5 shows that there was overlap at the 15% and 20% levels of increased volume when the background noise was 60 dB. But there was no statistically significant difference (p=1.000). Additionally, participants performed differently when background noises were 60dB and 80dB with the increase of volume of 15% (p=0.045). There was also the same result with an increase of volume of 20% (p=0.011). However, the post-hoc test revealed that performances of all background noises were not different with the increase volume to the 25% level.

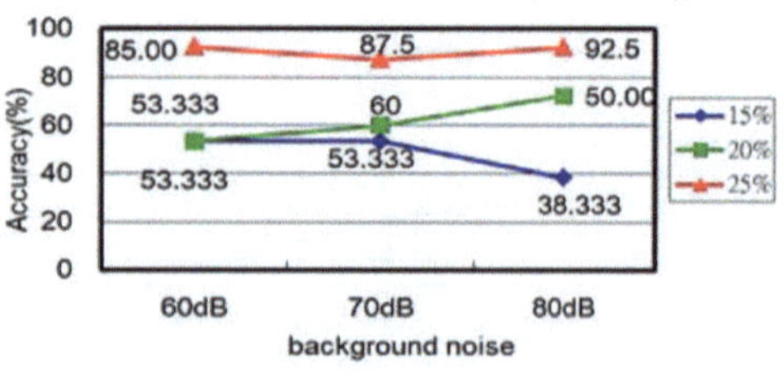

Fig.5. Interaction between background noise and increment in the volume level (overall)

B. Different vision status: Late-blind and congenitally blind

We further aimed to see whether vision status would affect the performance of the auditory-words perception ability. The blind group was divided into two groups: congenitally blind and late-blind. We conducted a 2 (vision status) x 3 (increasing level) x 3 (background noise) factor ANOVA.

Although the performance of congenitally blind individuals were better than late-blind individuals, there was no significant difference (p=0.185) between these two groups in auditory-words perception tasks. The interaction of increasing volume levels and background noise was significant (p=0.048). In all background noise conditions (60dB, 70dB and 80dB), the performance of congenitally blind individuals was better than late-blind individuals. A LSD post-hoc test revealed that performance of congenitally blind individuals did not significantly differ for background noise conditions. Late-blind individuals also performed the same behavior. A significant performance difference of performance did appeared for the 70dB condition in the congenitally blind group and both 60dB (p=0.035) and 70dB (p=0.014) conditions in the late-blind group.

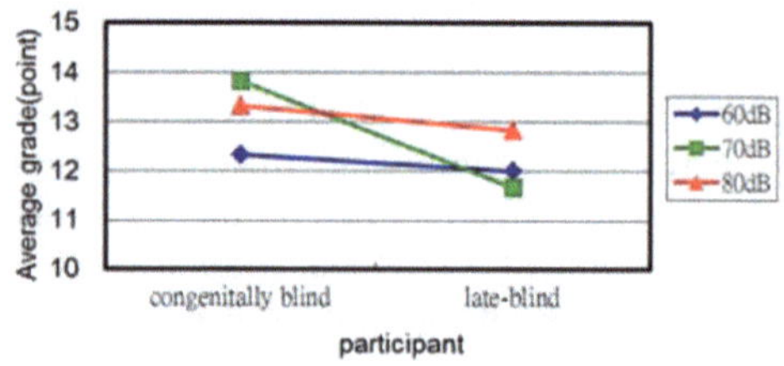

Fig.6. Interaction between vision status (congenitally blind and late-blind) and background noise

Additionally, the main effect of the increase in volume levels was significant (p=0.000). It means that blind individuals had better performance with auditory-words perception tasks with incremental increase in volume levels.

IV. CONCLUSION

The purpose of the present study was to explore the differences of ability on auditory-words perception between sighted and blind individuals. In the all conditions of the auditory-words perception task, the blind individuals performed better than did the sighted individuals. Rokem & Ahissar (2009) [14] also found that whether the condition was a quite or noise environment, blind individuals outperformed sighted individuals on auditory-word perception tasks.

Overall on auditory-word perception tasks, all participants did better with each incremental increase in volume of the test words. This result is consistent with the findings of Bonnel et al. (2003) [13] and Wan et al. (2010) [3]. They showed that participants could discriminate the sound when the diversity of volume increased to a certain level. Sighted individuals had an accuracy of 0.8333 when the volume was increased by 25%. However, blind individual could obtain similar accuracy (0.722) with only a 15% increase in the level of volume. This result implies that blind individuals exhibited cross-modal compensation. Rokem and Ahissar (2009) [14] also proposed that blind individuals had better hearing threshold compared with sighted individuals.

The interaction of increasing volume levels, background noise and visual-status were significant. The results showed that sighted individuals could perform over 0.8 of accuracy under the 60dB, 70dB and 80dB of background noise when volume was increased by 25%.

However, blind individuals could reach to over 0.7 of accuracy under all background noise conditions when the volume was increased by only 15%. The results further showed that blind individuals could perceive the sound of word when the diversity of the target volume and background noise was over 15%. Patterson (1982) [12] suggested that the sound of a signal should at least be over 15dB of the background sound but it would lead to interference or distraction when it was over 25dB. To avoid hearing distraction, the stimuli were controlled from 15dB to 25dB over the background noise volume level in the present study.

The interaction of background noise and visual-status were not significant. However, Muller-Gass et al. (2001) [9] explored hearing ability by using two consonant-vowel syllable combinations as their stimuli. Participants had to discriminate the unusual syllable under four different background noise conditions (quiet, 65dB, 70dB and 75dB). The results indicated that accuracy would be lower when the volume of the background noise was incrementally. But in the present study, the results were not consistent. The task of present study was more difficult than in their study. Participants had to pay more attention on the auditory-words perception task.

Performances of blind and sighted individuals were poorer when the volume of the background noise was adjusted from 60 dB to 80 dB under 15% of the increased volume level. The results further showed that the 15% of increased volume level condition was not loud enough for participants to successfully complete their perception task. Most participants had clear understanding of the sound when the volume level was increased by 20%.

We also investigated performance of congenitally blind and late-blind individuals on the discrimination of aurally presented words in advance. The results indicate that there was no difference between the two groups for any of increased volume levels and background noise conditions. Vision status of the congenitally blind and late-blind did not influence on the performance of the participants and the ability of discrimination of aurally presented words was similar for the two groups. However, the interaction of background noise and visual-status were significant. There was no difference in the ability to ability to discriminate aurally presented words with different background noises between two groups. A post-hoc test revealed that the significant difference of performance only appeared between the conditions which 70dB in the congenitally blind group and both 60dB and 70dB in late-blind group. The reason was that late-blind individuals had the poorest performance for 70dB of background noise level of all background noise conditions.

REFERENCES

[1] Gougoux, F., Lepore, F., Lassonde, M., Voss, P., Zatorre, R.J., Belin, P. (2004). Neuropsychology: pitch discrimination in the early blind. Nature, 430, 309.

[2] Mahon, B.Z., Anzellotti, S., Schwarzbach, J., Zampini, M., Caramazza, A. (2009). Category-specific organization in the human brain does not require visual experience. Neuron, 63, 397-405.

[3] Wan, C.Y., Wood, A.G., Reutens, D.C., Wilson, S.J. (2010). Early but not late-blindness leads to enhanced auditory perception. Neuropsychologia, 48, 344-348.

[4] Després, O., Candas, V., Dufour, A. (2005). Auditory compensation in myopic humans: Involvement of binaural, monaural, or echo cues? Brain Research, 1041, 56-65.

[5] Lewald, J. (2002). Vertical sound localization in blind humans. Neuropsychologia, 40, 1868-1872.

[6] Voss,P., Lassonde, M., Gougoux, F., Fortin, M., Guillemot, J.P., Lepore, F. (2004). Early- and late-onset blind individuals show supra-normal auditory abilities in far-space. Current Biology, 14, 1734-1738.

[7] Kozou, H., Kujala, T., Shrytov, Y., Toppila, E., Starck, J., Alku, P., Naatanen, R. (2005). The effect of different noise types on the speech and non-speech elicited mismatch negativity. Hearing Research, 199, 31-39.

[8] Martin, B.A., Sigal, A., Kurtzberg, D., Stapells, D.R. (1997). The effects of decreased audibility produced by high-pass noise masking on cortical event-related potentials to speech sounds /ba/ and /da/. Journal of the Acoustical Society of America, 101, 1585-1599.

[9] Muller-Gass, A., Marcoux, A., Logan, J., Campbell, K.B. (2001). The intensity of masking noise affects the mismatch negativity to speech sounds in human subjects. Neuroscience Letters, 299, 197-200.

[10] Sperry, J.L., Wiley, T.L., Chial, M.R. (1997). Word recognition performance in various background competitors. Journal of the American Academy of Audiology, 71, 71-80.

[11] Stuart, A., Philips, D.P. (1997). Word recognition in continuous noise, interrupted noise, and in quiet by normal-hearing listeners at two sensation levels. Scandinavian Audiology, 26, 112-116.

[12] Patterson, R. (1982). Guidelines for auditory warning systems on civil aircraft (CAA Paper 82017). London: Civil aviation authority.

[13] Bonnel, A., Mottron, L., Peretz, I., Trudel, M., Gallun, E., & Bonnel, A. M. (2003). Enhanced pitch sensitivity in individuals with autism: A signal detection analysis. Journal of Cognitive Neuroscience, 15, 226-235.

[14] Rokem, A., & Ahissar, M. (2009). Interactions of cognitive and auditory abilities in congenitally blind individuals. Neuropsychologia, 47, 843-848.

[15] Hugdahl, K., Ek, M., Takio, F., Rintee, T., Tuomainen, J., Haarala, C., Hämäläinen, H. (2004). Blind individuals show enhanced perceptual and attentional sensitivity for identification of speech sounds. Cognitive Brain Research, 19, 28-32.

[16] Röder, B., & Rösler, F. (2003). Memory for environmental sounds in sighted, congenitally blind and late blind adults: evidence for cross-modal compensation. International Journal of Psychophysiology, 50, 27-39.

Revisiting Lean Manufacturing Process with Vendor Managed Inventory System

C. L. Wen*, H. M. Wee, Simon Wu

Department of Industrial and Systems Engineering, Chung Yuan Christian University, 200 Chung-Pei Rd. Chungli, Taiwan 32023

(*chinlin@cycu.edu.tw, weehm@cycu.edu.tw, wusimon@cycu.edu.tw)

Abstract - **Due to global competitive, countries striving for economic growth and competitive advantage are focusing on manufacturing process and supply chain with vendor managed inventory policy. The aims are to reduce costs, eliminate waste, shorten lead time and enhance competitiveness. This study applies lean manufacturing with VMI (Vendor Managed Inventory) to shorten lead time and reduce inventories. The study also applies TOC (theory of constraints), 5W1H (who, when, where, which, what, how much) techniques to identify bottleneck and suggest continuous improvements in production, sales and supply chain management. The objective is to develop a framework to improve overall efficiency and ensure profitable business performance.**

Keywords - **Lean production, supply chain management, theory of constraints, vendor managed inventory**

I. INTRODUCTION

Due to the application of Toyota Production Systems and lean manufacturing, Toyota has overtaken General Motors (GM) to become the world's top car manufacturers [5]. However, while many companies have implemented lean production, many of them cannot achieve the same result as Toyota [15], [4]. The study seeks to look into the factors that affect the successful implementation of lean manufacturing. We investigate the cause using theory of constraints and problem-solving processes and techniques.

Theory of Constraints (TOC) was proposed by Goldratt who suggested the idea of bottleneck. He said that the strength of a system depends on the weakest link. Therefore, we must start with the weakest link in order to improve the efficiency of the entire chain. In the process, five questions are asked. They are: What to Change? Which direction? How to Cause the Change? What is the execution plan of action? How much resource should be allocated? Goldratt stated that the ultimate goal of business is to make profit, reduce inventory and operational costs while improving the throughput. Effective assessment should be based on customer demand to reflect the overall business indicators, enterprise value and market acceptance. However, the core in the problem solving process is solving the bottlenecks to find out the real problem that affect the overall goal of the company. Consolidated thinking process framework is suggested to maximize capacity, develop effective strategies monitor implementation to assessment the performance and problem-solving abilities. In this study, consolidated new problem-solving flowchart sorting is summarized in Fig.1.

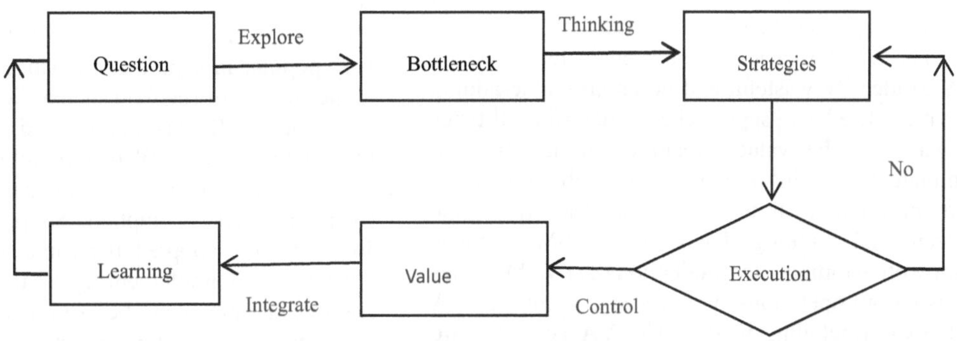

Fig.1. Consolidated new problem-solving flowchart

To ensure maximum output, holistic transformation and evaluation are helpful in achieving the business objectives. TPS (Toyota Production System) is a tool to eliminate waste and improve efficiency. The 5W1H (who, when, where, which, what, how much) is used to find the root cause and minimize lead time. Banerjee [2] proposed the concept of economic batch size in pursuit of minimum inventory as the ultimate goal. Liao and Shyu [19] proposed lead time as decision variables in the stochastic stock model. Logistics should comply with VMI (Vendor Managed Inventory) operation mechanism. Electronic information platform and efficiency IT will share the information immediately to the central plant and suppliers. The flow of information must be disclosed immediately, so that

E. Qi et al. (eds.), *Proceedings of the 5th International Asia Conference on Industrial Engineering and Management Innovation (IEMI2014)*, Proceedings of the International Asia Conference on Industrial Engineering and Management Innovation 1, DOI 10.2991/978-94-6239-100-0_26, © Atlantis Press and the authors 2015

suppliers and central plant can establish quick response mechanism with VMI. The e-commerce framework is shown in Fig.2.

VMI e-commerce framework

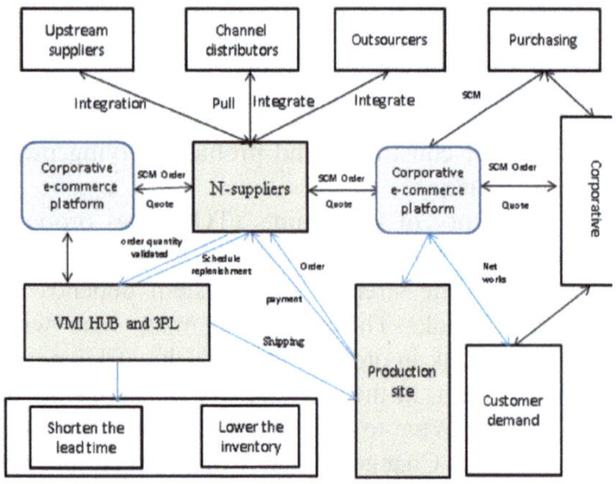

Fig.2. VMI e-commerce framework

II. RESEARCH REVIEW

A. Lean Production & Supply Chain

The term "lean" means a series of activities or solutions to eliminate waste, reduce non-value added (NVA) operations, and improve the value added (VA). The word "Lean" was first used in the Future Car Investigation by MIT professors; it represents Japan's new production system different from mass production [8], [20] "Waste" is defined as anything that interferes with the smooth production flow and does not add value [3], [22]. The eight wastes highlighted in TPS are overproduction, waiting, conveyance, over processing, excess inventory, movement, defects and unused employee creativity, and the biggest one being overproduction [27], [13].

The VSM is a lean supply chain tool used by TPS to identify wasteful and necessary value-adding activities. The "lean supply chain" identifies all types of waste in the value stream chain and tries to eliminate them; this is a major contribution of the lean production system [3] Different from the conventional thinking, Toyota uses VSM to focus mainly on avoiding over production [27] VSM begins by listing all operations, and classifies them into VA and NVA (including waste). The VA activities are those that customers are willing to pay money for, either tangible goods or intangible functions. The NVA work includes the eight wastes of TPS [3].

The application of VSM in the TPS identifies VA/NVA activities for waste elimination, and the status of their lead time in the supply chain [20], [13].

B. Theory of Constraints:

The TOC is a problem solving model developed by Eliyahu M. Goldratt [7], [29], [21]. Since mid-1970s, Goldratt has used scientific methods to develop management concept which have great value to industry. Goldratt also presented the structured TOC thinking processes using common sense. The problem solving process applies the Theory of Constraints thinking process and five Theory of Constraints logic tree diagrams [7], [25].

The TOC is a set of management principles that helps to identify obstacles to one's goal(s) and influence the changes essential to eliminate them [25], [26]. In essence, the Theory of Constraints is about change and how best to influence change.

In order to induce effective output from TOC thinking processes, Goldratt used three basic questions namely (1) What to Change? (2) What to change to? (3) How to cause the change? They are linked with five logic trees [7], [25], [26]. The detailed linkage is shown in Table I below:

TABLE I
TOC THINKING PROCESSES & LOGIC TREES

Key Quest	Intermediate Objectives	Logic Trees
What to change?	Identify core problems and find out root causes.	1. Current Reality Tree (CRT)
What to change to?	Eliminate the core problem starting with an objective to search for solution.	2. Evaporating Cloud Tree (for use with conflicts) 3. Future Reality Tree (FRT)
How to cause the change?	Identify obstacles to implement and devise detailed plans for overcoming these obstacles.	4. Prerequisites Tree (PRT) 5. Transition Tree (TT)

C. Vendor Managed Inventory (VMI)

VMI is an effective way of mitigating the bullwhip effect. Potentially VMI offers two possible sources of bullwhip reduction. Firstly, there is a reduction in decision-making level due to direct contact. Secondly, we reduce lead time by reducing information flow time delays [11].VMI has become very popular in supermarkets since 15 years ago due to the success of retailers such as Wal-Mart [11].

The VMI was popularized by the successful partnership between Wal-Mart and Proctor & Gamble in 1985. Since then, other companies such as Shell Chemicals, HP, Campbell Soup, and Johnson & Johnson have adopted the same approach [9]. The advantages of implementing VMI program are very significant and can be summarized as reduced inventory costs, better response to market changes, reduction in demand uncertainty, and more flexibility in production planning and distribution [9]

III. DISCUSSION

VMI enables suppliers to better control the inventory in the supply chains, reduces bullwhip effect and provides risk sharing. In order to focus on their core competencies, businesses should engage third-party logistics providers (3PL). Many 3PLs provide a range of services, including warehousing, distribution freight forwarding and contract manufacturing. VMI can benefit large-scale business company such as the *Hon Hai Enterprise* from Taiwan. Fig.3 shows a lean production research framework.

IV. CONCLUSION

This study develops a lean manufacturing with VMI (Vendor Managed Inventory) framework to shorten lead time and reduce inventories. The study also applies TOC (theory of constraints), 5W1H (who, when, where, which, what, how much) techniques to identify bottleneck and suggest continuous improvements in production, sales and supply chain management. The objective is to develop a framework to improve overall efficiency and ensure profitable business performance.

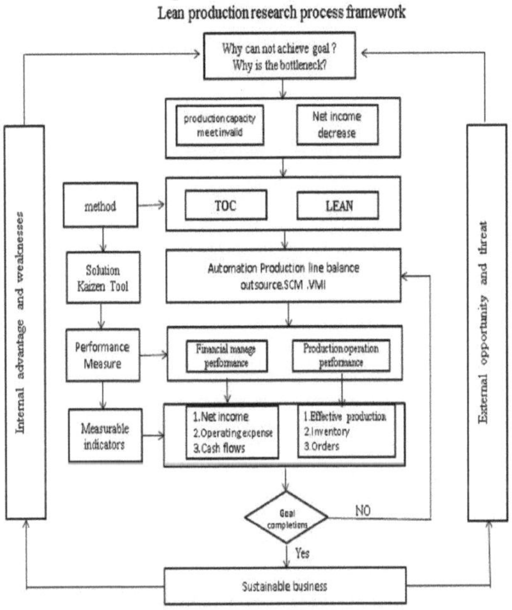

Fig.3. Lean production research framework

REFERENCES

[1] Aamodt, A. & Plaza, E., 1994, "Case-Base Reasoning: Foundational issues, methodological variations and system approaches". *The European journal on Artificial Intelligence, AICAM*, Vol. 7, No. 1, pp. 39-59

[2] Banerjee (1986) A joint economic-lot-size model for purchaser and vendor. *Decision Science* 17:292-311

[3] Bozickovici, R., Radosevic, M., Cosic, I., Sokovic, M. & Rikalovic, A., 2012, Integration of Simulation and Lean Tools in Effective Production Systems - Case Study, *Journal of Mechanical Engineering*,Vol.58, No.11.pp.642-652

[4] Brown, A., Eatock, J., Dixon, D., Meenan, B. J. and Anderson, J., 2008,"Quality and continuous improvement in medical device manufacturing", *The TQM Journal*, Vol. 20, No. 6, pp. 541 – 555.

[5] Bunkley, N., 2009, "Toyota Ahead of G.M. in 2008 Sales", *The New York Times*, Jan., 21, 2009

[6] Cetinkaya, S., Lee, C.-Y. 2000. Stock replenishment and shipment scheduling for vendor-managed inventory systems. *Management Science* 46 (2), pp.217–232

[7] Chou,Y.C., Lu, C.H., Chang, P.L., 2010, Using Theory of Constraints to Find the Problem about High Level Inventory in the *Aerospace Industry*, 978-1-890843-21-0/10©2010 IEEE

[8] Cottynab, J., Landeghemb, H.V., Stockmana, K. & Derammelaere,S., 2011, A method to align a manufacturing execution system with Lean objectives, *International Journal of Production Research*, Vol. 49, No. 14, 15 July 2011, pp.4397–4413

[9] Darwish, M.A., Odah, O.M., 2010, Vendor managed inventory model for single-vendor multi-retailer supply chains, *European Journal of Operational Research*, Vol.204, No.3, pp.473–484

[10] DB Schenker. 2012. Contract Logistics (http://www.dbschenker.com)

[11] Disney, S.M. & Towill, D.R., 2003, The effect of vendor managed inventory (VMI) dynamics on the Bullwhip Effect in supply chains, Int. J. *Production Economics*,Vol.85, pp.199–215

[12] Handfield, R.B., Nichols Jr, E.L. 1999. *Introduction to Supply Chain Management*. Prentice Hall, Upper Saddle River, NJ

[13] Hassan, K. & Kajiwara, H., 2013, Application of Pull Concept-based Lean Production System in the Ship Building Industry, *Journal of Ship Production and Design*, Vol. 29, No. 3, pp.105–116

[14] Hertz, S. and Alfredsson, M. 2003.Strategic development of third party logistics providers. *Industrial Marketing Management* Vol. 32, pp. 139-49.

[15] Holweg, M., 2007, "The genealogy of lean production", *Journal of Operations Management*, Vol. 25, No. 2, pp. 420-437

[16] Lee, J. and Peccei, R., 2008, "Lean production and quality commitment: A comparative study of two Korean auto firms", *Personnel Review*, Vol. 37, No. 1, pp. 5-25.

[17] Lee-Mortimer, A., 2008, "A continuing lean journey: an electronic manufacturer's adopting of Kanban", *Assembly Automation*, Vol. 28, No. 2, pp. 103-112.

[18] Lee, G. H., 2008, "Rule-based and case-based reasoning approach for internal audit of bank", *Knowledge-Based Systems*, Vol. 21, No. 2, pp. 140-147

[19] Liao and Shyu., 1991, An analytical determination of lead time with normal demand *International Journal of Operations & Production Management*, pp. 72–78

[20] Ramesh, V. & Kodali, R., 2012, A decision framework for maximizing lean manufacturing performance; *International Journal of Production Research*, Vol. 50, No. 8, pp.2234–2251

[21] Sobreiro, V.A. and Nagano, M.S., 2012, A review and evaluation on constructive heuristics to optimize product mix based on the Theory of Constraints, *International Journal of Production Research*, Vol. 50, No. 20, 15 October 2012, pp.5936-5948

[22] So, S. & Sun, H., 2011, An extension of IDT in examining the relationship between electronic-enabled supply chain integration and the adoption of lean production, *International Journal of Production Research*, Vol. 49, No. 2, pp.447-466

[23] Southard, Peter B., Swenseth, Scott R. 2008,

Evaluating vendor-managed inventory (VMI) in non-traditional environments using simulations. International Journal Production Economics, Vol.116, pp. 275-287.

[24] Tezuka.K., 2011. Rationale for utilizing 3PL in supply chain management: *A shippers economic perspective. IATSS Research*

[25] Tulasi, C.H.L., Rao, A.R., 2012,REVIEW ON THEORY OF CONSTRAINTS, *International Journal of Advances in Engineering & Technology*, Vol. 3, No., pp. 334-344

[26] Urhuogo, I., Williams, V., Hart, D. W.,2013, Root Cause Analysis: Using Theory of Constraints Application to Measure the Impact of Employee Demographics on the Adoption of Information Technology Equipment, *Journal of Information & Knowledge Management*, Vol. 12, No. 1, pp.1-15

[27] Vinodh, S. & Joy, D., 2012, Structural Equation Modelling of lean manufacturing practices, *International Journal of Production Research*, Vol. 50, No. 6, pp.1598–1607

[28] Yao, Y., Evers, P.T. Dresner, M.E., 2007, Supply chain integration in vendor-managed inventory. *Decision Support Systems*, Vol.43, pp. 663–674

[29] Wu, H.H., Tsai, S.H., Tsai,C.H., Tsai,R. and Liao, M.Y., 2011, A study of supply chain replenishment system of theory of constraints for thin film transistor liquid crystal display (TFT-LCD) plants, *African Journal of Business Management* Vol. 5, No.21, pp.8617-8633

Risk Assessment for International EPC Projects

Jing-nan Yang, Wei-ning Cai*, Wei-guo Fang
School of Economics and Management, Beihang University, Beijing 100191, P R China
(Cweining1985@163.com)

Abstract - **Risk assessment of international EPC project is the foundation of international EPC project management. This paper studies the risk assessment of international EPC projects in the light of their special characteristics. This paper tries to build an index system for risk assessment of international EPC projects from business operation of project, engineering construction and external environment. The Analytic Hierarchy Process method is applied to determine the weight of different indices and Fuzzy comprehensive evaluation method is used to evaluate the risk of international EPC project. In the end, this paper describes the application of evaluation system and assessment methods through an example of international EPC project risk assessment.**

Keywords - **Fuzzy Comprehensive Evaluation, International EPC Project, Risk Assessment**

I. INTRODUCTION

With the past economic development and globalization process, it becomes more competitive in the international market and the larger scale cooperation is required. The traditional practice of separating the design and engineering construction becomes outdated and it gradually evolves to a unified sub-contracting mode. The client requires higher quality services, more and more different modes have been adopted, e.g. EPC (Engineer-Procure-Construct), EP (Engineer-Procure), BOT (Build-Operate-Transfer), PPP (Public-Private Partnership). Based on the estimation from Design-Build Institute of America (DBIA), the market share of the EPC mode would reach 55% in 2015 [1].

EPC stands for Engineering Procurement and Construction. Under an EPC contract, the contractor will design the installation, procure the necessary materials and construct it as well as the quality assurance, construction safety, construction schedule and budget. There are two main characteristics in this type of contracts:

1) The structure of the contract between the client and contractor are succinct and there is clear division on the responsibility of both parties; there are sub-contracts between the sub-contractors and the main contractor, the main contractor is responsible for the overall process.

2) The most common pricing method is that the main contract has a fixed given number which mostly is higher than the prevailing prices in other models, this increases the profitability of the contractors; this usually stimulates main contractors to better organize and plan the overall schedule to ensure a higher quality delivery.

Since 1980s, with the increased number of the international projects and the evolvement of the contracting models, there are widespread interests in the study of the risk assessment of EPC projects across international communities. T M Williams proposed the basic principles of engineering risk management, he considered it as a dynamic and proactive process in 1993 [2]. Kliem worked on the theory and application of identifying, analyzing and managing risks with a systematic way [3]. In recent years, with more participation to international projects in China, there is significant progress in the study of the project risk management. Z Wang gave a summary on different risk management approaches, with a comprehensive case study on several aspects of risk analysis for the civil engineering project [4]. Lei Shengqiang [5], Qiu Wanhua [6] and Deng Tiejun [7] studied risk assessment and management for civil engineering projects and proposed several different theoretical frameworks and methods.

N G Bunni focused on the detailed studies for the shift of risks traditionally on the client/owner to the contractors by applying EPC [8]. P Galloway worked on the high risks for EPC contractors [9]. X Meng et. al. compared and contrasted the traditional approach and EPC approach from several aspects, including the applicable domains, bidding process, engineering management process and risk management; they proposed a collaborated way of risks and contingencies between clients/owners and contractors [10,11].

There is relatively little research on the risk management for international EPC projects compared with the thriving markets on such engineering projects in China. Usually the management passively responds to risks, without any prior analysis and evaluation on engineering risks, which would have guided the project execution. Therefore, a comprehensive study of the risk management in international EPC projects in this work explores the new territory of the research domain and will provide theoretical guidance to real-world engineering projects and applications.

II. MEASUREMENT GUIDANCE OF INTERNATIONAL EPC PROJECT RISK ASSESSMENT

It is the properties associated with the EPC contracts and the special elements in the international projects, which mainly contributes to risks in the international EPC projects. Based on the sources of risks, it can be analyzed from the following three aspects: operating risks, execution risks and external risks.

A. Operating Risks

Operating risks mainly include below items:

1) *Bidding decision risks*

The bidding from the contractors is based on the research and investigation of the construction

E. Qi et al. (eds.), *Proceedings of the 5th International Asia Conference on Industrial Engineering and Management Innovation (IEMI2014)*, Proceedings of the International Asia Conference on Industrial Engineering and Management Innovation 1, DOI 10.2991/978-94-6239-100-0_27, © Atlantis Press and the authors 2015

environment for the project, which includes the general information of the country where the project is built, the client/owner information, potential competitors information as well as accurate self-evaluation, as well as the bid-asking documents from the client/owner. The contractors face risks from inaccurate evaluation of the actual cost and inappropriate pricing strategy. The mis-price mainly comes from inaccurate estimation of the risks and/or the misunderstanding of the bid-asking documents. The essence of the EPC contracts is that the client/owner would pay contractors a premium to transfer the potential risks to the contractor's side. However, if multiple contractors compete for one project and offer a over-low price to earn the contract; it eliminates or significantly reduced the premium for risk-taking, which would eventually weaken the actual capability to provide necessary contingency. It is inappropriate/malicious competition to expect the profit on the assumption that no contingency plan would be actually used.

2) *Contract signing risks*

Most international EPC contract would use the standard EPC contract clauses, however, there is no single standardized template adopted. Client/owner modifies some clauses in the contract based on the specific characteristics of the project in most cases. The contract risks exists in the complete contract documents, including all clauses and all appendices. Sometimes, the client/owner takes the advantage of drafting the initial contract to gain upper-hand in the negotiation process, and modify the EPC contract substantially to their own advantage. This leads to contract-manipulation risks for contractors.

3) *Business operation risks*

Business operation is a key component in international EPC projects, which involves the government regulation, banking system, insurance, customs from multiple countries (owner/client country, contractor country and supply/raw materials providers' countries). Due to the great diversity and complexity in government regulation and law enforcement across different countries, there are risks from the inexperience in the complex business operation process. The main risk factors include: company registration, labor permit, equipment import/export, insurance and banking system etc.

B. Execution Risks

Execution risks include the following items:

1) *Design risks*

Design is the most fundamental part of a project, its execution quality and if it proceeds on schedule dictates the usability and functionality of the project, and is the most influential factor for the quality, cost and schedule of the project. Under EPC contracts, contractors are responsible for the design phase. They target to the most effective and efficient design to achieve the expected goal, and to maximize the profitability margin. However, no matter who is responsible to the design phase, either the client/owner or the contractor, there are many delicate risk factors during the design management, e.g. the communication and transfer of the technical specifications, the coordination during the design and contraction phases as well as some later stage change on the design. All of above directly impact the expected profit margin for contractors.

2) *Procurement risks*

In an international EPC project, procurement is usually the most significant component in the contractor's cost composition. The quality of the equipment and raw materials purchased and the time taken to delivery have huge influence on the schedule and quality of the project. The procurement risks permeate the complete procurement process, including the initial stage of making the purchase plan, picking and managing the right suppliers for equipments and raw materials, getting the right equipment with high quality assurance, as well as the law enforcement regulation from the supplier's country and monitoring the timely delivery of the equipments and materials.

3) *Construction risks*

The core part of an EPC project is the construction phase, which is the execution to achieve the specified goal of the project. There are several risks involved during this phase. For example, the discrepancy on the quality of execution between the main contractor of the project and the sub-contractor who is doing the construction work; issues in the coordination and communication of the project management between main contractor and sub-contractors; some unexpected issues in the infrastructures; all of these may cause delay of the overall project delivery, impacts on the quality and an increased cost.

4) *Post-construction risks*

Even after the contractor finishes the construction and delivers it to the client, the contractor still has responsibility for the on-going maintenance work. In most international EPC project, it becomes more popular that both parties would agree on a second phase collaboration in the future, if the client is satisfied with the quality of the service in the first phase. However, if the client set high standard and expectation on the first phase, the contractor tends to include the potential profits from the future phases of collaboration (which is not guaranteed yet) or significantly increases the current phase execution quality. These may cause the reduced profitability of the contractors.

C. External Risks

External risks include the following items:

1) *Political risks*

Political risks is an important factor in an international EPC project. Constructing in a political unstable country or its periphery region would not only have higher chances for an accident, but also potentially huge lost if it happens. It may include fatality in the project crew, huge financial lost or even cease of the project half-way completed. There are several aspects in political risks: the governing power/party is changed with the initial chaos of the new power setup; conflicts between races and/or nations, or the existence of the anti-government military; the political tension between the construction country and its neighbors or other countries, which may escalate to higher level of confrontation actions.

2) *Economic risks*

The construction is done at the specified country, its overall economic environment is crucial to the contractor, such as inflation, fiscal policy changes etc.. Besides the economic conditions in the construction country, the country of the suppliers as well as the global economic fluctuation also plays a big role in the expected profit and completion timeline.

3) *Social risks*

Social risks come from all aspects in the construction country, which involves different domains and industries, these may have direct impacts on the normal construction process or lead to a higher rental cost. The social risks include: (i) religion and local traditions; (ii) safety of the neighborhood; (iii) the development level of the society, this includes if there is existence of some discrimination policy on international contractors, the effectiveness of government operation, and other related parties such as media and labor union's attitude towards the international contractors; (iv) the quality of local labor, including their technical skills, educational level and work attitude; (v) the law enforcement situation, if it is mature, fair and stable, it is advantageous to the international contractors, otherwise it may cause loses.

4) *Natural risks*

Natural environment has huge and direct impacts on the project construction. The contractor usually has very limited control over the natural risks. It may come from following aspects: (i) natural disasters, such as earthquake, tsunami, hurricane and flooding etc.; (ii) adverse climate condition, such as extended period of the rain season, the drought sprout, heat waves, winter storms and sand storms; such unfavored condition would not only delay the project schedule, but also have negative impacts on the usability of equipments; (iii) adverse geographical condition on the construction site, e.g. riftzone, quicksand, karst cave, as well as some underground infrastructure or barriers.

5) *Related parties risks*

The key parties involved in an international EPC project include: client/owner (investor), client representatives (engineers), agents, sub-contractors, suppliers, transporters, insurance companies, banks (for loans and financing) etc., their unexpected behaviors may cause potential risks to the contractor. For example, owner (investor) has difficulty to pay for the project as agreed; the technical and management level of sub-contractors and suppliers is below expectation and cannot deliver as promised; owner representatives (engineers) and agents may have issues in attitude, professionalism, effectiveness etc.; banks may have a low-rate loan and financing services.

Based on above analysis, this work proposed a two-tier assessment system for risks in international EPC projects, as shown in Table I.

TABLEI
INDEX SYSTEM OF INTERNATIONAL EPC PROJECT RISK ASSESSMENT

First Class indexes	Second Class Indexes
Operating risks	Bidding decision risks Contract signing risks Business operation risks
Execution risks	Design risks Procurement risks Construction risks Post-construction risks
External risks	Political risks Economic risks Social risks Natural risks Related parties risks

III. RISK ASSESSMENT METHOD

There are some difficulties to provide accurate quantitative measurement for some of the risks listed in above proposed risk assessment system. We use fuzzy evaluation methods to give estimation. This work proposes a fuzzy comprehensive approach for the risk assessment. The mathematical model of this approach is illustrated as follow:

1) Determine the evaluation factor set

$$U = \{u_1, u_2, \cdots, u_m\}$$

where, $u_i (i = 1,2,\cdots,m)$ are evaluation factors, m is the number of factors. This set constitutes the evaluation framework. For example, in this paper, the factor set corresponding to the first level index is U ={ operating risks, execution risks , external risks }.

2) Determine the remark set

$$V = \{v_1, v_2, \cdots, v_n\}$$

where $v_j (j = 1,2,\cdots,n)$ are evaluation consequences, n is the number of remark grades. This set stipulates the selection scope of evaluation consequence for an evaluation factor. The element v_j can be either a qualitative description or a quantitative score. For instance, the five-scale remark set V = {very serious, serious, moderate, mild, very mild} is completely qualitative.

3) Determine the weight vector

$$A = \{a_1, a_2, \cdots, a_m\}$$

where a_i $(i = 1,2,\cdots,m)$ represents the importance of factor u_i. The weight vector must satisfy nonnegative and normalization conditions, i.e.,

$$0 \le a_i \le 1, \ \sum_{i=1}^{m} a_i = 1$$

4) Establish the membership matrix. For the ith factor u_i, a single factor evaluation is conducted and a fuzzy vector $R_i = \{r_{i1}, r_{i2}, \cdots, r_{in}\}$ is constructed, where r_{ij} ($0 < r_{ij} < 1$) means the extent to which factor u_i belongs to remark grade v_j. The evaluation result of m factors constitutes an $m \times n$ matrix, called the membership matrix R.

$$R = \begin{bmatrix} r_{11} & r_{12} & \cdots & r_{1n} \\ r_{21} & r_{22} & \cdots & r_{2n} \\ \vdots & \vdots & \vdots & \vdots \\ r_{m1} & r_{m2} & \cdots & r_{mn} \end{bmatrix}$$

5) Through a fuzzy synthetic operation of the weight vector with the membership matrix, get a comprehensive assessment result. This is shown as follows.

$$B = A \circ R = \begin{pmatrix} a_1 & a_2 & \cdots & a_m \end{pmatrix} \circ \begin{bmatrix} r_{11} & r_{12} & \cdots & r_{1n} \\ r_{21} & r_{22} & \cdots & r_{2n} \\ \cdots & \cdots & \cdots & \cdots \\ r_{m1} & r_{m2} & \cdots & r_{mn} \end{bmatrix}$$
$$= \{b_1, b_2, \cdots, b_n\}$$

where, b_j $(j = 1,2,\cdots,n)$ is produced by an operation of A with the jth column of R. It represents the membership degree of the evaluated object as a whole belonging to remark grade v_j.

When the evaluated system is more complex where many factors should be considered, if using the above elementary evaluation model, one has to compare numerous factors at the same level, then the weight assignment is very difficult to implement. In this case, we need to classify the evaluation factors into different classes according to their attributes. For example, the evaluation factors in the risk assessment index system established in this paper are divided into three categories according to Operating risks, Execution Risks and External risks. By this way, we built a two-level index system. In multilevel evaluation, first we conduct a comprehensive evaluation for each category, and then conduct a high-level evaluation for the categories located at the same level, until the highest level is reached.

Let $U_1, U_2, \cdots$ be subsets of set U, $R_1, R_2, \cdots$ and $A_1, A_2, \cdots$ be fuzzy membership matrices and weight vectors corresponding to each subset respectively, and then the two-level fuzzy integrated evaluation model is described as

$$B = A \circ R = A \circ \begin{bmatrix} A_1 \circ R_1 \\ A_2 \circ R_2 \\ A_3 \circ R_3 \\ \vdots \end{bmatrix}$$

Presently there are several models for the synthetic operation $A \circ R$. We adopt the most often used model $M(\wedge, \vee)$, i.e., $b_j = \bigvee_{i=1}^{m} (a_i \wedge r_{ij}), j = 1,2,\cdots,n$.

In addition, we adopt AHP to determine the weight vector A of evaluation factors.

IV. A CASE STUDY

A large Chinese contractor wants to participate the bidding of an international famous company's EPC project at North Africa in September 2013. The project is planed by the international manufacture based on the company's developing strategy and the local economic development and marketing needs. The main contract uses the EPC projects standard, only parts of contract can be changed by the client and there is a possibility for the second phase in the future. We will use the fuzzy evaluation methods established in this paper to demonstrate the risk assessment for this project from beginning to the end.

First, determine the fuzzy membership degrees of evaluation factors with respect to remark set. Suppose the expert used a five-scale remark set V={very serious, serious, moderate, mild, very mild}, to evaluate the different factors under operating risks, execution risks and external risks.

Table II gives the fuzzy membership degrees of all second-level factors with respect to the remark set.

TABLE II
FUZZY MEMBERSHIP DEGREES OF FACTORS TO REMARK SET

	Evaluation factors	Remarks				
		very serious	serious	moderate	mild	very mild
operating risks	bidding decision risks	0.1	0.1	0.7	0.1	0
	contract signing risks	0.1	0.2	0.7	0	0
	business operation risks	0.1	0.1	0.6	0.2	0
execution risks	design risks	0.1	0.1	0.7	0.1	0
	procurement risks	0.2	0.6	0.1	0.1	0
	construction risks	0.0	0.3	0.4	0.3	0
	post-construction risks	0.1	0.2	0.3	0.4	0
external risks	political risks	0.2	0.5	0.2	0.1	0
	economic risks	0.1	0.4	0.4	0.1	0
	social risks	0.1	0.5	0.4	0	0
	natural risks	0	0.2	0.2	0.6	0
	related parties risks	0.1	0.5	0.4	0	0

The AHP method is used to determine the factor weights. Assume the expert made pairwise comparisons to the three factors of bidding decision risks, contract signing risks and business operation risks under the operating risks. The comparisons form the following judgment matrix.

$$C_1 = \begin{bmatrix} 1 & 1 & 3 \\ 1 & 1 & 3 \\ 1/3 & 1/3 & 1 \end{bmatrix}$$

The element c_{ij} of judgment matrix represents how many times factor *i* is more important than factor *j*.

By calculating we get the maximum eigenvalue of this judgment matrix $\lambda_{\max} = 3.0000$, and the eigenvector $W_1 = (0.6882, 0.6882, 0.2294)$.

The consistency ratio $\mathrm{CR} = \dfrac{CI}{RI} = \dfrac{(3-3)/2}{0.58} = 0 < 0.1$, showing the consistency of the judgment matrix is acceptable. By normalizing W_1, we obtain the weight vector of social impact factors

$$A_1 = (0.4286, 0.4286, 0.1428)$$

Assume the expert made pairwise comparisons to the four factors of design risks, procurement risks, construction risks and post-construction risks under the execution risks. The comparisons form the following judgment matrix.

$$C_2 = \begin{bmatrix} 1 & 3 & 3 & 5 \\ 1/3 & 1 & 1 & 3 \\ 1/3 & 1 & 1 & 3 \\ 1/5 & 1/3 & 1/3 & 1 \end{bmatrix}$$

By calculating we get the maximum eigenvalue of this judgment matrix $\lambda_{\max} = 4.0435$, and the eigenvector $W_2 = (0.8720, 0.3337, 0.3337, 0.1304)$.

The consistency ratio $\mathrm{CR} = \dfrac{CI}{RI} = \dfrac{(4.0435-4)/3}{0.89} = 0.0163 < 0.1$, showing the consistency of the judgment matrix is acceptable. By normalizing W_2, we obtain the weight vector of execution risks factors

$$A_2 = (0.1429, 0.4286, 0.4285)$$

Assume the expert made pairwise comparisons to the five factors of political risks, economic risks, social risks, natural risks and related parties risks under the external risks. The comparisons form the following judgment matrix.

$$C_3 = \begin{bmatrix} 1 & 3 & 5 & 1 & 1/2 \\ 1/3 & 1 & 3 & 1/3 & 1 \\ 1/5 & 1/3 & 1 & 1/5 & 1/3 \\ 1 & 3 & 5 & 1 & 1 \\ 1/2 & 1 & 3 & 1 & 1 \end{bmatrix}$$

By calculating we get the maximum eigenvalue of this judgment matrix $\lambda_{\max} = 5.1153$, and the eigenvector

$$W_3 = (0.6579, 0.2728, 0.0025, 0.5855, 0.3755)$$

The consistency ratio $\mathrm{CR} = \dfrac{CI}{RI} = \dfrac{(5.1553-5)/4}{1.12} = 0.0257 < 0.1$, showing the consistency of the judgment matrix is acceptable. By normalizing W_3, we obtain the weight vector of external risks factors

$$A_3 = (0.0551, 0.1290, 0.1290, 0.3434, 0.3434)$$

The judgment matrix of pairwise comparisons to the three factors of operating risks, execution risks, and external risks at the first level is as following

$$C = \begin{bmatrix} 1 & 3 & 1/2 \\ 1/3 & 1 & 1/4 \\ 2 & 4 & 1 \end{bmatrix}$$

By calculating we get the maximum eigenvalue of this judgment matrix $\lambda_{\max} = 3.1083$, and the eigenvector $W = (0.4881, 0.1862, 0.8527)$. The consistency ratio $\mathrm{CR} = \dfrac{CI}{RI} = \dfrac{(3.1083-3)/2}{0.58} = 0.0993 < 0.1$, showing the consistency of the judgment matrix is acceptable. By normalizing W, we obtain the weight vector.

$$A = (0.3197, 0.1219, 0.5584)$$

After having determined the weight and membership degree of each factor, we can get a comprehensive assessment result by fuzzy synthetic operation. From Table 2, the fuzzy membership matrix of operating risks factors is

$$R_1 = \begin{bmatrix} 0.1 & 0.1 & 0.7 & 0.1 & 0 \\ 0.1 & 0.2 & 0.7 & 0 & 0 \\ 0.1 & 0.2 & 0.6 & 0.1 & 0 \end{bmatrix}$$

According to $B_1 = A_1 \circ R_1$, calculate the fuzzy membership degree vector of the operating risks factor as a whole to the remark set

$$B_1 = (0.4286, 0.4286, 0.1428) \circ \begin{bmatrix} 0.1 & 0.1 & 0.7 & 0.1 & 0 \\ 0.1 & 0.2 & 0.7 & 0 & 0 \\ 0.1 & 0.2 & 0.6 & 0.1 & 0 \end{bmatrix}$$
$$= (0.1000, 0.1571, 0.6857, 0.0571, 0)$$

Through normalization, it becomes to $(0.1000, 0.1571, 0.6858, 0.0571, 0)$

Similarly, we obtain the fuzzy membership degree vector of the execution risks factor

$$B_2 = A_2 \circ R_2 = (0.1000, 02477, 0.4889, 0.1634, 0)$$

And the fuzzy membership degree vector of the external risks factor

$$B_3 = A_3 \circ R_3 = (0.1036, 03988, 0.2759, 0.2217, 0)$$

By assembling B_1, B_2, B_3, construct the fuzzy membership matrix of the three first-level indexes as below

$$R = \begin{bmatrix} B_1 \\ B_2 \\ B_3 \end{bmatrix} = \begin{bmatrix} 0.1000 & 0.1571 & 0.6858 & 0.0571 & 0 \\ 0.1000 & 0.2477 & 0.4889 & 0.1634 & 0 \\ 0.1036 & 0.3988 & 0.2759 & 0.2217 & 0 \end{bmatrix}$$

Finally we compute the comprehensive assessment result

$$B = A \circ R = (0.1020, 0.3031, 0.4329, 0.1620, 0)$$

Normalization yields

$$(0.1020, 0.3031, 0.4329, 0.1620, 0)$$

Further, in order to make a comprehensive assessment on the risk level of this project, the five remark grades are assigned a score between 0 and 100 respectively as shown in Table III.

TABLE III
ASSIGNED SCORES OF REMARKS

Remark grade	very serious	serious	moderate	mild	very mild
Score assigned	100	80	60	40	20

Now we calculate the comprehensive score S of the risk in accordance with Table III.

$$S = 0.1020 \times 100 + 0.3031 \times 80 + 0.4329 \times 60 + 0.1620 \times 40 + 0 \times 20 = 66.902$$

Define the corresponding relation between comprehensive score and risk level as in Table IV. According to Table IV, the risk level of this project is "high".

TABLE IV
CORRESPONDENCE BETWEEN COMPREHENSIVE SCORE AND RISK LEVEL

Comprehensive score	$S > 80$	$60 < S \le 80$	$40 < S \le 60$	$20 < S \le 40$	$S \le 20$
Risk level	very high	high	moderate	low	very low

V. CONCLUSION

The risk assessment of international EPC project is an essential part in its risk management process, and is the important element to the decision making process of the engineering management. This paper presented the risk assessment for international EPC projects, focused on their specific characteristics. This work set up a two-tier evaluation system on risk assessment for international EPC projects from three aspects: project operation risks, construction execution risks and external risks. It used a layered analysis to decide the weights for each element, and utilized fuzzy comprehensive evaluation to provide a complete risk evaluation case study. The future research direction could be: expand and fine-tuning the proposed two-tier risk assessment system to a three-tier one, and the third tier evaluation process includes simple and easily execution on more quantitative and objective measurement.

REFERENCES

[1] Y. B. Zhang, Y. Q. Chen, *Contract Management for International EPC / Turnkey Projects* (in Chinese). Xi'an: China Electric Power Press, 2009

[2] T. M. Williams. "Risk Management Infrastructures", *International Journal of Project Management* 1993, vol.11, no.1, pp. 5-10

[3] R. L.Kliem, I. S. Ludin, *Reducing Project Risk.* Surrey, UK: Gower, 1997

[4] Z. F. Wang, *Risk Management of Engineering Projects - theory, method and application* (in Chinese). Beijing: China Water Power Press, 2003

[5] S. Q. Lei, *Risk Management and Insurance of International Engineering (2nd ed.)* (in Chinese). Beijing: China Architecture & Building Press, 2002

[6] W. H. Qiu, *Modern Project Risk Management: Method and Practice* (in Chinese). Beijing: Science Press, 2003

[7] T. J. Deng, *Engineering Risk Management* (in Chinese). Beijing: China Communications Press, 2004

[8] N. G. Bunni, *The FIDIC Forms of Contract (3rd ed.).* Blackwell, 2005

[9] G. Patricia, "Design-Build / EPC Contractor's Heightened Risk—Changes in a Changing World", *J. Legal Affairs Dispute Resolution Eng. Constr.* Vol.1, no.2, pp.7-15

[10] X. H. Meng, C. R. Dunzhu, Q. Zhao, "Comparison of the EPC General Contract Mode and Traditional Mode" (in Chinese), *International Economic Cooperation,* 2004, no.11, pp. 49-50

[11] X. H. Meng, "The owner and contractor's risk sharing and countermeasures with EPC mode" (in Chinese), *International Economic Cooperation*, 2005, no.12, pp. 45-46

Specification Test for Segmented Time-varying Diffusion Models

Jing-wei Cai [1,2,*], Ping Chen[1]

[1] Department of Statistics and Financial Mathematics, Nanjing University of Science and Technology, Nanjing, China

[2] Department of Basic Courses, Jiangsu Polytechnic College of Agriculture and Forestry, Zhenjiang, China

(*Jingwei_cai@126.com, prob123@mail.njust.edu.cn)

Abstract - **In this paper, a generalized residual goodness of fit test is proposed as an improvement on an existing test method for one-dimension diffusion models. By comparative analysis, we find that the generalized residual goodness of fit test is an effective test method. When diffusion models are used to describe stock prices, interest rates and exchange rates, the parameters of their drift function and diffusion function will change over time rather than be constants due to changeful economic environment. Based on this idea, we propose a diffusion model with segmented time-varying parameters. The test proposed above is used to examine the segmented time-varying characteristic of the model parameters.**

Keywords - **Diffusion models, generalized residual goodness of fit test, segmented time-varying**

I. INTRODUCTION

Diffusion models have been widely used to describe the dynamic properties of basic financial variables such as interest rates, exchange rates and stock prices. With the aid of the interpretation for the economic background in a particular aspect, people have put forward all kinds of parametric models to describe these variables. For example, people describe the changes of the instantaneous interest rates and the exchange rates by the Cox-Ingersoll-Ross model [1] (abbreviating Cox-Ingersoll-Ross model as CIR model in the rest of the paper) and describe the stock prices by the Geometric Brownian motion, and so on. They have also derived the closed form solutions for a variety of derivative securities prices in the specifications of some simple models. For example, they have derived the European option pricing formula based on the Geometric Brownian motion and interest rate derivatives pricing formula based on the CIR model. Because derivative securities pricing and risk measures are both associated with specific forms of model specification, the model misspecification may lead to major errors of pricing, hedging and risk management. Therefore, the specification test of continuous time financial model is very important. Recently increasing literatures begin to concern about the test issues of diffusion class models [2, 3].

Pioneering work for the nonparametric specification test of diffusion models is proposed by Aït-sahalia [4]. He pointed out that the marginal density of a diffusion model was completely determined by their drift function and diffusion function, so the test statistic could be obtained by comparing the model marginal density estimator with its nonparametric counterpart obtained by discretely sampled data. He also proved that the statistic had the asymptotic normality. The advantage of this test method is that the structure of the test statistic is simple, and the method has a good result for the test of model with the same marginal distribution. However, the marginal density function cannot capture completely the dynamic characteristics of the underlying variable process, and it cannot distinguish two diffusion models which have the same marginal density function but do not have the same transition density function. Additionally, owing to the lasting dependency of sample data of diffusion process and the slow convergence rate of kernel density estimator, the finite sample properties of this test are not so good [5, 6, 7].

The test [4] was based on the transition density rather than the marginal density. Transition density contains all the information of a diffusion process for its Markov property. Based on this idea, Hong and Li [8] proposed a nonparametric test (Q test method) based on generalized residuals obtained by a dynamic probability integral transform. Its basic idea is, if the model is specified correctly, the series of generalized residual defined by the transition density function of the data should be i.i.d.. Given a certain time lag between the two states, if the joint kernel density estimator was close to 1, they judged the generalized residual series should be i.i.d.. This test method has some advantages. Firstly, the use of the transition density makes test statistic contain more comprehensive information about the underlying variable. It effectively compensates for the lack of the test [4] which may loss those alternatives that have the same marginal density with the null hypothesis. Meanwhile, because the generalized residual series have no dependency when the model is specified correctly, their nonparametric density estimators have a better performance in finite sample. Secondly, because the test gives regular condition to the transition density rather than the stochastic differential equation of the underlying process, it can be used in wider fields. In addition to single variable homogeneous diffusion models, for more continuous time or discrete time models, such as the non homogeneous, stochastic volatility, jump diffusion and multivariate diffusion models, the test can be applied well. The method provides a uniform framework under which people can determine if a model is correctly specified by means of an examination whether the generalized residual series are i.i.d..

However, some of the simulation analysis of the specification tests for the Geometric Brownian motion and the CIR model show that the test based on conditional density by Hong and Li has a trend to accept the null hypothesis excessively to a certain extent. The method produces a big probability of judging by mistake some of the data belong to the null hypothesis, in fact, those data

E. Qi et al. (eds.), *Proceedings of the 5th International Asia Conference on Industrial Engineering and Management Innovation (IEMI2014)*, Proceedings of the International Asia Conference on Industrial Engineering and Management Innovation 1, DOI 10.2991/978-94-6239-100-0_28,

belong to another model which is just similar to the null hypothesis. The reason may be investigated from the structure of the test statistic. It should be noted that, the major factor of (a test statistic in Hong and Li [8], includes the nonparametric estimator of two-dimensional joint density . Though the consistency of statistic has been proved theoretically, the influence of the "dimensional disaster" cannot be ignored in finite sample.

In the financial field, economic conditions are subject to change, so it is necessary to consider that the instantaneous expected return and instantaneous volatility of an asset are not only related to time but also related to the specified state variable such as the price level of stock or bond. Naturally, the underlying state variable should meet time-varying diffusion processes. At present, the time-varying diffusion models have been paid more and more attention to [9, 10, 11]. The representative work on the study of testing the model with time-varying parameters is the quasi likelihood ratio test [12]. They constructed a quasi likelihood test statistic. In a simulation analysis for 1461 sample data, the test reflected a good performance.

In this paper, we propose a specification test for diffusion models. The test's principle is same as Hong and Li, we still determine that a model is correctly specified if the generalized residual series are i.i.d. , but we use goodness of fit to examine the generalized residual series by contingency tables instead of the nonparametric estimator of the joint probability density. We will call this improved method a generalized residual goodness of fit test which effectively avoids the defect of excessive acceptance. A diffusion model with time-varying parameters is also proposed. The proposed model with time-varying parameters is different from the past, its parameters are not always in change. In a relatively short time, we assume that the parameters of the time-varying model are constants, while in a long time, they change with time, namely, parameters are mutative piecewise, so the proposed model is more practical. The generalized residual goodness of fit test method is used to test the segmented time-varying characteristic of a model.

II. GENERALIZED RESIDUALS GOODNESS OF FIT TEST

As the generalized residual goodness of fit test is most closely related to Hong and Li's Q test [8], so we follow this test first.

Suppose a state variable X_t follows a continuous-time diffusion model with dynamic state represented by *Itô* stochastic differential equation:

$$dX_t = \mu(X_t)dt + \sigma(X_t)dB_t, \qquad (1)$$

where $\mu(\cdot)$ and $\sigma^2(\cdot)$ are, respectively, the drift function and diffusion function, and B_t is a standard Brownian motion. Suppose the parametric specification of model (1) is:

$$dX_t = \mu(X_t,\theta)dt + \sigma(X_t,\theta)dB_t, \qquad (2)$$

let $\mu_0(\cdot)$ and $\sigma_0^2(\cdot)$ are, respectively, the true drift function and diffusion function. It is often assumed that they belong to a certain parametric family:

$$\Im = \{(\mu(\cdot,\theta),\sigma^2(\cdot,\theta)) \big| \theta \in \Theta\}, \qquad (3)$$

where $\Theta \subseteq R^d$, d is a integer. We want to know whether parameters exist to make the assumption of the model (2) is acceptable and $(\mu_0(X_t),\sigma_0^2(X_t)$ is in the above parametric family.

Let $\{X_{t_i}, i = 1,2,\cdots,n\}$ be a discrete sample of the diffusion model observed in a period of time $[0,T]$, and $\Delta = T/n$ be the sample interval. The transition density of X_t can capture the all dynamic characteristics of the model. Let $p_0(x,t|y,s)$ be the transition density of X_t , namely, for any $t\in[0,T]$, the conditional density of X_t given $X_s = y$. For the diffusion model (2), given the drift coefficient $\mu_0(X_t,\theta)$ and diffusion coefficient $\sigma_0^2(X_t,\theta)$, the transition density $p(x,t|y,s,\theta)$ has a closed expression. The hypothesis we want to test is:

$$H_0 : p(x,t|y,s,\theta_0) = p_0(x,t|y,s), \qquad (4)$$

for some $\theta_0 \in \Theta$ and almost all $t\in[0,T]$, versus the alternative hypothesis:

$$H_A : p(x,t|y,s,\theta) \neq p_0(x,t|y,s), \qquad (5)$$

for all $\theta\in\Theta$ and some $t\in[0,T]$.

Hong and Li use the following probability integral transform to convert the sample $\{X_{\tau\Delta}, \tau = 1,2,\cdots,n\}$:

$$Z_\tau(\theta) = \int_{-\infty}^{X_{\tau\Delta}} p(x,\tau\Delta | X_{(\tau-1)\Delta},(\tau-1)\Delta,\theta)dx, \tau = 1,\cdots,n. \qquad (6)$$

$\{Z_\tau(\theta)\}_{\tau=1}^n$ are called the generalized residuals of the model $p(x,t|y,s,\theta)$. According to Rosenblatt's theory (Rosenblatt 1952), the generalized residual series $\{Z_\tau \equiv Z_\tau(\theta_0)\}_{\tau=1}^n$ are i.i.d. $U[0,1]$ under the null hypothesis. Hong and Li gave a centered and scaled test statistic $\hat{Q}(j)$. If the model was correctly specified, given each of the lag j , then $\hat{Q}(j) \xrightarrow{d} N(0,1)$.

Unlike Hong and Li's idea of constructing test statistics, in this paper, we consider the goodness of fit test. It is still the basis of the test whether or not the generalized residual series of the specified model are i.i.d. $U[0,1]$. However, we would use the goodness of fit test by contingency tables to investigate whether $\{Z_\tau, Z_{\tau-k}\}_{\tau=k}^n (\forall k \geq 1)$ meet a uniform distribution on $[0,1]\times[0,1]$ rather than consider whether the generalized residual series $\{Z_\tau\}_{\tau=1}^n$ are i.i.d. directly. We will call this test method a generalized residual goodness of fit test. Accordingly, the test for the null hypothesis (4) and the alternative hypothesis (5) will be carried out as follows.

Step 1: Use an estimator which meet a property of $\sqrt{n}(\hat{\theta}-\theta^*)=O_p(1)$ to estimate the parameters of the specified model (4).

Step 2: Calculate the generalized residual series $\{\hat{Z}_\tau = Z_\tau(\hat{\theta})\}_{\tau=1}^n$ according to the equation (6). (The first two steps are similar to Hong and Li's Q test).

Step 3: Given a lag k, implement the goodness of fit test by contingency tables to investigate whether $\{Z_\tau, Z_{\tau-k}\}_{\tau=k}^n (\forall k \ge 1)$ meet a uniform distribution on rectangle $[0,1]\times[0,1]$. Concretely, divide the rectangle $[0,1]\times[0,1]$ into m^2 equal parts (one may choose $m=[\sqrt{n/10}]$, the size of m is determined by the sample size.), and denote $n_{ij}^{(k)}$ as the frequency number that sample points $\{Z_\tau, Z_{\tau-k}\}_{\tau=k}^n$ fall in the grid D_{ij} which located by line $(i-1)$, line i, column $(j-1)$ and column j. We may get $P\{\{Z_\tau, Z_{\tau-k}\}\in D_{ij}\}=1/m^2$ under the assumption that $\{Z_\tau, Z_{\tau-k}\}_{\tau=k}^n$ is the uniform distribution on $[0,1]\times[0,1]$. According to the Pearson Theorem, there are approximately:

$$K(k)=\sum_{j=1}^{m}\sum_{i=1}^{m}\frac{(n_{ij}^{(k)}-n/m^2)^2}{n/m^2} \sim \chi^2(m^2-1). \qquad (7)$$

So, if $K(k)$ is greater than $\chi^2(m^2-1,1-\alpha)(\forall k \ge 1)$, then we will reject the null hypothesis under the significance level α.

III. COMPARATIVE ANALYSIS OF THE HL'S METHOD AND THE GENERALIZED RESIDUAL GOODNESS OF FIT TEST

A. Geometrical Brownian motion model is the true model, the null hypothesis is that the sample data meet a Geometric Brownian motion model

The true model is supposed as the following Geometrical Brownian motion:

$$GB: dX_t = 0.5X_t dt + 0.1X_t dW_t,$$

where W_t is a standard Brownian motion.

The sizes of sample that we will choose are 200, 500, 1000 respectively, and for each size of sample, we will simulate 100 times. The listed values of statistics are the average values of 100 tests respectively. For the generalized residual goodness of fit test, if $K(1)\ge \chi^2(m^2-1,1-\alpha)$, we will reject the null hypothesis. Let $m=10, \alpha=0.05$, then $\chi^2(99,0.95)\approx 77.0$. We will test the Geometrical Brownian motion hypothesis by the HL's method and our method respectively.

From the Table I, we can conclude, for the null hypothesis that is accepted by the HL's method, by our method, the average values of the test statistic are all less than the critical value and most of the results in simulations accept the null hypothesis. So the model that can be tested by the HL's method also can be tested by our method.

TABLE I
GEOMETRICAL BROWNIAN MOTION MODEL IS THE TRUE MODEL, THE NULL HYPOTHESIS IS THAT THE SAMPLE DATA MEET A GEOMETRIC BROWNIAN MOTION MODEL

Sample size	$\hat{Q}(j)$	Test results	$K(1)$	Test results
n=200	-68.07	78%accepted	76.30	76%accepted
n=500	-62.15	77%accepted	63.62	74%accepted
n=1000	-64.83	82%accepted	65.74	86%accepted

B. Geometric Brownian motion model is the true model, the null hypothesis is that sample data meet a CIR model

Now let the sample data meeting a CIR model be the null hypothesis, we will simulate 200, 500, 1000 data from the Geometrical Brownian motion model, and we will simulate these data 100 times respectively.

TABLE II
GEOMETRICAL BROWNIAN MOTION MODEL IS THE TRUE MODEL, THE NULL HYPOTHESIS IS THAT THE SAMPLE DATA MEET A CIR MODEL

Sample size	$\hat{Q}(j)$	Test results	$K(1)$	Test results
n=200	-30.43	47% rejected	109.59	69% rejected
n=500	-14.60	30% rejected	118.16	65% rejected
n=1000	-68.31	65% rejected	104.53	72% rejected

Table II unveils quite satisfactory performances for both tests. When the Geometrical Brownian motion model is the true model, the average values of the test statistics are all larger than the critical value by the generalized residual goodness of fit test. The results of 70% of experiments reject the null hypothesis, so our method improves the test efficiency relatively in contrast to the HL's method.

C. CIR model is the true model, the null hypothesis is that the sample data meet a Geometric Brownian motion model

The true model is supposed as the following CIR model:

$$dX_t = k(\alpha - X_t)dt + \sigma\sqrt{X_t}dW_t,$$

where $(k,\alpha,\sigma)=(0.89218, 0.090495, 0.032747)$, and W_t is a standard Brownian motion.

TABLE III
CIR MODEL IS THE TRUE MODEL, THE NULL HYPOTHESIS IS THAT THE SAMPLE DATA MEET A GEOMETRIC BROWNIAN MOTION MODEL

Sample size	$\hat{Q}(j)$	Test results	$K(1)$	Test results
n=200	-64.34	53% rejected	100.10	55% rejected
n=500	-51.82	56% rejected	93.24	61% rejected
n=1000	-27.08	58% rejected	79.40	60% rejected

Analogously, we simulate 100 sample trajectories for the above CIR model, the sizes of sample are 200, 500, 1000 respectively. Let the sample data meeting a Geometrical Brownian motion model be the null hypothesis, we test the hypothesis by the HL's method and our method respectively.

Table III, we can see that the average values of the statistics based on the generalized residual goodness of fit test are all larger than the critical value, and in the 100 simulation tests, there are only 40% accept the null hypothesis, so our method is more effective more or less than HL's method which accept the Geometrical Brownian motion hypothesis more easily when the CIR model is the true model.

IV. SEGMENTED TIME-VARYING TEST

In terms of the modeling of financial data, one issue should be paid more attention to. If data meet a stochastic differential equation, the equation's drift and diffusion terms may be constants or only the functions of the state variable, may have time-varying parameters as well. It determines which kind of model we will use to describe the data whether or not the drift and diffusion coefficients have time-varying characteristic. Therefore, the time-varying test of a model is necessary. A diffusion model with time-varying parameters no longer has invariable one-dimensional marginal density, so a test based on marginal density does not apply to a model with time-varying parameters, while a test based on transition density does do.

If we divide a sample which contains n data into l parts and each part has $n_1,\cdots,n_l$ data respectively, then $\sum_{i=1}^{l} n_i = n$. The model we will consider is such a model that, in a relatively short period of time, the parameters of the model are constants, while in the whole sample time, they are mutative piecewise. This kind of model can be better in accorded with the changes of the real economic variables.

Suppose a state variable X_t follows a diffusion model with time-varying parameters:

$$dX_t = \mu(t, X_t, \theta_{1,t}, \theta_{2,t}, \cdots)dt + \sigma(t, X_t, \theta_{1,t}, \theta_{2,t}, \cdots)dB_t, \quad (8)$$

where $\mu(t, X_t, \theta_{1,t}, \theta_{2,t}, \cdots)$ and $\sigma^2(t, X_t, \theta_{1,t}, \theta_{2,t}, \cdots)$ are the drift and diffusion functions respectively, B_t is a standard Brownian motion, and $(\theta_{1,t}, \theta_{2,t}, \cdots)$ are the time-varying parameters. For convenience, here we assume that there are only two time-varying parameters, then the diffusion model with time-varying parameters may be represented by the following differential equation:

$$dX_t = \mu(t, X_t, \alpha_t, \beta_t)dt + \sigma(t, X_t, \alpha_t, \beta_t)dB_t, \quad (9)$$

$\alpha_t = \sum_{i=1}^{l} \alpha_i I\{t_{i-1} < t \le t_i\}, \beta_t = \sum_{i=1}^{l} \beta_i I\{t_{i-1} < t \le t_i\}$ and $I(\cdot)$ is the indicator function. Given $i \in \{1, 2, \cdots, n\}$, α_i, β_i are constants respectively. The proposed model takes into account that the whole sample period $[0,T]$ is divided into l segments: $[0,t_1],\cdots,(t_{l-1},t_l]$. A remarkable characteristic of the model is that its parameters change piecewise, accordingly, the transition density of X_t change piecewise. Let

$$p_i(x, t_{i-1} + \tau\Delta | X_{t_{i-1}+(\tau-1)\Delta}, t_{i-1} + (\tau-1)\Delta)$$

(where $\Delta = (t_i - t_{i-1})/n_i, \tau = 1,\cdots,n_i$) be the transition density of X_t in $(t_{i-1}, t_i]$.

Next, we will examine the segmented time-varying characteristic of the model parameters. Firstly, we use the generalized residual goodness of fit test to examine if the overall sample data meet the model with determined parameters by the model (4). If the null hypothesis (4) is accepted, then the parameters do not have time-varying characteristic. Regard the overall sample meeting the model with segmented time-varying parameters as the null hypothesis:

$$H_0 : p(x, t_{i-1} + \tau\Delta | X_{t_{i-1}+(\tau-1)\Delta}, t_{i-1} + (\tau-1)\Delta, \alpha_{i,0}, \beta_{i,0}) = p_i(x, t_{i-1} + \tau\Delta | X_{t_{i-1}+(\tau-1)\Delta}, t_{i-1} + (\tau-1)\Delta), \quad (10)$$

for some $(\alpha_{i,0}, \beta_{i,0})$, for all $\tau = 1,\cdots,n_i (i = 1,\cdots,l)$, versus the alternative hypothesis:

$$H_A : p(x, t_{i-1} + \tau\Delta | X_{t_{i-1}+(\tau-1)\Delta}, t_{i-1} + (\tau-1)\Delta, \alpha_i, \beta_i) \ne p_i(x, t_{i-1} + \tau\Delta | X_{t_{i-1}+(\tau-1)\Delta}, t_{i-1} + (\tau-1)\Delta), \quad (11)$$

for all (α_i, β_i), some $\tau = 1,\cdots,n_i (i = 1,\cdots,l)$.

We may calculate the generalized residuals $\{Z_\tau(\theta)\}$ of a segment $\{X_{t_{i-1}+\tau\Delta}, \tau = 1,\cdots,n_i, \Delta = (t_i - t_{i-1})/n_i\}$ just like the equation (6), and get the corresponding test statistic $K(k)$. If the overall sample data accept the null hypothesis (10), namely, each of the l segments meets the model with corresponding parameters, then the overall sample data meet the diffusion model with segmented time-varying parameters.

In a word, if one wants to test the segmented time-varying characteristic of a model, he or she need to examine that the test statistic $K(k)$ of the overall sample is greater than $\chi^2(m^2 - 1, 1-\alpha)$ for some $k(\ge 1)$ and the test statistic $K(k)$ is less than $\chi^2(m_i^2 - 1, 1-\alpha)(i = 1,\cdots l)$ for all integer $k(\ge 1)$ for each of the segments. If and only if the two conditions are both satisfied, we can predicate that the model belongs to the diffusion model with segmented time-varying parameters.

V. CONCLUSION

There are some deficiencies for using diffusion models with determined parameters to describe the interest rates, exchange rates and stock prices. Some literatures consider to using the diffusion models with time-varying parameters to describe interest rates, exchange rates and so on. In this paper, we propose the generalized residual goodness of fit test and use it to examine the diffusion models with segmented time-varying parameters. We can also extend other diffusion models to the circumstances with segmented time-varying parameters, such as the Geometric Brownian motion with segmented time-varying parameters, the Vasicek model with segmented time-varying parameters, and so on.

ACKNOWLEDGMENT

This research is partially supported by the National Natural Science Foundation of China (NSFC, Grant No. 11271189 and 11201229).

REFERENCES

[1] J. C. Cox, J. E. Ingersoll, and S. A. Ross, "A theory of the term structure of interest," *Econometrica*, vol. 53, pp. 385-407, 1985.

[2] Y. Aït-sahalia, and J. Q. Fan, "Nonparametric transition-based tests for jump diffusions," *Journal of the American Statistical Association*, vol.104, pp. 1102-1116, 2009.

[3] S. X. Chen, J. T. Gao, and C. Y. Tang, "A test for model specification of diffusion processes," *The Annals of Statistics*, vol. 36, pp. 167-198, 2008.

[4] Y. Aït-sahalia, "Testing continuous time models of the spot interest rate," *The Review of Financial Studies*, vol. 9, pp.385-426, 1996.

[5] F. Bandi, and P. C. B. Phillips, "Fully nonparametric estimation of scalar diffusion models," *Econometrica*, vol. 71, pp. 241-284, 2003.

[6] Z. W. Cai, J. Q. Fan, and Q. W. Yao, "Functional-coefficient regression models for nonlinear time series," *Journal of the American Statistical Association*, vol. 95, pp. 941-956, 2000.

[7] J. Q. Fan, and C. M. Zhang CM, "A reexamination of diffusion estimators with applications to financial model validation," *Journal of the American Statistical Association*, vol. 98, pp. 118-134, 2003.

[8] Y. M. Hong, and H. T. Li, "Nonparametric specification testing for continuous-time models with applications to term structure of interest rates," *The Review of Financial Studies*, vol. 18, pp. 37-84, 2005.

[9] R. Becker, W. Enders, and S. Hurn, "A general test for time-dependence in parameters," *Journal of Applied Econometrics*, vol. 19, pp. 899-906, 2004.

[10] P. Chen, and J. D. Wang, "Wavelet estimation of the diffusion coefficient in time dependent diffusion models," *Science in China Series A: Mathematics*, vol. 50, pp. 1597-1610, 2007.

[11] S. Orbe, E. Ferreira, and J. Rodriguez-Poo, "On the estimation and testing of time varying constraints in econometric models," *Statistica Sinica*, vol. 16, pp. 1313-1333, 2006.

[12] J. Q. Fan, J. C. Jiang, C. M. Zhang, and Z. W. Zhou, "Time-dependent diffusion models for term structure dynamics," *Statistica Sinica*, vol. 13, pp. 965-992, 2003.

Multidimensional R&D-Price Level Game Model Analysis of Duopoly Group

Xin Zhang [1,*], Jing-han Zhi [1], Cheng-ke Zhang [2]
[1]School of Management, Guangdong University of Technology, Guangzhou, 510520
[2]School of Economics & Commence, Guangdong University of Technology, Guangzhou, 510520
(*Damixiao@126.com, zhangck@gdut.edu.cn)

Abstract - **For meeting the need of development of enterprises integration and the trend of collaborative technological innovation, this paper establishes two dimensional game models of R&D cooperation and non-cooperation, according to the situation that there are two duopoly groups with two types of products. Based on the conditions of knowledge spillover and absorptive capacity, specifically analyze the impacts of each case of the knowledge spillover, the absorptive capacity of R&D investment, profits and product price level. Game analysis and numerical simulation results, considering from the level of R&D investment and equilibrium price point view, show that cooperation case could gain more profits than semi cooperation and competition case, which also precisely validate that collaborative innovation will be achieved when the cluster enterprises achieve maximum profit; and It is beneficial for R&D and pricing of oligopoly.**

Keywords - **Duopoly group, game model, price level**

I. INTRODUCTION

To promote innovation capability of enterprises by gather resources and maintain the core competiveness of enterprise and expand the market share, more and more companies on the way of integration. At the same time, enterprises are increasingly focused on a variety of similar institution conduct collaborative innovation for gain maximum benefit with pay at least. Enterprises R&D collaborate with different external institute, especially competitors, to get a variety of development opportunities and exchange information, knowledge or other resources to achieve complementary advantages, resources sharing knowledge, risk sharing and benefit sharing for sustainable development. Therefore, development of enterprises showing collectivize trends (ie, develop and produce various types of technology-related products simultaneously) and increasingly focused on competition and cooperation collaborative innovation with competitors. Traditional theory model is on the premise of oligopoly to produce a single product as a hypothesis. So it can not accurately analyze the evolution process of collaborative innovation about multi-product production and R&D investment level between competition and cooperation in cluster enterprises [1~4]. Multidimensional game model is an appropriate research analysis tool [5]. Liu studies the correlative problem of enterprise investment and market competition, which gets the results analysis close to the actual situation [6]. But he did not study the situation of the cluster produce more than two kinds of products in the R&D competing about input and output [5]. In this paper defining the R&D output including R&D input level, product pricing and profit based on the one-dimensional game of DJ model, and divide R&D cooperation between enterprises into three scenarios [7]. The following section is to establish a two-dimensional game model respectively according to the situation that there are two kinds of products in two enterprises, and to explore the Impact of two types of knowledge spillovers, absorptive capacity input to R&D, profits and product price levels in each of these cases.

II. TWO DIMENSIONAL PRICE-R&D LEVEL GAME MODEL OF DUOPOLY

The model assumptions established in this paper are standard, which are mainly quote from literature [6], and hence it is beneficial to comparing the results between this article and existing literature. To suppose two firms decision variable price is $p^{(i)}=\begin{bmatrix}p_{i1}\\p_{i2}\end{bmatrix}$, and input levels of R&D is $x^{(i)}=\begin{bmatrix}x_{i1}\\x_{i2}\end{bmatrix}$, The demand functions of the jth products of ith enterprise is:

$$q_{ij}=b_j-p_{ij}+r_{3-i,j}p_{3-i,j}+\beta_j(p_{i,3-j}+q_{3-i,3-j}).i,j=1,2 \quad (1)$$

Where: r_j denotes correlation coefficients of demand that is the relation between the $3\text{-}j^{th}$ product and the j product (complementary or alternative), $a_j>0$, $r_j\in[-1,1]$. The effective research and development contribution levels for the jth of enterprise i is:

$$h_{ij}=x_{ij}+\theta_j x_{i,3-j}+kg_i(x_{3-i,j}+\theta_j x_{3-i,3-j}),\quad k,g_i\in[0,1] \quad (2)$$

Where: θ_j indicates economic coefficient of economies of scope that reflects the economic development contribution levels of the j th and internal product of investment in research and development on $3-j$ th products. k denotes knowledge spillover coefficients, and the constant g_i denotes knowledge absorptive capacity [8~9] for the enterprise i ; kg_i is known as knowledge absorption coefficient of enterprise i, $kg_i\in[0,1]$. The marginal cost of the jth products of ith enterprise can be

E. Qi et al. (eds.), *Proceedings of the 5th International Asia Conference on Industrial Engineering and Management Innovation (IEMI2014)*, Proceedings of the International Asia Conference on Industrial Engineering and Management Innovation 1, DOI 10.2991/978-94-6239-100-0_29, © Atlantis Press and the authors 2015

expressed:

$$C_{ij} = c_{ij} - h_{ij} = c_{ij} - [x_{ij} + \theta_j x_{i,3-j} + kg_i(x_{3-i,j} + \theta_j x_{3-i,3-j})] \quad (3)$$

So that building the two dimensional game model of duopoly with output and R&D input level as follows: Game player: duopoly; The strategy of game participants: there are two kinds of the products expression of strategy vector selected by the firm i (i=1,2), described as $(p_{i1},\ p_{i2},\ x_{i1},\ x_{i2}) \geq 0$, $(p_{i1},\ p_{i2},\ x_{i1},\ x_{i2}) \in P_{i1} \times P_{i2} \times X_{i1} \times X_{i2}$, where P_{ij} and X_{ij} denote the strategy space of price and R&D level of the jth products of ith enterprise respectively. Payoff function of participants: the firm i's payoff function is profit function as fllow:

$$\begin{aligned} U_1 = & -b^T C^{(1)} + b^T P^{(1)} + b^T \theta x^{(1)} + kg_1 b^T \theta x^{(2)} - \tfrac{1}{2} m x^{(1)^T} I x^{(1)} - P^{(1)^T} \\ & \beta^T P^{(1)} - P^{(1)^T} \beta^T \theta x^{(1)} + kg_1 P^{(1)^T} \beta^T \theta x^{(2)} + C^{(1)^T} \beta P^{(1)} + P^{(2)^T} \\ & A_{21} P^{(1)} + P^{(2)^T} A_{21} \theta x^{(1)} + kg_1 P^{(2)^T} A_{21} \theta x^{(2)} - C^{(1)^T} A_{21}^T P^{(2)} \end{aligned} \quad (4)$$

$$\begin{aligned} U_2 = & -b^T C^{(2)} + b^T P^{(2)} + b^T \theta x^{(2)} + kg_2 b^T \theta x^{(1)} - \frac{1}{2} m x^{(2)^T} I x^{(2)} - P^{(2)^T} \\ & \beta^T P^{(2)} - P^{(2)^T} \beta^T \theta x^{(2)} + kg_2 P^{(2)^T} \beta^T \theta x^{(1)} + C^{(2)^T} \beta P^{(2)} + P^{(1)^T} \\ & A_{12} P^{(2)} + P^{(1)^T} A_{12} \theta x^{(2)} + kg_2 P^{(1)^T} A_{12} \theta x^{(1)} - C^{(2)^T} A_{12}^T P^{(1)} \end{aligned} \quad (5)$$

Where $b = \begin{bmatrix} b_1 \\ b_2 \end{bmatrix}$. $C^{(1)} = \begin{bmatrix} c_{11} \\ c_{12} \end{bmatrix}$. $C^{(2)} = \begin{bmatrix} c_{21} \\ c_{22} \end{bmatrix}$. $\theta = \begin{pmatrix} 1 & \theta_1 \\ \theta_2 & 1 \end{pmatrix}$ (6)

$$\beta = \begin{pmatrix} 1 & -\beta_1 \\ -\beta_2 & 1 \end{pmatrix}. \quad A_{12} = \begin{pmatrix} r_{11} & \beta_2 \\ \beta_1 & r_{12} \end{pmatrix}. \quad A_{21} = \begin{pmatrix} r_{21} & \beta_2 \\ \beta_1 & r_{22} \end{pmatrix}$$

A. Competition case

Using converse inductive method of dynamic game, the analysis will begin from the second stage, which is stage of price competition [10~12].

(1) Stage 2: the price competition

Both enterprises choosing their own pricing variables $p^{(1)}$ and $p^{(2)}$ independently, which is beneficial to maximize their corporate profits; according to competition equilibrium principle of Symmetrical Knott, the problems in this phase can be expressed as following:

$$\max_{p^{(i)}} U_i(p^{(1)},\ p^{(2)},\ x^{(1)},\ x^{(2)}) \quad (7)$$

the problem of Nash equilibrium condition is described as follows:

$$\begin{cases} \dfrac{\partial U_1}{\partial p^{(1)}} = b + \beta^T C^{(1)} - (\beta + \beta^T) p^{(1)} + A_{21}^T p^{(2)} \\ \qquad - \beta^T \theta x^{(1)} + kg_1 \beta^T \theta x^{(2)} = 0 & (8) \\ \dfrac{\partial U_2}{\partial p^{(2)}} = b + \beta^T C^{(2)} - (\beta + \beta^T) p^{(2)} + A_{12}^T p^{(1)} & (9) \\ \qquad - \beta^T \theta x^{(2)} + kg_2 \beta^T \theta x^{(1)} = 0 \end{cases}$$

Denote $C_1 = b + \beta^T C^{(1)}$, $C_2 = b + \beta^T C^{(2)}$, $A = \beta + \beta^T$

The equations of Solving this variable $p^{(1)}$ and $p^{(2)}$ is

$$\begin{cases} p^{(1)} = E_0 + E_1 x^{(1)} + E_2 x^{(2)} & (10) \\ p^{(2)} = F_0 + F_1 x^{(1)} + F_2 x^{(2)} & (11) \end{cases}$$

$$F_0 = \left[A(A - A_{12}^T)^{-1} (A - A_{21}^T) + A_{21}^T \right]^{-1} \left[A(A - A_{12}^T)^{-1} (C_1 + C_2) - C_1 \right]$$

where: $F_1 = \left[A(A - A_{12}^T)^{-1} (A - A_{21}^T) + A_{21}^T \right]^{-1} \left[I - (1 - kg_2) A(A - A_{12}^T)^{-1} \right] \beta^T \theta$

$$F_2 = \left[A(A - A_{12}^T)^{-1} (A - A_{21}^T) + A_{21}^T \right]^{-1} \left[(kg_1 - 1) A(A - A_{12}^T)^{-1} - kg_1 I \right] \beta^T \theta$$

$$E_0 = (A - A_{12}^T)^{-1} (C_1 - C_2) - (A - A_{12}^T)^{-1} (A - A_{21}^T) F_0$$

$$E_1 = (A - A_{12}^T)^{-1} \left[(A - A_{21}^T) F_1 + (1 - kg_2) \beta^T \theta \right]$$

$$E_1 = -(A - A_{12}^T)^{-1} \left[(A - A_{21}^T) F_2 + (1 - kg_1) \beta^T \theta \right]$$

Put equations (10) and (11) into (8) and (9) we can obtain the profit functions of the two firms:

$$U_1 = U_1(x^{(1)},\ x^{(2)}) \overset{\Delta}{=} U_1\left(p^{(1)}(x^{(1)},\ x^{(2)}),\ p^{(2)}(x^{(1)},\ x^{(2)}),\ x^{(1)},\ x^{(2)}\right)$$

$$U_2 = U_2(x^{(1)},\ x^{(2)}) \overset{\Delta}{=} U_2\left(p^{(1)}(x^{(1)},\ x^{(2)}),\ p^{(2)}(x^{(1)},\ x^{(2)}),\ x^{(1)},\ x^{(2)}\right)$$

(2) Stage 1: R&D competition

When each of the two enterprises anticipates their own choosing price $p^{(1)}$ and $p^{(2)}$ the profits of corporation were $U_1(x^{(1)},\ x^{(2)})$ and $U_2(x^{(1)},\ x^{(2)})$; in order to maximize their profits $U_i(x^{(1)},\ x^{(2)})$, enterprises independently choose their own level of R&D $x^{(i)}$. According to Knott's symmetrical principle of solving of game model, Nash equilibrium problem in this stage is:

$$\max_{x^{(i)}} U_i(x^{(1)},\ x^{(2)})$$

Using the vector functions of compound seeking differential rule, Nash Equilibrium condition is:

$$\begin{cases} G_{11} p^{(1)} + G_{12} p^{(2)} + G_{13} x^{(1)} + G_{14} x^{(2)} = G_{10} \\ G_{21} p^{(1)} + G_{22} p^{(2)} + G_{23} x^{(1)} + G_{24} x^{(2)} = G_{20} \end{cases}$$

Where

$$G_{11} = F_1^T A_{21} - E_1^T A - \theta^T \beta$$

$$G_{12} = E_1^T A_{21}^T + \theta A_{21}^T = (E_1^T + \theta) A_{21}^T$$

$$G_{13} = (F_1^T A_{21} - E_1^T \beta^T) \theta - mI$$

$$G_{14} = kg_1 \left[E_1^T \beta^T + F_1^T A_{21} \right] \theta$$

$$G_{10} = F_1^T A_{21} C^{(1)} - E_1^T C_1 - \theta^T b$$

$$G_{20} = E_2^T A_{12} C^{(2)} - F_2^T C_2 - \theta^T b$$

$$G_{21} = (F_2^T + \theta) A_{12}^T$$
$$G_{22} = E_2^T A_{12} - F_2^T A - \theta^T \beta$$
$$G_{23} = kg_2 \left[E_2^T A_{12} + F_2^T \beta^T \right] \theta$$
$$G_{24} = \left(E_2^T A_{12} - F_2^T \beta^T \right) \theta - mI$$

Put equation（10）and（11）into:

$$\begin{cases} H_{11}x^{(1)} + H_{12}x^{(2)} = H_{10} & (12) \\ H_{21}x^{(1)} + H_{22}x^{(2)} = H_{20} & (13) \end{cases}$$

Where:

$$H_{11} = G_{13} + G_{11}E_1 + G_{12}F_1$$
$$H_{12} = G_{14} + G_{11}E_2 + G_{12}F_2$$
$$H_{21} = G_{23} + G_{21}E_1 + G_{22}F_2$$
$$H_{22} = G_{24} + G_{21}E_2 + G_{22}F_2$$
$$H_{10} = G_{10} + G_{11}E_0 + G_{12}F_0$$
$$H_{20} = G_{20} + G_{21}E_0 + G_{22}F_0$$

Combing equations (12) and (13), Nash equilibrium solution of R&D Input levels of two products is

$$\begin{cases} x^{(1)^*} = \left(H_{11} - H_{12}H_{22}^{-1}H_{21} \right)^{-1} \left(H_{10} - H_{12}H_{22}^{-1}H_{20} \right) \\ x^{(2)^*} = \left(H_{22} - H_{21}H_{11}^{-1}H_{12} \right)^{-1} \left(H_{20} - H_{21}H_{11}^{-1}H_{10} \right) \end{cases}$$

Two firms select cooperation when they stay on the stage of R&D level, and select competition on price under this situation.

(1) Phase 2: to compete completely on price

By the theory of Nash equilibrium in complete information static game, the situation about the two enterprises solving process of Nash equilibrium price is similar to competition case. The equilibrium price is

$$\begin{cases} p^{(1)} = E_0 + E_1x^{(1)} + E_2x^{(2)} & (16) \\ p^{(2)} = F_0 + F_1x^{(1)} + F_2x^{(2)} & (17) \end{cases}$$

(2) Phase1: to cooperate on the stage of R&D level two enterprises collude setting their R&D level $x^{(1)}$ and $x^{(2)}$ so as to maximize the coalition profit when they select their price $p^{(1)}$ and $p^{(2)}$.

$$\max_{(x^{(1)},\ x^{(2)})} U\left(x^{(1)},\ x^{(2)}\right) = U_1\left(x^{(1)},\ x^{(2)}\right) + U_2\left(x^{(1)},\ x^{(2)}\right)$$

The first order of the above problem is:

$$B_{11} = \left(F_1^T + kg_2\theta^T \right) A_{12}^T - E_1^T A - \theta^T \beta$$
$$B_{12} = \left(E_1^T + \theta \right)^T A_{21}^T - F_1^T A + E_1^T A_{12} + kg_2\theta^T \beta$$
$$B_{13} = \left(F_1^T A_{21} + kg_2 E_1^T A_{12} \right)\theta - \left(E_1^T - kg_2 F_1^T \right)\beta^T\theta - mI$$
$$B_{14} = \left(E_1^T A_{12} + kg_1 F_1^T A_{21} \right)\theta - \left(F_1^T - kg_1 E_1^T \right)\beta^T\theta$$
$$B_{10} = F_1^T A_{21}C^{(1)} + E_1^T A_{21}C^{(2)} - E_1^T C_1 - F_1^T C_2 - (1 + kg_2)\theta^T b$$
$$B_{21} = \left(F_2^T + \theta \right) A_{12}^T + F_2^T A_{21} + kg_1\theta^T \beta - E_2^T A$$
$$B_{22} = \left(E_2^T + kg_1\theta^T \right) A_{21}^T - F_2^T A - \theta^T \beta + E_2^T A_{12}$$
$$B_{23} = \left(F_2^T A_{21} + kg_2 E_2^T A_{12} \right)\theta - \left(E_2^T - kg_2 F_2^T \right)\beta^T\theta$$
$$B_{24} = \left(E_2^T A_{12} + kg_1 F_2^T A_{21} \right)\theta - \left(F_2^T - kg_1 E_2^T \right)\beta^T\theta - mI$$
$$B_{20} = F_2^T A_{21}C^{(1)} + E_2^T A_{12}C^{(2)} - E_2^T C_1 - F_2^T C_2 - (1 + kg_1)\theta^T b$$

Then conditions transfer to

$$\begin{cases} B_{11}p^{(1)} + B_{12}p^{(2)} + B_{13}x^{(1)} + B_{14}x^{(2)} = B_{10} & (18) \\ B_{21}p^{(1)} + B_{22}p^{(2)} + B_{23}x^{(1)} + B_{24}x^{(4)} = B_{20} & (19) \end{cases}$$

Put equation (16) and (17) into (18) and (19) we can obtain

$$\begin{cases} B_{11}\left[E_0 + E_1x^{(1)} + E_{(2)}x^{(2)}\right] + B_{12}\left[F_0 + F_1x^{(1)} + F_2x^{(2)}\right] + B_{13}x^{(1)} + B_{14}x^{(2)} = B_{10} \\ B_{21}\left[E_0 + E_1x^{(1)} + E_{(2)}x^{(2)}\right] + B_{22}\left[F_0 + F_1x^{(1)} + F_2x^{(2)}\right] + B_{23}x^{(1)} + B_{24}x^{(4)} = B_{20} \end{cases}$$

where

$$L_{11} = B_{11}E_1 + B_{12}F_1 + B_{13}$$
$$L_{12} = B_{11}E_2 + B_{12}F_2 + B_{14}$$
$$L_{21} = B_{21}E_1 + B_{22}F_1 + B_{23}$$
$$L_{22} = B_{21}E_2 + B_{22}F_2 + B_{24}$$
$$L_{10} = B_{10} - B_{11}E_0 - B_{12}F_0$$
$$L_{20} = B_{20} - B_{21}E_0 - B_{22}F_0$$

The equilibrium result of R&D level is

$$\begin{cases} x^{(1)^*} = \left(L_{11} - L_{12}L_{22}^{-1}L_{21} \right)^{-1} \left(L_{10} - L_{12}L_{22}^{-1}L_{20} \right) & (22) \\ x^{(2)^*} = \left(L_{22} - L_{21}L_{11}^{-1}L_{12} \right)^{-1} \left(L_{20} - L_{21}L_{11}^{-1}L_{10} \right) & (23) \end{cases}$$

B. Cooperation case

In this case, the two enterprises maximize the formed coalition benefits as code of conduct in R&D stage and market pricing stage.

(1) stage 2: the choice of coalition price

The two companies jointly determine the price of two kinds of products $p^{(1)}$, $p^{(2)}$and maximize the coalition profits, the question is:

$$\max_{(p^{(1)},\ p^{(2)})} U\left(p^{(1)},\ p^{(2)}\right) = U_1\left(p^{(1)},\ p^{(2)}\right) + U_2\left(p^{(1)},\ p^{(2)}\right)$$

By the function maximum principle, optimal conditions are:

$$\begin{cases} Ap^{(1)} - \left(A_{12} + A_{21}^T\right)p^{(2)} = C_1 + (kg_2A_{12} \\ -\beta^T)\theta x^{(1)} + \left(kg_1\beta^T + A_{12}\right)\theta x^{(2)} - A_{12}C^{(2)} \\ -\left(A_{21} + A_{12}^T\right)p^{(1)} + Ap^{(2)} = C_2 - A_{21}C^{(1)} + \\ (kg_2\beta^T + A_{21})\theta x^{(1)} + \left(kg_1A_{21} - \beta^T\right)\theta x^{(2)} \end{cases}$$

That are

$$\begin{cases} Ap^{(1)} - \left(A_{12} + A_{21}^T\right)p^{(2)} = C_1 + (kg_2A_{12} - \beta^T) \\ \theta x^{(1)} + \left(kg_1\beta^T + A_{12}\right)\theta x^{(2)} - A_{12}C^{(2)} \\ -\left(A_{21} + A_{12}^T\right)p^{(1)} + Ap^{(2)} = C_2 - A_{21}C^{(1)} \\ +(kg_2\beta^T + A_{21})\theta x^{(1)} + \left(kg_1A_{21} - \beta^T\right)\theta x^{(2)} \end{cases}$$

Taking the symmetry position of the two companies into account, set $C^{(1)}=C^{(2)}=C^{(0)}$, $C_1=C_2=C_0$, $A_{12}=A_{21}=A_0$, $g_1=g_2=g_0$, $x^{(1)}=x^{(2)}=x/2$ when they cooperation. There is

$$p^{(1)^*} = p^{(2)^*} = M_0 + M_1x$$

In this equation,

$$M_0 = \left(A - A_0 - A_0^T\right)^{-1}\left(C_0 - A_0 C^{(0)}\right)$$

$$M_1 = \frac{1}{2}\left(A - A_0 - A_0^T\right)^{-1}\left(kgA_0 - \beta^T + kg\beta^T + A_0\right)\theta x$$

Substitute them into jointly profit function.

$$U = -2b^T C^{(0)} + 2C_0 p^{(1)} + b^T\theta x + kgB^T\theta x - mx^T Ix$$
$$-2p^{(1)^T}\beta^T p^{(1)} - p^{(1)^T}\beta^T\theta x + kgp^{(1)^T}\beta^T\theta x + 2p^{(1)^T}$$
$$A_0 p^{(1)} + p^{(1)^T}A_0\theta x + kgp^{(1)^T}A_0\theta x - 2C^{(0)^T}A_0^T p^{(1)}$$

Let $\frac{dU}{dx} = 0$,

$$2M_1^T C_0 + \theta^T b + kg\theta^T b - 2mIx - 2M_1^T AM_1 -$$
$$M_1^T\beta^T\theta x - \theta^T\beta p^{(1)} + kgM_1^T\beta^T\theta x + kg\theta^T\beta p^{(1)}$$
$$+2M_1^T(A_0 + A_0^T)M_1 + M_1^T A_0\theta x + \theta^T A_0^T p^{(1)}$$
$$+kgM_1^T A_0\theta x + kg\theta^T A_0^T p^{(1)} - 2M_1^T A_0 C^{(0)} = 0$$

We get $M_2 - M_3 x + M_4 p^{(1)} = 0$,

then $x^* = -\left(M_4 M_1 - M_3\right)^{-1}\left(M_2 + M_4 M_0\right)$,

In this equation,

$$M_2 = 2M_1^T C_0 + \theta^T b + kg\theta^T b - 2M_1^T AM_1$$
$$+ 2M_1^T\left(A_0 + A_0^T\right)M_1 - 2M_1^T A_0 C^{(0)}$$

$$M_3 = 2mI + M_1^T\beta^T\theta - kgM_1^T\beta^T\theta - M_1^T A_0\theta - kgM_1^T A_0\theta$$

$$M_4 = kg\theta^T\left(\beta + A_0^T\right) + \theta^T\left(A_0^T - \beta\right)$$

III. NUMERICAL SIMULATION AND ANALYSIS

This section will use the equilibrium solution expression of game model under a variety of circumstances which by the results of theoretical analysis from previous section. To make use of the computer simulation by choosing different values of model parameters, trying to find how parameter changes affecting corporate profits changing trends by compare and analysis the results.

A. The simulation results in competition case

Assuming: $b_1 = b_2 = 10$, $m = 10$, $c_{11} = c_{12} = c_{21} = c_{22} = 0.2$, $kg_1 = kg_2 = 0.5$, $r_{11} = r_{12} = r_{21} = r_{22} = 0.1$, $\theta 1 = \theta 2 = \theta \in [0,1]$. In order to look for the trends rule, we make different the value of β and for simulation. For this we take $\beta = 0$(Due to limited space, the situation of β=0, 0.1, 0.2, 0.3, 0.4, 0.5, 0.6 is omitted), to analysis the impact of parameters θ on equilibrium price and the of R&D investment levels. The simulation results are shown as in Fig.1.

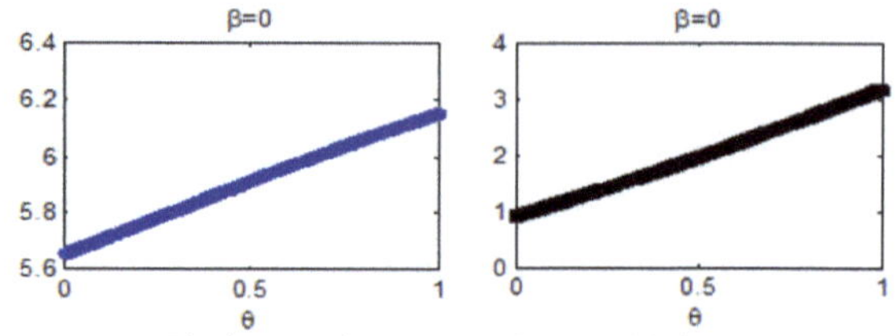

The impact of parameters θ on equilibrium price

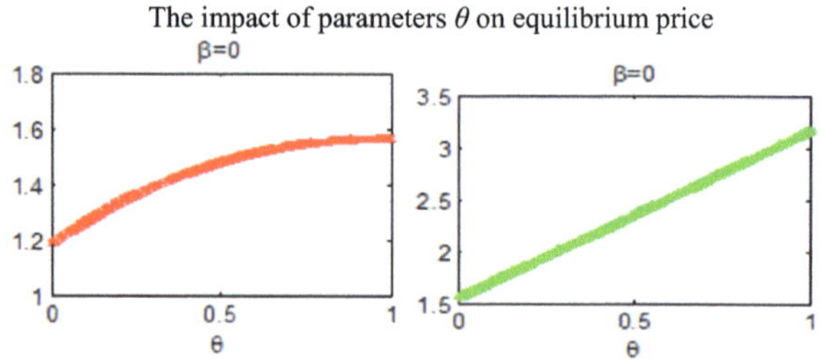

The impact of parameters θ on R&D investment levels

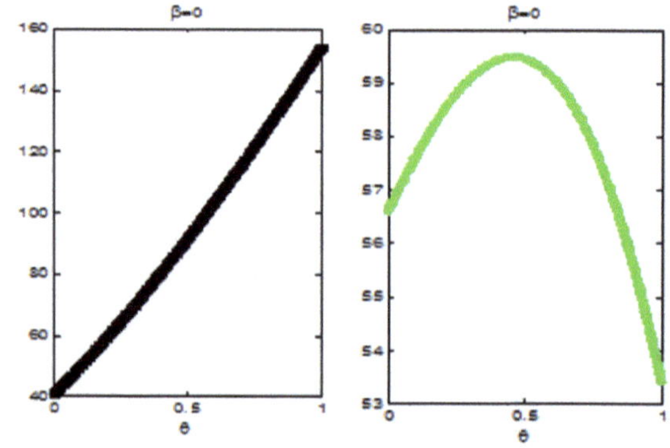

The impact of parameters θ on equilibrium profit

Fig.1. Simulation results 1

As can be seen from the simulation results, in the case of perfect competition, the effective range of β is 0. At this time, the correlation coefficient β is positively correlated with demand equilibrium price, equilibrium level of R&D investment and profit.

B. The simulation results in semi collusion situation

This section mainly analysis the affecting situation of θ on the equilibrium price and R&D when the β to a certain. For this we take $\beta = 0$ (Due to limited space, the situation of β=0, 0.1, 0.2, 0.3, 0.4, 0.5, 0.6 is omitted), simulation results are shown as in Fig.2.

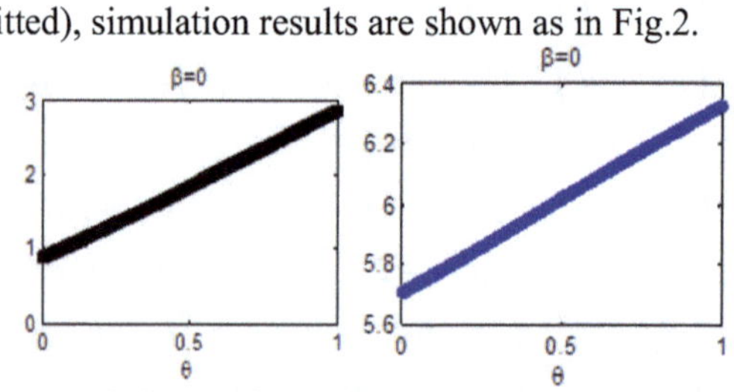

The impact of parameters θ on equilibrium price

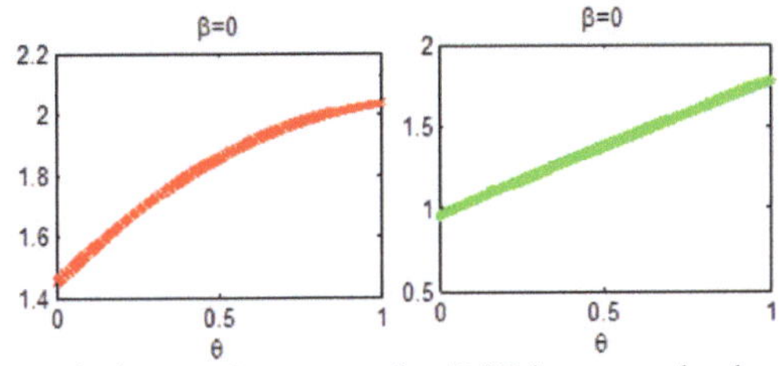

The impact of parameters θ on R&D investment levels

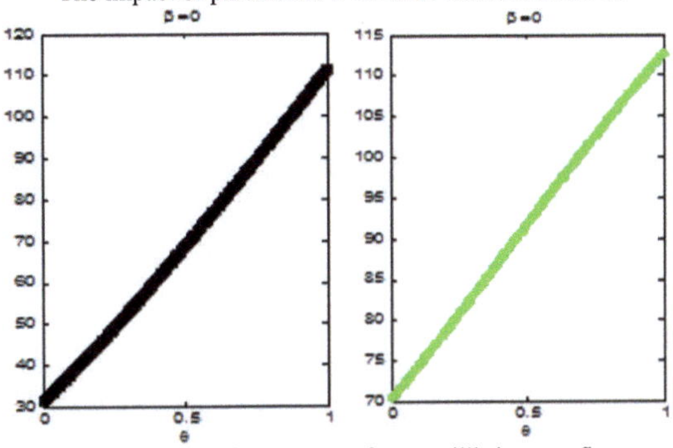

The impact of parameters θ on equilibrium profit

Fig.2. Simulation results 2

You can see from the results, the investment of enterprise 1and corporate profits of enterprise 2 are negative number when β exceeds 0.6, and this indicates that the effective range of the correlation coefficient β is [0,0.6].

IV. CONCLUSIONS AND POLICY RECOMMENDATIONS

In this paper, we consider two clusters of enterprises two kinds of products price $p^{(i)}$ and R&D investment levels $x^{(i)}$ as decision variables, while considering the basis of knowledge spillovers and the absorptive capacity[13~14] to establish two products price-R&D investment level in a two stages dimensional game model firstly. Secondly, analyzed and solve the models in a perfectly competitive, semi-competitive, full cooperation respectively, and given the equilibrium analytical expressions of these three circumstances by means of vector optimization techniques which lay the foundation for further theoretical analysis. Then, the equilibrium analytical expressions for the computer numerical simulation by choosing difference model parameters. We get the trend of equilibrium price and the equilibrium level of R&D investments and profits which be impacted by the changing of the parameters. These provide the inspirations for reference to company's R&D strategy.

V. ACKNOWLEDGEMENT

This work was supported in part by the National Science Foundation of China with grant No. 71171061, and Guangdong College of Humanities and Social Sciences project with grant No. 11WYXM022.

REFERENCES

[1] D'Aspremont, C., Jacquemin, A., "Cooperative and non-cooperative R&D in duopoly with spillovers", *The American Economic Review*, 78, 5, 1988, pp, 1133-1137.

[2] SONG, Z., SUN, Q., "Three Stage Game of Technical Innovation Firms-Based on the Conception of Absorptive Capacity", *Journal of Industrial Engineering Management*, Vol. 23, No.1, January 2009, pp. 112-138.

[3] QIN, Z, XIANG X., and CHEN T., "Game Analysis on spillovers, absorptive capacity and clusters technological innovation", *Shan Tou University Journal (Humanities &Social Sciences Bimonthly)* Vol.27, No. 2, pp. 57-60.

[4] ZHOU, K., ZHENG, Q., and PAN, Y., "The R&D Investment Analysis Based on Technical Differences and Absorptive Capacity", *Journal of Jilin Province Economic Management Cadre College*. Vol.26, No. 2, 2012, April, pp. 24-29.

[5] TAN, D., "Multidimensional game theory". Southwest Jiao tong University, Chengdu, Sichuan, China, Doctoral Thesis, 2004.

[6] Liu, J., "Analysis of Multi-Dimensional Game on R&D Investment and Market Competition", Southwest Jiao tong University, Chengdu, Sichuan, China, Master Thesis, 2009.

[7] Audretsch, D. B., Feldman, M. P., "R&D spillovers and the geography of innovation and production", *The American economic review*, Vol. 86, No. 3, 1996, PP. 630-640.

[8] Tramontana, F., "Heterogeneous duopoly with isoelastic demand function" *Economic Modelling*, vol. 27, 2010, pp. 350–357.

[9] Wulf, J., Singh, H., "How do acquirers retain successful target CEOs? The role of governance", *Management Science*, vol.57, 2011, pp. 2101–2114.

[10] Liao, Yu-ling, and Qian-lin Hong. "Survey of option game theory of R&D investment", *E-Business and E-Government (ICEE), 2011 International Conference on.* IEEE, 2011.

[11] Arsenyan, Jbid, Gülçin. Büyüközkan, and Orhan Feyzioğlu. "Modelling Collaboration Formation with a Game Theory Approach" *Proceedings of the World Congress on Engineering*, Vol. 1, 2011.

[12] Hofbauer, J., Schlag, K., "Sophisticated imitation in cyclic games", *Journal of Evolutionary Economics*, vol. 10. no. 5, 2000, pp. 523–554.

[13] Puranam, P., Singh, H., Zollo, H., "Organizing for innovation: managing the coordination-autonomy dilemma in technology acquisitions", *Academy of Management Journal*, vol. 49, 2006, pp. 263–280.

[14] Maksimovic, V., Phillips, G., "Post-merger restructuring and the boundaries of the firm", *Journal of Financial Economics*, vol. 102, 2011, pp. 317–343.

Linear Quadratic Zero-Sum Differential Games for a Class of Linear Systems with Markov Jump Parameters

Gong-liang Zhang [1,*], Cheng-ke Zhang [2], Huain-nian Zhu [2]
[1]School of Management, Guangdong University of Technology, Guangzhou 510520, China
[2]School of Economics & Commence, Guangdong University of Technology, Guangzhou 510520, China
(*zhangdry1984@hotmail.com; Zhangck@gdut.edu.cn; huainian258@163.com)

Abstract - **A class of two person zero-sum differential games for continuous-time linear systems which are subjected to the sudden changes in parameter values is studied in the paper. With using the algorithm which is formally similar to Kushner's stochastic maximum principle, the optimal control strategies are derived by means of the coupled differential Riccati equations.**

I. INTRODUCTION

A stochastic model of a control system provides one means to express the uncertainty about the way in which the controlled process will evolve in time. When an important and unpredictable variation causes a discrete change in the plant characterization at the isolated points in time, a Markov chain with a finite state space is a natural model for the plant parameter processes, such models were called Markov-jump models, namely, ones where there are jumps in the problem parameters modeled by a continuous-time Markov chain (see [1], [2], [3], [4], and also [5], [6], [7] and references therein).

On the other hand, the merge of differential games stems from a wide range of applications in network problems, complex systems, and financial engineering. Many problems arising in, for example, pursuit-evasion games, queuing systems in heavy traffic, risk-sensitive control, and constrained optimization problems, can be formulated as two-player stochastic differential games [8], [9], [10], [11], [12]. In light of these developments, it is natural to consider such differential games in which system parameters are modulated by Markov jumps.

In this paper, zero-sum games for linear quadratic systems governed by linear differential equations with Markov jumps are addressed, such class of games has wild applications, such as option pricing of the modern financial engineering. We will show that the coupled differential equation playing a keying role in determining the optimal feedback control strategies, which can be viewed as a direct generalization of the existing results of [13].

For convenience, we will make use of the following notations in this paper:

A': transpose of a matrix or vector A; A^{-1}: inverse of a matrix or vector A; $A>0$ ($A\geq 0$): positive definite (positive semi-definite) symmetric matrix A; $\mathbb{R}^n$: space of all n-dimensional real vectors with usual 2-norm $|\cdot|$.

II. PROBLEM FORMULATION

It will be assumed that the system to be controlled is described by the linear differential equation

$$\begin{cases} \dot{x}(t) = A(t)x(t) + B(t)u(t) + C(t)v(t),\, 0 \leq t \leq T \\ x(t_0) = x_0 \end{cases} \tag{1}$$

Where: $x(t)\in \mathbb{R}^n$ is the state variable, $u(t)\in \mathbb{R}^m$ and $v(t)\in \mathbb{R}^k$ are two control variables by player 1 and player 2. It is intended that (1) serve as a model for a system subject to abrupt changes in parameter values perhaps because of component failures or sudden shifts in environment. If the number of possible parameter values is finite, it is natural to model the parameter with a discrete state Markov jump process. Therefore, assume that each element of the random matrix $[A(t), B(t), C(t)]$ is a separable Markov process, which has the following transition probability:

$$\Pr\{[A(t+\Delta), B(t+\Delta), C(t+\Delta)] = [A_j, B_j, C_j] | [A(t), B(t), C(t)] = [A_i, B_i, C_i]\}$$
$$= \begin{cases} q_{ij}\Delta + o(\Delta), & i \neq j \\ 1 + q_{ii}\Delta + o(\Delta), & i = j; i, j \in \varphi = \{1, 2, \cdots, l\} \end{cases} \tag{2}$$

Where $q_{ij}\geq 0$ for $i\neq j$ and $q_{ii}=-\sum_{i\neq j} q_{ij}$. Fix $(t_0, x_0) \in [0, T] \times \mathbb{R}^n$. Let $U[0, T]$ and $V[0, T]$ be the set of the $\mathbb{R}^m$ and $\mathbb{R}^k$ -valued, square integral processes. Associated with each $(u, v) \in U[0, T] \times V[0, T]$ is a quadratic cost $J(u, v; x_0, r(t_0))$:

$$J(u,v;x_0,r(t_0)) = E\Big\{\int_{t_0}^{T} L\big(\tau, x(\tau;t_0,x_0,u,v), u(\tau; x(\tau;t_0,x_0,u,v), r(t_0)), v(\tau; x(\tau;t_0,x_0,u,v), r(t_0))\big) d\tau \,\big|\, t_0, x_0, r(t_0)\Big\} = E\Big\{\int_{t_0}^{T} [x'(\tau)Q(\tau)x(\tau) + u'(\tau)R_1(\tau) u(\tau) + v'(\tau)R_2(\tau)v(\tau)] d\tau \,\big|\, t_0, x_0, r(t_0)\Big\} \tag{3}$$

Where $R_1(t)>0$, $R_2(t)<0$, $Q(t)\geq 0$. $E\{\cdot\}$ represents the expectation of the enclosed random variable. In this paper, we want to find $u(t)$ and $v(t)$ composed of feedback strategies of the form

$$u(t)=u(t, x(t), r(t)),\ v(t)=v(t, x(t), r(t)) \tag{4}$$

E. Qi et al. (eds.), *Proceedings of the 5th International Asia Conference on Industrial Engineering and Management Innovation (IEMI2014)*, Proceedings of the International Asia Conference on Industrial Engineering and Management Innovation 1, DOI 10.2991/978-94-6239-100-0_30, © Atlantis Press and the authors 2015

Now, we consider the following zero-sum differential game problem.

Problem 1. Given a system described by (1) and (2), find $(u^*, v^*) \in U[0, T] \times V[0, T]$ such that the inequality $J(u^*, v; x_0, r(t_0)) \leq J(u^*, v^*; x_0, r(t_0)) \leq J(u, v^*; x_0, r(t_0))$ holds for all $(u, v) \in U[0, T] \times V[0, T]$.

That is, there are two players for the differential game. Player 1 chooses control u^* to minimize the objective J while Player 2 chooses control v^* to maximize J. In the next section, we will give our main results of this paper.

III. MAIN RESULTS

In this section, we shall use a technique similar to the maximum principle presented in [13] to solve the problem described in the preceding section, and we will show that a coupled differential Riccati equations yield the optimal feedback gains.

Fix $(t_0, x_0) \in [0, T] \times \mathbb{R}^n$. Define a vector valued random variable $p(t_0, x_0)$ as the solution to the following ordinary differential equation: Fix $(t_0, x_0) \in [0, T] \times \mathbb{R}^n$. Define a vector valued random variable $p(t_0, x_0)$ as the solution to the following ordinary differential equation:

$$\frac{d}{d\tau} p\left(\tau, x(\tau; t_0, x_0, u^*, v^*)\right) = -[A(\tau) + B(\tau)u_x^*(\tau, x(\tau; t_0, x_0, u^*, v^*), r(\tau))) \frac{-b \pm \sqrt{b^2 - 4ac}}{2a} + C(\tau)v_x^* (\tau, x(\tau; t_0, x_0, u^*, v^*), r(\tau))]' \cdot p\left(\tau, x(\tau; t_0, x_0, u^*, v^*)\right) + [L_x + L_u u_x^*(\tau, x(\tau; t_0, x_0, u^*, v^*), r(\tau)) + L_v v_x^*(\tau, x(\tau; t_0, x_0, u^*, v^*), r(\tau))], p(T, x) = 0, \quad \text{for all } x \tag{5}$$

As was noted in [13], in (5) the anomalous points have measure zero and the equation may be integrated forward in time. It can be, in fact, under some mild conditions, a solution to (5) exists and is unique.

If $(u, v) \in U[0, T] \times V[0, T]$, define the Hamiltonian at the point (t_0, x_0) to be If $(u, v) \in U[0, T] \times V[0, T]$, define the Hamiltonian at the point (t_0, x_0) to be

$$H(t_0, x_0, u, v) = p'(t_0, x_0) \left[A(t_0)x_0 + B(t_0)u(t_0, x_0, r(t_0)) + C(t_0)v(t_0, x_0, r(t_0))\right] - L(t_0, x_0, u(t_0, x_0, r(t_0)), v(t_0, x_0, r(t_0))). \tag{6}$$

Then the following is true.

Theorem 1. If $(u, v) \in U[0, T] \times V[0, T]$ and if $E\{p'(t, x)[A(t), B(t), C(t)] | t, x, r(t)\}$ is measurable on $[0, T] \times \mathbb{R}^n \times \varphi$, then the set on which $E\{H(t, x, u, v) | t, x, r(t)\} > E\{H(t, x, u^*, v^*) | t, x, r(t)\}$ has measure zero.

Excluding the measure theoretic qualifiers, the theorem simply states that the optimal control strategies optimize the conditional expectation of the Hamiltonian under almost any feedback condition. A proof is similar with theorem 1 in [13], so we omitted the detailed proof here.

As a consequence of Theorem 1, we have

$$u^*(t, x, r(t)) = \frac{1}{2} E\left\{R_1^{-1}(t)B'(t)p(t, x) | t, x, r(t)\right\}$$
$$v^*(t, x, r(t)) = \frac{1}{2} E\left\{R_2^{-1}(t)C'(t)p(t, x) | t, x, r(t)\right\} \tag{7}$$

To complete the description of the controller, an explicit relation between $p(t, x)$ and $(t, x, r(t))$ must be obtained. It will be supposed that the information available to the controller is sufficient to determine the instantaneous values of $A(t)$, $B(t)$ and $C(t)$, i.e., $r(t)$ is a vector with components equal to the elements of the $[A(t), B(t), C(t)]$ matrix. While the sample functions of the elements of $A(t)$, $B(t)$ and $C(t)$ are not continuous, they are independent of the control rules. Hence the equation for the optimal controller becomes

$$u^*(t, x, r(t)) = \frac{1}{2} R_1^{-1}(t)B'(t)E\left\{p(t, x) | t, x, r(t)\right\}$$
$$v^*(t, x, r(t)) = \frac{1}{2} R_2^{-1}(t)C'(t)E\left\{p(t, x) | t, x, r(t)\right\} \tag{8}$$

From (5) it is evident that

$$\frac{d}{d\tau} p\left(\tau, x(\tau; t, x, u^*, v^*)\right) = -[A(\tau) + B(\tau)u_x^*(\tau, x(\tau; t, x, u^*, v^*), r(\tau)) + C(\tau)v_x^*(\tau, x(\tau; t, x, u^*, v^*), r(\tau))]' \cdot p(\tau, x(\tau; t, x, u^*, v^*)) + 2Qx + 2R_1 u(t) \cdot u_x^* (\tau, x(\tau; t, x, u^*, v^*), r(\tau))' + 2R_2 v(t) v_x^*(\tau, x(\tau; t, x, u^*, v^*), r(\tau))' + 2R_2 v(t) v_x^*(\tau, x(\tau; t, x, u^*, v^*), r(\tau))',$$
$$t \leq \tau \leq T, p(T) = 0 \tag{9}$$

To determine the unique solution to (9), it will be supposed that $p(t, x)$ has the form

$$p(t, x) = 2K(t)x, \; 0 \leq t \leq T \tag{10}$$

Where $K(t)$ is a random process independent of x when conditioned on $(t, x, r(t))$ and differentiable everywhere.

From (8) and (10) it follows that

$$u^*(t, x, r(t)) = R_1^{-1}(t)B'(t)E\left\{K(t) | t, r(t)\right\} x$$
$$v^*(t, x, r(t)) = R_2^{-1}(t)C'(t)E\left\{K(t) | t, r(t)\right\} x$$

and hence from (9) and (10)

$$-2[A'(t)+E\{K'(t)|t,r(t)\}(B(t)R_1^{-1}(t)B'(t)+C(t)R_2^{-1}(t)C'(t))]K(t)x+2Qx+2B'(t)E\{K(t)|t,r(t)\}xR_1^{-1}(t)B'(t)E\{K(t)|t,r(t)\}+2C'(t)E\{K(t)|t,r(t)\}xR_2^{-1}(t)C'(t)E\{K(t)|t,r(t)\}=2\dot{K}(t)x+2K(t)A(t)x+2K(t)\left(B(t)R_1^{-1}(t)B'(t)+C(t)R_2^{-1}(t)C'(t)\right)E\{K(t)|t,r(t)\}x,K(T)=0.$$

Since $K(t)$ is conditionally independent of x and is symmetric with the indicated boundary conditions, it follows that

$$\dot{K}(t)=-A'(t)K(t)-K(t)A(t)-K(t)(B(t)R_1^{-1}(t)B'(t)+C(t)R_2^{-1}(t)C'(t))E\{K(t)|t,r(t)\}(B(t)R_1^{-1}(t)B'(t)+C(t)R_2^{-1}(t)C'(t))K(t)+E\{K(t)|t,r(t)\}(B(t)R_1^{-1}(t)B'(t)+C(t)R_2^{-1}(t)C'(t))E\{K(t)|t,r(t)\}+QK(T)=0 \quad (11)$$

It is desirable at this time to introduce some additional notation. At any time t, the augmented matrix $[A(t), B(t), C(t)]$ must be in one of a finite number of different states.

The event that $[A(t), B(t), C(t)]=[A_i, B_i, C_i]$ will be denoted by $(t, r(t))\in[j]$. Define

$$E\{K(t)|(t,r(t))\in[j]\}=\bar{K}_j(t)$$

If at time t, (t, r(t))∈[j], then

$$E\{\dot{K}(t)|(t,r(t))\in[j]\}=-A_j'\bar{K}_j(t)-\bar{K}_j(t)A_j-\bar{K}_j(t)\left(B(t)R_1^{-1}(t)B'(t)+C(t)R_2^{-1}(t)C'(t)\right)\bar{K}_j(t)+Q \quad (12)$$

From (11), it follows that $\dot{K}(t)$ is bounded on $[0, T]$ and consequently

$$E\{\dot{K}(t)|(t,r(t))\in[j]\}=\lim_{\Delta\to 0}\frac{E\{K(t+\Delta)|(t,r(t))\in[j]\}-E\{K(t)|(t,r(t))\in[j]\}}{\Delta}$$

The matrices $A(t)$, $B(t)$ and $C(t)$ are Markova, and hence

$$E\{K(t+\Delta)|(t,r(t))\in[j]\}=\sum_{i=1}^{l}\bar{K}_i(t+\Delta)\Pr((t+\Delta,r(t+\Delta))\in[i]|(t,r(t))\in[j])$$

Therefore, from (3)

$$E\{K(t+\Delta)|(t,r(t))\in[j]\}=\bar{K}_j(t+\Delta)+\sum_{i=1}^{l}\bar{K}_i(t+\Delta)q_{ji}+o(\Delta) \quad (13)$$

Finally, combining (12) and (13) and taking the indicated limit,

$$\dot{\bar{K}}_j(t)=A_j'\bar{K}_j(t)-\bar{K}_j(t)A_j-\sum_{i=1}^{l}q_{ji}\bar{K}_j(t)-\bar{K}_j(t)\left(B_jR_1^{-1}B_j'+C_jR_2^{-1}C_j'\right)\bar{K}_j(t)+Q,\dot{\bar{K}}_j(T)=0 \quad j\in\varphi \quad (14)$$

and

$$u^*(t,x,r(t))=R_1^{-1}B_j'\bar{K}_j(t)x,\ v^*(t,x,r(t))=R_2^{-1}C_j'\bar{K}_j(t)x, \text{ if } (t,r(t))\in[j]. \quad (15)$$

Equations (14) and (15) give the solution to this stochastic control problem. The gain matrices $\bar{K}_j(t)$, $j\in\varphi$ may be obtained by a standard numerical integration of (14). The stochastic nature of this problem places itself in evidence in the inter-coupling of Riccati equations identical to those derived in stationary deterministic problems.

IV. CONCLUSIONS

This paper has investigated the linear quadratic zero-sum differential games for a class of linear systems with Markov jump parameters in finite-time horizon, by using the technique similar to the maximum principle, we obtained the results that a coupled differential Riccati equations yield the optimal feedback gains which extend the existing results of [13] from LQ optimal control for linear systems with jump parameter to zero-sum differential game of two players.

ACKNOWLEDGEMENTS

This work was supported in part by the National Science Foundation of China with grant No. 71171061; Guangdong College of Humanities and Social Sciences project with grant No. 11WYXM022; China postdoctoral Science Foundation No. 2014M552177; Science and technology project of department of transportation of Guangdong Province with grant No.13ZK0303.

REFERENCES

[1] Di, Masi. G. B., Kabanov, Y. M., and Runggaldier, W. J., "Mean variance hedging of options on stocks with Markov volatility", *Theory of Probability & Its Applications*, 1994,vol. 39, no. 1, pp. 172–182.

[2] Ghosh, M. K., Arapostathis, A., and Marcus, S. I., "Optimal control of switching diffusions with application to flexible manufacturing systems", *SIAM Journal on Control and Optimization*, 1993, vol. 31, no. 5, pp. 1183–1204.

[3] Yin, G., Liu, R. H., and Zhang, Q., "Recursive algorithms for stock liquidation: A stochastic optimization approach", *SIAM Journal on Control and Optimization*, 2002, vol. 13, no. 1, pp. 240–263.

[4] Zhou, X. Y., and Yin, G., "Markowitz mean-variance portfolio selection with regime switching: A continuous-time model", *SIAM Journal on Control and Optimization*, 2003, vol. 42, no. 4, pp. 1466–1482.
[5] Blair Jr, W. P., and Sworder, D. D., "Feedback control of a class of linear discrete systems with jump parameters and quadratic cost criteria", *International Journal of Control*, 1986,vol. 21, no. 5, pp. 833–841.
[6] Ji, Y., and Chizeck, H. J., "Controllability, stabilizability, and continuous time Markovian jump linear quadratic control", *IEEE Transactions on Automatic Control*, 1990, vol. 35, no. 7, pp. 777–788.
[7] Mao, X., "Stability of stochastic differential equations with Markovian switching", *Stochastic Processes and Their Applications*, 1999, vol. 79, no. 1, pp. 45–67.
[8] Başar, T., and Bernhard, P., H_∞-*Optimal Control and Related Minimax Problems: A Dynamic Game Approach*, Springer, 2008.
[9] Elliott, R. J., & Kalton, N. J., *Existence of value in differential games*, volume 126 of Memoirs of the American Mathematical Society, AMS, 1972.
[10]Fleming, W. H., and McEneaney, W. M., "Risk-sensitive control on an infinite time horizon", *SIAM Journal on Control and Optimization*, 1995, vol. 33, no. 6, pp. 1881–1915.
[11]Baser, T., and Olsder, G. J., *Dynamic Noncooperative Game Theory*, Academic press, 1995.
[12]Dockner, E., Jorgensen S, Long, N. V. and Sorger, G., *Differential Games in Economics and Management Science*, Cambridge University Press, 2000.
[13]Sworder, D., "Feedback Control of a Class of Linear Systems with Jump Parameters", *IEEE Transactions on Automatic Control*, 1969, vol. 14, no. 1, pp. 9–14.

Research of Heterogeneous Product in R&D Investment and Output Strategies Based on Game Theory

Ming CAO[1,*], Cheng-ke ZHANG[2], Zeng-yang LIU[1]
[1]School of Management, Guangdong University of Technology, Guangzhou 510520, China
[2]School of Economics & Commence, Guangdong University of Technology, Guangzhou 510520, China
(caomig_ming@163.com)

Abstract - **This paper studies the R&D investment and output strategies of two enterprises produced heterogeneous products. In reference on the results of research by predecessors, the author set a model, and numerical analysis. It is concluded that the two enterprises should be cooperation and increase the knowledge absorptive capacity in order to get more profits.**

Keywords - **Output, profit, R&D investment**

I. INTRODUCTION

Nowadays, in order to obtain new knowledge and technology that are exploited to improve production efficiency and product quality, enterprises in the market will increase R&D investment for the increasing competition, such as expanding the competitive advantage and reducing costs. D'Aspremont and Jacque analysised cooperation forms by game theory to the technology R&D among enterprises for the first time. They established a two-stage duopoly game model to explored the completion of the enterprises in the production and marketing [1]. Bischi added the spillover to the game model about technology R&D. In the aspects of heterogeneous products [2], Zhong and Luo analysised permission strategies in professional technical of enterprises [3]. Hu and Chen explored the issue of what is the best technology content of technology innovation allowing companies to maximize profits [4]. This paper build the game model among two enterprises with taking spillovers, enterprise learning ability, the price sensitivity of two heterogeneous products and substitution factor into consideration, and then analyzing and comparing the equilibrium output, R&D investment and profit of enterprises under the two conditions of cooperation and non-cooperation. Finally, it is a conclusion that the best strategies to the two enterprises are improving learning ability to promote cooperation.

II. MODEL FORMULATION

Assumption: there are two enterprises in the same region, named enterprise 1 and enterprise 2, which produce two kinds of products with the characteristic of heterogeneity and certain substitutability respectively; and all products will be sold in this region and monopolize the market for such products. Let q_i stands for the production of enterprise i, C_i is for the cost function. The inverse demand function is $P_i = A-\beta q_i-\theta q_j$. In this equation, A represents for total demand of the market, β is the price sensitive coefficient, θ represents the substitution coefficient of heterogeneous products. In the real production activities, products between enterprises exist substitutive, the smaller is θ, the higher is the substitution between the two products. On the contrary, the lower is the substitution. $\theta \in [0,1]$, i, q=1, 2, $i \neq j$. Specially, when θ=0, both products are homogeneous products that can completely replace each other, when θ=1, both products irrelevant and no substitution.

R&D behavior is one of the effective methods for an enterprise to reduce production costs, and improve economic efficiency. In this paper, h_i is the effective contribution level of R&D of enterprise i. For enterprise i, effective contribution level of R&D depends not only on the level of R&D investment x_i, while relying on their ability to absorb knowledge $g_i(x_i)$ and R&D investment level in enterprise j because of the spillover in R&D activity. $h_i = \sqrt{x_i} + kg_i(x_i)\sqrt{x_j}$, k is a constant

E. Qi et al. (eds.), *Proceedings of the 5th International Asia Conference on Industrial Engineering and Management Innovation (IEMI2014)*, Proceedings of the International Asia Conference on Industrial Engineering and Management Innovation 1, DOI 10.2991/978-94-6239-100-0_31, © Atlantis Press and the authors 2015

represent knowledge spillover coefficient which means the extent of enterprise's own benefit through learning and absorbing from research results of other companies, and $kg_i(x_i)$ called as knowledge absorption coefficient of enterprise i [5], $k, g_i(x_i) \in [0,1]$. We assume c_i is the initial marginal costs of enterprise i, δ is a constant factor, which means that an increased cost in unit product of the enterprises in R&D product process due to technical depreciation, technology depreciation rate represents the rate of enterprise's technology updates. After R&D, the marginal cost of the enterprise's products is expressed as

$C_i = c_i - h_i + \delta = c_i - \sqrt{x_i} - kg_i(x_i)\sqrt{x_j} + \delta$ so we can get corporate profit function:

$\Pi_i = (P_i - C_i)q_i - x_i = (A - \beta q_i - \theta q_j - c_i + \sqrt{x_i} + kg_i(x_i)\sqrt{x_j} - \delta)q_i - x_i$

The following will analyze R&D investment, equilibrium output as well as equilibrium profit of two enterprises in the non-cooperative game and competing game under the condition of complete information.

III. MODEL ANALYSIS

Enterprise 1and enterprise 2 determine their own productions and R&D investment to maximize their profits. Strives for the first order derivative to profit function Π_i on q_i and x_i of enterprise i respectively,and make them equal to zero.

$$\frac{\partial \Pi_i}{\partial q_i} = A - c_i + \sqrt{x_i} + kg_i(x_i)\sqrt{x_j} - \delta - 2\beta q_i - \theta q_j = 0$$

$$\frac{\partial \Pi_i}{\partial x_i} = \frac{q_i}{2\sqrt{x_i}} - 1 = 0$$

Set $c_1 = c_2 = c, g_1 = g_2 = g, \sqrt{x_i} = X_i, \sqrt{x_j} = X_j$, we can get Nash equilibrium and equilibrium profit.

$$\begin{bmatrix} q_i \\ X_i \end{bmatrix} = \begin{bmatrix} \dfrac{2(A-c-\delta)}{4\beta + 2\theta - kg - 1} \\ \dfrac{A-c-\delta}{4\beta + 2\theta - kg - 1} \end{bmatrix}$$

$$\Pi_i = \frac{(A-c-\delta)^2(4\beta - 1)}{(4\beta + 2\theta - kg - 1)^2}, i = 1, 2$$

Now, we discuss this situation: two companies jointly negotiate to determine R&D investment and production in order to maximize the common interests, and this named concurrence of two-dimensional game. Strives for the first order derivative to total profit ($\Pi_1+\Pi_2$) on q_i and x_i, and make them equal to zero.

$$\frac{\partial(\Pi_1 + \Pi_2)}{\partial q_i} = (A - c - \delta) - 2\beta q_i - 2\theta q_j + \sqrt{x_i} + kg\sqrt{x_j} = 0$$

$$\frac{\partial(\Pi_1 + \Pi_2)}{\partial x_i} = \frac{1}{2}(q_i + kgq_j)\frac{1}{\sqrt{x_i}} - 1 = 0$$

We can get Nash equilibrium and equilibrium profit under cooperation of two-dimensional game.

$$\begin{bmatrix} q_i^c \\ X_i^c \end{bmatrix} = \begin{bmatrix} \dfrac{2(A-c-\delta)}{4\beta + 4\theta - (1+kg)^2} \\ \dfrac{(A-c-\delta)(1+kg)}{4\beta + 4\theta - (1+kg)^2} \end{bmatrix}$$

$$\Pi_i^c = \frac{(A-c-\delta)^2}{4\beta + 4\theta - (1+kg)^2}, i = 1, 2$$

Then, put the production and R&D investment with cooperative game and non-cooperative game in comparison and analysis.

$$q_i = q_i^c, \frac{2(A-c-\delta)}{4\beta + 2\theta - kg - 1} = \frac{2(A-c-\delta)}{4\beta + 4\theta - (1+kg)^2}$$

$$kg = \frac{\sqrt{1+8\theta} - 1}{2}$$

$$x_i = x_i^c, \frac{A-c-\delta}{4\beta + 2\theta - kg - 1} = \frac{(A-c-\delta)(1+kg)}{4\beta + 4\theta - (1+kg)^2}$$

$$kg = \frac{\theta}{2\beta + \theta}$$

$$\Pi_i = \Pi_i^c, \frac{(A-c-\delta)^2(4\beta - 1)}{(4\beta + 2\theta - kg - 1)^2} = \frac{(A-c-\delta)^2}{4\beta + 4\theta - (1+kg)^2}$$ when $kg = \dfrac{\sqrt{1+8\theta} - 1}{2}$, $q_i = q_i^c$; when $kg = \dfrac{\theta}{2\beta + \theta}$, $x_i = x_i^c$.

Here to discuss the impact of kg on equilibrium output, equilibrium R&D investment and equilibrium profit respectively through numerical analysis. We have a kind of hypothesis that A=2, C=0.2, δ=0.3, β=1, θ=0.5, and the simulation results shown in Fig.1, Fig.2, Fig. 3.

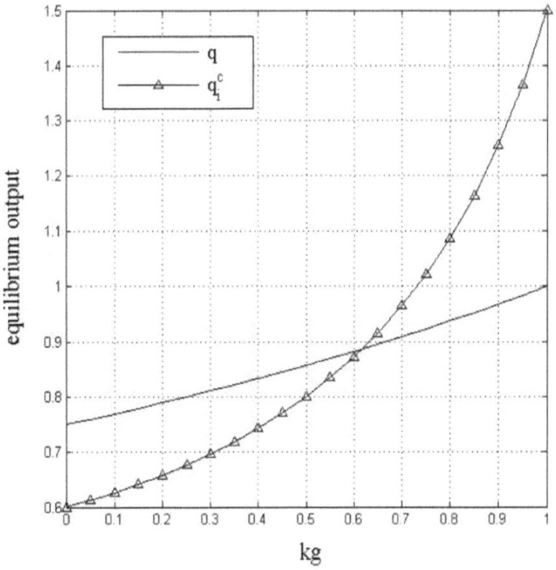

Fig.1. Comparative analysis of the impact on the equilibrium output of *kg*.

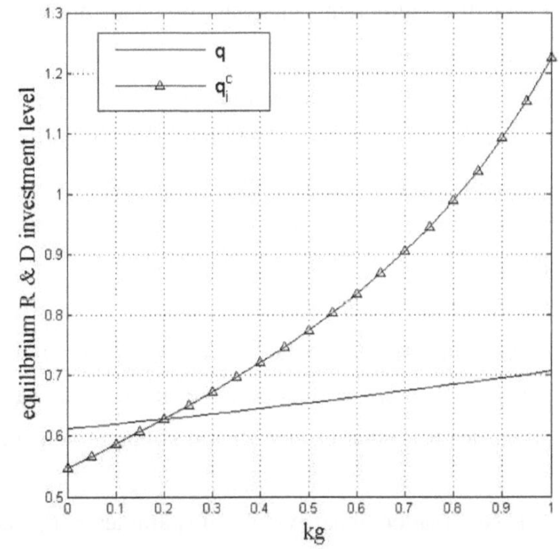

Fig.2. Comparative analysis of the impact on the equilibrium R&D investment level of *kg*.

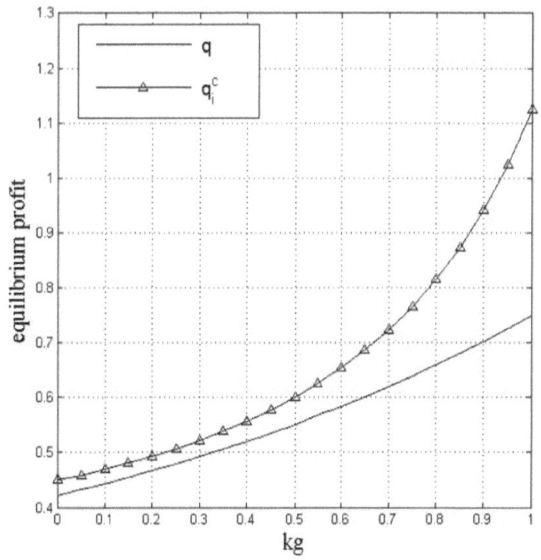

Fig.3.Comparative analysis of the impact on the equilibrium profit of *kg*.

From the numerical analysis the following conclusions can be drawn.

(1) Given the other parameters, in both the cooperative game and non-cooperative game situation, *kg* on equilibrium output, equilibrium R&D investment and equilibrium profit of the two enterprises are positively correlated.

(2) Given the other parameters, when *kg* to a certain extent, the equilibrium output of enterprise in cooperative game is greater than in non-cooperative game, and R&D investment level is greater than in non-cooperative game also.

(3) Under the condition of cooperative game, the equilibrium output, equilibrium R&D investment and its profit of an enterprise are growing faster than the non-cooperative situation.

(4) No matter what value of parameter *kg*, the equilibrium profit in cooperative game is always greater than that in non-cooperative game, and the greater *kg* is, the faster of equilibrium profit growth in cooperative game situation.

Substitution coefficient θ is also an important influencing factor. We have a kind of hypothesis that A=2, C=0.2, δ=0.3, β=1, kg=0.2. The following will discuss the impact of θ on equilibrium output, equilibrium R&D investment and equilibrium profit respectively through the way of numerical analysis. And the simulation results shown in Fig.4, Fig.5, Fig.6. when $\theta = \frac{(1+2kg)^2 - 1}{8}$, $q_i = q_i^c$;

when $\theta = \frac{2\beta kg}{1-kg}$, $x_i = x_i^c$.

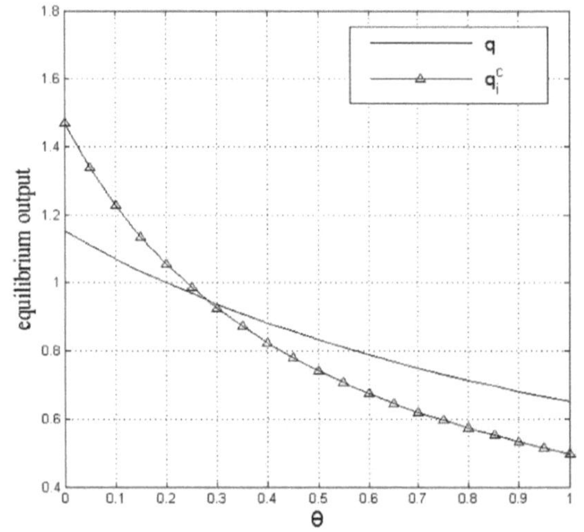

Fig.4. Comparative analysis of the impact on the equilibrium output of θ.

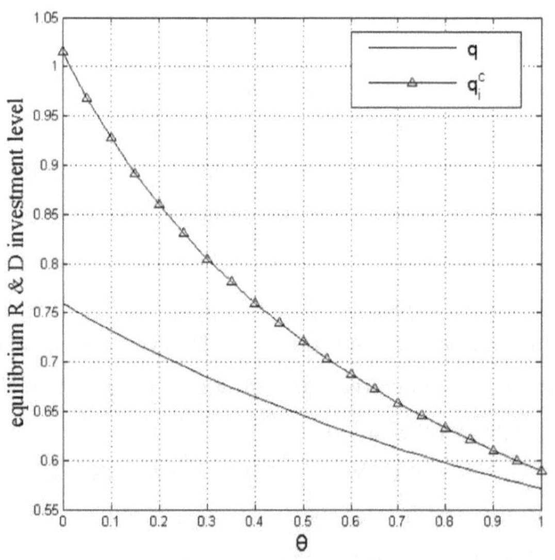

Fig.5. Comparative analysis of the impact on the equilibrium R&D investment of θ.

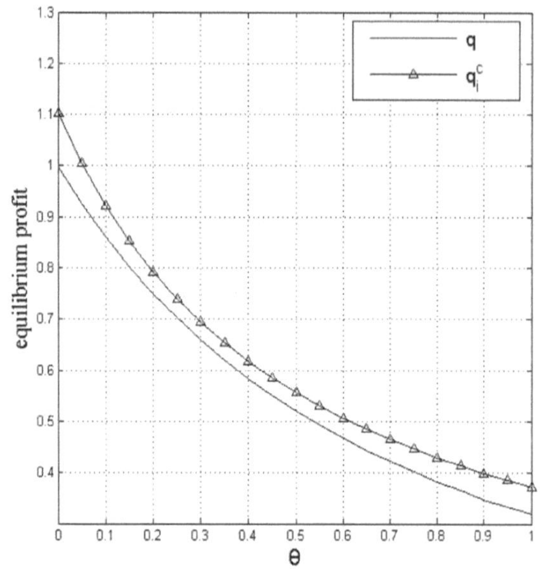

Fig.6. Comparative analysis of the impact on the equilibrium profit of θ.

From the numerical analysis we can draw the following conclusions.

(1) Given the other parameters, in both the cooperative game and non-cooperative game situation, θ on equilibrium output, equilibrium R&D investment and equilibrium profit of the two enterprises are negative correlated.

(2) Given the other parameters, when θ to a certain extent, the equilibrium output of enterprise in cooperative game is less than in non-cooperative game, and R&D investment level is smaller than in non-cooperative game also.

(3) No matter what value of parameter θ, the equilibrium profit in cooperative game is always greater than that in non-cooperative game, and the more augment of θ, the greater difference of equilibrium profit.

IV. CONCLUSION

In order to obtain more profits, the two companies should choose cooperation, and increase the value of knowledge absorptive capacity (*kg*). Cooperation is the first choice that will increase profits, and to improve the knowledge absorptive capacity (*kg*) [6] is the next step. When *kg* increasing close to 1, the greater difference of equilibrium profit between cooperative game and non-cooperative game, the same as equilibrium output and equilibrium R&D investment in two cases of game [7-9]. Complete competition among enterprises will no longer be a way out, cooperation can expand the competitive advantage and achieve win-win situation. Otherwise, it will fall into the "prisoner's dilemma" that behavior of corporate self-interest as the goal causes both sides to get a relatively inferior profits [10-13]. Given the value of *kg* and other parameters, regardless of substitution of products, the two companies will get higher profits if they cooperate [14, 15].

ACKNOWLEDGMENT

This work was supported in part by the National Science Foundation of China with grant No. 71171061, and Guangdong College of Humanities and Social Sciences project with grant No. 11WYXM022.

REFERENCES

[1] D'Aspremont, C., and Jacquemin, A, "Cooperative and noncooperative R&D in duopoly with spillovers", *The American Economic Review*, vol. 78, no. 5, 1988, pp. 1133-1137.

[2] Bischi, G. I., and Lamantia, F., " Nonlinear duopoly games with positive cost externalities due to spillover effects", *Chaos, Solitons & Fractals*, vol. 13, no. 4, 2002, pp. 701-721.

[3] De-qiang ZHONG , Ding-ti LUO , Wei-Jun ZHONG, Hui LIU," Licensing strategies in a differentiated Cournot oligopoly", *Systems engineering Theory & Practice*, vol. 9, no. 9, 2009,pp. 24-37.(Chinese)

[4] Rong HU, , Qin CHEN, and Peng YUAN, "Game analysis of the differentiated duopoly enterprises in R&D technical conten", *Science & Technology Progress and Policy*, vol. 26, no. 17, 2009, pp. 82-84.(Chinese)

[5] Jun LIU. *Analysis of Multi-Dimensional Game on R&D Investment and Market Competition*, Southwest Iiaotong University, Chengdu, Sichuan, China, Master Thesis, 2009.(Chinese)

[6] Cohen, W. M., and Levinthal, D. A., "Innovation and learning: the two faces of R&D", *The Economic Journal*, vol. 99, no. 397, 1989, pp. 569-596.

[7] Feldman MP,Kelley MR. The exante assessment of knowledge spillovers: Government R&D policy, economic incentives and private firm behavior[J].*Research Policy*, vol. 35, no. 10, 2006, pp.1509-1521.

[8] Kotar Suzumura. Cooperative and noncooperative R&D in an oligopoly with spillovers[J]. *The American Economic Review*,, vol. 82, no. 5, 1992, pp.22-25.

[9] Qiang WANG, WeiHong- LI, Rong HU, Game analysis for cooperative and noncooperative R&D in duopoly with price competition. *Science & Technology Progress and Policy*, vol. 27, no. 16, 2010, pp. 82-84. (Chinese)

[10] Pei-jun HUO, Jian CHEN, Ji-xiang CHEN.Optimal spillover for cooperative and noncooperative R&D in duopoly. *Chinese Journal of Management Science.* vol. 10, no. 6, 2002, pp. 92-96. (Chinese)

[11] AMIR R. Modelling imperfectly appropriable R&D via spillovers.*Internation Journal of Industrial Organization*, vol. 18, no. 7, 2000, pp. 1013-1032.

[12] Petit M L,Sanna-Randaccio F.Endogenous R&D and foreign direct investment in internaltion oligopolies. *International Journal of Industrial Organization.* vol. 18, no. 2, 2000, pp. 339-367.

[13] Shy O,Industrial Organization:Theory and Applicationa. Cambridge, MA: MIT Press,1995.

[14] Coe D T, Help man E. International R&D spillovers. *European Economic Review*, vol. 18, 1995, pp. 859-887.

[15] MOLTO, Cooperative R&D with endogenous technology differentiation. *Journal of Economics& Management Strategy,* vol. 14, no. 2, 2005, pp. 461-476.

Assessment of Fire Risk in Central Business District - Taking Yujiapu of Tianjin City as Example

Feng-ying Yan[1, 2, *], Qi Zhang[2], Ze-nan He[2]

[1] School of Architecture, Tianjin University, Tianjin, 300072, China

[2] School of Civil Engineering, Tianjin University, Tianjin, 300072, China

(fengying@tju.edu.cn)

Abstract - **Extra attention should be paid on the fire prevention for the safety of city since the central business district (CBD) is in a stage of rapid development. To assess the fire risk of CBD accurately, this paper gives an index system of assessment of fire risk which structured by AHP based on the Yujiapu of Tianjin city. Different grades of fire risk of some indicators were established by GIS technology. Also the fire risk indicators were quantitative scored based on the grade criterion, and classification of fire risk of the target area was given finally. The results show that the area in this study belongs to low fire risk (with about 78.6% of all are safe or low risk) and, the high population density, high-rise building, and far away from the fire station still lead to that nearly 4.8% area belong to high fire risk area.**

Keywords – AHP, Central Business District, fire risk, GIS

I. INTRODUCTION

The concept of central business district (CBD) appears in the United States during 20s in the last century. CBD in nowadays relates to the specific area of the metropolises, which concentrates abundant auxiliary facilities such as commercial, financial, culture, service and apartments. The developed modern infrastructures like convenient transport networks and communication devices benefit the process of the large-scale business activities at CBD [1, 2].

On the other hand, the complex buildings, enterprise groups and elites live in these surroundings become the major elements in CBD. Therefore, CBD usually becomes the economic center of the city, and locates the golden zone which owns best transportation condition and business trade opportunities. Also intensively land use and high-strength exploration reflect that the CBD plays the role of a material carrier, and the population density is higher in this area along with the developed business [3].

Take these characters into consideration comprehensively, the high concentration of facilities and the compact space increase the possibility of fire hazard, which will lead to more serious consequence than common area. So it is necessary to conduct a city fire risk assessment to avoid the property loss and personal casualties. Fire risk assessment of a city can conducts the scientifically planning and the reasonable stationed fire fighting force. Fire risk assessment is done based on the development of modern fire research, risk assessment and risk managements, and resulting in reducing the risk and threatens from the fire accident.

The main objective of this study is to establish an index system of fire risk assessment based on Analytic Hierarchy Process (AHP) and formulate a reasonable index score criterion for Yujiapu CBD area in Tianjin city. Geographic Information System (GIS) was employed to process the model of assessment, and evaluate the fire risk. Also the fire risk indicator was quantitative scored based on the grade criterion. At last, results of the fire risk assessment of Yujiapu CBD area were given and several proposals for planning were provided to improve the safety level on the current basis.

E. Qi et al. (eds.), *Proceedings of the 5th International Asia Conference on Industrial Engineering and Management Innovation (IEMI2014)*, Proceedings of the International Asia Conference on Industrial Engineering and Management Innovation 1, DOI 10.2991/978-94-6239-100-0_32,

II. FIRE RISK ASSESSMENT SYSTEM

A. *Theory of Fire Risk Assessment*

City fire risk assessment aims to predict the possibility and the degree of the fire accident via analyzing the factors relates to conflagration. According to the level of risk of conflagration, the configuration of fire fighting force, also the construct of fire control system and the planning of the area could be reorganized and be more reasonable. Literatures have been published about the theories of fire risk assessment [4, 5], and can be classified into three categories, namely qualitative method, semi-quantitative method, and quantitative method. The AHP method used in this study combines qualitative and quantitative method.

B. *Objects of Fire Risk Assessment*

The objects of the fire risk assessment can be classified into three based on their size. First is the monomer building, which has been researched by Shen [6] using the AHP, the indicator set and cumulative weights of the fire risk assessment were given. Second kind is the object like enterprise architectures mostly in Chemical industry. Dong [7] established the assessment system for a fertilizer plant, and evaluate the fire risk using comprehensive method include both AHP and fuzzy recognition method. The third category takes an area as an objective, and provides a scientific basis for regional fire management. Studies done by Yang [8] formulated two different index system of assessment for an area in Beijing, and gave final results of the fire risk based on the analysis method of AHP.

Both assessment index system and model was necessary for any kinds of objects. Although much works have been done in this field, the comprehensiveness of the assessment index system and the operability of the assessment model still could be optimized. On the other hand, the model of assessment depends on computer program more and more nowadays [9, 10]. And the platform of GIS has been invited into the field of fire risk assessment. So GIS technology was employed in this study for the convenient assessment.

III. ESTABLISHMENT OF FIRE RISK ASSESSMENT INDEX SYSTEM

A. *Characters of Yujiapu CBD Area*

Yujiapu is a peninsula, located in the core area of Tianjin Binhai New Area, the entire block east, west and south facing the Haihe River, covers an area of 3.86 square kilometers. The total construction area is 9.5 million square meters, of which 500 square meters is office space, 2 million square meters of commercial residential, 2.5 million square meters of schools, hospitals, shops and other public facilities. The rich three-dimensional structure of underground space and a larger-scale complex structure is the most unique character of this area. The total construction area of about 425,000 square meters, all three layers of underground space are connected with the surrounding buildings, to achieve the subway, underground parking, business and all through. The Yujiapu CBD is highly concentrated with economic, technological and cultural components, at the same time many public gathering places and large amount of high-rise buildings locates here. Fire hazard would be more possibility with the complex surroundings, and the loss would be immeasurable if fire happens in this area. So it is important to establish a comprehensive and reasonable index system for this kind of area.

B. *Determination of the Index System*

Index system is established according to the interaction and complex coupling between the main fire safety factors, and be established based on the analysis of the characteristics of the fire risk. Therefore, the index system is the key to fire risk assessment. Too many indicators will result in interference and increase the computational difficulty; while a small amount of factors cannot reflect the factors that affect the fire risk assessment roundly, and may results certain one-sidedness assessment. Therefore, to establish a scientific and reasonable index system should obey principles of scientific, integrity, completeness, simplicity, practical, operational, and comparability [11].

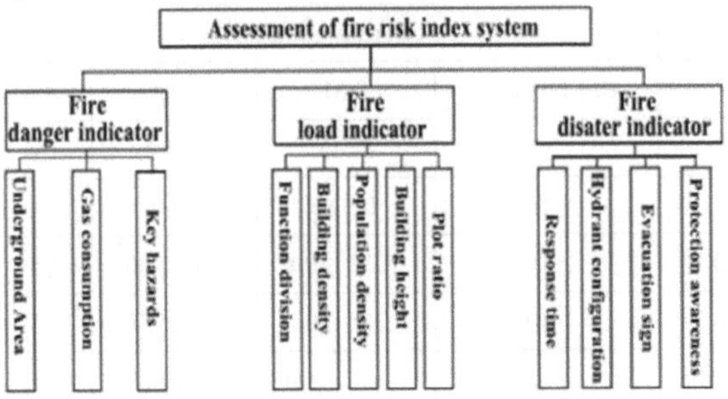

Fig.1. Index system of the fire risk assessment

Three main components of the index system were determined: fire danger indicator, fire load indicator and fire hazard indicator. The fire danger indicator relates to the possibility of the fire in the region, and fire load indicator means the extent of the damage brought by the fire accident. While on the other hand, fire hazard indicator refers to how to reduce the fire loss to minimum. According to the regional characteristics of the Yujiapu CBD district, index system has been established as shown in Fig.1, both the primary and secondary indicators factor were included.

C. *Determination of the Index Weight*

AHP is a method combined the quantitative and qualitative, and can be used to systematize, quantize and modeling the thinking process of the analysts. In this paper, the AHP 1-9 scale method was employed to compare each level indicator and process the results numerically, and then the weight of each factor was given. This approach is more conform to the vagueness and uncertainty of the fire risk assessment. Finally, the consistency test is necessary, and judgment matrix should be modified until the consistency is acceptable [12].

The importance of fire load indicator for urban area fire danger indicator was determined by the questionnaires, results show that about 75% of the survived people insist that the median is 3, and the importance of the fire load indicator to fire danger indicator is 3. At the same time, the judgment matrix of the primary indictors as shown in Table I.

According to the calculation, the consistency of the judgment matrix is 0.0088, which is lower than 0.1, indicates that the matrix is available. Also the weight calculation for the indicators of the Yujiapu CBD area were shown in Table II based on the calculate method mentioned above.

TABLE I

JUDGMENT MATRIX OF THE PRIMARY INDICATOR

	Load indicator	Danger indicator	Hazard indicator	Weight
Load indicator	1	2	3	0.54
Danger indicator	0.5	1	2	0.29
Hazard indicator	0.67	0.5	1	0.17

D. *Quantify and scoring guidelines*

Not all of the factors can be quantified in the index system, so the quantitative and qualitative indicators have been processed by different risk classification methods and scoring guidelines [13], and detail information is shown in Table II.

IV. BASIC DATA PROCESSING AND APPLICATION OF GIS

A. *Data acquisition*

Based on the survey of Yujiapu, the underground space configuration data, fire stations, petrol stations, gas stations, electrical substation sites distribution, traffic routes, and all the sub-region covered area, total land for construction, population size and so on are obtained. With the sorted data being input into the ArcGIS, the whole Yujiapu was divided into 126 sub-regions according to the traffic routes, namely every sub-region surrounds roads. The property information was attached to each region, such as the functional zoning, building density, population density, and plot ratio. Finally the database was established.

Then the assessment system of the corresponding infrastructure status, including the building height, functional zoning, building density, gas consumption, plot ratio, was obtained with the combination of the relevant statistics data, figuration information. All of the data is collected and analyzed by GIS platform.

TABLE II

WEIGHTS OF FIRE RISK ASSESSMENT INDEX SYSTEM AND SCORING RULES

Weights of indicator				Scoring guidelines			
Primary		Secondary		10 Marks(M)	8 M	6 M	4 M
Fire load indicator	0.54	Population density	0.14	≥3	2~3	1~2	≤1
		Plot ratio/%	0.30	≥10	5~10	1~5	≤1
		Building height/m	0.10	≥200	100~200	50~100	≤50
		Building density/%	0.21	≥30	20~30	10~20	≤10
		Function division	0.25	Administrative	Residential	Public	Others
Fire danger indicator	0.17	Underground area/m^2	0.49	≥1×10^4 (10 M)		<1×10^4	0
		Gas consumption/ 10^{6}m^3/day	0.32	≥0.90	0.36~0.90	0.13~0.36	≤0.13
		Key hazards/NO.	0.19	2	1	0 (4 M)	
Fire hazard indicator	0.29	Response time/min	0.43	≥5	5	4	≤3
		Hydrant configuration	0.26	< national standard (NS) (10 M)		=NS	> NS
		Evacuation sign	0.19	< NS (10 M)		=NS	> NS
		Protection awareness	0.12	Weak (10 M)		Best	Better

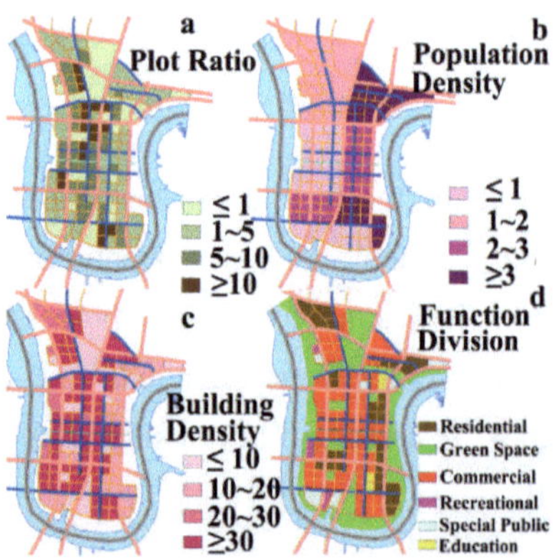

Fig.2. Basic information of Yujiapu

As is shown in Fig.2, the efficient traffic route in this district is typical feature and necessary function of CBD. The Plot Ratio is shown in Fig.2a. It can be seen that the maximum value of Plot Ratio is obtained which is located in the central district, south-east sub-region, and north-west region. Besides, the peak of population density can be observed in the north-east, south-east, while the density in middle-east region is also relative large (Fig. 2b). At the same time, the buildings of Yujiapu are mostly built in the central district, most of which are Ultra High-rise Buildings as the CBD (Fig. 2c). From the functional division map (Fig. 2d), it can be seen that the financial region is mainly located in the central road and the middle west and south region, while the residence part are distributed in the east and north-west region.

B. *Classification of fire risk*

After calculating the weighting value of each indicator through AHP, then the corresponding model of the actual situations are applied to obtain the degree of these indicators, finally we will have the fire risk classification of this region by combing the model and GIS. Firstly, the score field of the assessment model was added to the layer properties of the sub-region in the map database. Then the assessment results of each sub-region were stored into the corresponding assessment results field according to the only identification code in field calculator. At the last step, we will do the statistical analysis based on the Mathematical statistics function.

GIS can finish both the fire risk classification work automatically and presentation of the assessment results, which combines effectively the assessment results and the geographic information. Fig.3 shows the fire risk classification result through the analysis work of database and GIS technology. Based on the final assessment score of each sub-region, all these regions can be categorized into 4 levels: the score is not larger than 4.50 can be treated as the relative safe region, which is the green region in the map; the low risk region usually has the score between 4.50 and 6.50, as is the light green region; the orange region in the map is the middle risk region, which has the score between 6.50 and 7.50; and the region which has score between 7.50 and 8.14 is the relative high risk region, which is the red region.

Statistic works have been done on the assessment results from GIS, and results show that almost 13.5% of the sub-regions belong to safety area, and about 65.1% refer to low fire risk area. This result indicates that the Yujiapu CBD is in low fire risk level in general. Still 16.6% sub-regions here are under moderate risk level, and nearly 4.8% are classified into high risk level.

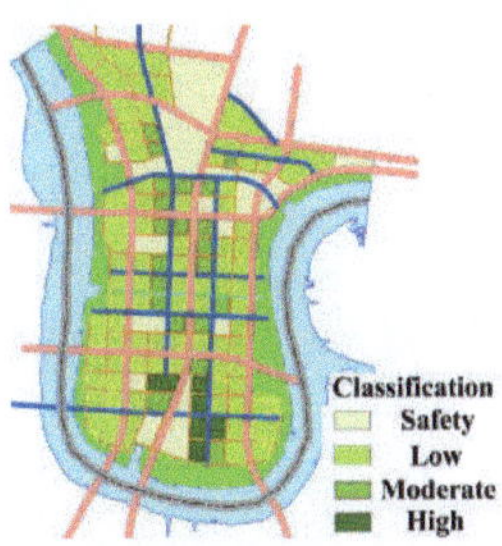

Fig.3. Classification of fire risk

Results in Fig.3 show that sub-regions those belong to moderate risk level mainly distribute around the central avenue of Yujiapu CBD, where has the highest building density (Fig.2c). On the other hand, high risk level sub-regions concentrate in the south of Yujiapu CBD. The highest population density, building density and plot ratio may contribute to this result. So these sub-regions should be key points of the fire fighting work in future.

C. *Planning proposals*

The class classification and the assessment score is not the final target of the assessment of the fire risk, but the planning proposals based on the results is valuable. Based on the results shown in Fig.3, safety evacuation area, fire emergency radio and fire refuge should be configured around the central avenue and south part of the Yujiapu CBD area. On the other hand, fire fighting skills, evacuation and fire protection awareness should be enhanced, especially in the high risk level area. Also the configuration of fire alarm interlocks and improvement of fire extinguishing device density can lower the fire loss and improve the safety level. Details configuration of planning proposals is shown in Fig.4. About seven refuge evacuation sites and some fire hydrants and fire extinguishers are configured on the green land near the central avenue. On the other hand, an apron for the rapid evacuation is configured in the southern part of Yujiapu CBD area.

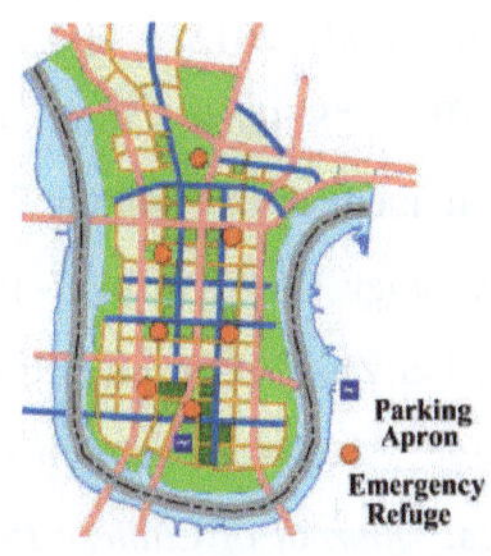

Fig.4. Proposals for planning

V. CONCLUSIONS

The fire safety condition is very important for Yujiapu, a new CBD area in Tianjin city. Hence, a comprehensive fire risk assessment index system was established based on Yujiapu area. The weight of each indicator was determined based on AHP method, and the fire risk level was scored and presented by the GIS technology.

Results show that Yujiapu CBD area belongs to low fire risk level in general. Also the assessment model has been certified be practicality, and can supply important reference for urban planning, rational allocation of fire fighting sources. Moreover, the assessment model is available for the fire assessment of other cities, especially for the CBD area.

ACKNOWLEDGEMENTS

This work was supported by the Nature Science Foundation of Tianjin (13JCYBJC19700).

REFERENCES

[1] Deci Z., "A General Introduction to the Development of Central Business District," *South Architecture*, 4, pp. 8-12, 2008.

[2] Ministry of Construction, GB/T50280-1998 Urban planning basic terminology standards: *Standards Press of China*, 1998.

[3] Bingze F., "*A Study on Development and Implementation Strategy for the Central Business District in Binhai New Area*," Master Thesis of Tianjin University, 2008.

[4] Juntao Y., Ye Ch., "City Regional Fire Risk Assessment Technology," *China Police Daily·Fire Weekly*, 2013, p. 004.

[5] Weijun W., "Summarization on fire risk evaluation method of building," *Fire Science and Technology*, (27) 7, pp. 477-481, 2008.

[6] Shen L., Lei Zh., "On a Fire Risk Assessment on Large Scale Comprehensive Public Buildings," *Journal of Chinese People's Armed Police Force Academy*, 12, pp. 35-38, 2011.

[7] Dongdong Zh., Zunze H., Bo L., Xiaoxing C., "Application of Synthetic Assessment Based on Analytical Hierarchy Pricess and Fuzzy Recongnition to Chemical Corporation Fire Risk Assessment," *Proceedings of 2006 (Shenyang) International Colloquium on Safety Science and Technology*, 2006.

[8] Rui Y., Zunze H., "Researches for the method on city zone fire safety evaluation," *Journal of Chinese People's Armed Poloce Force Academy*, 2003.

[9] Chunhong M., Kun Sh., Yanfei G., "A GIS based urban regional fire risk assessment system," *Science of Surveying and Maping*, (36)001, pp. 150-152, 2011.

[10] Xingwang L., Daming Zh., "Application of GIS to Fire Risk Assessment in Urban Community," *Geospatial Information*, (7) 5, pp. 78-80, 2009.

[11] Mei L., Jun L., "Study on Comprehensice Assessment Index System of Urban Fire Risk in Beijing City," *Fire Technique and Products Information*, 07, pp. 11-14, 2007.

[12] Xue D., Jiaming L., Haojian Z., Junyang Ch., Junfeng Zh., "Research on Computation Methods of AHP Wight Vector and Its Applications," *Mathematics in Practice and Theory*, (24)7, pp. 93-100, 2012.

[13] Rong J., Lizhi W., Xilin D., Minxue Zh., Kaoding Sh., "Urban Residential Fire Risk Evaluation," *Fire Technique and Products Information*, 2, pp. 5-8, 2005.

5th International Asia Conference on Industrial Engineering and Management Innovation (IEMI 2014)

Production Throughput Improvement for Plant Streamline N

Wei-hua GAN*, Peng-fei CHE, Jin-ning YOU
Department of Logistics Management and Industrial Engineering, East China Jiaotong University, Nanchang, China
(*weihuagan@163.com, 1043133504@qq.com, 1004981550@qq.com)

Abstract - **The objective of this research is to improve the throughput of the plant streamline N based on the methods of industrial engineering, e.g. job analysis, fishbone diagram and work study, for increased productivity. In this case study, bottlenecks diagnose and continuous productivity improvement is studied. The new production system has been designed and compared with the present throughput. The results show that new throughput of plant streamline N significantly increases from 140 to 202 per day.**

Keywords - **Bottlenecks diagnose, job analysis, production process, throughput improvement**

I. INTRODUCTION

Company N is a large transmission technology joint-venture plant. Its establishment is less than one year by the end of 2013 in Nanchang city, China. The equipment and labor are still in the introductory period. At present the production throughput is 140 per day, far away from the goal 180. It is urgent and necessary to improve the production throughput.

Since the conception of the assembly line and the following development of the Toyota Production System (TPS), efficiency has been a central objective of manufacturing [1]. Larger firms are more likely to implement lean manufacturing practices [2]. With rapid increasing of demand in production, Company N needs to increase its potentials in production and effectiveness to compete against the competitors. At the same time, the production process needs to be equipped with the ability to have lower cost with higher effectiveness. Therefore, the way to solve the problem about the production is very important.

Rachna Shah [3] postulate four "bundles" (JIT, TQM, TPM, HRM) of inter-related and internally consistent practices on the operating performance of plants concluding JIT/continuous flow production and quick changeover methods are included most frequently. Adriaan Van Horenbeek [4], in order to assist the maintenance manager on selection of the relevant MPI, an analytic network process (ANP) model and methodology is presented which is based on the designed maintenance performance measurement (MPM) framework, which decrease the failure rate of equipment. Krisztina Demete [5] use cluster and correlation analysis to separate manufacturers based on the extent of their leanness and to examine the effect of contingencies, improving turnover performance. Franci Pusavec [6] achieves production sustainability on a machining technology level.

II. DESCRIPTION OF PRESENT ASSEMBLY LINE N

An important task of empirical validation is to test the internal and external validity [7]. The assembly line in Company N is a typical process-oriented layout. There are 23 workers including 1 supervisor, 6 parts manufacturers, and 16 assembly workers. The manufacturing process is shown in Fig.1.

Assembly Line N consists of three sections, i.e. parts assembly section (OP1010-OP1080), loop-like general assembly section (OP2010-OP2130) and line testing (OP3010 and OP3020).

Fig.2 shows that OP1010-OP1080 belongs to the department of the external load, which are engaged in large-size components' assembly and components processing, OP2010-OP2130 belongs to loop-like general assembly line, which are responsible for parts manufacturer and associated assembly, OP2130 is shunting station.

OP1010, OP1020 and OP1060 is operated by one person. OP1050 means print label and glue seal, which is operated by one person, too. Within this station, Work-in-process is transferred to OP3010 and the trolley backflow to OP2010. OP3010 and OP3020 belongs to line testing. OP3030 assemble shifting yoke components. OP3040 is package. The final is off-line.

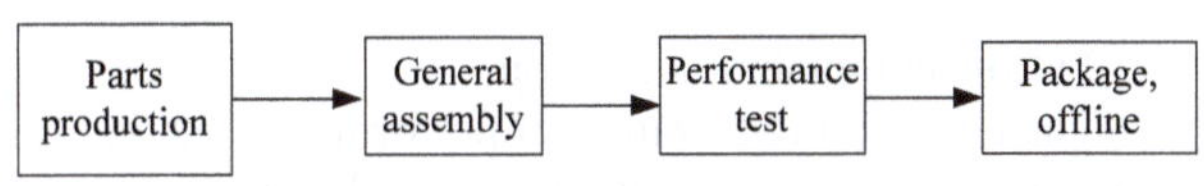

Fig.1. Production process of streamline N

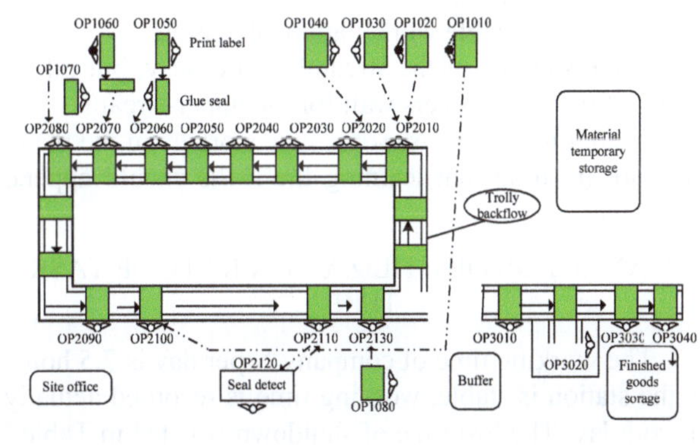

means parts material handling direction,
means conveyor direction,
means conveyor.

Fig.2. Assembly line N layout

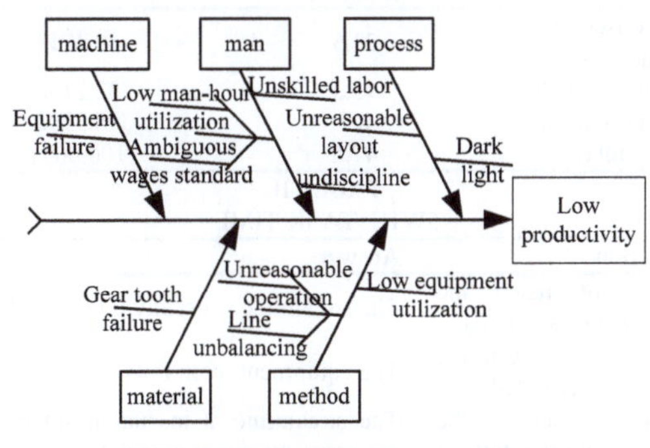

Fig.3. Fish-bone chart: low productivity

E. Qi et al. (eds.), *Proceedings of the 5th International Asia Conference on Industrial Engineering and Management Innovation (IEMI2014)*, Proceedings of the International Asia Conference on Industrial Engineering and Management Innovation 1, DOI 10.2991/978-94-6239-100-0_33,

III. CAUSE-EFFECT ANALYSIS OF LOW PRODUCTIVITY

Fish-bone diagram is used to analyze and classify factors related to the problems [8-9]. Now the problem is the throughput only 140, far away from the goal 180. The cause of low productivity is analyzed by fish-bone chart in Fig.3, from the perspective of man, machine, material, method, and process.

A. Low man-hour Utilization Rate

Operation time is lost because of low man-hour utilization rate and production line interruption. This result is mainly caused by two reasons as follow:

1). Inobservant discipline

Workers usually play mobile-phone and chat with each other during the working hours. The streamline is interrupted as a result of human factors.

2). Equipment failure

As conveyor belt often fails, assembly trolley cannot reach the next working station automatically. Therefore, workers should push the trolley. Moreover, glue seal machine is unstable. Workers should re-patch the glue because of leakage.

B. Unbalanced Streamline

1).Low equipment utilization

The equipment utilization of performance test is low. Operators have long idle time. It shows that man-machine matching is unreasonable.

2). Unreasonable work station allocation

Some stations of the streamline are busy, but others are idle. So some workers wait for the upper-stream.

In addition, there are some new workers in the plant. They are fresh so that learning the skills should append some time.

IV. ANALYSIS OF UTILIZATION RATE OF TASK TIME

The working time of company N per day is 7.5 hours. As the station is stable, working time is recorded actually on workday. The lost time of shutdown is listed in Table I. The data we collected is average time.

TABLE I
LOST TIME CAUSED BY SHUTDOWN

reasons of shutdown	lost-time (minute)	The proportion of lost time
inobservant discipline	23.5	49.47%
equipment failure	15.3	32.21%
other reasons	8.7	18.32%
total	47.5	100.00%

TABLE II
5W1H: TASK TIME

Question	Answer
Why not reach the expected productivity?	The streamline is usually interrupted.
Why is the streamline usually interrupted?	The equipment often fails.
Why does the equipment often fails?	The streamline is in running-in period and some machines are unstable.
How to decrease instability?	To increase the times of equipment examination
Why not adjust the examination in time?	Because equipment administrator regard the line as a mature one, and underestimate equipment failure rate

Calculate the utilization rate of task time:

$$\text{Loss Rate of Task Time} = \frac{\text{Lost Task Time}}{\text{Nominal Task Time}} \times 100\%$$

$$= \frac{47.5}{7.5 \times 60} \times 100\% = 10.56\%$$

$$= 1 - 10.56\% = 89.44\%$$

Then we use "5W1H" to find the solutions to the task time (Table II).

Based on 5W1H, we respond the reason to the equipment administrator, now the lost time decreases from 47.5m to 23.8m per day, and the utilization rate of task time is 94.71%.

$$\frac{23.8}{7.5 \times 60} \times 100\% = 5.29\%$$

$$1 - 5.29\% = 94.71\%$$

V. ANALYSIS OF LINE BALANCE RATE

High line balance rate does not only benefit for equipment but also for man-hour utilization. Meanwhile, resulting in the quality and safety of products [10-11]. Lean manufacturing represents a multifaceted concept that may be grouped together as distinct bundles of organizational practices [12]. We analyze the manufacturing processes into two sections: parts manufacturer and general assembly.

A. Balance Rate of Parts Manufacturer

Parts manufacturer station on the periphery includes 10 machines. Assume flexible rate is 1.15, operation time is shown in Table III.

Calculate the streamline balance rate: The status of assembly line balancing is mainly evaluated by balance delay (Bd) index. It means the percentage that the leisure time accounts for the total time on assembly line:

$$B_d = (N \times C - t) / N \times C$$

N—number of work station,

C—theory pitch time,

t—total work time.

∵ C=121.29s; N=6;

TABLE III
OPERATION TIME

Upper station	operation	Down station	manpower	Average (S)	Standard time(s)
OP 1010	Assemble odometer	OP2080	1/3	32.55	37.43
OP 1020	Assemble 1st shaft	OP2010	1/3	40.47	46.54
OP 1030	Assemble 2nd shaft	OP2010	1	95.83	110.20
OP 1040	Assemble intermediate shaft	OP2020	1	98.66	113.46
OP 1050	Front housing marking	OP2050	1	99.24	114.13
OP 1060	1st shaft end cover gland seal	OP2060	1/3	32.45	37.32
OP 1070	Assemble rear shell body	OP2070	1	95.39	109.7

OP 1080	Assemble console cover	OP2150	1	96.78	111.3

∴ t=(37.43+46.54+37.32)+110.2+113.46+114.13+109.7+111.3 =680.08s.

∴ *Bd*=(121.29*6-680.08)/(6*121.29)*100% = 6.55%

Balance Rate: η=1- *Bd* = 93.45%

It has reached the excellent level based on the line balancing criteria (Table IV), there is no need to improve.

B. Balance Rate of Assembly Line

The time of each process is definite by stop-watch. The record is shown in Table V. We can find there is big gap between some process time. As a result, production plan cannot be finished. The line productivity is limited because of uneven operation time especially bottlenecks. The insufficient productivity is made up by working long hours or working on weekend. Now calculate the assembly line balance rate.

$$B_d = \frac{176.21 \times 16 - 1958.49}{16 \times 176.21} \times 100\% = 30.53\%,$$

η=1-*Bd* =69.47%.

According to Table IV, 30.53%>20%, it is estimated worse. The line needs to be improved.

$$Target\ Pitch\ Time = \frac{Nominal\ task\ time \times utilization\ rate}{target\ throughput}$$

$$= \frac{7.5 \times 3600 \times 94.71\%}{180} = 142s$$

As seen in Table V, the standard time of OP2010, OP2110 and OP3020 are more than target pitch time 142s. Now, improvement will finish respectively.

TABLE IV
CRITERION OF BALANCING RATE

Value of Bd	Evaluation grade
$B_d \leqq 10\%$	excellent
$10\% < B_d < 20\%$	good
$B_d \geqq 20\%$	bad

TABLE V
RECORDS OF ASSEMBLY OPERATION TIME

Name of station	Process	Manpower	Measure (s)	Standard time /s
OP2010	final assembly on line	1	144.63	166.32
OP2020	assemble Intermediate shaft	1	89.79	103.26
OP2030	Assemble reverse gear shaft on second shaft	1	100.6	115.69
OP2040	Assembly parts on intermediate shaft	1	99.17	114.06
OP2050	Assemble shifting yoke mobile shaft	1	99.50	114.23
OP2060	Assemble front housing	1	78.33	90.08
OP2070	Assemble front housing shaft end cover	1	116.21	133.64
OP2080	Assemble interlayer, takedown base	1	89.50	102.93
OP2090	assemble rear shell body	1	106.67	122.67
OP2100	Assemble odometer check	1	112.40	129.26
OP2110	measure air tightness	1	132.00	151.80
OP2130	Assemble console cover, product shunting	1	86.01	98.91
OP3010	Hand lubrication	1	90.10	103.62
OP3020	Performance test of assembly	1	153.23	176.21
OP3030	Oil pumping, disengaging fork lever	1	103.05	118.45
OP3040	Packaging and warehousing	1	102.08	117.39

VI. DIAGNOSIS OF BOTTLENECKS

Developing the necessary position, process and trajectory may provide operations management systems with a more dynamic strategic management capability [13], however, bottlenecks is the key.

A. Improvement of OP3020

The line can be optimized by improving the bottleneck process [14-15]. The time of OP3020 is the maximum. Therefore, ideal rhythm C=176.21s. In reality production, there is only one channel in test station. Machine and operator is idle on account of feeding path detouring. In order to solve the problem, one exit is added in Fig.4.

The simplest and most direct way to increase streamline balance rate is decreasing the bottleneck time continuously using process analysis, time analysis and operation analysis [16]. To solve the problem of long waiting, man-machine Process Analysis together with ECRS is used in Table VI.

In the Table VI, P is periodic, WT is working time, IT is idle time, U is utilization.

Improved Stand Time of Work Station =

*Average Total Task Time of Improved*Comparison Coefficient** Flexible Rate

=110.62*1*(1+0.15)

=127.21s

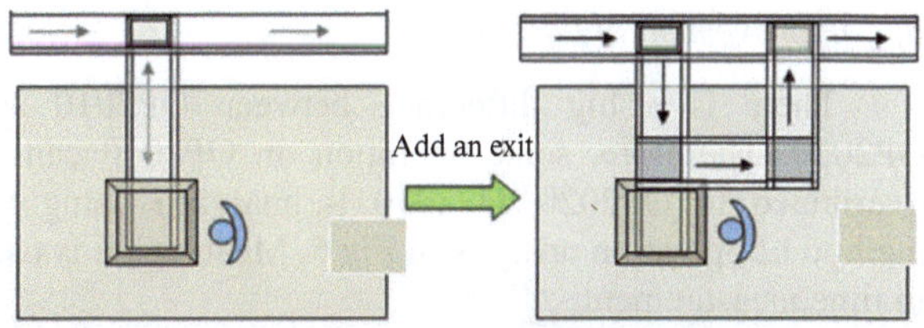

Fig.4. Feeding on OP3020

TABLE VI
MAN-MACHINE PROCESS ANALYSIS ON OP3020

Before improvement			time/(s)		After improvement		
transmission on belt	Work-er	Test-er			transmission on belt	Work-er	Test-er
adjust saddle	idle	idle	4.82	13.41	Adjust saddle	assemble and adjust	assemble tester and adjust
to the test station			21.34	88.69	To the test station	test	test
Idle	assemble and adjust	Assemble and adjust	34.75		waiting		
	test	test	11				

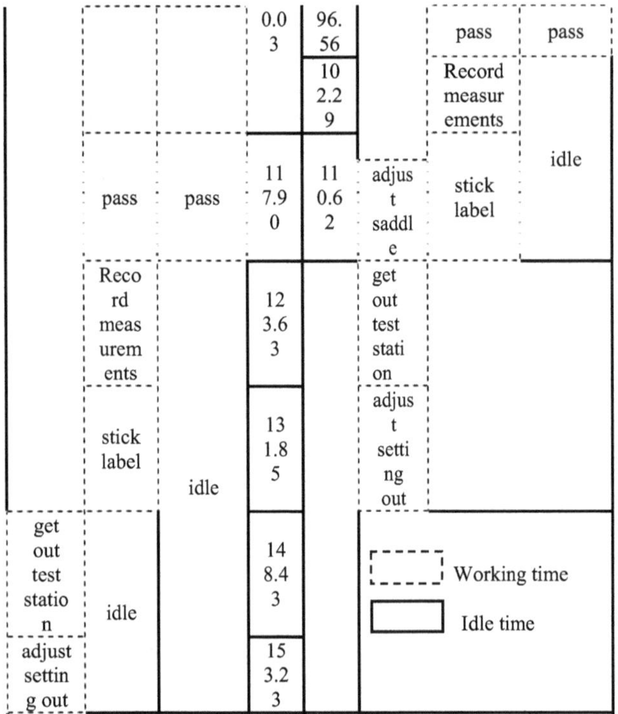

Statistics before improvement					Statistics after improvement				
project	P	WT	IT	U	U	IT	WT	P	project
man	153.23	110.62	42.61	72.19%	100%	0	110.62	110.62	man
machine	153.23	96.56	56.67	63.02%	87.30%	14.06	96.56	110.62	machine

B. Improvement of OP2010

There is a big difference between OP2010 and OP2020. Therefore, some operation on OP2010 can be distributed to OP2020. OP2010 is analyzed using the method of operation analysis in Fig.5. Mean value is used in time measurement.

*Improved Standard Time of Work Station =Average Total Task Time of Improved*Comparison Coefficient** Flexible Rate =109.31 s*1*1.15=125.71s

Considering that some work attribute to OP2010 has been arranged to OP2020, and it leads to working hour change about OP2020, it needs to retest the task time/standard time. Improvement of OP2020 can be seen in Fig.6.

Improved Standard Time=130.74s<*Target Pitch Time*=142s, although the work station operation increased, it has no influence on the whole pitch time.

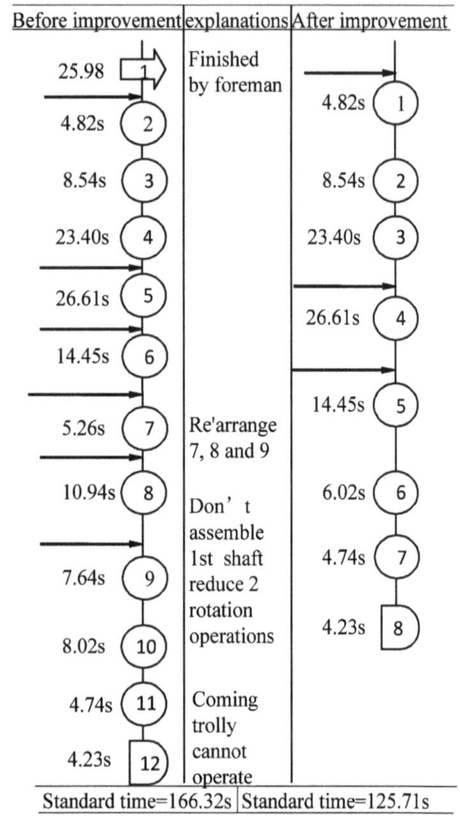

Fig.5. Job analysis of OP2010

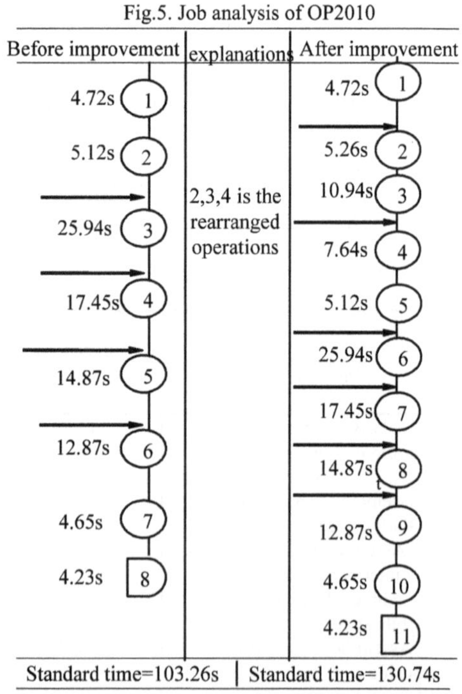

Fig.6. Job analysis of OP2020

C. Improvement of OP2110

OP2110 focuses on air tightness test for the ontology assembly in front. Similarly, with the improvement idea of OP3020 and OP2010, the standard time (ST) decreases to 125.58s.

D. Effect Evaluation

Now measure each station's time again (Table VII).

Balance rate: η=1- *Bd*=1-14.33%=85.67%

It has reached the good level based on the line balancing criteria. So, the *improved Team Capacity= Task Time*Utilization Rate of Task Time/Pitch Time*

=60 s/m*60 m/h*7.5h*94.71%/133.64s

=202 Units

TABLE VII
OPERATION TIME AFTER IMPROVEMENT

sequence number	operation	man power	Average(s)	standard time /s
OP2010	final assembly on line	1	109.31	125.71
OP2020	assemble Intermediate shaft	1	113.69	130.74
OP2030	assemble reverse gear shaft on second shaft	1	100.6	115.69
OP2040	assemble parts on intermediate shaft	1	99.17	114.06
OP2050	assemble shifting yoke mobile shaft	1	99.50	114.23
OP2060	assemble front housing	1	78.33	90.08
OP2070	assemble front housing shaft end cover	1	116.21	133.64
OP2080	assemble interlayer, takedown base	1	89.50	102.93
OP2090	assemble rear shell body	1	106.67	122.67
OP2100	assemble odometer check	1	112.40	129.26
OP2110	measure air tightness	1	109.20	125.58
OP2130	assemble console cover, product shunting	1	86.01	98.91
OP3010	hand lubrication	1	90.10	103.62
OP3020	performance test of assembly	1	110.62	127.21
OP3030	oil pumping, disengaging fork lever	1	103.05	118.45
OP3040	packaging and warehousing	1	102.08	117.39

VII. Conclusion

This paper tries to improve the throughput of plant streamline N based on IE methodologies. Although the balance rate increases to 85.67%, new bottlenecks will occur again. Actually, the bottlenecks exist all the time. Therefore the company should improve efficiency endless.

ACKNOWLEDGEMENT

This work was supported in part by Jiangxi 12th-five-year social science project in 2011 and the funding from Jiangxi provincial education department.

REFERENCES

[1] Holweg, M., 2007. The genealogy of lean production. Journal of Operations Management 25 (2), 420–437.

[2] Shah R., WardP.T., 2003. Lean manufacturing: context, practice bundles, and performance. *Journal of Operations Management* 21 (2), 129–149.

[3] Rachna Shah, Peter T. Ward, Lean manufacturing: context, practice bundles, and performance, *Journal of Operations Management* 21 (2003) 129–149.

[4] Adriaan van, horenbeek Liliane pintelon, Development of a maintenance performance measurement framework-using the analytic network process (ANP) for maintenance performance indicator selection, *Omega* 42(2014) 33-46.

[5] Krisztina, Demeter ZsoltMatyusz, The impact of lean practices on inventory turnover, *Int. J. Production Economics* 133(2011)154–163

[6] Franci Pusavec, Peter Krajnik, Janez Kopac, Transitioning to sustainable production-Part I: application on machining technologies, Journal of Cleaner Production 18 (2010) 174–184.

[7] Ma Ga(Mark)Yang, PaulHong, SachinB.Modi, Impact of lean manufacturing and environmental management on business performance: An empirical study of manufacturing firms, *Int. J. Production Economics* 129 (2011) 251-261.

[8] Xing Wenying, Basic Materials of QC Team [M]. Beijing: China Society Press, 2004: 125-127.(Chinese)

[9] Zheng Zhaoning, Wu Yuying. The Enterprise Diagnosis combined with Fishbone Diagram and AHP [J]. China Soft Science, 2001, 16(1): 118-121.(Chinese)

[10] Sun Jianhua, Research on Line Balancing Methods [J]. Group Technology and Production Modernization, 2004, 21(4): 34-36. (Chinese)

[11] Xu Xuejun, A Study of Relationship Between Production Line Balancing and Corporate Profits [J]. Industrial Engineering, 2009. 12(4):41-45. (Chinese)

[12] McLachlin, R., 1997. Management initiatives and just-in-time manufacturing. *Journal of Operations Management* 15 (4), 271–292.

[13] Edson Pinheirode Limaa, Sergio E. GouveadaCosta, Jannis Jan Angelis, Juliano Munikb, Performance measurement systems: A consensual analysis of their roles, *Int. J. Production Economics* 146 (2013) 524–542.

[14] Mentian Anhong, The New Toyota Production System[M]. Baoding: Hebei University Press, 2001.

[15] Jin jing Zheng ming, Site Improvement: Low Cost Management Method [M]. Beijing: China Machine Press, 2000. (Chinese)

[16] Li Ruiqin, The Importance of Process Balance in Assembly Line [J]. Inner Mongolia, Technology and Economy, 2008, 160 (6): 337-339. (Chinese)

Vertical Cooperative Advertising Game Models between Members in Supply Chain

Zhu Ying[1,*], Zhang Chengke[2], Liu Zengyang[1]
[1]School of Management, Guangdong University of Technology, Guangzhou, 510520
[2]School of Economics & Commence, Guangdong University of Technology, Guangzhou, 510520
(*zhuyingsom@163.com; zhangck@gdut.edu.cn)

Abstract - **This paper considers the vertical cooperative advertising tactics in a two-echelon supply chain system, which is composed of single manufacturer and retailer. In view of the situation that market demand is influenced by the retail price and advertising investment, a non-cooperative and cooperative game models are established respectively, and a model of income distribution is given based on Shapley value. Comparing with the equilibrium of two game models, we conclude that the total revenue of the supply chain, in the case of non-cooperative game, is less than that of cooperative game, as well as in the case of cooperative game, advertising investment is higher than that of non-cooperative game.**

Keywords - **Game models, supply chain, Shapley value, vertical cooperative advertising**

I. INTRODUCTION

With rapid development of logistics industry and E-commerce, companies can increase market awareness of their products and stimulate market demand by advertising on an E-commerce platform. As an effective way of advertising, cooperative advertising can be divided into two types: horizontal and vertical cooperative advertising [1]. The vertical cooperative advertising in the supply chain system can be seen as a coordination mechanism, which refers to two vertical companies who join together to advertise to co-promote their products, and which has interactive relationship between manufacturers and retailers. In order to promote the sale of a product in the supply chain, manufacturers and retailers can conduct national and local advertising. The advertising costs can be paid solely by manufacturers or retailers. The manufacturer can also pay part of the advertising cost as subsidies to the retailers.

At present, there are a large number of theoretical literatures on vertical cooperative advertising. Cooperative advertising in the supply chain system has become a very popular form of members' cooperation in the supply chain. Jørgensen, Taboubi and Zaccour studied dynamic advertising and promotional tactics of marketing channels in the supply chain. Retailers promoted manufacturers' products. Manufacturers built product goodwill by advertising. They got the optimal strategies for the manufacturers and retailers to build goodwill and cooperation in the non-cooperative Nash game and Stackelberg game. They also studied the impact of goodwill on marginal revenue. The research findings indicated that, regardless of the impact of goodwill on marginal revenue, cooperative advertising in marketing channels was a mechanism for coordination and participants would get higher income [2]. Huang, Li and Mahajan used game theory to study the brand investment and the role of local advertising cost and the function of advertising costs sharing in the supply chain cooperative advertising [3]. Xie and Neyret established three noncooperative game model (Nash game, retailers based Stackelberg game and manufacturers based Stackelberg game) and a cooperative game model to discuss cooperative advertising and product pricing of two-echelon supply chain of the manufacturers and the retailers [4]. He, Prasad and Sethi used Stackelberg game model to study cooperative advertising and pricing of dynamic stochastic supply chain system, and presented a method for coordinating supply chain channels after a comparison with advertising and retail pricing strategy of the vertical integration channels[5]. Javid and Hoseinpour used game model to study the coordinating of cooperative advertising decisions in the supply chain with only one supplier and one retailer [6]. Aust and Buscher considered advertising and pricing decisions based on supply chain with only one supplier and one retailer, established three non-cooperative game under information symmetry or information asymmetry as well as a cooperation game model which pursued overall benefits maximization, got a conclusion that the overall benefits of the supply chain was the highest when participants cooperated [7]. Wu, Zhao and Liu took full account of the uncertain market demand, established a game analysis model of vertical cooperative advertising in the supply chain. They explored bilateral optimal advertising and cooperation strategies under the Stackelberg and Nash Equilibrium according to different position of manufacturers and retailers in the supply chain. They got the relationship between the retailer's inventory levels and advertising cooperation and level. They concluded that the retailer's inventory level was independent from and advertising cooperation and advertising level [8]. Fu and Zeng studied two-echelon supply chain structure based on newsboy type of product. By introducing order quantity as a variable, they investigated the optimal advertising strategy of manufacturers and retailers and the optimal ordering policy of retailers under Stackelberg game and Nash game. And they explored the inhibition from randomness of the market

E. Qi et al. (eds.), *Proceedings of the 5th International Asia Conference on Industrial Engineering and Management Innovation (IEMI2014)*, Proceedings of the International Asia Conference on Industrial Engineering and Management Innovation 1, DOI 10.2991/978-94-6239-100-0_34, © Atlantis Press and the authors 2015

demand to advertising investment level of manufacturers and retailers [9].

As can be found from the studies, the retailers' or the manufacturers' market share can be influenced not only by their advertising investment but also by the advertising investment of other members in the supply chain [10~12]. So, it is vital for retailers and manufacturers to decide how to arrange advertising investment under the support of manufacturers and how to arrange the subsidy policy. Inspired by the research results, the paper considers that the market demand is influenced by both the retail price and advertising investment. The demand function is the multiplication of the retail price and advertising investment. It uses game theory to research vertical advertising model and cooperative game for manufacturers and retailers in the supply chain which have the same decision-making authority. And it gives income distribution model based on cooperative game by Shapley value method. It expects the resulting conclusions can provide scientific decision-making theory for the retailers to budget the advertising spending and the manufacturers to make strategy selection of advertising subsidies.

II. NOTATION AND MODELING ASSUMPTIONS

The research of this paper is simply from a supply chain with two echelons, which are manufacturer and retailer, and only considers the situation that there is a single product and firm in each echelon. Decision variables for manufacturer are advertising expenditure, wholesale price, and the cooperative advertising compensation. Retailer's decision variables are advertising expenditure and retail price. In this paper, a and A denote the retailer's local advertising and the manufacturer's national advertising investment respectively, and the proposed models are based on the following assumptions.

Assuption 1: The market demand function $D(p, a, A) = g(p)\,h(a, A)$ depends on the retail price p, the advertising level a and A in this case; $g(p) = \alpha-\beta p$ is linearly decreasing with respect to p, and $h(a, A) = K - Ba^{-\gamma}A^{-\delta}$ is the function that Huang and Li(2001) proposed to model advertising effects on sales in a static way. K is the sales saturate asymptote.

Assuption 2: The following notation is used throughout the chapter:

α, β, B, γ, δ are positive constants, γ is local advertising elasticity and δ is national advertising elasticity;

w denotes unit wholesale price(the manufacturer's transfer price to the retailer);

p denotes unit retail price;

c denotes manufacturer's unit production cost;

d denotes the retailer's unit cost incurred in addition to the purchasing cost;

t denotes the fraction of the local advertising expenditure, which is the percentage the manufacturer agrees to share with the retailer;

Π_m, Π_r, Π_{m+r} denotes profit function of manufacturer, retailer, and total supply chain, correspondingly.

Under these assumptions, the profits of the manufacturer, the retailer and the whole supply chain can be expressed as follows, respectively:

$$\Pi_m = (w-c)(\alpha-\beta p)(K - Ba^{-\gamma}A^{-\delta}) - ta - A \quad (1)$$

$$\Pi_r = (p-w-d)(\alpha-\beta p)(K - Ba^{-\gamma}A^{-\delta}) - (1-t)a \quad (2)$$

$$\Pi_{m+r} = (p-c-d)(\alpha-\beta p)(K - Ba^{-\gamma}A^{-\delta}) - a - A \quad (3)$$

In order to handle the problem in an equivalent but more convenient way, we apply an appropriate change of variables as shown below:

$$\alpha' = \alpha - \beta(c+d) > 0\ ,\quad p' = \frac{\beta}{\alpha'}(p-(c+d)) > 0\ ,$$

$$w' = \frac{\beta}{\alpha'}(w-c) > 0\ ,\quad B' = \frac{\alpha'^2}{\beta}B\ ,\quad K' = \frac{\alpha'^2}{\beta}K\ ,$$

$$A' = \frac{A}{\sqrt[\gamma+\delta+1]{B'}}, a' = \frac{a}{\sqrt[\gamma+\delta+1]{B'}}, \Pi' = \frac{\Pi}{\sqrt[\gamma+\delta+1]{B'}}.$$

After final change of variables leads to the following expressions for the manufacture's, the retailer's and the whole system's profits:

$$\Pi_m = w'(1-p')\left(\frac{K'}{\sqrt[\gamma+\delta+1]{B'}} - \frac{1}{a'^{\gamma}A'^{\delta}}\right) - ta' - A' \quad (4)$$

$$\Pi_r = (p'-w')(1-p')\left(\frac{K'}{\sqrt[\gamma+\delta+1]{B'}} - \frac{1}{a'^{\gamma}A'^{\delta}}\right) - (1-t)a' \quad (5)$$

$$\Pi_{m+r} = p'(1-p')\left(\frac{K'}{\sqrt[\gamma+\delta+1]{B'}} - \frac{1}{a'^{\gamma}A'^{\delta}}\right) - a' - A' \quad (6)$$

III. TWO FORMS OF RETAILER-MANUFACTURER RELATIONSHIP

A. *The non-cooperative Nash game (N)*

Under the Nash game situation, we assume a symmetrical distribution of power between manufacturer and retailer. The situation can be modeled by means of Nash equilibrium, where both players take their decisions simultaneously and non-cooperatively. The objective of them is to design their move in such a way as to maximize their revenue, and hence the decision problem of the manufacturer can be formulated as following:

$$\underset{A,t}{Max}\,\Pi_m = w(1-p)\left(\frac{K}{\sqrt[\gamma+\delta+1]{B}} - \frac{1}{a^{\gamma}A^{\delta}}\right) - ta - A \quad (7)$$

$$s.t.\ 0 \le t \le 1,\ A \ge 0$$

The decision problem of the retailer is:

$$\underset{a}{Max}\Pi_r = (p-w)(1-p)\left(\frac{K}{\sqrt[\gamma+\delta+1]{B}} - \frac{1}{a^\gamma A^\delta}\right) - (1-t)a \quad (8)$$

We solve problem (7) and (8) by derivation and set the first order equations to zero. Therefore, we can obtain the results:

$$\frac{\partial \Pi_m}{\partial A} = \delta w(1-p)(a^{-\gamma} A^{-\delta-1}) - 1 = 0 \quad (9)$$

$$\frac{\partial \Pi_r}{\partial a} = \gamma(p-w)(1-\mathrm{p})a^{-\gamma} A^{-\delta-1} - (1-t) = 0 \quad (10)$$

$$\frac{\partial \Pi_r}{\partial p} = \left(\frac{K}{\sqrt[\gamma+\delta+1]{B}} - \frac{1}{a^\gamma A^\delta}\right)(1-2p+w) = 0 \quad (11)$$

It is obvious that the optimal value of t is zero because of its negative coefficient in the objective (7). Then

$$p^N = \frac{2}{3},\ w^N = \frac{1}{3},\ a^N = \left[\frac{\gamma}{9}\left(\frac{\gamma}{\delta}\right)^\delta\right]^{\frac{1}{\gamma+\delta+1}},\ A^N = \frac{\delta}{\gamma}\left[\frac{\gamma}{9}\left(\frac{\gamma}{\delta}\right)^\delta\right]^{\frac{1}{\gamma+\delta+1}} \quad (12)$$

Thus manufacturer, retailer and total supply chain profits respectively:

$$\Pi_m^N = \frac{1}{9}\left\{\frac{K}{\sqrt[\gamma+\delta+1]{B}} - \left(\frac{\gamma}{\delta}\right)^\delta\left[\frac{\gamma}{9}\left(\frac{\gamma}{\delta}\right)^\delta\right]^{\frac{\gamma+\delta}{\gamma+\delta+1}}\right\} - \frac{\delta}{\gamma}\left[\frac{\gamma}{9}\left(\frac{\gamma}{\delta}\right)^\delta\right]^{\frac{1}{\gamma+\delta+1}} \quad (13)$$

$$\Pi_r^N = \frac{1}{9}\left\{\frac{K}{\sqrt[\gamma+\delta+1]{B}} - \left(\frac{\gamma}{\delta}\right)^\delta\left[\frac{\gamma}{9}\left(\frac{\gamma}{\delta}\right)^\delta\right]^{\frac{\gamma+\delta}{\gamma+\delta+1}}\right\} - \left[\frac{\gamma}{9}\left(\frac{\gamma}{\delta}\right)^\delta\right]^{\frac{1}{\gamma+\delta+1}} \quad (14)$$

$$\Pi_{m+r}^N = \frac{2}{9}\left\{\frac{K}{\sqrt[\gamma+\delta+1]{B}} - \left(\frac{\gamma}{\delta}\right)^\delta\left[\frac{\gamma}{9}\left(\frac{\gamma}{\delta}\right)^\delta\right]^{\frac{\gamma+\delta}{\gamma+\delta+1}}\right\} - \left(1+\frac{\delta}{\gamma}\right)\left[\frac{\gamma}{9}\left(\frac{\gamma}{\delta}\right)^\delta\right]^{\frac{1}{\gamma+\delta+1}} \quad (15)$$

B. *Cooperative game (CO)*

In this section, we apply a cooperative game approach [13] between the manufacturer and retailer in the supply chain, in order to find out whether both parties can increase their profit if they work cooperatively. Using this approach, the manufacturer and the retailer work together to determine *a* and *A*. According to the whole supply chain profits maximization, they decide the retail price and local/national advertising investment. A cooperation relationship between manufacturer and retailer, which target is to maximize the profit, is characterized by the following expression:

$$\underset{A,a}{Max}\,\Pi_{m+r} = p(1-p)\left(\frac{K}{\sqrt[\gamma+\delta+1]{B}} - \frac{1}{a^\gamma A^\delta}\right) - a - A \quad (16)$$

$$s.t.\ A \geq 0,\ a \geq 0$$

We solve the problem (16) by derivation and set the first order equations to zero. Therefore, the results obtained can be described as following:

$$p^{CO} = \frac{1}{2},\ a^{CO} = \left[\frac{\gamma}{4}\left(\frac{\gamma}{\delta}\right)^\delta\right]^{\frac{1}{\gamma+\delta+1}},\ A^{CO} = \frac{\delta}{\gamma}\left[\frac{\gamma}{4}\left(\frac{\gamma}{\delta}\right)^\delta\right]^{\frac{1}{\gamma+\delta+1}} \quad (17)$$

We set equations (17) into equation (16), hence we derive for the total supply chain profit:

$$\Pi_{m+r}^{CO} = \frac{1}{4}\left\{\frac{K}{\sqrt[\gamma+\delta+1]{B}} - \left(\frac{\gamma}{\delta}\right)^\delta\left[\frac{\gamma}{4}\left(\frac{\gamma}{\delta}\right)^\delta\right]^{\frac{\gamma+\delta}{\gamma+\delta+1}}\right\} - \left(1+\frac{\delta}{\gamma}\right)\left[\frac{\gamma}{4}\left(\frac{\gamma}{\delta}\right)^\delta\right]^{\frac{1}{\gamma+\delta+1}} \quad (18)$$

Gains allocation is a critical issue in the supply chain after the formation of cooperation coalition. Multiplayer cooperative game is also known as the coalitional game. To maintain a stable coalition, a major premise is that each party of the coalition gets more gains than that they can get under non-cooperation. Therefore, in the following text, Shapley value method is adopted to solve the allocation problem of the supply chain under the cooperative game.

IV. REVENUE DISTRIBUTION MODEL OF COOPERATIVE GAME BASED ON THE SHAPLEY VALUE METHOD

The definition of Shapley value implies that the participants outside coalition S form another coalition N-S against S. Under this assumption, the coalition S's gain is certain. As Shapley participants are assumed to join a particular coalition randomly and are subject to uniform distribution, which is inconsistent with the reality. That is why the distribution of gains applying Shapley value needs further consideration [10, 11]. Let N be the set of the participants of the supply chain. N={manufacturer, retailer}={1, 2}. Assume V(ø)=0. The following considers the non-empty set of coalition.

TABLE I
MANUFACTURE M'S PROFIT ALLOCATION RESULTS

S	(1)	(1, 2)
$V(S)$	Π_m	Π_{m+r}^{CO}
$V(S\|1)$	0	Π_r
$V(S)-V(S\|1)$	Π_m	$\Pi_{m+r}^{CO}-\Pi_r$
$\|S\|$	1	2
$u=\frac{(n-\|S\|)!(\|S\|-1)!}{n!}$	1/2	1/2
$u[V(S)-V(S\|i)]$	$\frac{1}{2}\Pi_m$	$\frac{1}{2}(\Pi_{m+r}^{CO}-\Pi_r)$

According to Tabl.,the manufacturer's revenue is

$$\Pi_m^{CO} = \frac{1}{8}\left\{\frac{K}{\sqrt[\gamma+\delta+1]{B}} - \left(\frac{\gamma}{\delta}\right)^\delta\left[\frac{\gamma}{4}\left(\frac{\gamma}{\delta}\right)^\delta\right]^{-\frac{\gamma+\delta}{\gamma+\delta+1}}\right\} + \left[\left(1-\frac{\delta}{\gamma}\right)\right.$$

$$(\frac{\gamma}{9})^{\frac{1}{\gamma+\delta+1}}-(1+\frac{\delta}{\gamma})(\frac{\gamma}{4})^{\frac{1}{\gamma+\delta+1}})](\frac{\gamma}{\delta})^{\frac{\delta}{\gamma+\delta+1}} \tag{19}$$

The similar method as tab1 can get retailer's revenue

$$\Pi_r^{co}=\frac{1}{8}\left\{\frac{K}{\sqrt[\gamma+\delta+1]{B}}-(\frac{\gamma}{\delta})^{\delta}[\frac{\gamma}{4}(\frac{\gamma}{\delta})^{\delta}]^{\frac{\gamma+\delta}{\gamma+\delta+1}}\right\}-(1-\frac{\delta}{\gamma})[(\frac{\gamma}{\delta})^{\delta}\frac{\gamma}{9}]^{\frac{1}{\gamma+\delta+1}} \tag{20}$$

V. EQUILIBRIUM ANALYSIS

Comparing with two game models decisions variables and revenue the corresponding theorems are given:

Theorem1: Comparisons on (15) and (18) it shows that the supply chain revenue under non-cooperative game is less than cooperative game.

Theorem 2: Our analysis in (12) and (17) shows that the retail price is always higher than cooperative game retail price.

Theorem 3: National advertising and local advertising expenditure under the cooperative game is higher than non-cooperative Nash game. As stated above, the highest advertising expenditures are received when the players cooperate.

VI. NUMERICAL EXAMPLE

To verify the correctness of these three theorems, a numerical example is given below. Suppose the parameters is $\gamma = [0, 2]$, $\delta = [0, 2]$, $K = 10$, $B = 1/2$.

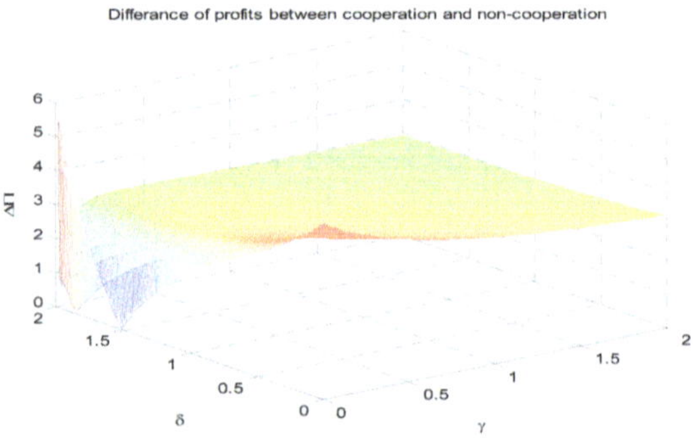

Fig.1. Relative difference of profits between cooperation and non-cooperation in the supply chain

The most important performance measure of the distribution supply chain deals with profits. It can be seen that whole supply chain profits under cooperative game is situation higher than non-cooperative game. Fig.1 presents the values of relative difference between cooperation and non-cooperation the total profits of the supply chain is positive, as the parameters γ and δ increase. Therefore, enterprises in the supply chain should set the whole supply chain profits maximization as the target when they make decisions.

VII. CONCLUSIONS

In the foregoing analysis, this paper have shown that a supply chain system based on game theory is composed of a manufacture and a retailer [14~15], which researches the problem of manufacturer and retailer's advertising expenditure, revenue, and gains allocation based on Shapley value with cooperative game, and then discusses the equilibrium results and profits of players under the situation of advertising sharing and pricing respectively. Obtain: First, the participation ratio (t) of manufacturer is zero if manufacturer and retailer implement non-cooperative game. It means in cooperation game manufacturer will share the advertising expenditure of retailer. Second, manufacturer and retailer are willing to cooperate when it is profitable. Third, it will better coordinate supply chain when players implement on cooperative game, since the whole supply chain gain more.

ACKNOWLEDGEMENTS

This work was supported in part by the National Science Foundation of China with grant No. 71171061, and Guangdong College of Humanities and Social Sciences project with grant No. 11WYXM022, Science and technology project of department of transportation of Guangdong Province with grant No. 13ZK0303.

REFERENCES

[1] Nie, J. J., and Xiong, Z. K., "Differential game models of vertical cooperative advertising with multiple advertising media", *Journal of Management Science In China*, 2010, vol. 13, no. 5, pp. 1-10.

[2] Jørgensen, S., Taboubi, S., and Zaccour, G., "Cooperative advertising in a marketing channel", *Journal of Optimization Theory and Applications*, 2001, vol. 110, no. 1, pp. 145-158.

[3] Huang, Z., Li, S. X., and Mahajan, V., "An Analysis of Manufacturer-Retailer Supply Chain Coordination in Cooperative Advertising", *Decision Sciences*, 2002, vol. 33, no. 3, pp. 469-494.

[4] Xie, J., and Neyret, A., "Co-op advertising and pricing models in manufacturer-retailer supply chains", *Computers & Industrial Engineering*, 2009, vol. 56, no. 4, pp. 1375-1385.

[5] He, X., Prasad, A., and Sethi, S. P., "Cooperative advertising and pricing in a dynamic stochastic supply chain: Feedback Stackelberg strategies", *Production and Operations Management*, 2009, vol. 18, no. 1, pp. 78-94.

[6] Javid, A. A., and Hoseinpour, P., "A game-theoretic analysis for coordinating cooperative advertising in a supply chain", *Journal of Optimization Theory and Applications*, 2011,vol. 149, no.1, pp. 138-150.

[7] Aust, G., and Buscher, U., "Vertical cooperative advertising and pricing decisions in a manufacturer–retailer supply chain: A game-theoretic approach", *European Journal of Operational Research*, 2012, vol. 223, no. 2, pp. 473-482.

[8] Fu, Q., and Zeng, S. Q., "Game analysis of cooperative advertising and ordering strategies in a supply chain under demand uncertainty", *System Engineering Theory & Practice*, 2008, vol. 3, no.3, pp. 56-63.(Chinese)

[9] Huang, S., Yang, C., and Zhang, X., "Pricing and cooperative advertising decision models in dual-channel supply chain", *Computer Integrated Manufacturing Systems*, 2011, vol. 17, no.12, pp. 2683-2692.

[10] Jorgensen, S., and Zaccour, G., "Equilibrium pricing and advertising strategies in a marketing channel", *Journal of Optimization Theory and Applications*, 1999, vol. 102, no. 1, pp. 111-125.

[11] Kargin, V., "Uncertainty of the Shapley Value", *International Game Theory Review*, 2005, vol. 7, no. 4, pp. 517-529.

[12] D'aspremont, C., Jacquemin, A., "Cooperative and noncooperative R&D in duopoly with spillovers", *The American Economic Review*, 1988, vol. 78, no. 5, pp. 1133-1137.

[13] Li, S. X., Huang, Z., "Cooperative advertising, game theory and manufacturer- retailer supply chains", *Omega*, 2002, vol. 30., no. 5, pp. 347-357.

[14] Li, X., Wang, Q., "Coordination mechanisms of supply chain systems", *European Journal of Operation Research*, 2007, vol. 179. No. 1, pp. 1-16.

[15] Jorgensen, S., Tiboubi, S., & Zaccour, G., "Retail promotions with negative brand image effects: Is cooperation possible?", *European Journal of Operationa Research*, 2003, vol. 150, No. 2, pp. 395-405.

Total Quality Situation and Analysis in China 2013: Empirical Research Based on Quality Observations

Hong Cheng[1], Dan-dan Li[1], Lian-fa Luo[2,*]
[1]Institute of Quality Development Strategy (IQDS), Wuhan University, Wuhan City, Hubei Province, China
[2]Corresponding Author, Institute of Quality Development Strategy (IQDS), Wuhan University, Wuhan City, Hubei Province, China
(398105109@qq.com)

Abstract - **In this article, a Total Quality Index evaluation model and data from quality observations are adopted to do an empirical research on total quality situation in China from four dimensions, quality safety, quality satisfaction, public service on quality and citizen concept on quality. It is implied by data that the fluctuations in total quality index in China 2013 is less than that in 2012, and product safety is in a period of rise; market competition promotes the satisfaction on products quality; consumers' demands on products quality have exceeded their demands on products safety; variations in products quality are in consistent with variations in economy. Empirical analysis above simultaneously proved that the four dimensions of total quality index evaluation model have internal stability and reasonability.**

Keywords - **Empirical research, quality observation, quality evaluation**

I. INTRODUCTION

Problems on quality in China are concerned by all people, and as time goes, these concerns increasingly turn into anxiety and dissatisfaction[1]. Chinese government has taken many measures to solve quality problems in recent years, but still do not reverse these emotions. Reasons for situations above are multidimensional, among which, the most basic one is the lack of a comprehensive and rational evaluation on quality problems in China [1]. Only if the current situation of total quality has been known well, can the problems be dealt with effectively [2].

Integrated with research done by some experts, the definition of quality is modified as the degree that inherent performance could satisfy needs [3][4][5], by *Quality Control and Quality Assurance Terminology* published by ISO in 2000. In other words, a certain condition of quality can be resulted from many factors, but it can eventually be reflected by consumers' selection. Based on this, evaluations from consumers are equivalent to overall situation of quality developments in an area. According to Total Quality Index evaluation model proposed by Cheng Hong etc. [6], a large-scale questionnaire survey, quality observations, has been conducted every year since 2012. Quality observations is a process of long-term interviews, investigations, statistics, measurements and analysis facing consumers, carried out in the form of participatory surveys and focuses on quality situation and characteristics [7].

What on earth is the real version of quality in China? This article will answer this question based on data from quality observations, empirically analyze general characteristics of total quality situation in China 2013 and main changes in quality situation compared to 2012, so as to provide basic data support to scientific decisions and effective management measurements on quality in China.

II. METHODOLOGY

A. Evaluation model

According to Total Quality Index evaluation model proposed by Cheng Hong etc. [6][8], four dimensions *quality safety, quality satisfaction, public service on quality and citizen concept on quality* compose an total quality situation evaluation system facing consumers, as shown in Table I. The meanings of these four dimensions respectively are:

◎*Quality Safety* refers to consumers' evaluations on the risk and harms resulted from poor quality.

◎ *Quality Satisfaction* refers to the degree to which consumers' needs or expectation on quality could be satisfied.

◎ *Public Service on Quality* refers to various public services and resources provided by different government departments or social organizations for quality management, as well as a series of supervisions and managements on all kinds of microeconomic units, which aim at safeguarding consumers' rights on quality and promoting the elevation of quality level in an area.

◎ *Citizen Concept on Quality* refers to consumers knowledge, morality and education level about quality, and consciousness on rights safeguarding towards quality problems.

There are 100 questions in the survey in total, designed based on the four dimensions. Every question can be scored from 1 to 10, where the higher the score, the better the performance of that sub-item is.

TABLE I

TOTAL QUALITY INDEX EVALUATION SYSTEM

	Observation dimension	Weight	Content of investigation
Total Quality Index (TQI)	Quality Safety	30%	Safety investigations on four fields, products, services, projects and environment. Products include foods, grains, oil, meats, dairies, household appliances, drugs, computers, consumer goods,

This study was supported by China's Ministry of Science and Technology Public Industry Special Research grants(201210117, 201310202), the key project of National Social Science Fund in China (11&ZD158) and the 'Twelfth Five Year' National Science and Technology program (2011BAK06B06). The authors wish to thank the anonymous reviewers for helpful comments and suggestions.

E. Qi et al. (eds.), *Proceedings of the 5th International Asia Conference on Industrial Engineering and Management Innovation (IEMI2014)*, Proceedings of the International Asia Conference on Industrial Engineering and Management Innovation 1, DOI 10.2991/978-94-6239-100-0_35, © Atlantis Press and the authors 2015

			clothing, lifts and agriculture supplies
	Quality Satisfaction	30%	Safety investigations on four fields, products, services, projects and environment. Industries involved in the field of products are the same as those in *Quality Safety*.
	Public Service on Quality	20%	Overall government image, investment in quality, consumption environment, quality early warning and prevention, information on quality, education on quality and consumer relief
	Citizen Concept on Quality	20%	Quality consciousness, quality knowledge, quality activities

B. Data sources and samples

6217 consumers from 92 cities, 29 provinces in China were involved in the quality observations of 2013 [7]. Stratified sampling was adopted in terms of the population in each province. In each area, samples were randomly selected in stratified way according to career, education, income and urban-rural dimension. Therefore, samples with consistent portions with population were obtained, and 824,000 records were collected. Considering regional representation and sample size, both the quality observations in 2013 and 2012 meet the requirements of large sample analysis.

TABLE II
SAMPLE STATISTICS OF QUALITY OBSERVATIONS

Year	Quantity of Provinces	Quantity of Cities	Quantity of Questionnaires
2012	26	48	2815
2013	29	92	6217

III. RESULTS

A. The total quality index is above the qualified line

The total quality index (TQI) in 2013 is 63.74, which is above the qualified level overall. As to the structural indexes, the quality safety index is highest, with a score of 65.89; the citizens' concept on quality ranks the second, with a score of 64.51; the quality satisfaction ranks the third, with a score of 64.51; only the public service in quality index is below the qualified level, with a score of 57.82. These results show that China's overall quality develops steadily.

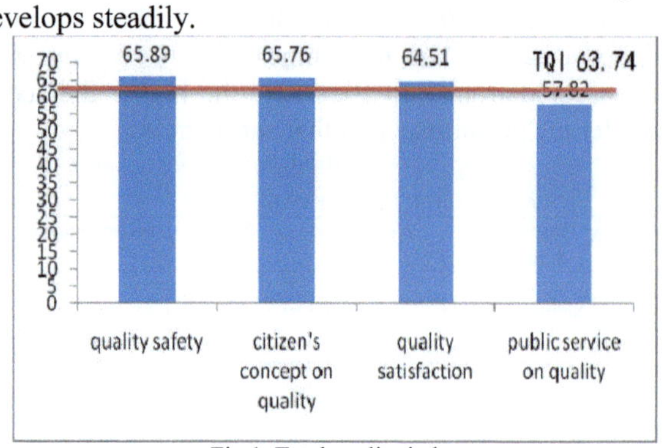

Fig.1. Total quality index

B. The quality safety and quality satisfaction develops steadily and with some increases.

Compared with 2012, the quality safety index is almost the same as 2012, with only a 0.09% decline. The quality satisfaction increase by 2.49 or 4.02% by percentage. As is shown that, the quality safety is steady, while the quality satisfaction has increased to some degree. The relatively steady results between 2012 and 2013 also verifies that the quality observation model is valid in evaluating China's quality.

Fig.2. The yearly change of quality safety and quality satisfaction

C. Product quality satisfaction index decreases slightly compared with 2012

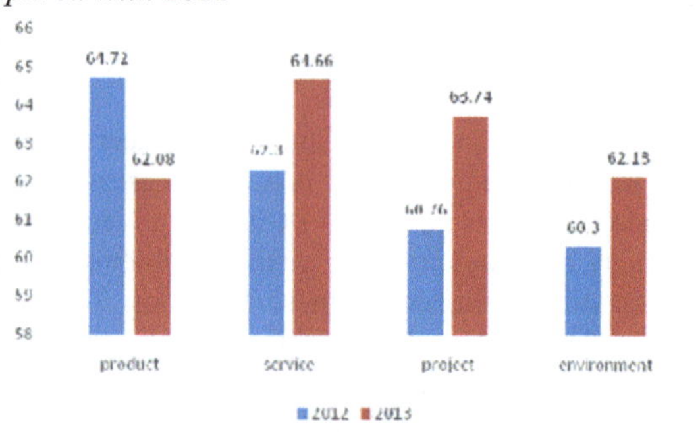

Fig.3. The quality index of four main fields

In the four quality evaluation fields, the service quality satisfaction has a biggest increase, which increase by 3.79% from 62.3 to 64.66; the project quality satisfaction increase by 4.9%; the environment quality satisfaction increase by 3.03%. But the product quality satisfaction decreases by 4.08%. This illustrates that the overall quality is on the path of developing while product quality is under fluctuation.

D. Household appliances score the highest while food score the lowest in the quality safety evaluations

Fig.4. Quality safety indexes of different products

According to the data, quality safety evaluations of goods from all the 11 surveyed categories have (we limit the range to these that are sensitive to safety problems) passed over the line of 60, and scores of home appliances and computer are above 70, ranking the top. Quality safety evaluation of food, cosmetics and children's products' scores were 61.37 points, 63.27 points and 63.85 points respectively, ranking the last three.

E. Home appliances, computer and mobile phone rank the top three in the quality satisfaction evaluation

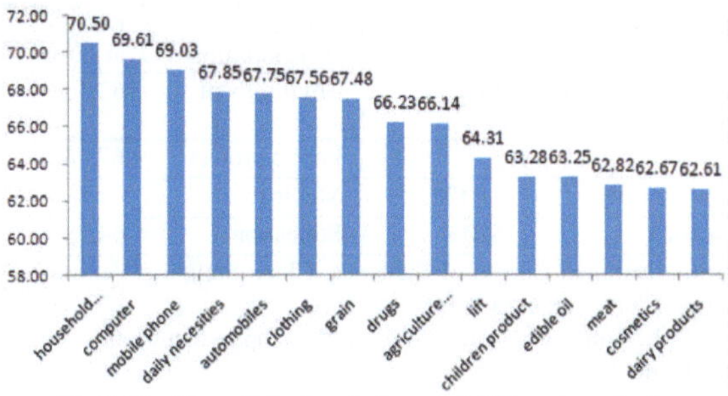

Fig.5. Quality satisfaction indexes of different products

In the product quality satisfaction evaluation, indexes of goods from all the 15 surveyed categories are over the line of 60. Home appliances, computer and mobile phone rank top three, while dairy products, cosmetics and meats are the last three. In the 15 categories only home appliances' score of quality satisfaction is above70. With respect to quality and safety, satisfaction products' quality in China is relatively weak.

F. Public services on quality is weakest in the four quality dimensions

In general, consumers' trust and satisfaction to government's public services on quality is low. All the scores of government's public services on quality does not reach the qualified level of 60. Among the quality public services, the overall quality investment, government image and consumption environment are the last three, scores being 56.83 points, 56.83 points and 57 points respectively[9][10].

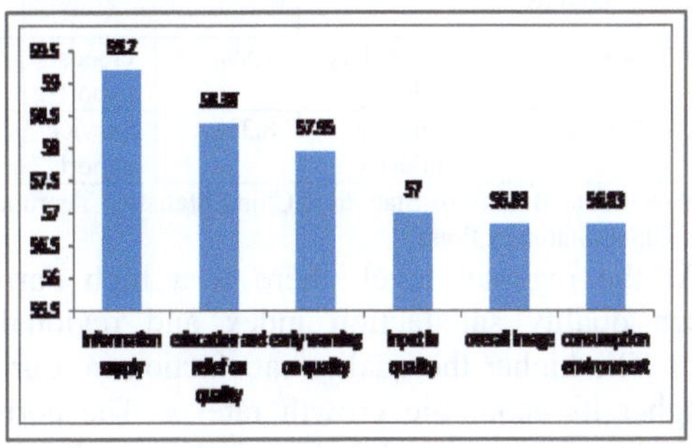

Fig.6. Indexes for government public service in quality

G. Both consumer' knowledge in quality and ability to react upon quality issues are weaker than their quality consciousness.

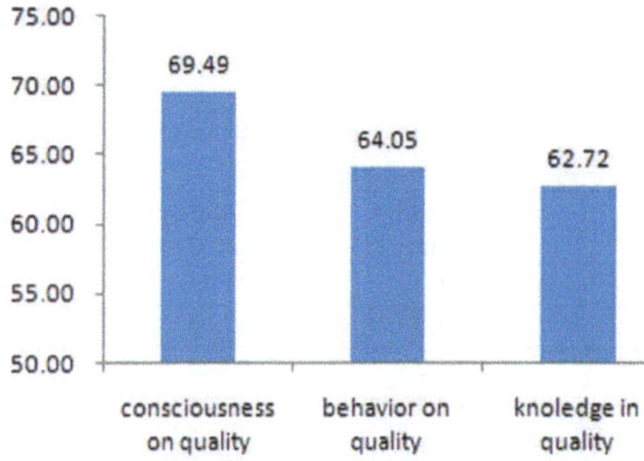

Fig.7. Indexes of Citizen Concept on Quality

In the structure of citizen concept on quality, consumers' quality consciousness score is higher than their quality actions and quality knowledge, and consumers' quality action is higher than quality knowledge. This shows that Chinese consumers' quality consciousness goes ahead of their quality knowledge and quality behaviors, but they are in lack of corresponding supports in behavior and knowledge [11].

IV. DISCUSSION

A. China's product quality safety is entering a steady upward development stage

According to data from the quality survey, the ratio of Chinese consumers who are hurt by product quality only counts 13.41% (including minor injuries) in their life since 2012. Product quality safety index is 66.58 in 2012 and 66.52 in 2013 respectively, almost being the same. Especially, quality safety index of food, household appliances and consumer products remain stable annually, while quality safety index of drug has increased greatly from 64.09 points in 2012 to 66.67 points in 2013. This shows that, the bottom of quality safety has been solidly built in China, and it has generally entered a steady upward development stage.

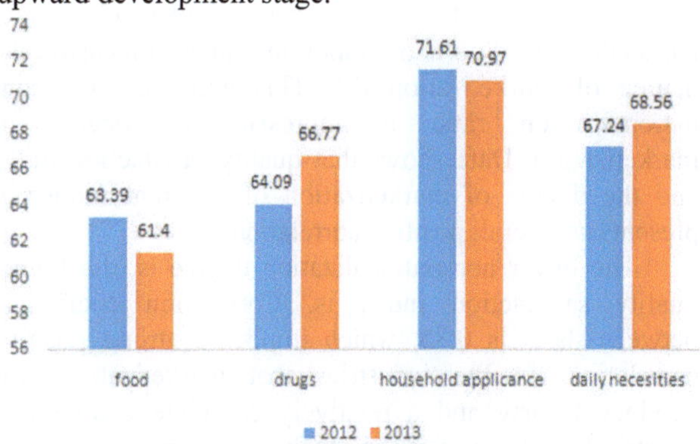

Fig.8. Yearly comparison of product quality safety indexes

B. Consumers' demands for product quality shifts to demands for satisfaction

As shown in Table III, in the 15 surveyed categories of products that are most relevant to consumers' eating, clothing and using, quality safety index is more than 60 for all products. This shows that, consumers do not feel dissatisfied with the quality safety of products, and think that daily consumption products meet their basic requirements.

TABLE III
COMPARISON OF QUALITY SAFETY AND QUALITY SATISFACTION

Items	Quality safety (1)	Quality satisfaction (2)	(1)-(2)
Food	61.37	61.44	-0.07
Grain	68.29	67.48	0.81
Edible oil	64.25	63.25	1.00
Meat	63.29	62.35	0.94
Dairy	62.90	62.45	0.45
Household appliance	70.97	70.04	0.93
Drugs	66.77	65.68	1.09
Computer	70.43	69.61	0.82
Daily necessities	68.56	67.14	1.42
Cosmetics	63.27	61.95	1.32
Children's product	63.85	63.28	0.57
Clothing	68.54	67.56	0.98
Automobiles	68.21	67.28	0.93
Lift	65.37	64.31	1.06
Agriculture supplies	66.58	65.71	0.87

Only for food, consumers' quality safety evaluation index is slightly less than quality satisfaction index. All other industries quality satisfaction indexes are significantly higher than the quality safety index. The highest gap between quality satisfaction index and quality safety index is in the category of daily goods (such as toothbrush and toothpaste etc.). Even for household appliances and computer whose quality safety index is ranked the first and the second, their quality satisfaction index is lower than their safety index by 0.93 points and 0.82 points respectively.

C. Marketization has promoted quality satisfaction of Chinese manufacturing

According to the existing research, non-nationalization ratio is an important index to measure the degree of marketization [2]. This paper choose non-nationalization ratio to represent the degree of marketization. Data shows that quality satisfaction index and the degree of marketization of different industries presents an overall positive correlation.

The lower non-nationalization degree is, the higher quality satisfaction index is. Correlation coefficient between them is 0.85, which implies a strong positive correlation. In the industries that marketization was conducted early and a relatively complete competition market has been formed such as home appliances, automotive, real estate and so on, their non-nationalization employment proportion is above 70%, and its quality satisfactory level at or near the 70. Meanwhile in the industries that mainly are operated by the government sector such as public projects, health, education and other industries, their non-nationalization employment proportions are below 20%, and their quality satisfaction indexes are significantly lower. These results show that the more fully competition at the industries, the higher quality satisfaction is [12] [13].

TABLE IV
QUALITY SATISFACTION INDEX OF DIFFERENT INDUSTRIES

Industries	Non state-owned ratio	Items in questionnaire	Quality satisfaction index
Manufacturing	91.3%	Household appliance	70.03
		Automobile	67.28
Real estate	82.9%	House	67.88
Finance	71.2%	Finance	66.26
Information	70.4%	Telecom	65.59
Transportation	37.2%	Public transportation	64.22
Environment and water supply	14.3%	Environment	62.13
		Public project	67.06
Social security	11.1%	Medical care	61.84
Education	5.2%	Education	64.73

Sources: National Bureau of Statistics: China Statistics Yearbook 2013, Beijing: China Statistics Press.
Notes: The non-state owned refers to the employment ratio of non-state owned enterprises.

D. Product quality satisfaction index variations are consistent with economy fluctuations

In 2013, the growth rate of third industry (service industry) is 8.3% in China, 0.5 percentage higher than the second industry growth rate, and is also 0.6 percentage higher than the overall growth rate. Meanwhile the growth rate of service export reached 10.6%, 3.3 percentage higher than the goods export growth rate. At the same time, the main industries in the first industry such as cereals, dairy products have a decline in 2013 to a certain degree. Consistently, agriculture industry increases by a rate of 7.7%, the lowest rate of industry growth among all industries. The above results show that the products' quality is still not stable enough to sustain Chinese economy development.

TABLE V
QUALITY SATISFACTION CHANGE AND INDUSTRIES GROWTH

Items	Quality satisfaction change	Industries	Growth rate	Export	Growth rate
Product	-4.08%	Secondary industry	7.8%	Goods export	7.3%
Service	3.79%	Tertiary industry	8.3%	Service export	10.6%

Sources: National Bureau of Statistics: China Statistics Yearbook 2013, Beijing: China Statistics Press.

At the regional level, there is a high correlation between quality satisfaction index and regional GDP growth. The higher the quality satisfaction in a province, the higher its economic growth rate is. The correlation coefficient between quality satisfaction index, quality safety index and regional GDP growth rate were 0.18 and 0.12 respectively. The correlation between the quality level and economic growth can be verified at the regional level.

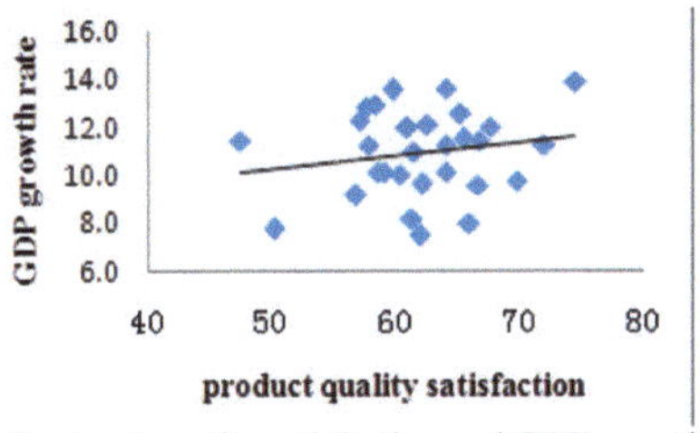

Fig.9. Regional quality satisfaction and GDP growth rate

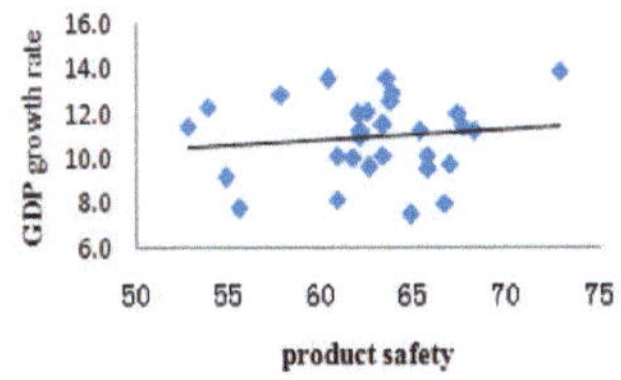

Fig.10. Regional product quality safety and GDP growth rate

E. The Increase in micro product quality promotes the economic development quality in China

The regional quality safety and satisfaction index is positively correlated with the consumption rate, with a correlation coefficient of 0.152. If we correlates the regional macro quality, represented by the product safety index, with the consumption rate, a positive correlation will be derived. The coefficient is 0.11. In addition, the public quality service also positively correlates with the consumption rate (Fig.11). The public service in quality promotes the regional consumption by bettering the consumption environment. The characteristics above show that, the quality can stimulates the consumption propensity, thus the regional consumption rate will be increased [14][15].

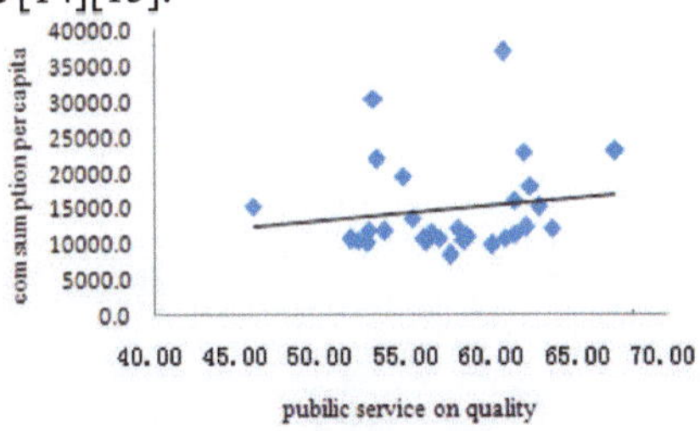

Fig.11. The government's public service on quality and the consumption per capita

V. CONCLUSION

Based on Total Quality Index evaluation model proposed by Cheng Hong etc. [6] and data from quality observations, overall quality situation in China 2013 have been empirically analyzed. The major conclusions are: total quality index in China 2013 is 63.74, above the qualified line; safety status of products is on the path of going up; consumers' demands on products quality have exceeded their demands on products safety; market competition promotes the elevation of satisfaction on products quality; variations in products quality are highly consistent with fluctuations in economy.

Above results proved that the quality observations based on consumers' evaluations have relatively strong explanatory power on total quality phenomenon in China and fluctuations in macroeconomic developments. In future research, macro quality observations based on consumers should be further enhanced, and the logic relationships between products quality from micro perspective and quality of economic developments from macro perspective should be established.

REFERENCES

[1] Cheng Hong, Chen Xinzhou, Luo Lianfa, "Research on Some Important Issues Concerning the Strategy of Prospering the Nation with Quality," Journal of Macro-Quality Research, 2013, 3: pp.1-14.

[2] Cheng Hong, "The general situation of China's quality in 2012: consumer perception and model construction." Journal of Macro-quality Research, 2013, 1:33–48

[3] Cheng Hong, Li Dandan, Fan Hanbing, Macro Quality Statistics and Analysis, BJ: Peking University Press, 2011, pp.45-52.

[4] Deming WE, Deming on quality management, SY: Hainan Press, 2003, pp. 89-93.

[5] Juran JM, Godfrey BA, Juran Quality Manual, BJ: China Renmin University Press, 2003, pp.110-118.

[6] Cheng Hong, Li Dandan, Luo Lianfa "The Chinese perception of quality: model building and analysis based on consumers' perception," Journal of Chinese Management 2014, 1:pp.1-22

[7] Chinese quality observation research group of Institute of quality development strategy of Wuhan University, Report on Chinese quality observation research 2012, BJ: China Quality Press, 2013, pp.1-32.

[8] Cheng Hong, Li Qingquan, "A Study on the System and the Test and Appraisement of the Index Model of the Quality of China's Total Regions," Management World, 2009, pp.2-9.

[9] Li Han, "Customer assessment and influence factors on Chinese government responsibility of quality and security-Based on the empirical research of national survey of 2012," Journal of Macro Quality Research. 2013, 1: pp.118-128.

[10] Luo Lianfa, "Does urban–rural duality of quality exist in China," Journal of Macro-quality Research, 2013, 1, pp.107–117.

[11] Zhang Jihong, "Analysis and comparison on factors that influence regional quality satisfaction," Journal of Macro Quality Research. 2013, 1: pp.68-75.

[12] Cheng Hong, Fan Hanbing, Luo Ying, "Research on quality management institution of the United States government and its implications." China Soft Science, 2012,12:pp.1–15

[13] Frank B, Enkawa T., "How economic growth affects customer satisfaction," Asia Pacific Manage Review 2008, 13(2), pp.531–544.

[14] Kotler P., Marketing management, NJ: Prentice Hall, 2011, pp.8-12.

[15] Li Dandan, "Study on the influence factors of government quality regulatory customer satisfaction index," Journal of Macro-quality Research, 2013, 1, pp.76–88

Application Research on the Cooperate Quality Monitoring Model in the Internet Environment
- An Empirical Study on the Case of Mengniu Product

Hong Cheng*, Wei Xu
Institute of Quality Development Strategy, Wuhan University, Wuhan, China
(16661021@qq.com)

Abstract - **In Internet environment, consumers can deliver various quality information of corporate quality by themselves, and such information is completely open, accessible and identifiable. This article collect, statistically process and analyze corporate quality information that consumers exposed in Internet, taking Mengniu, a company listed in HK Stock Exchange for instance, to perform empirical study on the Internet information related to quality with the basic classification theoretical methods of quality evaluation "inherent performance "and "requirement satisfaction" .**

Keywords - **Internet information, monitoring model, quality analysis**

I. INTRODUCTION

There is a great quantity of product quality data in Internet; more and more consumers posted their comments regarding product quality or service quality on web. The discussions of product quality over Internet expose the positions and views of netizens on current product quality phenomenons and hot issues. They have become the barometer of product quality in the market, exerting in-depth impact on the shopping decision of consumers, public image and profit of enterprises. Tracing back the events of product quality taking place in recent years, we find quite a lot of ones were first exposed in Internet, then became hot topics attracting leading medias to follow up, and further give rise to the burst of an quality safety network event[1]. It has become an imperative issue, how to monitor the information of product quality risk in Internet, and perform analysis on the tendency of quality risks based on consumers' evaluation information, to monitor product quality risk information through Internet.

What this paper studies is the quality information that consumers post in Internet related to enterprises, especially related to quality safety of enterprises. Quite a number of consumer quality safety information appearing in Internet are posted by micro-blog, a self-media, so it is impossible for us to get these consumer quality information which accounts for a high percentage just with common search engine, we need more professional quality information collection platform to perform analysis over consumer quality information[2]. The article backs on Quality Safety Online Information Monitoring and Pre-alert Platform (hereinafter referred to as "Quality Online Monitoring Platform") of Quality Development Strategic Institute of Wuhan University as the tools to acquire data. The Quality Online Monitoring Platform applies quality risk theory and value engineering theory in combination, utilizes the integrated innovations of a new generation of network information technologies including search technologies, cloud calculation, data mining, semantic analysis, quality evaluation system, to collect at real-time a great quantity of quality information that consumers are aware from public Internet, upon in-depth mining, refining, analyzing, processing these information, the platform will then be able to provide enterprises, consumers, governmental authorities with valuable quality information services. The online monitoring of product quality based a great quantity of consumer Internet information provides an effective method and technology for network-based monitor of product quality.

II. MODEL CONSTUCTION AND HYPOTHESIS

The radical reasons for the issue of quality safety lie in asymmetry of quality information. The development of Internet technology makes it possible to obtain the information of similar products which is previously unavailable due to the constraints of information exchange cost, consumers more and more often disseminate and share the evaluations on product quality of a company. The classification model of quality information built on Internet consumer evaluation information must obey the general theories of quality evaluation; moreover, it must reflect the features of consumption quality evaluation information in Internet environment.

Traditionally, the evaluation and analysis on quality is based on the observation over two dimensions, i.e.: object (products) and subject (user). First, quality evaluation should be the observation on the inherent characteristics of products things, and the analysis according to the conformity to standards identified through tests [3]. On the other hand, quality evaluation should be the observation on the ability of things to satisfy customer expectations, that is to say, it is an analysis over quality status based on the event to which customer requirements are satisfied. Therefore, in International Standardization Organization, ISO, 2000, quality is defined as "the degree of a group of inherent features and abilities to satisfy requirement" [4]. There are two methods mostly widely used in the classification evaluation of corporate quality risk from the dimension of inherent characteristics. One is ISO 9000 quality management system, which sets forth the methods of evaluating corporate quality using 4 different classification indicators just like management responsibilities, resource management, product realization

E. Qi et al. (eds.), *Proceedings of the 5th International Asia Conference on Industrial Engineering and Management Innovation (IEMI2014)*, Proceedings of the International Asia Conference on Industrial Engineering and Management Innovation 1, DOI 10.2991/978-94-6239-100-0_36, © Atlantis Press and the authors 2015

as well as measurement, analysis and improvement. The other is Criteria for Performance Excellence, which is also used to analyze and observe quality status of enterprises with 7 different classification indicators, i.e.: leading, strategy, customer and market, resources, process management, measurement analysis and improvement, results. The classification evaluations of corporate quality risk from the perspective of requirement satisfaction have been studied by some scholars and some countries in both theoretical and application fields. According to the definition that Kotler (1991) made, customer satisfaction means the senses like joy or disappointment compared to his/her expectations. American Customer Satisfaction Index (ACSI, 1994) developed SCSB from Sweden, broadened customer satisfaction indexes to six indicators, e.g.: customer expectation, quality of perception, value of perception, customer satisfaction, consumer's complaint and customer loyalty [5]. European Customer Satisfaction Index (ECSI, 1999) nearly borrowed ASCI in entirety, except that the indicator "corporate image" is introduced in lieu of the indicator "customer complaint".

TABLE I
CLASSIFICATION MODEL OF CORPORATE QUALITY SAFETY BASED ON INTERNET INFORMATION

Dimensions for Observation	Classification Indicators	Contents of Information
Inherent characteristics	Product performance	Reliability, easiness of use, sensory evaluation
	Safety	Physical injuries, change of properties, sense of unsafely
	Service quality	Convenience, service environment, employees, customer service system
Requirement satisfaction	Economical efficiency	Cost-effective, product diversity, using cost
	Operation quality	Management proficiency, human resources, humanistic environment
	Public images	Social responsibility and environment protection, image of managers, corporate reputation, external communication

By integrating various literatures, Cheng Hong summarized 6 classification indicators according to some new features of consumer quality information emerging in Internet age, and identified the contents of Internet information under each of the six indicators [6]. As a result, a classification model of corporate quality safety information under Internet environment comes into being. Backs on existing quality evaluation theories, the classification model attaches more importance to those quality information classification indicators reflecting the characteristics of Internet age. It not only acquires quality evaluation information of consumers to the most extent, but also embodies the features of consumer quality safety information dissemination over Internet. Based on the theoretical fundaments of quality classification "inherent characteristics" and "requirement satisfaction", the classification model not only reflect more regarding the indicators of safety, operation quality and product performance for "inherent characteristics", but also not only reflect more regarding the indicators of cost-effectiveness, service quality and public image for "requirement satisfaction". See Table I for its framework.

In addition, for those Internet quality information of products of a company neither suitable for performance dimensions nor suitable requirement satisfaction, they will be put in the comprehensive category.

III. DATA COLLECTION AND PROCESSING

Quality safety of dairy has become a hot topic drawing strongest concern of consumers [7]. This article chose the Internet information on product quality of Mengniu as study object [8]. As one of leading manufacturers of milk, yoghourt and dairy in China mainland, Mengniu has become the milk product firm with the second sales volume in China, its output of liquid milk and ice cream even rank No. 1 in home. It's holding company, Chinese Mengniu Dairy Company Limited. (Stock No. of HK Stock Exchange: 2319) is an industrial firm listed in HK Stock Exchange. During Jan. 1, 2012 – Dec. 31, 2012, the product quality Internet monitoring platform mined, processed, screened and classified the Internet information on Mengniu's dairy quality, thus obtained 11,778 pcs of valid data, and built up a Internet information database for product quality, by sequence of time, the paper performed empirical analysis on the data.

A. Design of realizing method

In recent years, Internet text classification technologies based on natural language processing technologies have achieved great progress; quite a lot of technical methods have become available for reference. Classification of Internet texts means the process of judging a text to be one or more established types based on the information contents of given Internet text [9]. As the basic way of technical realization in this paper, this method can be used to identify and classify oceans of wide-spectrum complex safety information quickly and accurately [10]. This article proposed an algorithm of quality text classification oriented to online consumers, namely, it classifies text information based on support vector machine (SVM) algorithm, and identifies the classification of the contents of the text information. SVM was first invented by Vapnik (1982), it is a model identification method based on statistics learning theory [11]. On basis of the principle of structural risk minimization, the method pool together and compress the original data set in SVM set, learns to get classification decision function. Its basic idea is: for a given learning task with a limited number of training samples, how to find out an optimum ultra-plane in a high-dimension space as the partition between two categories, so as to minimize the error [12]. SVM has obvious advantages in the identification of small-sample, nonlinear and high-dimension model texts of consumer quality information, besides, it features higher classification accuracy and recall rate.

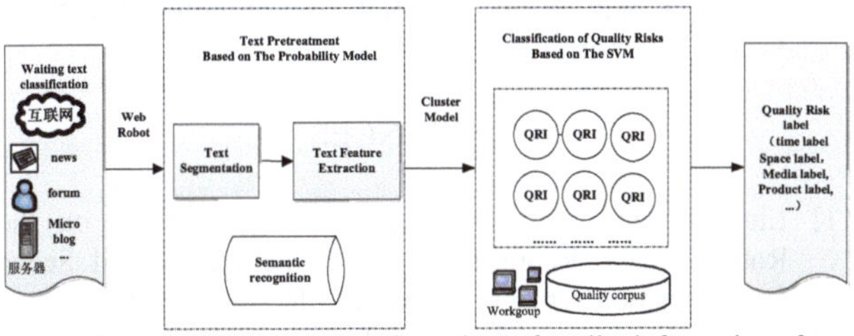

Fig.1. Classification process flow chart of quality information of network text

B. Data collection

These information primarily come from Micro-blogs, forums, complaint platform, blogs as well as leading network news media's, cover more than 90% of top 2000 Chinese Internet medias elected by Alex[13]. The data are acquired with web crawler technology, to continuously acquire data from designated Internet media source. The collected data is to be filtered by company name and product name, thus pick out all the data related to the company [14]. Then, all the data regarding corporate quality risks is marked means of AI automatic judgment by nature of industry, to create classified information source accordingly.

The platform is a professional quality information monitoring platform. The acquisition of consumer text data from Internet is based on the only content "Quality" without incorporating the text data of any other content [15]. In order to improve the accuracy of acquiring Internet information on quality, the quality monitoring platform developed corpus especially for dairy industry, which is able to identify the quality meanings of the texts, and accurately distinguish the texts on the quality of dairy quality from various information released by net citizens [16]. In order to ensure the accuracy of quality information classification models and its corpus, the platform also employs precision (P), recall rate (R and harmonic mean (F1 as the evaluation method for making tests, so as to improve the accuracy of system platform.

IV. DISCUSSION

A. 6-dimension analysis on internet quality information

According to statistics of Internet information of Mengniu products' quality, There are 11778 such messages per annum, Speaking from the dimension of quality risk, there are 7871 messages of safety, which account for the highest percentage of 66.83%, next, there are 2930 message of operation quality, which account for 24.88%; next, there are 367 messages of product performance, which account for 3.12%; next, there are 350 messages of public image, which account for 2.97%; next, there are 214 messages of service quality, which account for 1.82%; next, there are 46 messages of product cost-effectiveness, which account for 0.39%; the rest are comprehensive messages, totally 8202.

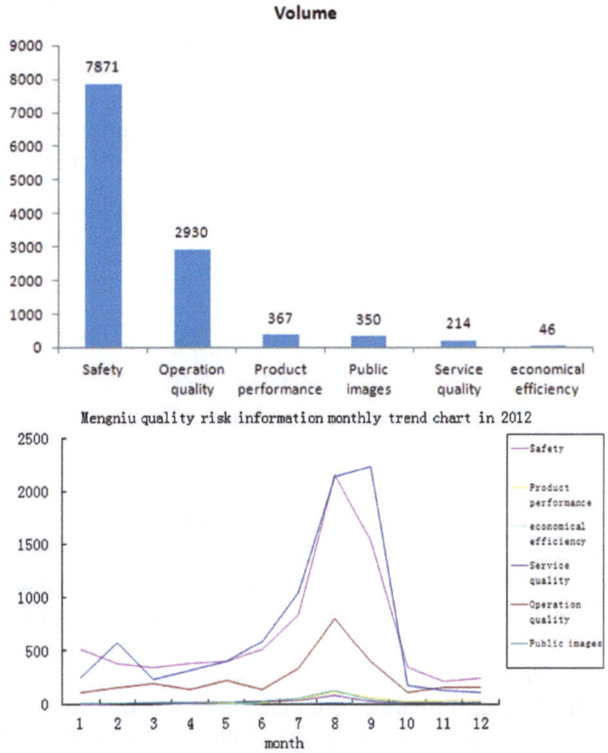

Fig.2. 6-dimension Statistics on Classified Internet Information of Mengniu Products' Quality

In a year, the third quarter occupies peak in the annual curve of dairy quality problem. There are 2807 pcs of Internet Information about Mengniu dairy quality recorded in Spring 2012, accounting for 14.36% of total, this figure in Summer, Autumn, Winter is 3238, 12315 and 1800 respectively. The value in the third quarter is the highest, this is partly due to the seasonal features of dairy quality, and however, so many quality safety events taking place in the third quarter have made Mengniu the focus of concern in China.

B. Safety of dairy is still the strong concern of consumers

The safety of dairy quality exerts significant impact on the perception of consumers toward total quality of a company; especially it is closely related to the evaluation of consumers on the operation quality of the company. After a year's statistical analysis, we found there exist certain relevance between corporate safety and operation quality. This article divides all Internet information on Mengniu quality into 36 segment, then test the relevance between safety and operation quality with SPSS. Results showed that Pearson correlation coefficient is 0.723, indicating there is relevance between safety and operation quality. Safety is a big concern of consumers on dairy quality; the issue of safety usually affects the evaluation of consumers on the operation quality of a company.

To a wider extent, the perception of consumers toward the safety of product quality comes from a sense of unsafety due to absence of trust; it is not an actual degree of quality safety. In the first-level safety dimension of Mengniu quality information, there are 3 underlying second-level dimensions, i.e.: sense of unsafety, change of properties, physical injuries. As shown in the pie graph of safety dimension: there are 5783 pcs sense of unsafety, accounting for 57% in first-level safety dimension; 1791 pcs property change index, accounting for 14% of safety performance dimension; 291 pcs physical injuries index, accounting for no more than 4% of safety dimension.

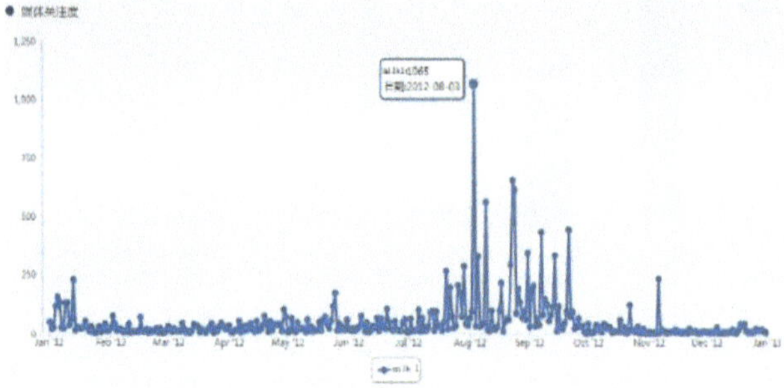

Fig.3. Yearly Information Distribution Curve of Mengniu

The year-around statistics show that, there are 1065 messages released on Aug. 3, 2012, which is the highest of the year in terms of daily release number? Tracing back these messages, we found that the surprising product quality safety event "Date Alteration" of Mengniu was also exposed on Aug. 3, 2012. According to the circular that Zhejiang Provincial Administration for Industry and Commerce posted on Aug. 20, without consulting Head Company, the manager of Mengniu in Yiwu, Zhejiang extended the production date of 3000 boxes of Mengniu Milk whose shelf life has expired to 6 months later, and sold them at the price for normal products. Zhejiang Provincial Administration for Industry and Commerce has handed over the case to the police. The alteration of production date happening to Mengniu not only hurt consumers' physical and mental health, but also damaged its own image; as a result, its stock price dropped 0.4% on the same day.

To sum up, this article studies the quality safety information coming from Internet, applies existing Internet quality information classification model, makes data collection and sorting, and then performs empirical study on Internet quality information of Mengniu. Certainly, the article need further study the following two aspects: First, build corpus and classification indicator of Internet quality safety information for wider range of industries and products. Second, perform study on change tendency of stock market based on the accumulation of quality text information.

ACKNOWLEDGMENT

This article is the key project "Study on Quality Safety Evaluation and Internet Pre-alert Method of Our Country" (11&ZD158) under National Social Science Fund, a project under National S&T Support Program for 12th Five-year Plan (No. 2011BAK06B06), also the achievement of Supporting Study on Quality Regulatory Technology and Safety Risk Information System (No. 201210117), a non-profit industrial scientific research project supported by MOST (Ministry of Science & Technology), supported by "the Fundamental Research Funds for the Central Universities".

REFERENCES

[1] Liu Jie, Du junping，Wang Xiaoru. Research on the Robust Image Representation Scheme for Natural Scene Categorization. Chinese journal of Electronics, 2013(2).

[2] Young, K., Mincheol, K., Goh, B., et al. A comparison between first-timers and repeaters at a food event [J]. Journal of Culinary Science & Technology, 2010, 7 (4): 239-249.

[3] Anderson, E. W., Fornell, C., Lehmann, D. R. Customer satisfaction, market share, and profitability: findings from Sweden [J]. The Journal of Marketing, 1994: 53-66.

[4] Fornell, C. A national customer satisfaction barometer: The Swedish experience [J]. Journal of marketing, 1992, 56 (1).

[5] Mitchell, V.-W. Consumer perceived risk: conceptualisations and models [J]. European Journal of marketing, 1999, 33 (1/2): 163-195.

[6] Cheng hong , Fan hanbing. The theoretical framework of enterprise quality safety risk effective governance [J]. Management World, 2012, 12 :29-36.

[7] Ting, S., Tse, Y., Ho, G., et al. Mining logistics data to assure the quality in a sustainable food supply chain: A case in the red wine industry [J]. International Journal of Production Economics, 2013.

[8] Van Boxstael, S., Habib, I., Jacxsens, L., et al. Food safety issues in fresh produce: Bacterial pathogens, viruses and pesticide residues indicated as major concerns by stakeholders in the fresh produce chain [J]. Food Control, 2013, 32 (1): 190-197.

[9] Zou, W., Tang, H., Zhao, W., et al. Data mining tools for Salmonella characterization: application to gel-based fingerprinting analysis [J]. BMC bioinformatics, 2013, 14 (Suppl 14): S15.

[10] Baeta, A., Niquil, N., Marques, J. C., et al. Modelling the effects of eutrophication, mitigation measures and an extreme flood event on estuarine benthic food webs [J]. Ecological Modelling, 2011, 222 (6): 1209-1221.

[11] Åberg, M. Food safety and contamination assessment using metabolomics [J]. Hot topics in Metabolomics: Food and Nutrition, 2013: 41.

[12] Garcia, A. B. The use of data mining techniques to discover knowledge from animal and food data: Examples related to the cattle industry [J]. Trends in Food Science & Technology, 2013, 29 (2): 151-157.

[13] Hadjigeorgiou, A., Soteriades, E. S., Philalithis, A., et al. National Food Safety Systems in the European Union: A Comparative Survey [J]. IJFS, 2013, 2 (2013): 105-117.

[14] Hurst, W., Harris, L. Integrating Hazard Analysis Critical Control Point (HACCP) and Statistical Process Control (SPC) for safer nut processing [J]. Improving the safety and quality of nuts, 2013: 119-147.

[15] McIntyre, L., Vallaster, L., Wilcott, L., et al. Evaluation of food safety knowledge, attitudes and self-reported hand washing practices in FOODSAFE trained and untrained food handlers in British Columbia, Canada [J]. Food Control, 2013, 30 (1): 150-156.

[16] Van Driel, M. A., Bruggeman, J., Vriend, G., et al. A text-mining analysis of the human phenome [J]. European journal of human genetics, 2006, 14 (5): 535-542.

An Empirical Study on Food Safety Early-warning Based on Internet Information

Hong-wei Yu
Institute of Quality Development Strategy, Wuhan University, Wuhan China
(yuhongwei928@163.com)

Abstract - **It is a new perspective to use consumer information on the Internet for food safety warning. The IQDS (Institute of Quality Development Strategy) builds a platform named Shendu Company based on Internet information for quality security warning,. This paper describes the basic framework of early warning analysis platform, including early warning information classified into three categories, namely, information on body injury, feeling of insecurity and characters changes, as well as the risk of all types of information classification. Then this paper selects the 2008 "Sanlu milk powder incident" as a case, to achieve early warning information gathering and through the platform. The result shows that the early warning platform can issue "yellow warning" signal three months before the outbreak of the event, which can play an effective role for food safety early warning.**

Keywords - **Early-warning model, food safety, internet information**

I. INTRODUCTION

It is a new perspective of food safety governance in network era to use the quality information that consumers released on the Internet for food safety pre-warning. According to statistics, by the end of June 2013, the internet users in China have amounted to 591 million, with 464 million of mobile net citizen, including 401 million of blog and personal space users, and micro blog users are also 331 million (CNNIC, 2013) [1]. The customers as net citizen, who can spread food safety risk assessment information through the internet directly to the society, have become the potential providers and communicators for food safety risk information. By means of information technology, through the real-[1]time collection for huge amounts of food risk information released by network consumers, we can know the food security situation more comprehensively and deeply that the consumers face. Therefore, pre-warning based on food risk information of network consumers is a new idea for food safety management in the big data era [2-4].

At present, there are rare studies directly focused on food safety network precaution, while more studies focus on discussing the network validity for quality safety precaution by using existed information systems (Rotais et al, 2010 [5]; Tang Xiaocun, 2008 [6]; Zhang Dongling, 2010) [7] or discussing the network system structure of quality safety precaution system. IQDS of Wuhan University has devoted to studies on food safety network precaution since 2008. The Institute believes that to realize quality safety precaution based on Internet information, three problems need solving in particular: firstly, classify the network quality risk information [8, 9]. According to the general precaution principle to define the scope of precaution information and choose the comments related to quality safety and freely released by consumers, and to make reasonable classification of quality information in different forms and contents, so as to enable the precaution information to realize its value in unified analysis paradigm; Secondly, grade the risk of network quality information that under classification, i.e. establish risk precaution grade of information in different classification according to the quality information risk degree showed by information presentation forms and specific contents; Thirdly, integrate the product quality information in different classifications and risk grades. Information uncertainty and imperfection should be fully considered in integration; therefore, it can consider to use relevant methods of artificial intelligence [10-12].

This thesis will take the 2008 Sanlu milk powder incident as the analysis object, and to propose the general analysis framework of food safety network precaution in aspects of network information classification, risk grades, alert classification, etc., realize data collection through existed quality safety network monitoring and precaution platform in the Institute, and make empirical precaution analysis for the cases by using intelligent information integration methods, and verify the applicability and feasibility of the network information precaution.

II. PRE-WARNING FRAMEWORK

Cheng Hong et al (2012) [13], researchers of the Institute, constructed risk classification indicators in 6 dimensions, covering public image, economical efficiency, service quality, operation quality and other information categories relevant to enterprise operation management and social responsibility. However, food safety precaution discussed in this thesis particularly focus on the feature of safety, so according to the research conclusions of Cheng Hong et al (2012), this thesis divide the Internet food safety information into three categories: physical injury, insecurity, character change.

1) Physical Injury: Physical injury refers to bodily injury resulted by unbearable power in a certain time imposed on human body when consumers contact or use the product. This is the most direct injury caused by potential quality and safety hazard. Generally, forms that caused injuries are physical injuring chemical injuring and biological injuring and so on (GB/T22760-2008), such as

This article is supported by the National Social Science Fund (11&ZD158), National S&T Support Program for 12th Five-year Plan (No. 2011BAK06B06), and non-profit industrial scientific research project supported by MOST (Ministry of Science & Technology) (No. 201210117), also supported by "the Fundamental Research Funds for the Central Universities"(No.639274134) and "the China Postdoctoral Science Foundation".

E. Qi et al. (eds.), *Proceedings of the 5th International Asia Conference on Industrial Engineering and Management Innovation (IEMI2014)*, Proceedings of the International Asia Conference on Industrial Engineering and Management Innovation 1, DOI 10.2991/978-94-6239-100-0_37, © Atlantis Press and the authors 2015

death, poisoning, skin allergy and diarrhea, etc. National Electronic Injury Supervisory System (NEISS) of America, the Rapid Alert System for Non-Food Products (RAPEX) of EU, and National Consumption Information Network System (PIO-NET) of Japan all use the collected information of quality physical injury on consumers as the important basis to monitor the quality safety risk and make corresponding precaution [14, 15]. According to practical experiences of each country, this thesis classifies the risks of physical injury in Internet quality information into 5 grades: (1) fatal, which can lead to catastrophic injury, like death, vegetative being, and high paraplegia, etc.; (2) very severe, which can lead to irreversible injury, and cause serious negative effects on human body, like physical disability, large area facial scaring, etc.; (3) severe, which are injuries that will recover through emergency ward treatment or regular hospitalization; (4) moderate, which just need to visit the doctor or treat in the outpatient clinics, and the impact on human body is slight; (5) minor, which need no doctor treatment, and can handle at home by oneself, and the impact on human body is small with uncomfortable feeling to a certain extent.

2) Insecurity: Insecurity refers to the psychological feeling of uneasy and fear of the consumers brought by some characters of the product quality. Although no physical injuries in the users, it will produce a sense of worry and fear in application because of the psychological changes in the consumers. Such kind of psychological feeling description information is ubiquitous in the monitored network text data of researches of Cheng Hong et al (2013) [16]. Sources of this security may come from the quality injury that once borne or seen, or quality injury that predicted through quality character changes. It reflects the possibility of suffering the similar quality injuries felt by the consumers subjectively which is expressed by the consumers' feeling indirectly. Referring to Likert Five-Point Scale, this thesis classifies the insecurity grades of the consumers into 5 grades in sequence of strong to weak, and also constructs 5 risk grades of insecurity information: inevitable, very likely, likely, negligible and unlikely, by combining the risk occurrence possibility described in *General Principle for Consumption Goods Safety Risk Evaluation (GB/T22760-2008)*.

3) Character Change: Character change refers to the changes of physical and chemical character of the product quality in comparison with that in factory delivery state. This change can be felt through eyes observation or simple test and inference methods made by the consumers. Generally, consumers' description for character change does not directly reflect the quality injury suffered, but it can really reflect the fact of existence of potential quality safety. These potential quality safety hazards will become the potential risks injurious to consumers' health and consumption rights. Moreover, the larger the information number is, the bigger the hidden risk becomes. Consumers' description information about quality character changes on Internet is very common, and quality risk information including words like foreign materials, out of date, mildewed, out of order and so on are very ubiquitous in BBS, blogs and microblogs. Similar to the quantity changes of public opinion information on the Internet, this thesis considers to use information total quantity to represent the risk grades of information in character change category, and constructs 5 grades of frequency characteristic: very frequent, frequent, often, sometimes and rarely, to classify the risk grades of character changes.

4) Alert Classification: This paper has classified the consumers' network quality information into physical injury, insecurity and character change, which also are respectively classifies into 5 risk grades. As per this method, this thesis will use the alert situation in 5 grades: no alert, slight alert, medium alert, serious alert and severe alert, to display the quality safety network precaution results, and display the grade of alert situation in sequence of high to low by colors of red, orange, yellow, blue and green respectively. To be particularly emphasize, green alert is the lowest risk grade which just needs no precaution measure but requires continuous monitoring in a special manner; while the state higher than that displayed by green indicate that the quality safety problem is relatively prominent which need to take precaution and response measures corresponding to the alert situation.

5) Precaution Framework: Through research on network quality information classification, information risk grading and alert grading, this thesis constructs the general analysis framework for quality safety network precaution as shown in Fig.1, which can realize the input of a large deal of quality information and output of final precaution results.

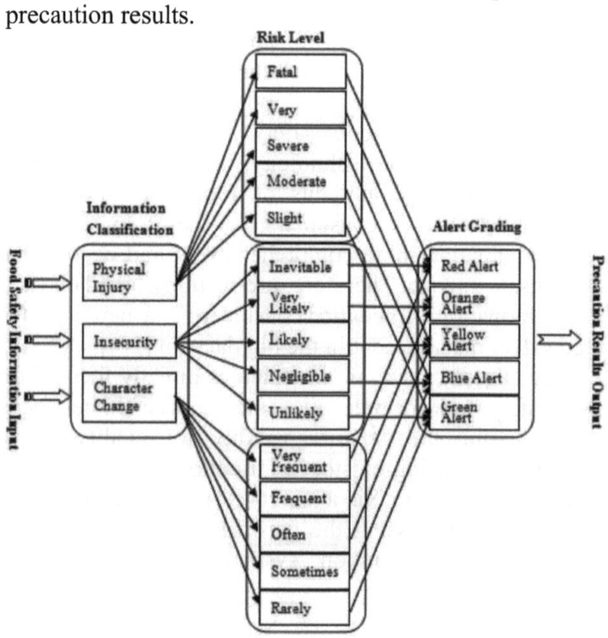

Fig.1. Analysis Framework for Food Safety Network Precaution

III. DATA

2008 Sanlu milk powder incident is the most severe food safety incident in China, resulting that over 40,000 infants and young children were hospitalized for drinking milk powder containing melamine, among which there are more than 1,600 kids suffered unrecoverable injury and 4 kids died, which made a huge impact at home and abroad. This thesis will use deep web to collect Internet information about Sanlu milk powder incident happened in 2008. The deep web is the scientific research transformation platform of Wuhan University Institute of Quality Development Strategy. Through web crawlers, semantic recognition, linguistic corpus and other information technologies, it can realize the real-time acquisition of food safety information on the Internet and numerous information oriented capture according to time, areas, product brands and incidents [17].

Firstly, use the key words of "milk powder" "Sanlu", "melamine" to search relevant network information happened during January to December of 2008. The search scope includes various blogs, BBS, communities and news website reports, etc., and the results are shown as Fig.2.

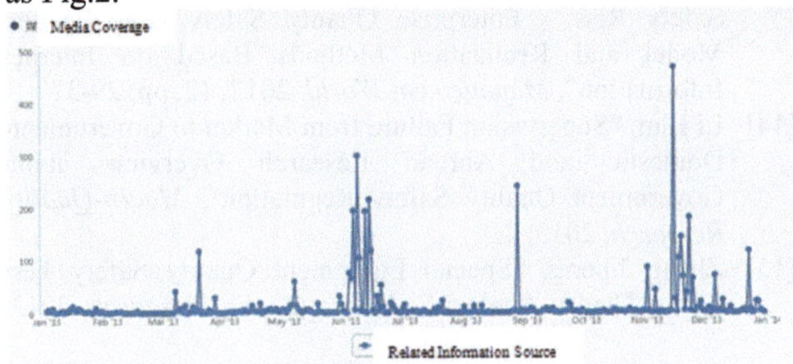

Fig.2. Information Search Results of Sanlu Milk Powder Incident during January to December 2008

As shown in Fig.2, from the preliminary search, it can be found that Internet attention to Sanlu incident obviously took place during March to April, May to June and September to November 2008. Make further particular analysis for the network information in these three time periods and acquire the hierarchical classification situation of these three time periods respectively through using the depth web, finally it can discover the result as shown in Table I.

TABLE I
FOOD SAFETY RISK INFORMATION HIERARCHICAL CLASSIFICATION RESULT

Information Category	Risk Level	Search Result		
		Mar.-Apr.	May-Jun.	Sep.-Oct.
Physical Injury	Fatal	0	0	117
	Very Severe	0	12	376
	Severe	3	35	510
	Moderate	8	257	768
	Slight	18	87	34
Insecurity	Inevitable	0	23	983
	Very Likely	5	226	1349
	Likely	13	167	587
	Negligible	30	103	35
	Unlikely	21	43	37
	Character Change	173	957	2364

IV. RESULT

To realize the output of the final precaution results, it needs to make effective integration for the above-mentioned classified food safety risk information. Considering the magnanimity, uncertainty and imperfection of the network quality information, this thesis will use the evidence theory method to proceed the integration of network food risk information.

Firstly, confirm the relative weight of each kind of information respectively is: physical injury information W_1 =0.4, insecurity information W_2 =0.4, and character change information W_3=0.2.

Then, confirm that the risk grade of each food safety information is evaluated as the probability value β^i_{jk} of the alert grade V_k ($k=1,\cdots,5$), and construct the evaluation matrix of food safety risk information as shown in Table II.

TABLE II
FOOD SAFETY RISK INFORMATION EVALUATION MATRIX

Risk Information	Weight	Alert Grade			
		V_1	V_2	......	V_5
Physical Injury	W_1	β^1_{11}	β^1_{12}	......	β^1_{15}
		......	......	......	......
		$\beta^1_{J_1 1}$	$\beta^1_{J_1 2}$	......	$\beta^1_{J_1 5}$
Insecurity	W_2	β^2_{11}	β^2_{12}	......	β^2_{15}
		......	......	......	......
		$\beta^2_{J_2 1}$	$\beta^2_{J_2 2}$	......	$\beta^2_{J_2 5}$
Character Change	W_3	β^3_{11}	β^3_{12}	......	β^3_{15}
		......	......	......	......
		$\beta^3_{J_3 1}$	$\beta^3_{J_3 2}$	......	$\beta^3_{J_3 5}$

Now, the reliability of the food safety information can be represented:

$$m^i_{jk} = W_i \beta^i_{jk},\ m^i_{j\phi} = 1 - \sum_{k=1}^{5} m^i_{jk}$$

Finally, to use the recursive computation method of Dempster compound principle, it can work out the final precaution output value.

As per this process, respectively calculate the precaution result in the time period of March to April, May to June and September to November, as shown in Table III.

TABLE III
PRECAUTION RESULT

Precaution Stage	Precaution Result				
	V_1	V_2	V_3	V_4	V_5
Mar.-Apr.	0.431	0.395	0.157	0.017	0.000
May-Jun.	0.110	0.216	0.573	0.093	0.008
Sep.-Oct.	0.034	0.045	0.074	0.339	0.508

From the depth web data collection and precaution result, it can be found that: during March to April, probability of Sanlu milk powder safety risk in state of "green alert" and "blue alert" is respectively 0.431 and 0.395, which should be given emphasis by relevant departments and be advised to take corresponding measures; during May to June, the probability of the safety risk in state of "blue alert" and "yellow alert" is respectively 0.216 and 0.573; at this moment, it should send precaution signal in a definite way, and notify relevant parties to take countermeasures; during September and November, the probability of the safety risk in state of "orange alert" and "red alert" is respectively 0.339 and 0.508; at this moment, the food safety crisis has already broken out, and it should take remedial measures positively.

V. CONCLUSION

IQDS has built a quality safety monitoring platform based on internet information. This platform is a real-time access to food safety information, and we can directionally obtain information by time, region, brands and events through this platform. In this paper we introduce the general analytical framework of the food security early warning platform from information classification, risk stratification and alert grades. Using this platform, we collect consumer information about "Sanlu milk powder incident" between January 2008 to December 2008.The data shows that during three sections of time information online is most concentrated, i.e. March to April, May to June and September to November. This paper selects data during three sections for early-warning computational analysis, and the results shows that the platform can issue a "yellow alert" through the platform between 5-6 months, which will be 3 months earlier than the occurrence of the crisis. This platform can play an effective role for food safety early warning.

ACKNOWLEDGMENT

I would like to thank Professor Hong Cheng for his Guidance and inspiration. I also thank participants at the Biweekly Academic Seminar in IQDS, especially for Doctor Fan Yu and Wei Xu's supports. All the data are form http://www.shendu.info.

REFERENCES

[1] China Quality Observation Research Group of Wuhan University Institute of Quality Development Strategy, "2012 China Quality Development Report", *Macro-Quality Research,* 2013,1,pp:1-10.

[2] Adrie J.M. Beulens, Yuan Li, Mark R. Kramer, Jack G.A.J. Vander Vorst, 2006: "Possibilities for Applying Data Mining for Early Warning in Food Supply Network" [EB/OL].

[3] Peter Karim Ben Embarek, "Information sharing between national and international authorities", *Senior Regional Food Safety Authority Response Coordination Roundtable*, 2007, 11(7).

[4] Tom Sidebottom, "Food Emergency Response Network (FERN)",*http://www.fda.gov/ohrms/dockets/ac/03/slides/4001s1_07_Sciacchitano. ppt.*, (5), 2009: pp:01-04.

[5] Rortais A, Belyaeva J, Gemo M, et al. "An early-warning system for the detection of (re-) emerging food-and feed-borne hazards". *Food Research International*, 2010, 43(5), pp: 1553-1556.

[6] Tang Xiaochun, "Food Safety Precaution System in Multi-Perspectives", *China Soft Science,* 2008,6.

[7] Zhang Dongling, Gao Qisheng, Yang Zehui, "Agricultural Products Quality Safety Risk Evaluation and Precaution Model: Example-Vegetable Export Demonstration Base in Shandong Province", *Systems Engineering-Theory & Practice*, 2010, 6.

[8] Kong Fantao, *Animal Products Quality Safety Precaution Theory and Methods, Economic Press·China.* 2009.

[9] *Consumption Goods Safety Risk Evaluation Principle* (GB/T22760-2008) Appendix A: Consumption Goods Injury Type.

[10] Huang Guansheng, Linwei, "General Theoretical Research of Risk Precaution System", *China Standardization Press,* 2006, 3.

[11] Maslow,A. H., "A Theory of Human Motivation", *Psychological Review*, 1943,50,:pp.370-396.

[12] Fan Ruimin, "Consumer Evaluation for Quality Social Supervision Function of China Consumers Association", *Macro-Quality Research,* 2013, 2.

[13] Cheng Hong, Fan Hanbing, Xiao Yu, "Theoretical Framework to Effectively Govern Enterprise Quality Safety Risk - Enterprise Quality Safety Classification Model and Realization Methods Based on Internet Information", *Management World,* 2012, 12, pp: 29-37.

[14] Li Han, "Supervision Failure from Market to Government-Domestic and Abroad Research Overview about Government Quality Safety Regulation", *Macro-Quality Research*, 2013, 2.

[15] Zhang Jihong, "Special Equipment Quality Safety Test Game Theory Analysis", *Macro-Quality Research*, 2013, 2.

[16] Cheng Hong, Liu Yun, 2013: "Standard Theory Framework and System Innovation of Benefit Unification-Case Study Based on "EU Standard", *Macro-Quality Research*, 2013, 2.

[17] Yu Hongwei, Yu Fan, Xu Wei, 2013: "Quality Safety Network Precaution and Model Realization", *Macro-Quality Research*, 2013, 3.

Part II

Industrial Engineering Practice

Tackling Message Congestion in Agent-Based Manufacturing Control

W L Yeung
Department of Computing and Decision Sciences, Lingnan University, Hong Kong
(wlyeung@ln.edu.hk)

Abstract - **Agent-based distributed task allocation based on a contract-net negotiation protocol is known to be susceptible to the problem of message congestion which could seriously degrade the performance of agent negotiation process. Despite that, while many studies on agent-based manufacturing control involve such a protocol, few of them attempt to address to the problem. This paper reviews these attempts and suggests ways of further investigation into tackling the problem.**

Keywords - **Audience restriction, Control Net Protocol, focused addressing, publish/subscribe**

I. INTRODUCTION

Agent technology has been developed for various areas of the manufacturing industry over the past few decades [1]. For manufacturing control at the shop-floor level, agent-based distributed task allocation has received much attention for its promises to reduce development cost, increase maintainability and enhance fault tolerance [2], [3]. While studies have shown promising results in agent-based manufacturing control, there are also some concerns with the capacity and efficiency of software agents in handling substantial amount of workload.

The contract net protocol [4] first found its applications in manufacturing control in the 1980's [5], [6]. Real-life manufacturing applications included the Production 2000+ system [7] and the Holomibles [8] assembly system in the automotive industry. In a typical application scenario, instead of relying on a centralised control program for monitoring and controlling an overall manufacturing system consisting of various elements including machines, tasks, sub-tasks, people, etc., control of each element is distributed among software agents that are programmed to interact with each other in determining the allocation of tasks among machines. The design of these distributed control programs is guided by a negotiation metaphor exemplified in the Control Net Protocol (CNP) [4] wherein resource (machine) agents "bid" for available tasks and task agents "award" tasks to the best bidders according to criteria such as production cost and expected finishing time. This involves agents signalling their intentions/decisions (to announce a task, to bid, to award, etc.) to one another and the programming of agents to act in response to received signals (e.g. evaluate task announcement upon an announce signal). Signals are typically conveyed by means of coded messages delivered through buffered communication channels. Message buffering is required to give agents leeways to handle signals that arrive simultaneously. However, with message buffering, there is also the issue of message congestion which could undermine the ability of software agents under certain workload conditions.

This paper reviews the literature on the message congestion problem and some major approaches to its solution. Suggestions for further work are also discussed. The paper is organised as follows. The next section reviews the relevant literature on the message congestion problem. Section 3 presents some previous proposals to tackle the problem. Section 4 provides suggestions for further work.

II. MESSAGE CONGESTION

Smith [4] discussed the issue of message congestion in the original proposal of CNP in a sensor network environment where tradeoffs had to be made in the negotiation scheme in order to avoid generating excessive volume of messages. There was even provision for task agents to stop announcing tasks for bidding during times of heavy system loading; instead, managers simply award tasks to suitable volunteers. Smith and Davis [9] reiterated the importance of minimising communication among nodes especially when a large number of fast processors are collaborating in a task and the nature of collaboration demands broadcast communications as in the case of CNP.

In an agent-based vehicle routing application, [10] observed that bidders for delivery tasks could be overwhelmed by task announcements. Tilley [11] conducted a comprehensive simulation study of a CNP-based multi-agent manufacturing system. The results show that *task evaluation time* (TET) incurred by resource agents is the main factor affecting negotiation performance. Long TET undermines resource agents' ability to cope with a large influx of task announcements which in turn lead to even more task re-announcements and, hence, message congestion.

Coudert et al [12] applied CNP to the scheduling of production and maintenance activities. They founded that re- source agents always submitted bids in response to relevant task announcements and this resulted in lengthy negotiation processes. Wong et al [13] found similar results in their simulation study of an agent-based integrated process planning and scheduling application. They pointed out that negotiation time had a significant impact on the performance of CNP- based negotiation.

III. SOLUTIONS

A. Focused addressing

In tackling the message congestion problem, Smith

E. Qi et al. (eds.), *Proceedings of the 5th International Asia Conference on Industrial Engineering and Management Innovation (IEMI2014)*, Proceedings of the International Asia Conference on Industrial Engineering and Management Innovation 1, DOI 10.2991/978-94-6239-100-0_38, © Atlantis Press and the authors 2015

[4] suggested the use of node (resource) available announcements in an extension to the negotiation scheme wherein resource agents announce themselves when they become available for work. Each task agent maintains a list of resource available messages. For each task, it tries to identify a suitable resource from this list and, if one is available, it sends a direct award message to the associated resource agent without going through the proper negotiation process. This extension to the CNP, known as *focused addressing*, is meant for individual task agents when they sense heavy loading on the network, although the issue of determining the network loading is not discussed in [4].

B. Audience restriction

Parunak [6] adapted CNP in a manufacturing control application. In this version of CNP, each task agent identifies a list of suitable resource agents and conducts negotiation with them only. The list is updated after each round of negotiation: those agents who have responded stay in the list while others are dropped. If the list is reduced to only one resource agent, the task agent simply awards the next task to it. Furthermore, AR allows resource agents to submit *null bids* for relevant tasks even when they are not available. Parunak [6] called this adaptation of CNP *audience restriction* (AR) and examined the effectiveness of AR in manufacturing control applications *analytically* and suggested that it could help reduce the volume of messages in reasonable factory- like settings under heavy loading, especially with relatively stable machines and specialised tasks.

Cantamessa [14] considered AR as an important feature for agent-based manufacturing systems and highlighted a potential side-effect, namely, *self-organizing capability* due to audience restriction. In a simulation experiment, task agents keep track of "favourable" resource agents based on the latter's track records on accomplishing tasks and/or winning bids. The results show that, starting with a strictly hierarchical link-structure, AR helps agents develop more direct links among themselves. However, Cantamessa [14] did not report on the impact of AR on message congestion nor the efficiency of negotiation.

Sugawara et al [15] considered audience restriction in large- scale multiagent systems and experimented with audience restriction. They considered issuing task announcements selectively to randomly selected agents. However, their simulation model assumes an application domain with thousands of agents with high level of parallel- and over-bidding—typically Internet-based e-commerce applications. Furthermore, message congestion is not a main issue addressed in the study.

C. Publish/Subscribe

Yeung [16] adapted CNP with *distributed bid evaluation based on publish-subscribe messaging* for a manufacturing control application in which resource agents avoid over- bidding by simply bidding one task at time, i.e. a resource agent always waits for the result of its last bid before attending the next task announcement. A resource agent with a backlog of task announcements increases its clearing rate if the waiting time for bidding results can be shortened. Instead of waiting idly for an explicit result sent by the task agent (or a timeout), *publish-subscribe* messaging helps disseminate bid information shared voluntarily among all bidders of the same task. If a bidder sees a better bid from others, it can stop waiting and attend to the next task announcement. Only the highest bidder keeps waiting until the arrival of an award message. While this adaption of CNP generates additional bid information messages among biding agents, simulation results show that it can effectively alleviate task announcement backlogs and hence reduce problematic task re-annoucements.

D. Miscellaneous

Sandholm [10] tackled message congestion in two ways. First, tasks are bundled together for announcement and bid- ding. Secondly, backlogs of task announcements received by resource agents are simply shortened by dropping the older ones. The rationale for the latter is that, as time passes, the validity of a task announcement becomes increasingly likely to expire. Sandholm and Lesser [17] suggested that agents should be charged for communication usage and monitor each other for possible "spamming".

IV. CONCLUSION AND FURTHER WORK

Despite the previous proposals outlined in the previous section, there has been little empirical evidence of a satisfactory solution to the message congestion problem in agent- based manufacturing control. On the other hand, while focused addressing and audience restriction are potentially promising means of reducing message congestion, they also fundamentally weaken the optimality of CNP-based task allocation by limiting the potential number of bids available for each task. Hence, the effectiveness of a solution cannot be seen only in terms of reduction in the volume of messages; it must also be judged by its impact on the overall optimality of task allocation in terms of operational metrics such as makespan, flowtime, work-in-progress, etc.

In order to tackle the message congestion problem which presents a major obstacle to the adoption of agent-based manufacturing control, the following threads of further work can be considered:

1) Simulation studies provide useful insights into the positive and negative as well as intended and unintended impact of potential solutions to the problem. Experiments designed for realistic workload conditions are needed.

2) The design of contract net protocol has often been tailored for addressing various manufacturing conditions and environments, with message congestion not being a primary concern. A more systematic design methodology, perhaps supported by an analytical mode

3) Previous proposals can perhaps be integrated to provide a more robust solution to the problem.

Finally, any future solutions to the message congestion problem need to be evaluated for their impact on the optimality of task allocation

REFERENCES

[1] M. Caridi and S. Cavalieri, "Multi-agent systems in production planning and control: an overview," *Production Planning & Control*, vol. 15, no. 2, pp. 106–118, 2004.

[2] N. Duffie and R. Piper, "Nonhierarchical control of manufacturing systems," *Journal of Manufacturing Systems*, vol. 5, no. 2, pp. 137–139, 1986.

[3] "Applications of Distributed Artificial Intelligence in Industry," in *Foundations of Distributed Artificial Intelligence*, G. M. P. O'Hare and N. R. Jennings, Eds. New York, NY, USA: John Wiley & Sons, Inc., 1996, ch. 4, pp. 139–164.

[4] R. G. Smith, "The Contract Net Protocol: High-Level Communication and Control in a Distributed Problem Solver," *IEEE Transactions on Computer*, vol. 29, pp. 1104–1113, 1980.

[5] M. J. P. Shaw, "The design of a distributed knowledge-based approach for the intelligent manufacturing information system," Ph.D. dissertation, Purdue University, W. Lafayette, IN, 1984.

[6] H. V. D. Parunak, "Manufacturing Experience with the Contract Net," in *Distributed Artificial Intelligence*, M. N. Huhns, Ed. Pitman, London, 1987, pp. 285–310.

[7] K. Schild and S. Bussmann, "Self-organization in manufacturing oper- ations," *Communications of the ACM*, vol. 50, no. 12, p. 79, 2007.

[8] S. Bussmann and J. Sieverding, "Holonic control of an engine assembly plant: an industrial evaluation," in *Proc. IEEE International Conference on Systems, Man, and Cybernetics*, T. Bahill, Ed., 2001, pp. 3151–3156.

[9] R. G. Smith and R. Davis, "Frameworks for Cooperation in Distributed Problem Solving," *IEEE Trans. on Systems, Man, and Cybernetics*, vol. 11, no. 1, pp. 61–70, January 1981.

[10] T. Sandholm, "An Implementation of the Contract Net Protocol Based on Marginal Cost Calculations," in *Proc. 11th National Conference on Artificial Intelligence (AAAI-93)*, B. Hayes-Roth and R. E. Korf, Eds. John Wiley & Sons Ltd., 1993, pp. 256–262.

[11] K. J. Tilley, "Machining task allocation in discrete manufacturing systems," in *Market-Based Control: A Paradigm for Distributed Resource Allocation*, S. H. Clearwater, Ed. River Edge, NJ, USA: World Scientific Publishing Co., Inc., 1996, ch. 9, pp. 224–252.

[12] T. Coudert, B. Grabot, and B. Archimede, "Production/ aintenance cooperative scheduling using multi-agents and fuzzy logic," *International Journal of Production Research*, vol. 40, no. 18, pp. 4611–4632, 2002.

[13] T. Wong, C. Leung, K. Mak, and R. Fung, "An agent-based negotiation approach to integrate process planning and scheduling," *International Journal of Production Research*, vol. 44, no. 7, pp. 1331–1351, 2006.

[14] M. Cantamessa, "Agent-based modeling and management of manufacturing systems," *Computers in industry*, vol. 34, no. 2, pp. 173 – 186, 997.

[15] T. Sugawara, T. Hirotsu, S. Kurihara, and K. Fukuda, "Performance variation due to interference among a large number of self-interested agents," in *Evolutionary Computation, 2007. CEC 2007. IEEE Congress on*. IEEE, 2007, pp. 766 – 773.

[16] W. L. Yeung, "Agent-based manufacturing control based on distributed bid selection and publish-subscribe messaging: A simulation case study," *International Journal of Production Research*, vol. 50, no. 22, pp. 6339–6356, 2012.

[17] T. Sandholm and V. Lesser, "Issues in Automated Negotiation and Electronic Commerce: Extending the Contract Net Framework," in *Proc. First International Conference on Multi-Agent Systems*, V. R. Lesser, S. Conry, Y. Demazeau, and M. Tokoro, Eds., 1995, pp. 328–335.

Development and Evaluation of a Computer Adaptive System for Enhancing Hands-on Skills

Shu-Hsuan Chang[1], Ai-Chiao Ku[2,*], Li-Chih Yu[3], Bor-Chen Kuo[4], Hong-wei Chen[5]

[1,2]Department of Industry Education and Technology, National Changhua University of Education, Changhua, Taiwan

[3]Department of Management Information Systems, National Chengchi University, Taipei, Taiwan

[4]Graduate Institute of Educational Measurement and Statistics, National Taichung University of Education, Taichung, Taiwan

[5]Department of Electroinics, National Taichung Industrial High School, Taichung, Taiwan

(g1h2c3@gmail.com)

Abstract - **This research develops a computer adaptive instruction system for enhancing hands-on skill (CAIS), which could assist students to clarify misconceptions and conduct remedial learning by themselves in fostering hands-on skills. The subjects were 35 students from grade 2 of the electronics department in a senior vocational school, Taiwan. Meanwhile, one-group pretest-posttest design was applied in the research. The results showed almost 60% of learners were able to correct their misconceptions by utilizing CAIS. At the same time, there was significant improvement in the interactions of conceptual and procedural knowledge as well as the hands-on skill performance. Moreover, the research examined the subjects' attitude with the In-depth interviews. Over 50% of learners considered CAIS was easy to operate and useful in clarifying their misconceptions as well as helpful to their hands-on skill learning.**

Keywords - **Computer adaptive instruction, Hands-on skill, Remedial learning**

I. INTRODUCTION

Hands-on skills (practical skills), which were defined as "an ability to use the techniques, skills and modern engineering tools necessary for technology practice", have been acknowledged as the core competences in technology education in many countries [1-2]. Technology educators divided technology knowledge into two categories, Conceptual Knowledge (CK), which is also named as 'knowing that', and Procedural Knowledge (PK), which is also named as 'knowing how' [3]. Partly because the history of technology education was regarded as "a craft, skills-oriented school activity" [4] and partly because procedural knowledge appeared to be easier to define than conceptual knowledge, technology education curriculum focused on procedural knowledge while conceptual knowledge was neglected [3]. It is true that students need to acquire certain procedural knowledge and technical skills that are utilized during their design and/or problem solving tasks. In addition to conceptual knowledge, procedural knowledge and technical skills have to be taught in action and have to be modeled by the teacher or an expert or a fellow student [5]. However, there should be no disjunction between theory and practice for technology education, because it is conceptual knowledge that makes possible the use of procedural knowledge [3]. Therefore, if teaching ignore students' comprehension of conceptual knowledge, it is highly probable that students cannot solve problems independently. It makes learner become inflexible to deal with novel problems, which is harmful to skill transfer [6]. The interactions of conceptual knowledge and procedural knowledge (the interactions of CK and PK) play an important role in skill learning and many researches have focused on the interactions between thought and action [7-11]. However, action, especially reflective practice, is an issue to enhance students' hands-on skills. Reflective practice is a concept that relates to the continuous process of learning from experience, and involves the individual considering critical incidents in his/her life experiences, asking questions about what we know and how we came to know it, and learning to learn [12]. Even if instruction

E. Qi et al. (eds.), *Proceedings of the 5th International Asia Conference on Industrial Engineering and Management Innovation (IEMI2014)*, Proceedings of the International Asia Conference on Industrial Engineering and Management Innovation 1, DOI 10.2991/978-94-6239-100-0_39,

occurs in rich contexts and involves interacting with peers while working on various activities, quality learning will not take place unless there is reflective introspection [13]. Therefore, it is important that teachers provide students with feedback to foster reflective practice in skill learning [11]. An effective feedback tool, which could provide the levels of learners' conceptual understanding and development status of their knowledge structure, is essential to hands-on skill learning. Then, according to the feedback tool, remedial instructions could be developed to help students overcome their learning difficulties.

With the vigorous development of computer and internet technology, development of instructional/learning systems refocus on adapting users' features, such as goals/tasks, knowledge, background, hyperspace experience, preference, and interests, to meet the individual needs of learner [14-16]. This research develops a computer adaptive instruction system for enhancing hands-on skill (CAIS), which could assist students to clarify misconceptions and conduct remedial learning by themselves in elevating the efficiency of hands-on skills. The development of CAIS concentrated on compiling and testing questions required for troubleshooting knowledge of electrical department students. Based on the results, learners could understand their misconceptions in hands-on skills and do the remedial learning by themselves. Moreover, in-depth-interview, a qualitative research approach, was used to gather an in-depth understanding of students' behavior and the reasons that govern such behavior. The study addressed the following research issues: (1) The effect of CAIS upon the interactions of CK and PK; (2) The effect of CAIS upon enhancing hands-on skills performance.

II. DEVELOPMENT OF CAIS

The CAIS was developed based on the KSAT system [17]. Item of CAIS was chosen according to the comparison results of expert and student knowledge structure to test various ability levels of learners. CAIS helped learners diagnose and clarify their misconceptions through adaptive remedial learning. Learners are expected to be able to enhance the interactions of CK and PK as well as hands-on skills after accomplishing KSATEHS. Furthermore, learners not only achieve learning transfer, but also strengthen their learning self-efficacy.

A. System architecture

The Client-Server architecture was applied to build CAIS. The server adopted MySQL as the database with PHP programming language to access client data between databases, such as the items database, algorithm of item selection, student portfolios, and user management. The procedure for developing CAIS starts from developing the expert knowledge structure and then constructs the item banks for pretest. After analyzing the data of pretest, the system created the student knowledge structure. The student knowledge structure was used to select item which is most suitable to the learner. Finally, CAIS integrated student and expert knowledge structures to expand the appropriate structure for remedial learning. The related knowledge structure was explained as follows.

B. Expert knowledge structure

Expert knowledge structure is defined as "the knowledge structure is constructed by teachers or experts according to teaching/learning principles and experiences." Teachers or experts should analyze the instructional goals, instructional materials, and learners' learning portfolios to design learning units, learning sequence and test items [17]. In the study, the expert knowledge structure and the test items of CAIS were constructed by 20 industrial electronics experts. They selected and arranged the important concepts for the interactions of CK and PK in each leaning unit according to the learning objective of "electronics meter operation and measurement skill", Furthermore, "the test items for diagnosing learning difficulties" based on each knowledge node were developed to clarify the following issue: the relation and suitability of each node in knowledge structure. Fig.1 illustrates the expert knowledge structure of "electronics meter operation and measurement skill". The lower nodes represent the fundamental concepts; the upper nodes, which are more complicated concepts than lower ones, are the integrated/interaction concepts of lower ones. According to the knowledge structure, to achieve the learning objective of "1 Measurement of multimeter (ACV)", students have to complete the learning unit in the following sequence, starting with "1-1" and then "1-2". Similarly, to accomplish the learning objective of "1-1", students need to complete the sub-skill learning units in the following sequence, "1-1-1" and then "1-1-2". The items were created by considering testing the students' possible misconception types.

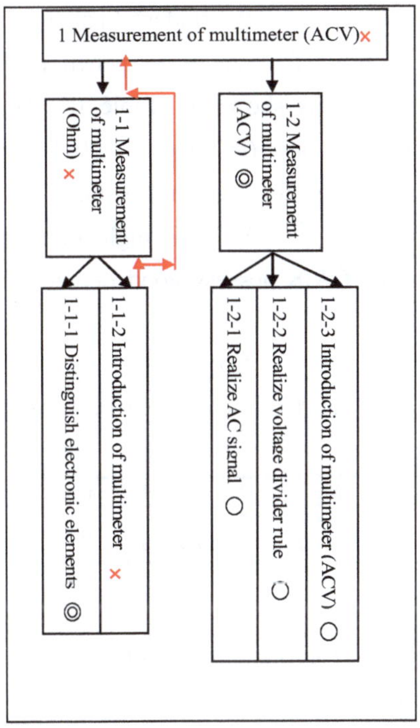

Fig.1. Student A's remedial path

C. Student knowledge structure

The student knowledge structure was constructed according to the 215 students' responses to the item bank via CAIS in accordance with the expert knowledge structure. The relation of the sequence in each level was computed by the item structure algorithm. First, the reliability and validity of the item bank should be analyzed; meanwhile, the parameters of Item Response Theory (IRT), such as difficulty, discrimination, and estimation, should be calculated for the reference of item selection and revision. The Cronbach α for the whole test is 0.772, which meets the minimal requirement (over 0.7) for internal consistency of test proposed by DeVellis [18] and Nunnally [19]. Kuo et al. [17] utilized Ordering Theory (OT) [20] and Item Relationship Structure analysis (IRS) [21] as the strategy to choose items for CAIS. This method is often used to define the relations in the item structure. The theory is described as follows. Define X (=X_1, X_2, … X_m) as a vector, including m binary item variables, which every student answers m items and each item gets a 0 or 1 vector. Table I lists the joint and marginal probability of item j and k.

TABLE I

THE CONJUNTIONAL AND MARGINAL PROBABILITY OF ITEM j AND K

		Item $_k$		
		$X_k = 1$	$X_k = 0$	Total
Item $_j$	X_j=1	$P(X_j=1, X_k=1)$	$P(X_j=1, X_k=0)$	$P(X_j=1)$
	X_j=0	$P(X_j=0, X_k=1)$	$P(X_j=0, X_k=0)$	$P(X_j=0)$
	Total	$P(X_k=1)$	$P(X_k=0)$	1

In Ordering Theory (OT), researchers define $\varepsilon_{jk}=P(X_j=0, X_k=1)<\varepsilon$, $0.02\leqq\varepsilon\leqq0.04$, ε_{jk} indicates the probability of violating lower item j to the higher item k, and ε indicates a threshold [20]. If item j is the fundamental item of item *k*, it is presented $X_j \rightarrow X_k$. Takeya [21] discovered a contradiction in the correlation coefficient from OT between the item and item structure. Therefore, he presented the Item Relationship Structure (IRS) to analyze and used another coefficient r_{jk} of measurable item sequence structure for revising the defect of OT. r_{jk} is defined as follows:

$$r_{jk}=1-(P(X_j=0,X_k=1) / P((X_j=0)P(X_k=1))$$

If $r_{jk}\geqq r$, we define item *j* as the fundamental one of item k and denote $X_j \rightarrow X_k$, where r is a threshold and the value is set to 0.5. This research invited 215 students from the electronics department to implement the pre-test expert knowledge structure. CAIS adopted the algorithm proposed by Airasian and Bart [20] and Takeya [21] to compute the pre-test result, to build the student knowledge structure, and to confirm the precision of the expert knowledge structure. The precision of the pre-test result reached 95%. Therefore, CAIS could accurately examine students and shorten the test time.

D. Remedial learning structure

Fig.1 illustrate CAIS used the student knowledge structure to proceed with adaptive assessment by gaining student A's responses to describe the process of test and remedial learning. The sequence of assessment from the upper to lower concepts is, 1→1-1→1-2→1-1-1→1-1-2→1-2-1→1-2-2→1-2-3. If students answer the upper items precisely (without misconception), they do not have to answer the lower items. Nonetheless, if they make errors on the upper items, they should have some misconceptions for the lower items. CAIS adopted the student knowledge structure to diagnose student's misconception and personalized the remedy learning from the lower missing nodes to the upper ones. As shown in Fig.1, student A had the misconception of "1-2-2", which led to the upper node error. Student A's remedial learning path should be 1-2-2→1-2→1.

III. METHOD

The one-group pretest-posttest design was applied to evaluate students' understanding improvement of the instrument operation and measurement principles after using CAIS. The subjects were 35 grade 2 students in the vocational high school. The independent variable was to use CAIS for remedial learning. The dependent variables in this research were the learning efficiency on the interactions of CK and PK and hands-on skills.

Evaluation tools

According to the aforementioned research structure, the evaluation tools of the study were Paper-Pencil pretest and posttest and the implementation of "electronics meter operation and measurement skill", as shown in Table II.

TABLE II

EVAULATION TOOLS

Objective of evaluation	Evaluation tool	Description
Knowledge	Paper-Pencil Pretest and Posttest	To examine student' understanding of the interactions of conceptual and procedural knowledge
Skills	Implementation of "electronics meter operation and measurement skill"	To evaluate the ability level of hands-on skill

IV. RESULTS

The evaluation for CAIS in the study consists of two parts: (1) the effect of CAIS upon the interactions of CK and PK; (2) the effect of CAIS upon enhancing hands-on skills performance. The evaluation results and discussion were addressed as follows.

(1) The effect of CAIS upon the interactions of CK and PK

After using CAIS, the students' paper-pencil pretest mean score was 75.7; the paper-pencil posttest mean score was 83.9, as shown in Table III. There was a significant efficiency ($t=7.076$, $p<.001$) between the scores of paper-pencil pretest and posttest. Moreover, the estimated effect ($\eta^2=59.6\%$) is high, meaning that the application of CAIS possess good degrees of practical significance.

TABLE III

THE LEARNING EFFICIENCY OF THE INTERACTIONS ON CAIS STATISTICS

	N	Mean	SD	SE	df	t value	p	η^2
Pre-test	35	75.7	9.2	1.56	34	7.076	*** .000	.596
Post-test	35	83.9	9.9	1.69	34			

***$p<.001$

(2) The effect of CAIS upon enhancing hands-on skills performance

One-way ANOVA is applied to compare the hands-on skills performance of students in high, medium and low score groups which were divided according to the posttest, as shown in Table IV. There was a significant efficiency ($p=0.01< 0.05$) between the high and low score group of posttest; meanwhile, there was a significant efficiency ($p=0.01<0.05$) between the medium and low score group of posttest. However, there was no significant efficiency ($p=0.995>0.05$) between the high and medium score group of posttest. As a result, after using CAIS, the students in high and medium score group, who have achieved well interactions of CK and PK, will have better hands-on skill performance. The effect of CAIS upon enhancing hands-on skills performance is partly supported.

TABLE IV

ANOVA FOR HANDS-ON SKILLS PERFORMANCE OF THE HIGH, MEDIUM AND SCORE GROUPS

Groups	N	Mean	SD	F	p	Post–Hoc (Scheffé)
1.High score	10	96.0	0.00	7.123*	.003	1,2>3
2.Medium score	17	95.1	2.66			
3.Low score	8	60.0	49.69			

*$p < .05$

V. DISCUSSION AND CONCLUSION

According to the evaluation results, almost 60% of learners were able to correct their misconceptions by utilizing CAIS. At the same time, there was significant improvement on the interactions of CK and PK and helpful in hands-on skills learning. One-group pretest-posttest design is applied in this study. A comparison of the experimental group and control group in quasi-experimental design should be provided for further study. CAIS is able to assist learners to improve their skill development for the interactions of CK and PK. (59.6% of learners showed a positive effect.). Further development of CAIS includes: (1) In addition to the test, we need to strengthen the interactive interface design between CAIS and learners as well as reinforcing learner motivation. (2) Regarding the learners with low learning achievement, if they fail in the active learning, CAIS gives them the appropriate assistance. (3) Remedial materials may provide adaptive materials and teaching strategies in relation to individual differences for improving skill development. (4) Establish an expanded skill training database to make CAIS a better expert system.

REFERENCES

[1] D. Barlex, "Assessing capability in design and technology: The case for a minimally invasive approach", *Design Tech. Educ. Int. J.,* vol. 12, no. 2, pp. 49-56, 2007.

[2] S. V. McLaren, "An International overview of assessment issues in technology education: Disentangling the influences, confusion and complexities", *Design Tech. Educ. Int. J.*, 12, 2, 2007, pp. 10-15.

[3] R. McCormick, "Conceptual and procedural knowledge", *Int. J. Tech. Design Educ.,* vol. 7, pp. 141-159, 1997.

[4] M. J. De Vries, and A. Tamir, "Shaping concepts of technology: What concepts and how to shape them", *Int. J. Tech. Design Educ.,* vol. 7, pp. 3-10, 1997.

[5] M. M. Sidawi, "Teaching science through designing technology", *Int. J. Tech. Design Educ.,,* vol. 19, no. 3, pp. 269-287, 2009.

[6] E. D. Gagne, C. W. Yekovich, and F. R. Yekovich, *The cognitive psychology of school learning* (2nd Ed). New York: HarperCollins College Publishers, 1993.

[7] R. A. Kimbell, K. Stables, T. Wheeler, A. Wozniak, and A. V. Kelly, *The assessment of performance in design & technology.* London: School Examinations and Assessment Authority (SEAC), 1991.

[8] P. Cobb, "Where is the mind? Constructivist and sociocultural perspectives on mathematical development", *Educ. res.,* vol. 23, no. 7, pp. 13-20, 1994.

[9] J. D. Bransford, and D. L. Schwartz, "Rethinking transfer: A simple proposal with multiple Implications", *Rev. Res. Educ.,* vol. 24, pp. 61-100, 1999.

[10] A. Albers, N. Burkhardt, and M. Ohmer, "The constructivist aspect of design education in the Karlsruhe education model for industrial product development KaLeP", In *Proc. of the 2nd international engineering and product design education conference* , Delft: TU Delft, faculty of Industrial Design Engineering, pp. 119-126, Sep 2-3 2004,

[11] Barak, M., "Motivating self-regulated learning in technology education", http://www.springerlink.com/content/wwt233w54760514p/, 2009

[12] D. A. Scho¨n, *Educating the reflective practitioner: Toward a new design for teaching and learning in the professions.* San Francisco: Jossey-Bass, Inc, 1996.

[13] S. Johnson, "Learning technological concepts and developing intellectual skills", *Int. J. Tech. Design Educ.,* vol. 7, pp. 161-180, 1997.

[14] T. C. L. Kinshuk, and S. Graf, "Considering cognitive traits and learning styles to open web-based learning to larger student community", ICTA 2007.

[15] N. Stash, Incorporating "Cognitive/learning styles in a general-purpose adaptive hypermedia system". Ph.D dissertation, Eindhoven: Technische Universiteit, Eindhoven, Netherlands, 2007.

[16] F. Lazarinis, S. Green, and E. Pearson, "Creating personalized assessments based on learner knowledge and objectives in a hypermedia Web testing application", *Comput. Educ.*, vol. 55, no. 4, pp. 1732-1743, 2010.

[17] Kuo, B. C., Liu, H. C., Sheu, T. W., Pai, C. H., Ko, L. W., Yang, J. M., and Lin, W. C, "A computerized adaptive testing system based on structure theories". In *Proc. of international conference on computer-based and standard-based assessment. Tainan*, Taiwan March 26-27, 2004.

[18] R. F. DeVellis, Scale Development: theory and applications. In Applied Social Research Methods Series (Vol. 26). California: Sage, 1991.

[19] Nunnally, J. C., *Psychometric Theory*. (2nd Ed.) New York: McGraw-Hill, 1978.

[20] Airasian, P. W., and Bart, W. M., "Ordering Theory: A new and useful measurement model", *J. Educ. Tech.*, vol. 5, pp. 56-60, 1973.

[21] Takeya, *New* item *structure theorem*. Tokyo: Waseda University, 1991.

Application on Determination of Key Factors of Hot Metal Silicon Content Based on Multivariate Statistics

Wen-wen Liu[1,2,*], Rui Miao[1], Bing-yu Tang[1], Qiang Zhang[1]
[1]School of Mechanical Engineering, Shanghai Jiao Tong University, 800 Dongchuan Road, Shanghai, China, 200240
[2]Shanxi Taigang stainless steel CO., LTD, 2Jiancaoping Street, Taiyuan, China, 030003
(*liuwenwen1239@163.com, miaorui@aliyun.com, mellotina@163.com, 740098255@qq.com, zq6670418@sjtu.edu.cn)

Abstract - **This paper use multivariate statistics to analyze and research the influencing factors of the hot metal silicon content, and apply principal component analysis to determine the key factors. According to blast furnace refining mechanism and production practice, determine 16 relevant operating parameters of hot metal silicon content, set up matrix and model, get the contribution rate of each variable to the principal, find out six key factors and use T2 statistics monitor hot metal silicon content changes in the actual production situation.**

Keywords - **Hot Metal Silicon Content, Multivariate Statistical, T2 statistics Chart**

I. INTRODUCTION

In practical production, many factors relate hot metal silicon content. Any one factor changes will affect the hot metal silicon content [1, 2]. Many scholars and technicians had research prediction mode of the metal silicon content [3, 4]. From several actual monitoring parameters, the purpose of this article is to find the key factors that influence on the metal silicon content, and improve the effectiveness on the molten iron quality control. Base on the field data, use principal component analysis and multivariate statistical process monitoring theory to get the more important factors on influence hot metal silicon content.

II. RELEVANT MONITORING PARAMETERS ANALYSIS ON INFLUENCE HOT METAL SILICON CONTENT

A. The reaction mechanism of silicon in the blast furnace

Reduction process of silicon in the blast furnace is step by step in accordance with the order of SiO2 - SiO - Si. The total reactive as follows:

$$SiO_2\,(l) + 2[C] = [Si] + 2CO\,(g) \qquad (1)$$

According to above formula, get the silicon content equation in blast furnace is

$$[\%Si] = \frac{K \cdot a_{SiO_2}}{f_{Si} \cdot P^2{}_{CO}} \qquad (2)$$

Where K is equilibrium constant, a_{SiO_2} isSiO2 activity of slag, P_{CO} is pressure of CO, and f_{Si} is activity coefficient of [Si].

From the formula (2), the influencing factors of [Si] content in molten iron are: the equilibrium constant K (mainly depends on the temperature T), the SiO_2contentof original fuel ash, a_{SiO2} and the P_{CO}.

B. Relevant monitoring parameters determination

Base on reduction mechanism of silicon in blast furnace, judge monitoring parameters which are the related factors on influencing the ultimate hot metal silicon content in the actual blast furnace. In the furnace monitoring parameters, blast volume (BV), blast speed, pressure difference (ΔP), permeability index (K value), rich oxygen (O_2), and furnace heat load (Q) have direct relationship with the blast furnace gas permeability, and influence the blast furnace gas volume. The blast furnace production (Yield), the batch number is on behalf of the volume of solid and liquid, these parameters influence the total pressure in the furnace. Blast temperature (BT), blast humidity (BH), and theoretical combustion temperature decide the change of the temperature field in the furnace. Silicon load, coke ash and pulverized coal ash content and fuel ratio (FR) is sources of silicon in blast furnace. The blast furnace gas utilization directly affects P_{CO}.

III. KEY FACTORS DETERMINATION

Principal component analysis establishes as little as possible new parameters during the related parameters. The new factors are on behalf of original hot metal silicon content information.

A. The principal component model establish

(1) The original data standardization.
(2) Calculate the correlation coefficient matrix.
(3) Calculate the eigenvalues and corresponding eigenvectors.
(4) Solve the cumulative contribution rate, and determine the number of principal components [5-16].

B. Key factors analysis

The calculation results are shown in Table I. From the table, the biggest variable correlation matrix characteristic value is 5.8960. The first principal component contribution rate is 36.9%. The second principal components contribution rate is 20.7%. By analogy, the variance of the principal component 5 contribution rate is 6.2%; the cumulative contribution rate reach 82.6%, suggesting that the first five principal components can provide plenty original data information. It also suggests that new five control factors can replace before 16 factors.

E. Qi et al. (eds.), *Proceedings of the 5th International Asia Conference on Industrial Engineering and Management Innovation (IEMI2014)*, Proceedings of the International Asia Conference on Industrial Engineering and Management Innovation 1, DOI 10.2991/978-94-6239-100-0_40, © Atlantis Press and the authors 2015

TABLE I
PRINCIPAL COMPONENT CHARACTERISTIC VALUE, RATIO AND ACCUMULATIVE CONTRIBUTION RATE

Number	1	2	3	4	5	6	7	8
Characteristic Value	5.8960	3.3115	1.7699	1.2458	0.9992	0.8666	0.5231	0.4875
Ratio	0.369	0.207	0.111	0.078	0.062	0.054	0.033	0.030
Accumulative Contribution Rate	0.369	0.575	0.686	0.764	0.826	0.881	0.913	0.944
Number	9	10	11	12	13	14	15	16
Characteristic Value	0.3186	0.1635	0.1486	0.1181	0.0760	0.0428	0.0272	0.0057
Ratio	0.020	0.010	0.009	0.007	0.005	0.003	0.002	0.000
Accumulative Contribution Rate	0.964	0.974	0.983	0.991	0.995	0.998	1.000	1.000

From Fig.1, we will find the characteristic value is less than 1 from the fifth principal component and gradually close to zero, it also suggests that five principal component is sufficient to meet the requirement of the amount of principal component selection. The coefficient of the maximum absolute value factors in the first five principal components is sorted out one by one, as shown in Table II.

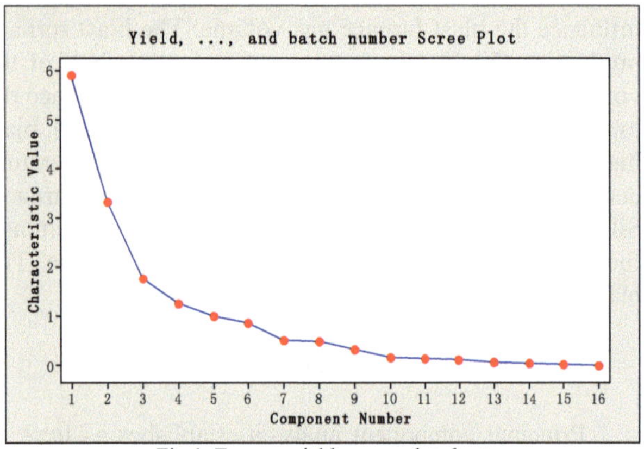

Fig.1. Every variable scree plot chart

TABLE II
MAIN COMPOSITION MAXIMUM ABSOLUTE VALUE AND CORRESPONDING CONTROL PARAMETERS

Principal Component	1	2	3	4	5
Maximum Absolute Value	0.397	0.503	0.512	0.467	0.621
Parameters	BV	BH	K	FR	SiliconLoad

Considering the maximum absolute value of control parameters and contribution, can be concluded that BV, BH, Q, K, FR and silicon load are the key factors on affecting blast furnace hot metal silicon content. Therefore, using principal component analysis can turn the above six factors as a control index of control hot metal silicon content, and carries on the adjustment of these factors to make hot metal components to meet customer needs.

IV. THE KEY FACTORS MULTIVARIATE T^2 STATISTICAL DISTRIBUTION

Base on above six factors the dates, analyze the keys factors multivariate T^2statistic analysis distribution. The multivariate T^2 control chart make up, as shown in Fig.2.

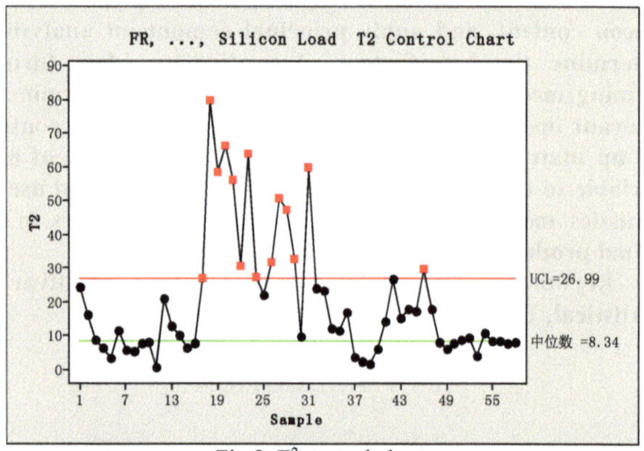

Fig.2. T^2 control chart.

In the Fig.2, we observe 14 statistics is beyond the scope of control, while the T^2 control chart point out that when the process is control, but not told which variables or which group is uncontrolled. However, we can decompose T^2 statistics to find out each variable's contribution. Use F distribution to test the significance of control points. Take Samplepoint18 to analysis. Table III shows the 18 sample P values of each significant variable.

TABLE III
SIGNIFICANT VARIABLES P VALUE

Variables	P value
FR	0.0000
BV	0.0000
Q	0.0025
BH	0.0171
K	0.0000

As you can see, 18 samples of T^2 statistics beyond control limit is mainly due to the FR, BV, Q, BH, and K. To analyze the five variables how to lead 18sample T^2 statistics beyond the control limit, we can observesingle value control chart of the five variables.

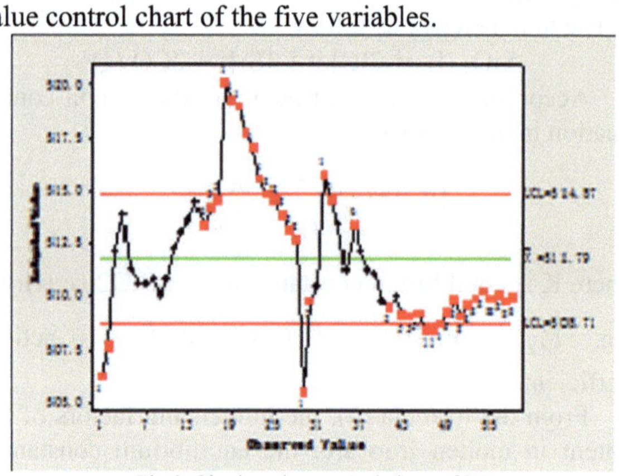
Fig.3. Single value control chart of FR

The measured value of FR at 18 samples is not in control, and is far with the center line. FR has made great contributions to the T^2 statistics in the sample.

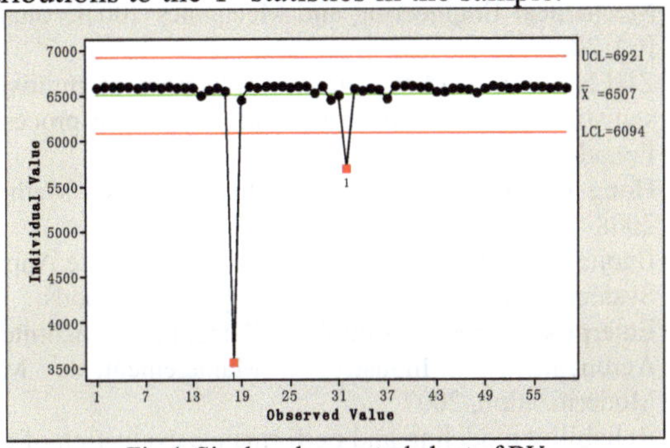

Fig.4. Single value control chart of BV

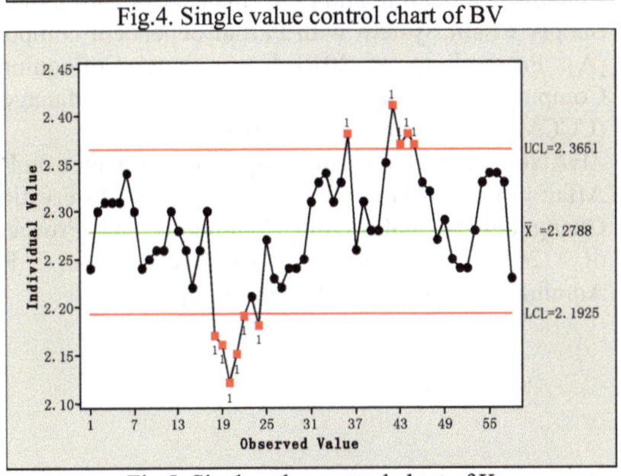

Fig.5. Single value control chart of K

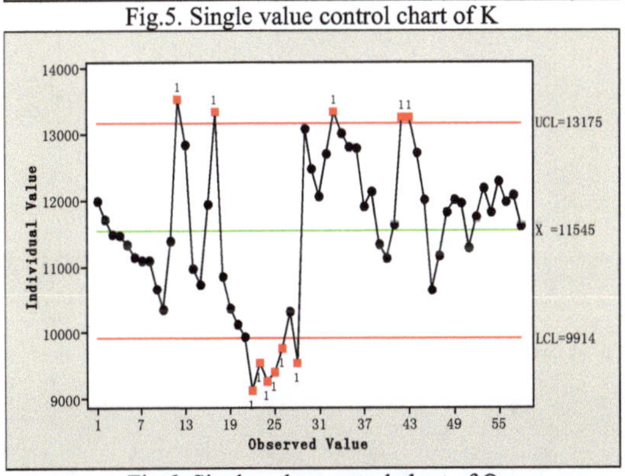

Fig.6. Single value control chart of Q

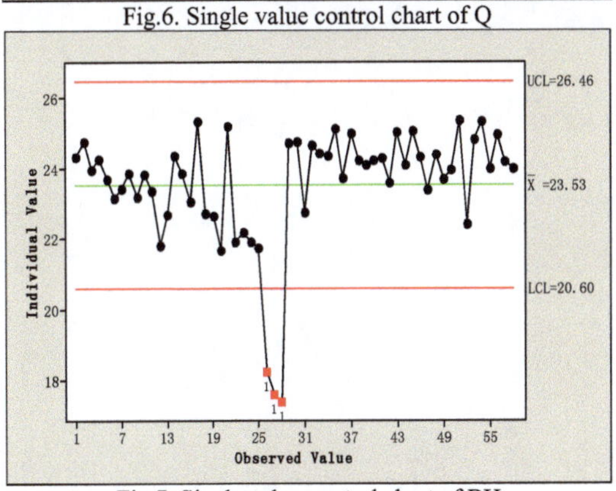

Fig.7. Single value control chart of BH

Similar, BV and K value at 18 samples are out of control, and their deviations are bigger than Q and BH, have made important contributions. From Fig. 6 and 7, Q and BH at 18 samples are in control, the contributions are smallest in five variables. Above analysis of an abnormal point, the T^2 statistics is a comprehensive reaction of various variables. By observing the value, can judge whether the whole process to control state, when find T^2 statistics is out control range, we should observe at this moment of each significant variable P values, if variable for the P value is less than 0.0001, we should focus on the variable, as to improve the efficiency of monitoring.

V. HOT METAL SILICON CONTENT AND T^2 STATISTIC CONTROL CHART RELATIONSHIP

Comparing Fig.2 and Fig.8, we can find the samples from 18 to 31 between T^2 control charts are in out of the control, the deviation of hot metal silicon content is also the largest. The sample standard deviation between 18 and 31 at the time is 0.156,58 overall samples standard deviation is 0.086.Hot metal silicon content was poor stability at this phase. At this point, the stability of the blast furnace smelting process is also poor. Therefore, in the blast furnace smelting process can use the T^2 control chart to monitor the state of the running.

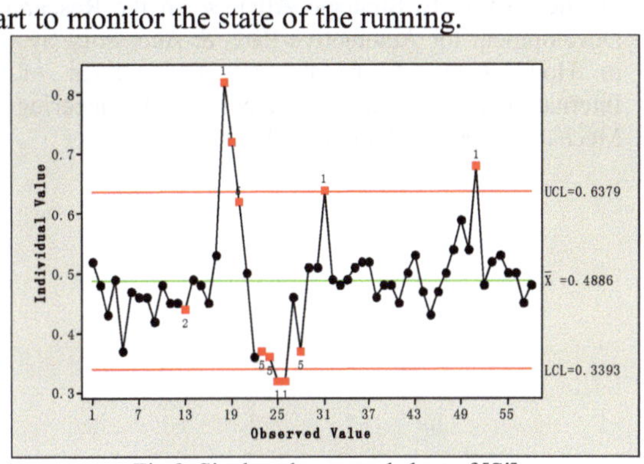

Fig.8. Singlc valuc control chart of [Si]

VI. CONCLUSIONS

In many practical influence factors about hot metal silicon content, by using principal component analysis method to find six key parameters respectively is: BV, BH, Q, K, FR and silicon load. And six key variables for multivariate T^2statisticanalysis are concluded that the T^2 statistics chart is better reflect actual changes of hot metal silicon content. In blast furnace production, therefore, can use the six key factors T2 chart to monitor the trend of hot metal silicon content, and reduce operating staff workload.

ACKNOWLEDGEMENTS

The authors gratefully acknowledge the financial support of the national natural science foundation, China (No.70932004) and the sixth innovation practice program of Shanghai (IAP6022).

REFERENCES

[1] MA Jin-fang, JIA Guo-li, Discussion on low silicon content smelting of BF [J], China Metallurgy, 2008, 18(9).

[2] ZHAO Xing-guo, TANG Shun-bing, Measures to realize low silicon hot metal smelting on a large blast furnace, Research and Iron&Steel [J], 2012, 40(6).

[3] LIU Hong-xin, ZHAO Jun, Practice of low silicon smelting of blast furnace [J], HeBei Metallurgy, 2012, 9.

[4] ZHOU Yong-ping, Simple analysis on low silicon hot metal production [J], China Metallurgy, 2009, 19(3).

[5] SUN Hui, LIU Yuan-yuan, Zhang Na-na, A case study of coal industrial competitiveness based on PCA [J], Resources & Industries, 2012, 14(1).

[6] Huang Qiong.Application of factorial analysis approach on elevation on school report of students from high medicine institute. Journal of Hubei Vocational Technical College . 2006

[7] Yu Jianying,He Xuhong.Statistic analysis and application of SPSS. 2004 [3] Xue Wei.Statistics package for social science and its application. . 2004

[8] Xue Wei.Statistics package for social science and its application. . 2004

[9] National Development and Reform Commission(NDRC). 11th Five-Year Plan for China National Economic and Social Development Report. . 2006

[10] Stephen S.F.LEE. Strategic Alliance on the Research & Development for Automotive Parts & Accessory Systems in Hong Kong & China [A]. Proceedings of the International Conference on Mechanical Engineering and Mechanics 2005(Volume 1) [C]. 2005

[11] Yiming SHAO. Evaluating Model and Empirical Analysis of Shipbuilding Industry Competitive Force [A]. Proceedings of the International Conference on Mechanical Engineering and Mechanics 2005(Volume 2) [C]. 2005

[12] ZHENG Yu, ZGANG Juan, Application of multivariate statistical process control in sand preparation process [J], Foundry, 2005, 54(8).

[13] Hong Liu. Discussion on Statistical Analysis Methods. . 2008

[14] Baojie Chen. Improving Employee Performance Appraisal System with Statistical Analysis Methods: Take Enterprises for Example Zhengzhou Institute of Aeronautical Industry Management. Market Modernization. 2007

[15] Reliability modeling and performance evaluation of Series Supply Chain System with Partial dependent components [A]. Proceedings of 2010 International Colloquium on Computing, Communication, Control, and Management (CCCM2010) Volume 4 [C]. 2010

[16] ZHANG Qiang,HAN Ying-ying. Co-governance on Public Affairs in Local Government - Empirical Observation on Guangdong No Fake Drug Community [A]. Proceedings of 2011 International Conference on Public Administration (7th) (Volume III) [C]. 2011

A Comprehensive Study on Fisher-General Motor: Outside Option and Hold-up Problem

Li-hua Hao [1,2]

[1] School of Economy and Management, Southwest Jiaotong University, Chengdu, China

[2] School of Archtecture and Cival Engineering, Xihua University, Chengdu, China

(lihuaxihua2011@hotmail.com)

Abstracts - **By introduction of effective outside options, the article analyzed classical Fisher-GM on a new perspective, and concluded: in the contract of Fisher-GM, because of a better mechanism design, Fisher didn't exist any hold-up motivation; but there were effective outside options, GM wanted to hold up Fisher by migration, while Fisher avoided it by refusal. GM had to purchase Fisher to reduce Fisher' outside option, and to warrant enduring, cheap closed-body provision.**

Keyword - **Fisher-General Motor, Hold-up, Outside Option**

I. INTRODUCTION

In the theory of transaction cost economics, Oliver Williamson (1985) talked about the hold-up problem in research of incomplete contracts, and argued that there exists suitability after specified assets invested. Benjamin Klein, Robert G. Crawford, Armen A. Alchain(1978) got the same conclusion, while they described the phenomenon as hold-up. Klein et al proved that there were three conditions to hold-up happened: incomplete contract, appropriate quasi-rent, and specified investment. Is there hold-up problem in presence of three factors? If yes, how to resolve it? Many scholars researched this problem, and get the different answers. I would review Klein's classical but controversial case of requisition of Fisher-General in literatures. In his literature he said that because of incomplete contract they signed, there is oscillation of closed body market, especially increasing demand for closed body, which made Fisher have more power of bargaining in contract implement, Fisher Body (later referred to as FB) hold-up General Motors (later referred to as GM) due to a lack of want of controlling cost [1]. Coase denied Klein's view it by his articles (2000, and 2006) [2][3]. He gathered enough evidences and investigated the whole process of requisition of Fisher-General, showed that there aren't enough evidences to prove invalid approach of cost-control being adopted by Fisher-body, and in the presence of hold-up problem. Therefore Klein "fraud" for confirmed theory but ignored the fact. Ronald H. Coase reputed the requisition of General-auto to get Fisher-body better into the overall business [2]. Freeland Robert (2000) recognized the existence of specified investment but hold-up problem, and denied that Hold-up problem could be resolved by vertical integration. He reputed that Fisher brothers had the opportunity to hold-up GM after being purchased [4]. Romon Casadesus-Masanell,Daniel F. Spulber(2000)showed that there was not specified investment and so not hold-up problem, the aim of requisition was to obtain coordination between production and investment and make use of Fisher brothers Human capital [5]. Yoshiro Miwa and J.Mark Ranseyer (2000) didn't recognize the existence of hold-up problem through the relation of A.O.Smith and GM [6]. Helper, MacDuffie and Sabel(2000)proved that it was not in accord with the actual operation in Japanese auto industry or FB [7].

Chinese Scholars researched the same problem. After sorting out the whole process of the case and analyzing it in the article of Assets specificity, Hold-up and Vertical Integration (2008), Nie Hui-hua and Li Jin-bo drew the following conclusion: there were assets specificity invested in FB and GM(no assets specificity, and no firms.); assets specificity might not induce hold-up problem, but environments of game might be. And they pointed it out:"If vertical integration could eliminate hold-up problems by the approaches of asymmetric bureaucratic arrangement, low efficiency incentive, and self government, like the prediction of transaction cost economics, we then said that assets specificity couldn't induce hold-up problem."[8]

While we find that above studies all ignored the fact-closed body was in short supply in the case and closed body brought super profit for car manufacturers. Thus, FB had many effective outside options, while GM didn't, even it could transact with any body manufacturers. I would consider it in the article to analysis the case again.

Outside options pointed to the alternative institution except that from partners, this could have influence on the form and performance of contract. The concept initially proposed was not to solve the problem.At the beginning of the theory conformation of transaction cost economy and contract, outside option was considered. Ronald H. Coase point out, the cause of firm formation was to reduce transaction cost, which compared government cost out of intra-firm to transaction cost of market transaction [9]. Oliver E. Williamson also said that specified investment, which was essential to foundation of firm, might induce applicability problem [10]. Klein reputed more explicitly that the one not to invest could have the motivation of hold-up the other one once specified investments were made. Basically, the cause of hold-up behavior came from outside option decreased even to zero, since parties to the contract could not predict all possibility in the process of contract implement, and could not allocate the surplus generated from specified investment and write in the contract. In general, Scholars argued that appropriate quasi-rent came from the part of value increasing but not being affected by market entry, just which came from price differences between contract outside and contract inside. So the nature of appropriate quasi-rent was external economies. However, outside option or external seldom was considered as a crucial factor in studies. For example, in requisition case of Fisher-GM, scholars focused on the cases of bilateral or unilateral monopoly. Of course, it still

E. Qi et al. (eds.), *Proceedings of the 5th International Asia Conference on Industrial Engineering and Management Innovation (IEMI2014)*, Proceedings of the International Asia Conference on Industrial Engineering and Management Innovation 1, DOI 10.2991/978-94-6239-100-0_41, © Atlantis Press and the authors 2015

was difficulty to square up outside option into bargaining problem in today's model.

Ken Binmore, Avner Shaked and John Sutton introduced outside option into game's model from strategic perspective-the strategy of carrot and stick, while which seemed to be identical with real market operation. Additionally, they proved it by an experiment: outside option was the only limitation in the scope of Nash bargaining efficiency solution; one part got the half of the surplus from the contract they signed, except that he allocated the other one less surplus because of absence of outside option, specially efficiency outside option; the threat of deal-me-out was credible when bargainer could obtain more benefit from outside option. The case studied how to form the contract price. But just as the experiment, each of outside options would not effect on the bargaining resolve [11]. We defined the efficient outside option as which had influence on the scope of bargaining resolve.

De Fraja(1990) suggested contract was able to solve the hold-up problem and it could provide almost incentive for specific investments when they exhibited sufficiently large direct externality under a mild condition[12]. Yeon-Koo Che got the Opposite conclusion [13]. David DE Meza and Ben Lockwood studied the effect of investment incentives using manager who had different asset as example. They deputed that the manager who make an important decision generally did not own the asset essential in work; independence asset should be owned simultaneously, strictly complementary asset should be owned by oneself [14].

Bake to our example, the assets of GM and FB was complementary, apparently should be owned by oneself. So, we could be persuaded by the conclusion if we ignored outside option. However requisition and divested business seemed to show that it was not true. We had to analyze the case of Fisher-GM again.

The secondly, I will describe the process of the case; thirdly, I would make a model in which outside option is introduced. Then we analyze it, and make my conclusion.

II. THE PROCESS OF THE COOPERATION OF FISHER-GM

I will have to sum up other scholar's description, and show it as follows, especially Ronald H. Coase's in 2000.

In 1919, the contract of investment and transaction was signed between GM and FB, which were to supply closed body during 10 years. To encourage FB for essential specified investment, it set the provision that GM must purchase closed body produced by FB, which decreased the probability of threatening FB after FB had made a corresponding investment. In addition, GM purchased 60% shares of FB; parties set up voting trust for 5 years, in which total 4 staff-two each side-managed it, and implemented one-vote veto. At the same time, the composition of the finance committee was 3:2 between GM and FB.

In 1923, for outside option of FB having gone up greatly, the closed body of FB produced increasingly met the needs of other auto manufacturers.

In 1925, GM requested FB manufactory to migrate near GM assemble plant. FB refused the recommendation because there were a large number of customers in original site.

In 1926, GM bought the remaining share of FB, and FB manufactory relocated.

But Ronald H. Coase did not follow closely external environment when he described and analyzed the process of the case. But the development and implementation of contract were all done in the market environment. Just as what Oliver Hart (2008) said, one part would deviate from contract when value or cost was not the general sense of "high" [15]. This may be the result of market fluctuations. So we described the market change in 1919 to 1926.

Before 1919, closed body existed as expensive product, which market share only had 10%. We compared the price of closed body cars to the price of non-closed cars in 1932, respectively 1200$ and 3000$. Importantly the demand to closed cars sharply rose. Just because of prediction of the upward trend, parties signed the contract and invested altogether in manufacturing closed metal body. During 1919-1922, the price of closed cars slowly went up in expectation, and sales of closed cars were small. The contract between GM and FB was executed well. Since 1923 the demand for closed cars increased largely, the price of closed cars remained high, which was not expected. Owing to price of procurement contract fixed from GM, which meant the stable revenue, FB would rather cater to the request of other car manufactures than GM.

Klein argued that Fisher had not the interest in controlling cost while transferring wealth because of plus-cost contract from GM, but which wasn't true [16]. On the one hand, GM controlled the finance committee and could monitor FB's behavior; on the other hand, FB sold closed body to other manufacturers and apparently had the incentive to control cost. Freeland Robert (2000) reputed Fisher Brothers hold up GM after requisition. While what Freeland Robert (2000) argued was not credible. FB would not like to hold-up GM after FB was bought. They would have the interest in attributing to the development of GM as the shareholders, and they indeed did so.

I would take account to the outside option in my model and analyze it again.

III. THE MODEL AND ANALYSIS

A. The model

Firstly, we assumed the final value of closed cars vt and manufacturing cost ct, vt≥ct≥0. because manufacturing apparatus of closed body were specified, at 0 data, GM and FB, who were all risk-neutral, signed an long-term purchase agreement to encourage FB investment, which lasted n. they adopted cost-plus contract, 17.6%. During the same time, the market price of closed body was $\bar{p}_t$ (which was average price accepted by all customers for FB.). We could easily obtain that $\bar{p}_t$ was higher than the price of GM to buy. And once the period of contract finished, GM had to buy closed Body at market price. And we defined the quantity of transaction between GM and FB as qt, the quantity of transaction between other car manufacturers and FB as $\bar{q}_t$.

During the performance of the contract, the unit revenue of GM was:

$$\pi_t^{GC} = v_t - 17.6\% \cdot c_t + 60\% \cdot \pi_t^{FC} \quad (1)$$

The first part of which was the revenue from GM itself business, the other part of which was investment income. The unit revenue of FB was:

$$\pi_t^{FC} = p_t - c_t \quad (2)$$

But FB had the payoff out of contract. We had to consider the transaction quantity. Now the whole payoff of FB was:

$$\Pi^F = \sum[17.6\% \cdot c_t \cdot q_t + (\bar{p}_t - c_t) \cdot \bar{q}_t] \quad (3)$$

The first part of payoff was from transaction with GM, and the other was from the transaction with other car manufacturers.

The whole payoff of GM was:

$$\Pi^G = \sum\{[v_t - 17.6\% \cdot c_t + 60\% \cdot 17.6\% \cdot c_t] \cdot q_t + 60\% \cdot [\bar{p}_t - c_t] \cdot \bar{q}_t\} \quad (4)$$

Throughout the period of performance of the long-term contract, the costs of Closed Body were not always same as a result of changes of material prices and labor cost. According to the theory of market economy, price would go up along demand increasing. So $\bar{q}_t' \succ 0$, and $\bar{p}_t' \succ 0$. The same, the production of lots of closed body would induce the cost rising, just is $c_t' \geq 0$, while marginal cost of closed body must be less than marginal price of closed body Auto. At the same time, $(\Pi^F{}_t)' \succ 0$ and $(\Pi^F{}_t)'' \geq 0$, that was Marginal revenue of FB went up owing to external revenue increasing, although the potential internal revenue might go down.

GM got benefit from efficient outside option increasing of FB. But its revenue enhancement came from its own business and investment in FB, and its own business revenue might decrease or go up not as its anticipation. This was described in the first part of (2), that is: $(v_t - 17.6\% c_t) q_t$. GM might focus more on the part than the other part, even investment in closed body of FB just was to obtain the more first part. That was,

$$(\Pi^G)' \succ 0 \text{ And } (\Pi^G)'' \geq 0 \quad (5)$$

B. Analysis

I would analyze the important events of requisition in this portion.

Firstly, I would analyze whether hold-up problem existed in the case or not.

Proposition1: If the whole revenue was considered, it seemed not hold-up happened. While outside option considered, FB took the opportunism behavior to GM.

Proof: Three condition of hold-up problem happened were incomplete contract, specified investment and appropriate quasi-rent. For the first two conditions obviously existing, we were concerned about the third condition. The traditional condition of hold-up problem was that the value of investment would decline if investment applied to alternative aspect after investment having been made. $(\Pi^G)' \succ 0$, so the hold-up problem seemed not to exist. But just as what I analyzed in the model, the whole revenue of GM increased because of increased sales of closed body in FB. This was not accord with the purpose of GM's investment. Just as what Oliver Hart and John Moore (2006) said: one part would strictly implement contract if he deemed to be treated well, where "to be treated well" meant he got the benefit set in advance [17]. Obviously in (4), GM did not realize for its purpose to increase the gain of the first portion, and could not realize it by alternative approach.

And FB maximized its revenue by virtue of no provision in incomplete contract. And more importantly, even though having set the provision in the contract, it was not verifiable by third-part, such as court.

Secondly, I would analyze why GM recommended FB relocation.

Just as what we dad described above, FB maximized its revenue by virtue of satisfaction with other auto manufacturers. One important cause was that FB was convenient to communicate with other auto manufacturers due to the factory of FB nearer to them. If the factory relocated near assembly plant of GM, parties would have more opportunity to communicate each other. This might improve the first part revenue and develop its own business, while FB's whole revenue perhaps declined. In some sense GM wanted to hold-up FB by virtue of the site migrated.

Thirdly, could vertical integration solve hold-up problem in the case?

Proposition 2: vertical integration could resolve hold-up problem when one part was seriously absent in outside option. And importantly FB merged into GM just because its outside option became decreasing.

Proof: this was not cooperative game. The theoretical basis was the property right perspective of Y. Barzel. He reputed: almost all transactions were restricted by the actual situation of parties in contract and were guaranteed by some mechanism. …and in general, who generate more revenue, who should bear the result of change in revenue [18]. We compared the price of closed body cars to the price of others, the price difference mainly laid in the price of bodies. Thus FB should obtain most of the benefits. After 1923 the competition in closed cars market was fiercer than before, so the profit from its own business declined. And GM had to obtain closed body at market price, which furthermore weakened the advantage of GM in closed cars after 1929.

There were three ways to resolve the problem: one was locate the side of FB manufacturer, of course, which had been refused. The second was to buy FB to enhance competitiveness of GM. The third just was to do nothing only to get the share revenue from FB. This meant what GM owned would be naught, which must be refused. We only focused on the second way.

$(\Pi^F)' \succ 0$ and $(\Pi^G)' \succ 0$, we could know that the whole revenue of GM must go up if GM acquired FB. And just as Yang Ruilong and Yang Qijing(2001) researched, the resource need and hard to replace in firm could obtain more corporate organization rent, which came from absence of outside options[19]. The theory of market economy supported it.

We had to focus on FB's willingness. In 1919, GM would want to requisite FB. While Fisher brothers noticed the prospects of closed body market, which was lack of competition, they would like to grow up their corporation by themselves. We could see $\bar{p}_t$ and $\bar{q}_t$ maintain at a high level. Until 1925 there were another closed body factories to build, which had seriously influence on $\bar{p}_t$ and $\bar{q}_t$, and especially outside option. That was $(\Pi^F{}_t)' \prec 0$. Fisher Brothers had to consider their development direction and accepted being merged into GM.

IV. CONCLUSION

The case of Fisher-General motors was important in the theory of firm, property right and contract. Perhaps only the case known clearly, study about these theories made progress. Then these theories would guide practices.

I tried to analyze the case of Fisher-General Motors by introduction of the notion of outside option. Although the study on the outside option was at the original stage, it attached market economy to institutional economy to some extend. And in the process of analyzing the case, I draw the following conclusion: the real cause of hold-up problem was absence of outside option; to some extend, vertical integration could resolve hold-up problem when one part was seriously absent in outside option, and one part would like to merge into the other part just when its outside option declined.

Of course, there were some shortages about this article, such as analyzing relatively simple, absence of enough market data about the case. I would like analyze the case and study outside option effect on contract theory.

ACKNOWLEDGEMENTS

This work was supported in part by Sichuan Circular Economy Research Center (XHJJ-1207).

REFERENCES

[1] Benjamin Klein,Robert G. Crawford, Armen A. Alchain, "Vertical Integration, Appropriable Rents and the Competitive Contracting Process," Journal of Law and Economics, vol. 21, no. 1, pp.297-326, 1978.

[2] Ronald H. Coase, "The Acquisition of Fisher-Body by General Motor," Journal of Law & Economics, vol. 43, no. 1, pp. 15- 31, 2000.

[3] Ronald H. Coase, "The Conduct of Economics: The Example of Fisher Body and General Motor," Journal of Economy & Management strategy, vol. 15, no. 2, pp. 255- 278, 2006.

[4] Freeland Robert, "Creating Holdup through Vertical Integration: Fisher Body revisited," Journal of Law & Economics, vol. 43, no. 1, pp. 33- 66, 2000.

[5] Romon Casadesus- Masanell, Daniel F. Spulber, "The Fable of Fisher Body," working paper, July 2000.

[6] Yoshiro Miwa, J. Mark Ramseyer, "Rethingking Relationship- Specific Investment: Subcontracting in the Japanese Automobile Industry," working paper, August 2000.

[7] Susan Helper, John Paul MacDuffie, Charles Sabel, "Advancing Knowledge While Controlling Opportunism," working paper, May 1998.

[8] Nie Huihua. 2008, "asset specification, hold-up and vertical integration-A comprehensive Study on Fisher-GM," Economist, no.4, pp. 44-49, 2008. (in China).

[9] Coase Ronald, "Nature of Firm," Economica, vol.4, pp. 386-405, 1937.

[10] Oliver E. Williamson, "the Vertical Integration: Market Failure Considerations," America Economy Review, vol. 61, no. 5, pp.112-123, 1971.

[11] Ken Binmore ,Avner Shaked, John Sutton,"An outside option experiment," The quarterly Journal of Economics, vol.104, no.4, pp. 753 – 770, 1989.

[12] De Fraja, After You Sir, Hold-up, "Direct Externalities, and Sequential Investment," Games Economy Behavior, vol. 26, no. 1, pp.22-39, 1999.

[13] Yeon-Koo Che, "Can a Contract Solve Hold- up When Investments Have Externalities? A Comment on De Fraja," Games Economy Behavior, vol.33, pp.195 – 205, 2000.

[14] David DE Meza and Ben Lockwood, "Does Asset Ownership Always Motivate Managers? Outside Options and the Property Rights Theory of the Firm," The Quarterly Journal of Economics, vol. 5, pp. 361-386, 1998.

[15] Oliver Hart, "Hold-up, Assets Ownership, and Reference Point," The Quarterly Journal of Economics, vol.2, pp.267-300, 2009.

[16] Benjamin Klein, "The Economic Lessons on the Fisher Body- General Motors," Int. J. of the Economics of Business, vol. 14, no. 1, pp.1-36, 2007.

[17] Oliver Hart, John Moore, "Contract as Reference Point," Discussion Paper, December 2006.

[18] Y. Baezel writing, Fei Fanf-yu, Duan Yi-cai translating, 2006,"Economic Analysis of Property Right," Shanghai joint Publishing House, Shanhai People's Publishing House, 2006(3), pp.17-35.

[19] Yang Rui-long, Yang Qi-jing. 2001, "specificity, exclusiveness and Enterprise Institution," studies of economy, no.3, pp. 3-11, 2001. (in China)

Using Kernel Density Function to Evaluate the Health Risk from Multi-Diseases

Chien-Chih Wang[1,*], Cheng-Deng Chang[2], Bernard C Jiang[3]
[1]Department of Industrial Engineering and Management, Ming Chi University of Technology, Taiwan
[2]Unimicron Technology Corporation, Taiwan
[3]Department of Industrial Management, Taiwan University of Science and Technology, Taiwan
(*ieccwang@mail.mcut.edu.tw, yabe.chang@gmail.com, bcjiang@mail.ntust.edu.tw)

Abstract - **Many previous studies have explored the risk factors and the prediction model for a specific disease. However, the focus of the preventive care is not only treating a disease, but also how to maintain health. This study developed a health risk index for the prevention of multi-diseases of the healthy people. The kernel density technique was proposed to estimate the distribution of the common risk factors of multi-diseases and to define the health risk index. A data set of the hypertension, hyperlipidemia and hyperglycemia from National Health Insurance Research Database in Taiwan was used to explain the proposed analysis process. The five common risk factors include Fasting plasma glucose (FPG), T-CHO, Triglyceride (TG), SBP, DBP were used to calculate the health risk index. When FPG, T-CHO, TG, SBP and DBP average reduced 0.2 mg/dl, 5.2 mg/dl, 13.4 mg/dl, 5.3 mmHg and 3.7 mmHg, separately. The analysis showed that the subjects could reduce 7.29% of health risk to suffer from the three diseases.**

Keywords - **Health risk, kernel density function, multi-diseases**

I. INTRODUCTION

Past reaearch have applied many feature selection methods to find key risk factors (physiological indicators) for specific diseases [1-2]. In practice, the doctor can use the value of the risk factor to judge the patient's disease. For example, hypertension patients are defined by systolic blood pressure (SBP) over than 140 mmHg or diastolic blood pressure (DBP) over than 90 mmHg. However, how to evaluate the degree of the healthy person when risk factors are falling into the reasonable range is rarely discussed. Reliability and maintainability are widely used in industry. They can be used in assessing the life of machinery or production system. Physical structures of the human body like a production system, how to transfer the concepts of reliability and maintainability used in the assessment of human health is the focus of this study.

Most of the relative studies focus on effect of maintenance interventions, like exercises or diet control. Fagard R. H. construct a one-year experiment to observe effects of exercise and diet on blood pressure [3]. SBP and DBP were average decreased 3.4 mmHg and 2.4 mmHg in 68 samples who only took exercise methods. They also find using exercises combined with diet control could get more significantly effect than only use one of the exercises or diet control method. Vaitkevicius P. Stewart K. J. *et al.* design three-times a week period of six week exercises maintenance plan [4]. Results show that exercise could help Improve physical fitness and decrease values of many physiological indicators related to cardiovascular diseases and diabetes. From the above discussion, we may know through exercises maintenance can reduce the value of the risk factors. However, the degree of health increase has not been estimated.

In this paper, a kernel density technique was used to develop health risk index by reliability concept. Then calculate the health risk index to estimate the degree of health for a healthy person, and the results can let people understand their health condition more easily.

II. METHODOLOGY

In this paper, the kernel density estimation method was used to fit the density curves of risk factor for disease. When all of the density curves were fitted, the probability of health state could be estimated. The estimator is called health risk index. The effect of maintenance could be calculated by health risk index using the change values of risk factor. The development of health risk index has two main methods: kernel density estimation and health risk calculation.

A. Kernel density estimation

The kernel density estimation approach was first provided by Rosenblatt [5] and Parzen [6], it is a nonparametric statistical method used to estimate the unknown probability distribution. Although the values of risk factor in large sample size generally assume a normal distribution, but lack of strong evidence to support that argument. Therefore, used kernel density estimation method to assess distributions of risk factor was suitable.

Let $x_1, x_2, \ldots, x_n$ be n^{th} independent and identical unknown distribution samples, the probability density function can be estimated by kernel density function as follows

$$\hat{f}_h(x) = \frac{1}{nh} \sum_{i=1}^{n} K\left(\frac{x - x_i}{h}\right)$$

E. Qi et al. (eds.), *Proceedings of the 5th International Asia Conference on Industrial Engineering and Management Innovation (IEMI2014)*, Proceedings of the International Asia Conference on Industrial Engineering and Management Innovation 1, DOI 10.2991/978-94-6239-100-0_42, © Atlantis Press and the authors 2015

The $K(\square)$ was a kernel function that was symmetric and was integrated into one. The h was a smooth function called bandwidth was used to determine the kernel function degree of smoothness. The common types of kernel function showed in Table I.

TABLE I
THE COMMON TYPES OF KERNEL FUNCTION

Types	kernel function
Uniform	$K(u)=\frac{1}{2}$ for $\lvert u\rvert \le 1$
Triangular	$K(u)=(1-\lvert u\rvert)$ for $\lvert u\rvert \le 1$
Biweight	$K(u)=\frac{15}{16}(1-u^2)^2$ for $\lvert u\rvert \le 1$
Triweight	$K(u)=\frac{35}{32}(1-u^2)^3$ for $\lvert u\rvert \le 1$
Epanechnikov	$K(u)=\frac{3}{4}(1-u^2)$ for $\lvert u\rvert \le 1$
Gaussian	$K(u)=\frac{1}{\sqrt{2\pi}}e^{-\frac{1}{2}u^2}$

Two parameters needed to be decided when using kernel density estimation method, one was which kernel function should be used, and the other one was how to setting bandwidth. For kernel function selection, this paper calculated probability density value of all values of risk factor, as was real density values. On the other hand, used each common type of kernel function showed in Table I, respectively, to estimate kernel density values for risk factor, as were estimated density values. Finally, according to use the minimum total difference between real and estimated density values technique to decide a suitable kernel function. The NDR0 method was used to estimate bandwidth [7]. The bandwidth can be calculated as follows.

$$\hat{h}=0.9\min\left(\hat{\sigma},IQR/1.34\right)n^{-1/5}$$

$\hat{\sigma}$ and IQR were the standard deviation and inter-quartile range of risk factor values.

B. Health Risk Index Calculation

Consider n kinds of multiple diseases are discussed, there is one normal state (is called healthy people), and the (2^n-1) kind combinations of suffering from different diseases. $f_{1,i}$ representative of healthy people in the ith risk factor of probability density function (from kernel density estimation approach). $f_{2,i},f_{3,i},\ldots,f_{2^n,i}$ represents the probability density function of different combinations of diseases (from kernel density estimation) under the ith risk factor. In this study, the definition of health risk $R_i(t)$ for healthy people of the ith risk factor is calculated as follows:

$$R_i(t)=\begin{cases} 0 & \text{for } t<x_i \\ \dfrac{f_{1,i}}{f_{1,i}+f_{2,i},\ldots,f_{2^n,i}} & \text{for } x_i\le t\le y_i \\ 0 & \text{for } t>y_i \end{cases}$$

t is measured value of the ith risk factor, the interval $[x_i,y_i]$ represents the range of values in the ith risk factors for healthy people. The human physiological system likes a series system. If any risk factors over specified values, then the person was suffering from disease. Therefore, the health risk index $R_h(t)$ for a healthy person can be presented as follows.

$$R_h(t)=\prod_{i=1}^{n}R_i(t)$$

III. ANALYSIS AND DISCUSSION

Hypertension, hyperlipidemia and hyperglycemia are three kinds of common diseases of metabolic syndrome. They are related to metabolic abnormalities. When suffered from those diseases, it was easier to lead more chronic diseases. In this study, a dataset comes from National Health Insurance Research Database in Taiwan was used to explain the proposed analytical procedures. There are 6496 subjects (3104 male and 3392 female) who were over 15 years old in Taiwan. According to the results of previous studies, we can get the five common risk factors: Fasting plasma glucose (FPG), T-CHO, Triglyceride (TG), SBP, DBP for Hypertension, hyperlipidemia and hyperglycemia.

The first step of analysis procedures was finding appropriate kernel function for each risk factor in every physiological state of Hypertension, hyperlipidemia and hyperglycemia. This study considered six kinds of commonly kernel functions, which includes Gaussian, Epanechnikov, Triangular, Uniform, Bright and Triweight, to estimate kernel density values for each risk factor. After calculating real density values of each risk factor, the suitable kernel function showed in Table II.

TABLE II
SUITABLE KERNEL FUNCTION

	FPG	T-CHO	TG	SBP	DBP
Normal	Gaussian	Gaussian	Gaussian	Gaussian	Gaussian
A	Triangular	Gaussian	Triangular	Gaussian	Gaussian
B	Triangular	Gaussian	Gaussian	Gaussian	Gaussian
B	Triangular	Gaussian	Triangular	Triangular	Triangular
A&B	Triangular	Triangular	Gaussian	Gaussian	Gaussian
A&C	Triangular	Triangular	Triangular	Gaussian	Gaussian
B&C	Gaussian	Gaussian	Rectangular	Triangular	Gaussian
A&B&C	Gaussian	Triangular	Gaussian	Triangular	Gaussian

(NOTES: A, B, C denotes Hypertension, Hyperlipidemia and Hyperglycemia, respectively.)

Table II showed when fitting density distribution at FPG in normal state, used Gaussian kernel function would be more suitable. But in hypertension state, the Triangular kernel function should be used. According to

Table II, each health risk curve of five physiological risk factors considered in eight states could be made as Fig.1.

From Fig.1, each health risk curve of risk factors was a monotonically decreasing curve. When FPG reaches 126 mg/DL, T-CHO or TG reach 200 mg/DL, SPB reach 140 mmHg and DBP reach 90 mm Hg, the health risk value becomes zero. That represented when any risk factor value over the threshold, probability of subject belongs to the normal physical state is 0. Multiplying of all health risk values of each risk factor could get the health risk index of the physiological system. Table III showed results of seven samples draw out from the dataset.

In Table III, the health risk index of physiological system, which was probability of physical state belong to normal, equal to the product of five risk factors in each sample. No matter how many diseases were suffered by a subject, the health risk index of physiological system of that sample becomes 0 because some values of risk factor over the clinical thresholds. Although there were four subjects had normal state of a physical system, but the reliability values of physiological systems were different. Sample 3 had a higher reliability value of physiological system than others, because he had better performance on each indicator. The study simulated the effect of these five risk factors after exercise maintenance. According to Stewart K. J. *et al.* [4], after six week exercise maintenance, FPG, T-CHO, TG, SBP and DBP average reduced 0.2 mg/dl, 5.2 mg/dl, 13.4 mg/dl, 5.3 mmHg and 3.7 mmHg, separately. Form the change value of risk factors calculated health risk index. The result showed that subjects reduce 7.29% of health risk to suffer from three diseases.

TABLE III
HEALTH RISK INDEX VALUES OF SEVEN SAMPLES

#	FPG	T-CHO	TG	SBP	DBP	Physiological state	Health risk index
1	89 (0.952)	177 (0.475)	153 (0.263)	141 (0.000)	90 (0.000)	Hypertension	0
2	88 (0.957)	185 (0.410)	109 (0.507)	124 (0.252)	92 (0.000)	Hypertension	0
3	86 (0.967)	119 (0.898)	56 (1.000)	97 (0.989)	60 (0.855)	Normal	0.7357
4	238 (0.00)	190 (0.360)	103 (0.559)	140 (0.000)	69 (0.675)	Hypertension Hyperglycemia	0
5	80 (0.979)	143 (0.718)	53 (1.000)	133 (0.066)	67 (0.743)	Normal	0.0347
6	105 (0.391)	167 (0.554)	99 (0.601)	92 (0.997)	60 (0.855)	Normal	0.1114
7	79 (0.980)	169 (0.537)	117 (0.437)	101 (0.978)	73 (0.499)	Normal	0.1128

(Notes: For example, FPG value in sample one is 89 (0.9527), means his SPG value is 89 and the reliability value of FPG=89 mg/dl is 0.9527.)

FPG T-CHO

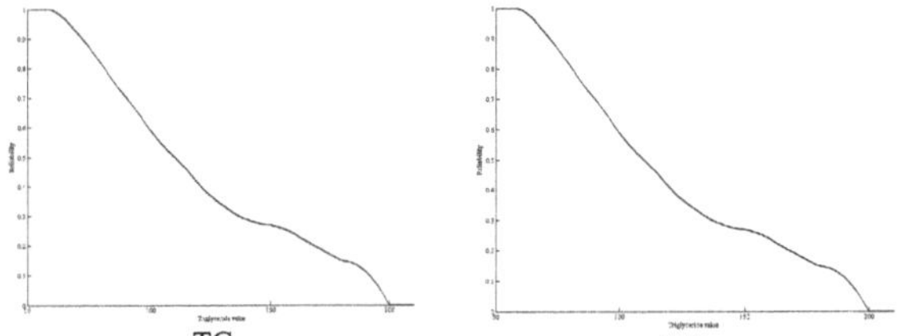

TG TG

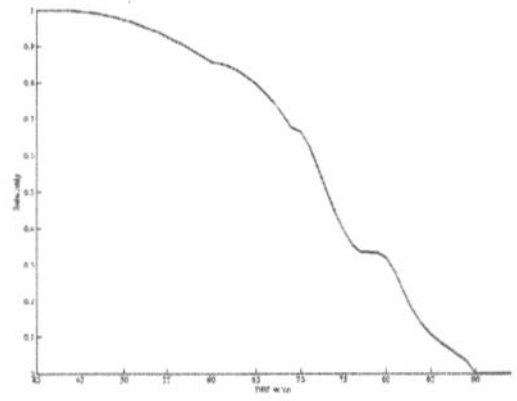

DBP

Fig.1. Health risk curves of five indicators

IV. CONCLUSION

This study used kernel density estimation method and proposed a novel health risk index to assess health degree of healthy person for multi-diseases. Because the human body system likes a complex tandem system, any risk factor value over the threshold, that subject was suffering from disease. Therefore, probability of physiological system belongs to normal state could be estimated by health risk index of all common risk factors. The research results can be used as an important reference index to quantify the effect of maintenance for a healthy people.

ACKNOWLEDGEMENTS

This work was supported in part by a grant from the National Science Council of Taiwan, Project Number NSC 101-2221-E-131-030-MY3.

REFERENCES

[1] Chang, C. D., Wang, C. C., and Jiang, B. C., "Using data mining techniques for multi-diseases prediction modeling of hypertension and hyperlipidemia by common risk factors", *Expert Systems with Applications*, 38, 2011, pp. 5507-5513.

[2] Jen, C. H., Wang, C. C., Jiang, B. C., Chu, Y. H., and Chen, M. S., "Application of classification techniques on development an early-warning system for chronic illnesses", *Expert Systems with Applications*, 39, 2012, pp. 8852-8858.

[3] Fagard, R. H., "Effects of exercise, diet and their combination on blood pressure", *Journal of Human Hypertension*, 19, 2005, pp.20-24.

[4] Stewart, K. J., Bacher, A. C.,Turner K., Lim J. G., Hees P. S., Shapiro E. P., Tayback M., and P. Ouyang.,"Exercise and risk factors associated with metabolic syndrome in older adults", *American Journal of Preventive Medicine*, 28,1, 2005, pp.9-18.

[5] Rosenblatt, M., "Remarks on some nonparametric estimates of a density function", *The Annals of Mathematical Statistics*, 27, 3, 1956, pp. 832-837.

[6] Parzen, E.,"On estimation of a probability density function and mode", *The Annals of Mathematical Statistics*, 33, 3, 1962, pp. 1065-1076.

[7] Liu, B., Yang, Y., Webb, G. I., and Boughton, J.," A Comparative Study of Bandwidth Choice in Kernel Density Estimation for Naive Bayesian Classification", *Lecture Notes in Computer Science*, 5476, 2009, pp.302-313.

Curve Monitoring for a Single-Crystal Ingot Growth Process

Liang Zhu[1,*], Chen-xu Dai[2], Yong-yue Sun[3], Wei Li[1], Ran Jin[3], Kai-bo Wang[2]
[1] The State Key Lab of Fluid Power Transmission and Control, Zhejiang University, Hang Zhou, China
[2] Department of Industrial Engineering, Tsinghua University, Beijing, China
[3] Department of Industrial and System Engineering, Virginia Tech., VA 24061, USA
(fyzhul@163.com)

Abstract - **In conventional statistical process control, the distributions of the monitoring statistics from a conforming process is assumed to remain unchanged. However, dynamic data stream of a variable over time are widely encountered for many complex manufacturing processes. In this paper, we study the ingot growing process, which exhibits a time-varying mean and variance even if the process is in-control. We propose an adaptive control chart for statistical monitoring. The adaptive control chart estimates the center line and the variability of the process in real time. Implementation of the newly proposed chart to the real engineering data shows that the chart is effective in detecting process failures in the growth curves.**

Keywords - **Profile, Statistical Process Control, SPC**

I. INTRODUCTION

Statistical process control (SPC) has been widely used in practice to investigate industrial processes and improve quality [1]. One of the major tools, control charts, is usually set up to monitor key process variables or quality characteristics, and trigger alarms when assignable-cause variation is detected. By removing such variation sources, the total variability of the process is expected to be reduced.

The control chart can be viewed as a graphical tool for quality control. When a sample becomes available, a charting statistic is calculated based on the sample, and a point that reflects the charting statistic is plotted on the chart. If the point falls outside the control limits, the process is concluded to be out-of-control. Then the procedure of searching for root causes shall be carried out.

In control chart implementation, one common assumption is that the distribution of the monitoring statistics from an in-control process remains unchanged, such as the mean and variance is assumed to be constant for a conforming process as shown in Fig.1. For this Shewhart control chart, the control limits are calculated based on the unchanged mean and variance from the in-control process. If the observed sample falls beyond the control limits, an out-of-control signal is triggered.

It has been widely studied when the process exhibit dynamic shift patterns. For example, in an inertia process, when a shift occurs, the resulting shift pattern follows a particular trend [2]. In a feedback-controlled process, a sudden shift in the process may lead to oscillated [3] or cyclical [4] signals. However, in these works, the mean of the process is always assumed constant.

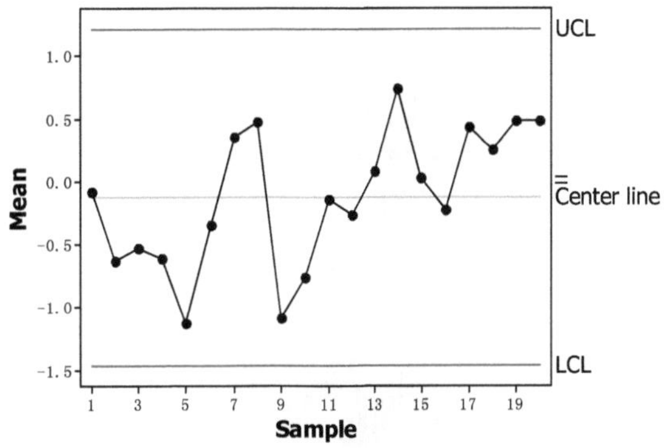

Fig.1. A typical control chart with constant mean and variance

There are cases in which the mean of the process is unstable. Reference [5] is one of the few works in the literature that has noticed this issue. The authors studied an etching process. Due to the accumulating effect, some parameters drift slowly in the normal operational runs and are reset when chamber cleaning happens. The variation of the parameters also changes over time. Reference [6] also proposed a regression tree based method to cluster and estimate multiple baseline distributions for an in-control process, and develop control chart for each cluster. In this paper, the authors proposed to use adaptive centering and scaling methods to incorporate such changes and designed a chart for trigging abnormal drifting effect, while the normal drift is allowed.

In this paper, we study an ingot growing process that is widely used in semiconductor manufacturing. We focus on the power consumption of the heater, which is one key process variable to affect ingot quality. The ingot growing process is a very complicated chemical-physical process. Fig.2 shows a photo of the equipment. The lower part of the equipment contains a quartz crucible, which is protected by insulation layers and used to melt raw silicon material. The upper part is the pulling chamber. In production, the raw silicon material is first melt; a crystal seed attached to a pulling thread touches the liquid surface and initiates the growth of the single-crystal silicon ingot. With the thread pulling up slowly, the single-crystal ingot grows. In the whole process, the power curve is one critical parameter that may determine the success or failure of the growing process.

Fig.3 shows the power over time from a conforming growing process. At the beginning of the growing process, the power increases sharply, but then decreases and oscillates after reaching an initial peak point. After several hours, the process becomes stable, and the power varies

E. Qi et al. (eds.), *Proceedings of the 5th International Asia Conference on Industrial Engineering and Management Innovation (IEMI2014)*, Proceedings of the International Asia Conference on Industrial Engineering and Management Innovation 1, DOI 10.2991/978-94-6239-100-0_43, © Atlantis Press and the authors 2015

slightly around a mean value. In the final stage of the process, the power increases slowly again until the process finishes. The sharp increase is because more power is required to melt the silicon crystal at the beginning stage of the process. Then the required heat gradually decreases as the ingot grows.

Fig.2. The single-crystal ingot growing furnace

In the crystal growing process, the thermal gradient in the furnace is critical to the success formation of the single crystal and quality of ingots. The thermal gradient is jointly determined by the power, the insulation layer, and the positions of the ingot and the crucible. As the ingot grows longer, the position of the ingot and the crucible is also changing. The furnace control system will automatically adjust the power to compensation the heat loss, and maintain the thermal gradient in the furnace. Therefore, this dynamically changing curve is considered normal in the process.

When some unexpected changes occur, it is possible that the process become nonconforming. Fig.4 shows an abnormal power curve from another furnace. Although the early and middle stage of the power curve shows a similar pattern as the one in Fig.3, the mean and variance of these two curves differ. This is mainly because the insulation layers in these two furnaces have different heat preservation effects due to different reliability conditions. Moreover, the power curve in Fig.4 has strong oscillation near the end of the process. The ingot becomes failure at this stage. The square mark shows the time point at which the process failure was identified by the operator.

Beyond the crystal growing process, such growing curves are widely encountered. A growing curve is a profile that extends/grows over time. It reflects the change of a quantitative variable in the time domain. For example, the wire slicing speed is slow at the beginning of the wafer slicing process, and becomes faster as the contact span increases. Different equipment will use different speed curves as the equipment maintenance conditions are not identical. Such growing curves share some common features: a) the in-control mean patterns shows a dynamic trend, i.e. a single baseline distribution is not adequate for the monitoring statistics, or the dynamic trend cannot be adequate modeled by a baseline model such as linear regression models or polynomial functions; b) the dynamic trend differs for different batches or equipment. This raises the challenge to control chart design in term of determining the in-control parameters, the center line and the control limits. We propose an adaptive control chart to address these challenges.

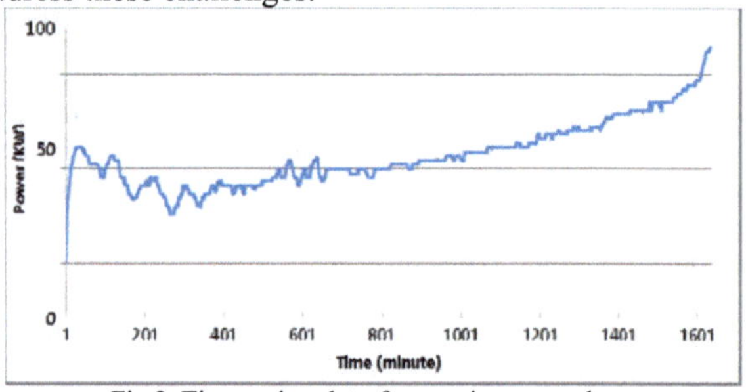

Fig.3. Time series plot of power in a normal run

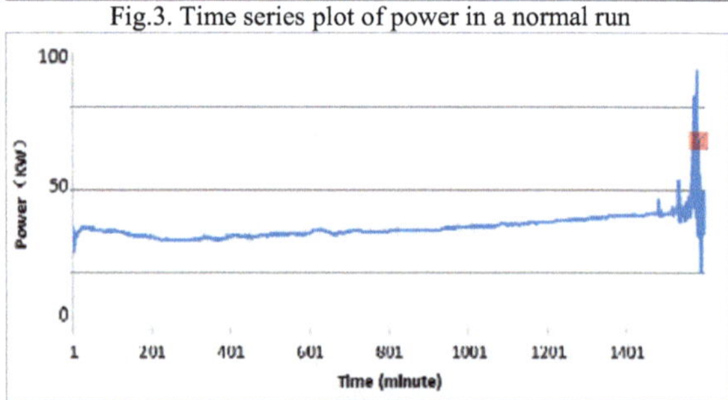

Fig.4. Time series plot of power in an abnormal run

The rest part of the paper is organized as follows. In Section 2, we first review relevant works on control charts. In Section 3, we propose a new adaptive chart to monitor such dynamic growth curves. The performance of the new chart is studied by applying it to samples collected from a real growth process in Section 4. Finally, conclusions and topics for future research are suggested in Section 5.

II. REVIEW OF RELATED CHARTING TECHNIQUES

There is a rich body of literature developed for control charts with a constant mean, such as the Shewhart chart [7], the cumulative sum (CUSUM) chart, and the exponentially moving average (EWMA) chart.

The EWMA chart is one of the most import charts for monitoring processes with small shifts. The common form of the EWMA chart is given by

$$z_t = \lambda x_t + (1-\lambda) z_{t-1}$$

where x_t is the observation, λ is a smoothing parameter that determines the weight assigned to historical observations, 0<<1. A small value of λ can help quickly detect small change in the mean shift and a larger value is helpful for quickly detecting the occurrence of large shifts.

It is difficult for a single EWMA controller to get a minimum ARL value for both small and large mean shifts [8, 9]. Reference [10] discussed the inertia problem of the EWMA scheme. When the charting statistic is close to the upper control limit, a sudden change in the opposite direction creates inconsistency in the process. Therefore the controller with smaller value of λ should overcome the inertia to react with sudden shift changes [11]. The inertia problem can also be solved by adding Shewhart limits to an EWMA control scheme. When the EWMA statistics falls outside of the EWMA limits, or the current observation is out of the Shewhart limits, an out of control signal is generated [9, 12].

The EWMA control scheme was further improved by [13], in which a new chart named the adaptive exponentially weighted moving average (AEWMA) control scheme was proposed. The AEWMA chart combines the EWMA and Shewhart in a smoother way to reduce the inertia. In contrast to the method introduce by [9], the AEWMA chart uses an adaptive smoothing parameter, which changes with observations. Therefore, the AEWMA chart is better for detecting shifts with an unknown magnitude of shifts. However, the AEWMA chart also assumes that the process mean is a constant. In this work, we will modify the AEWMA chart and borrow the algorithm for process updating.

The curve shown in Fig.3 and Fig.4 are typical functional profiles. In the literature, extensive research has been seen for profile monitoring. A comprehensive review is given in [14]. In profile monitoring, it is normally assumed that n (n>1) values of a response variable are measured along with the corresponding values of one or more explanatory variables. Reference [15] described the monitoring in two phases and discussed issues regarding profile monitoring.

Profile monitoring has normally been shown as a simple linear regression model. Reference [16] proposed Shewhart control charts for monitoring the slopes of the relationships between two or more measurement processes in order to assure their accuracy. Reference [17] considered using slope in profile monitoring in Phase I and Phase II. Reference [18] proposed a new method for monitoring simple linear regression profiles based on the multivariate control chart method of [19]. A self-starting method was proposed by [20] in which they avoided using distinction between the two phases of control. Refs. [21] and [22] considered change-point methods for detecting changes in the parameters of a simple linear regression model. These authors considered Phase I and Phase II applications, respectively. Reference [23] used a different approach of profile monitoring in which the estimated regression parameters is more effective in terms of the run length performances. However, although the curve considered in this work is a profile, the requirement for monitoring is quite different from the conventional scenarios. In the profile monitoring literature, each profile is treated as one sample; when calculating the charting statistic, the whole profile is known. In our work, the curve is a growing profile, the length of which extends with time; decisions must be made each time a new point is obtained from the profile. Thus, the conventional profile monitoring method cannot be used to this problem directly.

III. AN ADAPTIVE CONTROL CHART FOR GROWTH CURVES

As aforementioned, the power curve to be monitored has two distinct features. First, the curve shows a dynamic trend; the mean and variance of the curve are not constant with time goes by; second, the curve exhibit different patterns for different equipment. Therefore, a learning algorithm is needed to track the in-control parameter of each process.

Since the power curve exhibits a slow changing trend, a natural choice of the leaning algorithm would be the EWMA method. However, as the sample curve in Fig.3 shows, the power varies quickly in the initial stage, the EWMA algorithm shall prefer a large smoothing parameter so that the quick changing trend can be captured; nevertheless, in the later stage, the curve only changes slowly, and a small smoothing parameter shall be preferred. Therefore, the traditional EWMA algorithm with a fixed smoothing parameter is not favorable. Instead, we suggest to use the following adaptive EWMA algorithm suggested by [13]

$$\mu_t = \left(1 - w(e_t)\right)\mu_{t-1} + w(e_t)x_t \qquad (1)$$

where $w(e_t) = \phi(e_t)/e_t$, $e_t = x_t - \mu_{t-1}$ is the estimation error. The weighting function is given as follows:

$$\phi(e_t) = \begin{cases} e_t + (1-\lambda)k & \text{if } e_t < -k \\ \lambda e_t & ?\ |e_t| < k \\ e_t - (1-\lambda)k & \text{if } e_t > k \end{cases}$$

Since the process variability also changes over time, we update the standard deviation of the variable using the method proposed by [24]

$$\hat{\sigma}_t = \sqrt{(1-\gamma)\hat{\sigma}_{t-1}^2 \times + \gamma\left(x_t - \mu_{t-1}\right)^2}$$

where γ is a smoothing parameter. The magnitude of γ decides the speed of the parameter update.

Finally, the control chart for monitoring the dynamic growing curve is:

$$\begin{cases} UCL = \mu_t + 3\hat{\sigma}_t \\ CL = \mu_t \\ LCL = \mu_t - 3\hat{\sigma}_t \end{cases}$$

the center line, μ_t, is updated using the adaptive EWMA equation in (1). Whenever a new point becomes available and the chart is in-control, we update the center line; if the point becomes out-of-control, according to [5], the center line remains unchanged.

In this algorithm, the parameters λ decides the amount of weight assigned to new observations, and k is the threshold for estimation errors. If the estimation error e_t is large, $w(e_t)$ shall approach one, and the updated values is essential the new observation; if the estimation error is small, $w(e_t)$ approaches λ , the mean is then updated following the conventional EWMA equation. [13] noted that the former part is beneficial to quick changes, and the later part is beneficial to small changes. Therefore, the adaptive EWMA method combines the advantages of the Shewhart-type chart and the EWMA chart, and it is expected to perform well in tracking the dynamic mean of the growth curve.

It should be noted that the choice of λ is critical. If $|e_t|<k$, the effective smoothing parameter is λ; if $|e_t|>k$, the effective smoothing parameter is larger than λ.

IV. PERFORMANCE STUDY VIA SIMULATION

In this section, we study the performance of the proposed method and compare it with existing ones using Monte-Carlo simulation. A real example will be demonstrated in the following section. To mimic the dynamics of the ingot growth process, we construct the following two-stage model, which has a similar trend as the real process

$$y_t = \begin{cases} K + a\sin\left(\frac{t}{\omega}\right) + \varepsilon_{1t}, \varepsilon_{1t} \sim N(0,\sigma_t), if\ t \le T \\ y_T + b(t-T) + \varepsilon_{2t}, \varepsilon_{2t} \sim N(\mu_2,\sigma_t), if\ t > T \end{cases}$$

The first stage of the function changes as a sinusoidal wave, and the second stage follows an increasing linear trend. Each simulated process runs for 79 steps; correspondingly, each growing curve to be monitored contains 79 readings. When the process is out-of-control, it follows

$$y_t = \begin{cases} K + a\sin\left(\frac{t}{\omega}\right) + \varepsilon_{1t}, if\ t \le T \\ y_T + b'(t-T) + \varepsilon_{2t} + \delta, if\ t > T \end{cases}$$

In the in-control process model, the parameters are set as: K=0, α is randomly sampled from a uniform distribution within (5,10), b=0.75, $\sigma_t = 1.5\exp(-0.01t)$, ω=6, $T = 2\pi\omega$, and $\mu_2 = 0$. Three failure modes are studied, including variance shift, mean shift, and mean drift. When generating out-of-control samples, the variance shift, mean shift, and mean drift are added to the process at Step 45, 45 and 55, respectively.

Yu *et al.* (2012) proposed a method for Phase I profile outlier detection. When a curve of certain length is given, the method can be used to compare a new curve with a batch of existing ones. As the curve we encountered is a growing curve, we here modify the method proposed by Yu *et al.* (2012). Since the detection algorithm relies on a mean curve, and the curves in the simulation setting differ from each other, we first fit an autoregressive (AR) time-series model to each curve. The residual sequence, where the dynamic mean in the original variable is removed, are monitored using the method proposed by Yu *et al.* (2012). We name this method the *Profile-Residual chart.* In the following, we will compare this chart with the AEWMA chart introduced above.

To study the performance of the charts, 5000 in-control and 5000 output-of-control curves for each failure mode are generated and monitored by the charts. For a fair comparison, the control limits of all charts are adjusted so that they have an equal false alarm rate of 100 out of 5000 when the process is in-control. More specifically, the AEWMA chart uses $\lambda = 0.4, \gamma = 0.01, k = 1$, and the Profile-AR uses the threshold value of 51.975.

For the different out-of-control scenarios, two indicators are presented. The first is the number of alarms received. This indicator represents the number of curves that are identified as out-of-control from the total 5000 out-of-control curves. If the curve is reported out-of-control, one alarm is recorded (and the chart stops so that each curve can trigger one alarm at most). So the maximum value for this indicator is 5000. The second indicator is the average run length (ARL), which is defined as the average time to signal of all the profiles (if the curve never signals, the total length is treated as its time to signal). Since the curve is growing with time, if the curve signals, the exact time the alarm is triggered is also important. If the alarm is triggered before the change-point of the process, the alarm is treated as a false alarm. The simulation results are shown in Table I.

TABLE I
PERFORMANCE COMPARISON

Shift pattern	Shift size	AEWMA (# of alarms / ARL)	Profile-Residual (# of alarms / ARL)
Variation shift	$\sigma_{t>T} = 2e^{-0.01t}$	109/77.8	**734/77.5**
	$\sigma_{t>T} = 4e^{-0.01t}$	2294/66.3	4946/58.9
	$\sigma_{t>T} = 6e^{-0.01t}$	4652/51.8	**5000/52.6**
Mean shift	δ=5	**1898/65.6**	436/77.9
	δ=8	**4868/45.4**	1796/73.1
	δ=11	**5000/44.5**	4101/59.7
Mean drift	$b'=b$+0.7	**100/77.8**	49/78.9
	$b'=b$+0.9	**100/77.8**	63/78.9
	$b'=b$+1.1	**100/77.8**	79/78.8

Table I shows the number of alarms and ARL of the charts under different shift scenarios. For each shift scenario, the chart that gives the best performance (with the largest number of out-of-control alarms and the shortest ARL) is highlighted. It is learned from Table I that the AEWMA chart is insensitive to mean drift, but is sensitive to mean shifts. This happens since the slow drift may lead the adaptive EWMA procedure to a wrong directly, while a sudden shift is more prominent and cannot be eliminated by the EWMA procedure; while the Profile-Residual chart is good at detecting variation and mean shift, but is rather slow in detecting mean drift in the process. These findings provide useful guidelines for the practitioners to select charts that best suits their individual applications.

V. PERFORMANCE STUDY IN A REAL EXAMPLE

In this section, we apply the proposed adaptive chart to the real process and study its performance. Based on our analysis of the effects of the parameters, after trial and error, the following combinations give overall good charting performance: λ =0.4, γ =0.002, k =0.05. These settings are used to test all real samples. Since the initial stage of the process increases quickly, the first 50 steps is used to warm-up the control chart. So, no alarm is released in the first 50 steps in all charts.

Fig.5 shows the control chart for the power curve of a normal ingot. It can be seen that with the process goes on, the power curve (the curve in the middle) changes dynamically, while the control limits also vary with the power curve, which shows the adaptability of the control chart to the system dynamics. In addition, the interval between the two control limits is shrinking when time moves; this shows that the variability of the process is decreasing. This is consistent with the real situation since the curve becomes smoother in the later stage.

Fig.6 shows the estimation error, $e_t = x_t - \mu_t$, of the process. It is seen that the error gives rather random pattern; the main trend has been successfully captured by the adaptive EWMA algorithm.

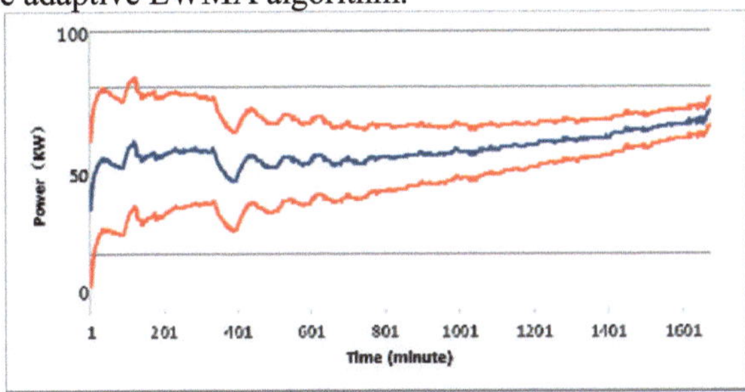

Fig.5. Control chart for the power curve of a normal ingot

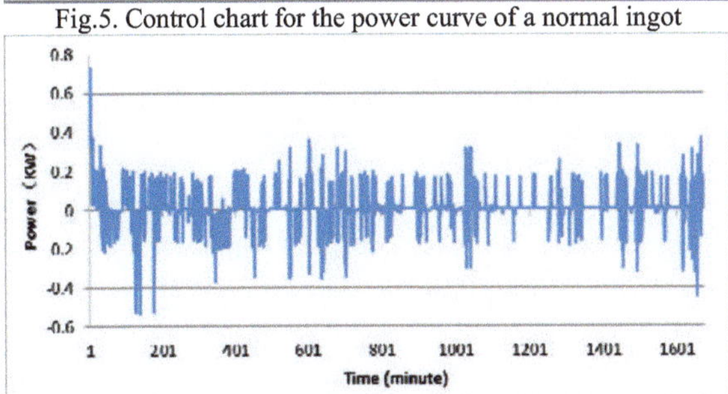

Fig.6. Prediction error of a normal ingot

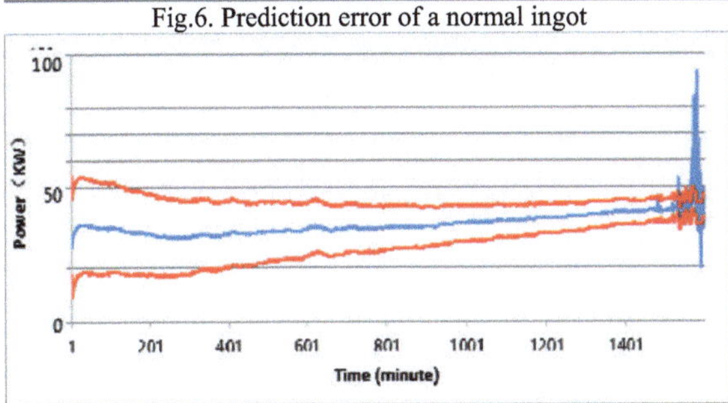

Fig.7. Control chart for the power curve of an abnormal ingot

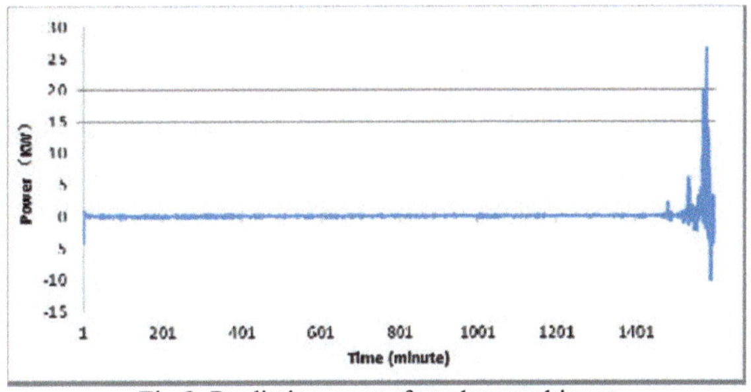

Fig.8. Prediction error of an abnormal ingot

Fig. 7 shows the control chart for an abnormal ingot. The ingot has been known from engineering knowledge to fail in practice. Fig.7 shows similar dynamic control limits as those in Fig.5. However, in the final stage, the power curve shows strong oscillation and the shift magnitude goes beyond the control limits. Further investigation shows that an engineering failure has happened, and consequently, the diameter of the ingot becomes abnormal. Therefore, the new chart can successfully capture the failure. Compared to the failure identified by the operator in Fig.4, the out-of-control signal is released nearly 90min earlier by the proposed chart. Fig.8 shows the prediction error of the variable, which shows random noise in the early stage and strong out-of-control signals in the later stage. This further confirms the ability of the new chart in tracking process dynamics and detecting abnormal failure signals.

VI. CONCLUSIONS

Traditional control charts usually assume that the monitoring statistics follows an unchanged distribution when the process is in-control. However, in some engineering processes, strong dynamics are seen and unstable process means are observed. In such cases, most conventional control charts cannot be applied directly.

In this work, we propose a new adaptive chart for such dynamic processes. The center line and the variability of the process are updated using an adaptive EWMA equation repeatedly. Consequently, the control limits of the chart are adjusted based on the updated process parameters. Implementation of the proposed chart to real data from an ingot growing process shows that the chart can successfully detect failure in the process.

The dynamics of the real process usually has complicated engineering background. If such domain knowledge is taken into consideration and incorporated in chart design, it is expected to further improve the charting performance. These are topics that deserve further research efforts.

REFERENCES

[1] W. H. Woodall, "Controversies and contradictions in statistical process control," *Journal of Quality Technology*, vol. 32, pp. 341-350, 2000.

[2] K. Wang and F. Tsung, "An adaptive T^2 chart for Monitoring Dynamic Systems," *Journal of Quality Technology,* vol. 40, pp. 109-123, 2008.
[3] K. Wang and F. Tsung, "Monitoring feedback-controlled processes using adaptive T2 schemes," *International Journal of Production Research,* vol. 45, pp. 5601-5619, 2007.
[4] F. Tsung and K. L. Tsui, "A mean-shift pattern study on integration of SPC and APC for process monitoring," *IIE Transactions,* vol. 35, pp. 231-242, 2003.
[5] G. Spitzlsperger, C. Schmidt, G. Ernst, H. Strasser, and M. Speil, "Fault detection for a via etch process using adaptive multivariate methods," *IEEE Transactions on Semiconductor Manufacturing,* vol. 18, pp. 528-533, Nov 2005.
[6] R. Jin and K. Liu, "Multimode Variation Modeling and Process Monitoring for Serial-Parallel Multistage Manufacturing Processes," *IIE Transactions,* vol. in-press, 2012.
[7] W. A. Shewhart, "Quality control charts," *Bell System Technical Journal,* vol. 5, pp. 593-603, 1926.
[8] S. V. Crowder, "A Simple Method for Studying Run–Length Distributions of Exponentially Weighted Moving Average Charts," *Technometrics,* vol. 29, pp. 401-407, 1987.
[9] J. M. Lucas and M. S. Saccucci, "Exponentially weighted moving average control schemes: properties and enhancements," *Technometrics,* vol. 32, pp. 1-12, 1990.
[10] E. Yashchin, "Some aspects of the theory of statistical control schemes," *IBM Journal of Research and Development,* vol. 31, pp. 199-205, 1987.
[11] W. H. Woodall and Maragah, "Discussion -- Exponentially Weighted Moving Average Control Schemes: Properties and Enhancements," *TECHNOMETRICS* vol. volume 32 pp. 17-18, 1990.
[12] S. L. Albin, L. Kang, and G. Shea, "An X and EWMA chart for individual observations," *Journal of Quality Technology,* vol. 29, pp. 41-48, 1997.
[13] G. Capizzi and G. Masarotto, "An adaptive exponentially weighted moving average control chart," *Technometrics,* vol. 45, pp. 199-207, 2003.
[14] W. H. Woodall, "Current research on profile monitoring," *Produção,* vol. 17, pp. 420-425, 2007.
[15] W. H. Woodall, D. J. Spitzner, D. C. Montgomery, and S. Gupta, "Using control charts to monitor process and product quality profiles," *Journal of Quality Technology,* vol. 36, pp. 309-320, 2004.
[16] T. C. Chang and F. F. Gan, "Monitoring linearity of measurement gauges," *Journal of Statistical Computation and Simulation,* vol. 76, pp. 889-911, 2006.
[17] J. Zhu and D. K. J. Lin, "Monitoring the slopes of linear profiles," *Quality Engineering,* vol. 22, pp. 1-12, 2009.
[18] Noorossana, "Monitoring Process Performance Using Linear Profiles " 2004a.
[19] J. D. Healy, "A note on multivariate CUSUM procedures," *Technometrics,* vol. 29, pp. 409-412, 1987.
[20] C. Zou, C. Zhou, Z. Wang, and F. Tsung, "A self-starting control chart for linear profiles," *Journal of Quality Technology,* vol. 39, pp. 364-375, 2007.
[21] M. A. Mahmoud and W. H. Woodall, "Phase I analysis of linear profiles with calibration applications," *Technometrics,* vol. 46, pp. 380-391, 2004.
[22] C. Zou, Y. Zhang, and Z. Wang, "A control chart based on a change-point model for monitoring linear profiles," *IIE Transactions,* vol. 38, pp. 1093-1103, 2006.
[23] S. Gupta, D. Montgomery, and W. Woodall, "Performance evaluation of two methods for online monitoring of linear calibration profiles," *International Journal of Production Research,* vol. 44, pp. 1927-1942, 2006.
[24] L. Huwang, Y. H. T. Wang, A. B. Yeh, and Z. S. J. Chen, "On the exponentially weighted moving variance," *Naval Research Logistics (NRL),* vol. 56, pp. 659-668, 2009.

Using Modified Value Multiplier to Measure the Effectiveness of Carbon Emission Relative Activities

Fang-jun Wang, Lin Wu, Li-xin Zhu*
School of Management, Xi'an Jiaotong University, Xi'an, China
(zhulixin.@mail.xjtu.edu.cn)

Abstract - **Based on the analysis of the relationship between activity cost and value creation, this paper integrates carbon emission factors into Activity Value Multiplier (AVM) model. By calculating the multiplier AVM, the modified AVM model can measure the efficiency and effectiveness of carbon emission activities, and help companies to recognize the structure of consumer value and identify the directions of carbon emission reduction. A discussion about four significant carbon emission relative activities in company is put forward to illustrating how modified AVM works and helps firms make decisions. Finally the conclusion is drawn.**

Keywords - **Activity Value Multiplier, activity cost management, carbon emission activity**

I. INTRODUCTION

The issue of the climate change caused by greenhouse gas emissions has increasingly aroused worldwide attention and developing low carbon economy has been recognized as an inevitable choice for every country in the world. Low-carbonization development of economy proposed higher requirements to companies. Chinese manufacturers face enormous pressures during the carbon emissions reduction process because of the laggard technology and the unreasonable energy structure. Carbon emission cost has become an important factor which must be considered seriously in production management and strategic decision.

In recent years, Activity-Based Costing (ABC) has been widely applied to manufacturing industry since ABC proposed activity and activity cost driver which makes overhead costs can be traced reasonably and accurately to products or services. Based on the calculable cost information, Activity-Based Costing Management (ABM) provides more tools such as activity value-added analysis, value multiplier analysis and activity efficiency analysis to measure and estimate activities and cost. Nowadays, ABM has become one of the most important management approaches which helps to make decisions.

Conventional approach assesses a carbon emission relative activity by measuring the relation between input and output. For example, new energy is input and carbon emission is output in a carbon emission reduction activity, so the efficiency and effectiveness of this activity are estimated by contrasting the new energy cost and final emission. Actually, the energy cost generally increases and final emission decreases if the new energy takes place of the traditional energy, however, we cannot confirm if the reduction of carbon emission cost exceeds the increase of energy cost.

Based on the principle that products consume activities and activities consume resources and focus on activities rather than input-output, the paper propose that it is the conduction of carbon emission activities that result in the consumption of resource and emission of greenhouse gas. So the efficiency of carbon emission activities should be measured based on activity analysis. The paper integrates carbon emission factors into Activity Value Multiplier (AVM) model to analyze the efficiency and effectiveness of carbon emission relative activities.

II. LITERATURE REVIEW

Input-output analysis is not only a powerful tool for economic study, but also a useful instrument for environmental analysis [1]. Since Leontief proposed environmental Input-output model, most environment management relative researches focus their attention on either input or output [2-4].

While as with the widely application of Activity-Based Costing in manufacturing companies, environmental cost relative research introduced ABC into environmental management. These studies focus on activities and provide a new perspective for environmental relative analysis [5-7].

According to whether activities could create value for customers or organizations, Activity-Based Management (*ABM*) divides them into value-added or non-value-added categories [8], which makes 'activity' become the tie of binding firms' resource costs and value creation. By identifying the environmental costs of all activities and analyzing all activities, a company can attempt to eliminate, or at least minimize the environmental costs of non-value-added activities [9]. But it is only limited to qualitative analysis and it usually highlights eliminating non-value-added activities and making value-added activities more efficiently, while ignores measuring the ability and effectiveness of activities to create customers' value.

Based on Value Creation Model (*VCM*), McNair [10] expanded the relationship between cost and value. The VCM model defines the firms' cost structure in terms of customer value-added activities, organization-value-added activities, as well as waste. McNair aligns the cost structure with product attributes, replaces value with market price, and makes use of value multiplier (*VMr*) to reflect the relationship between the value which customers place on product attributes and activity cost that the firm occurs in order to provide these attributes. *VMr* is comparable and can assess the effectiveness of the product attribute when combined with customer satisfaction. However, *VMr* can neither measure the contribution of a single activity to value, nor can it make comparison between the contributions of different

E. Qi et al. (eds.), *Proceedings of the 5th International Asia Conference on Industrial Engineering and Management Innovation (IEMI2014)*, Proceedings of the International Asia Conference on Industrial Engineering and Management Innovation 1, DOI 10.2991/978-94-6239-100-0_44, © Atlantis Press and the authors 2015

activities. Furthermore, *VMr* does not take the efficiency of activities into account. Wang fangjun [11] proposed Activity Value Multiplier (*AVM*) to identify and measure the relationship between activity cost and market value, and then evaluate the first-class measurement *AVM* and second-class measurement activity efficiency and activity effectiveness to provide essential information for cost control.

Based on the previous research, this paper integrates carbon emission factors into Activity Value Multiplier (*AVM*) to analyze the efficiency and effectiveness of carbon emission relative activities.

A. Development of Carbon Emission AVM Computation Model

Modified carbon emission AVM involves three factors, which are the value created by a carbon emission activity, the decreased value caused by carbon emission and the cost consumed by the activity or activity group. It identifies target activities that should be improved through analyzing the relationship between activity cost and value creation. The detailed computation procedures are as follows:

(1) Define product attributes through market analysis and customer survey, and then evaluate the value of each product attribute;

(2) Analyze contribution of each activity to the value of product attributes and identify the decreased value of each activity by measuring the carbon emission created by each activity.

(3) Calculate the resource costs consumed by the activity, and then calculate AVM of each activity.

B. Defining Product Attributes and Evaluating the Value and Decreased Value

Income is defined as a valid approximation of the value that firms create to satisfy consumers, so attribute value are created by multiplying income by the weight of each product attribute, and then the income derived from the specific product is allocated to each product attribute.

Suppose the income a firm obtains at a certain period is S, according to what described above: S can be regarded as a valid approximation of value U that the firm creates to satisfy its consumers; n kinds of attributes are defined according to market analysis and customer survey, and the product attributes set is $\{A_1, A_2, \ldots, A_n\}$. The weights of attribute $A_i (i=1,2,\ldots,n)$ is μ_i, obviously $\mu_1+\mu_2+\ldots+\mu_n=1$, thus value U_i of attribute A_i can be identified as:

$$U_i = S \times \mu_i \tag{1}$$

C. Identifying Activity' Contribution to Product Attribute and Evaluating Decreased Value Created by Carbon Emission.

In general, it is easy for the engineers and technicians to identify the linkage between product attributes and activity groups. Interviews with appropriate managers and operators about 'activities contribution to product attributes' are required. We can get the information about direct material costs related to each product attribute from financial department. We define the value reduced by carbon emission as "carbon emission value" which can be equivalent to the value or price of carbon emission credits bought from carbon emission trading market.

Suppose direct material costs of attribute A_i (i =1, 2, …, n) is DM_i and m kinds of activities happened in the firm's operation. Through investigating and analyzing, the weight of activity $VA_j (j = 1,2,\ldots,m)$contributing to attribute A_i is v_{ij}, so we can get $v_{i1}+v_{i2}+\ldots+v_{im}=1$. Define V_{j1} as the value created by activity VA_j and V_{j1} is computed as follow:

$$V_{j1} = \sum_{i=1}^{n}[(U_i - DM_i)\times v_{ij}] = \sum_{i=1}^{n}[(S\times\mu_i - DM_i)\times v_{ij}] \tag{2}$$

Define V_{j2} as the value reduced by activity VA_j because of carbon emission. Suppose the carbon emission value reduced by per unit activity is v_{j2} and the activity driver volume of activity VA_j is y_j. V_{j2} is computed as follow:

$$\begin{aligned} V_{j2} &= \text{Reduced Value of Carbon Emission} \\ &= \text{Carbon Emission Credits Price} \\ &= v_{j2} \times y_j \end{aligned} \tag{3}$$

The final value V_j created by activity VA_j is identified as follow:

$$\begin{aligned} V_j &= V_{j1} - V_{j2} \\ &= \sum_{i=1}^{n}[(S\times\mu_i - DM_i)\times v_{ij}] - v_{j2}\times y_j \end{aligned} \tag{4}$$

Obviously, supposing V_2 is the total carbon emission value, total value V created by all activities can be calculated as follow:

$$\begin{aligned} V &= \sum_{j=1}^{m} V_j = \sum_{j=1}^{m}\sum_{i=1}^{n}[(U_i - DM_i)\times v_{ij}] - \sum_{j}^{m} v_{j2}\times y_j \\ &= S - \sum_{i=1}^{n} DM_i - V_2 \end{aligned} \tag{5}$$

D. Calculating Resource Cost and AVM

Supposed rj is the activity driver rate of activity $VA_j (j=1, 2\ldots, m)$ and r_j is composed of r_{j1} and r_{j2}. r_{j1} represents the non-carbon emission or non-energy consumption part of activity driver rate and r_{j2} is the carbon emission or energy consumption part of activity driver rate. The activity driver volume is y_j. Thus the resource cost TC_j that activity VA_j consumes and total cost TC of all activities can be calculated as follows:

$$TC_j = (r_{j1} + r_{j2}) \times y_j \tag{6}$$

$$TC = \sum_{j=1}^{m} TC_j = \sum_{j=1}^{m}(r_{j1} + r_{j2}) \times y_j \tag{7}$$

AVM can clearly reflect every activity's ability to create customer value, income and profit. So AVM_j of activity VA_j can be calculated with formulae (4, 6), and the average multiplier $\overline{AVM}$ of all activities can be calculated with formula (5, 7). They are as follows:

$$AVM_j = \frac{V_j}{TC_j} = \frac{V_{j1} - V_{j2}}{(r_{j1} + r_{j2}) \times y_j}$$

$$= \frac{\sum_{i=1}^{n}[(S \times \mu_i - DM_i) \times v_{ij}] - V_{j2}}{(r_{j1} + r_{j2}) \times y_j} \quad (8)$$

$$\overline{AVM} = \frac{V}{TC} = \frac{S - \sum_{i=1}^{n} DM_i - \sum_{j}^{m} V_{j2}}{\sum_{j=1}^{m} (r_{j1} + r_{j2}) \times y_j} \quad (9)$$

Assuming $\frac{V_{j1} - V_{j2}}{y_j} = \alpha_j$, formula (8) can be rewritten as follow:

$$AVM_j = \frac{V_{j1} - V_{j2}}{y_j} \times \frac{1}{r_{j1} + r_{j2}} = \frac{\alpha_j}{r_{j1} + r_{j2}} \quad (10)$$

The formula (10) shows AVM of every activity could be divided into two parts. The economic meaning of α_j is customer value created by per unit activity driver after deducting carbon emission value, reflecting the effectiveness of this activity during the course of customer value creation. The lower part $r_{j1}+ r_{j2}$ is the activity driver rate which respectively means the non-carbon emission cost and the carbon emission cost consumed by per unit activity driver, reflecting the processing efficiency of this carbon emission relative activity.

III. ANALYSIS OF CARBON EMISSION AVM COMPUTATION MODEL

A. Analysis of the First-class Index AVM

The modified AVM takes environment costs caused by carbon emission into consideration, so it reflects the true value that the activity could finally provide to customers .The larger the multiplier is, the higher the value created by the activity with the same resource cost is. By comparing an activity's AVM multiplier to the average $\overline{AVM}$ of the company or the industry, the company can estimate that whether the activity is the key activity and have competitive advantages over competitors.

B. Analysis of the Second-class Index α, Namely Activity Effectiveness

If $\alpha=0$, V_1- V_2=0. It indicates that the value created by an activity is offset by the value of carbon emission during the activity, which means that the activity is an essential activity or non-value-added activity. If $\alpha>0$, V_1-V_1>0. It means the activity is still value-added after deduct carbon emission value.

Companies should not only pay their attention to increase the V1, but also take measures to decrease the value of carbon emission to improve the index α.

C. Analysis of the Second-class Index r_1+r_2

Here index r_1+r_2 represents activity driver rate. The higher r_1+r_2 is, the lower efficiency of the activity is. The purpose of analyzing r_1+r_2 is to identify the inefficient value-added activity and try to make improvement by reducing resource costs per unit activity consumed.

IV. DISCUSSION OF CARBON EMISSION RELATIVE ACTIVITIES AND PRACTICES IN COMPANIES

The existence of carbon emission costs sharply cut down companies' profit and the low-carbon economy has been increasingly recognized and emphasized in companies. Either the society or customer think more and more about the environmental protection attributes of products. In practice, companies have taken some measures to reduce carbon emission and carbon emission costs so as to increase their profit and satisfy their customers.

A. Replacing Traditional Energy with New Energy

Most companies believe that replacing traditional energy with new energy reduces the carbon emission and therefore reduces the carbon emission costs and increases the profit and product value. However, different conclusions can be achieved by using the carbon emission AVM model.

(1) V_1 increases because the usage of new energy adds environmental attributes value to products. While, V_2 does not always also decline taken other activities which support the usage of new energy into consideration. It is possible that the increase of carbon emission value balances the reduction created by activities using new energy. Furthermore, it is also uncertain whether the decrease of V2 exceeds the increase of V_1, which indicates that the change of index α is uncertain.

(2) Generally, using new energy will increase the energy consumption cost and thus increase the index r_2.

(3) According to formula (10), qualitative analysis cannot give an exact judgment about whether the activities' efficiency and effective are improved after replacing traditional energy with new energy. By using carbon emission AVM model, companies could take multiple complex uncertain factors into consideration and accurately calculate the integrated multiplier AVM based on actual data which provide an exact and veritable activity estimate.

B. Carbon Emission Reduction Relative Activity

Nowadays, most companies have implemented a series of carbon emission reduction relative activities including waste treatment, green product design, equipment renewal and process reengineering, etc.

(1) Carbon emission reduction activities increase r_1 which include equipment purchase cost, labor cost, R&D costs, etc. It incurs energy costs r_2 since the implement of reduction activities also consumes energy.

(2) V_1 increases and the increment equals the reduction of other activities' carbon emission value since the purpose of carbon emission reduction activities is cuting down the carbon emission created by other activities. It is easy to overlook that carbon emission reduction activities also create carbon dioxide, thus the carbon emission value V_2 is not zero. We cannot simply identify the change of V_1- V_2 which depends on the quantity of reduction and creation of carbon emission.

(3) According to formula (10), r_1+ r_2 increases and V_1- V_2 is uncertain. *AVM* model is needed in accurate computation to identify the efficiency and effective of carbon emission reduction activities.

C. Purchase Decision of Carbon Emission Credits

Generally, companies possess a specified volume of carbon emission credits allocated by the government. Companies need to purchase carbon emission credits in market once the carbon emission exceeds the limit.

(1) Carbon emission value V_2 is defined as the total market value of carbon emission credits that belong to a company. It indicates that V_2 is fixed for total activities and the product value decreases whether the companies release carbon dioxide or not.

(2) Carbon emission credits remained when carbon emission under the limit allocated by government. We can suppose that the company may take actions to reduce carbon emission which certainly increase the cost of relative activities. As a resource of company, the reminded carbon emission credits are wasted, which indirectly increases the index r_1+ r_2 and decreases V_1 and finally slashes the multiplier *AVM*. Based on the analysis above, companies should control the carbon emission at an appropriate level rather than just attempt to reduce the carbon emission. However, considering the environmental problem and external economy, better choices for companies are selling their remained carbon emission credits in the market.

(3) Companies need to purchase carbon emission from the market if their emission exceeds the limit. It is important to compare the market value of carbon emission with the value created by carbon emission activities, which is comparing the increase of V_2 and the increase of V_1. Only when V_1> V_2, it is value-added to purchase carbon emission credits.

D. Outsourcing Carbon Emission Relative Activity

Sometimes, companies outsource their high emission activities and emission reduction activities by assessing and comparing the outsourcing costs and original costs.

Carbon emission value V_2=0 and carbon emission activity driver rate r2=0 when an activity is outsourced. Non-energy consumption activity driver rate r_1 equals the outsourcing costs divided by activity driver volume *y*, which that means once a company outsources a carbon emission activity, the activity' $AVM=V_1/\ r_1$ for this company. Compared to the original *AVM*， V_1- V_2 increases and the quantity change of r_1+ r_2 are uncertain because r_2=0 but r_1 may increase.

In conclusion, decision made about carbon emission relative activities in companies is a Multi-factor complex problem. Conventional simple qualitative analysis can be replaced by carbon emission *AVM* model which provides a new perspective and a computable and comparable method for company carbon emission management.

V. CONCLUSION

In general, firms devote themselves to eliminating non-value-added activities and reducing non-value-activity costs under Activity-Based Management approach, but they always neglect the problems: relationship between customer value and cost, and influence of environmental value. McNair (2001b) developed Value Creation Model (*VCM*)to dispel the so-called 'cost-value gap' and Fangjun Wang extended *VCM* and proposed Activity Value Multiplier (*AVM*) measure to interpret the relationship between costs consumed by an activity and customer value created by the activity.

Based on the previous research, the paper integrates carbon emission factors into *AVM* model to efficient and effective identification and measurement of carbon emission relative activities. Taking the value of carbon emission credits as the carbon emission value, the carbon emission *AVM* model divides customer value created by firms into two component： General value provided by products (V_1) and reduced value caused by carbon emission (V_2). The activity driver rate is also composed of the non-carbon emission activity driver rate r_{j1} and carbon emission activity driver rate r_{j2}.

The modified *AVM* model can help firms to recognize the structure of consumer value and cost considering the carbon emission, and provide new perspective for how to create customer value from the aspects of increasing product value and reducing carbon emission value. By computing the multiplier *AVM*, the model can assess the effectiveness of carbon emission relative activities and identify the directions of cost control and process improvement. With *AVM*, the firms could not only systematically analyze the possible critical success factors and potential abilities to reduce the resource consumption internally; they can also pay deep attention to external environmental and social value to allocate resource more rationally.

Finally based on the analytical framework, the paper analyzed four common and significant carbon emission relative activities in company practices, which illustrates how *AVM* works as a new performance evaluation tool in carbon emission relative activities.

ACKNOWLEDGMENT

This work was supported in part by a grant from the National Social Science Foundation of China (Research on Accounting Measurement, Reporting, Auditing and Assurance of Carbon Emission Information, No 11CGL062).

REFERENCES

[1] Leontief. W, Environmental repercussions and the economic structure: An Input-Output Approach, Review of Economics and Statistics, 1970, 52(3):262-271.

[2] Duchin. F, Lange. G, Input-Output modeling: development and the environment in indonesia, final report to the canadian international development agency, 1992.

[3] Pan Xiaoming, Kraines. Steven, Environmental input-output models for life-cycle analysis, environmental and resource economics 2001, 9: 61-72.

[4] Goldar. A, Bhanot. J, Shimpoc. K, Prioritizing towards a green export portfolio for india: an environmental input-output approach. energy policy, 2011, 39 (11), 7036-7048.

[5] Brooks. Philip. L, Davidson. Laura. J, Palamides. Jodi. H, Environmental compliance: you better know your ABCs occupational hazards 55.2, 1993 (2): 41.

[6] Emblemsvag. J, Bras. B, Activity-Based cost and environmental management, Kluwer Academic Publishers, Boston, MA, 2000.

[7] Sarkis. Joseph, Meade. Laura, Presley. Adrien, An activity based management methodology for evaluating business processes for environmental sustainability, Business Process Management Journal 12.6, 2006: 751.

[8] Arthur. Andersen. LLP, Implementing activity based management: avoiding the pitfalls, Institute of Management Accountants, Montvale, NJ, 1998.

[9] Kreuze. Jerry. G, Newell. Gale. E, ABC and life-cycle costing for environmental expenditures, Management Accounting, 1994, 2 : 38.

[10] McNair. C. J, Lidija Polutnik, Riccardo Silvi, Cost management and value creation: the missing link. European Accounting Review, 2001a, 10, (1) 33-50.

[11] Wang Fangjun, Chang hua, Cao Jianan, Activity value multiplier: measuring the efficiency and effectiveness of activity, Journal of Management Sciences, 2008, 21(3):24-30.

Improved Rolled Throughput Yield: A Novel Method for Performance Measurement

Jie Jin*, Yi-zhong Ma, Lin-han Ouyang, Xiao-guang Gu
School of Economics and Management, Nanjing University of Science and Technology, Nanjing, China
(ginger_jinjie@163.com)

[1]*Abstract* - **The quality level measurement of a given process is essential to six sigma process improvement. Indicators such as rolled throughput yield and six sigma level have been applied to estimate the efficiency of total process in a certain organization. As an effective indicator, however, rolled throughput yield has not thoroughly explored in scarce studies undertaken. In this paper, we develop an Improved Rolled Throughput Yield (IRTY) that takes the effect of appraisal cost and external failure cost of each sub-process on process weights into account. Moreover, the rework and scrap cost are also incorporated into the proposed approach. The uniqueness of the proposed approach is that the determination of process weights is based on the cost of poor quality. A numerical example is illustrated to present the procedure and the effectiveness of the proposed approach.**

Keywords - **Process weights, rolled throughput yield, six sigma, total process**

I INTRODUCTION

Nowdays, the global market is highly competitive and in order to survive, organizations need to produce products and services of high quality to achieve customer satisfaction and loyalty to stimulate top-line business growth [1-2]. The six sigma methodology focus on reducing variation, measuring defects and improving the quality of products and processes [3]. Originated in Motorola in 1987, started with manufacturing, six sigma has been adopted by many other global firms such as GE, Allied Signal, Sony, Microsoft and Samsung, and expanded its realm to variety of fields such as healthcare, banking, servicing [4-10].

Academic research such as Schroeder et al. have tried to determine which elements in six sigma make it effective [11]. The improvement procedure generally known as DMAIC (Define, Measure, Analysis, Improve, Control) is regarded as a novel and effective contribution to quality management [12]. In the all five phases, the Measure phase is crucial for discovering opportunities for improvement, which can be used to improve and enhance the quality level of corporation. Rolled Throughput Yield (RTY), as a new six sigma metric that measure the overall performance of process, plays a significant role in the performance measurement of an organization.

On the basis of summing up the computation of qualified rate in the past manufacturing industry, Graves (1998) proposed Total Process Yield that was the earlier state of RTY. Hereafter, he further explained how RTY was related to rework, scrap, warranty and customer satisfaction and described the use of RTY for monitoring [13-14]. Some scholars have studied the combination of RTY with other methods, e.g. Dasgupta (2003) integrated RTY into the proposed supply chain performance measurement by using six sigma metrics [15]. Chien Lin (2008) also applied RTY to measure supply chain process performance in the supply chain performance measurement framework [16]. Abdelhamid (2003) and Hwang (2006) demonstrated the application of RTY to the Last Planner System in Lean Construction and to projects in integrated manufacturing execution systems respectively [17-18]. In the above studies, however, the importance of each sub-process was treated equally, the difference of each sub-process in overall performance measurement were ignored. Since each sub-process is unique in nature in one way or another, and interdependent, the quality level of each sub-process at a time may not be the same [19]. Therefore, it is wise and necessary to take the process weights into consideration in overall performance measurement of an organization. Based on this consideration, Ravichandran (2006) assigned weights for all critical processes based on their importance and then determined the 'weight-based sigma level' of an organization at a given point of time by using process weights and defects per million together. Following his work, there remained a number of queries on how to determine the process weights. In 2007,he proposed an approach in which both internal and external performances of the products and processes in terms of costs involved were used to determine cost-based process weights [20] Furthermore, Abbas (2012) proposed a new six sigma-based performance measure called Enhanced rolled Throughput Yield (ERTY). Unlike all previous methods, ERTY paid particular attention to such factors as the difference between scrap and rework cycle, the cost of scrap and rework and the sequence of stages [21]. Unfortunately, the effect of appraisal cost and external failure cost of each sub-process on process weights is ignored. In order to gain a more plausible result, however, it is more appropriate to consider these factors in practice.

This paper presents a model called Improved Rolled Throughput Yield (IRTY) that not only pays attentions to the cost of rework and scrap but also to the appraisal cost and external failure cost of each sub-process to calculate the overall performance of an organization. From the perspective of cost of poor quality, an effective metric for six sigma organizations to assess the actual overall performance and find opportunities for six sigma improvement was designed based on the RTY index.

[1] This work is supported by the State Key Program of National Natural Science of China (No. 70931002), the Youth Program of National Natural Science of China (No. 71301075) and colleges and universities in Jiangsu Province plans to graduate research and innovation projects (No.CXZZ13-210).

The organization of this paper is given as follows. In Section 2, basic concepts and limitations of RTY are reviewed. Sections 3 gives an improved RTY method that considers the cost of poor quality. In Section 4, a numerical example is presented to illustrate the computation of IRTY. Finally, a few concluding remarks are presented in Section 5.

II ROLLED THROUGHPUT YIELD

In general terms, RTY estimates the probability that a unit, whether a service or physical product, pass through a process defect-free [14, 22]. It is fairly obvious that RTY is closely related to rework and scrap: an increase in rework or scrap units implies a reduction in the probability of a unit passing the process without failures. Similarly, an increase or decrease in the probability also has significant effect on customer satisfaction and warranty while customer satisfaction is the ultimate goal of every organization in global market. RTY which evaluates the performance of processes from the view of system rather than that of a single process is an important metric. Therefore, RTY can truly describe the overall performance of processes.

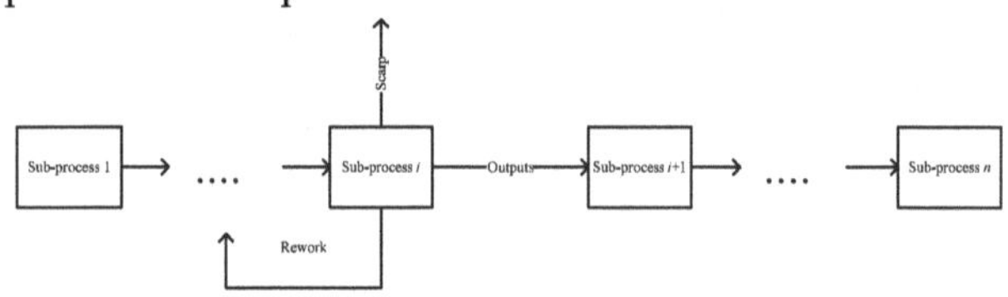

Fig.1. The typical diagram of a serial process

Suppose that a process consists of n sub-processes. Fig.1 shows the typical diagram of a serial process. For each sub-process, First Time Yield (FTY) refers to the probability of a unit completing the sub-process without reworking or scraping. Assuming that each sub-process is independent of each other, then the RTY of process could be easily obtained by multiplying the FTY of each sub-process:

$$\mathrm{RTY}=\prod_{i=1}^{n} P_i \tag{1}$$

where P_i denotes the FTY of sub-process *i*.

Indeed, the calculation result of formula (1) can reflect the capability of a given process to a certain extent, but it amplifies the effect of a single sub-process on overall performance. Suppose that there is a serial process consists of n sub-process. The FTY of sub-process i is only 50% while that of others is 100%, but all the unqualified output of sub-process i can be repaired through reworking. According to formula (1), RTY=0.5. However, the final yield is 100% which means no unqualified products are delivered to customer. It is the ignorance of process weights that lead to the deficiency in describing the overall performance. Although Abbas (2012) studied the process weights by considering the rework and scrap costs, he ignored the effect of appraisal cost and external failure cost. In order to incorporate the two costs into the construction of RTY, Section 3 gives an improved RTY method.

III IMPROVED ROLLED THROUGHPUT YIELD

Suppose that there is a serial process consists of n sub-process, the final yield of product depends on the performance of each sub-process while each sub-process has its own characteristic at a time. Namely, the importance of each sub-process is different. If the outputs of any sub-process do not meet the defined criteria, they will be clarified as unqualified outputs. If they are checked out to be unqualified, they should be either corrected or scrap. Otherwise, they should be considered in the next process and inflict losses. These losses caused by unqualified outputs are the 'invisible factory' which is exactly what RTY wants to reveal. With these in mind, this paper presents a method in the view of cost of poor quality.

Cost of poor quality refers to the losses caused by poor quality or the additional costs we have to pay in that we did not do it right at the first time. It includes prevention cost, appraisal cost and failure cost. Table I lists the composition of cost of poor quality.

TABLE I
THE COMPOSITION OF THE COST OF POOR QUALITY

Cost of poor quality	Cost of conformity	Prevention cost (No value added part)
		Appraisal cost (No value added part in pre-testing)
	Cost of nonconformity	Appraisal cost (Analyze the cause of failure)
		Failure cost (Internal and external failure)

Prevention cost refers to the no value added part in the upfront costs that we have to pay for prevention, including the costs of customer surveys, process control, product design, etc. Most of the prevention costs are usually paid for all the processes before the start of production rather than one single process except for the costs of process control which are paid during the whole process of production. Moreover, it seems difficult to clearly define the boundaries between non value-added part and value-added part in the prevention costs. Hence, prevention costs are not considered in determining the process weights in this paper.

Appraisal cost refers to the cost that is paid for testing whether the process output meet the specifications and finding the failure causes. In practice, there are a lot of inspection methods available used for testing and operators make decisions depend on many factors such as appraisal cost that have significant effect on the selection of inspection method. Since the losses caused by unqualified outputs that are produced in previous processes are affected by the selection of inspection method, the sub-process that has higher appraisal cost than others should have a higher priority during the process of optimization, namely, it is more important than others.

Failure cost refers to the losses caused by unqualified outputs, including internal and external failure cost. Assuming that the output of sub-process i do not meet the specification, if it is checked out before entering sub-process i+1, then it turns to rework or scrap that constitutes the internal failure cost. Otherwise, it goes to sub-process i+1 and causes a loss that constitutes the external failure cost. In a serial process, the probability of that the unqualified output of sub-process i goes through sub-process i and sub-process i+1 and has not been detected is extremely tiny so that we could exclude it beyond our consideration. Therefore, we suppose that the unqualified output produced by sub-process i must be unqualified after the production of sub-process i+1 if it escaped the test of sub-process i, and it will be detected in the test of sub-process i+1. Then we could measure the external failure cost by using the sum of the appraisal cost and process cost of sub-process i+1.

The notation is summarized below.

S_i	Unit scrap costs of sub-process i ;
R_i	Unit rework costs of sub-process i ;
C_i	Unit process costs of sub-process i ;
α_i	Proportion of scraps in unqualified outputs of sub-process i ;
β_i	Proportion of rework outputs in unqualified outputs of sub-process i ;
γ_i	Proportion of undetected outputs in unqualified outputs of sub-process i ;
F_i	Failure costs of sub-process i ;
T_i	Appraisal costs of sub-process i ;
w_i	Weight of sub-process i ;
P_i	FTY of sub-process i ;
n	Number of sub-process ;
X	External loss of process .

In this model, the importance of each sub-process is determined according to failure costs and appraisal costs. Failure costs of sub-process i can be written in the following form:

$$F_i = \alpha_i S_i + \beta_i R_i + \gamma_i (C_{i+1} + T_{i+1}) \quad , \quad i \in (1, n-1) \tag{2}$$

where $\alpha_i + \beta_i + \gamma_i = 1$, $F_n = \alpha_n S_n + \beta_n R_n$.

Formula (2) describes the sum of losses caused by the failure of sub-process i, including the cost of rework and scrap and external failure costs of sub-process i that constitute the architecture of unit failure costs of sub-process i. It is not too hard to aware of that the loss caused by a unit output of sub-process can only be one of three costs mentioned above. Meanwhile, we should understand that the importance of sub-process has nothing to do with the present yield of sub-process, so the high number of defects due to insufficient process capability has no effect on the determination of process weights. In order to avoid the repeated calculation, we take the failure proportion into account. There are some differences in calculation of failure costs between sub-process n and other sub-processes. The output of sub-process n is the product of all sub-processes that directly face to external customer, thus the external failure costs of sub-process n should be beared by all sub-processes rather than sub-process n. Based on this consideration, the external failure costs of sub-process n are assigned to each sub-process based on the process weights. Considering the above issues, the equation of the weight can be obtained:

$$w_i = \frac{w_i X + F_i + T_i}{X + \sum_{i=1}^{n} (F_i + T_i)} \tag{3}$$

where $\sum_{i=1}^{n} w_i = 1$

After simplification, the equation can be expressed as:

$$w_i = \frac{F_i + T_i}{\sum_{i=1}^{n} (F_i + T_i)} \tag{4}$$

thus, the proposed IRTY can be written in following form:

$$IRTY = \prod_{i=1}^{n} P_i^{w_i} \tag{5}$$

Weight of each sub-process is calculated by formula (5), then we can obtain the IRTY that describes the overall performance of process. The proposed method that pays particular attention to cost of poor quality provides a guideline for top managers to make strategic decisions, some opportunities for middle managers to improve existing process, a criterion for operators to strive to achieve. It should be noted that sub-processes in parallel also exist in actual production process. If we want to calculate the IRTY of a hybrid process, then we need to calculate the FTY of parallel process first and view it as the FTY of sub-process of serial processes, and then we can calculate the IRTY using the formula (5).

IV NUMERICAL EXAMPLE

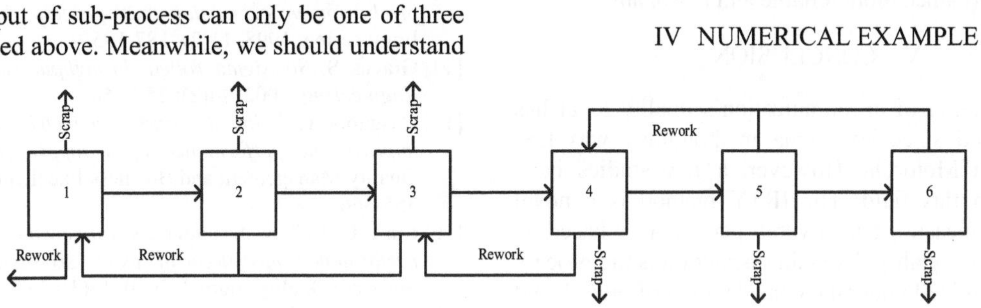

Fig.2. The diagram of a serial process

Fig.2 shows us a serial process consists of six sub-processes. If the outputs of sub-process 2 and 3 are tested to be unqualified and they could be repaired, then they will be returned to sub-process 1 for reworking. Similarly, if the unqualified outputs that can be repaired stem from sub-process 5 or sub-process 6 are tested to be unqualified, they will be returned to sub-process 4 for reworking. As for sub-process 4, it is sub-process 3 where it performs unqualified but repairable outputs will be returned to. Table II presents the parameter values of each sub-process and corresponding results calculated by using formula (2) and (4). In order to facilitate the analysis, the proportion of external failure costs of each sub-process is 5%. As can be seen in Table II, rework and scrap costs of each sub-process differ from each other and scrap cost is increasing from upstream to downstream process while rework cost do not show any regular changes. The failure proportion of sub-process 1 is the same as sub-process 2, but the rework and scrap costs of sub-process 2 are higher than that of sub-process 1. As a result, weight of sub-process 2 is larger. The difference between the weight of sub-process 1 and 2 reflects the effect of rework and scrap costs on process weight. It is also showed that although the rework and scrap costs of sub-process 3 are higher than sub-process 2, the weight of sub-process 2 and 3 turns out to be quite close in that the rework proportion of sub-process 3 is much lower than that of sub-process 2. Moreover, the failure cost of sub-process 2 is very close to that of sub-process 5, however, the results are different since the difference in appraisal costs.

TABLE II
THE PARAMETERS OF THE SERIAL PROCESS

	1	2	3	4	5	6
S_i	10	12	16	18	24	30
R_i	2	4	8	6	8	13
C_i	2	2	4	2	6	5
α_i	0.80	0.80	0.30	0.20	0.20	0.10
β_i	0.15	0.15	0.65	0.15	0.75	0.85
γ_i	0.05	0.05	0.05	0.05	0.05	0.05
F_i	8.435	10.475	10.525	4.9	11.17	14.05
T_i	1	0.7	1.5	2.5	2.0	2.4
w_i	0.136	0.16	0.173	0.106	0.189	0.236
P_i	95.3	98.5	97.5	99.0	98.5	99.5

According to the parameters given in Table II and the formulas (1) and (5), the results can be written as: $RTY=0.888 < IRTY=0.982$. Compared with RTY and ERTY, the proposed IRTY that takes process weights and cost of poor quality into account can describe the overall process performance more reliable and reasonable.

V CONCLUSION

The evaluation of an organization's quality level has been expanded since six sigma methodology was first introduced in Motorola. However, a few studies have been done in this field. The IRTY method is a novel approach for measuring the overall performance based on the cost of poor quality. Its main distinction is the process weights determined under the consideration of rework and scrap costs, appraisal costs and external failure costs. A numerical example is presented to illustrate the effectiveness and excellence of this proposed method. Comparing to the RTY and ERTY approaches, the proposed IRTY can give more reliable and reasonable results.

This research can be further extended to systems with more complicated process such as a hybrid process.

REFERENCES

[1] Maneesh Kumar, Jiju Antony, Christian N. Madu, Douglas C. Montgomery, Sung H. Park. *Common myths of six sigma demystified* [J]. International Journal of Quality and Reliability Management, 2008, 25(8):878–895.

[2] Chen C C. *An objective-oriented and product-line-based manufacturing performance measurement* [J]. International Journal of Production Economics, 2008, 112(1): 380-390.

[3] Furterer, S, Elshennawy, A. K. *Implementation of TQM and lean six sigma tools in local government: a framework and a case study* [J]. Total Quality Management, 2005, 16(10): 1179-1191.

[4] Lazarus, I R, Butler, K. *The promise of six sigma – part one* [J]. *Managed Healthcare Executive*, 2001, 11 (9): 22–26.

[5] Woodard, D T. *Addressing variation in hospital quality: is six sigma the answer?* [J], Journal of Healthcare Management, 2005, 50 (4): 226–236.

[6] Chassin M R. *Is health care ready for Six Sigma quality* [J]. *Milbank Quarterly*, 1998, 76(4): 565-591.

[7] Antony J, Kumar M, Madu C N. *Six sigma in small-and medium-sized UK manufacturing enterprises: Some empirical observations* [J]. International Journal of Quality and Reliability Management, 2005, 22(8): 860-874.

[8] Kwak Y H, Anbari F T. *Benefits, obstacles, and future of six sigma approach* [J]. Technovation, 2006, 26(5): 708-715.

[9] Immaneni A, McCombs A, Cheatham G, Ron Andrews. *Capital One banks on Six Sigma for strategy execution and culture transformation* [J]. Global Business and Organizational Excellence, 2007, 26(6): 43-54.

[10] Chakrabarty A, Tan K C. *The current state of six sigma application in services* [J]. Managing Service Quality, 2007, 17(2): 194-208.

[11] Schroeder R G, Linderman K, Liedtke C, Charles Liedtkeb, Adrian S. Choo. *Six Sigma: definition and underlying theory* [J]. Journal of operations Management, 2008, 26(4): 536-554.

[12] de Mast J, Lokkerbol J. *An analysis of the Six Sigma DMAIC method from the perspective of problem solving* [J]. International Journal of Production Economics, 2012, 139(2): 604-614.

[13] Graves S B. *Statistical quality control of a multistep production process using total process yield* [J]. Quality Engineering, 1998, 11(2): 187-195.

[14] Graves S. *Six sigma rolled throughput yield* [J]. *Quality Engineering*, 2002, 14(2): 257-266.

[15] Dasgupta T. *Using the six-sigma metric to measure and improve the performance of a supply chain* [J]. Total Quality Management and Business Excellence, 2003, 14(3): 355-366.

[16] Lin L C, Li T S. *An integrated framework for supply chain performance measurement using six-sigma metrics* [J]. Software Quality Journal, 2010, 18(3): 387-406.

[17] Abdelhamid T S. *Six Sigma in Lean Construction Systems: Opportunities and Challenges* [C]. Proceedings, 11th Annual Conference of the International Group for Lean Construction, 2003.

[18] Hwang Y D. *The practices of integrating manufacturing execution systems and Six Sigma methodology* [J]. The

International Journal of Advanced Manufacturing Technology, 2006, 31(1-2): 145-154.

[19] Ravichandran J. *Six-Sigma milestone: an overall Sigma level of an organization* [J]. Total Quality Management, 2006, 17(8): 973-980.

[20] Ravichandran J. *Cost-based process weights for DPMO and the overall performance of an organization* [J]. The TQM Magazine, 2007, 19(5): 442-453.

[21] Saghaei A, Najafi H, Noorossana R. *Enhanced Rolled Throughput Yield: A new six sigma-based performance measure* [J]. International Journal of Production Economics, 2012, 140(1): 368-373.

[22] Harry, M, Schroeder, R. *Six Sigma* [M]. New York: Currency, 2000.

Remarks on Dynamics-based Feedforward-Feedback Control of a Single–Link Flexible Arm

Kazuhiko Takahashi
Information Systems Design, Doshisha University, Kyoto, Japan
(katakaha@mail.doshisha.ac.jp)

Abstract - **This paper investigates a feedforward-feedback controller design, based on the dynamics of a distributed parameter system, which achieves the tip-position control of a single-link flexible arm. A mathematical model of the flexible arm is derived using Hamilton's principle. The feedforward control input is calculated by applying the numerical inversion of the Laplace transform to a inverse dynamic model of the flexible arm. A Lyapunov function related to the total energy of the flexible arm is considered and modified by potential energy shaping to derive the feedback controller. The computational simulations of the control of the flexible arm confirm the feasibility of the derived dynamics–based feedforward-feedback controller.**

Keywords – **Feedforward-feedback controller, flexible arm, inverse Laplace transform, Lyapunov function, passivity**

I. INTRODUCTION

The dynamics and control of flexible structures has gained increasing interest in various engineering fields because flexible structures frequently serve important roles in several important engineering applications, such as robotics, large-space structures and information instruments. A robot manipulator that incorporates elastic deformation is referred to as a flexible arm or flexible manipulator. In the last 25 years, because of the importance and usefulness of flexible arms, numerous studies on the dynamics, control and application of the flexible arm have been conducted worldwide [1-5]. These studies have improved the dynamic models and have adapted many techniques from the control theory to solve the control problems caused by the elasticity of the flexible arm. To analyse the flexible arms, different types of models have been employed, e.g. spring–mass discrete models, linear Euler–Bernoulli partial differential equations, Lagrangian equations and finite element decomposition. In control problems affecting the flexible arms, such as end-effector regulation and joint– or end–effector trajectory tracking, many types of control techniques have been investigated to achieve satisfactory control of the flexible arm, e.g. static–state feedback control, optimal control, robust control, adaptive control, neural network control and fuzzy control. A controller that is designed on the basis of the passivity of the system dynamics, which is used to investigate the input–output stability of nonlinear systems, has also been investigated [6, 7]. A passivity–based controller guarantees the stability of the flexible arm's control system using the output feedback. However, a control error is observed when the controller is applied to the tracking control of the flexible arm.

In this paper, a tip–position controller for a single–link flexible arm is discussed by considering both a feedforward control derived from the inverse dynamics of the flexible arm and a feedback control that is designed from the passivity of the flexible arm on the basis of a distributed parameter system. In Section II, the governing equations of the flexible arm are derived using the Hamilton's principle, while the nonlinear equations of motion are obtained using an unconstrained mode expansion method as the mathematical model of the flexible arm. In Section III, an inverse transfer function from the desired tip acceleration to the control torque of the flexible arm is utilized to design the feedforward control input. A Lyapunov function, which is related to the total energy of the distributed parameter system and is modified by the potential energy shaping, is introduced to derive a feedback control law for the flexible arm. Section IV describes the computational experiments for controlling the flexible arm using the proposed dynamics-based feedforward–feedback controller.

II. MODEL OF A SINGLE-LINK FLEXIBLE ARM

The single-link flexible arm, shown in Fig.1, is modelled as a continuous cantilever beam. Here, $\theta(t)$ is the angular position of the joint at time t, $y(x,t)$ is the elastic deflection in position x along the major axis of the cantilever beam, J_0 is the moment of inertia about the joint axis, ρ is the mass density of the flexible arm's material, A is the cross–section of the arm, L is the length of the arm, E is the Young's modulus of the arm's material, I is the area moment of inertia of the arm and m is the mass of the payload attached to the tip of the flexible arm. The flexible arm is driven in a horizontal plane by the input torque $\tau(t)$ that is applied to the joint.

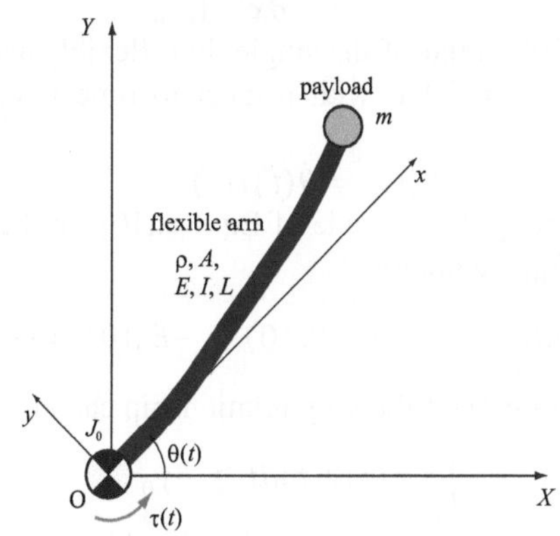

Fig.1. Single–link flexible arm driven in a horizontal plane

E. Qi et al. (eds.), *Proceedings of the 5th International Asia Conference on Industrial Engineering and Management Innovation (IEMI2014)*, Proceedings of the International Asia Conference on Industrial Engineering and Management Innovation 1, DOI 10.2991/978-94-6239-100-0_46,

The position of the flexible arm in the O-XY coordinate system $[X(t), Y(t)]$ is defined by

$$\begin{bmatrix} X(t) \\ Y(t) \end{bmatrix} = \begin{bmatrix} \cos\theta(t) & -\sin\theta(t) \\ \sin\theta(t) & \cos\theta(t) \end{bmatrix} \begin{bmatrix} x \\ y(x,t) \end{bmatrix}. \quad (1)$$

Considering the Bernoulli–Euler beam theory and small elastic displacements, and neglecting the internal damping forces, the kinetic and potential energies of the flexible arm and the work performed by the non-conservative actuating torque can be expressed as

$$T = \frac{1}{2}J_0\dot{\theta}^2(t) + \frac{1}{2}\int_0^L \rho A\left[\{x\dot{\theta}(t) + \dot{y}(x,t)\}^2 + y^2(x,t)\dot{\theta}^2(t)\right]dx + \frac{1}{2}m\left[\{L\dot{\theta}(t) + \dot{y}(L,t)\}^2 + y^2(L,t)\dot{\theta}^2(t)\right], \quad (2)$$

$$U = \frac{1}{2}\int_0^L EI\left\{\frac{\partial^2 y(x)}{\partial x^2}\right\}^2 dx, \quad (3)$$

$$W = \tau(t)\theta(t). \quad (4)$$

A dot denotes partial derivative with respect to time. Applying the Hamilton's principle yields the following governing equations and boundary conditions:

$$J_0\ddot{\theta}(t) + \int_0^L \rho A[x\{x\ddot{\theta}(t) + \ddot{y}(x,t)\} + 2y(x,t)\dot{y}(x,t)\dot{\theta}(t) + y^2(x,t)\ddot{\theta}(t)]dx + m[L\{L\ddot{\theta}(t) + \ddot{y}(L,t)\} + 2y(L,t)\dot{y}(L,t)\dot{\theta}(t) + y^2(L,t)\ddot{\theta}(t)] = \tau(t), \quad (5)$$

$$\int_0^L \rho A\{x\ddot{\theta}(t) + \ddot{y}(x,t) - y(x,t)\dot{\theta}^2(t)\}dx + \int_0^L EI\frac{\partial^4 y(x,t)}{\partial x^4}dx = 0, \quad (6)$$

$$m\{L\ddot{\theta}(t) + \ddot{y}(L,t) - y(L,t)\dot{\theta}^2(t)\} - EI\left.\frac{\partial^3 y(x,t)}{\partial x^3}\right|_{x=L} = 0, \quad (7)$$

$$\left.\frac{\partial^2 y(x,t)}{\partial x^2}\right|_{x=L} = 0. \quad (8)$$

Furthermore, the boundary condition resulting from the geometry of the cantilever beam is

$$y(0,t) = \left.\frac{\partial y(x,t)}{\partial x}\right|_{x=0} = 0. \quad (9)$$

The differential of the single–link flexible arm's total energy $E_n(t) = T + U$ with respect to time is calculated by

$$\dot{E}_n(t) = \dot{\theta}(t)\tau(t). \quad (10)$$

Integrating the left side of Eq. ([eq10]) in the interval from 0 to T_0, we obtain

$$\int_0^{T_0} \dot{E}_n(t)dt = E_n(T_0) - E_n(0) \geq -E_n(0) = -\gamma_0. \quad (11)$$

Therefore, the following relationship can be derived.

$$\int_0^{T_0} \tau(t)\dot{\theta}(t)dt \geq -\gamma_0 \quad (12)$$

This shows that the single-link flexible arm is passive with respect to the angular velocity of the joint.

The deflection $y(x,t)$ can be described by an infinite series of separable modes. The candidate mode frequencies and the corresponding associated shapes can be derived by using the Bernoulli–Euler beam theory with respect to the appropriate boundary conditions. Assuming normal-mode motions, the solution of the linearized homogeneous equations in (5) and (6) can be written as $\theta(t) = \phi(t) + pe^{j\omega t}$ and $y(x,t) = \psi(x)e^{j\omega t}$, where $\phi(t)$ is the rigid body mode function of the angle, p is the modal coefficient, ω is the modal frequency of the flexible arm and $\psi(x)$ is the mode shape function. By solving the eigenvalue problem yields the candidate mode frequencies and the corresponding mode functions. The modal frequencies comprise the solutions of the following frequency equation:

$$1 + \cos\xi\cosh\xi + \frac{2mL^2}{J_0\xi^2}\sin\xi\sinh\xi + \left(\frac{m\xi}{\rho AL} - \frac{\rho AL^3}{J\xi^3}\right)(\cos\xi\sinh\xi - \sin\xi\cosh\xi) = 0 \quad (13)$$

where $\xi^4 = \rho A\omega^2/(EI)$ and $J = J_0 + \rho AL^3/3 + mL^2$. The mode function corresponding to the i-th frequency ω_i $(i = 1,2,\cdots)$ is given in the following form:

$$\psi_i(x) = D_i\{\sin\frac{\xi_i}{L}x - \sinh\frac{\xi_i}{L}x - \frac{\sin\xi_i + \sinh\xi_i}{\cos\xi_i + \cosh\xi_i - 2\rho AL^3J_0^{-1}\xi_i^{-3}\sinh\xi_i} \times (\cos\frac{\xi_i}{L}x - \cosh\frac{\xi_i}{L}x + 2\rho AL^3J^{-1}\xi_i^{-3}\sinh\frac{\xi_i}{L}x - 2mL^2J_0^{-1}\xi_i^{-2}x)\} \quad (14)$$

where the parameter D_i is defined by the appropriate orthogonality condition of the mode functions

$$\int_0^L \psi_i(x)\psi_j(x)dx + \frac{m\psi_i(L)\psi_j(L)}{\rho A} - \frac{Jp_ip_j}{\rho A} = \frac{\delta_{ij}}{\rho A}, \quad (15)$$

$$\int_0^L \psi_i^{(2)}(x)\psi_j^{(2)}(x)dx = \frac{\xi_i^4}{L^4}\frac{\delta_{ij}}{\rho A}. \quad (16)$$

Here, δ_{ij} is Kronecker's delta. The coefficient p_i is defined to satisfy the following equation:

$$Jp_i + \rho A\int_0^L x\psi_i(x)dx + mL\psi_i(L) = 0. \quad (17)$$

Substituting $\theta(t) = \phi(t) + \sum_{i=1}^{\infty} p_i q_i(t)$ and $y(x,t) = \sum_{i=1}^{\infty} \psi_i(x)q_i(t)$ into the governing equations and using the orthogonality condition for the mode function, the following equations of motion can be obtained for the model of a single–link flexible arm:

$$\left\{J+\sum_{i=1}^{\infty}\sum_{j=1}^{\infty}(\delta_{ij}+Jp_ip_j)q_i(t)q_j(t)\right\}\ddot{\phi}(t)$$
$$+\left\{\sum_{i=1}^{\infty}\sum_{j=1}^{\infty}(\delta_{ij}+Jp_ip_j)q_i(t)q_j(t)\right\}\sum_{i=1}^{\infty}p_i\,\ddot{q}_i(t)$$
$$+2\left\{\sum_{i=1}^{\infty}\sum_{j=1}^{\infty}(\delta_{ij}+Jp_ip_j)q_i(t)\dot{q}_j(t)\right\}$$
$$\times\left\{\dot{\phi}(t)+\sum_{i=1}^{\infty}p_i\,\dot{q}_i(t)\right\}=\tau(t), \quad (18)$$
$$\ddot{q}_j(t)+\omega_j^2q_j(t)-Jp_j\ddot{\phi}(t)$$
$$-\left[\sum_{i=1}^{\infty}\{\delta_{ij}-m\psi_i(L)\psi_j(L)+Jp_ip_j\}q_i(t)\right]$$
$$\times\left\{\dot{\phi}(t)+\sum_{i=1}^{\infty}p_i\,\dot{q}_i(t)\right\}^2=0. \quad (19)$$

III. CONTROLLER DESIGN

In this section, a feedforward-feedback controller, enabling the tracking of the tip position of the flexible arm to the desired position, is designed on the basis of the dynamic model of the flexible arm.

The feedforward input to the system is generated by using an inverse model of the flexible arm. Applying the Laplace transform with respect to time to the linearized equations of motion derived by assuming small motion around the equilibrium point, a transfer function of the flexible arm from the control torque $\tau(t)$ to the tip position $z(t)=L\theta(t)+y(L,t)$ can be derived as follows:

$$Z(s)=\frac{n_t(s)}{d(s)}T_L(s)=G_t(s)T_L(s) \quad (20)$$

where $Z(s)$ and $T_L(s)$ are the Laplace transforms of $z(t)$ and $\tau(t)$, respectively, and

$$d(s)=(Js^2+\zeta_0 s)\prod_{i=1}^{\infty}(s^2+\omega_i^2). \quad (21)$$

$$n_t(s)=L\prod_{i=1}^{\infty}(s^2+\omega_i^2)$$
$$+Js^2\sum_{i=1}^{\infty}\left[\{Lp_i+\psi_i(L)\}p_i\prod_{j=1,j\neq i}^{\infty}(s^2+\omega_j^2)\right]. \quad (22)$$

The feedforward input torque can be calculated by applying the inverse Laplace transform to the inverse model, which is represented by $T_{L_d}(s)=G_t^{-1}(s)Z_d(s)$, where $Z_d(s)$ denotes the Laplace transforms of the desired tip position $z_d(t)$. Since the inverse transfer function $G_t^{-1}(s)$ is not proper function, the desired acceleration of the tip position $a_d(t)$ is considered to construct the inverse Laplace transform with the proper function given by

$$T_{L_d}(s)=G_t^{-1}(s)\frac{1}{s^2}A_d(s)=\frac{d(s)}{s^2n_t(s)}A_d(s). \quad (23)$$

where $A_d(s)$ is the Laplace transforms of $a_d(t)$. Similarly, the desired joint angle $\theta_d(t)$ can be obtained by the inverse model as

$$\Theta_d(s)=\frac{d(s)}{s^2n_a(s)}A_d(s). \quad (24)$$

where

$$n_a(s)=\prod_{i=1}^{\infty}(s^2+\omega_i^2)$$
$$+Js^2\sum_{i=1}^{\infty}\left[p_i^2\prod_{j=1,j\neq i}^{\infty}(s^2+\omega_j^2)\right] \quad (25)$$

and $\Theta_d(s)$ is the Laplace transforms of $\theta_d(t)$.

By considering the passivity of the single-link flexible arm, a feedback controller for the angular position of the flexible arm is designed. Assuming a point-to-point control of the angular position, the following function $V(t)$ is assumed to be a candidate for the Lyapunov function:

$$V(t)=\frac{1}{2}k_1J_0\dot{\theta}^2(t)+\frac{1}{2}k_3\{\theta_d-\theta(t)\}^2$$
$$+\frac{1}{2}k_2\int_0^L\rho A\left[\{x\dot{\theta}(t)+\dot{y}(x,t)\}^2+y^2(x,t)\dot{\theta}^2(t)\right]dx$$
$$+\frac{1}{2}k_2m\left[\{L\dot{\theta}(t)+\dot{y}(L,t)\}^2+y^2(L,t)\dot{\theta}^2(t)\right]$$
$$+\frac{1}{2}k_2\int_0^L EI\left\{\frac{\partial^2y(x,t)}{\partial x^2}\right\}^2dx \quad (26)$$

where k_i $(i=1,2,3)$ is a positive coefficient and θ_d is the desired angular position. The function $V(t)$ is positive definite and reaches a minimum value at $\theta_d-\theta(t)=0$. The differential of the function $V(t)$ with respect to the time is calculated as

$$\dot{V}(t)=\left[(k_1-k_2)J_0\ddot{\theta}(t)+k_2\tau(t)-k_3\{\theta_d-\theta(t)\}\right]\times\dot{\theta}(t). \quad (27)$$

The control torque is generated in the following form:

$$\tau(t)=\frac{k_3}{k_2}\{\theta_d-\theta(t)\}-\frac{K_r}{k_2}\dot{\theta}(t)-\left(\frac{k_1}{k_2}-1\right)J_0\ddot{\theta}(t), \quad (28)$$

where K_r is the control gain ($K_r>0$). Thus, the differential of the function $V(t)$ becomes

$$\dot{V}(t)=-K_r\dot{\theta}^2(t)\leq 0. \quad (29)$$

Therefore, $V(t)$ is the Lyapunov function. This result shows that the angular position control of the single–link flexible arm is stable for output feedback control that consists of the angular position of the joint, the angular velocity of the joint and the angular acceleration of the joint.

IV. COMPUTATIONAL EXPERIMENTS

To investigate the characteristics of the feedforward-feedback controller, the computational simulations of the control of the tip position of the flexible arm were conducted. In the experiments, the desired tip acceleration $a_d(t)$ comprised a bang–bang input defined: $a_d(t)=$

$0 (0 \le t < t_1, t \ge t_2); a_0 \left(t_1 \le t < \frac{t_1+t_2}{2}\right); -a_0((t_1 + t_2)/2 \le t < t_2)$ where a_0 is the maximum acceleration.The flexible arm dynamics were simulated by numerically solving the nonlinear simultaneous differential equations, which include the first five modes ($i = 1,2,\cdots,5$), with the implicit Runge-Kutta method in MathematicaTM (Wolfram Research, Inc.). The physical parameters of the flexible arm used in the computational experiments were set to $\rho = 2.7 \times 10^3$ [kg/m3], $A = 5 \times 10^{-5}$ [m2], $E = 6.5 \times 10^{10}$ [N/m2], $I = 4.17 \times 10^{-12}$ [m4], $L = 1$ [m], $J_0 = 0.0025$ [kg m2] and $m = 0.25$ [kg]. The natural–modal frequencies of the flexible arm are: $\omega_1 = 12.1938$, $\omega_2 = 32.1369$, $\omega_3 = 74.7949$, $\omega_4 = 149.912$ and $\omega_5 = 254.119$. In the equations of motion, the modal damping components were introduced for the asymptotic stability of the system and all damping factors were set to 0.001 in the experiments. To calculate the feedforward input torque and the desired angular position using the numerical inversion of the Laplace transform, the first three modes ($i = 1,2,3$) were considered.

First, the effect of the feedforward control input for controlling the flexible arm was examined. The top part of Fig.2 shows the tip position of the flexible arm, while the bottom part of Fig.2 shows the angular position of the joint. Here, only the feedforward control input was used as the input torque $\tau(t)$. Due to the nonlinear dynamics of the flexible arm and disregarding the higher modes ($i = 4,5$) in the inverse model, the tip position of the flexible arm cannot track the desired tip position that is generated by the desired tip acceleration.

Second, the control performance of the feedback controller was examined. Here, the feedforward control input was not utilized with the feedback controller. Thus, the flexible arm is controlled by enabling the angular position $\theta(t)$ to track the desired angular position. Fig.3 shows the response of the flexible arm, in which the parameters of the feedback control law were set to $k_1 = 1$, $k_2 = 1$, $k_3 = 5$ and $K_r = 0.1$. With these parameters, the feedback control law is equivalent to the proportional derivative (PD) control law. As shown in Fig.3, the angular position of the joint tracks the desired angular position, but the residual vibration caused by the elasticity of the link can be observed at the tip position.

Third, the proposed feedforward-feedback controller was investigated. Fig.4 shows the response of the flexible arm using the feedforward torque input and the feedback control. Although small residual vibrations with higher modes remain after the tip position of the flexible arm achieves the desired position, as shown in Fig.4, the feedforward-feedback controller can achieve the controlling task of tracking the tip position of the flexible arm to the desired tip position. This result indicates the feasibility and effectiveness of the feedforward-feedback controller.

Last, the control performance with respect to the variation of the mass of the payload was examined to investigate the robustness of the feedforward-feedback controller. Fig.5 shows the response of the flexible arm when the mass increases by 50%. To achieve high efficiency in the controlling task with the feedforward-feedback controller, the feedforward control input shouldbe recalculated and the parameters of the feedback controller should be fine–tuned according to the mass of the payload.

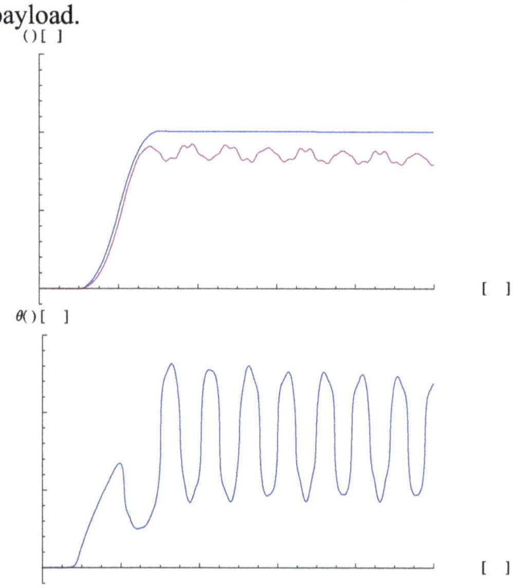

Fig.2. Response of the single-link flexible arm with feedforward control input (top: the red line indicates the tip position and the blue line indicates the desired tip position; bottom: joint angle)

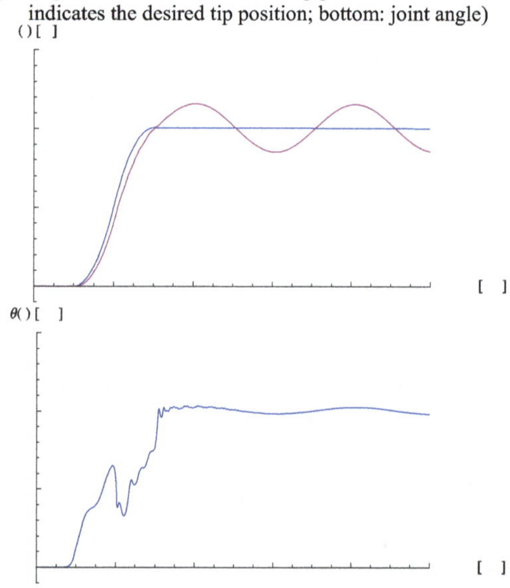

Fig.3. Response of the single–link flexible arm with feedback control, where $k_1 = 1$, $k_2 = 1$, $k_3 = 5$ and $K_r = 0.1$ (top: the red line indicates the tip position and the blue line indicates the desired tip position; bottom: joint angle)

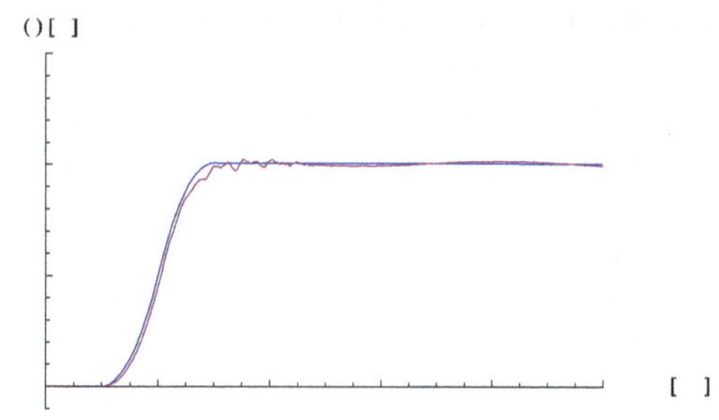

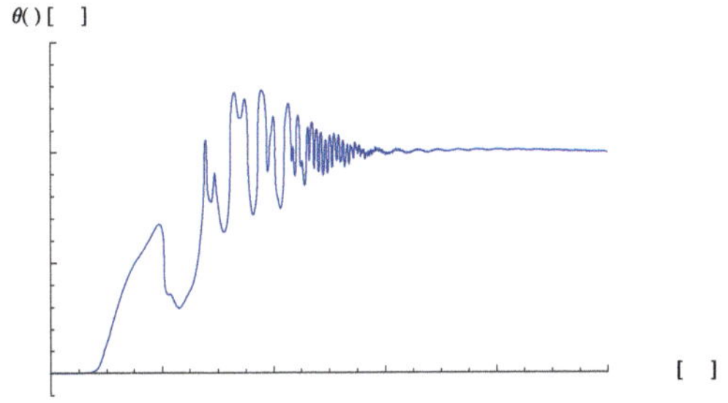

Fig.4. Response of the single–link flexible arm using feedforward–feedback control, where $k_1 = 1$, $k_2 = 5$, $k_3 = 5$ and $K_r = 0.1$ (top: the red line indicates the tip position and the blue line indicates the desired tip position; bottom: joint angle)

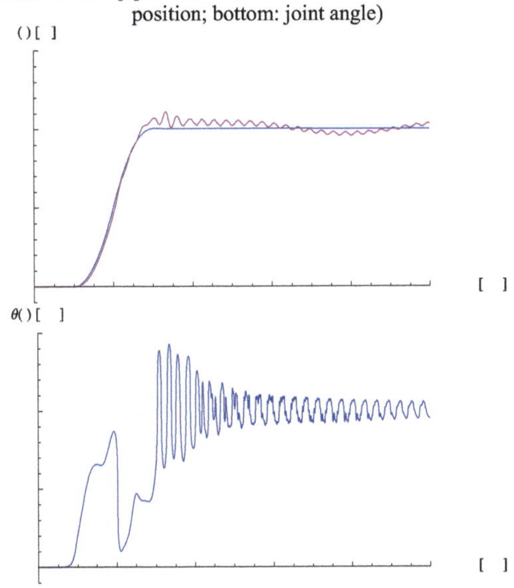

Fig.5. Response of the single–link flexible arm using feedforward-feedback control, in which the mass of the payload increased by 50% and the control parameters are $k_1 = 1$, $k_2 = 2$, $k_3 = 20$ and $K_r = 0.1$

V. CONCLUSION

This paper has investigated the tip–position control of a single-link flexible arm using a dynamics–based feedforward-feedback controller. The feedforward control input was obtained by applying the numerical inversion of the Laplace transform to the inverse transfer function that was derived from the linearized mathematical model of the flexible arm. The feedback controller was designed after considering the passivity of the flexible arm based on a distributed parameter system, in which a Lyapunov function was modified by potential energy shaping to achieve a point-to-point control of the angular position of the joint. The computational simulated control of the flexible arm confirmed the feasibility and characteristics of the proposed feedforward-feedback controller.

REFERENCES

[1] M. Benosman and G. Le Vey, "Control of Flexible Manipulators: A Survey", *Robotica*, Vol. 22, pp.533–545, 2004.

[2] S. K. Dwivedy and P. Eberhard, "Dynamic Analysis of Flexible Manipulators, A Literature Review", *Mechanism and Machine Theory*, Vol. 41, No. 7, pp.749–777, 2006.

[3] M. O. Tokhi and A. K. M. Azad (Eds.), Flexible Robot Manipulators: Modeling, Simulation and Control, Institution of Engineering and Technology, London, 2008.

[4] W. Xu, B. Liang and Y. Xu, Survey of Modeling, "Planning, and Ground Verification of Space Robotic Systems", *Acta Astronautica*, Vol. 68, pp.1629–1649, 2011.

[5] H. N. Rahimi and M. Nazemizadeh, "Dynamic Analysis and Intelligent Control Techniques for Flexible Manipulators: A Review", *Advanced Robotics*, Vol. 28, No. 2, pp.63–76, 2014.

[6] C. I. Byrners, A. Isidori, and J. C. Willems, "Passivity, Feedback-Equivalence, and the Global Stabilization of Minimum Phase Nonlinear Systems", *IEEE Transactions on Automatic Control*, Vol. 36, No. 11, pp.1228–1240, 1991.

[7] R. Ortega, A. Loria, R. Kelly, and L. Praly, "On Passivity-Based Output Feedback Global Stabilization of Euler-Lagrange Systems", *International Journal of Robust and Nonlinear Control*, Vol. 5, No. 4, pp.313–323, 1995.

Upstream Development Model Based on Business Priorities in Fuel Gas Business Application

Hong-fa Shi[1,*], Long Zhu[1], Song-tao Sun[1], Qiang Liu[1], Xiao-ping Xue[2], Jian-lin Gong[2], Chao Li[2]
[1]PetroChina Huabei Oilfield Company, Cangzhou, China
[2]PetroChina Kunlun Gas Energy Company, Beijing, China
(602839569@qq.com)

Abstract - **With the rapid development of the national economy, Long-term excessive consumption of coal has caused serious air pollution. Therefore, the Country has to accelerate the pace of Energy structure, the proportion of natural gas consumption has improved significantly. Facing the domestic city gas, The LNG, CNG rapid expansion of the market developments and intense competition in the market situation, Huabei Oilfield Company is to accelerate the development of the overall business gas business and innovative technology for the gas industry, reconstruct the leading product, derivative of new industrial chain and optimize the management structure in order to comprehensively improve the overall operating efficiency of the gas industry.**

Keywords - **Business Priorities, CNG, LNG, Natural gas**

I. INTRODUCTION

In recent years, the air pollution is worsening, environmental problems have caused people to pay close attention. The consumption ratio of clean, high quality, efficient natural gas is gradually increasing, the size of the domestic use of natural gas in primary energy consumption ratio will account for 10% by 2015. It brings a rare development opportunity and a great challenge for gas companies because of the huge demand of gas market [1]. But at the same time, Some foreign companies have also invested and built factories in the country, engaged in the natural gas exploration and development, construction and operation of gas projects in cities; Some strong momentum of development of private enterprises even has become an important force in the city gas market, the gas market competition is increasingly fierce [2].

II. ESTABLISH PRIORITY TO THE DEVELOPMENT OF THE UPSTREAM BUSINESS CONCEPT

A business to achieve industrial development has two main elements: the first is the upstream and downstream business should connect closely and form a complete industrial chain; the second, it has a product that should occupy a central position and play a leading role in the industrial chain. The industrial development of the gas business is the process of a gradual improvement and development, on the whole the industry chain of gas is formed at the beginning and simply described as:

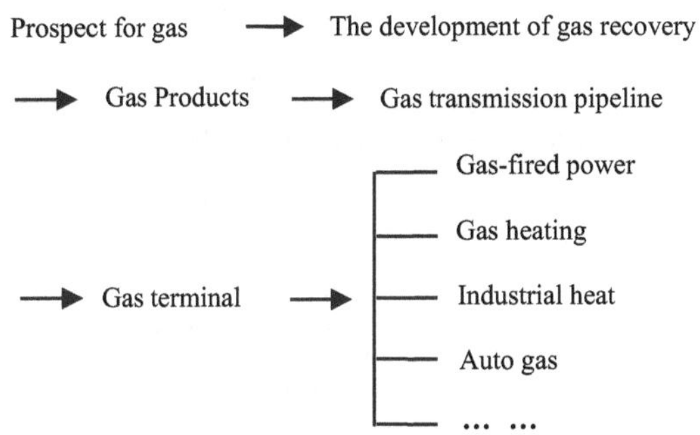

Among them: occupying the central position and the dominant product is natural gas, it has dominated and controlled the coordinated development of upstream and downstream businesses [3]. Natural gas is the main product of the early industrial chain, and it is in the upstream of industry chain [4]. With the continuous development of the natural gas business, the new industry chain has gradually improved as follows (See the next page):

The new industry chain from the initial single user mode directly supplied by pipeline gas has derived two natural gas products of CNG and LNG, formed two kinds of new ways of CNG and LNG supply and creased the two branched ends. It is the output of the upstream pipeline natural gas for deep cold treatment at one end of the branched-chain. It should convert to LNG product, then to storage and distribution, and finally supply for downstream users. In the process, it gradually forms LNG as the leading product of the industrial chain that is LNG industry chain. LNG is the basis for promoting the development of the whole industry chain, and it is also in the upstream of the branched chain. Similarly, in another CNG industry chain, CNG is leading products and is also in the upper reaches of the branched chain. The leading products in one of the main chain and two branched chain and each main products are all in the upstream of the business, therefore, it should be given priority to the development. But it belongs to Business areas of the oil and gas companies. As a business engaged in natural gas downstream gas business the only option is an innovative concept, focuses on the end of the branched chain gas industry, controls the two main products of LNG and CNG and as the upstream business branched priority as soon as possible so as to form industrial development.

E. Qi et al. (eds.), *Proceedings of the 5th International Asia Conference on Industrial Engineering and Management Innovation (IEMI2014)*, Proceedings of the International Asia Conference on Industrial Engineering and Management Innovation 1, DOI 10.2991/978-94-6239-100-0_47, © Atlantis Press and the authors 2015

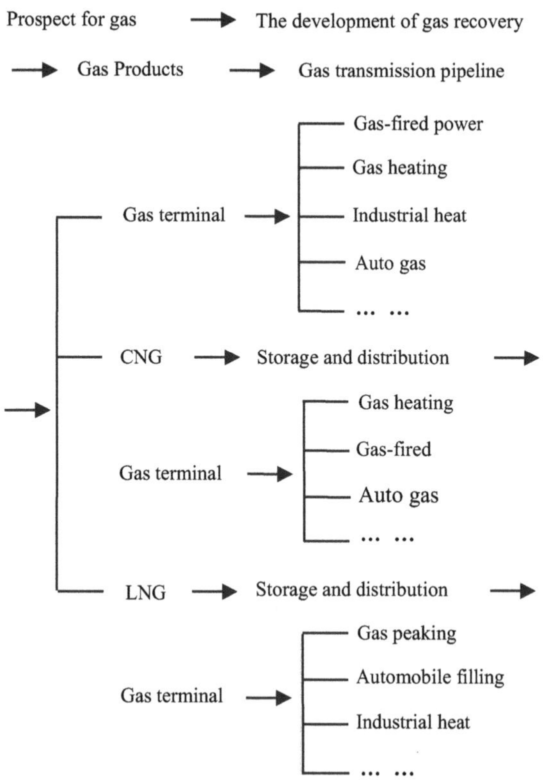

III. ACCORDING TO DEVELOPMENT PHILOSOPHY OF THE LEADING PRODUCT TO BUILD BUSINESS OPTIMIZATION DEVELOPMENT MODEL

As the two leading LNG and CNG products is deep processing of natural gas pipelines, they are high value-added products and have big profit space (see Table I), On the basis of ensure the supply of domestic gas and industrial gas, how to make limited gas pipe more and more be used in production of LNG and CNG, and make the overall management benefit be the largest. It needs to build a comprehensive structure optimization model:

TABLE I
THE CONTRAST OF GROSS PROFIT OF UNILATERAL PIPELINE GAS, CNG AND LNG

	Natural gas pipeline	CNG	LNG
Gross profit(yuan)	0.26	0.36	0.56
Increase in gross profit(yuan)	0	0.1	0.3

Assumes that the supply of total natural gas is T, among them: the direct supply for users is T_1, and the indicator for LNG production is X, the indicator for CNG production is Y. According to the different profit space of gas species, how to be reasonable arrangement of LNG and CNG supply, it is the core problem of maximum profit in order to obtain consolidated operating profit.

Re-established:

(1)The purchase price of gas is C_1;

(2)The sales price of pipeline gas, LNG and CNG are respectively P_1, P_2, P_3;

(3)The processing costs of LNG and CNG are respectively C_2, C_3;

(4)The pipeline fees of pipeline gas and the transportation costs per kilometer of LNG and CNG are respectively M_1, M_2, M_3, and the pipeline and transport distances are respectively S_1, S_2, S_3.

Then:

(1) Gas sales: $P=P_1T_1+P_2X+P_3Y$;

(2) Costs: $C= C_1T+C_2X+C_3Y+M_1S_1+M_2S_2+M_3S_3$.

Therefore, it can get profits:

$$Z=P-C=(P_2-C_2)X+(P_3-C_3)Y+P_1T_1-C_1T-M_1S_1-M_2S_2-M_3S_3$$

That is:

$$Y=\frac{C_2-P_2}{P_3-C_3}X+\frac{Z+C_1T+M_1S_1+M_2S_2+M_3S_3-P_1T_1}{P_3-C_3}$$

(Set it to linear L)

And X, Y satisfy the following conditions:

$$\begin{cases} 0<X+Y\le T-T_1 \\ a_1\le Y\le a_2 \end{cases}$$

(Among them a_1, a_2 are respectively the minimum supply and maximum supply of CNG)

To make maximum consolidated operating profit, the mathematical model is established as follows:

$$Max(Z)=(P_2-C_2)X+(P_3-C_3)Y+P_1T_1-C_1T-M_1S_1-M_2S_2-M_3S_3$$

X, Y satisfy the constraint conditions:

$$\begin{cases} 0<X+Y\le T-T_1 \\ a_1\le Y\le a_2 \end{cases}$$

X, Y will satisfy the constraints expressed in plane coordinate (see Fig.1). The parallelogram ABCD is the scope of X, Y of the area. The following will solve the above mathematical model graphically:

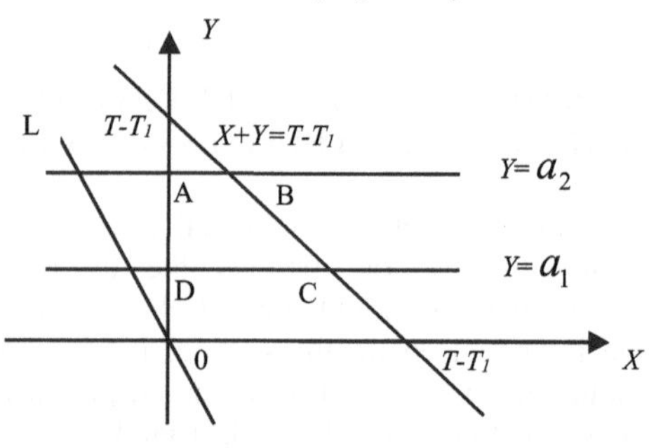

Fig.1. The schematic diagram of graphical method

(1)When the $\left|\frac{C_2-P_2}{P_3-C_3}\right|>1$, shift the straight line L to the upper right, When the straight line L through the point C, Z gets the maximum value, that is the most profitable. In this case, the X and Y values: $\begin{cases} X = T - T_1 - a_1 \\ Y = a_1 \end{cases}$;

(2)When the $\left|\frac{C_2-P_2}{P_3-C_3}\right|<1$, shift the straight line L to the upper right, When the straight line L through the point B, Z gets the maximum value, that is the most profitable. In this case, the X and Y values: $\begin{cases} X = T - T_1 - a_2 \\ Y = a_2 \end{cases}$;

(3)When the $\left|\frac{C_2-P_2}{P_3-C_3}\right|=1$, shift the straight line L to the upper right, When the straight line L and the straight line $X+Y=T-T_1$ coincides, Z gets the maximum value, that is the most profitable. In this case, the X and Y values are on the line segment BC.

Substituted into data: Data generations will get into in order to reach a conclusion.

It is known gas purchase price as follows: $C_1 = 2.5$ yuan/m^3; the processing costs of LNG and CNG are respectively: $C_2 = 0.69$ yuan/m^3, $C_3 = 0.32$ yuan/m^3; the sales price of pipeline gas, LNG and CNG are respectively: P_1=2.76yuan/m^3, P_2=5.21yuan/kg, P_3=3.18 yuan/m^3.

In addition, the pipeline gas per cubic can be processed into 0.72 kg of LNG, then can get $P_2 = 5.21 \times 0.72 = 3.75$ yuan/m^3, substituted into $\left|\frac{C_2-P_2}{P_3-C_3}\right|$, it can be obtained by $\left|\frac{C_2-P_2}{P_3-C_3}\right|=\left|\frac{0.69-3.75}{3.18-0.32}\right|=1.07>1$, Thus it satisfies the condition (1), Z to obtain maximum value at C point. Therefore, only when the supply of CNG to take the minimum and the supply of LNG to take maximum, it can make maximum profits obtained.

To this, after the implementation of the policy limit indicators supply for Gas terminal business enterprises by PetroChina, Huabei Oilfield Company to optimize the business structure of gas in order to improve the overall operational efficiency of gas business, and establishes the priority development LNG industry, preferential development of CNG business and the healthy development of city gas business philosophy. It has promoted the development of the gas business and improved the overall operating efficiency of the gas business [5].

IV. CARRY OUT THE OPTIMIZATION DEVELOPMENT MODEL OF GAS BUSINESS

A. Carry Out the Optimization Development Model of Gas Business

1) In LNG Vehicle Promotion as the Breakthrough Point, Fostering LNG Downstream Market at Full Blast. The first, Sources from the production and sale of vehicles start off LNG vehicle applications boom, establishing relations of cooperation with Dongfeng, huge Qimao group such as automobile manufacturers and dealers successively and promoting LNG heavy vehicle applications with their perfect sales network and market influence; the second, in urban public transport, logistics and other large and medium-sized passenger and freight enterprises as the breakthrough point, with using their own field to the construction of a Internal gas stations to kick-start the application of LNG vehicle and successively carry out the cooperation LNG vehicle applications with Zhenjiang Jiangtian group, Cangzhou bus group and more than other 30 passenger transport enterprises; the third, the total investment is 300 million yuan that to buy the 180 units of LNG tankers for huabei oilfield area transportation vehicles and other logistics companies to have a free try.

2) Take Containerized Refueling Station as the Starting Point and Rapid Flow of LNG Supply Chain Resources. The first, sign a strategic partnership agreement to purchase container and metering stations. At the early stage of the market development giving priority to the construction of containerized Refueling Station, when filling a certain scale we can build fixed stations in the transport routes and vehicles gathering place along the freight and passenger lines; second, purchase mobile charging cars to meet the demands of LNG gas customer timely. We can provide a vehicle for the LNG filling functions at any time, any place, make up for the deficiency early due to fewer gas stations influencing customer to add liquid and dismiss the concerns of potential customers at the same time. It's to guarantee the filling need of LNG vehicles timely and appropriately that greatly promotes the development of LNG vehicle market [6].

3) Perfect the LNG Application Technology and Further Widen the LNG Applications [7]. First, based on extensive market research for the Grand Canal, the Yangtze River we build LNG refueling stations in the Beijing-Hangzhou Grand Canal and the ferry port to charge for passing ships and break though the marine LNG ultimately; second, in motorway service areas, tourist attractions restaurants and hotels, industrial parks and other markets, we popularize LNG to replace diesel, coal and other fuels by Dewar, gasification station. There are 17 gasification stations that have been put into use in

Jiangsu, Beijing and other places and their daily sales are more than 50 tons; the third is to actively communicate with a drilling company in Bohai Drilling and a domestic diesel engine factory to carry out pilot applications of gas drilling machine, and further promote the use of natural gas in the oil rig power equipment [8].

B. Make the Upstream and Downstream to be Integrated, and Strengthen the Ability of Controlling Resources

1) Optimize the Layout and Choose Scientifically, Speed up the Construction of LNG Liquefaction Factory. Starting from the control of resources and focus on the layout of the market, the layout of LNG processing plant give full consideration to the source, location, price and other factors [9]. Relying on the regional market in Beijing Tianjin Hebei, Shanxi Inner Mongolia Henan, Suzhou Zhejiang Shanghai and other downstream businesses, LNG key determines three liquefaction factories. Relying on the seam gas resources of Huabei Oilfield in Shanxi, it plans to construct the Shanxi Jincheng Qinshui coalbed methane liquefaction project, to realize the in situ conversion of coal seam gas, and to further enhance the added value of the products. Relying on the planning and construction of the PetroChina Shanxi Beijing line four and Inner Mongolia coal gas resources, it plans to construct the Inner Mongolia Wulanchabu Xing and LNG factory project, and to highlight the price, regional and market advantages. Relying on the acquired second Shanxi Beijing Renqiu 34# distribution station pipeline gas resources, it plans to expand Renqiu north gate station liquefaction, storage, peak load capacity to ensure the formation of Beijing and Shanxi Hebei Inner Mongolia LNG gas network [10].

2) Improve the Station Construction and Build Actively CNG\LNG Filling Sales Network. The first is orienting the LNG vehicle promotion. It is based on the full investigation of passenger, freight vehicle route and the number, around the Su-Xi-Chang-Tong passenger line and the Inner Mongolia-Beijing-Tangshan, Shanxi-Hebei-Tianjin two Jin Meng coal transport line, to seize the high-quality market, rapid filling site layout. The second is the PetroChina sales sector cooperation. To use its sales network, optimize oil station, layout gas station. The third is through the strict examination, serious project approval, strict planning, approval, settlement and other aspects, to establish the "Atlas of the standardized design of the station and cost control", to realize the field station construction standardization, normalization, modularization, and to speed up the construction schedule, control the cost of the investment. By the end of 2013, it has been completed and put into LNG filling stations accumulated 160 points, day Dosing total of more than 700 tons, CNG filling stations are 31, and the gas network has begun to take shape, the operating efficiency initially has been apparent (See Fig. 2 and Fig.3).

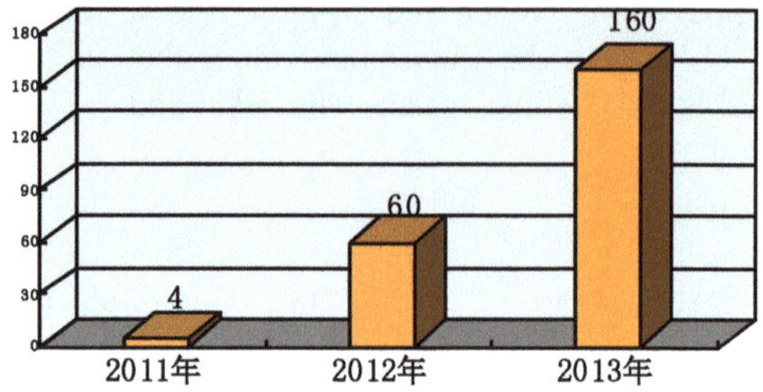

Fig.2. In 2011-2013 the construction of LNG refueling station (Unit: Tower)

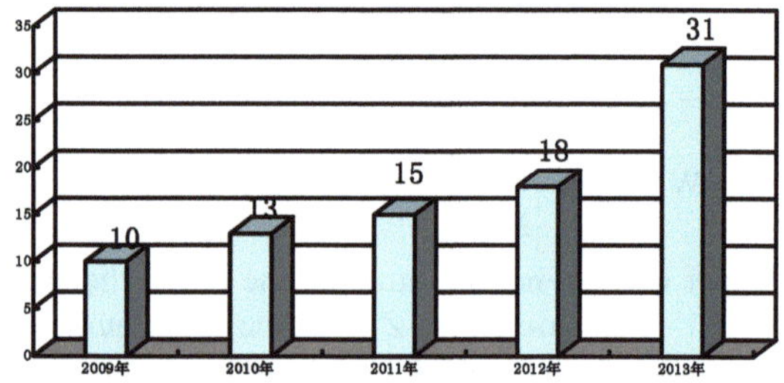

Fig.3. In 2009-2013 the construction of CNG refueling station (Unit: Tower)

3) Start from a Variety of Resources, Actively Try for Pipeline Gas Resources from the PetroChina. In 2013, we should actively try for Shanxi-Beijing forth-line located in Zhangjiakou fool-proof valve chamber and West distribution stations Lushan Henan Baofeng two openings right. The shanxi-beijing second-line 36# valve chamber and the shanxi-beijing first- line of the end canal project was completed with ventilation condition. From difference aspects and multi-channel we will report to the superior, in order to seek support guarantee gas demand in winter, which successively for the gas supply index amounted to 200000 cubic meters/day. 2013 winter, facing Renqiu urgent situation of LNG plant shutdowns and Jiangsu rudong pier limited supply, we should stock (LNG) liquid resources from Shaanxi, Inner Mongolia, Zhangjiakou, Ningxia and other places nine factories, so that we can realize the basic steady supply fluid with the transported to strengthen distribution.

C. Optimize the Management Structure and Accelerate the Development of Gas Industry

1) Allocate Resources Efficiently, Optimize and Adjust the Gas Structure. The key to ensure that adjusting the structure to increase business variety of LNG and CNG is adequate gas supply. To do this, the first is the policy guidance and increasing the additional indicators balance. Through the rational use of gas to guide users, thereby it can save gas indicators to facilitate optimal adjustment of business structure, and operate margin targets for LNG and CNG. The second is

scientific prediction to protect the index increment. It should predict the old and new gas market indicators increment and annual LNG, CNG gas indicators increment on time, actively sought to be secured gas supply to PetroChina so that for LNG, CNG business to attract more gas indicators in general. In 2013, compared with 2012, the LNG and CNG revenue increase 956 million yuan and 94 million yuan respectively, which effectively adjust the operating structure (See Table II).

TABLE II
THE COMPARISON OF GAS BUSINESS PROPORTION IN 2011, 2012 AND2013

	Natural gas pipeline	LNG	CNG	LPG
2011	71.714%	17.87%	8.77%	3.03%
2012	60.29%	24.3%	12.47%	2.94%
2013	46.5%	40.4%	11.38%	1.72%

2) Give Full Play to the Advantages of Technology and Speed up the Development of LNG and CNG Industry. Make a transformation to relatively scattered, smaller-demanded markets in the face of fierce market competition. The way that extending gas supply through a pipeline needs large investment and the cost of unilateral gas supply is relatively high. So we should give full play to the technology of LNG and CNG and the processing advantages by giving priority to the users of LNG device prizing stations that have the gas ability, CNG branch stations and relatively concentrated remote areas. Creatively implement "Station performance appraisal management approach" by combining the filling volume of LNG per day and the performance pay of the staff at the station as well as combining the filling volume of LNG per year with cadre examination in order to promote rapid increase of LNG sales. Also firstly open 400 free advice hotlines in the country and give the first domestic LPG station GIS system a test run to provide users with the LPG navigation as well as business advisory service. The above innovative service can booster the development of market, promote sales promotion and rapid expansion of LNG/CNG business.

3) Give Play to the Advantage of Gas resource and Give a Steady Development of Pipeline Gas. In line with the key that we will reasonably plan and construct high-pressure gas pipeline in Hebei and Shanxi oilfield area, we constantly expand the network coverage. Besides the existing 240 km main gas pipeline including such as Yongqing-Suqiao, Yongqing-Beijing and so on in the oil field, we newly construct more than 420 km high pressure gas pipeline. The total pipeline mileage reaches more than 660 kilometers and the transmission capacity of natural gas reaches 2.5 billion cubic meters per year. The operation ability of gas pipeline gets a constant improvement and the gas sales of pipeline reaches more than 940 million cubic meters per year (See Fig.4).

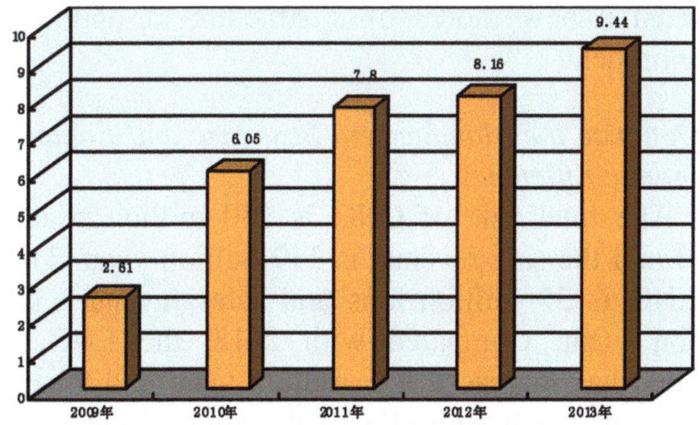

Fig.4. In 2009-2013 the situation of pipeline gas sale (Unit: one hundred million cubic meters)

4) Give Full Play to the Advantages of the Policy of Petrochina and Orderly Development of Civil Gas. It is based on the principle of serving the society and undertaking social responsibility. According to the principle of the new buildings development, the transformation of the old buildings of natural gas and taking into account some key rural users, to stable operation of city gas business. At the same time of residents completed oilfield gasification, it has been developed six counties markets, for example, Hebei RenQiu, Dacheng and so on, in order to steadily advance gasification local residents, and has piloted gasification eight villages in Renqiu. By the end of 2013, it has built a total of medium voltage network of more than 600 kilometers, and provided gas service for 17.8 million households and more than 800 businesses and households (See Fig.5).

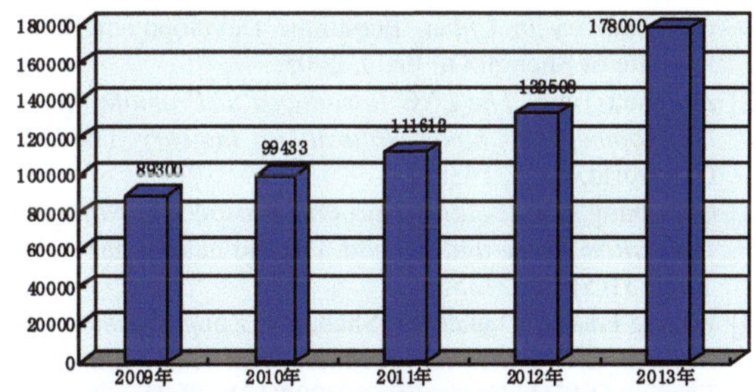

Fig.5. In 2009-2013 the total number of cases of city gas users (Unit: household)

V. IMPLEMEMENTATION EFFECT

A. Rich Industrial Chain Theory and Innovate Development Pattern

It has further enriched the theory of industrial chain, derived CNG and LNG two industrial branches in the natural gas industry chain and reengineered the two leading products CNG and LNG; It has innovated priority to the development of upstream business model, established the priority development LNG industry, preferential development of CNG business and the healthy development of city gas business philosophy, and selected a new model of a structure to optimize the development.

B. Optimize the Management Structure and Enhance the Economic Efficiency

The total sale of CNG is 150 million squares in 2013 and the sale revenue is 359 million yuan. The sale of LNG is 32 million tons and sales revenues is1.476 billion yuan. Compared with 2012, the LNG, CNG revenue has increased 956 million yuan and 94 million yuan respectively and effectively adjusted the operating structure. The annual sale revenue is 3.649 billion yuan, profit is 262 million yuan, the sale has been increased 1.327 billion yuan and the profit has been increased 35 million yuan compared to 2012.

VI. CONCLUSION

Through optimization of the operating structure of the gas business, Huabei Oilfield Company improves overall operational efficiency, while achieving a rapid development of the gas business. It is conducive to stable, healthy, sustained and rapid development of the company that establishing priorities LNG industry, meriting CNG business development and stabling development of city gas business philosophy.

REFERENCES

[1] Aiwen Ma. *The Reflection on the Development of Natural Gas Industry in China.* Economic Development Research Institute of Shengli Oil Field, 2005.

[2] Zhenhua Tang. *The LNG Introduced to Promote the Great Development of China's Natural Gas Industry.* The Oil and Gas world, 2006, 143(12).

[3] Gengliang Chen. *Status Quo and Prospect of Natural Gas Utilization Home and Abroad.* Oil and natural gas industry, 2002, 31(5): 231-235.

[4] Futang Huang, Diancheng Shen, Rui Zong. *Development of Natural Gas Utilization Technology at Home and Abroad.* Foreign Oilfield Engineering, 2002(12): 28-31.

[5] Qiang Wang, Yan Li. *Advantages of LNG in Automobile Industry Development.* Gas Industry, 2002, 22(5): 93-96.

[6] Ben Hua, Biao Xiong. *Accelerate the Development of LNG Industry Chain in China.* Chinese and Foreign Energy, 2007.

[7] Xia Wang, Xianliang Zhu. *Discussion on Clean Fuel LNG Alternative Motor Fuels the Superiority of Diesel.* Resource Saving and Rings Insurance, 2011, (6): 59-60.

[8] Luan Jia, Haiming Zhao. *LNG is the Best Choice of Motor Vehicle Gas Instead of Oil.* Shanxi Energy and Conservation, 2009, (2): 27-30.

[9] Xianglong Kong. *Development Status and Prospect of LNG Vehicles.* Value Engineering, 2011, 23:24-25.

[10]Guang Li, Shuhui Zhou. *Development Status and Direction of Natural Gas Vehicles in China.* International Petroleum Economics, 2008, 16 (7): 69-74.

An MCDM Model Based on KL-AHP and TOPSIS and its Application to Weapon System Evaluation

Jian-jun Li*, Li-wei Liu
School of Science, Nanjing University of Science & Technology, Nanjing, China
(*j_jli@139.com, lwliu@mail.njust.edu.cn)

Abstract - **In this paper, we propose a multiple criteria decision-making (MCDM) model based on the proposed KL-AHP and TOPSIS methods, where the KL-AHP method combines analytic hierarchy process (AHP) and Kullback-Leibler divergence (KLD) to determine the relative weights of decision criteria, and the technique for order performance by similarity to ideal solution (TOPSIS) is used to obtain the final rankings of the feasible alternatives. It is reasonable that both subjective judgment and objective information hidden in practical data should be taken into account in decision-making. Therefore, we use AHP to introduce the subjective judgment of decision makers and apply KLD to extract the information of practical data. Then, integrating the subjective judgment and objective information, a new method for determining the relative weights of decision criteria, called KL-AHP (Kullback, Leibler, and AHP), is proposed. For the rankings of the feasible alternatives, the rational and understandable TOPSIS method is employed. Finally, we construct a numerical example for evaluating fighter planes to illustrate our proposed model.**

Keywords - **Kullback-Leibler divergence, KL-AHP, MCDM, OR in military, TOPSIS**

I. INTRODUCTION

Evaluating weapon systems is a very important issue due to the fact that weapon systems have a significant impact on the efficiency of defense systems. Exactly as stated by Cheng [1-2], the performance evaluation of weapon systems has multi-level and multi-factor features, so it can be regarded as MCDM problem.

The analytic hierarchy process is an MCDM method. It was first proposed by Saaty [3-4]. Since its invention, AHP has been widely used in almost all the applications related with decision-making, such as banking sector evaluation [5], manufacturing system evaluation ([6] and [7]), software selection [8], project management [9], supplier selection [10], benefit–cost analysis [11], location allocation [12], resource allocation ([13] and [14]), forecasting ([15] and [16]), ranking ([17] and [18]).

Also, AHP was used in different weapon system problem. Cheng and Mon [19] evaluated weapon systems with AHP based on fuzzy scales. Mon et al. [20] used fuzzy AHP based on entropy weight to evaluate weapon systems. Cheng et al. [21] evaluated attack helicopters by AHP based on linguistic variable weight. Cheng [1-2] assessed naval tactical missile systems and weapon systems with fuzzy AHP and ranking fuzzy numbers, respectively. Dağdeviren et al. [22] developed an evaluation model based on AHP for the selection of optimal weapon in a fuzzy environment.

Saaty's AHP enables the decision-makers to structure the elements of a complex problem in a hierarchical form and to evaluate a large number of quantitative and qualitative elements in a systematic manner under conflicting multiple criteria. That is, Saaty's AHP provides a comprehensive and rational framework for analyzing MCDM problem. This is the reason why AHP is widely used.

However, AHP has some shortcomings [1-2], such as (1) Ranking of the AHP is not precise, (2) The subjective judgment, selection and preference of decision makers have large influence on the AHP result, i.e., if the judgment is in error, the decision is probably incorrect.

In this paper, for overcoming the above mentioned problems, we build an MCDM model, which is based on the proposed KL-AHP and TOPSIS methods, where the KL-AHP method combines AHP and KLD to reduce the impact of subjectivity on the results of decision-making. It is reasonable that both the subjective judgment and objective information hidden in practical data should be taken into account in decision-making. Therefore, we use AHP to introduce the subjective judgment of decision makers and apply KLD to extract the information of practical data. Then, integrating the subjective judgment and objective information, the KL-AHP method is proposed to determine the relative weights of decision criteria. Finally, we utilize the rational and understandable TOPSIS method to obtain the rankings of the feasible alternatives. More reasons of employing TOPSIS can be found in Dağdeviren et al. [22].

This paper is arranged as follows. In section 2, the KL-AHP method is proposed. Section 3 briefly presents the TOPSIS method. Then in section 4, the MCDM model for evaluating weapon system is proposed, and we apply the proposed model to a fighter plane evaluation problem in section 5. Finally, conclusion is given in section 6.

II. THE KL-AHP METHOD

In this section, we briefly describe Saaty's AHP method and KLD, and then propose the KL-AHP method.

A. The AHP method

AHP is characterized by three basic principles: hierarchical structure, the relative weights of decision criteria; and consistent judgment [23]. A complex decision problem is structured as a multilevel hierarchy firstly. The objectives, criteria, subcriteria, and alternatives are arranged in the hierarchical structure from the top down. Then the pairwise comparison is followed. It starts from the second level and finishes in the lowest level. In each level, the criteria are compared pairwise

E. Qi et al. (eds.), *Proceedings of the 5th International Asia Conference on Industrial Engineering and Management Innovation (IEMI2014)*, Proceedings of the International Asia Conference on Industrial Engineering and Management Innovation 1, DOI 10.2991/978-94-6239-100-0_48,

according to their levels of influence and based on the specified criteria in the higher level [22]. The results of the pairwise comparison can be summarized in a reciprocal matrix A in which every entry represents the relative importance of decision elements, represented as follows:

$$A = \begin{pmatrix} a_{11} & a_{12} & \cdots & a_{1n} \\ a_{21} & a_{22} & \cdots & a_{2n} \\ \vdots & \vdots & \ddots & \vdots \\ a_{n1} & a_{n2} & \cdots & a_{nn} \end{pmatrix}$$

$$= \left(a_{ij}\right)_{n\times n}, a_{ij} = 1/a_{ji},\ a_{ii} = 1 \qquad (1)$$

where the relative importance, a_{ij}, is expressed as a numerical scale shown in Table I.

The relative weights of decision elements can be obtained by normalizing the eigenvector corresponding to the largest eigenvalue, $\lambda_{\max}$, of A if A is consistent. To check the consistency, the consistency index (CI) and consistency ratio (CR) are needed. They are defined as follows:

$$CI = \frac{\lambda_{\max} - n}{n-1},\ CR = \frac{CI}{RI}, \qquad (2)$$

where n is the matrix size, i.e. the number of decision elements, and RI is the random consistency index provided by Saaty [24] as shown in Table II.

If the CR does not exceed 0.1 then the judgment is consistent; else judgments should be reviewed and improved.

The detailed algorithm about AHP is referred to Al-Harbi [9].

B. The Kullback-Leibler divergence

The Kullback-Leibler divergence is a measure of the distance between two distributions.

It is originally introduced by Kullback and Leibler [25]. Since its introduction, KLD has been applied in theory and practice. Song [26] used order statistics and KLD to create a new nonparametric goodness-of-fit test. Li and Wang [27] proposed a test for homogeneity based on KLD. Cabella et al. [28] presented applications of KLD for magnetic resonance imaging. Ye [29] applied a new method based on KLD to the fault diagnosis of turbine.

Definition Given two probability distributions p and q, the Kullback-Leibler divergence between p and q is defined as

$$D(p \| q) = \sum_{x\in\aleph} p(x)\log\frac{p(x)}{q(x)}$$

$$= E_p \log\frac{p(x)}{q(x)} \qquad (3)$$

with equality if and only if $p(x) = q(x)$ for all x.

TABLE I
PAIRWISE COMPARISON SCALE FOR AHP PREFERENCES

Definition	numerical rating
Equally important	1
Moderately more important	3
Strongly more important	5
Very strongly more important	7
Extremely more important	9
Intermediate values	2, 4, 6, 8

TABLE II
RANDOM CONSISTENCY INDICES FROM Saaty [4]

n	3	4	5	6	7	8	9	0
RI	0.58	0.9	1.12	1.24	1.32	1.41	1.45	1.49

Property 2 $D(p \| q)$ is convex in the pair (p, q), that is, if (p_1, q_1) and (p_2, q_2) are two pairs of probability distributions, then

$$D(\lambda p_1 + (1-\lambda)p_2 \| \lambda q_1 + (1-\lambda)q_2)$$
$$\le \lambda D(p_1 \| q_1) + (1-\lambda)D(p_2 \| q_2)$$

for all $0 \le \lambda \le 1$.

The proofs of above properties can be found in Cover [30].

C. The proposed KL-AHP method

In multiple criteria decision-making problem, decision makers always express their preferences on criteria, which will influence the result of decision-making. The preference information on criteria belongs to the subjective judgment of decision makers. It is reasonable that the subjective judgment should be reflected in decision-making.

On the other hand, the objective information which is hidden in practical data should not be neglected. For example, if the data of all alternatives are the same, the corresponding criterion can be neglected, that is, the relative weight can be equal to zero, but the subjective weight may be large. On the contrary, if the data is very different, the corresponding criterion should be paid more attention, that is, the relative weight is supposed to be larger, but the subjective weight may be very small. So, we should extract the information hidden in practical data to adjust the relative weight.

We use AHP to introduce the subjective judgment of decision makers and apply KLD to extract the objective information hidden in practical data. A new method for determining the relative weights of decision criteria, called KL-AHP, is proposed by integrating the subjective judgment and objective information. The steps of KL-AHP method are as follows:

Step 1: Run the AHP method. The subjective weights of decision criteria, $w_s = (w_{s1}, w_{s2}, \cdots, w_{sn})$, can be obtained by running the AHP method.

Step 2: Utilize the KLD to obtain objective weight.

a. Establish the decision matrix. The structure of the matrix can be expressed as follows:

$$D = \begin{array}{c} \\ A_1 \\ A_2 \\ \vdots \\ A_i \\ \vdots \\ A_m \end{array} \begin{array}{c} C_1 \; C_2 \cdots C_j \cdots C_n \\ \begin{bmatrix} f_{11} & f_{12} & \cdots & f_{1j} & \cdots & f_{1n} \\ f_{21} & f_{22} & \cdots & f_{2j} & \cdots & f_{2n} \\ \vdots & \vdots & \cdots & \vdots & \cdots & \vdots \\ f_{i1} & f_{i2} & \cdots & f_{ij} & \cdots & f_{in} \\ \vdots & \vdots & \cdots & \vdots & \cdots & \vdots \\ f_{m1} & f_{m2} & \cdots & f_{mj} & \cdots & f_{mn} \end{bmatrix} \end{array}, \qquad (4)$$

where A_i denotes the alternatives i, $i = 1,2,\cdots,m$; C_j represents jth attribute or criterion, $j = 1,2,\cdots,n$; and f_{ij} is a crisp value indicating the performance rating of each alternative A_i with respect to each criterion C_j.

b. Normalize the decision matrix. Let s_j, $j = 1,2,\cdots,n$, be the sum of the jth column, and f'_{ij} be the relative frequency $f'_{ij} = f_{ij} / s_j$, then we have

$$D' = \begin{bmatrix} \frac{f_{11}}{s_1} & \frac{f_{12}}{s_2} & \cdots & \frac{f_{1j}}{s_j} & \cdots & \frac{f_{1n}}{s_n} \\ \frac{f_{21}}{s_1} & \frac{f_{22}}{s_2} & \cdots & \frac{f_{2j}}{s_j} & \cdots & \frac{f_{2n}}{s_n} \\ \vdots & \vdots & \cdots & \vdots & \cdots & \vdots \\ \frac{f_{i1}}{s_1} & \frac{f_{i2}}{s_2} & \cdots & \frac{f_{ij}}{s_j} & \cdots & \frac{f_{in}}{s_n} \\ \vdots & \vdots & \cdots & \vdots & \cdots & \vdots \\ \frac{f_{m1}}{s_1} & \frac{f_{m2}}{s_2} & \cdots & \frac{f_{m1}}{s_j} & \cdots & \frac{f_{mn}}{s_n} \end{bmatrix} = \begin{bmatrix} f'_{11} & f'_{12} & \cdots & f'_{1j} & \cdots & f'_{1n} \\ f'_{21} & f'_{22} & \cdots & f'_{2j} & \cdots & f'_{2n} \\ \vdots & \vdots & \cdots & \vdots & \cdots & \vdots \\ f'_{i1} & f'_{i2} & \cdots & f'_{ij} & \cdots & f'_{in} \\ \vdots & \vdots & \cdots & \vdots & \cdots & \vdots \\ f'_{m1} & f'_{m2} & \cdots & f'_{m1} & \cdots & f'_{mn} \end{bmatrix} \qquad (5)$$

c. Calculate the Kullback-Leibler divergence. Let $f'_j = (f'_{1j}, f'_{2j}, \cdots, f'_{mj})$ and $e = (\frac{1}{m}, \frac{1}{m}, \cdots, \frac{1}{m})$, where f'_j and e are regarded as two probability distributions. The KLD between f'_j and e is

$$D(f'_j \| e) = \sum_{i=1}^{m} f'_{ij} \log \frac{f'_{ij}}{1/m} = \log m + \sum_{i=1}^{m} f'_{ij} \log f'_{ij}, \quad j = 1,2,\cdots,n. \qquad (6)$$

If $f_{ij} = 1/m$, $i = 1,2,\cdots,m$, then $D(f'_j \| e) = 0$; else $D(f'_j \| e) > 0$.

d. Determine the objective weight based on KLD. The objective weights of decision criteria are denoted by $w_o = (w_{o1}, w_{o2}, \cdots, w_{on})$, where

$$w_{oj} = \frac{D(f'_j \| e)}{\sum_{j=1}^{n} D(f'_j \| e)}, \quad j = 1,2,\cdots,n. \qquad (7)$$

Step 3: Obtain the overall weight. Since product of weights can produce much sharper judgment than their linear combination, we obtain the overall weight by the following formula:

$$w_{synj} = \frac{w_{sj} w_{oj}}{\sum_{j=1}^{n} w_{sj} w_{oj}}, \quad j = 1,2,\cdots,n. \qquad (8)$$

III. THE TOPSIS METHOD

TOPSIS is a practical and useful technique for tackling ranking problems in real situations. It was first developed by Hwang and Yoon [31]. The basic idea of TOPSIS is rather straightforward. According to this technique, the best alternative should have the shortest distance from the positive ideal solution and the farthest distance from the negative ideal solution. The detailed procedure is described in the following [32].

Step 1: Establish a decision matrix for the ranking. The structure of decision matrix is the same as (4) in section 2.3.

Step 2: Calculate the normalized decision matrix. The normalized value r_{ij} is calculated as:

$$r_{ij} = \frac{f_{ij}}{\sqrt{\sum_{i=1}^{m} f_{ij}^2}}, \quad i = 1,2,\cdots,m; \; j = 1,2,\cdots,n. \qquad (9)$$

Step 3: Calculate the weighted normalized decision matrix. The weighted normalized value v_{ij} is calculated as:

$$v_{ij} = w_j r_{ij}, \quad i = 1,2,\cdots,m; \; j = 1,2,\cdots,n, \qquad (10)$$

where w_j is the weight of the jth attribute or criterion, and $\sum_{j=1}^{n} w_j = 1$.

Step 4: Determine the positive ideal and negative ideal solution.

$$A^+ = \{v_1^+, v_2^+, \cdots, v_n^+\} = \{(\max_i v_{ij} \mid i \in I), (\min_i v_{ij} \mid i \in J)\}, \qquad (11)$$

$$A^- = \{v_1^-, v_2^-, \cdots, v_n^-\} = \{(\min_i v_{ij} \mid i \in I), (\max_i v_{ij} \mid i \in J)\}, \qquad (12)$$

where I is associated with benefit criteria, and J is associated with cost criteria.

Step 5: Calculate the separation measures, using the n-dimensional Euclidean distance. The separation of each alternative from the positive ideal solution is given as:

$$d_i^+ = \sqrt{\sum_{j=1}^{n} (v_{ij} - v_j^+)^2}, \quad i = 1,2,\cdots,m. \qquad (13)$$

Similarly, the separation of each alternative from the negative ideal solution is given as:

$$d_i^- = \sqrt{\sum_{j=1}^{n} (v_{ij} - v_j^-)^2}, \quad i = 1,2,\cdots,m. \qquad (14)$$

Step 6: Calculate the relative closeness to the ideal solution and rank the performance order. The relative closeness of the alternative A_i is defined as

$$RC_i = \frac{d_i^-}{d_i^- + d_i^+},\ i = 1, 2, \cdots, m, \qquad (15)$$

where $0 \le RC_i \le 1$. The larger the index value, the better the performance of the alternatives.

IV. THE PROPOSED MCDM MODEL

Our MCDM model is composed of KL-AHP and TOPSIS methods. It consists of three basic stages: (1) identify the criteria which will be used in the model, (2) calculate the relative weights of decision criteria by using the proposed KL-AHP method. (3) evaluate alternatives with TOPSIS and determine the final rank.

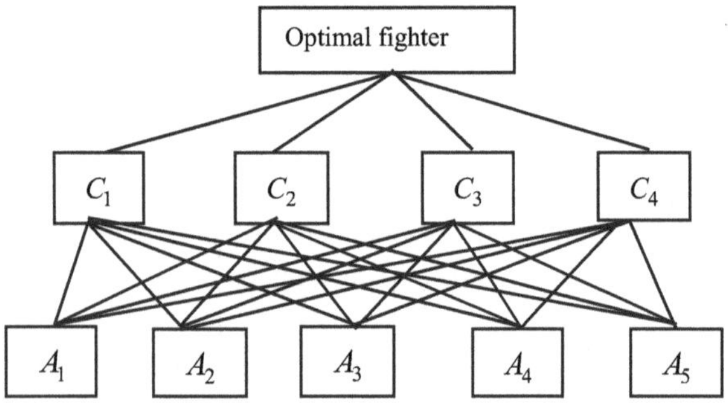

Fig.1. The structure model of evaluating five fighters

V. NUMERICAL EXAMPLE

The beyond visual range (BVR) air combat performance of fighter plane will be evaluated by the proposed model in this section. We select F-15C (A_1), F-16 (A_2), Mirage 2000-5 (A_3), Su-27 (A_4) and Mig 29M (A_5) fighters with specific equipment as our evaluating entities. The evaluation is based on the four criteria: situation awareness capability (C_1), combat capability (C_2), subsisting ability (C_3) and mobility (C_4). The hierarchy structure is constructed as in Fig.1.

After establishing the hierarchical structure, the subjective weights of decision criteria are calculated by using AHP method. The pairwise comparison matrix for decision criteria are presented in Tables III.

TABLE III
THE PAIRWISE COMPARISON MATRIX FOR CRITERIA

	C_1	C_2	C_3	C_4
C_1	1	1	5	7
C_2	1	1	5	7
C_3	1/5	1/5	1	2
C_4	1/7	1/7	1/2	1

TABLE IV
THE PERFORMANCE DATA OF FIGHTERS [33]

Alternatives	Carry scheme	Standardized performance data			
		C_1	C_2	C_3	C_4
A_1	4×AIM-120	0.6595	4299.75	0.0973	0.8731
A_2	2×AIM-7F	0.3228	735.60	0.9522	0.8396
A_3	4×MICA	0.6128	4146.19	0.9408	0.8230
A_4	2×AA-10	0.4841	3547.11	0.0792	0.9742
A_5	2×AA-10	0.3662	2604.60	0.3862	0.9638

Next, we determine the objective weights of decision criteria based on KLD. The standardized performance data was identified by Dong's study [33], as shown in Table IV. Then, we can establish the decision matrix, and obtain the objective weights of decision criteria by formula (4)-(7).

By formula (8), the synthesized weights can be obtained.

The resulting weights for AHP (w_s), KLD (w_o) and KL-AHP (w_{syn}) are shown in Table V.

TABLE V
WEIGHT-DERIVING PROCESS FOR CRITERIA

Alternatives	d_i^+	d_i^-	RC_i	rank
A_1	0.2041	0.2447	0.5452	2
A_2	0.2447	0.2084	0.4600	4
A_3	0.0126	0.3110	0.9610	1
A_4	0.2160	0.1906	0.4688	3
A_5	0.1818	0.1457	0.4449	5

TABLE VI
RESULTS OF TOPSIS ANALYSES

Criteria	w_s	w_o	w_{syn}
C_1	0.4254	0.0742	0.1630
C_2	0.4254	0.2284	0.5017
C_3	0.0934	0.6925	0.3340
C_4	0.0558	0.0049	0.0014

According to Table V, it is seen that the synthesized weights are very different from the subjective weights coming from AHP. Obviously, the objective weights have influenced the subjective weights. We can see that the subjective weight of C_1 is 0.4254, but the difference of the corresponding performance data of all alternatives is small, the objective weight of C_1 is only 0.0742, so the overall important weight of C_1 is much lower than the subjective weights. This is reasonable.

We use TOPSIS to obtain the final rank. The results of TOPSIS analyses are summarized in Table VI. Based on RC_i values, the rankings of the alternatives in descending order are A_3, A_1, A_4, A_2 and A_5. The results indicate that A_3 (Mirage 2000-5) is the best alternative with RC_i values of 0.9610.

VI. CONCLUSION

The performance evaluation of weapon systems is a multiple criteria decision-making problem. Several methods exist for MCDM. AHP is one of them. It is used to determine the importance weights of evaluation criteria usually. Subjective judgment of decision makers is considered in AHP, but the information hidden the practical data is neglected.

It is reasonable that both subjective judgment and objective information hidden in practical data should be taken into account in decision-making. Therefore, in this paper, we use AHP to introduce the subjective judgment of decision makers and apply KLD to extract the information of practical data. Then, integrating the subjective judgment and objective information, the KL-AHP method is proposed to reduce the impact of subjectivity on the results of decision-making. For the rankings of the feasible alternatives, the rational and understandable TOPSIS method is employed. Based on the proposed KL-AHP and TOPSIS methods, a more reasonable MCDM model is presented. From the illustrated example, we can see that the proposed model can efficiently handle the weapon system evaluation problems.

ACKNOWLEDGMENT

This work was supported in part by National Statistical Science Research Key Program of China (Grant No. 2013LZ45) and National Natural Science Foundation of China (Grant No. 11201229).

REFERENCES

[1] C. H. Cheng, "Evaluating naval tactical missile systems by fuzzy AHP based on the grade value of membership function", European Journal of Operational Research, vol. 96, no. 2, pp. 343–350, 1997.

[2] C. H. Cheng, "Evaluating weapon systems using ranking fuzzy numbers", Fuzzy Sets and Systems, vol. 107, no.1, pp. 25–35, 1999.

[3] T. Satty, "An eigenvalue allocation model for prioritization and planning", In: Working paper, Energy Management and Policy Center: University of Pennsylvania, 1972.

[4] T. Satty, "A scaling method for priorities in hierarchical structures", Journal of Mathematical Psychology, vol. 15, no. 3, pp. 234–281, 1977.

[5] N. Y. Seçme, A. Bayrakdaroglu, and C. Kahraman, "Fuzzy performance evaluation in Turkish banking sector using analytic hierarchy process and TOPSIS", Expert Systems with Applications, vol. 36, no. 9, pp. 11699–11709, 2009.

[6] A. Göleç, H. Taşkın, "Novel methodologies and a comparative study for manufacturing systems performance evaluations", Information Sciences, vol. 177, no. 23, pp. 5253–5274, 2007.

[7] C.-L. Yang, S.-P. Chuang, and R.-H. Huang, "Manufacturing evaluation system based on AHP/ANP approach for wafer fabricating industry", Expert Systems with Applications, vol. 36, no. 8, pp. 11369–11377, 2009.

[8] V. Lai, B.K. Wong, W. Cheung, "Group decision making in a multiple criteria environment: A case using the AHP in the software selection", European Journal of Operational Research, vol. 137, no. 1, pp. 134–144, 2002.

[9] K. M. Al-S. Al-Harbi, "Application of AHP in project management", International Journal of Project Management, vol. 19, no. 1, pp. 19–27, 2001.

[10] Z. Chen, W. Yang, "An MAGDM based on constrained FAHP and FTOPSIS and its application to supplier selection", Mathematical and Computer Modelling, vol. 54, no. 11-12, pp. 2802–2815, 2011.

[11] K. S. Chin, S. Chiu, and V. M. R. Tummala, "An evaluation of success factors using AHP to implement ISO 14001 based EMS", International Journal of Quality & Reliability Management, vol. 16, no. 4, pp. 341–361, 1999.

[12] M. Badri, "Combining the AHP and GP for global facility location–allocation problem", International Journal of Production Economics, vol. 62, no. 3, pp. 237–248, 1999.

[13] R. Ramanathan, L.S. Ganesh, "Using AHP for resource allocation problems", European Journal of Operational Research, vol. 80, no. 2, pp. 410–417, 1995.

[14] N. K. Kwak, L. Changwon, "A multi-criteria decision-making approach to university resource allocations and information infrastructure", European Journal of Operational Research, vol. 110, no. 2, pp. 234–242, 1998.

[15] J. Korpela, M. Tuominen, "Inventory forecasting with a multiple criteria decision tool", International Journal of Production Economics, vol. 45, no. 1–3, pp. 159–168, 1997.

[16] S. B. Kim, K. S. Whang, "Forecasting the capabilities of Korean civil aircraft industry", Omega, vol. 21, no. 1, pp. 91–98, 1993.

[17] Z. Babic, N. Plazibat, "Ranking of enterprises based on multi-criteria analysis", International Journal of Production Economics, vol. 56–57, no. 1–3, pp. 29–35, 1998.

[18] A. W. Lalib, G. B. Williams, and R. F. O'Conner, "An intelligent maintenance model (system): An application of analytic hierarchy process and a fuzzy logic rule-based controller", Journal of the Operational Research Society, vol. 49, no. 7, pp. 745–757, 1998.

[19] C. H. Cheng, M. L. Mon, "Evaluating weapon system by analytic hierarchy process based on fuzzy scales", Fuzzy Sets and Systems, vol. 63, pp. 1–10, 1994.

[20] D. L. Mon, C. H. Cheng, and J. C. Lin, "Evaluating weapon system using fuzzy analytic hierarchy process based on entropy weight", Fuzzy Sets and Systems, vol. 62, no. 2, pp.127-134, 1994.

[21] C. H. Cheng, K. L. Yang, and C. L. Hwang, "Evaluating attack helicopters by AHP based on linguistic variable weights", European Journal of Operational Research, vol. 116, no. 2, pp. 423–435, 1999.

[22] M. Dağdeviren, S.Yavuz, and N. Kılınç, "Weapon selection using the AHP and TOPSIS methods under fuzzy environment", Expert Systems with Applications, vol. 36, no. 4, pp. 8143–8151, 2009.

[23] J. Lee, S.-H. Kang, J. Rosenberger, and S. Kim, "A hybrid approach of goal programming for weapon systems selection", Computers & Industrial Engineering, vol. 58, no. 3, pp. 521–527, 2010.

[24] T. L. Saaty, "Multicriteria decision making: The analytic hierarchy process", Pittsburgh, PA: RWS Publications, 1990.

[25] S. Kullback, R. A. Leibler, "On information and sufficiency", Annals of Mathematical Statistics, no. 4, pp. 99–111, 1951.

[26] K. S. Song, "Goodness-of-fit tests based on Kullback-Leibler discrimination information", IEEE Transactions on Information Theory, vol. 48, no. 5, pp. 1103–1117, 2002.
[27] Y. Li, L. Wang, "Testing for homogeneity in mixture using weighted relative entropy", Communications in Statistics—Simulation and Computation, vol. 37, no. 10, pp. 1981–1995, 2008.
[28] B. C. T. Cabella, M. J. Sturzbecher, W. Tedeschi, O. B. Filho, D. B. de Araujo, and UP. C. Neves, "A numerical study of the Kullback-Leibler distance in functional magnetic resonance imaging", Brazilian Journal of Physics, vol. 38, no. 1, pp. 20–25, 2008.
[29] J. Ye, "Fault diagnosis of turbine based on fuzzy cross entropy of vague sets", Expert Systems with Applications, vol. 36, no. 4, pp. 8103–8106, 2009.
[30] Thomas. M. Cover, A. Thomas. Joy, "Elements of information theory", 2nd ed, John Wiley & Sons, Hoboken, New Jersey, 2006.
[31] C. L. Hwang, K. Yoon, "Multiple Attribute Decision Making Methods and Applications", Springer, Berlin Heidelberg, 1981.
[32] H. J. Shyur, H. S. Shih, "A hybrid MCDM model for strategic vendor selection", Mathematical and Computer Modelling, vol. 44, no. 7-8, pp. 749–761, 2006.
[33] Y. Dong, L. Wang, Z.Wang, and H. Zhang, "Air combat effectiveness assessment model based on operational pattern and analytic hierarchy process", Systems Engineering and Electronics, vol. 28, no. 6, pp. 885-888, 2006.

Sound Field Modelling and Multi-Objective Optimization for Workshop Layout

Ji-hong Yan*, Bo-han Lv, Li-zhong Zhao, Lai-chun Suo
Department of Industrial Engineering, Harbin Institute of Technology, Harbin, China
(*jyan@hit.edu.cn, lbh19881230@126.com, zhaolzh@hit.edu.cn, suolc@126.com)

Abstract - **Manufacturing enterprise mainly concerns makespan or efficiency of logistics in the process of workshop layout planning. However, environmental impact of noise which also has direct influence on staff productivity is not always considered to improve the productivity of enterprise. Guided by experience, the implementation is always based on pre-existing planning and layout principle, but do not provide quantitative analysis of environmental noise through scientific method. This paper proposes an effective approach to analyze environmental noise quantitatively. Mathematical model of noise on the condition of multiple sound sources is built. Then the numerical simulation and quantitative analysis to environmental acoustic field of workshop is presented. Finally, the double objective function of Flexible Assembly Workshop is established to obtain shortest logistic time and minimum noise, which is solved based on Genetic Algorithm. This work provides a new idea for workshop layout planning to reduce environmental impact of noise.**

Keywords - **Environmental Noise, Genetic Algorithm, Optimization, Quantitative Analysis, Sound Field**

I.INTRODUCTION

In the process of equipment operation, noise has enormous impacts both on internal and outer environment of the workshop. Therefore, quantitative computation method of environmental noise is essential for environmental impact analysis. In view of noise monitoring research, the common technical method is to detect noise intensity by utilizing acoustic measuring instrument [1-2]. The disadvantage is that the external shape and size of instrument might cause error in the process of measurement. Additionally, the handheld measuring method is not suitable for noise monitoring in some extreme environment, such as high temperature and high pressure. In order to solve those problems, some researchers proposed methods to detect environmental noise through developing noise monitoring system. Vilela et al. [3] established industrial machine noise online monitoring system. They designed data acquisition program in order to realize multi-channel noise signal acquisition. Although its measurement result is accurate, it could not be able to acquire noise intensity of all position coordinates in three-dimensional space. Cheng [4] analyzed acoustic characteristics of ship noise through spectral analysis method which provided basis of noise research.

In order to realize acoustic field's quantitative analysis, this paper proposes an effective approach to analyze environmental noise quantitatively combined with common acoustic index. Firstly, acoustic characteristics of the noise about acoustic superposition and attenuation are analyzed. Mathematical model of noise on the condition of multiple sound sources is built. Then the numerical simulation and quantitative analysis to environmental acoustic field of workshop is presented.

II.MULTIPLE SOURCES SOUND FIELD MODELLING FOR WORKSHOP

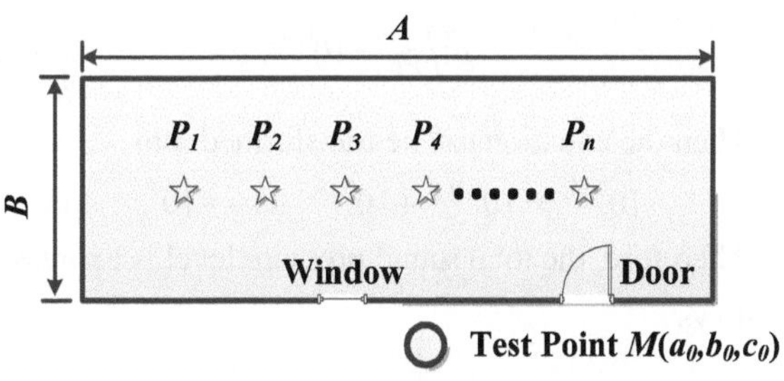

Fig.1. Physical Model of a Workshop

There are various kinds of equipment in a machine workshop. The outdoor sound field analysis should take multiple sound sources superposition into account [5-6]. A physical model of workshop is established with attenuation caused by walls and decreasing with propagation distance increased in air considered.

E. Qi et al. (eds.), *Proceedings of the 5th International Asia Conference on Industrial Engineering and Management Innovation (IEMI2014)*, Proceedings of the International Asia Conference on Industrial Engineering and Management Innovation 1, DOI 10.2991/978-94-6239-100-0_49,

A. Establishment of Physical Model

If there is a workshop with the size of L x W x H (length x width x height), and there are n equipment inside with the noise intensity $P_1,P_2,...P_n$, as shown in Fig 1.

Boundary condition:

All equipment is treated as point sound source in the workshop; ignoring the vibration of machine tools shell; the actual noise intensity will attenuate in the process of noise propagation in air.

B. Establishment of Multiple Sound Field Model

In the process of noise measurement and analysis, SPL is often utilized as a quantitative index in order to evaluate ears' subjective feeling to the intensity and frequency of the noise. The unit of SPL is Decibel which is logarithmic quantization. Therefore, the calculation of multiple noise intensity is not simple algebraic addition, but compound of logarithmic quantization.

If there are n noise sources with SPL $L_{p1}, L_{p2},..., L_{pn}$, then the complex noise intensity is expressed as

$$L_p = 20\times\lg\frac{p}{p_0} = 10\times\lg\frac{p^2}{p_0^2} \tag{1}$$

If the sound pressure of n noise sources are $p_1, p_2,..., p_n$, then the total noise pressure can be expressed as

$$p^2 = p_1^2 + p_2^2 + \cdots + p_n^2 \tag{2}$$

And the inverse logarithm of formula (1) is

$$p^2/p_0^2 = 10^{0.1L_p} \tag{3}$$

Then the equation can be transformed into

$$10^{0.1L_p} = 10^{0.1L_{p1}} + 10^{0.1L_{p2}} + \cdots + 10^{0.1L_{pn}} \tag{4}$$

Therefore, the total sound pressure level is expressed as follows

$$L_p = 10\times\lg\left(\sum_{i=1}^{n}10^{0.1L_{pi}}\right) \tag{5}$$

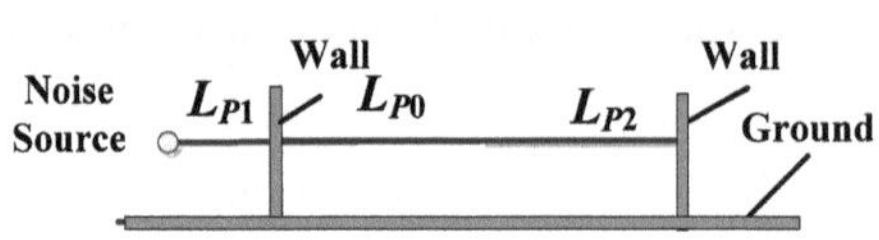

Fig.2. Estimation of indoor sound source's boundary noise value

When analyzing noise's environmental impact to the workshop, sound isolation behavior of wall should be considered. Since the material and structure of the wall are different, noise's insulation quality varies a lot. Some scholars studied the noise attenuation characteristics of the wall. We can know that sound insulation quality of the same material increases with its thickness.

In the process of noise propagation, besides attenuation caused by wall of the building, noise intensity also decreases with propagation distance increased in air. Ghering [7] has estimated the boundary noise value of indoor sound source. Estimation of indoor sound source is shown in Fig.2.

$$L_{p2} = L_{p1} - TL_2 + 10\times\lg S - 20\times\lg r - 14 \tag{6}$$

Where, L_{p1} represents indoor reverberation noise level (dB); L_{p2} represents boundary noise level (dB); TL_2 represents octave frequency band sound insulation quantity from reverberation field inside the building to the outside of the free field (dB); S represents surface area of the building (m^2); r represents the distance from the noise source to the wall (m).

If there are n noise sources with the coordinate

$$P_i = (x_i, y_i, z_i) \qquad i = 1,2,\cdots,n \tag{7}$$

where i represents the serial number of n machine.

The actual noise intensity of the n equipment is expressed as

$$Noise_i = k_i \qquad i = 1,2,\cdots,n \tag{8}$$

There is a test point with the coordinate $M(a_0,b_0,c_0)$. Then, the distance between M and n equipment can be calculated by formula (9)

$$Distance_i = \sqrt{(x_i - a_0)^2 + (y_i - b_0)^2 + (z_i - c_0)^2} \tag{9}$$

The actual noise of combination of n noise sources in point M is expressed as follows

$$LES_i = k_i - \Delta + 10\times\lg S - 20\times\lg(Distance_i) - 14 \tag{10}$$

where S represents the surface area of building exterior wall and Δ represents octave band sound insulation quantity from internal system to the outside free field.

With noise attenuation considered, we assumed that the sound reduction index of the wall is R_1, the sound reduction index of the window is R_2 and the sound reduction index of the door is R_3. Then, the actual noise intensity in the test point M caused by n sound source is expressed as

$$LESD_i = LES_i - \min\{R_1, R_2, R_3\} \quad i = 1,2,\cdots,n \qquad (11)$$

According to the theory of Sound Pressure Level calculation, the accumulative total noise intensity of n sound source in the test point M can be expressed as follows

$$LES = 10 \times \lg\left(\sum_{i=1}^{n} 10^{0.1 \times LESD_i}\right) \qquad (12)$$

C. Multiple Sources Sound Field Modelling and Simulation

Set Fig.2 as an example. If there are 8 sound sources in the workshop making up a production line, the distance between each sound source is 1 meter and actual noise intensity of the sound source is 100dB, we can obtain the final sound field through noise calculation as in Fig.3.

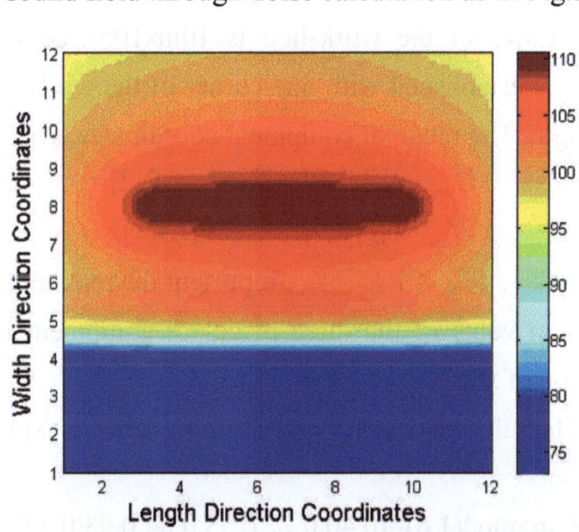

Fig.3. Multiple acoustic field intensity distribution

From Fig.3, we can obtain all the information about sound field of multiple sources and easily find out the location of the main noise source coordinates in the workshop. Besides, we can make effective prediction of noise impact before workshop layout planning. Sound field's quantitative analysis method will make contribution to analyzing the influence of high noise equipment in the noisy environment. Therefore, we can fully consider the influence of the equipment noise in the workshop layout planning, reasonably plan the work area within workshop and minimize noise interference among workshop in each work area.

III. MULTI-OBJECTIVE OPTIMIZATION FOR WORKSHOP LAYOUT

Currently, sustainable manufacturing has become a hot issue [8-9]. It requires enterprises not only simply increasing their productivity, but also considering the environmental impact [10-11]. However, a large amount of enterprises only concern the makespan without analyzing the workshop noise interference between each other from an overall perspective in layout planning. This kind of situation will lead to the high noise equipment causing interference to the outside environment and seriously affect the working efficiency of the work area around. On the contrary, if only considering the effect of high noise equipment in the process of workshop layout planning without concerning logistics relevance between the various devices, then transportation cost of production process will increase that violates workshop product-guided principle and reduce the production efficiency of the enterprise. Therefore, this paper quantitatively evaluates the relationship between the effects of environmental noise and the makespan.

A. Makespan Model

Based on the physical model of the workshop, we analyze its whole production efficiency and establish logistics time model. Firstly, processing time of n machine is expressed as

$$Machining_i = T_i \quad i = 1,2,\cdots,n \qquad (13)$$

where i represents the serial number of n machine.

The distance between upstream equipment and downstream equipment is expressed as

$$D_i = \sqrt{(y_{i+1} - y_i)^2 + (x_{i+1} - x_i)^2} \quad i = 1,2,\cdots,n-1 \qquad (14)$$

Average transportation time between upstream equipment and downstream equipment can be expressed as

$$transport_i = \sqrt{(y_{i+1} - y_i)^2 + (x_{i+1} - x_i)^2} \Big/ v_i \qquad (15)$$

where the average motion speed of work-piece between upstream equipment and downstream equipment is $v_i \ (i = 1,2,\cdots,n-1)$.

In the actual production process, the average waiting time due to the backlog of semi-finished products in the buffer of each work-piece is expressed as

$$Waiting_i = t_i \qquad i = 1, 2, \cdots, n \qquad (16)$$

Therefore, the total makespan to produce each product in the workshop can be expressed as

$$Makespan = \sum_{i=1}^{n} Machining_i + \sum_{i=1}^{n-1} transport_i + \sum_{i=1}^{n} t_i \qquad (17)$$

B. Multi-Objective Function

According to sound insulation quantity statistical analysis of different structure of the window from 'Acoustic Manual', sound insulation of wall R_1=43dB, sound insulation of glass steel window R_2=24 dB and sound insulation of wooden door R_3=28.5dB. Therefore, when concerning acoustic attenuation, the actual noise intensity in point M is

$$LESD_i = LES_i - 24 \qquad i = 1, 2, \cdots, n \qquad (18)$$

Then the actual noise intensity caused by 15 machines in point M can be expressed as

$$LES = 10 \times \lg\left(\sum_{i=1}^{n} 10^{0.1 \times LESD_i}\right) \qquad (19)$$

Based on makespan and multiple sound field model with target weight of each objective α、β considered (the target weight can be adjusted with different production requirement $\alpha + \beta = 1,\ \alpha, \beta \in (0,1)$), the dual objective function is finally established as follows

$$ObjectiveFunction = \alpha \times \left(\frac{LES - LES_{min}}{LES_{max} - LES_{min}}\right) + \beta \times \left(\frac{Makespan - Makespan_{min}}{Makespan_{max} - Makespan_{min}}\right) \qquad (20)$$

Constraints (with space between equipment, quantity and size of the machine considered):

(1) Equipment space coordinates cannot overlap and equipment need to keep a certain distance between each other.

(2) There should be enough room for handling equipment between equipment and the wall.

(3) If there are production lines, size and the distance between each device should be considered.

C. Optimization Based on Genetic Algorithm

This paper will solve this problem based on Genetic Algorithm. In view of the workshop layout optimization problems, the solution procedure of Genetic Algorithm will be provided as below.

Firstly, Chromosome Coding should be performed. The nature of the coding is map from the problem of the solution space to the chromosome coding space. A candidate solution (individual) is expressed with a string of symbolic representation [12-13]. The essence of workshop layout optimization is arranging the position of its internal mobile devices. Therefore, real number encoding will be used for this double objective optimization problem [14-15]. Every chromosome sequence contains position coordinates information of all the equipment. Also, each chromosome is a set of feasible solution of workshop layout planning. Chromosome coding method of workshop layout optimization will be described in detail.

If there are 3 pieces of equipment in the workshop and the size of the workshop is 10mx10m, coordinate system is established with one corner of the workshop as the origin. The range of equipment coordinates is

$$X_{\max} = [7, 7, 9];\ X_{\min} = [1, 2, 2]$$

$$Y_{\max} = [8, 9, 9];\ Y_{\min} = [1, 1, 2]$$

where X_{max}、X_{min}、Y_{max}、Y_{min} represent the maximum and minimum value of the three pieces of equipment in the direction of x and y respectively.

Then the length of a chromosome will be expressed as

Pop=rand(1,6)=[0.40 0.22 0.53 0.84 0.93 0.72];

Convert the array elements into the device coordinate space respectively.

$X(1)=X_{min}(1)+Pop(1).*(X_{max}(1)-X_{min}(1))$;

$Y(1)=Y_{min}(1)+Pop(1).*(Y_{max}(1)-Y_{min}(1))$;

Finally we can obtain a chromosome sequence representing device coordinate information.

Pop= [3.40 3.10 5.71 6.88 8.44 7.04];

IV.CASE STUDY

The implementation of our research focuses on Flexible Manufacture System (FMS) assembly workshop at Harbin Institute of Technology. The system model is built in Flexsim. Material flow is from M_1 to M_{15}, which represents the number of machine. It is easy to find out from CNC milling machine to the manipulator, its long haul distance leads to high transportation cost. Therefore, facility layout of this workshop is urgent to be improved.

Based on logistics time model and the computation model of multiple sound sources, study both noise and total logistics time with production efficiency first and noise impact properly considered. Therefore, the two target weights are $\alpha = 0.3$; $\beta = 0.7$ respectively.

Constraints:

$$\begin{cases} x_i \geq 0 \; y_i \geq 5 \; z_i \geq 0 \quad i = 1,2,\cdots,15 \\ |x_i - x_j| \geq 0.8 \; |y_i - y_j| \geq 0.8 \; i,j = 1,2,\cdots,15 \; i \neq j \\ z_i \leq 1.5 \quad i = 1,2,\cdots,15 \quad 0 < v_0 < 1 \quad v_1 \approx 1 \end{cases}$$

The parameters of genetic algorithm are finally set as: the weight of the logistics time $\beta = 0.7$, weight of the noise $\alpha = 0.3$, iterations are 200, population size is 20 and aberration rate is 0.01. The layout of workshop before and after optimization is shown in Fig.5 and Fig.6. The comparison of computation results before and after optimization is shown in TABLE I.

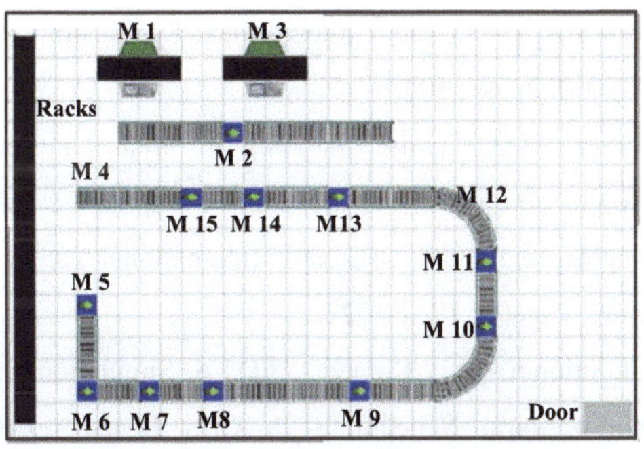

Fig.4. Flexible Manufacture System Model

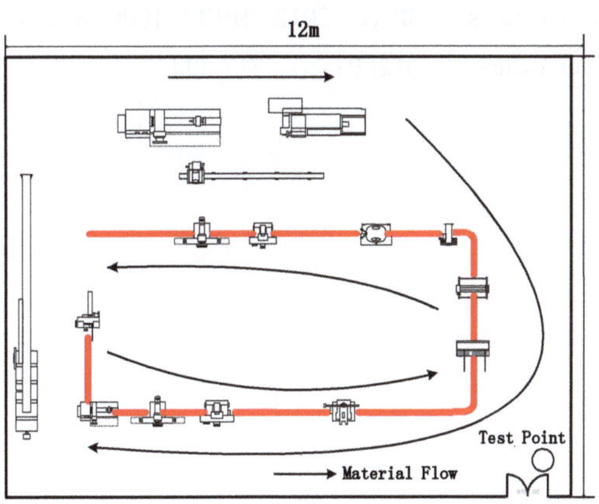

Fig.5. FMS workshop layout before optimization

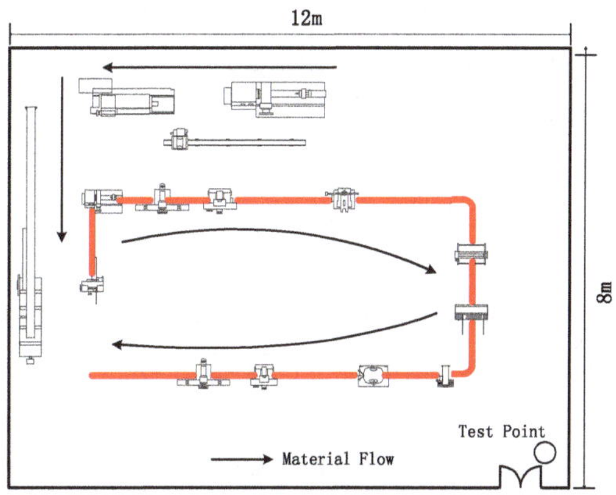

Fig.6. FMS workshop layout after optimization

TABLE I

COMPUTATION RESULTS BEFORE AND AFTER OPTIMIZATION

Workshop Layout Condition	Logistic Time (S)	Noise Intensify (dB)
Before Optimization	330	95
After Optimization	242	87

As shown in Table I, the indoor noise value (test point in Fig.6) is about 87dB and the total assembly time is 242 seconds. Compared with the result before optimization, the indoor noise value is about 95dB and the total assembly time is 330 seconds. The comparative results show that the layout of the workshop has been improved in the aspect of noise impact and makespan.

V. CONCLUSIONS

This paper proposes an effective approach to analyze environmental noise quantitatively combined with sound field superposition. Mathematical model of noise on the condition of multiple sound sources is built. Multi-objective function with minimum noise value and shortest makespan considered based on spatial coordinates of the equipment is established. After optimizing based on Genetic Algorithm, FMS workshop is improved in the aspect of both environmental noise impact and production efficiency. The sound field quantitative analysis and optimization method for workshop we proposed can help enterprises carry out more scientific and effective workshop layout planning. Furthermore, this method has important practical significance for long-term and stable development of enterprise under the background of sustainable manufacturing.

ACKNOWLEDGMENT

This work was supported in part by a grant from the National Science Foundation (#71271068).

REFERENCES

[1] Reza Banai-Kashani, “Discrete Mode-Choice Analysis of Urban travel Demand by the Analytic Hierarch Process”. Transportation; Jan 1, 1989. Page(s). 81.

[2] Kjetil K. Haugen, Arild Hervik (2003) “A Game Theoretic ‘Mode-Choice’ Model for Freight Transportation”. The Annals of Regional Science; (2004) 38: 469-484.

[3] Vilela RM, MetrGlho J C, Cardoso J C. Machine and Industrial Monitorization System by Analysis of Acoustic Signatures. IEEE MELECON 2004, May 12 - 15, 2004, Dubrovnik, Croatia. 277~279.

[4] Cheng Yijie. “spectrum analysis method for ship noise”. CHINA SCIENCE AND TECHNOLOGY INFORMATION. 2006, Page(s): 203..

[5] Caligiuri, L.M.; Sabato, A. Multiresolution analysis of environmental noise signals: an application to railway noise. Instrumentation and Measurement Technology Conference, 2004. IMTC 04. Proceedings of the 21st IEEE. 2004, Page(s): 1891-1895.

[6] Qi Wang. Application of grey relational analysis and hierarchical cluster analysis in regional environmental noise pollution. Grey Systems and Intelligent Services, 2007 IEEE International Conference on 2007, Page(s): 336-340.

[7] W.L. Ghering. "Noise control reference manual". Shanghai Science and Technology Press. 1982. Page(s): 23.

[8] Li Xiuchun “Evaluation on International Competitiveness of Coal Industry Based on Objective and Subjective Integrated Method” E-Business and E-Government (ICEE), 2010 International Conference on 2010 , Page(s): 4013 - 4017

[9] Amrina, E.; Yusof, S.M. Key performance indicators for sustainable manufacturing evaluation in automotive companies. Industrial Engineering and Engineering Management (IEEM), 2011 IEEE International Conference on 2011, Page(s): 1093-1097.

[10] Rachuri, S.; Sriram, R.D.; Sarkar, P. Metrics, standards and industry best practices for sustainable manufacturing systems. Automation Science and Engineering, 2009. CASE 2009. IEEE International Conference on 2009, Page(s): 472-477.

[11] Chengcheng Fan; Carrell, J.D.; Hong-Chao Zhang. An investigation of indicators for measuring sustainable manufacturing. Sustainable Systems and Technology (ISSST), 2010 IEEE International Symposium on 2010, Page(s): 1-5.

[12] Kamkar-Parsi, A.H.; Bouchard, M. "Improved Noise Power Spectrum Density Estimation for Binaural Hearing Aids Operating in a Diffuse Noise Field Environment". Audio, Speech, and Language Processing, IEEE Transactions on 2009, Page(s): 521-533.

[13] Mikhael, W.B.; Hill, P.D. "Acoustic noise cancellation in a multiple noise source environment". Circuits and Systems, 1988., IEEE International Symposium on 1988, Page(s): 2399-2402.

[14] Eldien, H.H. "Noise mapping in urban environments: Application at Suez city center". Computers & Industrial Engineering, 2009. CIE 2009. International Conference on 2009, Page(s): 1722-1727.

[15] Gkoutioudi, K. ; Karatza, H.D. A Simulation Study of Multi-criteria Scheduling in Grid Based on Genetic Algorithms. Parallel and Distributed Processing with Applications (ISPA), 2012 IEEE 10th International Symposium on 2012, Page(s): 317-324.

Theory Research of Intensive and Automatic Warehouse System Based on Lane-Switch Shuttles

Min-de Shen*, Lan-qiang Cong, Cheng-min Gao, Yun-na Xue
Qilu University of Technology, Key Laboratory of Advanced Manufacturing and Measurement and Control Technology for Light Industry in Universities of Shandong, Jinan Shandong, China, 250353
(smd.1130@163.com)

Abstract - **In this paper, the intensive and automatic warehouse system based on lane-switch shuttles is researched. This system can be classified into two types according to with transporting lanes or not. A lane-switch shuttle has the function of walking along a vertical and crossing path on the plane, which can be transferred between different storage channels by oneself, to improve the storage efficiency of system. And the scheduling method of walking for lane-switch shuttle has been introduced.**

Keywords - **Intensive and automatic warehouse system, lane-switch shuttle, theory**

I. INTRODUCTION

At present, Automatic Storage & Retrieval System (AS/RS system in abroad) has become the main technology and equipment of the logistics. With the cost of land for production increasing rapidly, people come up with higher requirement to improve the storage rate and reduce the storage cost, which prompts to research technology of intensive storage. From the developing trend of the current, intensive automatic warehousing system will become prevailing storage form of logistics industry in the future, which has higher storage density of goods and utilization of warehousing space than the traditional Automatic Storage & Retrieval System [1],[2] .

Intensive automatic warehousing system has a variety of forms. Early warehouse of put-in rack is one of the simplest intensive storage system. In the newly developed intensive automatic warehousing system, the gravity rack storage system with damping roller and the shelves storage system with shuttle plate (also called shuttle car) are most widely used [3],[4]. Comparing two forms of the intensive automatic warehousing system, the shelves storage system with shuttle plate has been widely recognized. It is becoming a hot spot of application and research because of the simple structure, the convenience of operation and maintenance, and low construction cost.

At present, the shelves storage system with shuttle plate being used on the market is equal to add a shuttle plate track to the bottom of each channel of the traditional put-in rack. Shuttle plate can travel along the rails to the low part of each unit-load, lift and travel along the channel direction. Then, the goods loading and unloading are realized. Because the shuttle plate can only reciprocate along a straight line (one-way), the in-out warehouse stacker or high forklift must be used to realize the transfer between different channels, which can inevitably affect the operational efficiency of the whole storage system.

In this paper, an intensive automatic warehousing system based on lane-switch shuttles is introduced. Lane-switch shuttles have the function of traveling along a vertical and crossing path on the plane, so don't have to take up the stacker or high forklift. The efficiency of storage system is improved. Combining lane-switch shuttles with elevator can take the place of stacker and high forklift operating in-out warehouse. It generates the simplified system structure, and decreases the construction cost.

II. SYSTEM COMPOSITION

Intensive automatic warehousing system based on lane-switch shuttles is composed of lane-switch shuttles, intensive put-in rack, shuttle plate rack, haulage roadway, crossing rail, loading and unloading elevator and stack, a complete set of electric control system and upper machine scheduling management system, etc. According to the different layout, not only stacker or forklift can replace in-out warehouse elevator, but also several forms of lifting equipments can be used at the same time. Fig.1 is schematic figure for an intensive automatic warehousing system based on lane-switch shuttles.

E. Qi et al. (eds.), *Proceedings of the 5th International Asia Conference on Industrial Engineering and Management Innovation (IEMI2014)*, Proceedings of the International Asia Conference on Industrial Engineering and Management Innovation 1, DOI 10.2991/978-94-6239-100-0_50, © Atlantis Press and the authors 2015

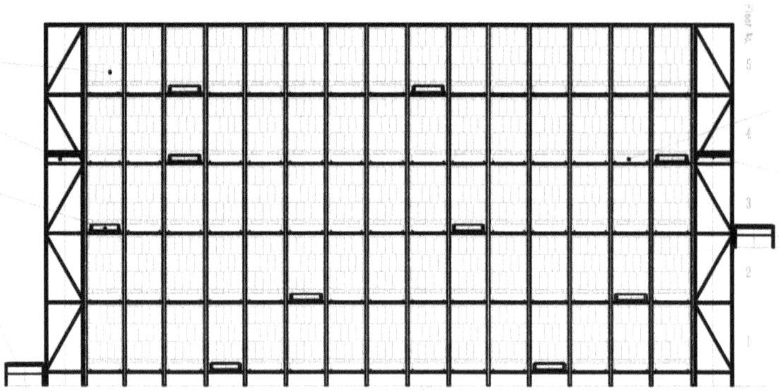

1-Loading cargo station, 2-Loading elevator, 3-Lane-switch shuttle, 4-Unit-Load, 5-Intensive put-in rack, 8-Unloading elevator, 9-Unloading cargo station

Fig.1. The structure of intensive automatic warehousing system based on lane-switch shuttle

III. WORKING PRINCIPLES OF LANE-SWITCH SHUTTLES MECHANISMS

Lane-switch shuttle is the key equipment of the system, which is the new model of logistics transport vehicles. Fig.2 is the structure of lane-switch shuttle. It possesses the functions to travel two-way on the perpendicular and crossing rail and carry goods [5]. Fig.3 is working principles of lane-switch shuttle mechanisms.

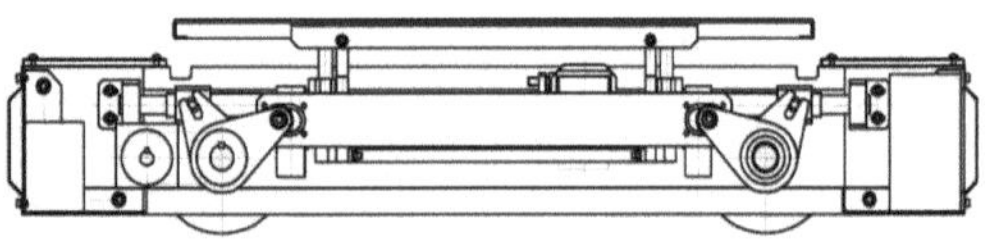

Fig.2. Schematic figure of lane-switch shuttle

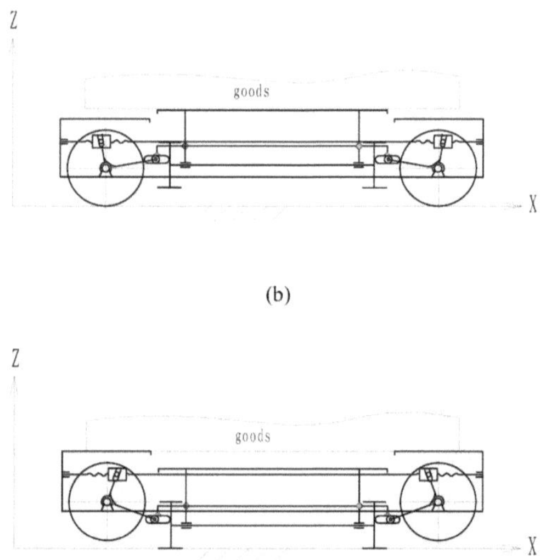
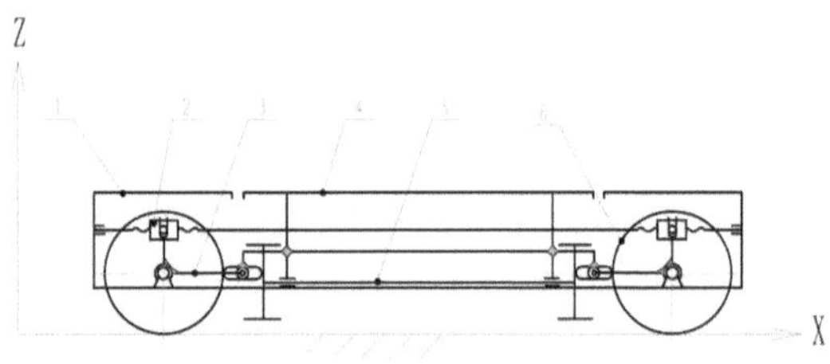

(a)

(b)

(c)

1-Shuttle plate's body, 2-Nut, 3-Shifting yoke, 4-Lifting frame, 5-Transverse traveling wheel, 6-Longitudinal traveling wheel

(a) Schematic of mechanism (b) Lift goods (c) Transfer transverse or longitudinal traveling mechanism

Fig3: Mechanic principles of lane-switch shuttle

Lane-switch shuttle is composed of three basic mechanisms, namely the lifting mechanism, longitudinal and transverse traveling mechanisms.

Working principle of the lifting mechanism: Start a set of

lifting mechanism of deceleration servo motor, then drive the connected worm shaft. The synchronous belt gear on the worm shaft rotates with worm shaft. Through the synchronous belt meshed with it and another synchronous belt wheel drives across another worm shaft to rotate synchronously; Four worm shafts at the group's both ends of the worm shaft rotate together with shaft, each worm drives worm gear meshed with it to rotate, worm gear drives the eccentric disk installed inside the sliding bearing to rotate. Spin to the configuration in the worm gear and worm pair makes four eccentric disks distributed on both sides of the shuttle plate's body into a mirror symmetry; Through the rotation of the eccentric disk can make a set of transverse traveling wheel (5) or longitudinal traveling (6) installed in the lifting frame (4) lift at the same time. The lifting height is precisely controlled by deceleration servo motor.

Working principle of transverse traveling mechanism: Drive transverse geared-motor, drive the drive-shaft to rotate, drive-shaft drives transverse traveling wheel (5), realize function of transverse traveling , through the motor positive or negative rotation and changing speed to realize transverse traveling forward, backward and traveling with variable speed.

Working principle of longitudinal traveling mechanism: Drive transverse geared-motor, drive universal transmission shaft to rotate, universal transmission shaft drives the Longitudinal traveling wheel (6) to realize function of longitudinal traveling, through the motor positive or negative rotation and changing speed to realize transverse traveling forward, backward and traveling with variable speed.

Lane-switch shuttle uses rechargeable battery to provide direct current. Direct current converts into industrial frequency current (ac) by current-source inverter to drive every motor. The control unit is used to control the motor's logic action, It communicates with the computer of shuttle's logistics system at the same time.

IV. THE BASIC WORKING PATTERN

There are two basic layout forms of intensive automatic warehousing system based on lane-switch shuttles: The first is a form with haulage roadway, as shown in Fig.4; The second is a form without haulage roadway form, as shown in Fig.5.

In the first kind of layout form, every layer of the racks densely arrays put-in channels to store the goods. There are haulage roadways, crossing vertically with put-in channels, as the roadway that shuttle plate carrying the goods in-out. Shuttle plate travels along the haulage roadway to the target channel, then reversing to travel into the channels to pick-up or place the goods. Haulage roadway is commonly decorated in central of the channel. Each layer of rack channel is divided into A and B regions .The length of two regions' channel can be equal or different. Each put-in channel's A and B segments are used to store one kind of goods. In the same channel, the goods allocation principles are followed: when loading, if region A is full, the goods will be stored in region B; When unloading, if region A is empty, the goods will be fetched in region B. This can make the region A and B are used alternately, to realize goods "first-in, first-out grouping" and in region A or B with the goods "first-in, last-out". This kind of "conditional first-in, first-out" way is suitable for short storage period, multi-varieties, and minority single-variety goods.

In the first form, the vertical and crossing rail, lying at the bottom of haulage roadway, connects haulage roadway with each put-in channel's A and B regions. Lane-switch shuttle can reverse to travel between haulage roadway and each put-in channel. Setting fixed elevator at two-side of haulage roadway, the lifting platform of it is equipped with adjustable plate fork which is similar to the fork of stacker. Adjustable plate fork combines with lane-switch shuttle to realize the goods in-out. In the low frequency and quantity of in-out, we can also set elevator at the single side of the haulage roadway and use it in-out at the same time.

Loading operation cycle with haulage roadway form: Shuttle plate waits for orders at a specified point; The central controlling computer system issues loading instructions which make shuttle plate travel along the roadway to arrive the pick-up point under the direction of a radio frequency; Elevator's fork sends the goods onto shuttle plate. Shuttle plate goods detection sensors to detect the goods, then the lift-up device will hold goods. Then the shuttle plate travels along the roadway to the opening target position. In according with the principle of the goods distribution, it reverses to drive into the channel section A or B. Arriving at the target position and put down the goods, then it back to the specified point waiting for new instructions. Unloading circulation process can be got similarly, go here.

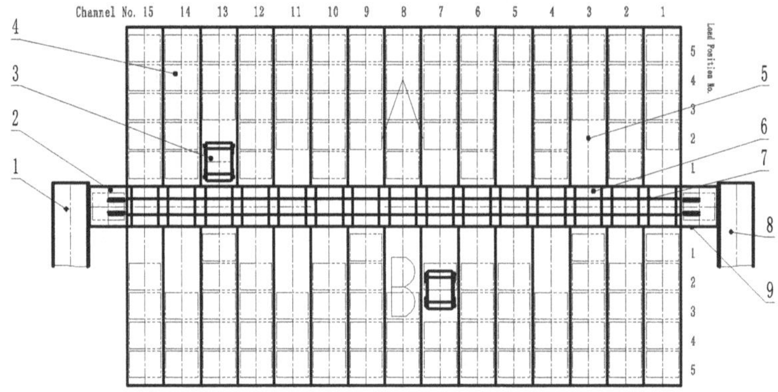

1-Loading cargo station, 2-Loading elevator, 3-lane-switch shuttle, 4-Unit-Load, 5-Goods channel, 6-Haulage roadway, 7-Cross rail, 8-Unloading Cargo Station, 9-Unloading elevator

Fig.4. With haulage roadway layout form

The second layout form without haulage roadway is similar to the traditional one-way shuttle plate intensive warehouse. The ends of put-in channels are loading and unloading sides, which use stacker or high forklift to realize in-out warehouse. In this form, the goods moving from the loading side to the unloading side is accomplished by shuttle plate in the channel. Unlike traditional one-way shuttle plate, lane-switch shuttle's transfer doesn't have to use the stacker or high forklift between the different channels, but through crossing rail laid on both sides of the bottom of the channel, which doesn't have to take up the stacker or high forklift's working time to improve the system loading and unloading efficiency. This kind of layout can achieve the full "first-in, first-out" working way.

Loading operation cycle without haulage roadway form: Shuttle plate waits for orders at a specified point, the central controlling computer system issues loading instructions. Then loading stacker send the goods waiting for loading to the specified position of the entrance of the channel; Shuttle plate travel along the roadway to arrive the downside of goods under the direction of a radio frequency .Shuttle plate goods detection sensors to detect the goods. Then the lift-up device will leg up and take the goods away from the entrance of the channel. Then the shuttle plate travels along the roadway to deliver the goods until arriving the storing position; When we need shuttle plate traveling to another channel to carry on loading task, the shuttle plate traveling along the current rail under of the channel arrives the side of the loading channel. Then it traveling along the cross rail arrives the loading side of the target channel to begin to execute new loading task. Unloading circulation process can be got similarly, go here.

In practical application, we can depend on storage conditions and requirements to composite other forms of working forms.

V. THE SCHEDULE OF LANE-SWITCH SHUTTLE

The advantage of the task scheduling is that the management of the process of the scheduling task is simpler. The load balance among the shuttle plates is easy to process[6].

Take the form of the first layout as example. According to different types of loading and unloading task, implementation process of shuttle plate can be decomposed into five steps of unloading or loading. When executing the unloading task: ① To determine target position of pick-up goods (position coordinates); ② Start traveling mechanism to reach target point; ③ Complete the pick-up goods; ④ Start traveling mechanism reach the side of unloading; ⑤ Complete unloading. When executing the loading task: ①To determine target position of stock (position coordinates); ② Start traveling mechanism to reach the side of loading; ③ Complete the pick-up goods; ④ Start traveling mechanism reach the point of stock; ⑤ Complete unloading.

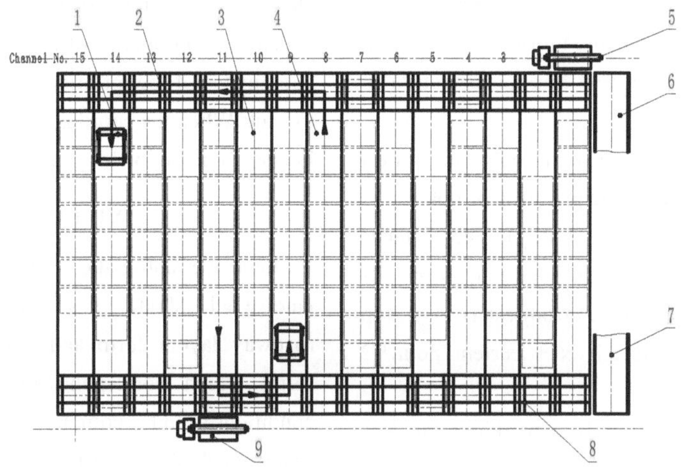

1-Lane-switch shuttle, 2-The cross rail at the bottom of the entrance of channel, 3-Goods channel, 4-Unit-Load, 5-Loading stacker (or forklift), 6-Loading cargo station, 7-Unloading cargo station, 8-The cross rail at the bottom of the exit of channel, 9-Unloading stacker (or forklift)

Fig.5. With no haulage roadway layout form

As shown in Fig.4, establish position coordinates of goods. Set X axis as the increasing direction of the channel number and Y -axis as the increasing direction of the channel number. Region A is positive and region B is negative. In Fig.4, stand-by point of the first plate is coordinate (9, 1), stand-by point of the second plate is coordinate (1, 0).

The principles of determining target point coordinates of loading and unloading:

The status information of the stock is indicated by "0, 1". 0 indicates the position is empty, 1 indicates the goods in stock; In the next place, region A has opposite properties to region B. That means, when region A is for pick-up region, region B is the stock region, vice versa. A specific position is chose as the target position of unloading and pick-up. The status information of position is 1 and satisfies the absolute value of the ordinate to be minimum. A specific position is chose as the target position of loading and pick-up. The status information of position is 0 and satisfies the absolute value of the ordinate to be maximum.

Algorithm for shuttle plate reaches target point from the current position: current_point i,j, current_point is the current position of shuttle plate, i indicates the number, j indicates the number of axis. Such as current_point 1,1 is coordinate of X axis for the car A, current_point 2, 2 is coordinate of Y axis for the car B.

Fig.4, the X axis is the number of c hannel, Y axis is the number of position, region A is positive, region B is negative.

The waiting position of the first car is (7,0), the waiting position of the second car is (3,0).

Fig.5, x axis is from 0 to 12, y axis is from 0 to 10.

The program to achieve target of Fig.2.

Set a target position is (x, y), judge current_point i, if 1 = x. If it is not, to the x axis. If X>current_point i,1 reverses forward.If X<current_point i,1 reverses back. Until x=current_point i,1, in the same way, y = current_point i,2, get the target point.

The program to achieve target of Fig.3.

Set a target position is (x, y),

1) Judge if current_point i,1=x.

A, if it is not. Judge if current_point i,2=0 or current_point i,2=11.

1. If it is, judge if current_point i,1>x.It is true to travel back. It is not to travel forward. Until current_point i,1 =x. In the same way, if current_point i,2 =y, get the target point.
2. If it is not, judge if current_point i,2>8.It is true to travel forward until current_point i,2=11. Otherwise, travel back until current_point i,2=0.Judge if current_point i,1>x.It is true to travel back. It is not to travel forward. Until current_point i,1 =x. In the same way, if current_point i,2 =y, get the target point.

Loading task is divided into ① reach target point (loading stacker) ② reach target point of picking up (loading position) ③ put the goods ④ execute the next mission or return to waiting point.

When a unit-load is loading or unloading only, the nearest car will work. The second car judges to give way or not.

The coordinate point of unloading is (x,y). The distance from car A to pickup point is a_distance = |current_point1, 1- x| + |current_point1,2 - y|. The distance from car B to pickup point is b_distance = |current_point2,1 - x| + |current_point1,2 - y|.

If (a_distance >= b_distance), car B is working, car A is waiting.

If (a_distance < b_distance), car A is working, car B reaches the point (4,1) to give way. When the task is finished, two cars return to the waiting point.

When only a unit-load is loading, just judge the distance to loading stacker.

When need many unit-loads to load or unload, two cars need to work together. The first task is assigned to the nearest car, following tasks are in turn. When need the car to avoid, the car travel to the nearest channel according the priority .

Calculation method of priority:

Priority of car A is a_priority. Priority of car B is b_priority.

a_priority = is_load*2 + 4*is_x + 1;

b_priority = is_load*2 + 4*is_x + 0;

Is_load indicates loading or unloading. If loading, it is 1.Otherwise it is 0;

Is_load indicates on X axis or not. If on X axis, it is 1.Otherwise it is 0;

The judgment and methods of avoiding:

Use the variable: current_point is the car's coordinate.Car's line (coordinates) of this task has not been finished and the coordinates of a series of points (assuming two channels).

When the coordinates of a series of points exist intersection, one of two cars must avoid.

The choice of avoiding way:

When the coordinates of high priority car include the coordinates of low priority car. The car need to enter the channel for avoidance.

The choice of entering the channel for avoidance:

It is not the channel for another car to enter; The forward channel has the nearest distance with the current coordinate; When the coordinates of high priority car do not include the coordinates of low priority car, the low priority car only need to wait; When the coordinates of a series of points do not exist intersection, the low priority car can continue to work.

VI. CONCLUSIONS

Intensive automatic warehousing system based on the lane-switch shuttle has two basic layout: with haulage roadway layout form and without haulage roadway layout form, which correspond to first-in, first-out method of conditions and completely first-in, first-out method. Lane-switch shuttle of the system is composed of three basic institutions, namely the lifting mechanism, longitudinal and transverse traveling mechanisms, which makes lane-switch shuttle has the function of traveling along a vertical and crossing path on the plane. In the schedule of lane-switch shuttle, implementation process of shuttle plate can be decomposed into five steps of unloading or loading; When many shuttle plates work together, operating management takes the method of comparative priority to decide which shuttle plate to avoid.

ACKNOWLEDGEMENTS

The work was supported by a project of Jinan Higher Education Independent Innovation Program (Grant No.201202074).

REFERENCES

[1] Zhang Bo, Stereoscopic Warehouse Concentrated Storage [J]. Logistics Technology, 2002, 11(12): 15-16

[2] Chen Zhentian, Intensive Storage System Construction and Application [J]. LOGISTICS & MATERIAL HANDLING, 2013, №2: 61-64

[3] Shen Bin, Luo Qilong, Gravity AS/RS [J]. Mechanical Manufacture, 2003, 41(467): 61-64

[4] Zhu Congmin, Wang Weiqing and YU Hansheng, The design and application of automatic shuttle plate [J]. Logistics Technology, 2007, 26(7): 96-98

[5] Shen Minde, Chen Zhaoqiang and Zhang Peng, A travel along the cross rail shuttle plate [P]. Chinese Patent: CN201310052063.7, 2013-05-15

[6] Meng Xianfu, Study of Task Scheduling and Load Balancing Models Based on Priorities [J]. MINI-MICRO SYSTEMS, 2005, 9(9): 2-5

Non Production Direct Workforce Work Measurement Challenges and Case Study

Thong Sze Yee*, Zuraidah Bt. Mohd Zain, Bhuvenesh Rajamony
Department of Manufacturing Engineering, Universiti Malaysia Perlis (UniMAP), Perlis, Malaysia
(*szeyee2@gmail.com, zuraidah@unimap.edu.my, bhuvenesh@unimap.edu.my)

Abstract - **Most of the operations in the production line consist of discrete work activities. With the systemic approach in work measurement methods, setting time standards for the works and deriving the DL (Direct-Labor) workforce requirements are achievable. However, advanced technology has changed the work nature of the operations with increased supporting activities to sustain routine tasks like quality assurance, machine changeover management, trouble-shooting, material replenishing, and administrative works. The combined activities have now become non-discrete, non-routine, non-repeatable in a predictable pattern, and the work process steps are now relatively loosely defined. Workers who perform these works are categorized as unique workforce namely NPDL (Non-Production-Direct-Labor). Setting time standards for NPDL is challenging due to the work nature and the work measurement methods used in this region are mainly stopwatch time study, basic PMTS (Pre-determined-Motion-Time-System), and shop floor self-recording systems. Without defining the time standards, the common practice to determine the NPDL workforce is based on past experience or estimated DL: NPDL ratio. This paper discusses case study of using a systemic approach to overcome the difficulties to determine NPDL workforce by defining steps of categorizing the NPDL work activities. The approach also contributes to measure the NPDL productivity objectively with the existing work measurement methods.**

Keywords - NPDL (Non-Production-Direct-Labor), PDL (Production-Direct-Labor), work measurement

I. INTRODUCTION

In order to get the most out of the factory production line, work measurement methods are used. The work measurement is carried out to set the machine and labor standard time, and thus to compute the machine capacity and labor resource required to support the factory production build plan. In general, the labor resource that supports the production line execution needs is called the "direct labor". As per the American National Standard (IIE, 2000) the direct labor means any labor whose cost is directly allocated to a product [1].

In the conventional factory set up, most of the direct labors' work in the production line has clear cut assembly work task. This group of workers is termed as PDL (Production-Direct-Labor) in this paper. However, the advanced technology has changed the work nature of the operations with increased supporting activities to sustain routine tasks like quality assurance, machine changeover management, trouble-shooting, material replenishing, and administrative work. Due to their work nature which is different from the "conventional stationed base labor work", they are categorized as unique workforce namely NPDL (Non-Production-Direct-Labor).

With the nature of the task activities whose cycle time are discrete, routine, repeatable in a predictable pattern as per the operating specification, there is no difficulties in modeling out the requirement of PDL with the conventional work measurement methods. However, NPDL's task activities are largely opposite to the PDL's work in many perspective. Their combined activities are non-discrete, non-routine, non-repeatable in a predictable pattern, and the work process steps are relatively loosely defined. Industrial Engineers face the challenges of modeling this type of non-conventional activities by using the methods which are based on the conventional activities. Without defining the time standards, the common practice to determine the NPDL workforce is based on past experience or estimated DL: NPDL ratio.

It is important to determine the NPDL requirement with the correct method from manufacturing overhead cost perspective because labor cost is part of it. The current practices for direct labor cost computation include workers from both PDL and NPDL. Moreover, due to the unique skillset and qualification, NPDL is generally at a higher salary scale.

This paper describes several of the more popular work measurement methods used, and discusses the challenges encountered during the execution of some of the methods in four separate NPDL groups.

II. LITERATURE REVIEW

A. *The Considering Factors*

Federick Taylor's scientific management for labor work began with the objective to set a fair day work (Salvery, 2001) pay with a systemic approach. The methodology breaks the task into small and simple actions and thus precise time can be determined for each of the action in the task scope [2, 3]. With the introduction of this time and motion study, labor wage is determined objectively. Over the decades, numerous theories and methods have been developed to measure labor work based on Taylor's principle of scientific management, widely known as work measurement. Work measurement is popularly defined as the application of techniques to establish time standards for a qualified worker to carry out a task at a defined performance level. Time Standards provide a means for the management to measure workers' performance,

E. Qi et al. (eds.), *Proceedings of the 5th International Asia Conference on Industrial Engineering and Management Innovation (IEMI2014)*, Proceedings of the International Asia Conference on Industrial Engineering and Management Innovation 1, DOI 10.2991/978-94-6239-100-0_51,

hire the appropriate workforce, install equipment and facilities capacity, and measure the utilization of its capital tools. In short, work measurement sets productivity level accordingly for an organization workforce and overall overhead cost target.

B. Types of Work Measurement Methods

There are two main categories of work measurement methods which are used to measure cycle time and set time standards for the task. These two categories are typically called 'Engineered' and 'Non-engineered' methods (Zandin, 2006 [4]). Non-engineered methods are usually non systemic, and often involve pure or best guess, and hence they are less accurate and inconsistent. In addition, data from past history is normally used in combination with experience within a particular area used as reference. Stopwatch time study and work sampling fall into non-engineered standards but they produce more accurate and consistent standards despite some drawbacks. On the other hand, engineered methods customarily use data sheets that comprise time values that use Predetermined Motion Time System (PMTS) as a basis for measurement. This system was developed as early as 1930 by the Gilbreth and others who originated the concepts of scientific management for work measurement. PMTS systems are based on the concept of analyzing human capabilities and hence determining the average rates at which a task can be accomplished. Some examples of the engineered methods are MTM (Methods Time Measurement), MSD (Master Data), MOST (Maynard Operation Sequence Technique) and MODAPTS (Modular Arrangement of Predetermined Time Standard). They are all called 'Pre-determined' because all the necessary time values have been predetermined based on very detailed time studies (Zandin, 2006 [4]). The drawbacks of PMTS methods are the activities steps and motions must be clearly defined and cycle time is discrete.

Table I outlines various types of non-engineered and engineered methods.

C. How Modern Technology Affects Work Measurement Methods

With more diverse types of manufacturing more diverse types of manufacturing industries blooming, production lines must also evolve to support operation needs accordingly. Some key driving factors for the revolution are product life span, reliance on IT in operating a production line, heavy use of intelligent machines, and the pattern and intensity of human-machine interaction. Apart from the product and machine perspective, the job market is also rapidly changing. The ageing population, workers' literacy levels, and the influx of female workforce have shaped the way standard time is established. Workers population is not 'homogeneous', and thus Industrial Engineers must be creative and resourceful in analyzing the contributing factors of their (the workers') capabilities to set the right standard for the population. Standard time must be established in a timely manner in order to meet its purpose despite frequent changes of products and processes in the production line.

Of late, the availability and implementation of IT systems such as electronic records of cycle time within the machine, efficient recording device camcorder, smart phone, PDA, and data processing software have inevitability changed the way data is captured for work measurement analysis. Traditional work measurement methods have been reengineered to accommodate and leverage the capabilities and convenience provided by electronic systems. Table I shows some commonly used examples of 'modified data sheet' or 'in-house developed' work measurement method, shown in the 'Hybrid' column.

TABLE I
WORK MEASUREMENT METHOD EXAMPLES

Commonly Used Work Measurement Methods in Factory Production Line			
Category			Method
Engineered	Non-Engineered	Hybrid	
√			MTM
√			MODAPTS
√			MOST
√			SGD (General Sewing Data)
	√		Stopwatch Time Study
	√		Work Sampling
√			Synthesis from standard data
	√	√	Estimation
	√	√	Analytical estimating
	√	√	Comparative estimating
	√		Shop Floor - Time Value Download
√	√	√	Self-developed System

D. Hybrid Methods

Hybrid methods are created because the current single systems do not entirely meet their needs in the business environment. They comprise combinations of 2 or more systems to complement the shortcomings of available single systems (combination of Stopwatch time study for labor activities with Shop Floor System Download for machine process cycle time). By grouping a few standard times into batches for a task, data can be retrieved for use in similar jobs. They meet the desired level of speed and data accuracy to derive the standard, provided the work analysis work is carried out by trained and experienced analysts.

In essence, the combinations of work measurement methods or hybrid methods are normally developed for in-house use and they are modified along the way for task time standards determination.

E. Selection of Suitable Measurement Methods

As discussed above, different methods are there for different work tasks, in order to meet different expected standard of setting the time standards. Some methods provide a higher accuracy whilst others are based on estimation but require less analysis time. Table II shows a comparison of MTM, Shop Floor System Download method, stopwatch time study, and work sampling method. Each of the methods has its strength and shortcomings in analyzing different types of work activities. Detail and accuracy are the trade offs against speed and cost of application. (Daniels, 1991 [5])

TABLE II
WORK MEASUREMENT METHODS AND THE CHARACTERISTICS

	Cycle Time Range				Task Process Step				Movement & Motion Repetation				Group / Individual Activities				Task Type		
Method	Short	Medium	Long	Very long	Highly mature	Simple step	Complicated step	>50 steps with no discrete start and end	Fixed	Mixture & predictable	Random	Random & unpredictable	Individual	1-2 person	3-5 persons (small group)	>5 person (Large group)	Physical	Physical motion & thinking	Judgmental decision making
PMTS - MTM	1					1			1				1				1	1	
Shop Floor System Download	1	1	1	1		1			1				1				1		
Work Sampling	1	1	1	1				1		1	1	1	1	1	1	1		1	1
Stopwatch Time Study	1	1			1	1			1				1				1		

Frederick Taylor, in his book 'Principle of Scientific Management' (Taylor, reprinted 1997), laid down the fundamental principles of large scale manufacturing through assembly-line factories. Standard time for each worker's daily task is dissected to be segmented according to the process steps. In another words, timd standards are established according to the specific defined job scope. This standard task and work order are repeatable and similar for every worker assigned to the same job in the production line. Their job performance is measured against the time standards that have been set. In the production line, this type of workforce is categorized under direct-labor workforce hired to support routine operation activities.

III. THE STUDY

This paper describes four case studies carried out in three electronics assembly and box build manufacturing factories under the same company. One of the factories is located in Eastern Europe and the other three located in two countries in Asia region. The head count is approximately five hundred workers for two factories respectively and 5000 for another one factory. In all three locations, there are resident Industrial Engineer whose job oversees direct labor planning using work study methods. The author shadowed the Industrial Engineers for planning the work measurement method, designing the template for data collection, carrying the observations during data collection stage, analyzing the data and was involved in the many discussion sessions related to the implementation of the work study measurements. In addition, the author also interviewed the workers in the area where the case studies taken place to understand any potential hiccup issues [6, 7].

The focus of the paper is on non-production direct labor, and it is limited to support crews who work in the production line. Office administrative workforce with the same job grade level was excluded. The number of non-production direct labor in this case study ranged from fifteen to thirty percent of the total direct labor workforce. Their primary job task is not working on the production line to directly produce products, but rather, they enable the production execution to be more efficient. The service provided by these groups are quality control audit, raw material supply, maintenance & trouble shooting of equipment, tools, & electronics systems, collection & verification of problematic products, finished goods, products transportation, assisting in administrative work, and facilitating & filling up the gap among work stations or working zones [8, 9].

IV. RESULTS AND DISCUSSION

A. *NPDL versus PDL*

When the work measurement methods were initially developed, they were primarily to measure standard work tasks with discrete cycle times in which the motion was describable and the time value was assigned accordingly. With highly repeatable work segments and clearly defined respective standard time, the workforce determination was done based on the standard. However, today, the activities involved in NPDL are not discrete, routine, repeatable in a predictable pattern, and the work process steps are relatively loosely defined. The characteristics of these types of work tasks do not match with the work measurement 'pre-requisite' conditions.

Data for motion and time collection studies are collected by visual observation in this conventional work measurement method. However, the ability to observe, record, and analyze for a full day, a full week or a full month work is not possible without resorting to routine and repeatable patterns. The next problem in the cycle time collection is that the sample size is usually small, because the NPDL work activities are not in a predictable pattern. Much resource is demanded for conducting field observations to record the task activities, motion, and time values. This will create concerns of inconsistency in the collection and production of compatible and comparable standard time for large use throughout the entire factory NPDL population. Table III shows the differences of (PDL) Production Direct Labor versus the (NPDL) Non Production Direct Labor task activities in motion.

The result of this scenario is that the NPDL standard time and workforce determination is excluded from the Industrial Engineer's work measurement scope. The negative impact will occur in the long run, because, without the standard time being defined, the common practice to determine the NPDL workforce is based on past experience or estimated DL: NPDL ratio. The consequence of this is that, the workforce assigned will either be too high or too loose in the factory operation. This can account for fifteen to thirty percent of labor costs. In addition, there might be the intangible impact of perceived 'uneven or unfair' work load distribution among the direct labor.

TABLE III
PDL AND NPDL ACTIVITIES COMPARISON

Comparison criteria	Production Direct Labor (PDL)	Non Production Direct Labor (NPDL)	Non Production Direct Labor (NPDL)
Work Type	Direct Labor - SMT Machine operator	Non Production Direct Labor - Warehouse operator	Non Production Direct Labor - 'Trouble shooting' operator
Cycle time range	Fixed amount of time to load a stack of products into machine	Varies depends on carton size and quantity	Varies depends on defect mode
Repetition of activities (random or fixed pattern)	Fixed pattern for every batch	Non routine pattern	Non routine pattern
Type of activities	Follow the operating documentred standard work	Load, unload, adminstrative and other work	Trouble shooting work on various defect modes
Group or Individual activities	Individual operator has a standard work list	Group activities	Individual operator with multiple trouble shooting guidelines
Process Steps	Process flow is defined in the operating standard work	No operating standard process step	Normally working guidelines, but not in detail process step
Driven factors - Motion driven	Physical work driven, minimum thinking	Physical work driven, minimum thinking	Physical work driven, with considerable amount of thinking process

B. Factory NPDL Work Measurement

Before any detailed work measurement study is started for NPDL, a basic field observation was carried out to scan through the work activities involved. A rating, which is measured against a set of criteria is given in Table IV. This scanning and rating process is important for a factory that is just about to embark on the NPDL work measurement. For example, starting work measurement from production line Quality Assurance (QA) operators would be easier than from a warehouse operator because there are more predictable, discrete, and repeatable activities here, based on a standard work process. Hence, the analysis results would yield to a higher acceptance level.

Another area that deserves close attention is the target workers in the analysis. The NPDL's have never been measured for their activities, and thus they should be mentally prepared to avoid being perceived that their 'self-assumed reference standards' are challenged. Fig.1 summarizes the process flow used in the preparation stage for NPDL work measurement.

C. NPDL Work Measurement Process

Due to a wide range of work nature and activities involved in all sorts of NPDL work in the factory, work measurement methods are varied in order to get optimum analysis and results. Table IV shows a list of criteria with rating scale that objectively rates the characteristics of the NPDL works. Six rating criteria are categorized into two groups, namely i) activities discreteness and ii) activities occurrence frequency or pattern.

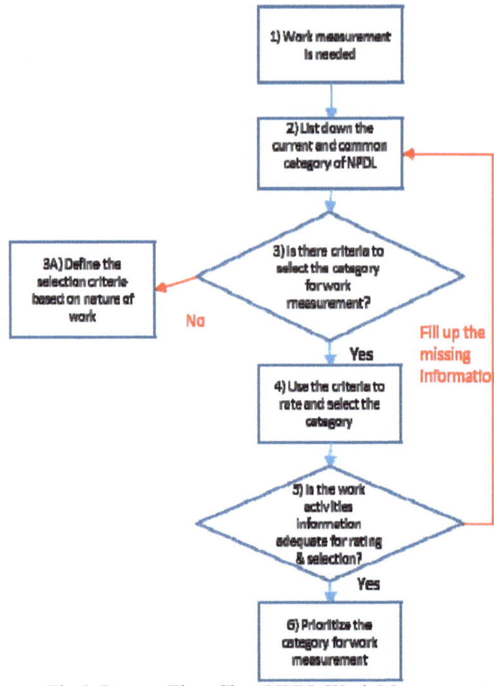

Fig.1. Process Flow Chart NPDL Work Measurement

TABLE IV
NPDL ACTIVITIES RATING CRITERIA

			Rating			
Category	Criteria #	Criteria description	1	2	3	4
Discrete	1	Cycle time range	Short	Mid	Long	Highly mixture
Discrete	2	Group or individual activities	Individual	1- 2 person	small group	Large group
Discrete	3	Steps	1-10 steps/ cycle	20-50 steps/ cycle	>50 steps / cycle	>50 steps without clear start and end
Discrete	4	Driven factors - Motion driven	Physical motion	Physical motion & thinking	Thinking, judgmental decision making	
Frequency & Pattern	5	Repetition of activities (random or fixed pattern)	Fixed	Mixture & predictable	Random	Random & unpredictable
Frequency & Pattern	6	Expectation of the activities	List of work steps available	Partially defined	Non defined	
D: Activities discreteness						
F: Occurrence frequency or pattern						

To demonstrate the use of the activities rating criteria, Table V shows an example using the Incoming Parts Quality Control (IPQC) Operator daily task.

TABLE V
IPQC ACTIVITIES RATING

Rating Criteria / Task Activities	Cycle time range	Group or individual activities	Number of steps / cycle	Driven factors - Motion driven	Repetition of activities (random or fixed pattern)	Expectation of the activities	Overall Rating
1) Start of shift audit (1x)	Short	Individual	1-10 steps/ cycle	Physical motion (occasionally thinking)	Fixed	List of work steps available	1
2) Product conversion audit (varies depends on production scheduling plan)	Short	Individual	1-10 steps/ cycle (occasionally extra steps)	Physical motion (occasionally thinking)	Mixture & predictable	List of work steps available	1
3) Hourly audit	Short	Individual	1-10 steps/ cycle	Physical motion (occasionally thinking)	Fixed	List of work steps available	1
4) 4 hourly audit	Short	Individual	1-10 steps/ cycle	Physical motion (occasionally thinking)	Fixed	List of work steps available	1
5) Production line interuption audit (occur on ad hoc basis whenever the productionn stop working due to any reason)	Short to Medium	Normally individual (occasionally 1- 2 persons)	20-50 steps/ cycle	Physical motion & thinking	Mixture & predictable	List of work steps available. Some abnormal cases are partially defined)	2
6) Out of spec limits control verification and report preparation	Short to Medium	Normally individual (occasionally 1- 2 persons)	20-50 steps/ cycle	Physical motion & thinking	Mixture & predictable	List of work steps available. Some abnormal cases are partially defined)	2

Based on the 6 daily tasks, task 1 to task 4 have the overall rate of one; task 5 and 6 are rated 2 for the 3 criteria namely criteria #3) steps, criteria #4) driven factors where there are more thinking process involved on top of physical motion. Because for task 5 and task 6 activities occur at random based on predicted yield and probability percentage, it scores 2 for criteria #5.

This kind of rating exercise is carried out for other types of NPDL tasks as well. The summary of the task is as shown in Table VI. Based on the rating score, IPQC is the lowest, which means the activities are more discrete and they occur at a more routine and predictable pattern, when compared to other tasks.

TABLE VI
JOB TASK PRIORITIZATION FOR WORK MEASUREMENT ANALYSIS

Criteria #	IPQC	Store hand	Material handler	Engineer helper	Admin clerk	Leader
1	1	2	2	2	3	4
2	1	2	1	1	2	1
3	1.5	1	1	3	4	4
4	1.5	1.5	1.5	2	3	3
5	2	2	2	2	2	3
6	1	2	2	2	2	3
Sum	5	6.5	5.5	8	12	12
Priority	**1**	**3**	**2**	**4**	**5**	**6**

D. *Case Study #1: IPQC*

The scanning of the IPQC operator's task shows that most of the activities are discrete. The seemingly challenging part of setting standard time for this group of job tasks is the frequency of occurrence. However, this frequency is based on the part quality yield or track record. In other words, the computation of resource for IPQC can be computed as follows:

<u>Sub Activity #1</u>

Out of spec check cycle time = 50 minutes per cycle

Frequency of occurrence = 2 time per week

Total time required / week = 100 minutes

Average time / shift (12 hours) = 7.14 minutes

<u>Sub Activity #2 to #n</u>

Following the same computation for other ad hoc tasks with the pre-defined or estimated yield percentage.

In summary, by adding all the sub activities, the total time required can be translated to the quantity of resource. In this case study, since the task can be segregated to discrete start and end of motions, the time study with stopwatch and the PMTS methods are sufficient to handle the work measurement analysis.

E. *Case Study #2: Trouble Shooting Operation*

Another case study, that involved the NPDL, is the trouble shooting cum repair operation worker. This type of work involved trouble shooting and repairing customer-returned units. The failure modes of each of the products can be varied, and so is the frequency of occurrence. Hence, the difficulty involved both activities discreteness and occurrence pattern and frequency.

To tackle the data discreteness concern, a recipe 'reference book' was created, whereby all past failure modes were listed. Using an advanced electronic recording system, past trouble shooting details were extracted from the trouble shooting station. The information included start and end time of each trouble shooting, failure modes, and the frequency of occurrence. However, the station records did not provide the details of the motion of each step in the operator's or technician's activities step. It was observed that there was a 'high level' motion time for the trouble shooting activities available and associated with the respective failure mode. To make the standard time analysis complete, the Industrial Engineer filled in the gap of recording other standard activities such as:

1) Walking time between unit storage area to trouble shooting stations
2) Spare parts retrieval time
3) Time for administrative work related to the task
4) Time for other standard activities

In summary, this case study demonstrated a good use of an electronics recording system that provided cycle time and frequency of occurrence. The record was not as accurate compared to the PMTS level, but it was good enough for headcount resource planning. The occurrence was not in a predictable pattern, but the average frequency from the electronics record was a good estimation. Besides, the cycle time for trouble shooting and repairing different types of failure mode can be retrieved from the record as well. The operation could fine tune the variables of routine-ness with the failed or breakdown machines owners from the profiling records.

F. *Case Study #3: Machine Maintenance*

Not all the trouble shooting works were equipped with electronic recording systems, or carried out with power supply on. In the absence of the convenience of an electronics system, conventional methods could be used. Another case study scenario was a group of technicians whose routine activities were working in the production line to clear machine stoppage and other trouble shooting jobs individually or in small groups. The task also involved in entering data at computer stations, attending briefings, and participating in improvement projects. The time spent on task activities were driven by the workers' skill set, and the occurrence was influenced by internal and external factors that beyond the worker's sphere of influence.

There was no discrete activity and fixed occurrence frequency in this case study. The measurement process started with finding the proportion of time spent of activities mentioned above. As shown in Table II, the best tool used for this type of group activity work measurement is Work Sampling (Ray, 2006). Fig.2 is the sample template which was used in this case study. One Industrial Engineer was assigned to collect the 'raw state' data anytime within an hour for 8 hours every day for minimum 2 weeks for each production line. Multiple production lines were included in this work sampling data collection to find the technical utilization rate.

Date: 05-May-13
Station: Technician #1 Note:

8:10am	9:30am	0.05am	1:45am	2:30pm	Time	1:23pm	2:40pm	3:18pm	4:02pm	5:30pm
D1	D2	D3	D4	D5	**State**	D6	D7	D8	D9	D10
				2	Fine tuning parameters	1	1	1	1	1
	2		2		Setting Up and Testing					
1					PM / Cal					
1		2			Jam clearing					
					Idling	1	1	1	1	1
					Others					

Technician #2

8:11am	9:35am	0.15am	1:50am	2:35pm	Time	1:25pm	2:40pm	3:20pm	4:05pm	5:34pm
D1	D2	D3	D4	D5	**State**	D6	D7	D8	D9	D10
					Fine tuning parameters	2	2	2	2	2
2	2	2	2	2	Setting Up and Testing					
					PM / Cal					
					Jam clearing					
					Idling					
					Others					

Technician #3

8:10am	9:33am	0.10am	1:45am	2:30pm	Time	1:25pm	2:40pm	3:25pm	4:05pm	5:30pm
D1	D2	D3	D4	D5	**State**	D6	D7	D8	D9	D10
2	2	2	2	2	Fine tuning parameters	2			2	
					Setting Up and Testing					
					PM / Cal					
					Jam clearing			2		2
					Idling		2			
					Others					

Fig.2. Work Sampling Template

Fig.3 is an example of how the data can be analyzed for utilization rate for the machine maintenance crew. The proportion use of time for each activities was listed. Approximate fifteen percent of idling time and over nine percent was classified under 'others'. Nearly one quarter of nonproductive time each day was likely an indication that the existing pool of resource was underutilized or not efficiently used.

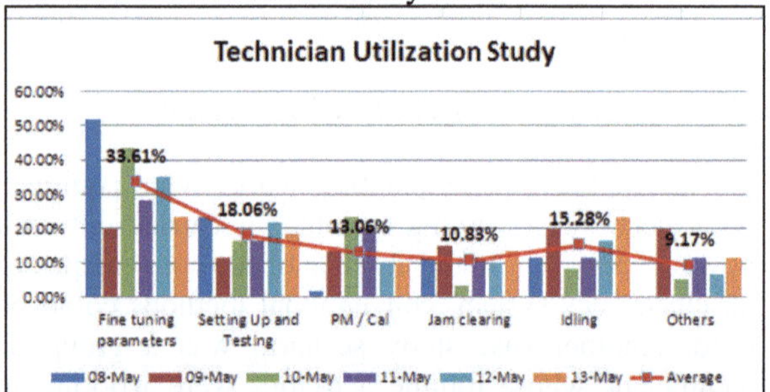

	08-May	09-May	10-May	11-May	12-May	13-May	Average
Fine tuning parameters	51.67%	20.00%	43.33%	28.33%	35.00%	23.33%	33.61%
Setting Up and Testing	23.33%	11.67%	16.67%	16.67%	21.67%	18.33%	18.06%
PM / Cal	1.67%	13.33%	23.33%	20.00%	10.00%	10.00%	13.06%
Jam clearing	11.67%	15.00%	3.33%	11.67%	10.00%	13.33%	10.83%
Idling	11.67%	20.00%	8.33%	11.67%	16.67%	23.33%	15.28%
Others	0.00%	20.00%	5.00%	11.67%	6.67%	11.67%	9.17%

Fig.3. Work Sampling Utilization Analysis

In summary, Work Sampling does not provide time standards based on motion, but for group activities in this case study, it is an ideal tool that is easy to apply, and it involves fewer resources for data collection. Additionally, cycle time can be obtained through time study for PM (Preventive Maintenance), machine conversion set up and Testing process because there are normally operating specification available for these types of activities. For the most optimum use of resource, another round of work sampling should be carried out after improvement / Kaizen activities. The study is to be repeated until the proportion of value-added activities reach the desired level. This way, the management shall have a higher confidence in the next resource adjustment planning.

G. Case Study #4: Warehouse and Store Crew

The factory from case study #3 also adopted the same approach for its warehouse store-hand crew. The state of work was modified to include the goods distribution work to multiple floors sub-stores (was called supermarket) within the same building. The store crew was responsible to work with the incoming trucks to verify the goods by sampling, arrange the goods into proper locations within the store, distribute the goods to sub-stores as and when the service was triggered by the sub-stores. They were responsible for all sorts of paper work for the inventory management. Despite each of the workers had their respective primary work scope, they were expected to assist others when situation called for.

2 Industrial Engineers were assigned to collect data for 2 weeks. Upon the data collection completed, the 'work state' analysis was carried out. The analysis also focused on each floor (a.k.a. functional area) store hand utilization performance as shown in Fig.4. This detail level allowed the management to benchmark the operation execution and narrow the gap from productivity perspective.

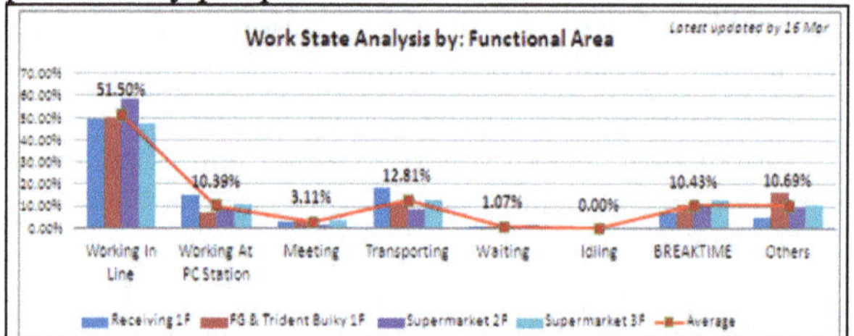

Work State	Receiving 1F	FG & Trident Bulky 1F	Supermarket 2F	Supermarket 3F	Average	Rank
Working In Line	49.55%	50.40%	58.72%	47.30%	51.50%	1
Working At PC Station	15.10%	6.93%	8.56%	10.96%	10.39%	5
Meeting	3.30%	3.93%	1.25%	3.97%	3.11%	6
Transporting	18.39%	11.33%	8.84%	12.67%	12.81%	2
Waiting	1.21%	0.50%	0.91%	1.66%	1.07%	7
Idling	0.00%	0.00%	0.00%	0.00%	0.00%	8
BREAKTIME	7.10%	10.66%	11.41%	12.56%	10.43%	4
Others	5.34%	16.25%	10.30%	10.88%	10.69%	3

VA Job	67.96%	61.25%	68.54%	62.23%	65.00%
NVA Job	19.60%	11.84%	9.75%	14.33%	13.88%
Ranking					
VA Job	2	4	1	3	Supermarket 2F at the top
NVA Job	1	3	4	2	Receiving 1F at the top
BREAKTIME	4	3	2	1	Supermarket 3F at the top
Others	4	1	3	2	FG & Trident bulky at the top

** VA job: value added job, eg. work in line, work at PC and meeting
** NVA job: non value added job, eg. transporting, waiting and idling

Fig.4. Work 'State' Analysis by Functional Area

During the work sampling data collection, the Industrial Engineers managed to quantify cycle time for some segments of work activities. These included cycle time that associated to different transportation route, loading and unloading cycle time and paper work processing time. However, the argument points were the incoming truck unload depended on too many variables such as supplier's delivery truck load level, material supplied to production line and sub-store was on ad hoc basis and the quantity were not standardized.

There was no doubt that the variables were not controllable by the warehouse crew, but the similar events could be classified in batch. For example, 'sending material to the second floor regardless to the exact production lines or exact location' was considered as one batch. The frequency of 'going to second floor' could be traced by records. Similar approach could be used for the incoming truck unloading process. With the assistance from the purchasing department and incoming loading bay crew, past record of truck arriving portfolio could be established.

In summary, the cycle time and the occurrence frequency seemed messy and without a predictable pattern. However, with systematic batching of activities and finding the respective past records is a way to associate the respective cycle time and the occurrence frequency. The amount of the required labor can be computed from the cycle time and frequency. Fine tune and update the computation model frequently to capture the change of truck and product profile is critical to ensure all the assumptions in the batch data does reflect the existing condition.

V. DISCUSSION & INSIGHTS FOR PRACTITIONERS

A. *Experience of Use*

The 4 case studies are mapped in Table VII based on the criteria discussed in section #4.3, Table IV on NPDL Activities Rating Criteria. Criteria #1, 2, 3 and 4 are cycle time related whilst criteria 5 and 6 are occurrence frequency and pattern related. The mapping indicates that despite the occurrence frequency is not fixed, but the randomness can be handled with track record reference. It might be challenging for data retrieval based on hard copy records in the past, but not anymore with the electronics system. Another area that deserves a closer attention is the cycle time determination. Non discrete cycle time, high mixture of cycle time length and steps due to undefined process steps are the most influential factors to set time standards.

Three out of four case studies show that the work task cycle time is non discrete in which the motion is not describable and thus time value is difficult to be assigned accordingly. To overcome situation as such, instead of measuring each motion, time value is assigned to a batch of activities in a high level context. The batch activities are then assigned with a referencing code or equivalent identification. Otherwise, time can be measured with engineered, non-engineered or hybrid method for the segments which are describable in 'motion' or steps. For any other task activities which are neither measurable with any methods aforementioned, the alternate verification method is work sampling. Work sampling may be carried out started from macro levels, then gradually into micro levels.

TABLE VII
COMPASITON OF APPLICATION METHODS

Criteria # / Worker type	IPQC	Repair Man	Maintenance	Store hand
Cycle time range	1	4	4	4
Group or individual activities	1	1	2	2
Number of steps / cycle	1.5	4	4	2
Driven factors - Motion driven	1.5	3	3	1
Repetition of activities (random or fixed pattern)	2	2	2	2
Expectation of the activities	1	2	2	2

Variables	Method			
Method / Worker type	**IPQC**	**Repair Man**	**Maintenance**	**Store hand**
Cycle time	Time Study	Electronic System; Time Study	Time study for activities which have process steps defined.	Time Study for discrete activities
Frequency	Track record; Yield trend	Track record; Product quality profile		Track record
Alternate Verification Needs	n/a	n/a	Work sampling	Work sampling

B. *Limitations and Concerns of Use*

The product design is changing rapidly whereby it drives the change for many aspects in manufacturing throughout the pipeline as well. Product quality performance, type of raw material needed, machines and tools types conversion may have to be taken into consideration. This kind of chain effect will affect the cycle time and frequency of task activities. To ensure the existing NPDL model's validity, new model validation must be carried out timely.

Traditionally, NPDL resource is justified through list of activities by the owners. Because the justification is only substantiated by high level task list but without the standard time and occurrence frequency and pattern in numbers, no verification by Industrial Engineer is carried out. Hence, it is a possible convenient way to add in buffer during resource planning. With the task activities clearly defined, it means the resource usage is transparent in this new NPDL model. Resistance from the owner groups may arise and thus refuse to be cooperative to support the new model. Management must ensure the continuous support so that documentation of standard work, cooperation in all the verification continues.

VI. CONCLUSION

It has been demonstrated in this paper that NPDL workforce can be derived from systematical and logical approach through work measurement methods. The computation of time standards and occurrence patterns are not similar to the conventional PDL's task, but they can be overcome with scientific approach, instead of pure guess, gut feel, past experience or estimation of DL: NPDL ratio in the past.

Different NPDL job task and situations warrant the use of different types of work measurement methods. Industrial Engineers may use one or combine more types of work measurement methods to establish time standards. Because of the breath of conditions apparent in the manufacturing shop floor, it is imperative that the right choice of work measurement method be carried out. Hybrid methods can be developed to meet special needs. Therefore, the Industrial Engineers in charge should have the necessary skills, wisdom and be creative to select and use the best method(s) in order to give data that contributes to measure the NPDL productivity objectively and for the organization's resource planning purposes.

VII. FUTURE RESEARCH

The future research will look into task which involved more complicated cycle time and event occurrence patterns. The challenges of those task activities will contribute to improving the methodology and overall as a field in Industrial Engineering for better labor resource utilization. Currently, there is no proper selection methodology available, that can be used as guidance for the novice Industrial Engineer. Hence, a selection framework should be developed appropriately.

REFERENCES

[1] IIE, "Industrial Engineering Terminology, Revised Edition", *ANSI Standard Z94.0-2000*, IIE Engineering and Management Press 2000, pp. 6-15.

[2] Gavriel Salvery, *Handbook of Industrial Engineering, 3rd Edition*, John Willey & Sons, Inc, 2001, pp.1411

[3] Frederick Winslow Taylor, *Chapter 2, The Principles of Scientific Management*, Dover Publications 1997

[4] Kjell Zandin *Chapter 4, Work Measurement in: Handbook of Manufacturing Engineering*, 2nd edition. CRC Press 2006/2005

[5] Barrie Daniels, (1991) *Choosing a work measurement system*. vol 40. Emerald Article 2007

[6] Paul S. Ray, The University of Alabama, *Chapter 7, Work Sampling,* CRC, Taylor & Francis Group, LLC, 2006

[7] Andrew Sherlock, "Work Measurement Notes", in *Mechanical Engineering with Management Techniques 4*: The University of Edinburgh, Scotland

[8] D.S. Barrett, (1997) "Standard data in batch production at Rank Xerox", in *Work Study Volume 46, Number 1*, 1997, pp. 9-19.

[9] Shigeyasu Sakamoto, "Work Measurement" in *Beyond World-Class Productivity: Industrial Engineering Practice and Theory*, Springer 2010

Empirical Research on Product Quality Management Based on PEM

Qing-hui DAI*, Zhi-yang FAN
School of Energy Power and Mechanical Engineering, North China Electric Power University, Hebei, China
(*dcba6789@126.com, 258407857@qq.com)

Abstract - **Performance excellence model(PEM) is not only used to review the quality management award, but also used for the performance of self evaluation, and make the organization achieve a great leap which is excellent. The paper introduces the basic principle and the framework of PEM. This article takes PEM into a water pump factory as an example, and analyzes six systems' construction process and implementation method. The six systems include corporate culture system, strategic management system, customer and market management system, the management system, process management system, performance measurement, analysis and improvement system. It sums up of the pump enterprise's achievements, existing problems and the matters needing attention. The example shows that bring in PEM is not only an effective method to improve the level of product quality management, but also the small manufacturing enterprise' striving direction, which is sustainable development and the pursuit of excellence.**

Keywords - **Comprehensive performance, empirical research, PEM, quality management**

I. INTRODUCTION

Performance excellence model is widely and internationally accepted, is an advanced model, which is about quality and organizational management, and it is derived from Baldrige Quality Award evaluation criteria [1]. It focuses on results of business by corporate, and uses quantitative scoring method to evaluate the company's performance. Enterprises can make a self-assessment with this standard, which is internationally known as the "guiding for organizational success"[2]. Many enterprises of our country have experienced five stages of imitation and application, which is QE, SQC, TQC, international quality management (ISO9000 certification), zero defect and 6σ management [3]. Recently, in order to improve the product quality, management level and comprehensive performance of enterprises, as a small and medium-level manufacturing enterprise, it is necessary to study PEM and implementation issues [4].

II. BASIC PRINCIPLE OF PEM

The connotation of PEM is to achieve the organization's performance excellence and to integrated organization system, conduct comprehensive and the continuous improvement method [5]. PEM review structure includes seven aspects: leadership, strategy, customers, markets, resources, process management, measurement, analysis and improvement, business results [6]. The framework model of PEM is shown in Fig.1.

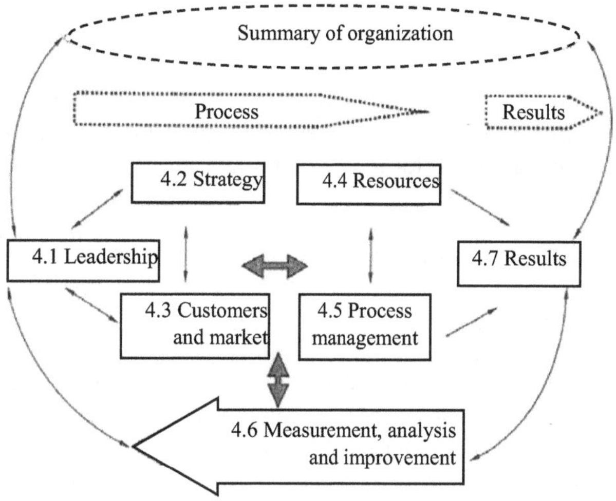

Fig.1. Framework model of PEM

Summary of organization includes the organization's environment, relationships and challenges, shows the key factors and background conditions of the organization operation. "Leadership" controls the organization forward direction, and pay close attention to the "result". "Leadership" and "strategic", "customer and market" form a triangle of "leading role", and it is the initiative, emphasis on leadership is focus on strategy and customers and the market; "Resources", "process management" and "results" constitute "resource, process and results" triangle, and it is driven, show the use of resources, through process management to achieve results [7]. The "measurement, analysis and improvement" is the foundation of organization operation, and is connected to the two triangles "chain", and turn the improvement and innovation of the organization [8].

III. CONSTRUCTION OF PEM

The pump factory was founded in twentieth Century 60 years. It has more than 4000 million RMB in fixed assets. Factory has all kinds of equipment more than 400 sets, including precision casting, precision machining, welding, heat treatment, machining center, tool manufacturing, chemical and physical test, metallographic analysis, nondestructive flaw detection, length and built thermal measurement and so on. It also has the domestic first-class large pump experiment center. After many years of running the enterprise ISO9000 system, product quality has been a corresponding protection goals and assessment system but the goals and assessment system were not got enough attention, it also has to be improved. The factory's main products boiler feed pump on the market reflect well, it was approved by most domestic and

E. Qi et al. (eds.), *Proceedings of the 5th International Asia Conference on Industrial Engineering and Management Innovation (IEMI2014)*, Proceedings of the International Asia Conference on Industrial Engineering and Management Innovation 1, DOI 10.2991/978-94-6239-100-0_52, © Atlantis Press and the authors 2015

international customers, but the products also exist the quality of wave phenomena occur, and the quality of stability needs being further strengthened. The pump factory has not established a comprehensive performance management system, now it just runs the performance measure which based on the requirements of ISO9001 quality management system.

In view of the above situation of the factory, it should be established a set of suitable scientific performance appraisal system by the introduction and implementation of PEM, to optimize the enterprise management, improve the operating performance, corporate strategic goals in earnest.

A. Build enterprise culture system

A good business must be built a good corporate excellence of culture, which has become the consensus of the majority of the business community [9]. From 47 years ago, a pump factory adhere to the people-oriented, the concept of sustainable development, and gradually formed its own unique corporate culture. Leaders set an example to the entrepreneurial spirit, open attitude, to create a pioneers, called "wise men" of the team, under the guidance of the corporate mission, vision and strategy, adhere to customer-centric, striving good corporate citizen as responsibility, and constantly promote the PEM, for the development of Chinese equipment manufacturing plants have made an outstanding contribution.

B. Building strategic management system

Strategies decide the development direction of the enterprise, it can ensure that companies do the right things and lead the companies go higher and farther [10].

(1) Develop strategy. We should pay attention to three points in the process of strategies development: Clear strategy formulation process, considering the key factors when collecting and analyzing data and information, clear key strategic objectives. The strategic deployment process is: According to the company's strategic goals, the factory office conduct business decomposition, develop business plans, decompose the strategic goals to every level of each business, and implement it to the annual business plan, make sure that the strategies are put into effect.

(2) Deployment strategy.

① Developing and deploying strategic Planning. There are two strategic priorities, the research and development of pumps and related products, strategic business, which can promote enterprise achieve rapid rise. The three strategic securities are organization management platform, high-leveled skilled labors and core values. Then according to the activities that reach the key performance indicators, we decompose and implement the indicators to all relevant sectors, establish a performance measurement system.

② Allocate resources rationally to ensure the implementation of the strategy. Configure for material resources, technology, market, environmental and other resources rationally.

③ The measurement method of key performance. According to the levels of finance, customers, process and learning, same performance measurement indicators were developed systematically. It covers all key strategic areas and stakeholders, and has been decomposed and implemented to all relevant sectors as to ensure coherence.

(3) Performance prediction. Conduct regular performance prediction according to determined key performance metrics. By predicting we can accurately grasp the level our corporate in the industry, which laid a good foundation for the deployment and implementation of the strategy.

C. Build customer and market management system

(1) Understanding customers and market. First, segment the market and customers. Second, establish the multichannel omni-directional information network, and understand the customer's demand and the expectation. Finally, ensure that customer needs and expectations of methods are suitable for strategic planning and development direction [11].

(2) Customer relationship and customer satisfaction. Customers can be divided into strategic customers, key customers, loyal customers, general customers and potential customers, different customers follow a different approach [12]. Perfect sales and service network provide convenience for the communication with customer. Then, in order to make sure that customers can visit the web site at any time and contact with the relevant departments and communication and get the user information in time in different aspects, companies establish enterprise website using the network technology, and set up official business mailbox address for enterprise all levels of departments. All of these are to ensure that complaints can be resolved in time by using the institutionalization of management, use the complaint information to improve services and further improve service satisfaction.

(3) Customer satisfaction measurement. It requires companies to establish customer satisfaction measurement. Establish the method of product, service quality tracking, collection, analysis and use, and acquire customer satisfaction information of rivals and benchmark enterprise.

D. Build a human-oriented management system

Building a human-oriented management system can make the company achieve the strategic security of "talent thriving enterprise plans", use many forms and multichannel to arouse the enthusiasm of the employees' work, develop the potential of employees, create elite team, and foster a sound environment for people can play their talents [13].

(1) Work system. It can be divided into two aspects. One is the work of organization and management. Strengthening the management of work and position, through multiple channels take the advice of employees and customers and sharing and taking effective ways of information communication and skill are needed. The other is the staff performance management system. Based

on the principle of balanced scorecard, employee performance appraisal system of four-dimensional is born. Further more, establishing many kinds of salary management system patterns which can coexist is also needed.

(2) Learning and development of employee. It can be divided into two aspects: one is strengthening personnel training; two is to create conditions for career planning, and give full play to personal potential.

(3) Survey of the rights and interests of employees and satisfaction. It is divided into two aspects: one is to establish a good working environment, which includes the provision of occupational health and safe working environment, pay attention to the interests of the staff and customers and carry out various forms of mass quality management activities. The other is to focus on staff feelings and meet the needs of different levels.

E. Build process management system

As the analysis show, there are four key processes which a pump product is produced: product development process, the procurement process, manufacturing process and marketing services. And three support processes: financial management, human resource management, equipment management.

There are four key processes: product development process which through the strengthening of research and development management system to establish a procedural and standardized product development process. The procurement process through the establishment of strategic cooperative relations with supplier, create value for the customers and stakeholders, essential to achieve corporate strategic goals and vision. The manufacturing process is that enterprises through the productivity gains to the continuous improvement of the production guarantee the continuous growth in sales revenue, product manufacturing enterprises to create a huge profit value. Marketing services, which supporting product implementations and the achievement of strategic objectives directly.

There are three support processes: financial management, as prescribed by national accounting standards and other regulations, the planning of the comprehensive budget management, fund management, financial management process, forming a perfect financial management framework. Human resources management, following the enterprise strategic planning and national regulations, combining with corporate culture background, concerning the competitor information and external labor market dynamics and external business partners to provide management information, designing and optimizing the management structure. Equipment management is based on the life cycle of equipment design, including "the equipment prophase management, equipment management, equipment management" three modules of the management process and the formation of two management closed loop: maintenance and fault information collection-repair and maintenance; maintenance and accident repair and maintenance information processing"[14].

F. Build a system of performance measurement, analysis and its improved Performance measurement.

(1) Performance measurement. It includes the indexes of scientific selection, comprehensive information collection, data classification, information system integration. Comparative data from each other and then learn from them to use it effectively. Analyze organized performance deeply. The results are showed by vertical transmission and horizontal communication.

(2) Information management. To collect and finishing relevant information by the responsible departments, prescribed from the process and responsibilities, make information more true and reliable [15]. Through the implementation of various measures, it is to ensure the integrity, timeliness, reliability, security, accuracy and confidentiality of enterprise information.

(3) Improvement. Management improvements: determine a goal, develop improvement plans; fully implemented, tracking; objective evaluation.

IV. IMPLEMENTATION OF PEM

There is a pump importing this mode within two years and constructing its self-evaluation system. It implements the improved management continuously, and improves the management of enterprises. During the process of PEM construction, senior leaders personally involved and constitute the "standard import team" with middle managers of each department. They implement hands-on guidance to promote the implementation of the whole plant, carry out generally excellent performance management activities in the enterprise, and organize staff forum to discuss so that the whole factory workers can fully understand the importance and necessity of importing advanced management mode; They set up a special working group and a clear labor division so that everything was in charge and implemented. During almost two years' exploration and practice, a unique pump PEM was established on basis of extensive research and careful analysis by each working group in enterprises.

A. Implementation results

After more than two years of hard work, the enterprise's strategic management, human resource management, marketing management, performance management and other management has embarked on a standardized, scientific road. Effect of pump PEM is shown in Table I.

TABLE I
COMPLETION OF 2010 -2012 MAIN INDEX

Key performance indicators	Units(RMB)	2010	2011	2012
The main business income	million	15446	21072	31036
Net profit	million	876	1292	2119
Sales revenue of new products	million	3519	5394	6711
International business sales revenue	million	2323	3330.1	4904.3
Overall labor productivity	million/person	8.43	9.50	10.32
Rate of return on net assets	%	1.89	2.12	3.05

Current assets turnover	%	64.46	69.99	73.92
Customer satisfaction	%	83.3	84.77	86.24
Market share	%	21.56	22.34	23.03
New product R&D program rate	%	107.8	115.64	117.6
New product feedback rate	%	28.42	31.36	36.26
Quality of feedback rate	%	11.76	12.74	14.7
To reduce the rate of purchasing cost	%	1.862	2.058	2.45
Employee satisfaction	%	70.17	73.89	77.22
Completion rate of training plan	%	100	100	100

From the table, significantly improve the economic efficiency of enterprises, expanding market share, enterprise management level has improved significantly, increased customer satisfaction, product and service to achieve results, more harmonious humanistic environment in enterprises, actively carry out public duties, drive the relevant parties to benefit from.

The company's products have been a number of national patents, has won several awards, such as "provincial excellent quality products", "provincial famous brand", "provincial hundred technology innovation", "top ten innovative power results", etc.. These products all over the country 29 provinces, city, autonomous regions, and exported to Turkey, India, Indonesia, Pakistan, Thailand, Vietnam, Cuba, Guatemala and other countries, thermal power plant and sales areas of power, petroleum, chemical and other system and power plant and the local power plant owned large enterprises.

B. Problem Analysis

(1) Enterprise training efforts is not enough. Some related people's understanding of performance excellence model is too simple, and someone even thinks that it has nothing to do with itself. The enterprise has its own problems. For PEM, which requires strong system thought, it's hard to make all of person of the enterprise to accept immediately.

(2) Comparing with peers, competitors and bench-marking is not enough. The plant belongs to the power pump industry that competition is intense. They are wary of each other, and the information is highly confidential. From what has been discussed above, compare and analysis are relatively lack.

(3) Integration of PEM and the other standard system is not enough in the enterprise. ISO9001, OHSAS18001, ISO14001 and other standards are run now for years before bring in PEM. But enterprise did not combine these standards very well at the beginning of the moment, when enterprise brings in PEM. It leads to spend more manpower and material resources for the enterprise so that it causes great waste.

C. Matters needing attention

The implementation of PEM in small and manufacturing enterprises, should pay attention to the following matters:

① Business leaders should investment actively and support the implementation of the PEM work for a long time, so as to get the results which want to achieve.

② Small-scale manufacturing industry personnel should to develop the habit of strategic thinking, work according to the requirements of enterprise strategy.

③ The implementation of the PEM involved everyone in the enterprise, if you want to achieve superior performance, you must have excellent talents.

④ For small-scale manufacturing industry, the enterprise staff only through training and examination, can they adapt to the request of the PEM gradually.

⑤ Improve the 6σ management, to create a execution culture atmosphere of practice what you preach and pay attention to the performance and the participation of cultural environment.

⑥ The formation of the PEM is a longer system process. Small manufacturing enterprises have the limited resources, should be promote from simple to complex.

V. CONCLUSION

The criterion of PEM represents the highest requirements of today's organization management. The essence of PEM is starting by point of product quality management. It is a detailed rule of the implementation of standardized, systematic and specific to achieve the overall management of the organization. In the increasingly fierce market competition environment, small manufacturing enterprise which wants to take the road of sustainable development, they must create new management model. With implementation of the method and idea of PEM, it may help small manufacturing enterprise learn about their situations, play their own advantages, establish and implement strategic objectives effectively, improve their cohesion and competitiveness. It can also improve customer's satisfaction, achieve better performance, and help them realize spanning from the good to great gradually.

ACKNOWLEDGEMENTS

This work was supported by Senior Engineer Huang Qi's a lot of help.

REFERENCES

[1] Shanmugaraja, Nataraj, Gunasekaran. "Total performance excellence: a model for successful implementation of Six Sigma," *International Journal of Procurement Management*, Vol.3, No.6, PP.297-328, 2013

[2] Yang QU, "Brief introduction of the world's three Quality Award," *China quality*, No.6, PP.13-15, 2001 (In Chinese)

[3] Lin MA, "PEM understanding and Implementation Guide," China Standards Press, PP.99-101, 2005 (In Chinese)

[4] Gorji, Mohammadbagher; Emami, Azize, "Self-assessment of organization's performance with regard to enablers criteria excellence model," *Research Journal of Applied Sciences, Engineering and Technology*, Vol.17, No.4, PP.3048-3055, 2012

[5] Xiang-wen YI, "Performance excellence leading enterprise development - Discussion on the implementation of performance excellence model in small and medium sized

enterprises," *China quality technology*, No.3, PP.51-53, 2013 (In Chinese)
[6] Xiao-wen FENG, "Comparison of PEM and ISO9000," *The mass standard*, No.S1, PP.18-20, 2008 (In Chinese)
[7] James R Evans, "Quality management and quality control," Renmin University of China press, PP.102-104, 2010 (In Chinese)
[8] Yahe FU, Yu-lin XU, "Performance management," Fudan University press, PP.45-48, 2008 (In Chinese)
[9] Zhen-yong LEI, Jin-ping BAI, Sha-ning CHEN, "The excellent performance management to business excellence in the practice of road," University of Electronic Science and Technology Press, PP.73-75, 2006 (In Chinese)
[10] Xinghai QU, "Study on performance management mode of electric power enterprises," *Managers*, No.5, PP.66-75, 2013 (In Chinese)
[11] Yihua ZHOU, "The establishment and implementation of the exploration and practice of evaluation of the effectiveness of the quality management system," *Quality and standardization*, No.12, PP.50-55, 2011 (In Chinese)
[12] Zhen-nian REN, "Modern hospital quality management excellence performance evaluation - background and summary in other countries," *Chinese hospital*, No.3, PP.48-52, 2006 (In Chinese)
[13] Dong-feng ZHANG, Min LIU, Dong-hong ZHANG, "The paradigm of excellent performance management," *Economic Forum*, No.10, PP.60-62, 2009 (In Chinese)
[14] Qiu-xiang ZHAO, Xin-chun ZHAO, "Performance excellence model and the method of quality management," *Technology Supervision in petroleum industry*, No.3, PP.42-46, 2010(In Chinese)
[15] Wei-chuan YE, Rong-xin JI, "Excellent performance-The Improvement of Quality Management," *The Railway Technical Supervision*, No7, PP.86-90, 2010 (In Chinese)

A New Problem-Solving Procedure based on TRIZ Methodology and QC Story

Sheng-Yuan, Hsu
(Department of Industrial Engineering and Management, Chienkuo Technology University, Taiwan)
(jackhsu@ctu.edu.tw)

Abstract - **In this study, "creative problem solving QC Story" (CPS-QC story) has been proposed for improving the traditional quality control process. Through the combination of traditional quality control process and systematic innovation (TRIZ), the concept of problem solving, innovation tools and methodologies of TRIZ will be included to QC story for strengthening the lack of problem solving techniques of traditional QC story. TRIZ is the core of CPS-QC story, for providing systematic, logical, and procedures of the ways to solve the problem. In an actual case studies, a mask company in Taiwan use CPS-QC story on the case of improvement of the environmental control system of mask exposure machines. The quality level of the improvement is equivalent to the traditional program, cost and delivery is only one-third of traditional programs. It's showed the CPS-QC story will be helpful for the photomask industry.**

Keywords - **QC Story, QCC, TRIZ**

I. INTRODUCTION

This study integrates TRIZ and the traditional process of the QC story. A creative problem-solving QC (CPS-QC) story is developed based on the combination of the analytical process of the QC story and TRIZ's tools for creative problem solving. The efficiency of problem solving for businesses and individuals will be enhanced.

The real case of a mask company in Taiwan is employed in this study. A CPS-QC story will be used to solve the environment improvement problem of mask exposure equipment. The new problem-solving process can provide rapid and valuable innovative solutions. The six performance indicators of the manufacturing operations, including cost, quality, and delivery (Miltenburg, 1965), will be used for evaluation.

II. MODELING THE CPS-QC STORY

The QC story is a common aspect of QCC activities or quality improvement projects. The output of these improvement activities can be significantly enhanced by a QC story (Noriaki, 1999 [1]). However, in the process of improvement activities, team or individual brainstorming or engineering analysis and experiments are critical tools, such as in "topic choice and status analysis," "factor analysis and countermeasures." Currently, businesses focus on fast, valuable, and innovative solutions, which create a significant gap. The TRIZ developed by the global patent analysis, which is a type of process analysis of problem solving based on scientific knowledge, can be used to provide a scientific, logical, and systematic approach to thinking that differs from brainstorming, which is psychological, illogical, and divergent (Nakamura, 2003a, b [2, 3]). The integration of the TRIZ with the QC story employs unique problem-solving tools of TRIZ. The main weakness - the lack of problem - solving tools - will be reinforced in the QC story using the TRIZ.

Sugiura and Shantian (2003) proposed the "problem-solving type of QC story," which will be considered as the foundation [4]. Four steps of the TRIZ problem (Mann, 2002) and the general TRIZ problem-solving flow (Domb, 1998a, b) will be considered for building the CPS-QC story [5, 6, 7]. Suitable tools of TRIZ will be selected to integrate TRIZ with the procedure of the problem-solving type of QC story. The CPS-QC story involves three phases. The first phase involves the integration of TRIZ's problem-solving tools with the topic choice and status analysis in the QC story. The second phase involves the integration of TRIZ's problem-solving tools with the factor analysis in the QC story. The third phase involves the integration of TRIZ's problem-solving tools with the creative countermeasure in the QC story.

Based on this three-phase approach, the tools of TRIZ will be integrated with the problem-solving type of QC story (Fig.1). TRIZ tools will be used in suitable steps in the story flow of the problem-solving-type of QC. All tools will enhance the problem-solving abilities and the creativity of the solution.

E. Qi et al. (eds.), *Proceedings of the 5th International Asia Conference on Industrial Engineering and Management Innovation (IEMI2014)*, Proceedings of the International Asia Conference on Industrial Engineering and Management Innovation 1, DOI 10.2991/978-94-6239-100-0_53,

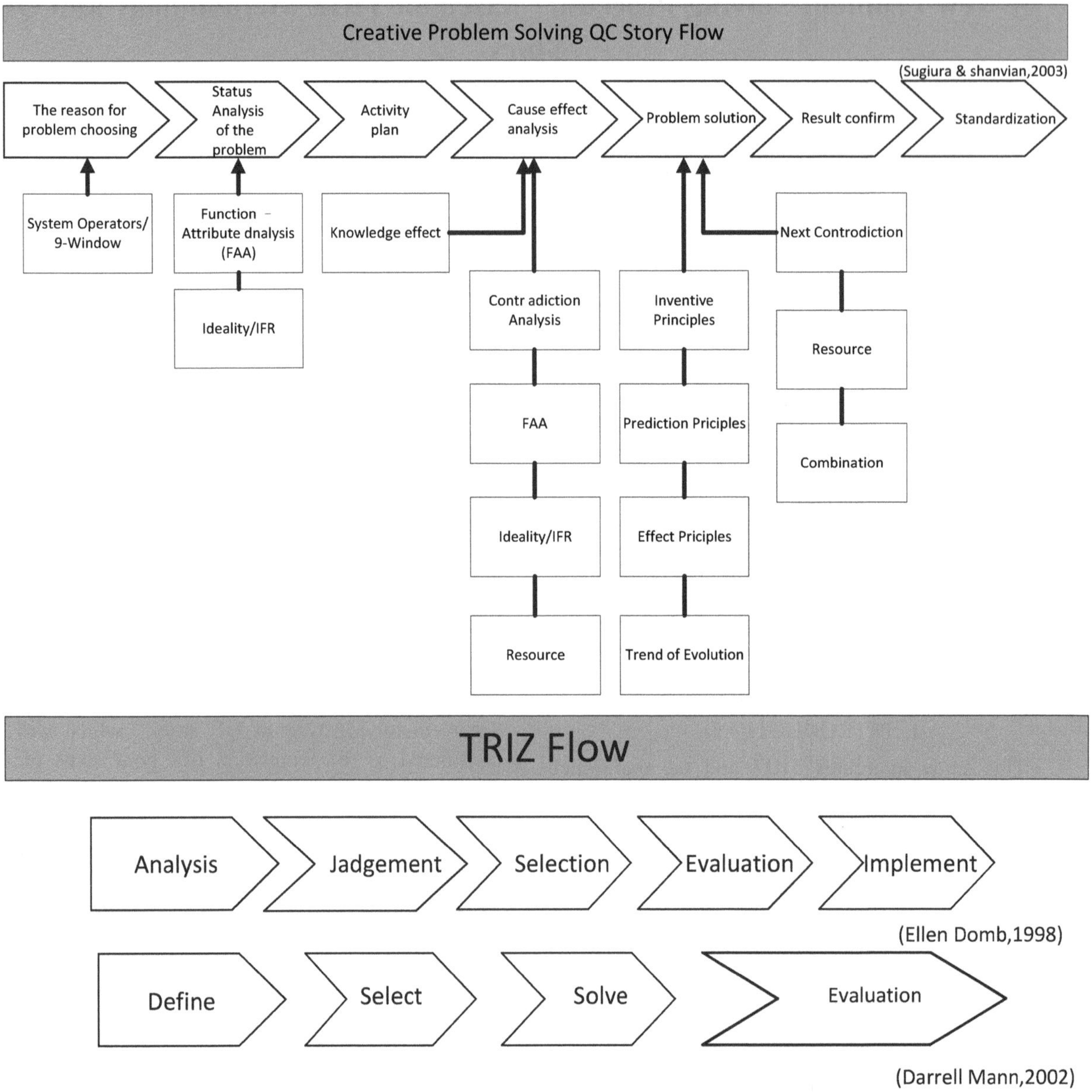

Fig.1. the structure of creative-problem-solving QC Story

III. CASE STUDY

A. Background

In mask companies, the exposure machine (writer) demonstrates the highest percentage (approximately 40% to 50%) of investment in production equipment. Similar to the high-level electronic beam (e-beam) writer, it costs more than ten million US dollars. Despite the purchasing and operating costs of machines, such as the "environment chamber" that exhibits a unique design and installation, it enables strict control of the environmental temperature, humidity, wind speed, cleanliness, chamber pressure, magnetic fields, noise, and other parameters. Mask companies must provide the perfect environment for e-beam lithography machines to produce high-end photomask manufacturing products. In addition to the construction cost of the central control room, the maintenance cost is also high. The improvement in management and maintenance of the "environment chamber" is an important issue. "Exposure" is critical in the front-end process of the photomask manufacturing process.

The exposure machine is the main process equipment. In the semiconductor industry, yield control directly determines the productivity. The primary reasons for yield loss are as follows: 1. parameter exceeds the specification limits or incorrectly defined parameters; 2. process abnormalities or equipment failure; 3. chemical contamination; and 4. particle pollution. Because the semiconductor photomask industry is similar to the lithography process, the yield control problem is the key to capacity (Levinson, 1999 [8]). As shown in Fig.2, the particles in the mask will cause particle defects during exposure. The repair costs in the backend process are exorbitant. If a fatal flaw (killed defect) is detected, the products should be discarded (reject), which will increase manufacturing costs, erode profits, and affect product delivery. Therefore, "reduce defects, improve yield" is an important goal of mask companies. The mission will serve as the primary goal of the production line, in which "fatal flaw exposure machine improvement" is an important project. This study will use the CPS-QC story to assist the case company. Using the structural process of a CPS-QC story, innovative, systematic, and comprehensive problem definition and analysis can be accomplished.

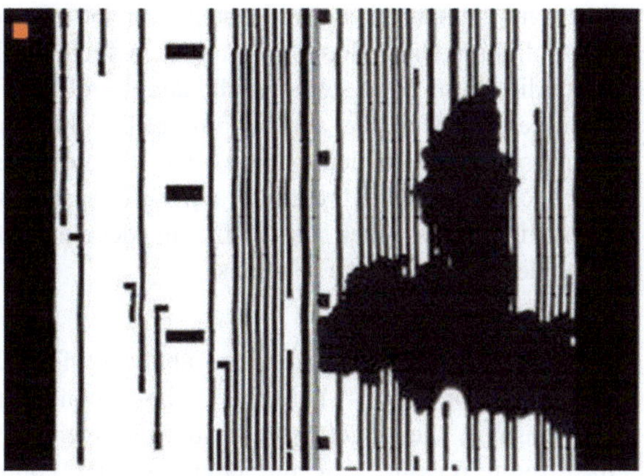

Fig.2. the killed defect of mask products

B. CPS-QC story

According to the CPS-QC story, the real problems of the case company are sequentially implemented and described as follows:

Step 1: The reason for problem selection

Five or six of the shop staff comprises a group for the establishment of the QCC. The activity matrix will be used to identify the master problem. The problem items B, F, G, and, H score more than 15 points with the highest priority. After discussions with the QCC members, problem H, which entails the improvement of the killed defect, will determine the activity topic. Using the TRIZ 9-window analysis, the killed defect affects the yield of the exposure machines and the future development of the advanced process. An internal subsystem will affect the maintenance, repair and maintenance costs of development, monitoring plans, and other internal machines. An external super system will affect the rate of delivery, customer satisfaction, and cost. Thus, the problem of "killed defect affect the exposure machine yield" must be effectively addressed. The consolidated results of the analysis reference 9-windows system indicate that the QCC's members selected the problem of the killed defect as the improvement theme.

Step 2: Status analysis of the problem

A Pareto analysis was applied to the scrap pieces of the exposure machines. Scrap materials, such as a killed defect, accounted for approximately 63.63% of the total quantity of scrap materials. First, the function/attribute analysis of TRIZ will be adopted to analyze the exposure machine systems and improve the mutual environmental relations within the component parts (subsystems). The functions and the problem to be improved have been clearly defined. The main function of the system is the exposure, the harmful function is the contaminant, and the inadequate function is the specific environment for the exposure process. The environmental control system includes hardware facilities, air-conditioning systems, magnetic systems and materials, as well as the tools used in the environmental control chamber, such as the product transmission box, clean clothing, and gloves. The environmental control system is unable to fulfill the requirement of special environmental conditions for exposure machines, which will result in the pollution of the exposure machines or killed defects. The ideal final results (IFR), which employ a series of questions, are used to establish the idealistic system, the idealistic level of the system and subsystems, and the requirements for solving the underlying problem. The IFR can be used to define the problem or solve the problem and facilitate the use of the knowledge effect and the contradictions problem.

Step 3: Activity plan

To improve the pollution problems of the environmental control system, develop an activity plan.

Step 4. Cause-and-effect analysis

Checklists are used for the cause-and-effect analysis. The items are confirmed by discussion with the QCC members about the project based on their practical experiences, including contamination of clothing worn by personnel, contamination of the operator's gloves, contamination of machine parts, contamination of the machine feeder box, quality variation in the main material (blank), dynamic operations pollution, static pollution of clean degree, dynamic pollution of clean degree, and deviation in the chamber operating pressure deviation. The verification items are classified according to 4M-1E and the evaluation or detection methods. Next, the possible causes of contamination in the environmental control system will be extended to the fishbone diagram. Based on the verification method for each item all problems will be checked sequentially. The main cause of the pollution in the environment control system will be identified. Based on the test results, item 9 (the deviation of operating pressure in air-conditioning (chamber) is recognized as the primary problem. The normal value of the pressure deviation is "+2" to "−2".

Although the test results of items 1–8 were normal, the negative pressure of the air-conditioning chamber will result in pollution in the chamber by the external space, which exhibits a lower level of cleanliness. The particles on the indoor table and the killed defects of the product were used to verify these findings. According to the TRIZ knowledge effect, "inertia change" is a main condition of the pressure deviation from the classification of attributes in the knowledge effect database. The inertia change can be checked by the inertia, chamber, and centrifugal force of the air-conditioning system. After a comprehensive examination of the cause-and-effect/function analysis and knowledge effect, the fan of the environmental control system was determined to run with a full load; however, the load is insufficient for maintaining positive pressure in the chamber.

Using the functional analysis of the TRIZ, the mutual relations for all components (subsystems) in the pollution and environmental control system were analyzed. The environmental control system cannot establish a sufficient positive pressure in the environment chamber due to insufficient air-conditioning equipment output, which will contaminate the mask products. The critical improvement issue is the environmental control system, which is air conditioned.

A traditional solution is the installation of larger capacity air-conditioning equipment, which requires a delivery time of approximately 30 days (needs to be imported). The construction period consists of approximately two days. The mask exposure machine must be shut down. The one-day value for the exposure machine output is approximately thirty thousand dollars. Updating the device costs approximately one hundred and twenty thousand dollars. The traditional solution requires a total of 32 days. The investment cost of the equipment and the loss in output is approximately one hundred and eighty thousand dollars. Traditional solutions will result in numerous contradictions and conflicts, such as cost (including the investment cost of update equipment and productivity losses) and the complexity of design and construction (including the removal or modification of old equipment, the shutdown of exposure machines, and construction planning and shop floor management). Using TRIZ contradiction analysis, conflicting relationships with the problem can be identified. To improve the room pressure problem of the environment chamber, the traditional solutions for updating air-conditioning equipment capacity produce major contradictions of cost and complexity. These issues are defined as technical contradictions. The improvement is the pressure of the environment chamber; the problem is the cost and system complexity.

In resource analysis, Domb (1998) defined resource as "All the things can be used, including parts, component, energy, information, functions, within the system or peripheral. Resources can be combined with the system or subsystem and used to improve or enhance system functionality." According to the classification of Mann (2002), six types of resources exist: resources in the environment, low-cost resources, material resources, special resources, manufacturing process-type resources, and resources associated with humans. QCC members will check and observe the related resource of the problem based on the six types of resources. Each resource will be analyzed for use or conversion to a useful type.

For the IFR of TRIZ, a contradiction has been identified as a "product is not a Killed Defect" and "the environmental control system cannot provide non-pollution shop floor" in step 2, which is the contradiction of the system and the target problem for improvement. For the analysis of TRIZ, this case comprises a "contradictory type" and "improvement type" of issue. This research will use systematic innovation tools, such as TRIZ, to develop fast, valuable, and innovative solutions.

Step 5. Problem solution

After using contradiction analysis for the essential problem between the conflicting relationships, the target problem of improvement is the "room pressure" of the environmental control chamber (environment chamber). A traditional solution consists of updating the capacity of the air-conditioning equipment, which will result in a substantial contradiction of air-conditioning equipment for the new purchase of the increase in total cost and the "complexity" from the construction process required to update the old equipment. The problem is a contradictory problem and comprises a technical contradiction. The five steps of technology contradiction can be applied to solve this problem (Mann, 2002).

After discussing the 31 types of evolutionary trends in TRIZ's technology forecast (Domb, 1998, Mann, 2002), the evolutionary trend "mono-bi-poly (similar)" can be applied to improve the environmental control system in the problem. The key concepts and reasons for jump is the quantity for improving the amount of useful and deliverable functions. It is a feasibility design concept that QCC members can use to improve the chamber pressure of the environmental control system. "Increase single group (or multiple groups) of small air-conditioning systems to deliver useful functions." After completing a site survey, QCC members considered the lack of shop floor space, which cannot be used for improvement. However, "mono-bi-poly (similar)" provided an excellent design concept for the QCC members: the initial system design for a single system should consider the planning and capacity for future expansion, including system space, function, installation, and the line of motions for installation. Using the substances field/76 standard solution (Su-Field/76 standards) analysis to improve the environmental control problem (Mann, 2002 [5], Xinjun, 2006 [9]), the problem is classified as "Insufficient/Excessive Relationships." In accordance with the proposed standard solution, to choose a new field for supporting the system: "Make use of a field that already exists in the environment around the system" (standard solution #5.2.2, Mann, 2002) provided field energy to S1 (the environmental control equipment). Searching the surrounding environment of the system revealed two types of fields that can be applied, including the static field (nonpollution outdoor air system), and the dynamic pressure source (other clean room space surrounding hydrostatic).

The IFR concepts for increasing benefit and reducing costs and damage are critical in the current and highly competitive environment. The IFR can be used to guide all feasible solutions of the QCC members to improvement the ideality of the solutions. The photo mask plant is one of the key components of the semiconductor industry, in which the main indicators of competition focus on quality, cost, and time-to-delivery. Therefore, based on the analysis of the IFR, "the use of the dynamic pressure source (other clean room space surrounding hydrostatic)" will be selected because it can be achieved without consuming resources outside the clean room and is a low-cost and highly efficient solution.

Based on various tools and methods of the TRIZ, all solutions obtained by the QCC members. Items 2 and 5 comprise the selection of solutions that are adapted by the QCC members. Item 2 consists of changing the volume of air-conditioning in the environment control room: add an adjustable air volume damper under the raised floor in the central control room for adjusting the air flow plate to control the volume of air conditioning. Item 5 consists of installing a field to the environment control room and utilizing the static pressure of other spaces surrounding the clean room space.

Step 6. Confirmation of results

Through measurement and certification of the cleanroom environment system, the static pressure of the environmental control chamber has been successfully improved from −2.0 pa to +2.1 pa to achieve the goal of +2.0 pa. (killed defect ratio also improved from 63.63% to 25%. These two quality indicators are the main goals.

Step 7. Standardization

After finishing the project of improving the environmental control system, standard operating procedures (SOP) should be established in accordance with the requirements of the case company. Standards for implementation and training for related members have been developed. The results of the operation will be listed for the key items in the shop floor auditing.

IV. CONCLUSION AND SUGGESTIONS

This paper proposed the CPS-QC story, which consists of the combination of the traditional QC story and TRIZ. TRIZ's innovative tools and its unique systematic problem solution concept were employed in the quality control process (QC story). The traditional QC story is based on the QCC and brainstorming for problem solving, which is always insufficient and nonsystematic. The CPS-QC story integrates the advantages of TRIZ in problem solving. The efficiency will be improved. Actual case studies, such as the case company, used the CPS-QC story of the environmental control system of a mask machine to improve its quality. The performance of a CPS-QC story is superior to the performance of a traditional program. The cost and delivery time of a CPS-QC story is only one-third the cost of traditional programs. The CPS-QC story can improve the competitiveness of the case company in the mask industry.

REFERENCE

[1] Noriaki, K., 1999, "Management topics reaching type QC-Story," Pioneer Enterprise Think Tank Press, Taiwan.

[2] Nakagawa, T., 2000, "Approaches to Application of TRIZ in Japan", The TRIZ Journal, Aug., 2000.

[3] Nakawaga, T., 2003, "Invention on Computer Keyboard A TRIZ base Analysis", The TRIZ Journal, Nov., 2003.

[4] Sugiura, Z., and Shanrian, J., 2003, "QC Story Utilization Method," Corporate Synergy Development Center Press, Taiwan.

[5] Mann, D. L., 2002, "Hands on systematic innovation", CREAX Press.

[6] Domb, E., 1998a, "Increase Creativity to Improve Quality", The TRIZ Journal, Dec., 1998.

[7] Domb, E., 1998b, "QFD and TIPS/TRIZ", The TRIZ Journal, June, 1998.

[8] Levinson, H. J., 1999, "Lithography Process Control", Vol. TT28. pp. 119-120, SPIE Optical Engineering Press, Bellingham, Washington.

[9] Xinjun, Z., 2006, "Develop New Kind of Plough by using TRIZ and Robust Design", TRIZCON2003, Philadelphia PA USA, March, 2006.

A Study of Critical Success Factors and Prioritization by Using Analysis Hierarchy Process in Lean Manufacturing Implementation for Thai SMEs

T. Kiatcharoenpol[1*], T. Laosirihongthong[2], P. Chaiyawong[1], C. Glincha-em[1]
[1]Industrial Engineering Dept., Faculty of Engineering, King Mongkut's Institute of Technology Ladkrabang, Thailand
[2] Department of Industrial Engineering, Thammasat University, Bangkok, Thailand
(tossapol_k@yahoo.com)

Abstract – **This study identifies and prioritizes critical success factors (CSF) based on the implementation of Lean Manufacturing for Thai SMEs. After review of International research papers, 12 CSFs were selected and categorized into four resource types from the lens of Resource-Based View (RBV) Theory. These are Organizational, Technology, Human and Financial Resources. Interviews and surveys with Lean experts experienced in SME consultation were carried out to answer the proposed research questions. Analytical Hierarchy Process (AHP) is then applied to prioritize the relative importance of the CSFs. The results show that "Technology resource" is the most important to enhance implementation of Lean in SME organizations. This indicates the knowledge of production technology is a prime key to support process improvement according to Lean methodology. This study contributes to the existing body of knowledge by evaluating and prioritizing CSFs reflecting by grounded theory and expected outcomes at firm level from experts' experience.**

Keywords – **Analytical Hierarchy Process (AHP), Lean manufacturing, Resource-Based View**

I. INTRODUCTION

Small and Medium Enterprises (SMEs) are importance economic foundation of the country. It make production, employment, income and development of the country competitiveness. Operation strategy ae introduction in Thai Industry in order to level up manufacturing performance. Lean is one of prime strategy encouraged among Thai SMEs. Lean manufacturing takes a holistic and multidimensional systems approach towards understanding and providing solutions for reducing waste, and thus develops close links between quality, cost, delivery, customer satisfactions and continual improvement. By implementing Lean, organizations could achieve breakthrough process improvement with a dramatic impact not only on financial benefits but also customer satisfaction and production capability.

While Lean has made a substantial impact on industry, academic research in this area is lacking, particularly research regarding what makes a successful Lean implementation for SMEs. The main purpose of this study is to identify and prioritize factors for successful Lean implementation by employing the Analytic Hierarchy Process (AHP) approach. This method allows us to define those success factors in a hierarchical structure of factors, evaluate factors in pairs, and quantify the relative importance of each factor to the successful implementation. Critical Success factors resulting from a number of publications are reviewed and selected Preference data from selected experts involved in consulting Lean for Thai SMEs in many years are utilized in this study to identify and prioritize the significant CSFs. The following section provides relevant literature in the field and identifies the factors for successful implementation of Lean. A detailed discussion on research methodology is given in below section, while data analysis and findings are shown in Section 4. Finally, managerial implications are discussed in the conclusion section.

II. LITERATURE REVIEW

A. Analytical Hierarchy Process (AHP) Methodology

In order to ascertain the management of SMEs to understand on factors that affect the successful Lean implementation at firm level, this study conducted an in-depth research in the Thai electronics components manufacturing industry using the AHP approach. AHP is a decision-making approach which integrates simultaneously qualitative and quantitative information for prioritizing alternatives when multiple criteria must be considered. According to Saaty [1], a decision making approach should have the following characteristics: be simple in structure, be adaptable to both group and individual decision making environments, and be natural to human intuition and general thinking,

The modeling process of AHP involves four steps [2]:

1) Assessment of success factors in Lean implementation,

2) Structuring the problem as a hierarchy and building the AHP model,

3) Collection and compilation of experts' opinions and application of the prioritization procedure, and

4) Determination of critical factors through the synthesis of normalized priority weights.

B. Resources to achieve implementation of Lean manufacturing in SMEs

The basic concept of RBV is that firms are collections of resources that are: (a) valuable to the firm; (b) rare to come by; (c) imperfectly mobile and not imitable by competitors; and (d) not substitutable [3]. The efficiently and effectively resources utilization leads the firms to enhance their organizational capabilities. Resources, in addition, consist of a bundle of potential services, whereas capabilities are intangible bundles of skills and accumulated knowledge exercised through organizational routines [4-6]. Applicability of RBV theory in developing the firm's capability has been explained by

E. Qi et al. (eds.), *Proceedings of the 5th International Asia Conference on Industrial Engineering and Management Innovation (IEMI2014)*, Proceedings of the International Asia Conference on Industrial Engineering and Management Innovation 1, DOI 10.2991/978-94-6239-100-0_54,

the idea of how firms can achieve sustainable competitive advantages by using firm-specific resources. To apply RBV theory in explaining implementation of Lean manufacturing, the resources could be considered as CSFs. What resources play important roles in enhancing success of implementing Lean in environment of Thai SMEs. The CSFs from literatures are arranged into four resources [7] views as follow in Table I and Fig.1;

1) Organization resource (strong leadership and commitment, clear vision and target deployment, project management, and continuous improvement culture),

2) Human resource (empowerment, training and skill building, internal expert, and external consultant),

3) Technology resource (production technology support, and Lean technique knowledge), and

4) Financial resource (financial support for Lean project, and reward).

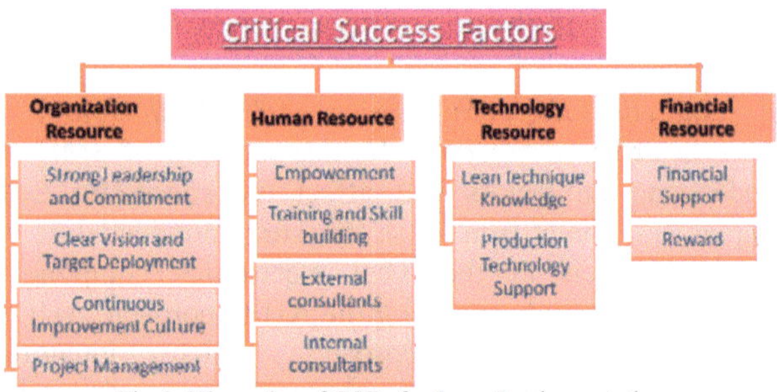

Fig.1. Hierarchy of CSFs for Lean Implementation

TABLE I
CRITICAL SUCCESS FACTORS OF LEAN IMPLEMENTATION: RESOURCE-BASED VIEW PERSPECTIVE

Type of Resource	Critical Factors	Ref.
Organization Resource: is a sets of management practices that could be used to establish the strategic direction of SMEs (for both policy and operations level). The strong leadership, well-established organizational policy, and organization culture play a vital role in supporting the successful of Lean implementation. This resource also includes the ability of project manager to prioritize all selected improvement projects, and regularly review the progression to ensure that all expected outcomes will be achieved within the proposal timeframe.	Strong Leadership and Commitment	[8-18]
	Clear Vision and Target Deployment	[10,12-16,19]
	Project Management	[8-9, 11,13-14,18]
	Continuous Improvement Culture	[8-11, 14,18-19]
Human Resource: is the SMEs capability to hire, maintain, and develop qualified personal in order to achieve the firm objectives .To achieved the Lean implementation, involvement and empowerment of team leader as well as leader is needed. In addition, training among team members in multi-skills is one of the most important enabling factors to being success.	Empowerment	[10,13,17]
	Training and Skill Building	[8-9, 11,13-14,18-19]
	Internal Expert	[9,11-12,14-15,18]
	External Consultant	[9,11,14]
Technology Resource: is the SMEs capability to maintain and control existing operation procedure by acquiring new process technology and to identify appropriate source of technology. Knowledge of Lean Tools/techniques are also included.	Production Technology Support	[9,13,18]
	Lean Technique Knowledge	[13-14]
Financial Resource: is the management budget allocating for Lean implementation. Reward for team member and sufficient financial support for each Lean project, could be considered as one of the most important success factors.	Financial Support	[8-9,11]
	Reward	[8,11]

III. METHODOLOGY

A combination of research methodology approaches has been employed in this research project. This comprises literature review and specialist interview. The literature review conducted extensively at the initial stages of this research to settle the CSFs for implementing Lean in Thai SMEs. These are categorized based on RBV perspectives. A survey was carried out among Lean experts who have been consultants for Thai SMEs. Most of them have more than 5 years experiences for training Lean Manufacturing topic and implement at shop-floor level. The data were processed in order to give interpreting priority of CSFs using the AHP technique selected because of the following reasons:

- It is an appropriate tool for determining priority of CSFs with respect to different dimensions,
- This technique does not involve statistics or probability theory, thus giving the user a better sense of reality, and
- It is the well established methodology to evaluate important factors in other research focus on operation strategy such as total quality management implementation, and Six-Sigma™ [7,20]

In this study, the structured questionnaire was developed for the eight experts to provide the important score for each CSF. It should be noted that these evaluation score is come from the specifically view point for Thai SMEs not for general firms.

IV. RESULTS

Numerical evaluation was carried out in two levels. Firstly, four resources are compared to determination of important weights to prioritize the significant resources for implementation of Lean Manufacturing. In next level, among CSFs are evaluated to determine of important weights on each CSFs. Fig.2. Show the weight calculated by AHP based model for each of resources categories. The top two highest weights are "Technology Resource" and "Organizational Resource", which are 0.364 and 0.232 respectively.

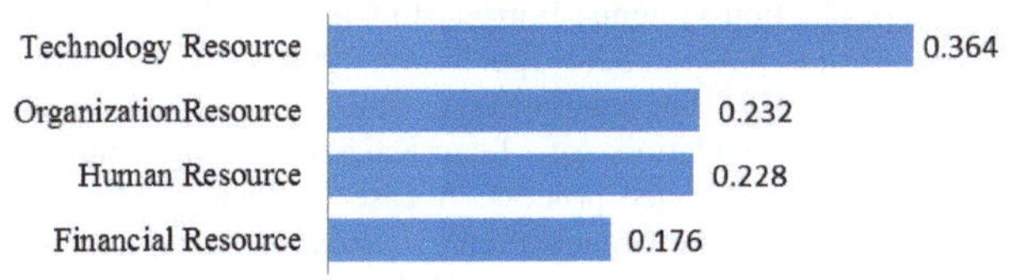

Fig.2. Priority weight of four Resources using AHP based model

In AHP weight respect to Technology Resources as shown in Fig.3, "Production Technology Support" is the highest weight which indicates the most significant CSFs to implement Lean in Thai SMEs. This factor means capability to maintain and control existing operation process by maintaining process knowledge, acquiring new process technology and identify appropriate source of technology. For weights respect to Organization Resources in Fig.4, the CSFs with the highest weight is "Empowerment". This enable Lean member to have enough authority and role to implement Lean effectively.

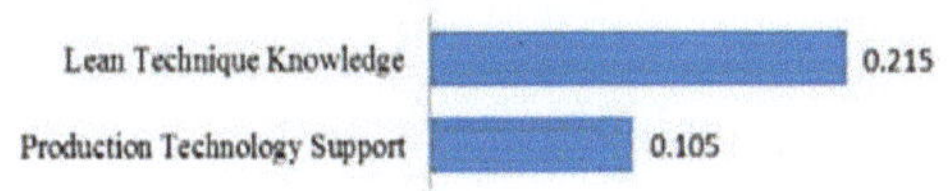

Fig.3. Priority weights of CSFs with respect to Technology Resource using AHP based model

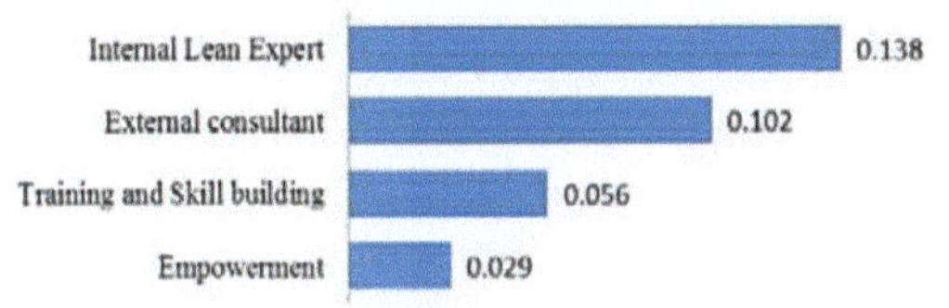

Fig.4. Priority weights of CSFs with respect to Organization Resource using AHP based model

When consider weight among all twelve CSFs as shown in Fig.5, the CSFs can be divided into two groups; the higher and lower groups. In the higher group, each success factor have weight more than the average weight value (0.083) which comprise Production Technology Support (0.215), Empowerment (0.138), Strong Leadership and Commitment (0.12), Lean Technique Knowledge (0.11) and Training and Skill Building (0.10). It should imply that Thai SMEs could play more focus on this priori group of CSFs in implementation of Lean in their firms.

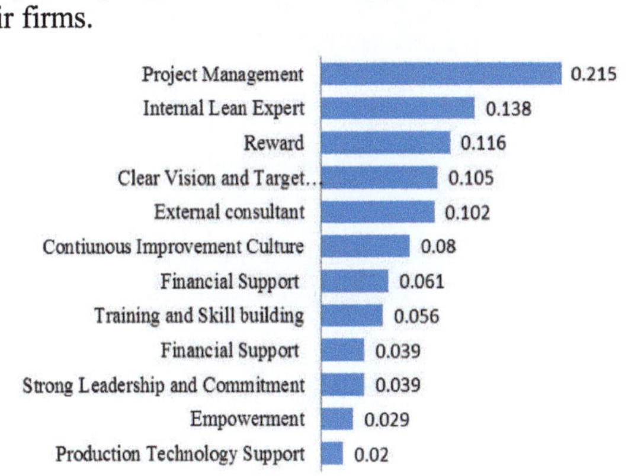

Fig.5. Priority weights of Critical Success Factors

V. DISCUSSIONS AND CONCLUSION

This study is focused on critical success factors in implement Lean in Thai SMEs. The comprehensive literature of published journals are investigated to construct CSFs for SMEs. Four different areas of resources are examined in RVB perspective. The survey opinion of Lean Experts who have experienced as a professional consultant were carried out to set the priority of resources and success factors that are significant to enhance implementing Lean in SMEs. Then, AHP techniques are used as numerical evaluation to interpret the priority which is the principle outcome of this study

The priority weights of four resources are compared in this evaluation. "Technology Resource" and "Organization Resource" are considered as the two top priorities in resources for achieving effective implementation of Lean Manufacturing. This finding suggests that Lean is Improvement strategy centric on process level, the role of Production Technology Support is prevalent, and when it is synergized with solid capabilities from all employees coming from Empowerment organizational culture, high performance will be realized. It is noted that "Financial Resource" gets the lowest priority among four resource types. The two main components of this resource are "Financial support" for budgeting Lean implementation program and "Rewards" for team members who achieve Lean improvement target. The low priority of Finance resource contradicts to the most belief that financial budget is key success factor to efficiently promote Lean in Thai SMEs environment.

This finding also indicates what is priority of CSFs in implementing Lean in SMEs environment. These top success factors are Production Technology Support, Empowerment, Strong Leadership and Commitment, Lean Technique Knowledge and Training and Skill Building. It can imply that Thai SMEs need knowledge of Production Technology as basically intrinsic base to improve firm performance. However, the implication of this study is limited to a selected country, Thailand, only and may reflect local cultural preferences/perceptions, which may not apply generally to other countries. Nevertheless, this work may have important learning points for other countries at the early stage of Lean implementation.

ACKNOWLEDGEMENT

The authors would like to thank Lean Manufacturing Experts from Technology Promotion Association (Thailand-Japan) who give opinion and fill data according to AHP based model questionnaire. This surveyed data are helpful to fulfill the research objective considerably. And we would appreciate to the faculty of engineering, KMITL for funding some important part of this research.

REFERENCES

[1] T. L. Saaty and L. G. Vargas, *The Logic of Priorities.* Kluwer Nijhoff Publishing, US, 1982.

[2] T. Laosirihongthong, S. Rahman, K. Saykhun, Critical success factors of six-sigma implementation: an analytic hierarchy process based study", *International Journal of Innovation and Technology Management*, Vol. 3, No. 3, pp. 303–319, 2006.

[3] B. Wernerfelt, "A resource-based view of the firm," *Strategic Management Journal*, vol. 5, pp. 171–180, 1984.

[4] G. F. Luo, and W. L. Huang, "A multi-stage model for evaluating new technology-based firms' innovation resource on the view of venture capital" in *Proc. 7th Conf. Machine Learning and Cybernetics*, Kunming, pp. 12-15, 2008.

[5] R. R. Nelson and S. G. Winter, "An evolutionary theory of economic change," Harvard University Press, 1982.

[6] D. J. Teece, G. Pisano and A. Shuen, "Dynamic capabilities and strategic management," *Strategic Management Journal*, vol. 18, pp. 509–533, 1997.

[7] T. Kiatcharoenpol, P. Punnakitikashem and T. Laosirihongthong, "Six-Sigma™ implementation: Resource-Based View Perspective", in *Proc. IEEE 18Th International Conference on Industrial Engineering and Engineering Management*, China, 3-5 Sept., 2011.

[8] P. Achanga, E. Shehab, R. Roy and G. Nelder,"Critical success factors for lean implementation within SMEs", Journal of Manufacturing Technology Management, 17(4), pp.460-471, 2006.

[9] S. Bhasin, "Performance of organizations treating lean as an ideology", *Business Process Management Journal*, 17(6), pp. 986-1011, 2011.

[10] V. Crute, Y. Ward, S. Brown, and A. Graves, "Implementing Lean in aerospace challenging the assumption and understanding the challenges", *Technovation*, 23, pp. 917-928, 2003.

[11] J. Czabke, E. N. Hansen and T. L. Doolen, "A multisite field study of lean thinking in US and German secondary wood products manufacturers", *Forest Products Journal*, 58(9), pp. 77-85, 2008.

[12] K. Goutam, and B. M. Murali, "Critical success factors for implementing Lean practices in IT support services", *International Journal for Quality Research*, 6(4), pp. 301-312, 2012.

[13] A. Grove, J. Meredith, J. Anjelis, and K. Neailey, "Lean implementation in primary care: health visiting services in National Healthy Service UK", *Quality and Safety in Health Care*, 19, pp. 1-5, 2010.

[14] J. P. Womack and D. T. Jones, *Lean Thinking*. Free Press, New York, 2003.

[15] Kumar, M., Antony, J. and Douglas, A., "Does size matter for six sigma implementation?-Findings from the survey in UK SMEs", *The TQM journal*, 21(6), pp. 623-635, 2009.

[16] R. N. Mefford, "Increasing productivity in global firms: The CEO challenge", *Journal of International Management*, 5(3), pp. 262-272, 2009.

[17] M. Scheer-Rathje, T.A. Boyle and P. Delforin, "Lean, take two! Reflections from the second attempt at the Lean implementation", Business Horizons, 52(1), pp. 79-88, 2009.

[18] K. L. Sim and J. W. Rogers, "Implementing lean production systems: Barriers to Change", *Management Research News*, 32(1), pp. 37-49, 2009.

[19] M. F. Sua'rez-Barraza and J. Ramis-Pujol, "Implementation of Lean-Kaizen in the human resource service process: A case study in a Mexican public service organization", *Journal of Manufacturing Technology Management*, 21(3), pp. 388-410, 2010.

[20] K. S. Chin, et al., "An AHP based study of critical factors for TQM implementation in Shanghai manufacturing industries," *Technovation*, vol. 22, pp. 707-715, 2002.

Risk Assessment of Aviation Maintenance Error Based on Set Pair Analysis and BP Neural Network

Ying Duan*, Ling-ling Yuan
Department of Mechanical Engineering, Chongqing University, Chongqing, China
(15923062685@163.com)

Abstract - **According to the complex association of the aviation maintenance error, using the SHEL model to divided the content of maintenance error into four subsystems including human-hardware subsystem, human-software subsystem, human-environment subsystem, human-human subsystem, and establishing a hierarchical model of ladder structure based on this. Considering the uncertainty and complexity of each subsystem, first establishing a set of risk assessment models of maintenance error through the set pair analysis (SPA), and dealing with date from same-indefinite-contrary points in order to acquire the degree of connection, and combining with AHP analysis method to determine subsystem's weight,. Then each component values of the connection's degree as input layer of BP neural network, simulating the maintenance error comprehensive risk by using BP neural network method. Finally, taking an aviation maintenance enterprise for an example, the network model is trained and tested from the practical data, and the quantitative results indicate the effectiveness and feasibility of the proposed method, which can provide a reasonable decision support for aviation maintenance management.**

Keywords - **Aviation maintenance error, BP neural network method, SHEL model, set pair analysis (SPA)**

I. INTRODUCTION

With the rapid development of the national economy, China's air transport industry has been rapid development. In closing the world's distance and providing convenient travel at the same time, the people's demand for air's security, reliability and maintainability is higher and higher. Aviation accidents rates caused by the mechanical fault is dropped to the original 1/4. But the risk of locomotive maintenance had an upward trend, therefore maintenance error management become more and more important for the quality of aviation maintenance. The risk assessment of aviation maintenance error is set as an important part of maintenance error management, which will be more and more important. In order to improve the quality of maintenance, this paper evaluates the risk of aviation maintenance error from the point of people.

At present, the methods of risk assessment used at home and abroad mainly include principal component analysis, grey comprehensive evaluation method, fuzzy analytic hierarchy process, set pair analysis method and a set of qualitative and quantitative integration method. These methods have been certainly applied in practice. But the method of principal component analysis need strict conditions, and the precision of evaluation is low; the calculation of grey comprehensive evaluation method is simple, but its implementation is difficult involving the large amount of data; establishing membership degree, the fuzzy AHP is subjective; set pair analysis methods may make objective assessment by analyzing the relationship and transformation between various influencing factors with dialectical perspective, but there are parameters problem; using neural network algorithm to simulate evaluation weight can reduce people's subjectivity randomness [1-9].

Aviation maintenance error is an open and complex giant system of a dynamic, multi-variable, the human factor played a leading role with uncertain factors and logic relations. According to the characteristics of the aviation maintenance error, this paper aggregates the risk factors from different levels, and establishes a hierarchical structure model of the influence of human factors. On the basis of the characteristics and structure of the model, a practical calculation method is put forward. This paper processes the index data using set pair analysis method from three aspects, and also determines system's weight combined with AHP analysis method. Finally using the BP neural network algorithm to simulate the weight of each component in the connection's degree, which can overcome the influence of the parameters in set pair analysis method. The example proves the feasibility and effectiveness of the method.

II. MAINTENANCE ERROR DESCRIPTION

Based on analysis of status, history and statistical data in aviation maintenance, this paper divides the risk of maintenance error into four subsystems with SHEL model [10]: man-hardware, man-software, man-environment, man-man. The four subsystems are graphically depicted in Fig.1.

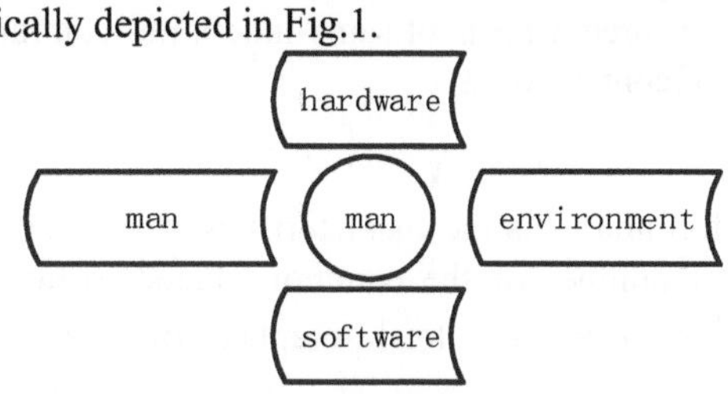

Fig.1. System model

SHEL model is centered on human factors, and studying on relationship between man, hardware, software, environment. The following is four subsystems:

Man-hardware subsystem: performing the interaction between man and work environment (the location of dangerous goods storage, airport facilities and equipment, ect) in maintenance, which decide man how to interact with physical work environment.

E. Qi et al. (eds.), *Proceedings of the 5th International Asia Conference on Industrial Engineering and Management Innovation (IEMI2014)*, Proceedings of the International Asia Conference on Industrial Engineering and Management Innovation 1, DOI 10.2991/978-94-6239-100-0_55, © Atlantis Press and the authors 2015

Man-software subsystem: performing the relationship between man and support system in working environment, such as regulations, operation rules, safety checklist, ect.

Man-environment subsystem: performing the relationship between man and internal, eternal environment. Internal environment contains temperature, weather, noise, vibration, air quality and enterprise culture, etc. External environment contains economic environment, political environment, etc.

Man-man subsystem: performing the interpersonal relationship in maintenance place, such as team members, quality inspector, etc.

Based on the analysis of maintenance error's risk in four subsystems, an index system is proposed and graphically depicted in Fig.2.

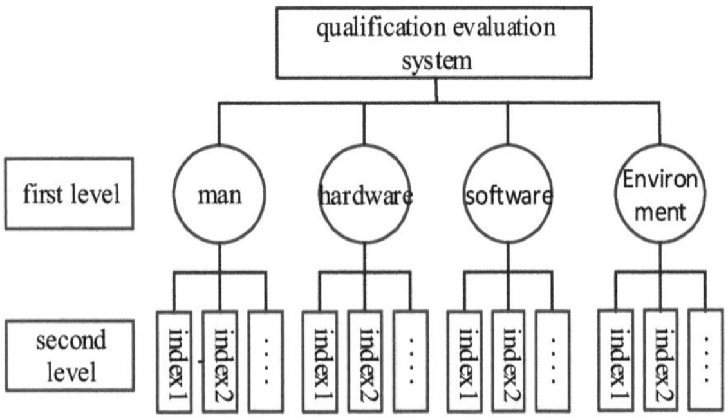

Fig.2. Index system

III. ALGORITHM DESIGN

In order to improve the accuracy and effectiveness of the assessment, two-phase algorithm are proposed to solve the risk assessment problems of aviation maintenance errors.

A. Set Pair Analysis

The core theory of set pair analysis [11-12] is to build set pair $H(\phi, \psi)$ from set ϕ and set ψ, and analyze the two sets from three aspects of same-indefinite- contrary. The degree of connection is,

$$\mu = \frac{\partial}{N} + \frac{\gamma}{N} i + \frac{\beta}{N} j \tag{1}$$

In the formula, N is the characteristics of set pair H ; ∂ is the total number of the common characteristics of set pair H ; β is the total number of the contrary characteristics of set pair H ; $\gamma(\gamma = N - \partial - \beta)$ is the total number of the other characteristics of set pair H ; i is indefinite coefficient, $i = [-1,1]$; j is contrary coefficient, and its normal value is -1.

$a = \frac{\partial}{N}$, $b = \frac{\gamma}{N}$, $c = \frac{\beta}{N}$ respectively denote the same degree, indefinite degree and contrary degree of set pair H under the specific problem, the simplified equation is

$$\mu = a + bi + cj \tag{2}$$

According to the classification of index grade, we can divide the bi into five dimension connection's degree $\mu = a + b_1 i_1 + b_2 i_2 + b_3 i_3 + cj$, and build a set pair by secondary index combined with the standard of corresponding classification. Among them, the same degree a corresponds the risk grade of 1; the indefinite degree b_1 corresponds the risk grade of 2; the indefinite degree b_2 corresponds the risk grade of 3; the indefinite degree b_3 corresponds the risk grade of 4; the contrary degree c corresponds the risk grade of 5.

1) Assumption of risk assessment

Assumption1 The first level $V = \bigcup_{i=1}^{m} V_i$ is consist of four subsystems, $i \in [1, m]$. And when any $j \neq i$, $V_j \cap V_i = \phi$. The second level $V_i = (\mu_{i1}, \cdots, \mu_{ik}, \cdots, \mu_{in})$, $k \in [1, n]$ is consist of assessment index in every system.

Assumption2 Every subsystem needs to build a standard evaluation of index set, including all the index in the subsystem. The set pair $H(V_i, V_i^0)$ is built by subsystem V_i and the corresponding standard subsystem V_i^0. The different subsystem may have different number of index, in order to ensure the same numbers, if the subsystem don't exist the index, the value is 0.

Assumption3 The component coefficient i, j is just to mark, not to consider their value range.

2) Assessment model of index

According to the above conditions, building the membership function of assessment index and normal assessment index, and the degree of connection is,

$$\mu_{mn} = \begin{cases} 1 + 0i_1 + 0i_2 + 0i_3 + 0j & x = 0 \\ \frac{s_1 - x}{s_1} + \frac{x}{s_1} i_1 + 0i_2 + 0i_3 + 0j & x \in (0, s_1] \\ 0 + \frac{s_2 - x}{s_2 - s_1} i_1 + \frac{x - s_1}{s_2 - s_1} i_2 + 0i_3 + 0j & x \in (s_1, s_2] \\ 0 + 0i_1 + \frac{s_3 - x}{s_3 - s_2} i_2 + \frac{x - s_2}{s_3 - s_2} i_3 + 0j & x \in (s_2, s_3] \\ 0 + 0i_1 + 0i_2 + \frac{s_4 - x}{s_4 - s_3} i_3 + \frac{x - s_3}{s_4 - s_3} j & x \in (s_3, s_4] \\ 0 + 0i_1 + 0i_2 + 0i_3 + 1j & x \in (s_4, +\infty) \end{cases} \tag{3}$$

In the formula, s_1、s_2、s_3、s_4 denotes the limited value of classification standard of index n ; x denotes the actual risk value.; m denotes the subsystems; n denotes assessment index.

3) Assessment model of subsystem

According to the operational rule of the connection's degree, getting average connection's degree of evaluation index in first level.

$$\overline{\mu_m} = \frac{1}{n}\sum_{k=1}^{n} \mu_{mn} \quad (4)$$

In the formula, $\overline{\mu_m}$ denotes the average connection's degree of subsystem m.

4) Determine the weight of subsystem

AHP [13] is a combination of qualitative and quantitative analysis, which is a kind of evaluation and decision method to express and process the human's subjective judgment with quantity form, the following is the main steps to determine the weight of subsystems:

1) According to the scaling theory, constructing judgment matrix $B = (b_{ij})_{n*n}$;

2) Normalizing the data of the judgment matrix B, $\overline{b_{ij}} = b_{ij} \Big/ \sum_{h=2}^{n} b_{hj}$, $j = 1,2,\cdots,m$;

3) Working out the summation of each line elements, $\overline{W_i} = \sum_{j=1}^{n} \overline{b_{ij}}$, $i = 1,2,\cdots,n$;

4) Normalizing $\overline{W_i}$, $W_i = \overline{W_i} \Big/ \sum_{j=1}^{n} \overline{W_i}$;

5) Checking the consistency and randomness, if not satisfied the conditions, modifying the judgment matrix until satisfied.

B. BP neural network

BP neural network [14-15] is a self-organizing neural network to meet the given direction of input and output relationship, which is one of the most mature and commonly used model. It can better solve the uncertain and complex systems problem. Using the BP neural network algorithm to simulate the weight of each component in the connection's degree.

1) Select the type of network structure

In the risk assessment of maintenance error, selecting the feed forward networks, mainly using its characteristics of non-linear mapping function to approach any space from the finite-dimensional space to another finite-dimensional space. Containing a principle: a hidden layer with a transfer function of Sigmoid and its output layer with a linear function, which can approximate any practical function. Thus, there is no meaning of too much intermediate layers, on contrary, it will increase the complexity of the calculations. So this paper adopts three-layer network structure including an input layer, an output layer and a hidden layer.

2) Decide the type of network structure

Data in the input and output layers are associated with some practical problems represented some practical significance. By the above calculation, each component value in the connection degree of maintenance errors will be regarded as the input layer of the neural network, and the comprehensive risk value of errors will be regarded as the output layer. The hidden layer is established by the requirements of easy degree and model problem, and the number of nodes g is the key of this method:

$$g = \sqrt{m + n} + a \quad (5)$$

In equation (5), the parameter m represents the number of nodes in the output layer; Parameter n represents the number of nodes in the input layer; Parameter a represents the constant between 0 to 10.

3) Risk classification of Maintenance error

Based on the established network structure, using the final output value to define the risk. The lower its value is, the lower the risk grade, the higher the security. At present there is no general standard to determine the risk level, so dividing the interval [0,1] into five-level by the equipartition principle: grade I corresponds to the interval [0,0.2); grade II corresponds to the interval [0.2,0.4); grade III corresponds to the interval [0.4,0.6); grade IV corresponds to the interval [0.6,0.8); grade V corresponds to the interval [0.8,1], which are described as pole-low risk, low risk, medium risk, high risk, pole-high risk level.

IV. CASE

A. Basic data

Based on collecting and analyzing the accident research report of aviation maintenance, the index is selected and classified, finally the remaining index use the method of questionnaire to investigate maintenance crews and primary level managers, and it proposes the evaluation TABLE I.

TABLE I
THE INDEX OF "MAN-ENVIROMENT-HARDWARE-SOFTWARE"

subsystem	index
man—man(S1)	communication skills(x_{11}), team consciousness(x_{12}), cooperation(x_{13}), emotion mediation (x_{14})
man—environment(S2)	emergency ability to environment (x_{21}), psychological adjustment competence to environment diversification(x_{22})
man—hardware(S3)	Be familiar with equipment performance (x_{31}), technical level (x_{32}), operation control competence (x_{33}), maintenance competence to equipment (x_{34})
man—software(S4)	safety awareness(x_{41}), sense of responsibility (x_{42}), education and training (x_{43}), observe operation flow (x_{44}), safe operant behavior (x_{45})

B. Basic data

This paper collects 10 evaluation cases of error risk in an aviation maintenance base, and establishes the expert group of risk evaluation. According to the actual probability and economic loss, the specialist team evaluate each index and give value-at-risk. An original value-at-risk is graphically depicted in TABLE II. Using the AHP analysis method to determine the weight of each subsystem, and 1~9 scaling method to construct judgment matrix B. The result is graphically depicted in TABLE III.

TABLE II
VALUE-AT-RISK OF EACH INDEX

Index	X_{11}	X_{12}	X_{13}	X_{14}	X_{21}	X_{22}	X_{31}	X_{32}	X_{33}	X_{34}	X_{41}	X_{42}	X_{43}	X_{44}	X_{45}
1	11	6	5	7	1	18	4	2	5	18	8	10	9	13	1
2	17	4	14	1	1	18	1	14	1	14	3	13	8	3	3
3	10	3	9	9	8	12	4	15	12	7	1	3	15	18	15
4	2	5	17	2	6	8	3	4	7	2	7	3	7	4	2
5	5	3	2	3	14	8	12	9	18	3	10	15	14	8	11
6	16	3	13	5	10	17	4	4	12	3	18	15	9	9	7
7	9	15	8	4	9	15	15	15	6	5	14	9	4	2	13
8	7	19	8	13	7	11	4	1	13	11	3	8	3	2	14
9	10	7	3	18	4	13	3	6	2	17	3	7	8	3	20
10	18	18	4	10	0	0	2	16	10	12	4	12	8	4	10

TABLE III
WIGHT OF SUBSYETEM

Judgment matrix B	S1	S2	S3	S4	$\overline{W_i}$	λ max	*CI / CR*
S1	1	4	3	2	0.4400	4.187	0.070<0.1
S2	1/4	1	1/2	1/4	0.0814		
S3	1/3	2	1	1/5	0.1196		
S4	1/2	4	5	1	0.3589		

C. Maintenance error risk evaluation

Input layer（5）Hidden layer（5）Output layer（1）

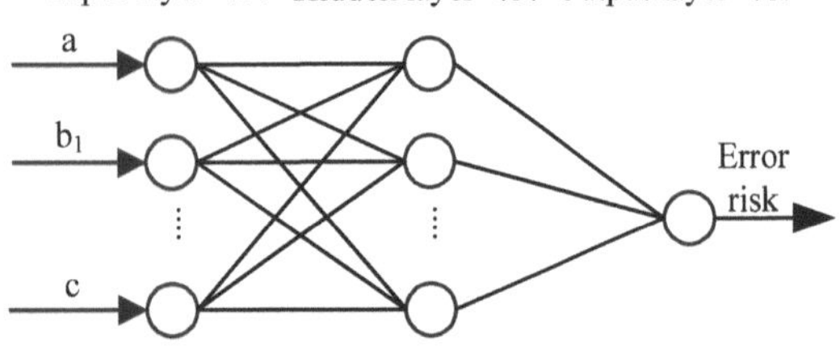

Fig.3. BP network structure

Parameter setting of BP neural network: μ denotes minimum training rate, its value is generally defined as 0.9; dynamic parameters is selected by experience, its value is generally defined as 0.7; ε denotes permissible error, $\varepsilon \in 0.001\text{-}0.00001$; T denotes the number of iterations, $T = 1000$; Sigmoid denotes the transfer function, its range is defined as [0.9~1.0]. A network structure is graphically depicted in Fig.3.

D. The Input and Output

A is the highest safety level and is consisted of no risk content, which connection degree is set as $\mu_A = 1 + 0*i_1 + 0*i_2 + 0*i_3 + 0*j$. The evaluation criteria is used to construct training samples of BP neural network, and thirteen connection degrees are given according to the safety degree of highest to lowest. Using linear interpolation method to equally distribute the range 0-1 of risk degree and giving the corresponding expected value-at-risk and grade. The connection degree of maintenance error is calculated by AHP method and set pair analysis method, whose value of each connection is set as input of BP neural network input layer, and the value-at- risk of maintenance error is set as output of BP neural network output layer. A network structure is graphically depicted in TABLE IV.

TABLE IV
BP NETURAL NETWORK SAMPLES

training sample							**test sample**							
	(Input)					(Output)		(Input)					(Output)	actual value
No	**a**	**b1**	**b2**	**b3**	**c**	**I**	**No**	**a**	**b1**	**b2**	**b3**	**c**	**I**	
1	1	0	0	0	0	0	1	0.069	0.250	0..361	0.211	0.109	0.315	0.309
2	0.5	0.5	0	0	0	1/12	2	0.212	0.324	0.072	0.139	0.253	0.489	0.470
3	0.5	0	0.5	0	0	1/6	3	0.100	0.192	0.283	0.224	0.201	0.369	0.359
4	0.5	0	0	0.5	0	1/4	4	0.187	0.485	0.218	0	0.110	0.233	0.238
5	0.5	0	0	0	0.5	1/3	5	0.118	0.325	0.216	0.201	0.140	0.624	0.615
6	0	0.5	0.5	0	0	5/12	6	0.035	0.265	0.209	0.187	0.304	0.712	0.710
7	0	0.5	0	0.5	0	1/2	7	0.036	0.255	0.299	0.198	0.212	0.413	0.419
8	0	0	1	0	0	7/12	8	0.094	0.219	0.312	0.194	0.181	0.348	0.357
9	0	0.5	0	0	0.5	2/3	9	0.086	0.329	0.278	0.085	0.222	0.517	0.509
10	0	0	0.5	0.5	0	3/4	10	0.076	0.188	0.299	0.098	0.339	0.786	0.793
11	0	0	0.5	0	0.5	5/6								
12	0	0	0	0.5	0.5	11/12								
13	0	0	0	0	1	1								

Using matlab to simulate the test, and the result of evaluation is graphically depicted in Fig.4. From the diagram, we can see the high degree of agreement between the prediction of BP neural network model and actual level evaluation, and the error in a reasonable range. The result shows that the BP neural network is feasible for the risk prediction of Maintenance error.

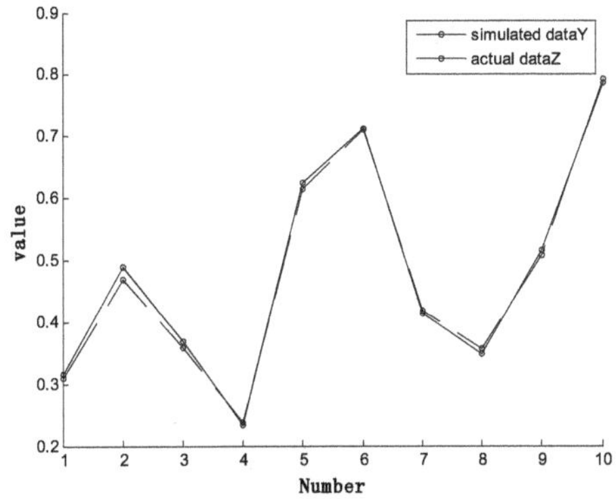

Fig.4. BP network structure

According to risk judgment standard of the maintenance error, there are three maintenance errors in the fourth grade, belonging to the higher risk, which need to adopt risk control measures to conduct a comprehensive management and avoid danger.

V. CONCLUSION

According to the complex, multivariable, uncertainty of aviation maintenance errors, this paper is considering various factors comprehensively from the system point of view, and is also modeling and simulating to approach the risk distribution approximation. Compared with other methods, this method has unique advantages. First, using SPA to establish the risk assessment model of maintenance errors, and combining with the AHP analysis method to determine the weight of each subsystem. Then using BP neural network to simulate the risk of maintenance error, which can avoid the direct computation of association degree of difference coefficient, but also make up for the deficiency of judging principle of the maximum connection's degree. It is better to solution the quantitative assessment of maintenance error risk, and the accuracy of evaluation result is scientific and significant. But this method depends on digging and improving the index data, through continuous accumulating, it can simulate better results, which need to in-depth study the index.

REFERENCES

[1] MENG Yu, ZHAO Shan-shan, WANG Shu-sheng, LI Lie-ping. Risk Evaluation Researching of Power-supply Company [J]. Safety and Environmental Engineering, vol.16, no.2, pp.82-85, 2009.

[2] WANG Su. Fire risk assessment of high-rise buildings based on grey relational analysis [J]. Journal of Safety Science and Technology, vol.9, no.8, pp. 83-89, 2013.

[3] JING Guo-xun, WANG Wei-min, WEN Hong-min. Risk Assessment of Fire Accidents in Henan Province of China Based on BP Neural Network[J].China Safety Science Journal, vol.17, no.8, pp.16-18, 2007.

[4] WANG Wen-sheng, JN Ju-liang, LI Yue-qing. Risk Degree Assessment of Natural Disaster Based on Set Pair Analysis Method [J]. Journal of Sichuan University (Engineering Science Edition), vol.41, no.6, pp.6-11, 2009.

[5] WANG Juan. Application of Fuzzy Analytic Hierarchy Process to the HSE Risk Assessment of Storage Tank Construction [J]. Safety and Environmental Engineering, vol.18, no.5, pp.96-99, 2011.

[6] Nieto-Morote, A, Ruz-Vila, F. A fuzzy approach to construction project risk assessment [J]. International Journal of Project Management, vol.29, no.2, pp.220-231, 2011.

[7] MerzB,Hall J, Disse M. Fluvial flood risk management in a changing world [J].Natural Hazards and Earth System Sciences, vol.10, no.10, pp.509-527, 2010.

[8] XU Hong-jie, GONG Bo, LI De-shun. Risk assessment of coal mine based on set pair analysis [J]. Journal of Safety Science and Technology, vol.8, no.6, pp.86-90, 2012.

[9] WEI Yong, XU Kai-li. Application of set pair analysis on safety evaluation of tailing dam stability[J]. Journal of Safety Science and Technology, vol.9, no.8, pp.101-105, 2012.

[10] WANG Qi-quan. Neural Network Safety Assessment Method for Aviation Enterprises Based on SHEL Model [J]. China Safety Science Journal, vol.20, no.2, pp.46-53, 2010.

[11] GONG Bo, XU Kai-li, LI De-shun. Risk assessment of building fire based on set pair analysis [J]. Journal of Safety Science and Technology, vol.9, no.8, pp.102-106, 2012.

[12] Ming-Wu Wang, Peng Xu, Jian Li, Kui-Yuan Zhao.A novel set pair analysis method based on variable weights for liquefaction evaluation[J]. Natural Hazards, vol. 70, no.2, pp 1527-1534, 2014.

[13] Grace K. L. Lee, Edwin H. W. Chan.The Analytic Hierarchy Process (AHP) Approach for Assessment of Urban Renewal Proposals [J]. Social Indicators Research, vol.89, no. 1, pp.155-168, 2008.

[14] WEI Kai, GUAN Zhi-chuan, WEI Jie-hong, FU Sheng-lin, ZHAO Ting-feng. Drilling Engineering Risk Evaluation Method Based on Neural Network and Monte-Carlo Simulation [J].China Safety Science Journal, vol.23, no.2, pp.123-127, 2013.

[15] Li fang Peng, Lingling Lai. A service innovation evaluation framework for tourism e-commerce in China based on BP neural network [J]. Electronic Markets, vol.24, no.1, pp 37-46, 2014.

Part III

Management Innovation

Research Framework of Leadership Behavior and Team Innovation Performance

Qiao-yun He[1,2,*], Chun-hua Chen[1]
[1]Department of Business Administration, University of South China University of Technology, Guangzhou, China
[2]Department of Economics & Management, Guangdong Institute of Science and Technology, Guangzhou, China
(*Tracy_hqy@163.com, bmchchen@scut.edu.cn)

Abstract - **Team leadership has been confirmed to be a key important factor in all the factors influencing team effectiveness. According to Heuristic model of team effectiveness, the research constructs the mediator pattern, team leadership - team process and team status - Team performance. Positive and Negative leadership behavior of team leader is a situational factor which has an influence on team process (team conflict) and team status (group affective tone) and has a final effect on team effectiveness (team innovation performance), and verify the conformity of the model through the following empirical study, for proposing leadership strategies promoting team innovation performance.**

Keywords - **Authentic leadership, abusive supervision, team conflict, Heuristic model of team effectiveness, team emotion atmosphere, team innovation performance**

I. RESEARCH BACKGROUND

With profound changes of the organization environment, more and more organizations apply team to complete the objectives of the organizations for increasing organizational effectiveness and improving the productivity. And in harmony with this, the role of leader is not only to lead and motivate the subordinates any longer, and how to lead and motivate the subordinates for achieving the team's objective should be included [1]. It is thought that innovation is the key to obtaining and keeping the dominance in the dynamic situation for organization [2, 3]. As an elementary cell of organization in the new situation, team is an important executor of team strategies. So more and more scholars pay attention to team innovation performance and its effect mechanism.

The research indicates that team innovation performance is influence by many factors. However, in recent years, leadership has achieved attention as an important factor affecting team innovation [4]. Leadership has always been a hot research topic attracting administration scholars. The research orientation on leadership can be roughly classified into 2 categories. One is to study effective leadership behavior and its positive effect on subordinates, team and organization from the positive perspective, and the research orientation has always been the mainstream [5-7]. And one is to study effective leadership behavior and its positive effect on subordinates, team and organization from the negative perspective, and the research orientation was ignored in previous studies on leadership and has attracted scholars' attention gradually in recent years such as abusive supervision proposed by Tepper in 2000, tyrant behavior proposed by Ashforth in 1997, authoritarianism leadership proposed by Zheng Boxun in 1995 and bullying behavior proposed by Einarsen in 1999, etc. Some achievements have been achieved by the study on negative leadership behavior at present, which indicates that the field also has the trend of vigorous development and expansion [8-12].

The threats of SARS, financial tsunami and Foxconn falls happened in recent years tests the faith of Chinese people on leaders. In Chinese corporation, the scandals created by leaders abusing public trust seem to be continuous [13]. Everyday the newspaper has the report that business executives are forced to resign and given conviction and sentencing, even be chained and thrown into prison because of violating ethics and morality, which has already been familiar with readers, such as Wong Kwong Yu incident, Wu Ying incident and Tang Jun educational background door incident [14]. So far, studies on positive and negative leadership behavior are centered in foreign countries, and the theories and empirical studies on positive and negative leadership behavior on the basis of organizational context in China are very deficient [15]. Therefore, it is necessary to study positive and negative leadership behavior on the basis of organizational context in China. Theories of positive and negative leadership behavior in Western countries can provide some ideas for the development of team innovation performance in enterprise and the construction of relationship between leaders and subordinates at the turning stage of Chinese economy.

II. METHODOLOGY CONCEPT DEFINITION

A. Authentic Leadership

In 2007, Endrissat and Muller and Baum pointed that authentic leadership was the origin of positive leadership styles including transformational leadership, moral leadership and charismatic leadership. The concept of authentic leadership originated in the events of leaders lacking morals such as Worldcom, Worldcom and Martha Stewart, and the increase of more and more social challenges such as 911 terrorist attacks, stock market fluctuation in USA, economic depression and SARS, which needs more positive leadership theories to restore people's confidence. Many researches claim that the existing frameworks of leadership theory can't cultivate the leaders needed in the future.

High moral standards is the principle for the actions of authentic leaders who are the models of organization members, possess high self awareness and psychological maturity and are full of hope, confidence, optimistic and high resilience. The employees can also identify and practice sincere behavior under the influence of authentic leaders, and exhibit more positive working emotion and behaviour. Through concluding the above definition on

E. Qi et al. (eds.), *Proceedings of the 5th International Asia Conference on Industrial Engineering and Management Innovation (IEMI2014)*, Proceedings of the International Asia Conference on Industrial Engineering and Management Innovation 1, DOI 10.2991/978-94-6239-100-0_56,

authentic leadership and developing process, Walumbwa proposed four factor model on authentic leadership on the basis of the past studies in 2008, that is, self awareness, relationship transparency, balanced information processing and internalizing morality.

B. Abusive Supervision

As a negative leadership behavior, abusive supervision is a growing research topic in recent years. It points hostile actions of leaders exhibiting on the subordinates continually including irony, taunt, threat, solation and corporal (physical) punishment except for body contact. Abusive supervision was firstly proposed by American scholar Bennett J.Tepper in 2000. Next, Tepper and his workmates made a series of studies on abusive supervision and acquired a great achievement. The studies of Tepper pointed out that common abusive supervision behavior consists of openly irony or accusing of subordinates, controlling information, coercing subordinates, violating the privacy of subordinates, breaking promises and lying to subordinates. Although these actions have hostility, they still belong to the social tolerance range.

In 2007, Tepper defined the concept of abusive supervision from 4 aspects, that is, (1) it expresses the hostility of managers on the particular subordinates, (2) it excludes infringement in some respects such as bodily injury, sexual harassment, etc. (3) it is an intentional behavior, (4) abusive supervisors don't have the intention of adverse consequences. The four dimensions can distinguish abusive supervision from other related concepts. From the concept of abusive supervision, we can summarize the following characteristics, subjectivity, constancy, hostility and non physical contact.

C. Team Conflict

As an important factor of team process, team conflict is an important variable for effective operation in a team and has important influence on improving team performance and forming team overall cognitive. Team conflict is generally considered a moderator variable among team environment, antecedent of team individual and result of team process. Researchers generally think that conflict has two structure types which are related evidently, one is cognitive conflict caused by different opinions of organization members on the task, the other is affective conflict caused by antagonistic sentiments. There is difference in cause and objective between two conflict types.

D. Group Affective Tone

Group affective tone points the emotion and feeling produced by emotional interactions and interpersonal relationships among team members. Researchers break group affective tone into two aspects, positive group affective tone and negative group affective tone. In this research, the connotation of group affective tone is also divided into two aspects, positive group affective tone and negative group affective tone. And exploring group affective tone. Next, analyzing and explaining the two aspects. 1. Positive group affective tone is good group affective tone which not only can condense the consensus among team members, but also can build harmonious and safe emotion, and the team members have identity and sense of belonging. 2. Negative group affective tone is bad group affective tone which not only can't condense the consensus among team members, but also has conflict and helplessness among teams, and the team members have no identity and sense of belonging.

E. Team Innovation Performance

Team innovation means some new ideas, processes, products or procedures for the team which are introduced and applied consciously to benefit individual, team, organization and even the whole society. Team innovation performance is effective output in the process of executing innovation mission for the team, which reflects the completion of innovation objectives.

III. RESEARCH FRAMEWORK

A. Deficiencies of Existing Studies

Team leadership has been confirmed to be an key important factor in all the factors influencing team effectiveness. There were not many researches in the past, and they only studied traditional leadership concept including transaction type and conversion type. Compared with the study of individual leve, the influence of positive and negative leadership behavior on team level results don't receive enough attention, and there have been few research results being published up to now. The inherent relationship between positive and negative leadership behavior and team innovation performance may be complicated, therefore, it needs more further researches on its intrinsic mechanism. Most literatures regard the leaders as a regulation variable in studying the influence of multiple variables on team innovation, which means that the leaders are as situational factor and is assumed to have an influence on the strength and direction of the relationship between other antecedent variables and innovation. However, because the leaders have an important and even decisive effect on cultural atmosphere, policy institution and goal setting, the research considers that if the leaders could be as the main antecedents variable influencing team innovation to study the influence of their behavior on team process and team state variables (such as team conflict and group affective tone) and to investigate the influence on team innovation, which may help us to recognize and understand team innovation from different aspects.

B. Theoretical Basis

In traditional researches relating to team, IPO (input-process- output) model was proposed by McGrath in 1964 and is the framework which is most commonly used. Input is a factor influencing team effectiveness, process is interactive process among team members and output means team effectiveness. In this model, team process is considered to be an important mechanism to connect team input variables and team output variables. In recent years, researches relating to team have surpassed traditional ideas only observing team process. And the concept reflecting team cognition, motivation or emotion and team status are incorporated into the researches on team input and team output so as to stretch the framework of IPO type.

In addition to the external environment, heuristic model of team effectiveness proposed by Cohen and Bailey in 1997 includes four important variables of team level. The fist variable is task design of team (including autonomy, interdependence and so on in a team), constitution (including size, formation age in a team) and situation (including leadership and award, etc.). The second variable is team process (including team conflict and communication). The third variable is mental characteristics of team (team status, including sharing mental model and group norm). The fourth variable is team effectiveness (including performance).

C. Research Model

According to Heuristic model of team effectiveness, positive and negative leadership behavior can be considered to be a situational factor of team, which has an influence on team process (team conflict) and team status (group affective tone) and has a final effect on team effectiveness (team innovation performance).

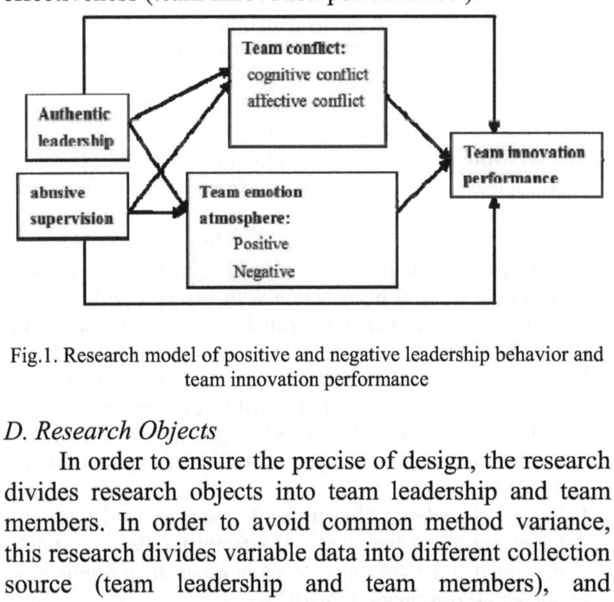

Fig.1. Research model of positive and negative leadership behavior and team innovation performance

D. Research Objects

In order to ensure the precise of design, the research divides research objects into team leadership and team members. In order to avoid common method variance, this research divides variable data into different collection source (team leadership and team members), and collecting data at T1andT2 with an interval of 3-5 months.

The questionnaire takes the mode of pair and is divided into two categories, team leadership questionnaire and team member questionnaire. In the first investigation (T1), team leadership questionnaire includes the background information of leader and team, team innovation performance, and team member questionnaire consists of the background information of individual, team conflict and group affective tone. After completing the questionnaire, respondents put the questionnaire into envelopes and send directly the questionnaire to researchers or give the questionnaire to Human Resources Department in every company to send to researchers intensively.

E. Research Instruments

1) Measurement of authentic leadership: The existing empirical studies about authentic leadership is few, and measurement of authentic leadership is not uniform. In 2003, Kemis pointed that sincerity included four aspects, self awareness, impartial information processing or balancing processing, sincere action and Sincere and transparent relations. Based on the study of Kemis and viewed by structure, Ilies observed authentic leadership model, the research results are similar to Kemis. The classification is the most widely used now. Fredo. Walumbwa developed measurement scale of authentic leadership on the basis of the classification.

This research took the measurement scale developed by Fredo. Walumbwa which includes 8 entries and each dimension has 2 entries, and applied Li Kete six point scale, in which 1 represents very disagree and 6 represents very agree, all items take positive score and the higher the score, the more satisfied with authentic leadership behavior.

2) Measurement of abusive supervision: Abusive supervision scale including 10 items in Chinese taken in studies by Aryee (2007) is adopted. This scale is constructed with 10 culturally neutral items of abusive supervision scale of 15 problems developed by Tepper in 2000. It covers irony, indifferent attitude, making employee embarrassed, violating privacy, lacking of trust, cheating, anger, rough and negative treatment in working process of leasers.

3) Measurement of team conflict: Measurement scale adopts conflict scale developed by Jehn in 2001, including task, process, relationship conflict and 9 measurement items. Many organizational environment including work group and senior management team have taken ICS, and researches verified the effectiveness of the measurement scale.

4) Measurement of group affective tone: Team members repot. Applying measurement scale developed by LIU which includes 4 subjects. The higher the score, the more positive group affective tone, for example, we feel vigorous in team, and every member in team is full of youthful spirit.

5) Measurement of team innovation performance: Measurement of team innovation performance in this research is reported by team leader and applies the measurement scale developed by LOVELACE. Investigating team innovation performance from the angle of innovation results. And conducting the measurement from innovation of products, the number of originality, overall technical performance and ability to adapt to change. There are 4 items, for example, the working achievement of this team involve novelty, and the team produces much originality and new ideas.

Previous studies indicate that background variables of leaders (age, sex and seniority) and background variables of teams (size and age) have an influence on team process, team status and team performance (Chen G, Kirkman B L, Kanfer R, Allen D, Rosen B., 2007). Therefore, the research regards age, sex and seniority of leaders and size and age of team as controlled variables to dispose.

6) Statistic and Analysis: In order to examine the discrimination validity of key variables including authentic leadership, abusive supervision, team conflict, group affective tone and team innovation performance, and the corresponding measurement parameter of every measurement scale, this research applies AMOS 7.0 to carry out confirmatory factor analyses. Making a contrast among five factor model, four factor model, three factor model, two factor model and single factor model.

The research adopts SPSS17.0 to process data. In addition to conventional statistical methods including average value and variance of variables and correlation coefficient, the research mainly takes CFA, CFA and SEM, and applies AMOS7.0 to prove the hypothesized model. (1) Influence on intervening variables (team conflict and group affective tone). Firstly, introducing controlled variables, then the effect of independent variables (authentic leadership and abusive supervision). (2) Inspection of mesmeric effect. Firstly, introducing controlled variables, then the effect of independent variables (authentic leadership and abusive supervision), lastly, introducing intervening variables (team conflict and group affective tone).

7) Discussion of Results: Combined with the research subject in this paper, through logical reasoning and theory analysis, after empirical analysis, analyzing and explaining empirical results in combination with related conclusions, and proposing leadership strategies to promote team innovation performance.

IV. INNOVATION

Compared with the research on individual level, the influence of positive and negative leadership behavior on variables of team level results don't receive enough attention, and there are few research results published up to now. The research make the influence of positive and negative leadership behavior on variables of team level results expand the team level.

The inherent relationship between positive and negative leadership behavior and team innovation performance may be complicated, and immanent action mechanism should be given further investigation. Considering team process and team status on the basis of heuristic model of team effectiveness can help to understand internal process of leadership behavior. Based on inspiration model of team effectiveness proposed by Cohen and Bailey in 1997, considering the intermediary role of team conflict and group affective tone on the relationship between positive and negative leadership behavior and team innovation performance, which has vital significance on revealing the relationship between positive and negative leadership behavior and team innovation performance.

This research model is on positive and negative leadership behavior in Chinese context. Taking originality of Chinese culture such as high power distance, collectivism, high uncertainty avoidance, long-term orientation and relationship standard into consideration, it is necessary to develop research of positive and negative leadership behavior in Chinese context (Ling Wenshuan, 2009; Zheng Boxun, 2010).

This research model is more rigorous in study design. The research adopts the study design of longitudinal tracking and data collection method with multiple sources (two surveys intervals 3-5 months and evaluate different variables by team leaders and team members), so homologous variance problems are avoided effectively, which helps to reveal the causation among variables more effectively.

REFERENCES

[1] D. J. Beebe, "Signal conversion (Book style with paper title and editor)," in *Biomedical Digital Signal Processing*, W. J. Tompkins, Ed. Englewood Cliffs, NJ: Prentice-Hall, 1993, ch. 3, pp. 61–74.

[2] M. Akay, *Time Frequency and Wavelets in Biomedical Signal Processing* (Book style). Piscataway, NJ: IEEE Press, 1998, pp. 123–135.

[3] G. B. Gentili, V. Tesi, M. Linari, and M. Marsili, "A versatile microwave plethysmograph for the monitoring of physiological parameters (Periodical style)," *IEEE Trans. Biomed. Eng.*, vol. 49, no. 10, pp. 1204–1210, Oct. 2002.

[4] V. Medina, R. Valdes, J. Azpiroz, and E. Sacristan, "Title of paper if known," unpublished.

[5] E. H. Miller, "A note on reflector arrays (Periodical style—Accepted for publication)," *IEEE Trans. Antennas Propagat.*, in press.

[6] T. Menendez, S. Achenbach, W. Moshage, M. Flug, E. Beinder, A. Kollert, A. Bittel, and K. Bachmann, "Prenatal recording of fetal heart action with magnetocardiography" (in German), *Zeitschrift für Kardiologie*, vol. 87, no. 2, pp. 111–8, 1998.

[7] J. E. Monzon, "The cultural approach to telemedicine in Latin American homes (Published Conference Proceedings style)," in *Proc. 3rd Conf. Information Technology Applications in Biomedicine, ITAB'00*, Arlington, VA, pp. 50–53.

[8] F. A. Saunders, "Electrotactile sensory aids for the handicapped (Presented Conference Paper style)," presented at the *4th Annu. Meeting Biomedical Engineering Society*, Los Angeles, CA, 1973.

[9] J. R. Boheki, "Adaptive AR model spectral parameters for monitoring neonatal EEG (Thesis or Dissertation style)," Ph.D. dissertation, Biomed. Eng. Program, Univ. Fed. Rio de Janeiro, Rio de Janeiro, Brazil, 2000.

[10] J. P. Wilkinson, "Nonlinear resonant circuit devices (Patent style)," U.S. Patent 3 624 12, July 16, 1990.

[11] T. Menendez, S. Achenbach, W. Moshage, M. Flug, E. Beinder, A. Kollert, A. Bittel, and K. Bachmann, "Prenatal recording of fetal heart action with magnetocardiography" (in German), *Zeitschrift für Kardiologie*, vol. 87, no. 2, pp. 111–8, 1998.

[12] J. E. Monzon, "The cultural approach to telemedicine in Latin American homes (Published Conference Proceedings style)," in *Proc. 3rd Conf. Information Technology Applications in Biomedicine, ITAB'00*, Arlington, VA, pp. 50–53.
[13] F. A. Saunders, "Electrotactile sensory aids for the handicapped (Presented Conference Paper style)," presented at the *4th Annu. Meeting Biomedical Engineering Society*, Los Angeles, CA, 1973.
[14] J. R. Boheki, "Adaptive AR model spectral parameters for monitoring neonatal EEG (Thesis or Dissertation style)," Ph.D. dissertation, Biomed. Eng. Program, Univ. Fed. Rio de Janeiro, Rio de Janeiro, Brazil, 2000.
[15] J. P. Wilkinson, "Nonlinear resonant circuit devices (Patent style)," U.S. Patent 3 624 12, July 16, 1990.

Research on Hierarchical Structure of Higher Education Performance Evaluation Based on GW Center

Ying-xin Wu , Xia-yuan Li*
Department of Management and Economic, Tianjin University, Tianjin, China
(dtwuyx@tju.edu.cn, *lixy@tju.edu.cn)

Abstract - **This paper introduces the concept of Generalized Work Center (GW Center) into the field of higher education, and divides the field of higher education into three levels, namely the higher education, the work center of higher education, and the GW center of higher education. A hierarchical structure of higher education performance evaluation based on GW center is also established to discuss the relationship between internal performance and external performance in the higher education, and explore the new area of higher education performance evaluation.**

Keywords - **Generalized Work Center (GW Center), Higher Education, Performance Evaluation**

I. INTRODUCTION

Research in higher education has always been one of the major concerns of domestic and overseas scholars, and performance evaluation attracts even more attentions on its importance and difficulties in organizational studies. Improving performance evaluation research of higher education in the theory and practice has been the inevitable choice and the developing direction which could strengthen the management system reform, improve the management and teaching quality, and promote the theoretical and practical development of management development and performance evaluation in higher education [1], [2]. Higher education performance evaluation is a type of measurement for the efficiency and effectiveness of input and output, and is researched according to the economic benefit, namely the input-output ratio [3], [4], [5]. The concept of GW center is introduced into the field of higher education by this paper, and the foundation of hierarchical structure of higher education performance evaluation based on GW center is also considered.

II. DEFINITION OF GENERALIZED WORK CENTER (GW CENTER)

In the traditional ERP system, production and manufacturing units are collectively called the "Work Center". The establishment of GW center has built a bridge between Manufacturing Resource and Operational Plan and Dispatch, combining the two organically. GW center is not only a basic unit of production scheduling, capacity calculation and cost calculation, but also a basic process unit of ERP system [6], [7]. The basic work center is the most basic atomic structural unit in the enterprise model, which combines parts of enterprise resources and organization structures. It could perform some living example of operation within the range of its ability property set, and produce resource consumption during the process of performing living example of operation [8], [9].

Aimed at coping with the problems in the ERP system application, Professor Congdong Li and his team came up with a series of solutions to improve flexibility and reconfiguration of the ERP system, in which GW center is one of the basic concepts and assumptions [10]-[11].

GW center expands the definition of the work center, making it possible to manage both structure unit and dynamic behavior in a unified way in each phase of the enterprise product full-lifecycle and the whole enterprise management process. Furthermore, GW center is composed of one or more sub work centers that can be the basic work centers or the GW centers. GW center is a tree structure with ladder levels, and the capability set of GM center is a union set of its sub work centers capability set. Meanwhile, GW center can execute business processes or sub processes according to its scope of capability set. Operations of the business processes or inferior sub processes are distributed to work sub centers which could meet the capacity requirements [12-15].

Virtual work center is one kind of GW center, namely the execution unit which does not belong to the inner-enterprise, integrating shared resources and functionality from the virtual enterprise partners, the suppliers, the customers and other partners. Virtual work center encapsulates the heterogeneous and distributed information systems of corporate partners, realizes the integration of shared resources and functionality. Structural model refactoring within and between enterprises could be executed in a unified way.

It can be seen that basic work center and GW center both belong to the basic unit and element in ERP system to analyze process of enterprise production activities. Production resources and corporate missions are linked closely by the inputs and outputs of GW center. The enterprise organization, the organization's hierarchical structure and the various organization roles are all described by GW center hierarchy. Resources consumption of GW center is added up by resources consumption of every work sub center on the basis of

E. Qi et al. (eds.), *Proceedings of the 5th International Asia Conference on Industrial Engineering and Management Innovation (IEMI2014)*, Proceedings of the International Asia Conference on Industrial Engineering and Management Innovation 1, DOI 10.2991/978-94-6239-100-0_57,

hierarchical structure. And the hierarchical structure of GW center determines the hierarchical relationships in process. The single top level in GW center list, namely the root node of the tree, is on behalf of this enterprise.

III. HIERARCHICAL STRUCTURE ANALYSIS OF HIGHER EDUCATION PERFORMANCE EVALUATION

In the area of the higher education, the basic work center is the most basic atomic structure unit, integrating parts of resources and organizational structures in higher education. In addition, it could perform some living example of operation within the range of its ability property set, and produce resources consumption during the process of performing living example of operation. The basic work centers are the entity sectors like the college, the department, the laboratory and so on. GW center in higher education is also composed of many sub work centers, and the sub work center can be the basic work center or GW center. The sub work centers of GW consist of the entity sectors like the college, the department and the laboratory , the project groups are also included such as the teaching working group, the scientific working group, the academic working group, the department, and the scientific research group and so on. In terms of resources allocation and operation coordination, the main function the basic work center and GW center is to achieve the optimal allocation in the internal higher education resources. Virtual work center in higher education is also a kind of GW center, and it belongs to the execution unit in the process of daily work in higher education, not the unit in internal higher education. Furthermore, the virtual work center could integrate the shared resources and functionality from the virtual higher education partners, the logistical support suppliers, the teaching and scientific research customers and other partners. GW center in higher education involves the whole process of operation, management and development including teaching, scientific research, social services and so on. The whole higher education could be regarded as a combination of a series of GW centers. As the basic object to perform the higher education functions, GW center is not only the center of teaching activities and scientific research activities, but also the center of management, cost and information.

One higher education has only one top layer of GW center called the higher education work center. The higher education work center is a gather of GW centers, including the basic work centers and the virtual work centers. From the view of definition and description, there are almost no differences in the different higher education work centers, as their basic work center and GW center are both the structural units based on each phase of higher education full-lifecycle and whole management process.

This paper quotes from concepts and methods of GW center, on this basis, it strengthens the hierarchy of GW center, divides the field of higher education into three levels, including the higher education, the work center of higher education, and the GW center of higher education, and sets up a performance evaluation model with corresponding hierarchy (see Fig.1). Compared with other GW centers, something needs to be identified that although higher education work center is an aggregate of one or more GW centers, it has some differences as it possesses higher decision-making power in higher education, and triggers activities of other GW centers.

Fig.1. Hierarchical structure of higher education performance evaluation based on GW center

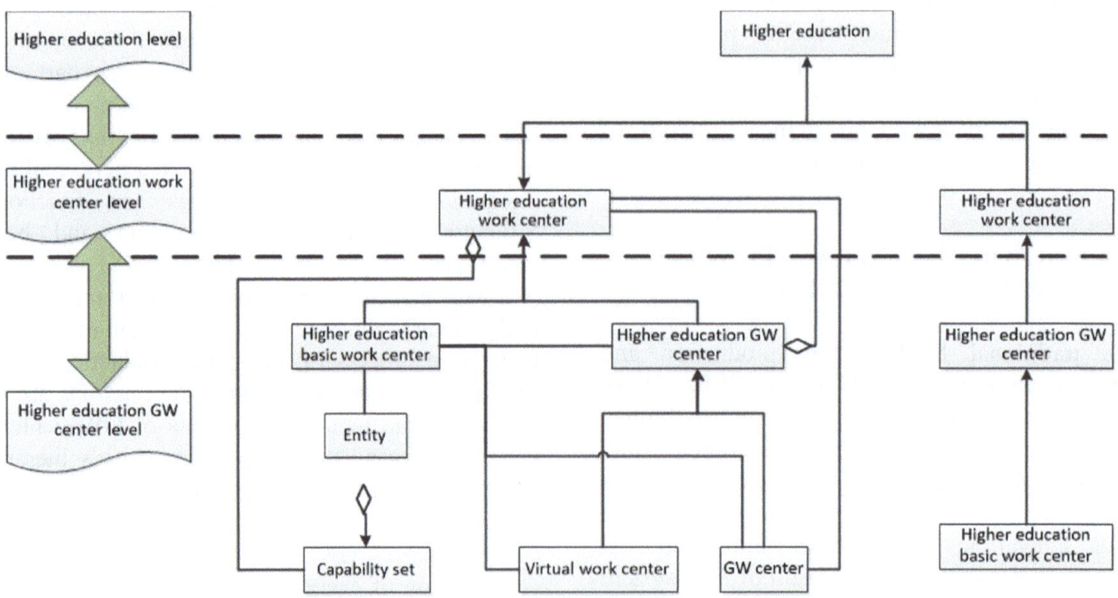

Thus it can be seen that, the field of higher education in the top of structure chart is the ultimate embodiment of the school performance, and the real school performance evaluation is reflected in the following two levels: one is the higher education GW center level (internal higher education). Relative to a certain education resources, the resources allocation and operation collaboration among many sub work centers in higher education, like achieving optimal resource allocation and forming a good resource allocation mechanism in different subject areas and different colleges, are realized under the coordination and decision-making of top level in the higher education work center. Furthermore, stakeholders in higher education like the teacher, the student and the administrative personnel, etc. have the accountability right and obligation to the running performance. Therefore, it is called the internal performance in higher education work center or internal higher education performance for short. The other one is the higher education work center. Under the supervision and accountability from all sectors of society including the government department, the cooperative enterprise, the donor, and the news media, as the agent, every higher education work center must publish the resources utilization information to the external stakeholders in order to get more education resources opportunities, as well as providing the decision-making basis for the whole social education resource allocation. Therefore, performance evaluation which does not belong to this kind of work center system is called external performance in higher education work center or external higher education performance for short. From the above, internal performance and external performance in higher education are the two components of higher education performance, as well as the two levels of realizing higher education performance, and lead to further research about the higher education performance evaluation comprehensively and systematically.

IV. THE RELATIONSHIP BETWEEN INTERNAL PERFORMANCE AND EXTERNAL PERFORMANCE IN HIGHER EDUCATION

The above analysis shows that internal performance and external performance in higher education belong to an organic whole, there are not only differences but close connections between the two.

Apparently, only if the students trained in higher education and the relevant outputs from teaching and scientific research both meet the needs of the development of higher education and economic society, the embodiments of high external performance can be showed. Meanwhile, only if the students have quality, innovation ability, practice ability and entrepreneurial spirit, the relevant outputs from teaching and scientific research could promote the social productivity development, and then better external performance could be formed. Moreover, only if the higher education carries out knowledge innovation, technology innovation and social services actively instead of only training students, higher internal performance could be formed. In other words, only if higher education forms good internal performance themselves under a certain amount of education resource inputs, it could form better external performance to the country and the society. Otherwise, if the internal higher education performance quality and the resources utilization rate are both low, the goal of talent training and the goal of the development in social and economic are disjointed seriously, the teaching content is outdated, the teaching quality is not high, the educatees are lack of innovation ability, practice ability, entrepreneurial spirit and dedicated spirit, teaching and research products run contrary to the social products, and the education outputs are little relative to the educational inputs, then it could not form high external performance. This inherent and inevitable positive correlation relationship between the internal performance and external performance is very important to the higher education performance evaluation research, as it proves that relationship between internal performance and external performance in higher education is homogeneous. Meanwhile, the higher education performance itself has consistency and continuity that could be studied and researched by the same theory and method, and the evaluation results could also be compared and analyzed accordingly.

Similarly, there are a lot of differences between the internal performance and the external performance in higher education showed in the following three aspects.

Firstly, the subjects to implement performance evaluation are different, and the differences among the subjects originate in the differences between internal and external stakeholders. The internal performance evaluations in higher education are mainly the evaluations and assessments to the subordinate departments on the basis of functions, work targets, regulations and requirements about resources consumption efficient from the superior departments. The subjects to implement performance evaluation are the internal departments from teaching, scientific research, finance and asset. The external higher education performance evaluations are mainly from the external stakeholders such as government. As a public welfare institution with financial allocation from the government, government departments measure, analyze and evaluate the function fulfillment situation, social benefit and economic benefit of higher education in certain ways. The subjects to implement performance evaluation are the government departments, social organizations and other nonprofit public organizations commissioned by the government departments.

Secondly, the objects to implement performance evaluation are different. The internal performance aims at every resources consumption unit in higher education work centers within the higher education, including every GW center including the college, the department, the laboratory, the agency, the affiliated institution and so on. The external performance evaluation aims at the performance evaluation among every higher education work center by the government departments and other social organizations, in fact, it is the performance evaluation among every artificial person of institution.

Thirdly, the index systems to implement performance evaluation are different. According to the actual situations themselves, the higher education sets up the assessment systems and methods adapting different characteristics of departments, professions and positions, and regards the assessment results as the main basis to verify and adjust the department performance appropriation. The date of evaluation index systems is from statistical data in each department, such as teaching, scientific research, finance, assets and so on. But government department, cooperative enterprise, donor, news media and other social organizations measure and analyze the efficiency of the resources consumption of higher education in different angles based on personal interests, and the data is from statistical yearbook of the state education departments, the public school information, and the public school reports as requested and so on. It is supposed to be noticed that a lot of the concrete evaluation indexes are consistent, only the caliber sources and the evaluation objects are different, such as the index of "undertaking appropriation expenditure" in the aspect of fiscal expenditure, its internal performance is based on the number of department expenditures from every college, department and so on, while its external performance is based on the financial accounts data which could evaluate the whole higher education. When it comes to the index of "number of the teaching and administrative staffs" in the aspect of human input, things are different. As its internal performance is based on the number of the staff at work from every school department, while its external performance is based on the number of staff from the whole school, so are some indexes such as "competitiveness research funding". These indexes all only have differences in the caliber sources and the evaluation objects. Obviously, there are many types of indexes which could be used in valuation, and the subjects could adopt different indexes into the evaluation system. But some specific evaluation indexes could not be commonly used between the internal performance and the external performance, such as the basic construction investment index. In external performance, its rights of decision-making and resource allocation belong to the decision-making level in higher education work center, as its investments to the basic construction are planned by school uniformly, and its most funds source are from fiscal appropriation and self-raised funds. But when it comes to the internal performance evaluation, fiscal appropriation in colleges does not include the basic construction. Thus the real internal performance evaluation system has no the basic construction index. Meanwhile, the teaching management institutions in the internal performance evaluation will carry out the satisfaction evaluation to one college or one lesson in the college in order to evaluate the teachers' professional spirit and teaching level, etc., which are regarded as the evaluation indicators for the position assessment and job title promotion. But when it comes to the external performance evaluation, the satisfaction evaluation will expand to the social reputation and overall satisfaction instead of being limited to one specific college or lesson.

The improvement of the internal performance will necessarily improve the external performance in the higher education. Macroscopically speaking, when the state analyzes the contribution that the higher education has made to the economic growth, in addition to the external performance analysis in higher education, the internal performance has become one important reference basis at the same time.

REFERENCES

[1] Hou Qiping, "The Application of Data Envelopment Analysis to Performance Evaluation of Research Oriented Universities", *R & D MANAGEMENT*, 17(1), 2005, pp.118-124.

[2] Shen Shuhai, Chen Xuean, *Fiscal Expenditure Effectiveness Evaluation,* China Financial and Economic Publishing House, 12, 2002.

[3] Wu Feng, *The Research on the Assessment of the Benefits from School-running and the Cost Inputs in Higher Education Institutions,* Third Military Medical University, Chongqing, Doctoral Thesis, 2007.

[4] Huang Chaofeng, *A dissertation submitted in satisfaction of requirements for the degree of PhD in Management Science and Engineering,* National University of Defense Technology, Anhui, Doctoral Thesis , 2005.

[5] Zeng Yuqing, Huang Chaofeng, "A Study of the Benefit of Running a University by DEA Evaluation Index System", *University Education Science*, 3, 2006, pp. 38-41.

[6] Li Congdong, Zhang Hongliang, "The ERP System Based on BOX and Feedforword Cost Control", *Computer Integrated Manufacturing Systems*, 5(10), 2004, pp.528-261.

[7] Li Congdong, Tang Yongli, Zhang Hongliang, et al., "ABC method based on generalized word centers and its application", *Computer Integrated Manufacturing Systems*, 11(2), 2005, pp.228-232.

[8] Huang huiyu, Li Congdong, "New networked manufacturing pattern based on bill of X &application service provider", *Computer Integrated Manufacturing Systems,* 3(13), 2007, pp.484-489.

[9] Tang Yongli, Li Congdong, Hu Xinyue, "Research on the Reconfigurable ERP System based on BOX", *Industrial Engineering Journal*, 1(10), 2007, pp.1-5.

[10] Tang Yongli, *Research on the Reconfigurable ERP System Based on BOX*, Tianjin University, Tianjin, Doctoral Thesis, 2005.

[11] Zuo Xiaoming, *Network Resources Planning System for Multi-Core Manufacturing Cluster Based on Bill of Hierarchies*, Jinan University, Guangzhou, Doctoral Thesis, 2009.

[12] Li Yongfan, *Knowledge Cost of Innovative Enterprise and Its Control Strategies Based on BOX System,* Tianjin University, Tianjin, Doctoral Thesis, 2006.

[13] Zheng Li, Zhang Wei, "the Discussion about Principle in Performance Evaluation of teachers in the higher education", Journal of Capital University of Economics and Business, 4, 2009, pp. 92-95.

[14] Zhen Zhiyong, Liu Chunqing, et al., "The Research and Practice of Teaching Quality Evaluation in the Internal Higher Education", *Water Conservancy Science and Technology and Economy,* 16(7), 2010, pp. 838-840.

[15] Meng Fang, "the Revelation from the Financial Performance Evaluation System of Foreign Higher Education", *Northern Economy and Trade*, 10, 2009, pp.166-167.

Comparison of International Incentive Policy of Green Building

Rui Liu*, Yan Xu

School of Economy and Management, North China Electric Power University, Beijing, China, 102206

(1696314936@qq.com)

[1]*Abstracts* - **With the increasingly grim issue of global climate change and the energy crisis, the development of green building and Green New Deal is becoming the best policy choices of governments around the world. In order to promote the green building to spread in the world, a great number of countries in the world have taken different incentives in succession. This paper carries on depth analysis on incentives of green buildings in west countries and China, compares the difference between supply-side and demand-side, and puts forward that our governments need to establish and improve incentives to promote the green building market prosperous.**

Keywords - **Demand-side, green building, incentive policy, supply-side**

I. INTRODUCTION

With the development of the construction industry and the improvement of the quality of people's lives, more and more people began to pay attention to the green building. Green building is able to provide people with a healthy, comfortable and safe place to live and work, which can efficiently use resources in the whole building life cycle. It not only can ensure the quality of people's lives, but also to some extent make tremendous contributions to the social environment. Therefore, many countries of the world have successively published corresponding incentive policies to promote the development of the green buildings. Comparing the various national green building incentive policies on different perspectives is better.

Green building incentive polices can be divided into two aspects, mandatory policies and economic incentive policies. The mandatory policies advantageously features direct function and quick results, disadvantageously features low individual enthusiasm and even contrary to the initial purpose when the individual income and value orientation conflict with policy. The economic incentives can be roughly divided into two types, tax policy and fiscal policy. Mandatory policies are more considerable than the economic policy in the interests of the participants of the green building. In the execution of the policy, whether for supply side and demand side of the green building has extensive applicability.

II. LITERATURE REVIEW

A. Abroad Research

Western developed countries started to focus on building energy conservation through 30 years of efforts from the energy crisis of 1973. The unit area energy consumptions of new buildings in western countries have been reduced to 1/3 to 1/5 of the original, where the incentive policies played a key role. In developed countries, the governments have put forward a variety of economic incentive policies to improve the building energy-saving system. For example, the EU have proposed a series of normative fiscal policy, such as energy tax, tax breaks, fiscal subsidies and establishing investment bank loans, etc. In the United States, the mainly incentive policies is implemented from two aspects of market and policy including the incentives of funds and material, tax breaks, special funds, carbon trading, voluntary projects and the Energy Start logo, etc. In addition, the Japanese government has encouraged to use the new energy and adopted the subsidy system of residential solar power. It also has established effective incentive measures, such as special funds, tax breaks, accelerating depreciation, rewarding the new technologies and so on to ensure the ecological space, efficiently use of energy, waste disposal and care for people's health, material selection, etc. Although the various countries' tax incentive policies have different starting point and incentive level, they have played a positive and effective role in promoting the

[1] Grant from: Scientific Research Foundation for Returned Scholars, Ministry of Education of China

E. Qi et al. (eds.), *Proceedings of the 5th International Asia Conference on Industrial Engineering and Management Innovation (IEMI2014)*, Proceedings of the International Asia Conference on Industrial Engineering and Management Innovation 1, DOI 10.2991/978-94-6239-100-0_58, © Atlantis Press and the authors 2015

development of energy-efficient construction.

B. Domestic Research

In recent years, with the increasing of green building support, the research of green building incentive policies gradually attracted widespread attention of many scholars. Domestic scholars have studied the theory of green building incentive policy, analyzed the green building incentives from different angles, compared the incentives policies between China and aboard, and then put forward the existing problems in our green building incentives and researched on how to solve them. All of those contribute to promote the development of the government incentives and perfect the green building industry. Under different incentive theories, there are many analyses on the green building incentive policies. Yan Jin (2008) form the perspective of the development of the green building incentive policies used Game Theory under the condition of limited rationality, through established the policy model of the development process of green building to conduct in-depth research on the asymmetric game of government and the developer community, finally took measures to improve the effectiveness of green building incentives [1].

Niu Yi and Yang Jie (2011) studied the economic incentive policies of building energy conservation on the developer's view [2]. Zhang Yingchun (2012) discussed the external incentive strategies on the economic development of green building [3]. Anna (2012) analyzed the demand-side incentive policies of green building using the game analysis method [4]. Li Shaolin (2011) studied the energy-saving economic incentive policies of green building of Shanxi province [5]. According to the general idea and the basic framework of the Beijing green building incentive policies in the period of "Twelfth Five-Year", Liu Yuming (2012) got into the government incentives of Beijing [6].

In contrast to Chinese and foreign green building incentive policy, Kai Yan et al. (2011) compared the internationalization of green building incentive policies performance between China and the United States [7]. Through analyzing the U.S. green building incentive policy, Chen Yan and Yue Xin (2010) studied how to publish effective incentives for developing green building in our countries [8]. By in-depth research the incentives in Singapore, Zhang Mingming summarized the experience in green building incentive policies [9]. On the other hand, Liu Xiaojuan and Wong Jianting compared the green building incentive policies between domestic and international [10].

At present, our incentive policies exist a lot of issues need to be solved, in this regard Yang Guangshen and Zhou Yingpu came up with four main factors restricting the development of Chinese green building, on this basis put forward the suggestion for the green building incentive policies [11]. Dai Xuezhi and Di Yanqiang (2011) raised establishing the comprehensive evaluation system for economic incentive policies of green building energy-saving [12]. Finally Liu Xiaotian, Ren Tao and Yang Jie (2008) pointed out the shortcomings of China's green building incentive policies and gave improvements by analyzing the foreign green building incentive policies [13].

Vigorously developing green building has become an important issue to the world, and also is an important task for the development of China's construction industry. In order to better promote the development of green building, establish sound and effective green building incentives, we should in-depth research and analysis the existing issues on government incentives, on this basis put forward a perfect and effective green building incentives. It is conductive to formation of green building development long-term mechanism, and has great significance for improving the efficiency of government regulation, reducing monitoring costs, increasing the level of social resource conservation and protecting the environment.

III. COMPARE DOMESTIC AND INTERNATIONAL GREEN BUILDING INCENTIVE POLICIES

A. Mandatory Green Building Incentive Policies is More Perfect in Developed Countries

On one hand, due to the development of the construction market in developed countries is relatively mature and the development background of regulatory regime and the market economy is relatively mature, which is better to establish a more comprehensive legal system. All of these are helpful to develop the green building standards, incentives, etc. and provide a strong legal protection for the development of green building. On the other hand, the developed countries has put more attention to the development of the green building in local markets, the proposed local policies are more targeted and more in line with the actual situation in various places. In the course of implementation, these are more effective to achieve the original goal.

Compared with developed countries, although China has a large number of green building-related policies, the regulations with practice are few. In terms of green building incentives, great majority regulations are made for green building demonstration project, which could not play good leaders, finally more and more regulations because of unrealistic been shelved.

B. Developed Countries' Green Building Incentive Policies Have Strong Maneuverability

By summarizing each country's green building incentive policies, the green building incentive methods can be roughly divided into: financial subsidies, tax incentives, preferential credit, research funding, support agencies and so on. In terms of incentive methods, China's incentive method of green building market mainly include financial subsidies, tax incentives and credit offers, other means are also involved, but in practice is not obvious. The incentive methods of western developed countries can realize the incentive effect on green building supply side and demand side, which is according to the behavior of the parties involved in the different markets from a different perspective. In terms of the incentive effect, in our case, seemingly incentive policies have a certain generality whether in our green building market supply side or the demand side, but because developed incentive policies are too broad, which led to policies in the implementation process cannot be specific applications, and the green building incentive policies ineffective.

C. Effect of Green Building Incentive Policies in Developed Countries is More Obvious

Green building incentive policies of developed countries not only can directly give financial compensation for green building projects, but also can give start-up capital for supporting the standards implement in the industry. In the aspect of fiscal subsidies and tax breaks there are clear provisions and clear rules on the preferential amount. In order to better implement green building, the government should give legislative support to green building rating system, offer strong protection for green building rating system, and then provide a certain amount discount for the green building developers and contractors.

China's green building incentive policy is in the policy promotion phase and limited to just support the government policies, the exemption amount of the comprehensive mandatory policies is not obvious and even includes penalty provisions. The green building credit incentive policies carrying out the situation in the major banks is uneven, cannot guarantee the owners of green building to enjoy the preferential policies at the first time and then reduce the enthusiasm of the green building owners. Incentive policies for green building consumers are ignored in the implementation process by the various departments of the government. The preferential policies which can fall into the consumers' hands are truly little. In addition, the property management companies can't provide reasonable management for green building and effectively protect the owners' interests. All of these eventually lead the result that consumers can't get the benefits from green building lose confidence in green building and give up to support the development of green buildings.

IV. SUGGESTION

A. To Strengthen the Practicability of Green Building Regulations System

Green building regulations system of our country's current existing has relatively wide coverage, but its design depth remains deficient, which needs China's central government and local governments offer more comprehensive interpretation. In the past few years, with the development process of green building incentives of our country, we had continuously learned advanced foreign policies, but in the process of learning our governments just paid attention to the policies' coverage blindly and ignored the policies' depth, which had leaded the incentives of green building just did work to demonstration project, but not had universal applicability.

B. Focus on the Effect of the Green Building Incentive Policies

The promulgation and implementation of the green building incentive policies aimed at promoting green building market to constantly improve and develop. However, our green building incentives enacted in recent years are not perfect, which lead to the implementation effect is not obvious. The implementation effects of the green building incentive policies is the best standards to detect the quality of green building incentives, only the incentive policies which has obvious effects should be should be implemented.

C. Strengthen incentive for the main demand subject of green building

Purchasers and users of green buildings are the key to develop the green building market, only when consumers are willing to spend more money to buy green building, the green building market can be really pushed to forward. However, it is not easy to let consumers spend more money to buy a commodity without immediate interests. This requires the governments to give consumers more immediate interests, allow consumers to realize the long-term interests that buying and residential green building, and to feel the real interests existing, so that it can encourage consumers to buy the products. So, the "supply driven demand" incentive model which blindly stimulate the supply side of green building to increase commodity reserves of the green building market, and ignore the desire of demand side of the green building market is not feasible.

D. The Amount of Green Building Tax Cuts and Subsidies Should Be Specific

At present, the amount of China's green building tax incentives and subsidies is obscure, which are decided by the government departments. Green building supply-side and demand-side cannot learn the discount amount they should get before pass approve, it will lead to many government departments deduct the preferential payments, and the real government compensation can't be fully carried out into the hands of green building supply-side or demand-side. On one hand, the specific preferential amount can make green building supply-side and demand-side more clear how much they should enjoy from the preferential policy, so that the beneficiaries can clearly see their own decision-making interests before making investment decisions. On the other hand, it also helps to standardize the government behavior for the green building market, reduce some governments' illegal behavior which will violates the green building beneficiary interests.

REFERENCE

[1] Yan Jin. "Research on the development of green building incentive policies", *China Henan Province: The Ping Dingshan Institute of engineering*, 10, 2008, pp. 119-03.

[2] Niu Yi, Yang Jie. "Based on the developers perspective research the building energy-saving economic incentive systems". *The Social and Science of Shan Dong*, 7, 2011, pp. 157-03.

[3] Zhang Yingchun. "Study on the External Incentives Strategies in Economic Development of Green Building", *Modern Business Trade Industry*, 22, 2012, pp. 70-01.

[4] An Na. "Game Analysis on the Demand Side of the Green Building Economic Incentives Policy", *Ecological Economy*, 2, 2012, pp. 107-110.

[5] Li Shaolin. "Research the Building Energy -Saving Energy-Efficient Economy Incentive Policies in Shanxi Province", *Chinese Academic Journal*, 2011.

[6] Liu Yuming. "Research on Incentive Policies for the Development of Green Building in Beijing", *The Journal of Beijing Jiaotong University*, 2, 2012, pp. 046-06.

[7] Kai Yan et al. "Comparison of the Incentive Policy Performance between China and the United States Green Building", *Housing Industry Forum*. 2011.

[8] Yan, Yue Xin. "The Revelation of the Policy System of the U.S. Green Building to China", *Environment and Sustainability Development*, 04, 2010, pp. 043-03.

[9] Zhang Mingming. "Singapore Green Building", *China Tianjin: Tianjin University*, 2011

[10] Liu Xiaojuan, Wang Jianting. "Comparative Study of Domestic and International Green Building Incentive Policy".

[11] Yang Gaosheng, Zhou Yingpu. "Research on the Government Incentive of Green Buildings in China",

China Nanjing: Hohai University.

[12] Dai Xuezhi, Di Yanqiang. "Discussion on Establishing the Comprehensive Evaluation System of Economic Incentive Policy in Building Energy Saving", *Housing Industry Forum*, 2011.

[13] Liu Xiaotian et al. "Discussion on Green Building Incentive Policy", 2008.

Research on Correlation between Chinese Listed Company Performances and Internal Control Quality

Jian-min Hang, Run-xiang Zhao*, Ying-xin Wu
Faculty of Management and Economics, Tianjin University, Tianjin, China
(zhaorunxiang@tju.edu.cn)

Abstract - **This paper uses the data of 2012 Shanghai and Shenzhen main board A-share listed companies to study the correlation between internal control quality and company performances and conducts robust tests with the data of 2011 Shanghai and Shenzhen main board A-share listed companies. ROE after deducting non-recurring gains and losses, ROE and Tobin's Q are chosen as dependent variables, internal control quality is chosen as independent variable and 7 control variables are chosen to conduct regression analysis and such results are gained: internal control quality is significant correlated to operating efficiency but the impact is small, while it is not significant correlated to company value; the promotion of internal control quality can improve company operating effectiveness, but internal control quality cannot be the basis of company evaluation for now.**

Keywords – **Company performances, correlation, Internal control**

I. INTRODUCTION

Since 2001, Enron event has made the investors begin to pay attention to the process of value creation when they invest, especially the authenticity. The American congress and government also set Sarbanes-Oxley Act because of the Enron event, requesting companies assess the effectiveness of internal control and hire internal control auditor to test the internal control and give professional opinions which should be compulsorily disclosed. Following the promulgation of Sarbanes-Oxley Act, China issued the "Basic Standards for Enterprise Internal Control" on June 28, 2008 and the "guidelines for Enterprise Internal Control" on April 26, 2010. The basic standards and the guidelines constitute the internal control system of the enterprise, which promote the building and improvement of internal control and mark the beginning of a more comprehensive and unified system of internal control standards and guidance. It has a huge impact on market regulators, companies and investors.

II. LITERATURE REVIEW

The separation of ownership and control has become the core features of modern enterprise management system. Since the appearance of this system, that how to ensure the manager serve shareholders' interests has become one of the most important issues of corporate governance. To solve this problem it is necessary to establish internal control system which could supervise the managers. By now the studies on internal control mainly focused on two aspects. One is about the factors that affect internal control quality. Goh et al suggested that efficient and independent board of directors and the audit committee have a significant impact on internal control deficiency correction [1]. Ashbaugh-Skaife et al found that companies that often disclose internal defects always have complex business projects, structure reorganization, existence of a higher audit risk, more frequent auditor replacement and almost have no resources available for internal control construction [2]. Broson et al found that a company with a larger scale, more frequent Audit Committee meeting, more investors and rapider income growth tends to disclose internal control report, while a company with a faster sales growth tends to not disclose internal control report[3]. The other one is about the economic consequence of internal control. Wang et al demonstrates that different quality of internal control has different impact on the annual report restatement [4]. Sun Guangguo et al obtained the result that high quality internal control can improve accounting information transparency [5]. Tong Yan et al found that high effective internal control can restrain earnings management behavior and improve earnings quality [6]. Yu Haiyun found company with higher quality internal control tends to disclose internal control and relevant information to deliver good news to market to get lower financing cost [7]. These studies focused on the correlation between accounting information quality, earnings quality, financing cost and internal control, while there is a lack of research on the correlation between internal control and corporate value. Zhang Xiaolan found that a company with a high quality internal control disclosure has a better performance [8]. But some other scholars found that internal control will lead to a decline in company value, such as Engel [9], and some scholars suggest that internal control had no significant effect on company value, such as Ogneva [10]. January 1, 2011, China officially implemented the "Implementation Guidelines", which marks a more completed and uniform norms that guide internal control system construction and means a new stage. Therefore the data of 2012 Shanghai and Shenzhen main board A-share listed company are selected to research the correlation between internal control quality and company performances and the data of 2011 Shanghai and Shenzhen main board A-share listed company are selected to conduct robust tests.

III. RESEARCH DESIGN

A. Hypothesis

According to the "Basic Standard for Enterprise Internal Control", the purpose of internal control is to make sure company's complying with the law and regulations, asset security, the authenticity and integrity of

E. Qi et al. (eds.), *Proceedings of the 5th International Asia Conference on Industrial Engineering and Management Innovation (IEMI2014)*, Proceedings of the International Asia Conference on Industrial Engineering and Management Innovation 1, DOI 10.2991/978-94-6239-100-0_59, © Atlantis Press and the authors 2015

financial reports and related information, improve operational efficiency and results, and fulfill enterprise development strategy. Internal control, as an indispensible part of company management, promote the company value mainly in three aspects: First, reduce the agency cost and alleviate agency conflicts by enhancing the company supervision; second, pay much more attention on value creation process to provide reasonable assurance of company value quality and authenticity; third, based on signaling theory, pass positive message to market to improve company value by gaining positive response of market.

However, Chinese internal control construction is still in its infancy stage and internal control efficiency and quality is difficult to guarantee, so Chinese internal control needs to be bettered under the guide of "basic standard" and "supporting guidelines" so it is debatable whether Chinese internal control can really improve company performances. As Engel's study found, internal control may impose an additional burden, increasing disclosure cost, which would lead to a decline in company value. Since Chinese internal control construction is still in its initial stage, the cost of internal control may exceed the revenue, which may have a negative impact on company performances. Therefore, this study uses ROE and ROE after non-recurring gains and losses to represent the company operating efficiency and Tobin's Q to represent the company value to verify the correlation between internal control quality and company performances from the views of efficiency and value. So, following hypotheses are put forward:

H_1: Internal control quality is positively correlated to operating efficiency;

H_2: Internal control quality is positively correlated to company value.

B. Internal Control Quality Measurement

If any one of the following situation appears then internal control quality is low: (1) the company or its directors, supervisors and senior manager is criticized or blamed by the CSRC; (2) annual report restatement; (3) being defendant involved in litigation and arbitration cases; (4) internal control audit report is not issued standard unqualified opinion; (5) disclosure of major or important internal control defect.

All the following conditions being meet, internal quality control is high: (1) No situation that suggests low quality appears; (2) hire the big 4 to audit annual report and the audit opinion is standard unqualified opinion; (3) no replacement of accounting company because of the company itself; (4) CEO duality.

Others are medium.

From the view of investors, such principles are followed to set up the above measurement: (1) the relative information is easy to be obtained outside the company and the cost is low; (2) not involved with complex math calculation and quantitative analysis; (3) objective and authentic; (4) cannot be easily manipulated by company managers; (5) to minimize the results deviation, factors which may be related to company performances are avoided when evaluate internal control quality.

C. Data Selection

Data are from CSMAR database, CCER database, and 2011 and 2012 A-share listed companies, deleting the following companies:(1) financial companies; (2) the new listed companies in 2011; (3) ST, PT, *ST in 2011 or 2012; (4) change of accounting company because of the accounting company in 2011 or 2012; (5) in the process of a major restructuring in 2011 or 2012; (6) issue stock in other market, such as B shares, H shares, etc at the same time; (7) abnormal data or lack of data. A sample of 566 companies is received.

D. Variable Definitions

Dependent variables include three: OROE, ROE and Q. OROE represents the rate of return on net assets after deducting nonrecurring gains and losses; ROE represents net asset income rates; Q is Tobin's Q. Independent variable is ICQ which represents internal control quality. Control variables include ASS, ALR, IR, TS, ATS, STA and C_i. ASS represents asset scale; ALR represents asset-liability ratio; IR represents increase rate of revenue; TS represents the largest shareholder ownership; ATS represents the average of the second to the fifth shareholder ownership; STA represents the style of ultimate controller; C_i represents INDj which according to CSRC is divided into 13 industries, including Agriculture, forestry, livestock farming and fishery, mining, manufacturing, electricity, utilities, construction, transportation, IT, wholesale and retail trade, finance and insurance, real estate, social services, communication and cultural industry and comprehensive. Financial and insurance companies are deleted. Variable definitions can be seen in Table I.

TABLE I
VARIABLE DEFINITIONS

variable	definition
OROE	rate of return on net assets after deducting nonrecurring gains and losses
ROE	Net asset income rates
Q	Tobin's Q, equal to [(stock value + liability)/asset]
ICQ	Internal control quality, if good, equal to 1; if medium, equal to 0; if bad, equal to -1
ASS	Asset scale, equal to Ln(asset)/100
ALR	Asset-liability ratio
IR	Increase rate of revenue
TS	The largest shareholder ownership
ATS	The average of the second to the fifth shareholder ownership
STA	Ultimate controller, if state-owned, equal to1, if not equal to 0
C_i	INDj, if a company belongs to industry i then C_i equals to 1, others equal to 0

E. Model Design

Model 1: $OROE=\alpha_0+\alpha_1 ICQ+\xi$

Model 2: $ROE=\alpha_0+\alpha_1 ICQ+\xi$

Model 3: $Q=\alpha_0+\alpha_1 ICQ+\xi$

Model 4: $OROE=\alpha_0+\alpha_1 ICQ +\alpha_2 ASS +\alpha_3 ALR +\alpha_4 IR +\alpha_5 TS^2+\alpha_6 ATS +\alpha_7 STA +\alpha_8 C_i +\xi$

Model 5: $ROE=\alpha_0+\alpha_1 ICQ +\alpha_2 ASS +\alpha_3 ALR +\alpha_4 IR +\alpha_5 TS^2+\alpha_6 ATS +\alpha_7 STA +\alpha_8 C_i +\xi$

Model 6: $Q=\alpha_0+\alpha_1 ICQ+\alpha_2 ASS+\alpha_3 ALR+\alpha_4 IR+\alpha_5 TS^2+\alpha_6 ATS+\alpha_7 STA+\alpha_8 C_i+\xi$

Model 1 to 3 is single factor regressions of company performances to internal control quality. The control variables are added in Model 4 to 6 on the basis of Model 1 to 3 respectively.

F. Descriptive Statistics Analysis

According to Table II the mean value of OROE is 0.07 and the Std. is 0.08. The mean value of ROE is 0.10 and the Std. is 0.08, which means the difference among the operating efficiency of different companies is small. The mean value of Q is 1.61which lies in the reasonable range. The mean value of ICQ is -0.20, which means the internal control quality of Chinese listed companies is at a lower middle level, which matches the situation that Chinese internal control is in its initial stage. There are409 companies (72%) whose internal control quality is at medium level. Descriptive Statistic results of the main variables can be seen in Table II.

TABLE II
DESCRIPTIVE STATISTIC RESULTS

variable	Min	Max	Mean	Std.	N
OROE	-0.29	0.45	0.07	0.08	566
ROE	0.00*	0.47	0.10	0.08	566
Q	0.66	6.57	1.61	0.84	566
ICQ	-1.00	1.00	-0.20	0.49	566
ASS	0.19	0.28	0.22	0.01	566
ALR	0.05	0.93	0.52	0.18	566
IR	-0.87	20.11	0.15	0.91	566
TS	0.04	0.84	0.36	0.16	566
ATS	0.00*	0.12	0.03	0.02	566
STATE	0.00	1.00	0.66	0.47	566

*: round off, the real number is not 0

According to the internal control quality, the sample is divided into 3 groups, low quality group (L group), medium quality group (M group) and high quality group (H group). By analyzing these three groups, such results can be found: for OROE and ROE, the mean value is highest in the high quality group and the lowest in the low quality group, but for Q there's no such results. Descriptive statistic results for different groups can be seen in Table III. By conducting Kruskal-Wallis test and Jonckheere-Terpstra test, the results of asymptotic significance show the same results. Nonparametric tests can be seen in Table IV.

TABLE III
DESCRIPTIVE STATISTIC RESULTS FOR DIFFERENT GROUP

	variable	Min	Max	Mean	Std.	N
	OROE	-0.00*	0.31	0.12	0.08	21
	ROE	0.01	0.31	0.13	0.07	21
	Q	0.88	2.41	1.37	0.47	21
	ASS	0.22	0.25	0.24	0.01	21
H	*ALR*	0.12	0.85	0.5	0.17	21
	IR	-0.52	0.37	0.05	0.2	21
	TS	0.08	0.73	0.47	0.18	21
	ATS	0.00*	0.09	0.03	0.03	21
	STATE	0	1	0.81	0.4	21
	OROE	-0.1	0.45	0.08	0.08	409
	ROE	0.00*	0.47	0.1	0.08	409
M	*Q*	0.66	6.57	1.63	0.87	409
	ASS	0.19	0.26	0.22	0.01	409
	ALR	0.05	0.92	0.51	0.18	409
	IR	-0.88	20.11	0.16	1.03	409
	TS	0.04	0.84	0.36	0.16	409
	ATS	0.00*	0.13	0.03	0.02	409
	STATE	0	1	0.65	0.48	409
	OROE	-0.29	0.37	0.06	0.08	136
	ROE	0.00*	0.37	0.08	0.08	136
	Q	0.74	5.82	1.61	0.8	136
	ASS	0.2	0.28	0.22	0.01	136
L	*ALR*	0.07	0.93	0.53	0.18	136
	IR	-0.41	2.74	0.13	0.48	136
	TS	0.09	0.76	0.35	0.15	136
	ATS	0.00*	0.1	0.03	0.02	136
	STATE	0	1	0.68	0.47	136

*: round off, the real number is not 0

TABLE IV
TYPE SIZES FOR CAMERA-READY PAPERS

test	asymptotic significance		
	OROE	*ROE*	*Q*
Kruskal-Wallis Test	0.00	0.00	0.36
Jonckheere-Terpstra Test	0.00	0.00	0.53

G. Regression Analysis

According to the regression results of Model 1 and Model 2, OROE and ROE are both positively correlated to ICQ, but the impact is small. Both of the models are positively significant, but R^2 and adjusted R^2 is both small which means the goodness of fit is not good enough. This may be caused by the following reasons: (1) different types of dependent variables and independent variable; (2) some other relative factors could be missed because they are too hard to be quantified. Furthermore, R^2 and adjusted R^2 are not absolutely authoritative. So the results support H_1 with reservations. According to Model 3, ICQ is not significantly correlated to Q; R^2, and adjusted R^2 and F are all small enough to suggest that H_2 is not supported.

Test the multicollinearity of Model 4 to 6 to make sure these three models are not severely multiple collinear respectively. According to the regression results of Model 4 to 6, the same results can be obtained. R^2 and adjusted R^2 of Model 4 and Model 5 are in a range from 0.1 to 0.3, which is the same with the previous researches. The signs of the coefficients of control variables correspond to reality. The regression results can be seen in Table V and Table VI.

TABLE V
REGRESSION RESULTS FOR MODEL 1 TO 3

	variable	coefficient	t	p
Model 1	intercept	0.08***	20.98	0.00
	ICQ	0.03***	3.53	0.00
Model 2	intercept	0.10***	29.01	0.00
	ICQ	0.02***	3.67	0.00
Model 3	intercept	1.61***	41.91	0.00
	ICQ	-0.03	-0.42	0.67

Model 1:R^2=0.02, adjusted R^2=0.02, F=12.43, p=0.00;
Model 2:R^2=0.02, adjusted R^2=0.02,F=13.45,p=0.00;
Model 3:R^2=0.00,adjusted R^2=0.00,F=0.178,p=0.674;
***:significant at level 1%, the same below

TABLE VI
REGRESSION RESULTS FOR MODEL 4 TO 6

	variable	coefficient	t	p
	intercept	-0.27***	-3.70	0.00
	ICQ	0.01**	2.06	0.04
Model 4	*ASS*	1.60***	4.59	0.00
	ALR	-0.08***	-3.69	0.00
	IR	0.01**	2.00	0.05

	TS^2	0.11***	3.64	0.00
	ATS	0.69***	4.58	0.00
	STA	-0.01	-1.04	0.30
	C_i	Control variable		
Model 5	intercept	-0.22***	-3.26	0.00
	ICQ	0.16**	2.44	0.02
	ASS	1.41***	4.37	0.00
	ALR	-0.04**	-2.21	0.03
	IR	0.01**	2.26	0.02
	TS^2	0.09***	3.23	0.00
	ATS	0.63***	4.56	0.00
	STA	-0.01	-1.83	0.07
	C_i	Control variable		
Model 6	intercept	7.25***	10.83	0.00
	ICQ	0.03	0.47	0.64
	ASS	-22.99***	-7.21	0.00
	ALR	-1.36***	-6.91	0.00
	IR	0.01	0.17	0.87
	TS^2	0.86***	3.18	0.00
	ATS	4.06**	2.97	0.00
	STA	-0.19**	-2.75	0.01
	C_i	Control variable		

Model 4:R^2=0.15,adjusted R^2=0.12,F=5.78,p=0.00;
Model 5:R^2=0.14,adjusted R^2=0.12,F=5.34,p=0.00;
Model 6:R^2=0.30,adjusted R^2=0.28,F=13.80,p=0.00;
**:significant at level 5%, the same below

IV. ROBUST TEST

To test the robustness of the above results, the data of 2011 and model 1 to model 6 are used to conduct robust tests. In the process of data selection, the company is deleted as long as the data of any year is not available. Therefore, the companies in the robust sample and the original sample are totally the same, making sure the consistency of company features.

The results of robust tests are almost the same with the above results. The only difference is that the coefficient of internal control quality becomes insignificant in Model 5, but still positive. So the above results are robust. The robust regression results can be seen in Table VII and Table VIII.

TABLE VII
ROBUST TEST REGRESSION RESULTS FOR MODEL 1 TO 3

	variable	coefficient	t	p
Model 1	intercept	0.10***	23.13	0.00
	ICQ	0.03***	3.18	0.02
Model 2	intercept	0.12***	30.27	0.00
	ICQ	0.02**	2.30	0.02
Model 3	intercept	1.72***	41.84	0.00
	ICQ	-0.01	-0.15	0.88

Model 1:R^2=0.02,adjusted R^2=0.02,F=9.49,p=0.02;
Model 2:R^2=0.01,adjusted R^2=0.01,F=5.27,p=0.02;
Model 3:R^2=0.00,adjusted R^2=0.00,F=0.02,p=0.88;

TABLE VIII
ROBUST TEST REGRESSION RESULTS FOR MODEL 4 TO 6

	variable	coefficient	t	p
Model 4	intercept	-0.31***	-3.83	0.00
	ICQ	0.02*	1.65	0.10
	ASS	1.73***	4.43	0.00
	ALR	-0.05**	-1.99	0.05
	IR	0.02***	5.35	0.00
	TS^2	0.09***	2.79	0.01
	ATS	0.48***	2.89	0.00
	STA	-0.00	-0.13	0.90
	C_i	Control variable		
Model 5	intercept	-0.19**	-2.42	0.02
	ICQ	0.01	1.19	0.24
	ASS	1.32***	3.46	0.00
	ALR	-0.03	-1.20	0.23
	IR	0.03***	6.01	0.00
	TS^2	0.06*	1.95	0.05
	ATS	0.57***	3.49	0.00
	STA	-0.00	-0.32	0.75
	C_i	Control variable		
Model 6	intercept	7.48***	10.19	0.00
	ICQ	0.04	0.55	0.58
	ASS	-23.94***	-6.84	0.00
	ALR	-1.54***	-7.10	0.00
	IR	0.10	2.53	0.12
	TS^2	0.86***	2.89	0.00
	ATS	4.65**	3.11	0.00
	STA	-0.18**	-2.36	0.19
	C_i	Control variable		

Model 4:R^2=0.18,adjusted R^2=0.15,F=6.97,p=0.00;
Model 5:R^2=0.16,adjusted R^2=0.13,F=5.97,p=0.00;
Model 6:R^2=0.31,adjusted R^2=0.29,F=14.63,p=0.00;
*:significant at level 10%

V. CONCLUSION

The data of 2012 Shanghai and Shenzhen A-share listed companies on the main board are used to study the correlation between internal control quality and company performances, meanwhile the data of 2011 Shanghai and Shenzhen A-share listed company on the main board are used to conduct robust tests. The following conclusions are obtained: internal control quality and company operating efficiency are significantly positive correlated but the effect is small, while internal control quality and company value are not significantly correlated. The reasons are as follows:

Internal control reduces the possibility of company fraud and information asymmetry between the managers and the shareholders by enhancing internal supervision, which makes managers operate the company more seriously and offers a reasonable guarantee for the process of value creation. But Chinese internal control construction is still in its initial stage. It's hard for it to play a strong role in company management, so it only has a small impact.

Company value is the result of market evaluation. Only when the market can precisely evaluate the company internal control quality and admit the evaluating results internal control quality can be the basis of company evaluation. With the constant improvement of internal control disclosure and the gradual transition to the mandatory disclosure in China, the market gradually comes back to its senses, realizing the possibilities of false information and misjudgments when evaluating internal control quality on the basis of disclosure or not, which lower the reliability of the way that assessing company value on the basis of the judgment of internal control quality.

However, with the improvement of the market information transparency and the standardization of internal control information disclosure in China, it is still possible that internal control quality becomes one of the standards of company value judgment. Yet, the situation that internal control quality and company value, in fact, are not correlated at all cannot be ruled out.

Through the robust tests, such phenomenon can be found: ICQ is not significantly correlated to ROE in 2011, but positively correlated in 2012, which means Chinese internal control is improving company operating efficiency with a better prevention and control of the effect from the outside market.

At last, due to the situations that the construction of Chinese internal control is still in its initial stage and there are many nonstandard aspects to be improved, which bring some limitations for this empirical research, the conclusions still need to be testified in the future.

REFERENCES

[1] B. W. Goh, "Audit committees, boards of directors, and remediation of material weaknesses in internal control*," *Contemporary Accounting Research*, vol. 26, no. 2, pp. 549-579, 2009.

[2] H. Ashbaugh-Skaife, D. W. Cpllins, and W. R. Kinney Jr, "The discovery and reporting of internal control deficiencies prior to SOX-mandated audits," *Journal of Accounting and Economics*, vol. 44, no. 1, pp. 166-192, 2007.

[3] S. N. Bronson, J. V. Carcello, and K. Raghumandamn, "Firm characteristics and voluntary management reports on internal control," *Auditing: A Journal of Practice & Theory*, vol. 25, no. 2, pp. 25-39, 2006.

[4] Y. F. Wang, "Internal control and financial quality: evidence from post-SOX restatement", *Accounting & Taxation*, vol. 5, no. 1, pp.19-28, 2013.

[5] Guang-guo Sun, and Jin-feng Yang, "Can high quality internal control improve accounting information transparency" (in Chinese), *Research on Financial and Economic Issues*, vol.7, pp. 77-85, 2013.

[6] Yan Tu, and Feng Xu, "Study on the dynamic relationship between internal control efficiency and earnings quality in Chinese listed companies" (in Chinese), *China Soft Science*, no.2, pp. 111-122, 2013.

[7] Hai-yun Yu, "Quality of internal controls, credit model and enterprise value: an empirical analysis by Shenzhen A-shared listed companies" (in Chinese), *The Theory and Practice of Finance and Economics*, vol.32, no.3, pp. 44-50, 2011.

[8] Xiao-lan Zhang, Hao-jie Shen, and Mo Yang, "The quality of disclosure of internal control and operational performance of companies: an empirical study based on panel data" (in Chinese), *Journal of Audit & Economics*, vol.2, pp. 64-73, 2012.

[9] E. Engel, R. M. Hayes, and X. Wang, "The Sarbanes–Oxley Act and firms' going-private decisions," *Journal of Accounting and Economics*, vol. 44,no.1, pp. 116-145, 2007.

[10] M. Ognvea, K. R. Subramanyam, and K. Raghunandan, "Internal control weakness and cost of equity: evidence from SOX Section 404 disclosures," *The Accounting Review*, vol.82, no.5, pp. 1255-1297, 2007.

Self-enhancement is not Restricted to Individualistic Cultures — The Evidence from Risk Decision Making

Wei-fang Song
Research Division of Railway Police College, Henan, China
(64621344@qq.com)

Abstract - **Whether self-enhancement motive is culturally relative or universal is a hot debated topic. The universalist proposed that the motive is pervasive all cultures, whereas the relativistic perspective posited that the motive is restricted to western cultures but absent in eastern cultures. The purpose of this study is to provide the evidence that self-enhancement is palculture by looking at individual's risk decision-making behaviors. A large number of studies have been documented to show that individuals in Asian culture (e.g., Chinese) are more risky than individuals in Western culture (e.g., American). Two studies were conducted in mainland China to explore the relation of self-enhancement and risk decision-making. Study 1 (N=158) shows that self-enhancement is a significant predictor of risk decision making (adjusted R^2=.039). Study 2(N=148) replicated the results by another scenario of risk decision-making relating one's self-esteem. These results of self-enhancement positively correlated with risk-taking behavior among Chinese participants provide an indirect evidence for self-enhancement is palcutural, because it is consistent with the universalist perspective.**

Keywords – **Pal culture, risk taking, self enhancement**

I. INTRODUCTION

Self-enhancement is known as the tendency to maintain a positive view of the self [1]. People think of themselves as better than others, for example, they believe they drive better than most other drivers [2], are fairer than most other people [3], and are smarter, warmer, more thoughtful, and more talented [4]. More than the expected 50% people evaluate themselves as above average on these desirable traits. Therefore, self-enhancement is also called as "above-average-effect" that denotes the drive to affirm the self [5]. Hundreds of studies have found that people behave in way that minimize shame and humiliation feeling and maximize feeling of pride [6]. As a basic motivating factor, researcher confirms that self-other bias beyond their own personality, they also believe their friends and loved ones are better than most other people, their romantic relationship are better than most other people's [7], and they also believed that the groups they belong to are better than the groups to which they do not ([8]. That is to say, anything that carries the designation "my" or "mine" tends to be evaluated in an overly positive manner. However, whether self-enhancement is culturally relative or universal is a hot debated topic [9, 10]. There mainly have two completely opposite viewpoints on this.

The purpose of the present study is to investigate whether self-enhancement is pan-cultural from the risk-taking perspective. Most literature above indicated that the self-enhancement tendency is not shared by all cultures, but is a unique phenomenon of western culture [11-13]. However, some studies on risk-seeking state that individuals in Asian culture were more risky than individuals in western culture. From this perspective, risk-seeking, self-enhancement and culture has such a progressive logic relationship that : 1) individuals in Eastern Asian are more risky than westerners; 2) people who have positive self-regard are generally more risky. We can make a conclusion that individual in Eastern Asian have high average positive self-regard (self-enhancement) [14-17]. If we can improve the casual relationship between self-enhancement and risk-seeking, we believed that this can provide an indirect evidence that people in east culture also is self-enhancement. We conducted two study to verify the hypothesis of the relationship between self-enhancement and risk-taking intention.

II. METHODOLOGY AND RESULTS

A. Study 1

Participants

Participants were recruited from several universities in Beijing. Announcements were made in various classes at these universities inviting them to participate this experiment. There are 158 undergraduate students involved the experiment at the end (52 male, 106 female; Mean age= 21.6 years, SD=4.8). Most of them from rural areas (107) and they don't from one-child family. They were all recruited on a voluntary basis and thanked by 10 Yuan after they finished it.

Procedure

There are two steps of the experiment. First, we asked the participants to complete a measurement of self-enhancement. After this they should accomplish an invest game. They filled in all their demographic information like age, sex etc at the end of the experiments.

Materials

Self-enhancement scale. The study first ask participants to indicate to what extent their traits higher than the average age college students with the same sex of other college students. The scale included 28 personal traits (positive: independent, creative, distinctive, attractive, talent, easygoing, cooperative, honest, courtesy, sincere, moral; negative: independent, uncreative, untalented, incompetent, uncooperative, discourtesy, inconsiderate). The question is: "Suppose you are a student in a common college, according the below personality traits, to what extent do you think you are better than the other students who has the same sex, same age to you?" The participants should sign the number on a 1-6 likert scale to indicate what extent do they agree with the description of the traits be in line with themselves.

E. Qi et al. (eds.), *Proceedings of the 5th International Asia Conference on Industrial Engineering and Management Innovation (IEMI2014)*, Proceedings of the International Asia Conference on Industrial Engineering and Management Innovation 1, DOI 10.2991/978-94-6239-100-0_60, © Atlantis Press and the authors 2015

Among that, 1 stand for "*absolutely higher*" and 6 means "*absolutely lower*". The individuals who gain more scores means that they are much more self-enhancement and the Cronbach alpha is .863 in this study.

Risk-seeking measurement. After the participant finished the self-enhancement scale, they should make a choice on two invest games expression. The stimuli were presented in booklets which measure individual's degree of risk-seeking read as follows: (Li, Fang, & Zhang, 2004):

Suppose that you are faced with 2 investment options, A and B.

If you choose OPTION A, you may either make 920 Yuan or lose 200 Yuan.

If you choose OPTION B, you may either make 110 Yuan or lose 10 Yuan.

Please indicate your choice by circling a number on the 7-point scale given below:

7	6	5	4	3	2	1
Definitely Choosing OPTION A					Definitely Choosing OPTION B	

Subjects were instructed that there were no right or wrong answers, and the experimenters were only interested in the subject's own thoughtful answer. After all of these finished, they were thanked by 10￥RenMinBi.

Results

Primary analyses

Primary analyses revealed that no statistically significant main effects or interactions effects of participant's gender and the type of family(whether he/she in the one-child family) on self-enhancement (t=.269, p=.789; t=1.01, p>.05). So we combined men and women participants in the further analysis. However, results shows that there is significant difference of birth place on self-enhancement (M_{city}=4.67>M_{rurul}=4.10, t=2.01, p<.01). For risk-seeking, there is only has significant difference of gender (M_{male}=5.68>M_{fmale}=4.69, t=2.554, p<.01), other geographic variables not. We also checked the descriptive of the responses to the 28 traits to explore an overall index of explicit positive self-presentation (a=.887). The mean score of self-enhancement is 4.55 in the 1-7 point linket scale that above the midpoint 4. The results states that participant in this study have a moderate higher level self-enhancement.

Regression analysis

We conducted the correlation analysis to explore the relationship between self-enhancement and risk-taking. The results show that there is a positive relationship between self-enhancement and risk seeking in this study by the self-esteem related scenario (r=.214, p<.001). We can make the conclusion from the results that people in Eastern culture may be have high level of positive evaluation because people are more risk-seeking in previous study. We conducted the stepwise regression analysis to further explore the causal relationship between the two. The results show that there is a significance predictive effect of self-enhancement on risk-seeking. After controlling the birth place and gender variables, the adjusted R^2 is .039 (p<.001) which indicated that people's self-enhancement is an important predictor variable on risk seeking. Individuals having higher level self-enhancement are more risk seeking from this results.

Discussion

Though gender and the type of family has no significant difference on self-enhancement, the birth place has. Specifically to say, individuals in city has higher level positive evaluation than rural one's. According to the gender, male participants are much more risk seeking than female ones. There is a significant r coefficient between self-enhancement and risk seeking. Results of the regression analysis indicated that self-enhancement is an important predictor of risk seeking which can explain .039 variation. This verified the causal relationship between the two that people who have higher level positive evaluation are more risk seeking. According to the existed study, people in Eastern culture context are more risky than western people that we can believed that Eastern people are more self-enhancement or at least equivalent with western people. This results confirmed the hypothesis that self-enhancement are pal-culture, not only belong to some special culture. However, the invest game used in this study is from economic exchange only. As an important concept of self, we try to use a self-esteem related paradigm of decision making to further verify our hypothesis.

B. Study 2

Participants:

The sample of the study is consisted of 148 college students who were recruited by announcements. There are 49 male (33%) and 99 female (67%) students whose mean age is 22.4 years. There are 79 participants (53.4%) from rural areas and their family has more than one child. They take part in the experiments on the voluntary basis and will receive 10 yuan RMN as payment after they finished the experiment.

Procedure

Participants in this study accomplished two tasks in sequence. Firstly, they were asked to complete the measurement of self-enhancement. Then the participants read a self-esteem related scenario about a joke which he/she will talk on the college graduation. After this, they should make the choice between the two to decide which joke they would like choose. They received the payoff after they finished all of the tasks and thanked by experimenters.

Materials

Self-enhancement scale: We used the same self-enhancement measurement as in the study 1 whose Cronbach alpha is .863.

Risk seeking scenarios: we ask the participants to read a scenario related to self-esteem that on the topic of jokes. They should imagine giving the commencement speech at their college graduation and having to choose a joke between two choices. Joke A *(high payoff/ high risk)*, which is hilarious and touching if successful but could fail completely, or Joke B *(low payoff/low risk)*, which is less impressive but has a 100% likehood of success. Participants were asked to decide between the jokes at 10 trials, where the likelihood of Joke A being successful decreases at each trial. For example, at Trial 1, Jokes A and B both have 100% likelyhood of success: it thus makes sense to choose Joke A at this trial (which all participants did). At trial 2, the likelihood of Joke A success if 90%, at Trial 3 is dwindles to 80%, and so forth. At trial 10, there is a 0% chance that joke A will be successful (as might be expected, all participants opted for Joke B at this trial). The scenario was read as below:

Imagine that you have been asked to give a speech at your graduation ceremony. Your are speaking to hundreds of your classmates. Now imagine that you can begin your speech with one of two jokes: joke A OR Joke B. if you tell Joke A, it can turn out to be extremely funny and touching. However, Joke A is a little complicated and there's a chance that no one will laugh. Joke B is mildly funny, but not as funny as joke A. however, Joke B is guaranteed to work - everyone will get it. The circles in the tables below represent your chances of successfully delivering each joke. If each box contains a circle, you chances of successfully delivering the joke are guaranteed, whereas if there are no circles, you have no chance of successfully delivering the joke. Imagine that for each numbered pair below, you must choose to deliver only Joke A OR Joke B. Note that you are offered varying chances of Joke A being successful, while your chances of Joke B being successful are always guaranteed. For each row, indicate which joke you would choose to tell by placing a checkmark in one of the two boxes.

Following completion of the decision making scenarios, participants were fully debriefed.

Results

Primary analysis

We first analyzed the difference of geographic variables on self-enhancement and risk-seeking. The primarily results revealed no statistically significant main effects or interactions effects involving participant gender (t=1.186, p>.05), so we combined men and women in all remaining analyses. Also, the number of child in a family also has no significant difference both in self-enhancement (t=.961, p>.05) and risk-taking (t=1.137, p>.05) respectively. However, there is significant difference of birth place on self-enhancement (M_{city}=4,70>M_{rural}=3.85, t=2.49, p<.01). And also there is a significant difference of gender on risky decision make. Specifically, male students are more risky than female students (M_{male}=6.05> M_{femle}=5.17, t=4.24, p<.001). We also checked the descriptive statistics of the self-enhancement and risk seeking. The mean score of self-enhancement scale is 4.831 which above the midpoint 4 in 1-7 linket. This results revealed that people have a moderate positive self evaluation. The mean choice of risk taking is 5.94 on the adventure choice of A which state that people are likely to choose the 60% risky answer.

Regression analysis

We first did the correlated analysis of self-enhancement and risky seeking to explore the relationship between the two. The correlation coefficient is positive significant (r=.247, p<.001). The results is the same as in the study 1 which reveals that individual who have a higher positive self evaluation are more risk seeking.

To further explore the causal relationship between risk seeking and self-enhancement we did the stepwise regression analysis. The result show that risk seeking has a significant predictive effect on self-enhancement. After controlled the birth place and gender, the adjusted R^2 of risk-taking on self-enhancement is .045, p<.001. This results verify the causality that self-enhancement is an important predictor of risk-seeking. Generally speaking, individual who have a high level self-enhancement always have high risk seeking intention.

Discussion

The results of this study verify the assumption t that individual in eastern culture also have positive evaluation of self from the risk seeking perspective. According to the two joke choice, most people are likely to choose the "choice A" which is at the point of risk level 60%. Adopt the self-related paradigm, the results indicated that people are moderate risk seeking above the midpoint. The correlation coefficient is significant in the correlation analysis of self-enhancement and risk seeking as in study 1. We also conducted the regression analysis to explore the causal relationship between the two and verify again that self-enhancement is a significant prediction of risk seeking. This study verified our hypothesis that individuals have positive regard motivation with the self-esteem related scenario in Eastern culture.

III. DISCUSSION

Whether the need for positive self-regard (i.e., self-enhancement motive) is culturally relative or universal is a topic of intense debate. We address this issue with a study that test the causal effect of risk-seeking on self-enhancement in eastern cultures. The results showed that there is a significant relationship between self-enhancement and risk-seeking intention, and individual's risk-seeking intention is an important prediction of self-enhancement. That is to say, individual's who risk-seeking are self-enhancement.

In the present world economy, China and United States, Chinese from China are more-risk-seeking than Americans from the United States, and Chinese were found to exhibit markedly higher degrees of overconfidence than Americans in their general knowledge and probability judgments. There are several cultural explanations for differences in risk preference between Chinese and American respondents. Some argue that in socially-collectivist cultures like China, family or other in-group members will step in to help out any group member who encounters a large and possibly catastrophic loss after selecting a risky option. In individualist cultures like the United States, on the other hand, a person making a risky decision will be expected to personally bear the (possibly adverse) consequences of their decisions. Collectivism thus acts as a cushion against possible losses, that is, as social diversification of the risks of risky options. In the same sense that the purchase of an insurance policy reduces risk, social diversification in paying for the downside of risky options quite objectively reduces the risks of risky options for the members. Cultural values and collective culture have a hand in influencing risk preference and self-enhancement.

Is Self-enhancement Restricted to Individualistic Cultures?

According the existed study, the existence or absence of a motive is difficult to study and is usually analyzed indirectly. The present study revealed such indirect findings that support the existence of self-enhancement motive in the East Asian culture of China from the risky decision making perspective. The results of some study believed that modesty plays in self-enhancement of agentic traits. When requirements for modesty are less internalized, that is to say, when modesty is lower, the self-enhancement motive can function more freely and higher self-enhancement is registered.

People in all culture have the motive striving to maintain and achieve positive self-regard. People have the same goal but may use different tactics to do so. In a similar vein, both individualistic and collectivistic cultures permit self-enhancement, but they do so through different norms. In the west, it is accepted or tolerated to flaunt one's successes. In the east, it is accepted or tolerated to expect reciprocity relying on the seniority rule. Both in the West and in the East, self-enhancement is sanctioned through upward mobility, status seeking, forms of artistic expression, the promotion of the self on dimensions that matter. Both in the west and in the East, people self-enhance tactically, strategically, and opportunistically by making the culture work form them - a feat that deserves to be seen as a tribute to human resourcefulness, flexibility, and adaptability.

Limitations and directions for tuture research

The present study support that self-enhancement is palculture that both Eastern and Western people have self-positive of themselveis from the perspective of risk-seeking. Howevere ,there are several question in this study and the future research should pay attention to and explore it.

There are many study indicated that in certain colletivist cultures self-enhancement is low. The common explanation is that the self-enhancement motive is not prevalent in East Asian culutres due to basic differences in their typical self-construal patterns. The independent self, dominant in individualist cultures, ia autonomous and self-contained, whereas the interdependent self typical of collecitivst cultures is part of a comprehensive social relationship and is partically defined by others in that relationship. Other scholars have suggested that cultural restrictions on the self in collectivest cultures are strong. One of the implications of such restrictions is a low need for uniqueness. In line with this model, it was suggested that the need for positive self-regard exists in East Asian cultures as well, but its manifestation is restricted by the culture and specifically by cultural demands for modesty. Several previous indirect findings have shown the relevance of cultural restrictions on the self to the manifestatoin of self-enhancement. comparisons of elf-enhancement levels for different types of traits have shown that when the self-enhancement of a tratit does not vilate cultural norms, the bias toward self-enhancing this trait increases. Self-enhancement in Singapore was foud to be stronger than in Israel for communal trait, as opposed to agentic traits. Another study condcted in Singapore showed that the degree of sensitivity of the self-enhancement measure to modesty norms affected the magnitude of measured self-enhancment. Thus the grater the sensititvity of the self-enhancment measure to self-presentation norms, the lower the measured self-enhancement level. It was also found that Japanese students described themselves less favorably in public on traits defined as self-profitable traits as compared to descriptions they gave in private context.

REFERENCES

[1] Sedikides, C., Gaertner, L., & Toguchi, Y. (2003). Pancultural selfenhancement. *Journal of Personality and Social Psychology, 84,* 60 –79.

[2] Svenson, O. (1981). Are we all less risky and more skilful that our fellow drivers? *Acta Psychologica, 47,* 143-148.

[3] Messick, D.M., Bloom, S., Boldizar, J.P. & Samuelson, C.D. (1985). Why we are fairer than others. *Journal of Experimental Social Psychology*, *21*, 480 - 500.

[4] Brown, J. D., & Smart, S. A. (1991). The self and social conduct: Linking self-representations to prosocial behavior. *Journal of Personality and Social Psychology, 60*, 368-375.

[5] Heine, S. J., Lehman, D. R., Markus, H. R., & Kitayama, S. (1999). Is there a universal need for positive self-regard? *Psychological Review, 106,* 766–794.

[6] Taylor, S. E., & Brown, J. D. (1988). Illusion and well-being: A social psychological perspective on mental health. *Psychological Bulletin,* 103,193–210.

[7] Paul A. M. Van Lange and Caryl E. Rusbult, My Relationship is Better than- and Not as Bad as -Yours is: The Perception of Superiority in Close Relations, *Pers Soc Psychol Bull* 1995; 21; 32

[8] Sedikides, C., & Strube, M. J. (1997). Self-evaluation: To thine own self be good, to thine own self be sure, to thine own self be true, and to thine own self be better. In M. P. Zanna (Ed.). *Advances in experimental social psychology* (Vol. 29, pp. 209–269). New York: Academic Press.

[9] Sedikides, C., Gaertner, L., & Vevea, J. (2007). Evaluating the evidence for pancultural self-enhancement. Asian Journal of Social Psychology, 10, 201-203

[10] Acker, D., & Duck, N.W., "Cross-cultural overconfidence and biased self-attribution", The journal of Socio-Economics, 377, 1815-1824, 2008

[11] Borkenau, P., Zaltauskas, K, & Leising, D., "More may be better but there may be too much: optimal trait level and self-enhancement bias", Journal of Personality, 77(3), 825-858, 2009

[12] Heine, S.J., Kitayama, S., & Lehman., "Cultural differences in self-evaluation - Japanese readily accept negative self-relevant information", Journal of Cross-

cultural Psychology, 32(4), 434-443, 2001
[13] Kitayama, S., "Culture and basic psychological processes-toward a system view of culture: comment on Oyserman et al. (2002)", Psychological Bulletin, 128(1), 89-96, 2001
[14] Kwan, V.S., John, O.P., Robins,R.W., & Kuang, L.L., "Conceptualizing and assessing self-enhancement bias: a componential approach", Journal of Personality and Social Psychology, 94(6), 1062-1077, 2008.
[15] Li, S., Chen, W.W., & Yu, Y. (2006). The reason for Asian overconfidence. Journal of Psychology, 140, 615-618
[16] O'Mara, E.M., Sedidides, L., C., Zhou, X.Y., Liu, Y.P. (2012). A longitudinal - experimental test of the panculturality of self-enhancement: self-enhancement promotes psychological well-being both in the west and the east. Journal of Research in Personality, 46(2), 157-163
[17] Brown, J.D, Albert, H., "My better half: partner enhancement as self-enhancement. Social Psychological and Personality Science, 3(4), 479-486, 2012

Purchasing Decision Rules of Airline Passengers under Revenue Management Circumstances

Chen ZHANG, Lan CHEN, Hui YANG*
School of Economics and Management, Nanjing University of Science and Technology, Nanjing, Jiangsu Province, China
(crg8578@126.com, chenlanlan_1990@126.com, *yanghui@njust.edu.cn)

Abstract - **In this paper, we present an empirical study on behavior characteristics of airline passengers based on prospect theory framework and Chinese market data. The goal is to find out the decision rule and utility measurement method of the passengers under revenue management circumstances. By analyzing the relationship between passengers' utility and their purchasing decision behaviors based on both expected utility theory and prospect theory, we find that prospect theory shows much more rationality than expected utility theory in explaining passengers' purchasing decision behaviors. By discussing different attitudes-to-risk settings, we verify that utility measurement for passengers with common risk attitudes is more consistent with the actual purchasing decision behaviors.**

Keywords - **Decision rule, expected utility theory, prospect theory, purchasing behavior, revenue management**

I. INTRODUCTION

Revenue management (RM) is one of the great success stories of Operations Research techniques applied to current management issues. An example is that airlines have been very successful at using dynamic pricing and capacity control to increase their profitability. In the airline industry, RM focuses on selling the right seat to the right customer at the right time for the right price. In other words, it controls the availability and/or pricing of travel seats across different booking classes with the goal of maximizing expected revenues or profit. Traditional RM models characterize customer demand exogenously. For example, market size is often represented by a demand distribution (as in EMSR model). The common feature among these modeling approaches is that customers are passive: they do not engage in any decision-making processes. However, with the development of online booking systems and e-commerce, customers can easily get information of inventory and historical prices to predict the price path. For many practical problems, neglecting these decision processes on the demand side may have significant repercussions [1]. Anderson(2003) shows that use of standard RM approaches to pricing by airlines can result in significantly reduced revenues (even approaches 7%) when buyers are using an informed and strategic approach to purchasing [2]. As a result, researches on customers' purchasing behaviors under RM situation have been pursued recently. Assuming that customers are absolutely rational, these researches explore optimal strategies and decision-making schemes for both customers and firms by seeking the equilibrium solutions of game.

In fact, decision behaviors of individuals, including passengers, are bounded rational - neither absolute short-sighted nor absolute rational. On one hand, passengers can predict price path according to their purchasing experiences. On the other hand, their predictions cannot be completely accurate due to the mental cost, the information cost and their cognitive abilities. Moreover, 'markets, decision changes to firms and the variation of customers' purchase intention also affect the selling process. Kahneman and Tversky(1979) presented a critique of expected utility theory (EUT) as a descriptive model of decision making under risk, and developed prospect theory (PT) [3]. Although this theory has been applied successfully in many fields, little researches associate it with RM. Shen and Su (2007) reviewed current models of customer behaviors in RM and auction literatures and suggested several future research directions [4]. Mathies and Gudergan (2011) investigated the effect of RM on customers and established utility measure systems of customers. They provided empirical support for the role of fairness in customer decision making and the pertinence of reference-dependent preference theory to explain customer choices, and demonstrated the applicability of this theory beyond price perceptions to a range of other attributes [5]. Bu and Xu (2012) proposed the marine shipping contract allocation and pricing optimization model, which showed that the optimal contact ordering quantity is affected simultaneously by the price of the contract and shipper's reference effects, while the carrier's optimal contract price will be less than that without reference effects [6]. However, these papers rarely verify the bounded-rationality of passengers' purchasing behavior. This paper focuses on illustrating airline passengers' bounded-rationality by empirically examining their purchasing behavior. The remainder of this paper is organized as follows. Section II describes the methodology employed in this paper. Section III analyzes passengers' utility function based on traditional EUT. Section IV analyzes passengers' value function and weight function based on PT under different attitudes-to-risk settings. Finally, section V offers concluding remarks.

II. METHODOLOGY

A. Prospect Theory

The key elements of PT are: a) gains and losses are defined relative to some neutral reference point; b) risk aversion for most gains and low probability losses, and risk seeking for most losses and low probability gains; and c) losses loom larger than gains. It presents a critique for EUT. Two modifications of this theory are: a) the carries of value are gains and losses, not final assets; and

E. Qi et al. (eds.), *Proceedings of the 5th International Asia Conference on Industrial Engineering and Management Innovation (IEMI2014)*, Proceedings of the International Asia Conference on Industrial Engineering and Management Innovation 1, DOI 10.2991/978-94-6239-100-0_61,

b) the value of each outcome is multiplied by a decision weight, not by an additive probability. These can be formulated with value function and weighting function. Let ω_i denote the asset one owns, ω_0 denote the neutral reference point. Gains or losses can be denoted as $\Delta\omega_i = \omega_i - \omega_0$. The two functions can be denoted by

$$v(\Delta\omega_i) = \begin{cases} \Delta\omega_i^{\alpha} & \Delta\omega_i \geq 0 \\ -\lambda(-\Delta\omega_i)^{\beta} & \Delta\omega_i < 0 \end{cases} \quad (1)$$

where $0 < \alpha < 1, 0 < \beta < 1, \lambda \geq 1$, and

$$w(p_i) = \frac{p_i^{\gamma}}{(p_i^{\gamma} + (1-p_i)^{\gamma})^{1/\gamma}} \quad (2)$$

where $0 < \gamma < 1$.

B. Questionnaire Design and Data Analysis

A questionnaire survey was carried out to obtain data about passengers' purchasing decision behaviors. The questionnaire displayed two alternative fare schemes for an air route: scheme A ¥500 - ¥1500 and scheme B ¥800 - ¥1200. The average fare of the two schemes are both ¥1000. But scheme A has higher variance than scheme B. In other words, passengers who choose scheme A take more risk than those who choose scheme B. Assume a situation that passengers need to make a choice between the two schemes under four cases of budget constraint: ¥1500, ¥1200, ¥1000 and ¥800. We developed an investigation among 480 people, distributing questionnaires to airport passengers, to netizens online through SO JUMP, and to MBA students in class. In all questionnaires, 300 of them are valid. Table I shows passengers' options under different budget constraints.

Passengers number for the scheme represents their preferences. From Table I, we can get several implications: First, passengers' preferences are not constant, they are affected by the budget constraint. Passengers gain when the budget constraint is higher than the average fare, otherwise they lose. In other words, the budget constraint is the reference point for passengers to measure their gains and losses which affect their purchasing decision behaviors. We calculate the correlation between the selected probabilities and budget constraints of each scheme. The result is -0.6895 for scheme A and 0.6895 for scheme B, which shows significant correlation between them. In addition, passengers prefer to scheme A when the budget constraint is lower than the average fare and prefer to scheme B when the budget constraint is higher than the average fare.

C. Notations

We assume that fares of each scheme are normally distributed and that 99% of all the probable fares fall inside the given range in each scheme. Based on these two assumptions, we can calculate the standard deviations of the two schemes. The result is 1.9411 for scheme A and 0.7764 for scheme B.

Let σ_i be the standard deviation of scheme i. Let μ be the average fare in both of the schemes. Let τ be the budget constraint. Let X_i be the varying fares which follows the normal distribution that $X_1 \sim N(10, 1.9411^2)$, $X_2 \sim N(10, 0.7764^2)$. Let U_i be the utility passengers get in scheme i. Let $[a_i, b_i]$ be the variable scope of fares of scheme i.

TABLE I
PASSENGERS' OPTIONS UNDER DIFFERENT BUDGET CONSTRAINTS

Budget Constraint	Number of Passengers Choosing A	Number of Passengers Choosing B	Total Number of Passengers
¥800	228	72	300
¥1,000	144	156	300
¥1,200	104	196	300
¥1,500	38	162	300

III. CALCULATION AND ANALYSIS BASED ON EUT

According to EUT, there is no reference point in passengers' utility measurement. Considering negative exponential utility function

$$u(x_i) = -e^{\alpha x_i} \quad (3)$$

(where α>0, denoting the coefficient of risk aversion), and linear utility function

$$u(x_i) = -x_i \quad (4)$$

we can obtain the expected utility as

$$U_i = \int_{a_i}^{b_i} \frac{u(x_i)}{\sqrt{2\pi}\sigma_i} \exp\left(-(x_i - \mu)^2 / 2\sigma_i^2\right) dx_i. \quad (5)$$

Table II shows the result of (5) with different coefficient of risk aversion. Fig.1 shows the value of negative exponential utility function for scheme A. It indicates that the utility function becomes steeper as the coefficient of risk aversion α increases from 0.05 to 0.1.

Table II illustrates that scheme B is always superior to scheme A, which violates the results in Table I. The discrepancy is even larger for linear utility function in which passengers are risk neutral.

TABLE II
UTILITY MEASUREMENT BASED ON EXPECTED UTILITY THEORY

Utility Function		Utility Value for Scheme A	Utility Value for Scheme B
Negative Exponential	α=0.1	-2.7383	-2.6986
	α=0.075	-2.1164	-2.0991
	α=0.05	-1.6394	-1.6334
Linear		-9.9000	-9.9000

Hence, EUT cannot explain passengers' purchasing decision behaviors. In addition, we find that the coefficient of risk aversion has a significant effect on utility measurement. This conclusion shows the significance of measuring passengers' risk attitudes.

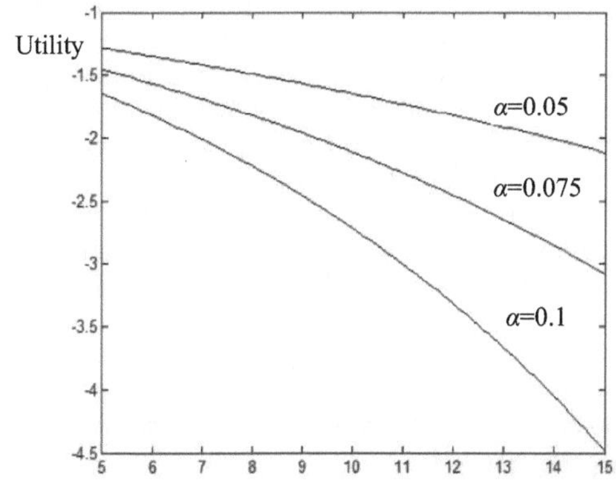

Fig.1. Negative exponential utility of scheme A with different coefficient of risk

IV. CALCULATION AND ANALYSIS BASED ON PT

Passengers' purchasing decision behaviors in our questionnaire survey can be qualitatively explained by PT. The budget constraint is the reference point to measure gains and losses. Passengers gain when the budget constraint is higher than the average fare, so they are risk averse and prefer to scheme A. Passengers lose when the budget constraint is lower than the average fare, so they are risk seeking and prefer to scheme B.

In this section we analyze passengers' purchasing decision behaviors based on PT. In order to illustrate the effect of passengers' risk attitudes to their decision behaviors, we consider two situations for comparison.

A. Passengers with Common Risk Attitudes

Let $u_{\tau g}(x_i)$ denote the utility of scheme i with budget constraint τ when the fare is lower than the budget constraint, let $u_{\tau l}(x_i)$ denote the opposite. Let P_{ig} denote the probability that the fare is lower than the budget constraint, let P_{il} denote the opposite.

PT considers value functions rather than utility functions,

$$u_{\tau g}(x_i) = \int_{a_i}^{\tau} \frac{(\tau - x_i)^{\alpha}}{\sqrt{2\pi}\sigma_i} \exp\left(-(x_i - \mu)^2 / 2\sigma_i^2\right) dx_i \quad (6)$$

$$u_{\tau l}(x_i) = \int_{\tau}^{b_i} \frac{-\lambda(x_i - \tau)^{\beta}}{\sqrt{2\pi}\sigma_i} \exp\left(-(x_i - \mu)^2 / 2\sigma_i^2\right) dx_i \quad (7)$$

and probabilities are replaced by decision weights,

$$w(p_{ig}) = \frac{p_{ig}^{\gamma}}{(p_{ig}^{\gamma} + (1 - p_{ig})^{\gamma})^{1/\gamma}} \quad (8)$$

$$w(p_{il}) = \frac{p_{il}^{\gamma}}{(p_{il}^{\gamma} + (1 - p_{il})^{\gamma})^{1/\gamma}} \cdot \quad (9)$$

We denote the utility of passengers for scheme i by

$$U_i = u_g(x_i) \times w(P_{ig}) + u_l(x_i) \times w(P_{il}) \quad (10)$$

We assume that passengers have common risk attitudes. In other words, they are common economic agents. We use the parameters estimated by Glenn, Harrison and Elisabet Rutström [7]: α=0.710, β=0.723, λ=1.380, γ=0.911. Table III shows that passengers choose the scheme which offers the higher utility.

PT can illustrate passengers' purchasing decision behaviors both qualitatively and quantitatively. Table III also verifies that passengers' attitudes of gains and losses approximately resemble common risk attitudes based on PT.

TABLE III

UTILITY MEASUREMENT FOR PASSENGERD WITH COMMON RISK ATTITUDES

Budget Constraint (times ¥100)	Utility Value for Scheme A	Utility Value for Scheme B
8	-1.7566(228)	-2.2194(72)
10	-0.1208(144)	-0.0603(156)
12	1.2423(104)	1.5931(196)
15	3.0533(138)	3.1069(162)

B. Passengers with Neutral Risk Attitudes

In order to verify the effect of risk attitudes on utility measurement, we use another group of parameters: α=β=0.99, λ=2.25, γ=0.99. In this situation, passengers' risk attitudes are approximate to neural risk attitudes. Table III shows the value of utility for each scheme [8, 9].

Table IV shows the consistence of passengers' utility with neutral risk attitudes and their purchasing decision behaviors. However, the difference of the utility between the two schemes is too small to reflect passengers' preference when τ=15. Hence, this group of parameters cannot explain passengers' purchasing decision behaviors as well as the former one.

TABLE IV

UTILITY MEASUREMENT FOR PASSENGERD WITH NEUTRAL RISK ATTITUDES

Budget Constraint (times ¥100)	Utility Value for Scheme A	Utility Value for Scheme B
8	-3.9385(228)	-4.3735(72)
10	-0.4588(144)	-0.1852(156)
12	1.7144(104)	1.9438(196)
15	4.8171(138)	4.8576(162)

C. Comparison and Discussion

Fig.2 shows the difference curves of the two schemes: the difference of the passenger number, the difference of the utility with common risk attitudes, and the difference of the utility with neutral risk attitudes. In order to compare them more intuitively, we take a 300-fold reduction on the value of the first curve [10-12].

As we can see in Fig.2, compared to the neutral risk attitudes, utility measurement for passengers with common attitudes is more consistent with the actual purchasing decision behaviors. This shows the necessity to set a reasonable level of attitudes-to-risk settings in studying airline passengers' purchasing decision behaviors [13].

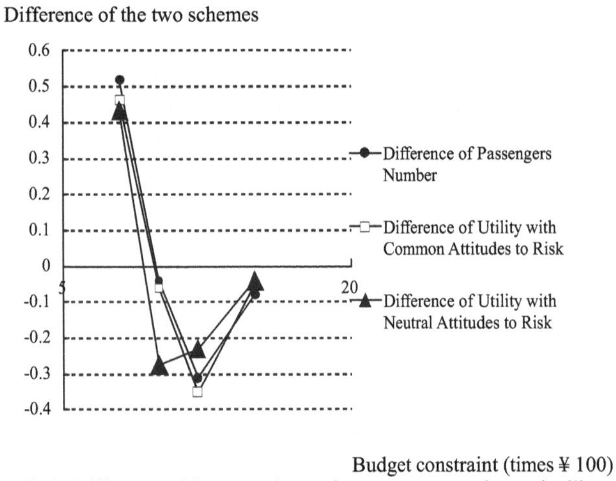

Fig.2. Difference of the two schemes in passengers number and utility

V. CONCLUSION

We have studied the purchase behavior rules of airline passengers by carrying out a questionnaire survey. Through calculation and analysis based on both EUT and PT, we come to the conclusion as follows:

a) PT shows much more rationality than EUT in explaining passengers' decision behaviors.

b) There is consistence of passengers' actual purchasing decision behaviors with utility measurement of common risk attitudes based on PT.

c) The level of passengers' risk attitudes has great influence on the utility measurement and it is very important to set a reasonable level of passengers' risk attitudes when passengers are common economic agents.

In order to improve the effectiveness of the RM strategy, further research will correct and improve traditional RM models with the hypothesis of bounded-rationality customer behavior .

ACKNOWLEDGMENT

This paper is substantially supported by research grants from the National Natural Science Foundation of China (Grant No. 71102073, 71172105).

REFERENCES

[1] J. Zhou and H. Yang, *Methods and Applications of Revenue Management, Beijing*: Scientific Press, 2009.

[2] Anderson C.K., Wilson, J.G. "Wait or buy? The strategic consumer: pricing and profit implications," *Journal of the Operational Research Society*, vol. 54(3), pp. 299–306, 2003.

[3] D. Kahneman and A. Tversky "Prospect theory: an analysis of decision under risk," *Econometrica*, vol. 47(2), pp. 263-291, 1979.

[4] Z. M. Shen and X. Su, "Customer behavior modeling in revenue management and auctions: a review and new research opportunities," *Production and Operations Management Society*, vol. 16(6), pp. 713–728, 2007.

[5] C. Mathies and S. Gudergan. "Revenue management and customer centric marketing - How do they influence travellers' choices?" *Journal of Revenue and Pricing Management*, vol. 6(4), pp. 331-346, 2011.

[6] X. Bu, L. Xu and Q. Zhao "Study on marine shipping contract allocation and pricing policy under shipper's pricing reference effects," *Journal of Management Sciences in China*, vol. 15(2), pp. 28-36, 2012.

[7] W. Glenn, E. Harrison and Elisabet Rutström. "Expected utility theory and prospect theory: one wedding and a decent funeral," *Experimental Economics*, vol. 12(2), pp. 133-158, 2009.

[8] Y. Levin, J. McGill and M. Nediak, "Optimal dynamic pricing of perishable items by a monopolist facing strategic consumers," *Production and Operations Management*, vol. 19(1), pp. 40–60, 2010.

[9] Q. Liu and G. J. van Ryzin, "Strategic capacity rationing when customers learn," *Manufacturing & Service Operations Management,* vol. 13(1), pp. 89–107, 2011.

[10] H. Li, Y. Zhang and W. Zhong, "Dynamic pricing policy under risk of strategical customers," *Journal of Management Sciences in China*, vol. 15(10), pp. 11-25, 2012.

[11] X. Xu, W. Chen and H. Peng, "Joint ordering and pricing policy considering consumer behavior and market segmentation," *Chinese Journal of Management Science*, vol. 20(6), pp. 78-86, 2012.

[12] G. Wu and R. Gonzlez, "Curvature of probability weighting function," *Management Science*, vol. 42(12), pp. 1676-1690, 1996.

[13] H. Xu, Y. Lou and J. Zhou, "A prospect-based user equilibrium model with endogenous reference points and its application in congestion pricing," *Transportation Research Part B*, vol. 45(2), pp. 311-328, 2011.

The Research on Customer Loyalty Assessment in Commercial Bank

Hui-juan YAN[1,2,*], Gui-shi DENG[1], Shi-min SHAN[3], Zi-rui LIU[4], Qi YU[4]
[1]Institute of Systems Engineering, Dalian University of Technology, Dalian, China
[2]City Institute, Dalian University of Technology, Dalian, China
[3]School of Software Technology, Dalian University of Technology, Dalian, China
[4]Department of Information Technology, Bank of Dalian, Dalian, China
(yanhuijuan0716@163.com)

Abstract - **Customer loyalty assesment is a very important issue of customer loyalty management. Building on typical and excellent literatures of customer loyalty assessment, this paper summarizes and compares the current assessment indicators and methods in commercial bank for the first time. It concludes that customer loyalty assessment model in commercial bank should be built via customer database data and data mining techniques. It also provides several future research references of customer loyalty in commercial bank.**

Keywords - **Assessment indicator, assessment method, commercial bank, customer loyalty, customer loyalty assessment**

I. INTRODUCTION

Commercial banks need more loyal customers under present intense competition so that loyalty assessment of each customer is very important to a commercial bank, and it can adopt targeted measures to improve customer loyalty and achieve economic benefit. In the process of assessment, it is necessary to build indicator system and assessment model. To construct suitable indicators and models, this paper reviews the connotation of customer loyalty, summarizes and compares the current assessment indicators and methods of typical and excellent literatures especially in commercial bank.

II. THE CONNOTATION OF CUSTOMER LOYALTY

In marketing, customer loyalty is researched mainly from three viewpoints: behavior dimension, attitude dimension and integrate dimension [1].

In the early study of customer loyalty, many researchers define and measure customer loyalty as same-brand purchase frequency [2-3], time of duration [4] and wallet share [5]. Tucker (1964) [2] considers that repurchasing is the best expression of loyalty and a customer who buys up to 3 times can be considered as a loyalty customer.

Following the development of the studies and enterprise practice, it is discovered that many consumers will recommend and advertise the product or service they have purchased and a true loyal customer not only purchases a company's product or service, but also keeps emotional attachment [6] to the company. Attitudinal loyalty is defined typically as word-of-mouth publicity, recommendation intention and purchase intention.

		Repeat Patronage: High	Repeat Patronage: Low
Relative Attitude	High	True loyalty	Latent loyalty
	Low	Spurious loyalty	No loyalty

Fig.1. Dick & Basu's loyalty model

Day (1969) [7] proposes a concept that customer loyalty should include both attitude dimension and behavior dimension. Dick & Basu (1994) [8] propose a theoretical framework based on repeat patronage and relative attitude, considering the direct antecedents to customer attitude and elements which may impact on customer attitude or behavior. They hold that only a repeat patronage with positive attitudinal intention is true loyal, and customers are divided into four categories (see Fig.1).

Many academics approve Oliver's definition of customer loyalty: a deeply held commitment to repurchase or repatronize a preferred product or service consistently in the future, thereby causing repetitive same-brand or same brand-set purchasing, despite situational influences and marketing efforts having the potential to cause switching behavior [9].

In summary, most academics intend to define customer loyalty from both behavior and attitude. A loyal customer will repurchase and recommend a preferred product/service, being not sensitive to the price while keeping longer relationship with a brand or enterprise.

III. INDICATORS SYSTEM OF CUSTOMER LOYALTY ASSESSMENT IN COMMERCIAL BANK

It is necessary to choose indicators which can measure customer loyalty before evaluating. According to above mentioned connotation of customer loyalty, there are indicators such as frequency of purchasing, time of duration, wallet share, word-of-mouth publicity, recommendation intention and purchase intention and so on. In following studies of customer loyalty assessment, empirical research in certain industry is applied extensively through questionnaire usually designed according to chosen indicators.

Beeli, Martin & Quintana (2004) [10] concludes that customer satisfaction and switching cost are the primary and direct drives to customer loyalty by empirical research in main retail banks. The research of Baumann, Burton & Elliott (2005) [11] shows that customer attitude is closely related to customer behavior intention and wallet share can be predicted. Wang, Zhang & Chi (2006) [12] add enterprise image into ACSI(America Customer Satisfaction Index) model and use customer satisfaction, customer complaints and enterprise image to evaluate a bank's overall customer loyalty. Qiao & Jiang (2007) [13] build indexes system based on ECSI(European Customer Satisfaction Index) and CCSI(Chinese Customer Satisfaction Index), and they reveal that a significant

E. Qi et al. (eds.), *Proceedings of the 5th International Asia Conference on Industrial Engineering and Management Innovation (IEMI2014)*, Proceedings of the International Asia Conference on Industrial Engineering and Management Innovation 1, DOI 10.2991/978-94-6239-100-0_62, © Atlantis Press and the authors 2015

positive correlation exists among relationship trust, switching cost and customer satisfaction of product attribute, service quality and bank image by empirical research in banking. Zou & Hao (2008) [14] build indexes from both attitude and wallet share, and verify a model of the two independent factors is better than that of one single factor.

With the growth of customer transaction data and the development of data mining techniques, many researches evaluate customer loyalty in view of the result of customer behavior. Facing with abundant database data, some literatures [15-17] primarily used the most important and popular R(Recency)-F(Frequency)-M(Monetary) indexes to calculate loyalty. Hosseini, Maleki & Gholamian add a indicator - period of product activity to RFM. Whereas Buckinx, Verstraeten & Van den Poel(2007) [18] choose more detailed variables including monetary spending, frequency of purchasing, recency of last purchase, length of the customer-company relationship, cross purchase time, returns of goods, purchase variety, promotion sensitivity, responsiveness on mailing and distance to the store to predict customer behavior loyalty. They find the most important indicator of behavioral loyalty consists of the variety of product previously purchased in retail store. Zhang, Yuan & Liu (2008) [19] builds index system of customer loyalty assessment based on objective attributes of credit card firstly in China. But it only refers to the data of credit card. Wang, Jin & Zhang (2012) [20] propose more comprehensive indexes including savings account, time savings account, bank card, loan and financing product of commercial bank, and the detailed indexes are built based on RFM model which lack of some attributes such as that of literature [19].

In short, indexes of customer loyalty include attitude indexes which are used to design questionnaire and behavior indexes which can be acquired from database data. Recency, frequency and monetary are used widely to build indexes. More indicators relative to bank business will be used in future.

IV. METHOD OF CUSTOMER LOYALTY ASSESSMENT IN COMMERCIAL BANK

There are many methods to evaluate customer loyalty. Here only summarizes and compares their characteristic and application researches especially in commercial bank so that a method can be adopted appropriately in a specific issue.

A. Model in Marketing Field

1) The Net Promoter Score (NPS) is a original method to measure customer loyalty by Fred Reichheld. By questionnaire, customers respond on a 0-to-10 point rating scale and are categorized as follows.

Promoters (score 9-10) are loyal enthusiasts who will keep purchasing and recommend others. The promoters' percentage is recorded as x.

Passives (score 7-8) are satisfied but unenthusiastic customers who are vulnerable to competitive offerings.

Detractors (score 0-6) are unhappy customers who can damage a brand and impede growth through negative word-of-mouth. The detractors' percentage is recorded as y.

Then a company's NPS can be calculated as following: NPS= x — y.

NPS reflects a company's overall customer loyalty and all targeted banks can be sorted by NPS. By Improving NPS, Tian [17] calculates loyalty score of every single customer to express customer attitude loyalty.

2) RFM analysis, proposed by Hughes (1994) [21] can analyze transaction data and predict customer behavior by three variables. Recency refers to the time span between the last purchasing time and present. Frequency refers to times of transaction in a particular period. Monetary refers to total transaction money in a particular period.

Classic RFM analysis procedure is that the three variables are sorted separately and divided into equal quintiles and these quintiles are assigned with the numbers from 5 to 1. Then the database will be divided into 125(5*5*5) roughly equal groups according to the three values. Customers with high scores are usually the most profitable. The literatures [15-17, 20], using RFM model, segment customers and calculate behavior loyalty score before clustering to obtain less number of ranks.

B. Statistical Analysis Method

Structural equation modeling (SEM), based on statistical theory, can construct measurement equation which describes the relationship between latent variables and prediction variables, and structural equation which describes the relationship among latent variables. Latent variables are not measured directly, but are estimated in the model from several measured variables. Factor analysis and regression analysis represent special cases of SEM and they usually research the relationship of customer loyalty and its antecedents. Beeli, Martin & Quintana (2004) [10] construct SEM of customer loyalty in retail banking and concludes that satisfaction and switching cost are direct antecedents of customer loyalty by empirical research. Qiao & Jiang (2007) [13] use factor analysis to research the relationship of every drive and customer loyalty, and achieve main drives of customer loyalty of banking. The model of literature [1] is constructed in view of information source and based on database data of telecommunications. Some literatures [12, 14] construct model to calculate customer loyalty score of the targeted banks based on the preceding model research.

C. Data Mining Method

In assessment of customer loyalty, clustering and classification of data mining techniques are applied widely. Clustering is to discover groups and structures in the data that are similar in some way, without using known structures in the data. Models can be built based on different theories such as customer social attributes, customer value, and customer behavior. Cheng & Chen

(2009) [15] category customers as 3/5/7 classes using RFM and k-means, then classify customers using the LEM2 algorithm(rough sets theory) not only to enhance classification accuracy but also to extract classification rules. Hosseini, Maleki, and Gholamian (2010)[16] propose a clustering procedure joining RFM to k-means. The important difference is that they evaluate the optimum k based on Davies-Bouldin Index with a detailed analysis and classification precision with decision tree and ANN.

Classification is to generalize known structure to apply to new data. Aside from literature [15], Buckinx, Verstraeten, & Van den Poel (2007) [18] and Zhang, Yuan & Liu (2008) [19] predict customer loyalty using classification method too. They use multiple linear regression and BP neural network to obtain higher predict precision than other classification methods separately. In addition, it is the same to Cheng & Chen (2009) [15] that literature [19] extracts loyalty classification rules too.

D. Analytic Hierarchy Process

Analytic Hierarchy Process (AHP) is a structured technique for organizing and analyzing complex decisions. The key of AHP is to determine weight of every hierarchy element through calculating maximum eigenvalue and eigenvector of judgment matrix, so AHP is popular to calculate weight of every element in assessment of customer loyalty.

There are abundant methods evaluating customer loyalty. It is significant that how to choose a suitable method in a specific application. So the primary methods in assessment of customer loyalty in banking are listed in detail (see Table I).

Usually, the target objects are each bank to obtain antecedents or to sort banks, so that the ones with lower ranking can take marketing measures or supply improved product or service to improve customer loyalty. With the producing of mass data, more data mining or machine learning methods are used to sort/group/predict each individual customer of commercial bank.

V. CONCLUSION

Summarizing and comparing the current assessment indicators and methods by reviewing typical and excellent literatures of customer loyalty assessment in commercial bank for the first time, this paper may help future research build more suitable indicators system and model. However, there are some problems and prospects as follows.

1) Assessment indicators systems are different in literatures, especially some related to product features of commercial bank, although indexes based on RFM is used frequently.

2) The research of customer loyalty is done more in view of antecedents than in view of consequence which means research is mainly based on customers' transaction data in database. Since huge amounts of data are produced in a short period of time, it is necessary to use data mining or machine learning techniques which can process mass data better and achieve more precise result than traditional assessment method.

TABLE I
COMPARISON WITH EVALUATION METHODS

	Method	Target	Function	Loyalty type	Literature	Data source
Marketing field	NPS	bank	sort	attitude	[17]	questionnair
	RFM	individual	sort/group	behavior	[15-17][20]	database
Statistical analysis	SEM	bank	antecedents	integrate	[10]	questionnair
		bank	antecedents/ sort	integrate	[12]	questionnair
		individual	sort	--	[1]	database
	regression analysis	bank	antecedents	integrate	[11]	questionnair
		individual	predict	behavior	[18]	database
	factor analysis	bank	antecedents	integrate	[13]	questionnair
		bank	antecedents/sort	integrate	[14]	questionnair
Clustering	K-means	individual	sort/group	behavior	[15]	database
		individual	sort/group	behavior	[16]	database
Classification	rough set(LEM2)	individual	predict	behavior	[15]	database
	BP neural network	individual	predict	behavioral	[19]	database

3) The target of traditional assessment of customer loyalty is to sort each bank or enterprise. But now the new target is to sort or predict each individual customer loyalty in order to supply personalized product or service and achieve the high value customers.

4) It's hard to acquire bank data which are top-secret and therefore unavailable by common researchers. Furthermore the structure of bank data is complex and massive so that it will take very long time to collect and process data.

5) With the development of e-commerce and e-bank, it is necessary and significant to develop the research on e-loyalty.

ACKNOWLEDGMENT

This research is supported by National Sci-tech Support Plan of China (no.:2013BAH01B03) and National Nature Science Foundation of China (no.:71372083, 61300087).

REFERENCES

[1] S. J. Yuan, "Customer loyalty model and management system building in view of information resource" (in Chinese), Ph.D. dissertation, Tongji University, Beijing, China, 2007.

[2] W. T. Tucker, "The development of brand loyalty", Journal of Marketing Research (JMR), vol.1, no.3, pp.32-35, Aug.1964.

[3] J. W. Newman, and R. A. Werbel, "Multivariate analysis of brand loyalty for major household appliances", Journal of Marketing Research (JMR), vol.10, no.4, pp.404-409, Nov.1973.

[4] K. L. Keller, "Conceptualizing, measuring, and managing customer-based brand equity", Journal of Marketing, vol.57, no.1, pp.1-22, Jan.1993.

[5] T. O. Jones, and W. E. Sasser, "Why satisfied customers defect", Harvard Business Review, vol.73, no.6, pp.88-99, Nov.1995.

[6] J. G. Barnes, "Closeness, strength, and satisfaction: examining the nature of relationships between providers of financial services and their retail customers", Psychology & Marketing, vol.14, no.8, pp.765-790, Dec.1997.

[7] G.S. Day, "A two-dimensional concept of brand loyalty", Journal of Advertising Research, vol. 9, no.3, pp.29–35, 1969.

[8] A. S. Dick, and K. Basu, "Customer loyalty: toward an integrated conceptual framework", Journal of Academy of Marketing Science, vol.22, no.2, pp.99-113, spring.1994.

[9] R. L. Oliver, "Whence customer loyalty?", the Journal of Marketing, vol.63, Special Issue,pp.33-44,Oct.1999.

[10] A. Beerli, J. D.Martin, and A. Quintana, "A model of customer loyalty in the retail banking market". European Journal of Marketing, vol.38,no.1/2, pp.253-275, 2004.

[11] C. Baumann, S. Burton, and G. Elliott, "Determinants of customer loyalty and share of wallet in retail banking", Journal of Financial Services Marketing, vol.9, no.3, pp.231-248, Mar.2005.

[12] J. J. Wang, Y. Zhang, and H. Chi, "The research on customer loyalty of Chinese commercial bank"(in Chinese). NanKai Business Review, vol. 9, no.4, pp.29-34, Aug.2006.

[13] J. Qiao, and Y. J. Jiang. "Individual customer loyalty evaluation and empirical study of commercial banks—a case study of bank of JiangSu branch", Management World, no.7, pp.94-101, Jul.2007.

[14] P. Zou, and Y. Y. Hao, "The model of customer loyalty measurement Based on Attitude and Wallet Share Two-dimensions"(in Chinese), Jounal of Management Sciences. vol.21, no.3, pp.58-66, Jun.2008.

[15] C. H. Cheng, and Y. S. Chen, "Classifying the segmentation of customer value via RFM model and RS theory". Expert Systems with Applications, vol.36, no.3, pp.4176-4184, Apr.2009.

[16] S. M. S. Hosseini, A. Maleki, and M. R. Gholamian, "Cluster analysis using data mining approach to develop CRM methodology to assess the customer loyalty", Expert Systems with Applications, vol.37, no.7, pp.5259-5264, Jul.2010.

[17] X. C. Tian, The research on the segmentation model of commercial banks' individual customer loyalty (in Chinese), Master's Thesis, Xi'an University of Architecture and Technology, Xi'an, China, 2012.

[18] W. Buckinx, G. Verstraeten, and D. Van den Poel, "Predicting customer loyalty using the internal transactional database", Expert systems with applications, vol.32, no.1, pp.125-134, Jan.2007.

[19] T. Zhang, B. Yuan, and W. H. Liu, "Predicting credit card customer loyalty using artificial neural networks", in Proceedings 11th Joint International Conference on Information Sciences(JCIC'08). Atlantis Press, 2008.

[20] W. X. Wang, Y. Jin, and D. B. Chen, "A study on personal customer loyalty based on RFM model"(in Chinese), Finance Forum, vol.195, no.3, pp.75-80, Mar.2012.

[21] A. M. Hughes, Strategic database marketing. Chicago: Probus Publishing Company, 1994.

Credible Certification Scheme Applies to Mobile E-Commerce Payment

Juan-juan Liang
Department of economy and management, Guangdong Polytechnic, FoShan, China
(ljjgdfy@163.com)

Abstract - **With the development of mobile Internet, e-commerce extensions make the payment from the Internet to the mobile Internet, it brings convenience to people, while it also brings security issues. In order to solve the trust of the mobile e-commerce payment terminals, this paper uses two-way trust authentications between the clients and the certification center to protect the trust of the payment terminals, while based on k-CAA problem in GDH group, this paper proposed an efficient authentication scheme. In the standard model, the identity authentication protocol can resist posing attack in parallel reset attack. Due to the simple interaction, low storage capacity, high efficiency and good safety performance, the identity authentication protocol is more applicable to mobile terminal equipments which havelimited computing and storage capacity.**

Keywords - **E-commerce, mobile, payment**

I. INTRODUCTION

With economic development and wide application of electronic information technology and the rapid spread of e-commerce with its fast, convenient, general, etc. become more and more recognition and trust of the community [1], electronic payment technology which is used as the core of e-commerce commercial transactions has been widely applied, and it has built an electronic payment system which is an important foundation for the development of electronic payment methods in the business, and the current inter-bank payment systems for various types of commercial bank money transfer system and within the (electronic funds transfer, online banking, bank cards, etc.) will provide e-commerce services [2]. However, due to the requirements of e-commerce environment, the need to develop appropriate payment instruments which can enable customers to securely access their bank accounts through the Internet. In addition, establishing the current payment system interfaces between the Internet and online commerce systems transmitting the payment information to the bank's payment system,to complete the appropriate transfer of funds and how to ensure the security of the transaction process are currently the electronic payment key issues [3].

Simultaneously with the development of mobile Internet, e-commerce expansion makes the paymet from the Internet to the mobile Internet, it brings convenience to people, and it also to brings security risks [4]. Because the opening performance of the mobile Internet and the rich resources, it has caused a network communication link security flaws which allow the unauthorized users to get the access to the system to intercept sensitive data and make malicious modification or insertion for the business data, while the computing power of mobile terminals and communication capability is relatively weak, so traditional security solutions are not suitable for application to a mobile terminal, and the current needs of the mobile e-commerce payment primary issue is the securely outcomes [5].

In order to solve the mobile e-commerce payment terminals in the issue of credibility, we use trusted computing technology [6-7] to protect trusted mobile payment terminals, and the traditional identity authentication strategies are more complicated, this paper proposes an authentication scheme suitable for the mobile terminals. The identity authentication protocol is based on the public key cryptosystem, the technology has an important role in the certification of electronic cash transfer and access control systems. The original identity authentication protocol was based on zero-knowledge proof ideas, because the zero-knowledge proof authentication scheme has a low efficiency. Later, based on the RSA problem Guillou-Quisquater (GQ) identity authentication protocol [8] and based on the discrete logarithm problem (DLP) of Schnorr identity authentication protocol [9] are considered to be efficient, but no security proofs were given, the two protocols are insecure under reset attacks [10]. Generally due to security requirements, based on the RSA or DLP-based identity authentication protocol requires a longer key and other parameters, so it has a lower efficiency. And based on elliptic curve cryptography scheme requires a smaller length of the key and the associated parameters to achieve the same degree of safety under the premise. Therefore, a simple and efficient authentication protocol on elliptic curve based on the discrete logarithm problem (ECDLP) is becomes hot topic in recent years. Kim M and Kim K made a scheme based on bilinear Diffie-Hellman problem (BDHP) in 2002 (referred to KK scheme [11]), compared with GQ programs and Schnorr schemes, it is more efficient, but KK is more complex and literature [11] has pointed out the security vulnerabilities. In 2004, Shao J, Cao ZF and Lu R proposed an identity-based authentication based on the strong Diffie-Hellman (SDH) problem [12] (referred to as SCL protocol). SCL protocol is more efficiency and higher security, but there is no proof for it. In order to ensure the identity authentication protocol can resist various attacks, it needs to use rigorous mathematical derivation to demonstrate the safety of authentication protocols. Based on k-CAA in GDH group, this paper proposed an efficient authentication scheme. In the standard model, we can prove that the identity authentication can resist posing attack in parallel reset attack. Due to the simple interaction, low storage capacity, high efficiency and good safety performance, the identity authentication is more suitable for the mobile terminal equipments which have the limited computing and storage capacity.

E. Qi et al. (eds.), *Proceedings of the 5th International Asia Conference on Industrial Engineering and Management Innovation (IEMI2014)*, Proceedings of the International Asia Conference on Industrial Engineering and Management Innovation 1, DOI 10.2991/978-94-6239-100-0_63,

II. THE TRUST BIDIRECTIONAL AUTHENTICATION PROTOCOL BETWEEN THE TERMINALS AND CERTIFICATION SERVER

Users will send the status information and configuration information of the mobile client platform to the server to prove whether they are trusted, if the terminal is trusted, it will be allowed to access the certification center, otherwise it will be refused to get the services, while mobile clients require the certification center to send its own state and platform configuration information to mobile client, the mobile client will check whether the configuration and status information certification center meets their security policies for mobile clients, the proof is a bi-proof certification. The server needs to measure the clients, while mobile clients need to measure the certification center, after two-way evaluation, the mobile client can access the authentication center. The following Fig.1 shows the process:

Fig.1. Double-sided credible proof

A. First Certification Center select a random number N_s , And transmitting information (1) to mobile clients, PCM_{needc}, SML_c , SML_{needc} Certification Center for the mobile client platforms need to provide mobile client configuration values and metrics, According to PCM_{needc}, SML_c mobile client need to collect configuration information.

$$m_1 = (N_s, PCM_{needc}, SML_c) \quad (1)$$

B. when the client receives mobile customers, Extracting, Extracting PCM_{needc}, SML_c , Certification required to determine whether the PCR_{need}, SML_c provided meets privacy and security policies of the mobile client, if the client does not meet the mobile access authentication center, and if so, move the client chooses a random number N_c , Then read configuration information PCM , from the local value, and then use the mobile client's private key to generate a signature (2), Then read the stored metrics log SML_c , Then it Sends (3) Certification Center after using the session key K encryption, PCM_{needs}, SML_s are the requested authentication configuration information provided by the center and measure logs for mobile clients.

$$S_c = Signm_1(PCM, N_c)_{SK} \quad (2)$$

$$m_2 = (SML_c, S_c, PCM_{needs}, PCM, N_c, AIK, SML_{needs}) \quad (3)$$

C. When the certification center receives (3), it will determine whether PCM_{need}, SML_{needs} meets the security policy and Certification Center, if not, the terminal will be denied, or the meets the security policy and Certification Center chooses a random number N_{cs} , Read its own PCM value, the it will use its private key to generate the signature (4), then it reads the stored measurement log SML_s , Certification Center uses the session key uses the K to encrypt (5) and sends it to the mobile client.

$$S_s = Signm_1(PCM, N_{cs})_{SK} \quad (4)$$

$$m_3 = (S_s, PCR, N_{cs}, SK, SML_s) \quad (5)$$

D. After moving a client receives (5), extracting PCM, SML_s , determine whether they meet the mobile client security policy, if they meet the security policy of the mobile client, then it has completed a two-way mobile clients trusted authentication and certification center, the mobile client access can use the services of the authentication center.

III. IDENTITY AUTHENTICATION

Authentication scheme have the parameter established phase, the key generation phase and user identity verification phase, the following is a detailed description of these three stages.

A. Parameters establishment

G_a, G_b are the P-order groups, P is a large prime, g is a generator of G_a , Bilinear mapping is (6), giving a hash function (7), Public parameters is (8).

$$e(G_a, G_a) \to G_b,\ e(g, g) = I \quad (6)$$

$$H\{0,1\}^* \to Z_p^* \quad (7)$$

$$G_a, G_b, g, P, e(g, g) = I \quad (8)$$

B. Key generation

The mobile client Firstly selects $x \in {}_R Z_P^*$, and computing (9), The (9) as the user public key, users register with the Certification Center using ID_i (9).

$$R = g^x \quad (9)$$

C. Verification

First, the user will be sent to the certification center (10), then Certification Center selects $y \in {}_R Z_P^*$, it will be sent to the mobile client, Mobile client will send (11) to the certification center. Certification Center will verify the (12) are established, If (12), the user's identity is legitimate.

$$PK = g^x \quad (10)$$

$$T = g^{\frac{1}{y+x}} \quad (11)$$

$$e(T, R^y) = I \quad (12)$$

IV. SECURITY ANALYSIS

Random oracle model (Random Oracle) is 1993 Bellare and Rogaway proposed random oracle model is a non-standardized calculation model, the model can be used in this way to prove the safety of the Statute of the agreement, if the attacker can prove with non-negligible probability to break cryptographic protocols, Then the attacker can use this kind of attack algorithm design, making this attack algorithm to calculate the non-negligible probability to break the difficult problems in number theory.

This paper describes the use of the random oracle actual system / ideal system (Real system / idea system) to prove credible pipeline safety protocols, anonymous authentication actual system (Real System) participants: user authentication center, there is also an attacker ***A***, and the environment ***E***, ***E*** provides participants input information, and interact with, the attacker ***A***, and finally to all participants attacker output. Ideal system (Ideal System), users, and Certification Center, the participants do not run directly cryptographic protocols, they put all the calculations request sent to an ideal trusted party (ideal all-trust part) ***T***, ***T*** have to calculate the output protocol based on the input of participants, the final result will be returned to the participant, if an agreement exists attacker simulator ***S***, and a practical system for each operating environment ***E*** as well as to control some of the participants, so that the environment cannot be distinction is run in the real system and interact with the attacker A, or in an ideal system, and run the simulator ***S***, it indicates that the password protocol is running safe.

Constructing an attacker ***A*** which communicates with ***C*** which is the owner of (13), then impersonate legitimate users prove the security of the proposed algorithm. Construct a performance function ***F***, the function is K-CCA attacker difficult problems, Performance function ***F*** is known in this set of data (14). Function puts the attacker ***A*** as its own subroutine, ***F*** simulates honest owners to ***A*** generated response.

$$PK = g^{\frac{1}{x}} \quad (13)$$

$$y_i^{'} \in \{h_1, h_2 .... h_k \in Z_p, g, g^x, g^{\frac{1}{x+h_1}}, g^{\frac{1}{x+h_2}} .... g^{\frac{1}{x+h_k}}\} \quad (14)$$

A. Interactive stages:

Function simulates the honest user $C_i \in \{C_1, C_2 .... C_n\}$, Interaction can be performed in parallel, all the owners of C_i have the same public-key (9), C_i independently initializes the required parameters, the attacker ***A*** disguises as verifier ***V*** in the interactive stage, selectively sending $y_i \in {}_R Z_p^*$ and $i \in 1, 2....n$ to C_i, the answer is A or $\perp$, if the answer is, ***C*** will sent (15) to the attacker ***A***, Otherwise putting $\perp$ return to ***A***.

$$T = g^{\frac{1}{y_i+x}} \quad (15)$$

Because of the attacker ***A*** knows the public-key of C_i, $e(T, Rg^y) = I$, ***A*** cannot distinguish between the real and the ideal environment for the environment, so this calculation protocol is safe, the attacker ***A*** can record (16) in order to new attack, then attacker re-attacking. All C_i own the same (9) which sending to the attacker , the attacker ***A*** randomly selects (17) and send it to C_i, Meet the conditions (14), C_i sends (9) and (18) to the attacker , Attacker in question calculates (19), Settlement results will not be x, so re-attacking on the program is invalid.

$$(R, y_i, T = g^{\frac{1}{x+y_i}}) \quad (16)$$

$$y_i^{'} \neq y_i \in {}_R Z_p^* \quad (17)$$

$$R = g^{\frac{1}{x+y_i^{'}}} \quad (18)$$

$$T / T_i = g^{x/y_i - x/y_i^{'}} \quad (19)$$

B. Verification stages:

A impersonate legitimate users C_i to attack, During this period, the performance function ***F*** play honest verifier, function sends (R, y_i, A_i) to ***A*** according y $(y \notin \{y_1, y_2 ..... y_n\})$, if ***A*** Successfully outputs r which making (20) is true, Description ***A*** successful then found a solution K-CCA, which in polynomial computing power is impossible, so in this stage protocol is secure.

C. Theorem:

If a computing capability is polynomial impostors re-attack *CR* in t, they must have ε superiority that can be success, so the $k-CAA$ problem on the group G_a can be solved in the time (21) according (21).

$$e(A, R^y) = I \quad (20)$$

$$t' = t \quad (21)$$

Since the *CR* attacks on anonymous authentication protocol strongest active attacks, passive attacks on the proposed scheme is invalid. Identification anonymous attacker eavesdropping session script (16), the attacker cannot obtain the private-key x information and cannot impersonate a legitimate user from the (16).However, the attacker can not obtain the private-key x and can not impersonate a legitimate user. The only adversary can do is continue to send (ID_i, $g^{\frac{1}{x+y}}$), A challenge value (22) in order to get the same value of its recorded session script (16), and then replay. Similarly birthday attacking, the same probability is approximately equal to (23), it is Negligible. In summary therefore proposed an anonymous authentication protocol is secure.

$$A = g^{\frac{1}{x+y}} \quad (22)$$

$$p \le 1/2^{80} \quad (23)$$

V. EFFICIENCY ANALYSIS

Communication and computation from the perspective of comparative comparative literature [13] authentication protocol efficiency programs and programs in this article. This article first program adds the private key extraction stage verification operation, inspection and certification center for the correctness of the generated key part to prevent dishonest camouflage operation of users, the security of further protection. Secondly, the user's private key authentication operation can be pre-computed and published the final identification of the user to verify the legitimacy of the actual need is a bilinear operator. Generally, the bilinear operator $\Pr$ is the most time-consuming, and exponential multiplication group M is also more time-consuming, other operators can not count, and traditional certification programs required computing and communication complexity comparison, in the security program under further protect the advance, the efficiency has significantly improved.

TABLE I
PERFORMANCE COMPARISON TABLE

	authentication scheme of Literature [12]	the article's proposal
Computational complexity	$\sum 2(M) + E(0) + \prod 2Me + 2\Pr$	$\Pr + 2Me$
Network traffic	$\|G_a\| + 2\|q\| + \|G_b\|$	$2\|G_a\| + \|q\|$
Security	Not resist re-attack	resist re-attack

VI. CONCLUSION

This paper analyzes the mobile e-commerce transaction process safety issues, and propose a suitable mobile e-commerce payment certification program, which firstly completes the two-way trusted authentication between the client and the authentication center, if it is trusted, the terminal will be allowed to access to authentication center, otherwise the center will refuse to provide services to users, and then the scheme based on the k-CAA problem on the GDH group, it proposes an more efficient identity authentication solution. In the standard model, the identity authentication protocol can resist posing attack in parallel reset attack. Due to the simple interaction, low storage capacity, high efficiency and good safety performance, the identity authentication protocol is more applicable to mobile terminal equipments.

REFERENCES

[1] Youqing Guan, Xiaojun Wang. E-Commerce Security Technology Beijing: Beijing University of Posts and Telecommunications Press, 2005.

[2] Qifeng Yang, Zhengwei Cheng, Ping Song, Research on Online Payment Mode Based On Internet Banking Payment Gateway. Communications of the ACM IEEE Computer Society, 2007, 11. ISBN:0-7695-3038-9.

[3] Qiang Zheng. Chinese e-commerce Blue Book: 2007 [M] Beijing: China Economic press, 2007.

[4] The Shanghai iResearch Co., Ltd.. IResearch-2006 China's online payment simplified version of the report .2006.

[5] Dengguo Feng. Rising e-commerce security issues. Information Systems Journal, 2009, 3(1):103-103

[6] Changxiang Shen Research and development of trusted computing, Chinese Science [J], 2010.2

[7] Huanguo Zhang, Jie luo, Gang jin. Trusted Computer Technology and Application. Wuhan University (Natural Science Edition) [J],2006.5, pp:513-518

[8] GUILLOU L S, QUISQUATER J J. A practical zero knowledge protocol fitted to security microprocessors minimizing both transmission and memory [A]. Advances in Cryptology-Eurocrypto'88[C]. Switzerland d, Springer-Verlag New Y0rk Inc, LNCS 330, 1988. 123-128.

[9] SCHNORR C P. Eficient identification and signatures for smart cards [A]. Advances in Cryptology-Crypto'89 [C]. Santa Barbara, Caledonia, USA, Springer-Verlag New Y0rk Inc, LNCS 435, 1990. 239-252.

[10] BELLARE M, PALACIOY A.GQ and Schnorr identification schemes: proofs of security against impersonation under active and concurrent auacks [A]. Advances in Cryptology-Crypto'02[C].Santa Barbara (California), USA, LNCS 2442, 2002.167-177.

[11] KIM M, KIM K. A new identification scheme based on the bilinear Diflie-Hellman group [A]. Information Security and Privacy: 7th Australasian Conference [C]. ACISP 2002 Melbourne, Australia, 2002.362-378.

[12] SHAO J, CAO Z E LU R X.A new efficient identification scheme based on strong Difie-Hellman assumption [A]. Proceedings of International Symposium on Future Software Technology (ISFST-2004) [C]. Xi'all, China, 2004.

[13] Brickell E, Camenisch J, Chen L. Direct Anonymous Attestation [C] // The Proceedings of the 11th ACM Conference on Computer and Communications Security. ACM Press, 2004:132-145

An Exploration of the Building of Enterprise Multi-project Management System

Hui-ming Wang*, Zhi-jian Yang, Yun-tao Guo
School of Management, Northwestern Polytechnical University, Xi'an, China
(wjswlw@163.com)

Abstract - **Project management has been put into practice by more and more enterprises, and the concept, "to implement management according to project", has been deepening gradually. Starting from the analysis of the existing problems of enterprise multi-project management, and taking Lanfei Company as an example, this paper puts forward an approach of building enterprise multi-project management system. If an enterprise intends to gain a competitive advantage in the fierce market competition, it must take market demand as the guidance, pursue technological progress and upgrade its products. This is the so-called "one generation's production, one generation's research, one generation's pre-research". At present, the application of Enterprise Project Management is quite common. Because in the process of an enterprise's scientific research and production, multiple projects (task) are conducted simultaneously, how to resolve the management problem of parallel multiple program, becomes the challenge that it cannot avoid and must meet.**

Keywords - **Enterprise project management, multiple projects, management system**

I. THE EXISTING PROBLEMS

With the rapid development of science and technology and the deepening of market economy, an enterprise faces that the cycle of product development is increasingly shortened, the requirements of quality are increasingly improved, cost control is increasingly fine, the conflict of resources are increasingly acute, parallel multiple projects often exists, and the difficulty of the organization, implementation coordination and management of projects is increasingly great. As a result, the original management system and approach of an enterprise faces great challenge, and many management problems are exposed [1, 2].

A. The problem of consistency between the project's goal and the corporate strategic goal

Corporate strategy is its long-term development goal, and the achievement of this strategy needs to rely on the implementation of the project. So in the implementation of project management, an enterprise needs to coordinate and reach an agreement between the project's goal and the corporate strategic goal, and ultimately achieves the corporate strategic goal through implementing a series of projects.

B. The problem of horizontal cooperation and communication of functional departments

With the increase of the types and quantities of enterprise projects, the implementation of many projects requires more and more collaboration among different departments, and it is difficult to cooperate and communicate between different functional departments and the information transfer is obstructed. After the problems arise within the project, each department often shirk its responsibility, making it difficult to achieve the overall objective of the project.

C. The problem of how to allocate the resources to avoid conflicts within the project

The number of parallel enterprise projects is enormous, and the working tasks increase sharply, but the technology, management personnel, equipment, factories and other resources don't increase at the same pace. The coordination and allocation of resources is difficult, and the phenomenon of competition for the progress and resources appear constantly, leading to the failure to finish some projects on schedule.

D. The problem of coordination between the processes and methods of project management and the original business processes and methods of enterprises

There are differences between processes and methods of implementing project management, and the original business processes and methods of enterprises (e.g. functional management style). Since two management processes and systems are in operation within the enterprise, it is essential to coordinate the two sets of management processes and methods, and deal well with the connection between the two interfaces.

II. THE SOLUTION

This paper takes Lanzhou Flight Control Company, Ltd (hereinafter referred to as Lanfei company) to propose a new enterprise multi-project management system [3].

A. Building an evaluation criterion of establishing projects and priority corresponding to the corporate strategy

Strategy builds the vision for the enterprise, and provides overall and guiding plan for fulfilling the goal. Projects are the carrier for achieving corporate strategy, clear criteria of establishing projects can provide the evidence for determining the annual project of Lanfei company. In the meantime, the hierarchical management ideas will be integrated into the project, which can ascertain the level of the project (important project, ordinary project) which can in turn serve as a basis for resource allocation and security.

E. Qi et al. (eds.), *Proceedings of the 5th International Asia Conference on Industrial Engineering and Management Innovation (IEMI2014)*, Proceedings of the International Asia Conference on Industrial Engineering and Management Innovation 1, DOI 10.2991/978-94-6239-100-0_64, © Atlantis Press and the authors 2015

According to the contents of the report of establishing project, every year Lanfei company makes an evaluation for each project in accordance with the indicators shown in Table I. The company organizes experts to score respectively for each project in a meeting of demonstrating the establishment of the project, and ranks these projects according to their scores. Projects with a score of 85 points or more (including 85 points) belongs to the company's important project, those between 65 to 85 points (including 65 points) are regarded as the ordinary project, and projects less than 65 points are not allowed to establish temporarily. The important projects should be guaranteed to have the priority to enjoy resources of the company.

TABLE I
THE EVALUATION CRITERION OF PROJECT ESTABLISHMENT AND ITS PRIORITY

indicator	weight W_i	Score a_i		
		1	3	5
income	20	operating income ≤2 million yuan	2 million yuan <operating income≤5 million yuan	operating income>5 million yuan
income/investment	15	(income/investment)≤1	1<(income/investment)≤2	(income/investment)>2
correlation with the group's focus	30	not belong to the group's focus		Belong to group's focus
correlation with the corporate strategy	20	basically uncorrelated	partially correlated	Key tasks to implement strategy
urgency of the project's time	15	extremely urgent	quite urgent	quite free

B. Establish a matrix-based project management organization

With the increase of projects, Lanfei company's original organization structure cannot effectively complete the allocation of resources, and a new organization structure is urgently needed to solve the contradictions faced by the company. Therefore, according to the characteristics of the established project, Lanfei company founded the project management organization structure based on company's original organization structure to improve the company's speed and ability to response to the rapid environment change. Lanfei company's project management organization can be divided into four layers: the decision-making layer of enterprise project management, the management layer of enterprise multi-project, the implementation layer of enterprise multi-project and single-project, as is shown in Fig.1.

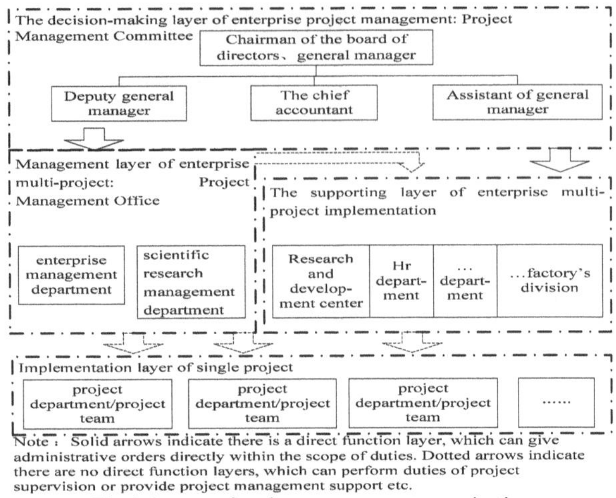

Fig.1. Layers of project management organization

C. Set up management system of the whole process of scientific project management

Lanfei sets up a standardized management system of the whole process of project management, focusing on the organization structure of the project operation. This system centers on the project to carry out its work, highlights the right of the project's manager for the project's decision-making, optimizes the information process, and reduces the link of information transfer and decision making. At the same time it quite properly solve the conflicts between the goal of the project team and the company's strategic goal, the conflicts among the goals, resources and interests of multi-project teams and the conflicts between the project teams and functional departments.

According to main management work of the project's entire life cycle, the project management process of Lanfei company can be divided into project establishment, project planning, project implementation and control and project closeout, the main work of each process is shown in Fig.2, 3, 4, 5. Through a clear input conditions and output results of each process, it emphasizes the standardized operation of the process, and guarantees consistency of the process and traceability of responsibilities.

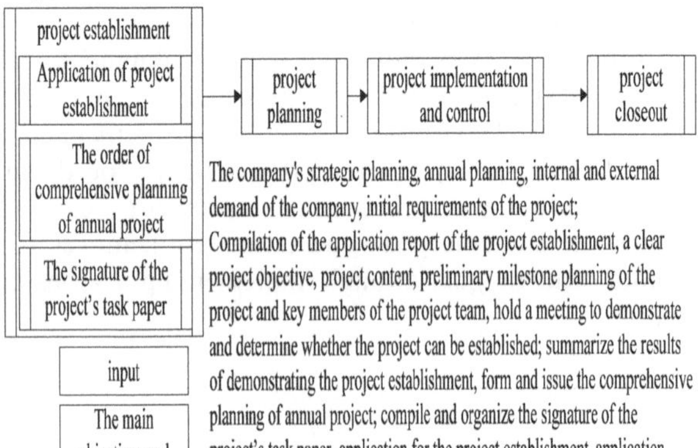

Fig.2. Schematic diagram of the project's main management work

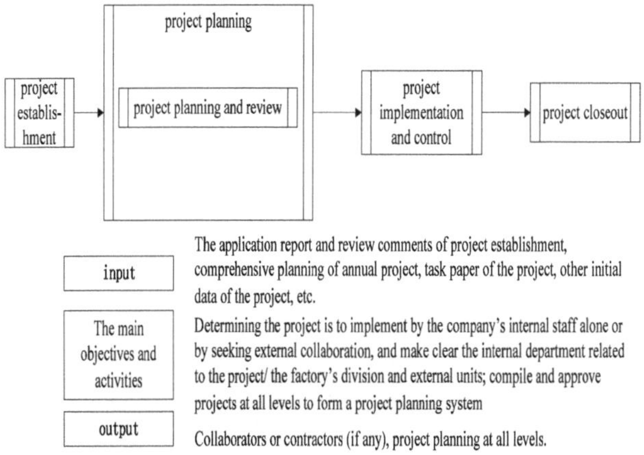

Fig.3. Schematic diagram of the main work of project planning

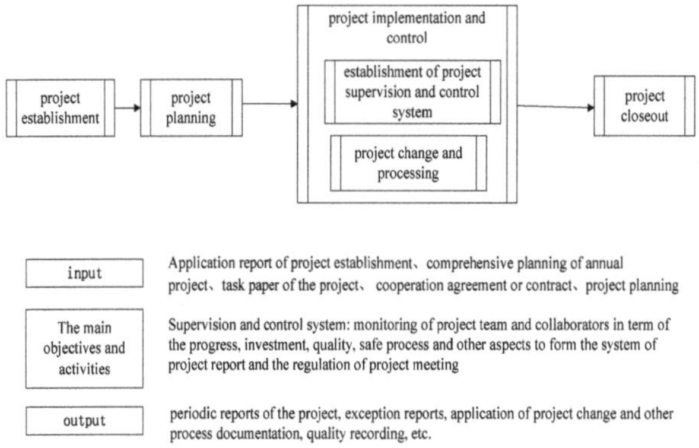

Fig.4. Schematic diagram of the main management work of the project's implement and control

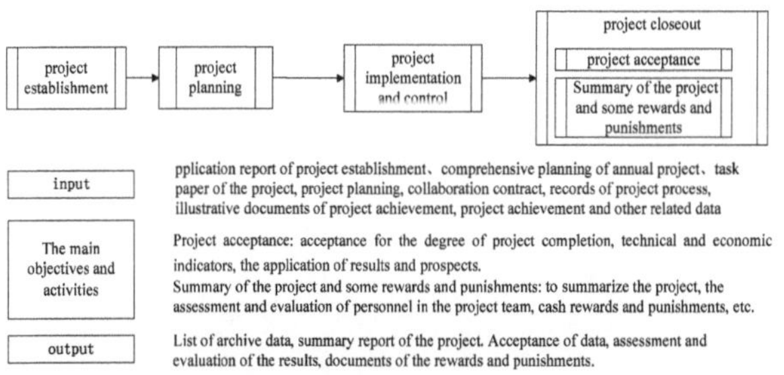

Fig.5. Schematic diagram of the main management work of the project closeout

D. Form the standardized multi-project management system

The core of the enterprise project management, "to manage according to the project", is a long-term organization management mode centering on the project. Specifically, to center on the project - to divide the enterprise business into a goal-oriented, resource-restricted process or process teams; to regard the project team as the main body of management - to make use of knowledge, tools, and methods of project management with the constraints of limited resources; to treat the enterprise project management system as the support - to provide resources, management environment, the normative documents behavioral consistency support, etc. for the project. The construction of project management system is the foundation for achieving the parallel enterprise multi-project management and the important means to rapidly improve the ability of enterprise project management. Lanfei company's project management regards the project process as the main line, and integrates the working content of project management into the business work to realize the combination of the project's process and management work to attain to the project's goal. Based on this point, the structure of project management system is shown in Fig.6.

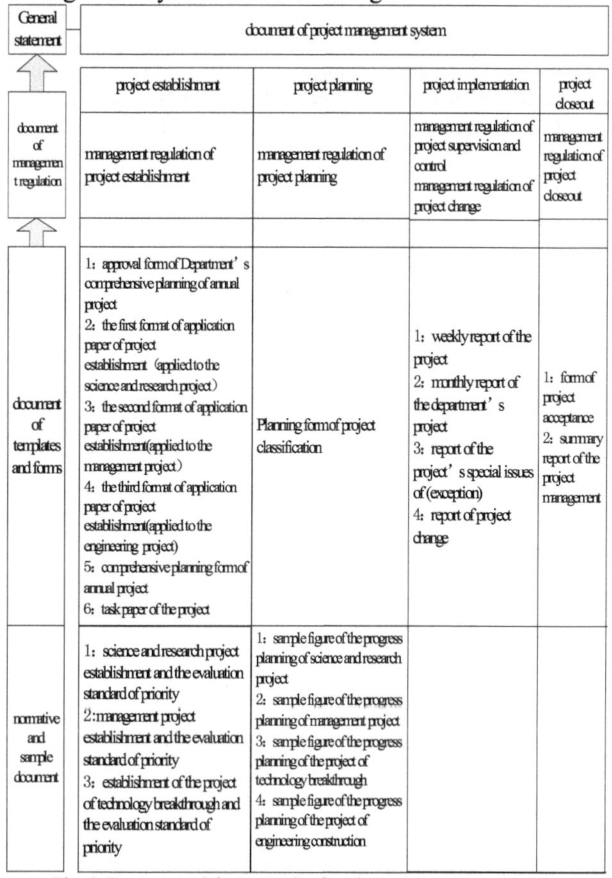

Fig.6. Documental framework of project management system

III. CONCLUSION

With the import and use of the concept of "to manage according to the project", more and more tasks in the enterprises can be managed in accordance with the project. The construction and practice of Lanfei company's multi-project management system has become important driving force to improve the company's management level and enhance its core competence. First of all, enterprise project management system optimizes, and reorganizes the enterprise's business process, reduces the links of information transfer, improves the speed of decision making and the enterprise's communication channels, and makes the case demonstration of scientific research project, project design and project type and other links connected tightly and carries out smoothly. Secondly, according to the project's priority and actual use of resources, the enterprise allocates the proper resources to the project with the most benefits so that those resources may play a maximum value. Thus, the enterprise effectively eases contradictions of limited resources and improves the efficiency of the utilization of its resources under the circumstances of arduous multi-project tasks, which serves as a good enlightenment and reference for similar enterprises.

REFERENCES

[1] Li Wen, Li Dan, Cai Jinyong. Practice of Enterprise project management, Beijing: Mechanical industry publishing house, 2010.
[2] Bai Sijun. General statement of modern project management, Beijing: Electronic industry publishing house, 2006
[3] Zhang Fumin, Wang Hongchen, Xu Jianglin. Office of efficient project management office. Beijing: Electronic industry publishing house, 2010

Using Galois Lattice to Represent and Analyze Information Security Policy Compliance

Xiao-long Wang*, Wen-li Li
Faculty of Management and Economics, Dalian University of Technology, Dalian, China
(michaelwangxl@mail.dlut.edu.cn)

[1] *Abstract* - **Employees' noncompliance with the information security policy results in a large number of information security incidents in organizations. The information security managers need to understand and manage the noncompliance behaviors of employees. The representation and pattern of the information security noncompliance or compliance will help managers to gain insights on, and to counter effectively the threats originated from the employees. This study proposes a Compliance Galois Lattice Diagram (CGLD) for visually representing and analyzing the employees' compliance patterns. Six compliance patterns, namely, compliance outlier, compliance core and peripheral, compliance subgroup, compliance partition, multiple compliance containment and compliance equivalence, have been obtained from the CGLD. A comparative analysis of these patterns and the structural features identified from the network generated by the UCINET software reveals that fairly good consistency has been reached between them.**

Keywords - **Compliance, Galois lattice, information security management, information security policy, two-mode network**

I. INTRODUCTION

A huge number of information security incidents are perpetrated by employees in the organizational setting [1-4], and the anthropogenic incidents caused even by one employee can be potentially devastating [4-6]. In order to protect the information resources and assets, strict information security policies have been formulated in a wide variety of organizations [3, 4, 7, 8]. However, these information security policies are only effective to the extent that employees comply with them [7]. There is an obvious need for the information security managers to understand and then to counter the threats originated from the employees' noncompliance with these policies [3, 7].

Empirical methods have been used to investigate the information security policy compliance or noncompliance of employees in many recent studies [3-10]. In the existing standards of information security metrics, some simple statistical indicators, e.g., percentage, have been used to measure the compliance performance of organizations [11, 12]. Actually, the information security compliance is concerned with two kinds of different objects, i.e., the employees and the clauses of the information security policy, hence a two-mode network [13-18] can be used to represent and analyze compliance data. In particular, a two-mode network can be constructed by the employee nodes, the clause nodes and the lines linking the two different kinds of nodes. By taking advantage of the structural duality of the two-mode network [14, 19], three kinds of compliance structures can be explicitly analyzed: the employee-to-employee structure, the clause-to-clause structure, and the employee-to-clause structure. Some insights into the employees' compliance or noncompliance with the information security policy may be gained from these structures and their derivatives.

In the present study, a Compliance Galois Lattice Diagram (CGLD) is proposed for a visual representation of the information security policy compliance of employees. Some compliance patterns have been obtained from the CGLD. These compliance patterns are expected to be useful for the managers to understand and manage the information security compliance of employees.

II. COMPLIANCE GALOIS LATTICE

Consider a triple (E, A, C), where E and A are two finite nonempty sets of employees and information security policy clauses, respectively, and $C \subseteq E \times A$ is a binary relation denoting the employees' compliance with the clauses. For example, $E = \{e1, e2, e3, e4, e5, e6, e7, e8, e9, e10, e11, e12, e13, e14\}$ and $A = \{a1, a2, a3, a4, a5, a6, a7\}$ are involved in the hypothetical two-mode compliance data matrix in Table I, and in the matrix, an entry of 1 in a cell indicates that the clause designated by the column has been followed by the employee designated by the row, whereas an entry of 0 means that the employee designated by the row has not complied with the clause designated by the column. Consider two lattices $S(E) = \{E_1, E_2, E_3, ...\}$ and $S(A) = \{A_1, A_2, A_3, ...\}$ as the collections of subsets of E and A. The relation C can be used to define the following two mappings:

1) The mapping $\uparrow : B \rightarrow B\uparrow$ from $S(E)$ to $S(A)$, here $B\uparrow = \{ a \in A \mid (e, a) \in C \text{ for all } e \in B, B \subseteq E \}$, where $(e, a) \in C$ means that the clause a has been complied with by the employee individual e. In Table I, the clause subsets complied with by the corresponding employee subsets can be derived from this mapping. For instance, the clause subset $\{a_1, a_2, a_3\}$ corresponds to the employee subset $\{e_3, e_5, e_6, e_{14}\}$.

2) The mapping $\downarrow : D \rightarrow D\downarrow$ from $S(A)$ to $S(E)$, $D\downarrow = \{e \in E \mid (e, a) \in C \text{ for all } a \in D, D \subseteq A\}$. The employee subsets which have complied with the corresponding clause subsets can be obtained from this mapping. For

[1] This work was supported in part by a grant from the National Natural Science Foundation of China (Nos. 70972058/G0211 and 71272092/G0211).

instance, $\{e_9, e_{10}, e_{14}\}$ is the employee subset which has complied with the clause subset $\{a_4, a_5, a_6, a_7\}$.

Let $T(E) = \{E_1\uparrow, E_2\uparrow, E_3\uparrow, \ldots\}$ be the collection of images of $\uparrow$, and $T(A) = \{A_1\downarrow, A_2\downarrow, A_3\downarrow, \ldots\}$ the collection of images of $\downarrow$. Note that $E_j\uparrow = A_j$ and $A_j\downarrow = E_j$ for each subscript j. The subsets of $T(E)$ form a lattice under inclusion, as do the subsets of $T(A)$. The two lattices are dual inverse, i.e., $A_j\downarrow \subseteq A_k\downarrow \Leftrightarrow E_j\uparrow \supseteq E_k\uparrow$, which can be represented by a single lattice $T(A) \times T(E)$. Thus, $(A_j\downarrow, E_j\uparrow) \leq (A_k\downarrow, E_k\uparrow) \Leftrightarrow (A_j\downarrow \subseteq A_k\downarrow) \& (E_j\uparrow \supseteq E_k\uparrow)$, where the binary relation $\leq$ is reflexive, anti-symmetric and transitive. $(A_j\downarrow, E_j\uparrow)$ in this dual lattice is a lower bound of $(A_k\downarrow, E_k\uparrow)$ when $A_j\downarrow \subseteq A_k\downarrow$ or $E_j\uparrow \supseteq E_k\uparrow$. This kind of dual lattice, in which each element is a pair, is named the Galois lattice [20-25] concerning with the employees' information security policy compliance.

A compliance Galois lattice can be displayed by a labeled line diagram, i.e., CGLD. In a CGLD, two labels are assigned to each node, one of them is specified by one or several elements belonging to $S(E)$, and the other by one or several elements of $S(A)$. These nodes are linked by ascending and descending lines to show their partial ordering. In this work, the ConExp1.3 software [26] is used to generate the CGLD. The CGLD representing the compliance data in Table I is shown in Fig. 1. Each node in the CGLD with a black lower half circle indicates that an employee subset is associated with it, and the node with blue upper half circle indicates that it is associated with a clause subset. Consider two nodes in such a diagram. The first node represents both the employee subset E_m and the clause subset A_n complied with by the E_m, likewise, the second one represents both E_p and A_q. The two nodes are linked by a descending line from the first to the second if $E_p \subseteq E_m$ and $A_n \subseteq A_q$. For instance in Fig. 1, the employee and clause subsets represented by the node denoted with the upper label $a2$ and the lower label $e4$ are $\{e3, e4, e5, e6, e14\}$ and $\{a1, a2\}$, respectively. The employee subset $\{e3, e5, e6, e14\}$ and the clause subset $\{a1, a2, a3\}$ are represented by a node labeled with $a3$ and $e3$, $e5$, $e6$. In the diagram, the two neighboring nodes are linked by a descending line.

III. COMPLIANCE PATTERNS

Specific compliance patterns with respect to the information security policy of an organization can be obtained from the CGLD. In what follows, six such patterns are introduced, and these patterns are expected to be useful for the information security managers to understand employees' compliance behaviors.

A. Compliance Outlier

There probably exist two kinds of compliance outliers in a CGLD. The first kind of compliance outlier is the employee outlier, which refers to the employee who has not complied with all of the clauses of the information security policy. In the CGLD shown in Fig.1, the top-most node indicates that $\boldsymbol{e2}$ and $\boldsymbol{e12}$ are employee outliers. The second kind of outlier is the clause outlier, which refers to the clause that has not been complied with by all of the employees. The two-mode compliance data in Table II are illustrated by the CGLD in Fig. 2. In the CGLD, the bottom-most node labeled by $\boldsymbol{a8}$ is a clause outlier.

TABLE I
THE HYPOTHETICAL TWO-MODE COMPLIANCE DATA: EXAMPLE I

	a1	a2	a3	a4	a5	a6	a7
e1	1	0	0	0	0	0	0
e2	0	0	0	0	0	0	0
e3	1	1	1	0	0	0	0
e4	1	1	0	0	0	0	0
e5	1	1	1	0	0	0	0
e6	1	1	1	0	0	0	0
e7	0	0	0	0	0	1	0
e8	0	0	0	1	0	1	1
e9	0	0	0	1	1	1	1
e10	0	0	0	1	1	1	1
e11	0	0	0	1	1	1	0
e12	0	0	0	0	0	0	0
e13	0	0	0	0	1	1	0
e14	1	1	1	1	1	1	1

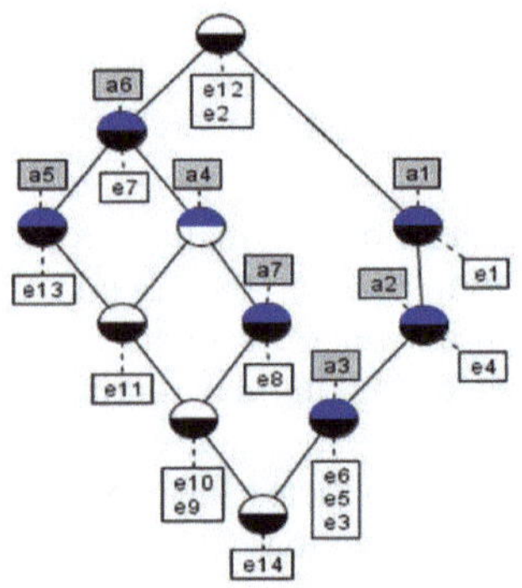

Fig.1. The CGLD of the compliance data in Table I

TABLE II
THE HYPOTHETICAL TWO-MODE COMPLIANCE DATA: EXAMPLE II

	a1	a2	a3	a4	a5	a6	a7	a8	a9
e1	1	0	0	0	0	0	0	0	1
e2	0	0	0	0	0	0	0	0	1
e3	1	1	1	0	0	0	0	0	1
e4	1	1	0	0	0	0	0	0	1
e5	1	1	1	0	0	0	0	0	1
e6	1	1	1	0	0	0	0	0	1
e7	0	0	0	0	0	1	0	0	1
e8	0	0	0	1	0	1	1	0	1
e9	0	0	0	1	1	1	1	0	1
e10	0	0	0	1	1	1	1	0	1
e11	0	0	0	1	1	1	0	0	1
e12	0	0	0	0	0	0	0	0	1
e13	0	0	0	0	1	1	0	0	1
e14	1	1	1	1	1	1	1	0	1

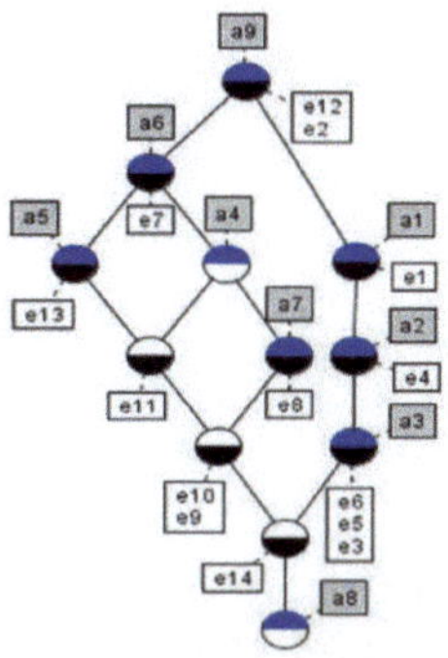

Fig.2. The CGLD of the compliance data in Table II

B. Compliance Core and Compliance Peripheral

There probably exist compliance cores and compliance peripherals in a CGLD. Here, a compliance core is defined as an element of the employee subset in which each employee has complied with all of the clauses, and each element in a clause subset adhered by all of the employees is also defined as a compliance core. A compliance peripheral is defined as an element of the employee subset in which each employee has adhered to not more than three clauses, and an element of the clause subset complied with by not more than three employees is also defined as a compliance peripheral. In Fig.1, for instances, the node with the lower label $e14$ indicates that the employee $e14$ is a compliance core; the node labeled with $a1$ and $e1$ indicates that $e1$ is a compliance peripheral. In Fig.2, the node labeled with $a9$ and $e1$, $e12$ indicates that the clause $a9$ is a compliance core; The compliance peripheral $e1$ is indicated by the node labeled with $a1$ and $e1$.

C. Compliance Subgroup

Two kinds of compliance subgroups may be found from the labels and the descending or ascending lines in a CGLD:

1) The first kind of subgroup is the employee subgroup, which refers to an employee subset complied with a specific clause subset. In Fig.1, for example, an employee subgroup $\{\boldsymbol{e3}, \boldsymbol{e4}, \boldsymbol{e5}, \boldsymbol{e6}, \boldsymbol{e14}\}$ is identified from the two linked nodes which are labeled with $\boldsymbol{a1}$ and $\boldsymbol{e1}$, and $\boldsymbol{a2}$ and $\boldsymbol{e4}$, respectively. This means that all of the employees belonging to the subset $\{\boldsymbol{e3}, \boldsymbol{e4}, \boldsymbol{e5}, \boldsymbol{e6}, \boldsymbol{e14}\}$ have adhered to the clause subset $\{\boldsymbol{a1}, \boldsymbol{a2}\}$.

2) The clause subgroup is the second kind of subgroup, which refers to a clause subset complied with by a specific employee subset. In Fig. 1, a clause subgroup $\{\boldsymbol{a4}, \boldsymbol{a6}\}$ is determined by three different nodes, namely, the node labeled with $\boldsymbol{e8}$, the one with $\boldsymbol{e9}$ and $\boldsymbol{e10}$, and the one with $\boldsymbol{e11}$. The subgroup $\{\boldsymbol{a4}, \boldsymbol{a6}\}$ means that all of the clauses belonging to $\{\boldsymbol{a4}, \boldsymbol{a6}\}$ have been complied with by the employees in the subset $\{\boldsymbol{e3}, \boldsymbol{e4}, \boldsymbol{e5}, \boldsymbol{e6}, \boldsymbol{e14}\}$.

TABLE III
THE ACTUAL TWO-MODE COMPLIANCE DATA

	*a*1	*a*2	*a*3	*a*4	*a*5	*a*6	*a*7	*a*8	*a*9	*a*10	*a*11	*a*12	*a*13	*a*14	*a*15	*a*16
*e*1	1	0	0	0	0	0	0	0	0	0	0	0	0	0	0	0
*e*2	0	0	0	0	0	0	0	0	0	0	0	0	0	0	0	0
*e*3	1	1	1	1	1	1	0	1	1	0	0	0	0	0	0	0
*e*4	0	0	0	1	0	0	0	0	0	0	0	0	0	0	1	0
*e*5	1	1	1	0	1	1	1	1	0	0	0	0	0	0	0	0
*e*6	0	0	0	0	0	0	0	1	1	1	0	1	1	1	1	1
*e*7	0	0	0	0	0	0	1	1	1	1	0	1	0	0	1	0
*e*8	1	1	1	1	1	1	1	1	1	0	0	0	0	0	0	0
*e*9	1	1	1	1	1	1	1	1	1	1	1	1	1	1	1	1
*e*10	1	0	1	1	1	1	1	1	0	0	0	0	0	0	0	0
*e*11	0	0	1	1	1	0	1	0	0	0	0	0	0	0	0	0
*e*12	0	0	1	0	1	1	0	1	0	0	0	0	0	0	0	0
*e*13	0	0	0	1	1	1	1	1	0	0	0	0	0	0	0	0
*e*14	0	0	0	0	0	0	0	0	0	0	0	0	0	0	0	0
*e*15	0	0	0	0	0	1	0	1	1	0	0	0	0	0	0	0
*e*16	0	0	0	0	1	0	1	1	0	0	0	0	0	0	0	0
*e*17	0	0	0	0	0	0	0	0	0	0	0	1	0	1	0	1
*e*18	0	0	0	0	0	0	1	1	1	0	0	0	0	0	0	0
*e*19	0	0	0	0	0	0	0	1	1	1	0	1	0	0	1	1
*e*20	0	0	0	0	0	1	1	0	1	1	1	1	1	1	0	0
*e*21	0	0	0	0	0	0	1	1	0	0	1	1	0	0	0	0
*e*22	0	0	0	0	0	0	0	0	0	0	0	0	0	1	0	1
*e*23	0	0	0	0	0	0	0	0	0	0	0	0	0	1	0	0

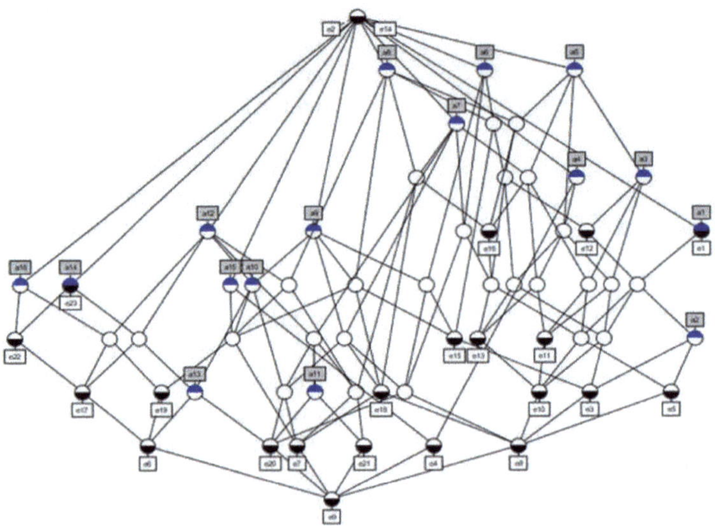

Fig.3. The CGLD of the compliance data in Table III

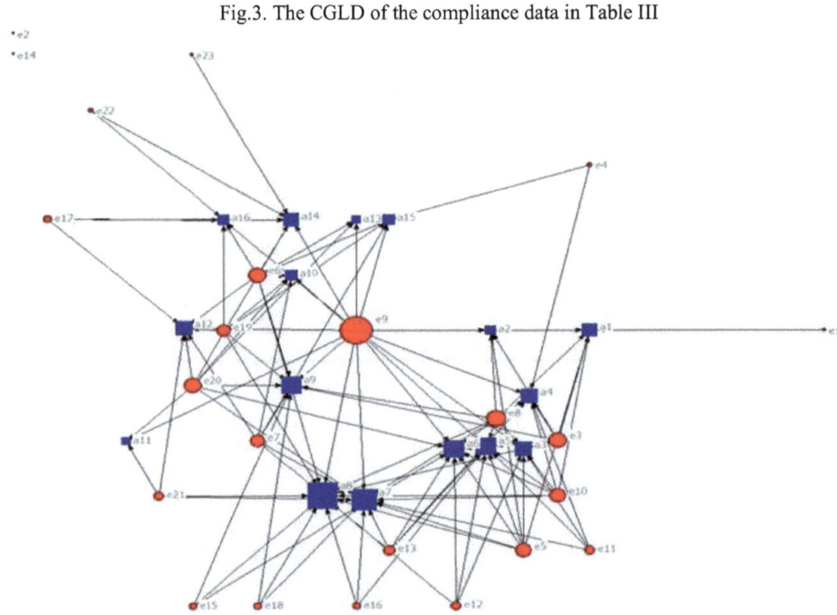

Fig.4. The network generated by the UCINET software

D. Compliance Partition

The compliance partition refers to the formation of subareas by the clustering of nodes in a CGLD. In Fig.1, two compliance partitions are formed without considering the top-most and the bottom-most nodes. The right-hand side compliance partition is formed by three different nodes: the node with labels $\boldsymbol{a1}$ and $\boldsymbol{e1}$, the one with labels $\boldsymbol{a2}$ and $\boldsymbol{e4}$, and the one with labels $\boldsymbol{a3}$ and $\boldsymbol{e3}$, $\boldsymbol{e5}$, $\boldsymbol{e6}$. The left-hand side partition is composed of the remaining nodes.

E. Multiple Compliance Containment

A CGLD can exhibit multiple containment structure with respect to both employees and clauses. In Fig.1, the following multiple compliance containment patterns exist: $\{e14\} \subseteq \{e9, e10, e14\} \subseteq \{e9, e10, e11, e14\} \subseteq \{e9, e10, e11, e13, e14\} \subseteq \{e7, e9, e10, e11, e13, e14\}$ and $\{a1, a2, a3, a4, a5, a6, a7\} \supseteq \{a4, a5, a6, a7\} \supseteq \{a4, a5, a6\} \supseteq \{a5, a6\} \supseteq \{a6\}$. The two compliance containments show that the employees belonging to each of the subsets $\{e14\}$, $\{e9, e10, e14\}$, $\{e9, e10, e11, e14\}$, $\{e9, e10, e11, e13, e14\}$, $\{e7, e9, e10, e11, e13, e14\}$ have complied in order with the clauses belonging to $\{a1, a2, a3, a4, a5, a6, a7\}$, $\{a4, a5, a6, a7\}$, $\{a4, a5, a6\}$, $\{a5, a6\}$, $\{a6\}$. It is also seen that the employee subsets get larger along the upward direction in the CGLD, while the clause subsets become smaller.

F. Compliance Equivalence

There probably exists structural equivalence [18] in a CGLD. In Fig. 1, the three employees $e3$, $e5$ and $e6$ are said to be structurally compliance equivalent for they have complied with the identical clause subset $\{a1, a2, a3\}$.

IV. COMPARATIVE ANALYSIS

In order to verify the availability of these compliance patterns derived from a CGLD, actual two-mode compliance data have been collected by means of survey questionnaires and interviews from a data service company in Shanghai. The compliance data of 23 internal employees with 16 information security policy clauses are presented in Table III. A CGLD generated from these data is shown in Fig.3. The six compliance patterns derived from this CGLD are compared with the corresponding structural features of the network generated by the UCINET software [27-30] given in Fig.4. The results of the comparative analysis are given as follows:

1) *Compliance outlier*: In Fig.3, the employees $e2$ and $e14$ are two employee outliers. They are also identified as outliers in Fig.4.

2) *Compliance core and compliance peripheral*: In Fig.3, $e9$ is a compliance core, and $e1$, $e4$, $e22$ and $e23$ are compliance peripherals. By calculating the degree centrality of each of the nodes in Fig.4, it is found that the values of the degree centralities of employees $e9$ and $e8$ and clauses $a8$ and $a7$ are higher, and the values for the elements belonging to $\{e1, e23, e22, e4\}$ and $\{a13, a11, a2\}$ are lower. That is, compliance cores and peripherals observed in Fig. 3 also appear in Fig.4.

3) *Compliance subgroup*: As for the employee subgroups, in Fig. 3 for instance, it is found that the two employee subgroups $\{e6, e17\}$ and $\{e8, e10, e11\}$ have complied with $\{a12, a14, a16\}$ and $\{a3, a4, a5, a7\}$, respectively; as for the clause subgroups, the two clause subgroups $\{a9, a10, a12\}$ and $\{a1, a3, a4\}$ have been complied with by $\{e6, e7, e9, e19, e20\}$ and $\{e3, e8, e9, e10\}$, respectively. Similar subgroup features are also seen in Fig.4.

4) *Compliance partition*: Two large employee partitions can be determined in Fig.3, namely, $\{e6, e7, e17, e18, e19, e20, e21, e22, e23\}$ and $\{e1, e3, e4, e5, e8, e10, e11, e12, e13, e15, e16\}$ Similarly, $\{a9, a10, a11, a12, a13, a14, a15, a16\}$ and $\{a1, a2, a3, a4, a5, a6, a7, a8\}$ are two large clause partitions. The similar structural partition feature can be seen in Fig.4.

5) *Multiple compliance containment*: $\{e9, e20\} \subseteq \{e6, e9, e20\} \subseteq \{e6, e7, e9, e17, e19, e20, e21\}$ and $\{a6, a7, a9, a10, a11, a12, a13, a14\} \supseteq \{a9, a10, a12, a13, a14\} \supseteq \{a12\}$ are found to be two multiple containments in Fig.3. In Fig.4, the three areas surrounded respectively by $\{e9, e20\}$, $\{e6, e9, e20\}$ and $\{e6, e7, e9, e17, e19, e20, e21\}$ are gradually expanded, the other three areas determined by $\{a6, a7, a9, a10, a11, a12, a13, a14\}$, $\{a9, a10, a12, a13, a14\}$, and $\{a12\}$ being gradually shrunk. This can be regarded as a kind of structural multiple containment, which is similar with the multiple containment pattern shown in Fig.3.

6) *Compliance equivalence*: In Fig.4, the calculation results of the Archimedes distances among $e1$, $e2$, $e14$ and $e23$ with the UCINET software indicate that the four employees are approximately structurally equivalent. Fig.3 shows that these employees have the similar structural compliance equivalence.

V. CONCLUSION

Compliance Galois lattice diagram has been proposed for a visual representation and analysis of employees' compliance with the information security policy of an organization. Six specific compliance patterns have been derived from the compliance Galois lattice diagram generated from the compliance data. These patterns are compared with the structural features identified from the network generated by the UCINET software using the same data. Fairly good consistency has been reached between the compliance patterns and the structural features. It should be mentioned that the compliance Galois lattice diagram gets very dense and complicated when the numbers of employees and clauses are very big, and it is hard to recognize explicitly the compliance patterns. For such a complex system, improved methods should be developed to represent and analyze the compliance Galois lattice diagram, and to obtain the compliance patterns.

REFERENCES

[1] W. Baker, M. Goudie, A. Hutton, C. D. Hylender, J. Niemantsverdriet, and C. Novak, et al., "Version 2010 data breach investigations report," 2010.

[2] R. Richardson, "2010/2011 CSI computer crime & security survey," 2011.

[3] R. E. Crossler, A. C. Johnston, P. B. Lowry, Q. Hu, M. Warkentin, and R. Baskerville, "Future directions for behavioral information security research," *Comput. Secur.* vol. 32, pp. 90–101, Feb. 2013.

[4] R. Willison and M. Warkentin, "Beyond deterrence: An expanded view of employee computer abuse," *Manage. Inf. Syst. Q.*, vol. 37, no. 1, pp. 1–20, Mar. 2013.

[5] R. Willison and M. Siponen, "Overcoming the insider: Reducing employee computer crime through situational crime prevention," *Commun. ACM*, vol. 52, no. 9, pp. 133–137, Sep. 2009.

[6] M. Warkentin and R. Willison, "Behavioral and policy issues in information systems security: The insider threat," *Eur. J. Inf. Syst.*, vol. 18, no. 2, pp. 101–105, Apr. 2009.

[7] M. Warkentin, D. Straub, and K. Malimage, "Featured talk: Measuring secure behavior: A research commentary," in *Proc. Annu. Symp. Information Assurance & Secure Knowledge Management, ASIA & SKM'12*, Albany, NY, pp. 1–8.

[8] K. H. Guo, "Security-related behavior in using information systems in the workplace: A review and synthesis," *Comput. Secur.*, vol. 32, pp. 242–251, Feb. 2013.

[9] A. C. Johnston and M. Warkentin, "Fear appeals and information security behaviors: An empirical study," *Manage. Inf. Syst. Q.*, vol. 34, no. 3, pp. 549–566, Sep. 2010.

[10] B. Bulgurcu, H. Cavusoglu, and I. Benbasat, "Information security policy compliance: An empirical study of rationality-based beliefs and information security awareness," *Manage. Inf. Syst. Q.*, vol. 34, no. 3, pp. 523–548, Sep. 2010.
[11] E. Chew, M. Swanson, K. Stine, N. Bartol, A. Brown, and W. Robinson, "Performance measurement guide for information security," *NIST Special Publication 800-55 Revision 1*, Jul. 2008.
[12] R. Barabanov, S. Kowalski, and L. Yngström, "Information security metrics: State of the art," *DSV Report Series No.11–007*, Mar. 2011.
[13] A. Davis, B. B. Gardner, M. R. Gardner, and W. L. Warner *Deep South: A Social Anthropological Study of Caste and Class*. Chicago: University of Chicago Press, 1941.
[14] R. L. Breiger, "The duality of persons and groups," *Soc. Forc.*, vol. 53, no. 2, pp. 181–190, Dec. 1974.
[15] S. Wasserman, and K. Faust, *Social Network Analysis: Methods and Applications*, Cambridge: Cambridge University Press, Nov. 1994.
[16] S. P. Borgatti and M. G. Everett, "Network analysis of 2-mode data," *Soc. Netw.*, vol. 19, no. 3, pp. 243–269, Aug. 1997.
[17] L. C. Freeman, "Finding social groups: A meta-analysis of the southern women data," in *Dynamic Social Network Modeling and Analysis*, R. Breiger, K. Carley, P. Pattison, Ed. Washington, D.C.: The National Academies Press, 2003.
[18] S. P. Borgatti, "2-Mode concepts in social network analysis," in *Encyclopedia of Complexity and System Science*, R. A. Meyers, Ed. Springer Link, 2009, pp. 8279–8291.
[19] P. J. Carrington, J. Scott, and S. Wasserman, *Models and Methods in Social Network Analysis*. Cambridge: Cambridge University Press, Feb. 2005.
[20] G. Birkhoff, "Lattice theory," *Amer. Math. Soc. Colloq. Publ.*, vol. 25, American Mathematical Society, New York, 1940.
[21] M. Barbut and B. Monjardet, "Order et classification: Algébre et combinatoire," Hachette Université, Paris, 1970.
[22] R. Wille, "Restructuring lattice theory: An approach based on hierarchies of concepts," in *Ordered Sets, NATO Advanced Study Institutes Series*, Vol. 83, I. Rival, Ed. Springer Netherlands, 1982, pp. 445–470.
[23] R. Wille, "Line diagrams of hierarchical concept systems," *Int. Classif.*, vol. 11, pp. 77–86, 1984.
[24] V. Duquenne, "Contextual implication between attributes and some representation properties for finite lattices," in *Proc. 11th Conf. Formal Concept Analysis, ICFCA'13*, Dresden, Germany, pp. 1–27.
[25] V. Duquenne, "The core of finite lattices," *Discr. Math.*, vol. 88, no. 2–3, pp.133–147, Apr. 1991.
[26] SOURCEFORGE, Concept explorer, *Http://sourceforge.net/projects/conexp*, 2013.
[27] L. C. Freeman and D. R. White, "Using Galois lattices to represent network data," in *Sociological methodology*, P. V. Marsden Ed. Oxford: Blackwell, 1993, pp. 127–146.
[28] S. P. Borgatti, M. G. Everett, and L. C. Freeman, *UCINET for Window version 6*, *Http://www.analytictech.com/ucinet*, 2013.
[29] B. A. Davey and H. A. Priestley, *Introduction to Lattices and Order*. Cambridge: Cambridge University Press, 2002.
[30] L. C. Freeman, "Cliques, Galois lattices, and the structure of human social group", *Soc. Netw.*, vol. 18, no. 3, pp. 173–187, Aug. 1996.

Empirical Research of Costs Stickiness Behavior in Chinese Manufacturing Listed Firms

Lu-ying Xu, Fang-jun Wang*, Yi-han Hu
Management School, Xi'an Jiaotong University, Xi'an, China
(wangfangjun@mail.xjtu.edu.cn)

Abstract - **This paper examines the existence of cost stickiness in total costs, operating costs, selling expenses and general & administrative expenses by using our modified model based on the ABJ model. With the analysis of accounting data during 2008 to 2011 from 1192 Chinese A-share listed companies, we find: (1) Total costs, operating costs and general & administrative expenses are sticky; (2) Selling expenses are anti-sticky; (3) The cost stickiness would converse in the subsequent period, but whether they would fall in the long time window is unclear.**

Keywords - **Cost behavior, cost stickiness, manufacturing listed companies**

I. INTRODUCTION

Traditional cost behaviour models in accounting literature distinguish between fixed and variable costs with respect to changes in sales volume, indicating that variable costs respond the same to upward and downward changes in sales activity. Recent research documents that many costs are "sticky"—they decline less in response to sales decrease than they rise for equivalent sales increase. These findings are inconsistent with the traditional models, and suggest an alternative theory of cost behavior that is based on deliberate managerial decisions.

Up to now, there has been a rich literature on costs stickiness, the measurements mainly include the ABJ model proposed by Anderson, Banker and Janakiraman (2003), DW model proposed by Dan Weiss (2010) and BCM model proposed by Banker, Byzalov, Ciftci and Mashruvala (2012), among which the ABJ model is the most widely used one. ABJ model with ratio and log forms alleviates heteroscedasticity, enhances the comparability between companies, and also offers a comparative good explanation for the coefficients of the regression model. However, it fails to reveal the complete relationship between costs and sales volume for it only describes the changes of slope without considering the changes in intercept. In order to fill the research gap, this paper modifies the ABJ model, and takes the manufacturing industry as an example to examine the existence of cost stickiness.

II. THEORY AND HYPOTHESIS

Traditional model of fixed and variable costs envisions a mechanistic symmetric relation between costs and sales. However, Banker and Johnston (1993) find that selling expenses respond differently to upward and downward changes in sales activity in US airline industry. Noreen and Soderstrom (1997) also offer the same evidences in the relation between general & administrative expenses and sales revenue in hospitals. Until 2003, Anderson, Banker and Janakiraman introduce the concept of expenses stickiness, they find that selling expenses and general & administrative expenses behave asymmetrically, falling less in response to sales decrease than they rise for the equivalent sales increase. Subramanian and Weidenmier (2003) add operating costs, together with selling expenses and general & administrative expenses, to the total costs, and their results indicate all the total costs and its components are sticky. Subsequent research has demonstrated that cost stickiness is pervasive across different cost categories, datasets and countries, and has explored the implications of sticky costs for both financial and cost accounting (Sun Zheng and Liu Hao, 2004; Banker and Chen, 2006; Calleja et al., 2006; Kong Yusheng et al., 2007; Weiss, 2010; Chen et al., 2012; Dierynck et at., 2012; Kama and Weiss, 2013).

Generally speaking, when sales revenue decrease, managers need to predict whether the sales revenue will increase in the future. Even it is necessary to cut down the costs, it will also take time to fulfill the process. In addition, according to the "agency theory", managers are reluctant to reduce the resources of their own. Collecting these arguments, we have:

H1: Total costs are sticky.

Operating costs are mainly consist of direct material, direct labor and manufacturing expenses. When there is a declare in sales revenue, most components, especially expenses related to assets and labor, are hard to reduce timely. We have:

H1a: Operating costs are sticky.

With the decrease in sales revenue, even selling expenses are easier to be adjusted comparing to operating costs, managers may still regard it as a temporary phenomenon and are not willing to lose control for the resources. For the reasons above, we have:

H1b: Selling expenses are sticky.

General & administrative expenses refer to all the expenses that occur in operating activities for administration department and are greatly influenced by managers' willing. As is stated above, the more

E. Qi et al. (eds.), *Proceedings of the 5th International Asia Conference on Industrial Engineering and Management Innovation (IEMI2014)*, Proceedings of the International Asia Conference on Industrial Engineering and Management Innovation 1, DOI 10.2991/978-94-6239-100-0_66, © Atlantis Press and the authors 2015

resources managers own, the more interests they will get, which may lead to expenses on the high-level even sales revenue decrease. We have:

H1c: General and administrative expenses are sticky.

III. METHOD AND EMPIRICAL MODELS

A. Sample Selection

In this paper, we use annual CSMAR data of all the listed manufacturing firms in the A-shares from Shanghai and Shenzhen Stock Exchange during the period of 2008 to 2011. The total initial sample size is 4494, and we exclude samples by the following principles : (1) firms with incomplete data during 2008-2011; (2) firms with industry change (no longer belongs to manufacturing industry) during 2008-2011. (3) firms with singular value (operating revenue is 0). Only 3847 samples of 1192 firms left for the single-period model analyses. To make the conclusion more convincing, we also make the robustness checks by the single-period with one-year lag model and the two-period model. The final sample in these two models consists of 2409 observations for 803 firms. In extension analyses, we examine the components of total costs (TC), including operating costs (OC), selling expenses (SC) and general & administrative expenses (GC).

B. Empirical Models

Traditional standard methodology examines cost stickiness by the following ABJ model:

$$In\frac{Cost_{i,t}}{Cost_{i,t-1}} = \beta_0 + \beta_1 \times In\frac{Revenue_{i,t}}{Revenue_{i,t-1}} + \beta_2 \times D_{i,t} \times In\frac{Revenue_{i,t}}{Revenue_{i,t-1}} + \varepsilon_{i,t} \quad (1)$$

Because the ABJ model only describes the slope change of a straight line between the cost and sales without considering the changes in intercept, this study proposes a modified measure of cost stickiness. We refine the ABJ model by using the following single-period model with a dummy item:

(Model A)

$$In\frac{Cost_{i,t}}{Cost_{i,t-1}} = \beta_0 + \beta_1 \times In\frac{Revenue_{i,t}}{Revenue_{i,t-1}} + \beta_2 \times D_{i,t} \times In\frac{Revenue_{i,t}}{Revenue_{i,t-1}} + \beta_3 \times D_{i,t} + \varepsilon_{i,t} \quad (2)$$

We also define a single-period with one-year lag model (Model B) and two-period model (Model C) to enhance the robustness of results.

(Model B)

$$In\frac{Cost_{i,t}}{Cost_{i,t-1}} = \beta_0 + \beta_1 \times In\frac{Revenue_{i,t}}{Revenue_{i,t-1}} + \beta_2 \times D_{i,t} \times In\frac{Revenue_{i,t}}{Revenue_{i,t-1}} + \beta_3 \times D_{i,t} + \beta_4 In\frac{Revenue_{i,t-1}}{Revenue_{i,t-2}} + \beta_5 \times D_{i,t-1} \times In\frac{Revenue_{i,t-1}}{Revenue_{i,t-2}} + \beta_6 \times D_{i,t-1} + \varepsilon_{i,t} \quad (3)$$

(Model C)

$$In\frac{Cost_{i,t}}{Cost_{i,t-2}} = \beta_0 + \beta_1 \times In\frac{Revenue_{i,t}}{Revenue_{i,t-2}} + \beta_2 \times D_{i,t} \times In\frac{Revenue_{i,t}}{Revenue_{i,t-2}} + \beta_3 \times D_{i,t} + \varepsilon_{i,t} \quad (4)$$

where:

$In\frac{Cost_{i,t}}{Cost_{i,t-j}}$ = log form in a certain cost of firm i in year t relative to year t-j, including the log form in total costs (TC), operating costs (OC), selling expenses (SC) and General & administrative expenses (GC) of firm i in year t relative to year t-j;

$In\frac{Revenue_{i,t}}{Revenue_{i,t-j}}$ = log form in sales revenue of firm i in year t relative to year t-j;

$D_{i,t}$ = dummy variable, 1 if sales revenue of firm i decreased in year t relative to year t-1, zero otherwise.

IV. EMPIRICAL RESULTS

The descriptive statistics of single-period model are presented in Table I. The average ratio of sales revenue is 1.1758 with its minimum value 0.0003 and the maximum value 27.1798, indicating an increasing trend in operating income during 2008-2011. Among the four types of costs in our model, the average ratio of general & administrative expenses (GC) is the highest with the value of 1.2149, suggesting this particular expense grows faster than the other three. While the ratio of selling expenses (SC) has the greatest range between the maximum and minimum, as well as the the highest standard deviation of 0.9294, which means firms have different attitudes in terms of sales activities.

TABLE I
DESCRIPTIIVE STATISTICS

	Mean	Maximum	Minimum
$\frac{Revenue_{i,t}}{Revenue_{i,t-1}}$	1.1758	27.1798	0.0003
$TC_{i,t}/TC_{i,t-1}$	1.1712	8.0513	0.0275
$OC_{i,t}/OC_{i,t-1}$	1.1744	9.7146	0.0204
$SC_{i,t}/SC_{i,t-1}$	1.1955	48.6337	0.0269
$GC_{i,t}/GC_{i,t-1}$	1.2149	8.9060	0.1006

The estimates for our main single-period model (Model A) are presented in Table II. According to the model, when sales revenue increase 1%, costs increase β_1%; when sales revenue decrease 1%, costs decrease ($\beta_1+\beta_2$)%. Based on the theory of cost stickiness, if a cost is sticky, then $\beta_1>\beta_1+\beta_2$, which means $\beta_2<0$, and the smaller coefficient β_2, the stickier costs will be. As expected, β_2 is negative in the interaction item of total costs, operating costs and general & administrative expenses, but it turns out to be positive in selling expenses. These evidences indicate the existence of cost stickiness in total costs, operating costs and general & administrative expenses, consistent with prior studies and our H1, H1a, H1c respectively as well. As for selling expenses, it reveals the pattern of significant anti-stickiness (costs decline more in response to sales decrease than they increase for equivalent sales increase) in the case of a prior sales decrease, so our H1b is not supported.

TABLE II
ESTIMATES FOR THE SINGLE-PERIOD MODEL

		Coefficient	Sig.	R^2_{adj}	F
	β_0	0.054**	0.000		
TC	β_1	0.661**	0.000	0.692	2884**
	β_2	-0.318**	0.000		
	β_3	-0.122**	0.000		
	β_0	0.046**	0.000		
OC	β_1	0.719**	0.000	0.715	3216**
	β_2	-0.312**	0.000		
	β_3	-0.118**	0.000		
	β_0	0.073**	0.000		
SC	β_1	0.509**	0.000	0.302	555**
	β_2	0.164**	0.000		
	β_3	-0.030*	0.042		
	β_0	0.133**	0.000		
GC	β_1	0.256**	0.000	0.112	163**
	β_2	-0.137**	0.000		
	β_3	-0.054**	0.000		

**significant at 1 percent level
* significant at 5 percent level

More specifically, when sales revenue increase, $\frac{\Delta Cost_{i,t}}{Cost_{i,t-1}} \approx e^{\beta_0} \times (1+\beta_1 \times \frac{\Delta Revenue_{i,t}}{Revenue_{i,t-1}})-1$; when sales decrease, $\frac{\Delta Cost_{i,t}}{Cost_{i,t-1}} \approx e^{\beta_0+\beta_3} \times [1+(\beta_1+\beta_2) \times \frac{\Delta Revenue_{i,t}}{Revenue_{i,t-1}}]-1$, if the costs are sticky, then,

$$e^{\beta_0} \times (1+\beta_1 \times \frac{\Delta Revenue_{i,t}}{Revenue_{i,t-1}})-1 >$$

$$-\{e^{\beta_0+\beta_3} \times [1+(\beta_1+\beta_2) \times \frac{\Delta Revenue_{i,t}}{Revenue_{i,t-1}}]-1\} \cdot$$

The result indicates that not until $\frac{\Delta Revenue_{i,t}}{Revenue_{i,t-1}} > 1.01\%$ can the total costs be sticky, so are the operating costs (>1.98%) and general & administrative expenses (>53.29%). As is stated above, selling expenses are anti-sticky, and the $\frac{\Delta Revenue_{i,t}}{Revenue_{i,t-1}}$ for selling expense turns out to be greater than -9.58%, indicating that selling expenses are anti-sticky at any small change in sales revenue.

In extension analysis, we estimate single-period with one-year lag model (Model B) and two-period model (Model C) for TC and the its main components (OC, SC and GC) to check whether the results are robust across samples. The estimates are presented in Table III. By the two new models, we can also check whether the cost stickiness of four types would converse in the subsequent period, and when extending the accounting period, how would the costs behave in the long run. Both of the new models show that the results are robust. In Model B, for most of these cost categories (except SC), the negative β_2 is changed into positive β_5 and is significant at 1 percent level, indicating the property of cost stickiness converses in the next period. As for SC, even its negative β_2 is not significant, the signal of β_5 is also inversed into positive at 5 percent level. Generally speaking, the estimates support our hypothesis. According to the Model C, the larger differences between β_1 and $\beta_1+\beta_2$, the stickier costs will be. Comparing the absolute value of β_2 in Model C with Model A, GC turns to be stickier than before, while TC and OC are the opposite, and the coefficient of SC is no longer significant. In all, whether the costs become stickier with extend period is up to the condition.

V. CONCLUSION

In this paper, we modified and examined the standard measure of sticky costs by an additional dummy item in manufacturing industry. Based on ABJ's intuition of deliberate managerial decisions in the presence of resource adjustment costs, our model also indicated that there is a complex pattern of cost asymmetry which combines two distinct processes : cost stickiness and cost anti-stickiness both conditional on a prior sales decrease. The empirical results suggested that the total costs, operating costs and general & administrative expenses are sticky, while the selling expenses are anti-sticky. We also proposed two new models—single-period with one-year lag model and two-period model— to check whether the results are robust across samples. Empirical estimates showed that the results were robust, and the expense stickiness would converse in the subsequent period, but whether they would fall in the long time window is unclear based on our sample size.

Our main findings are consistent with prior

studies and our predictions except that the selling expenses are anti-sticky. The results may be mainly influenced by the property of selling expenses as well as the financial crisis in 2008 and the economic bubble before. Selling expenses are directly related to sales volume, which may change with sales income automatically, indicating that the selling expenses are not sticky. In addition, managers' pessimism under the financial crisis may be reflected on the decrease in selling expenses. Before the financial crisis, most firms have high demand for investment, resulting in a large number of idle resources. When there is decrease in sales revenue, selling expenses are more likely to be reduced.

Our new model aims to conduct a more accurate measure of cost behavior not only for cost accounting but also for financial accounting topics such as earnings forecasts and earnings management. Further research is expected to build on this approach in exploring multi disciplinary accountings topics. Integrating management and financial accounting research is likely to benefit both disciplines.

ACKNOWLEDGMENT

This work was supported in part by a grant from the National Social Science Foundation of China (11CGL062) and the Fundamental Research Funds for the Central Universities of China (2012jdhz17).

REFERENCE

[1] Anderson, M. C., Banker, R. D., and Janakiraman, S. N., "Are selling, general, and administrative costs "sticky"?", Journal of Accounting Research, 1, 41, 2003, pp. 47-63.

[2] Balakrishnan, R., M. J. Peterson, and N. Soderstrom, "Does capacity utilization affect the‘stickiness’ of costs?", Journal of Accounting, Auditing and Finance, 19, 2004, pp. 283-299.

[3] Banker, R, D., Byzalov, D., and Ciftci, M., et al. "The moderating effect of prior sales changes on asymmetric cost behavior", Working paper, 2012.

[4] Banker, R, D and L. Chen, "Predicting earnings using a model based on cost variability and cost stickiness", The Accounting Review, 81, 2006, pp.285-307.

[5] Banker, R. D. and Johnston, H. H., "An empirical study of cost drivers in the US airline industry", Accounting Review, 1993, pp.576-601.

[6] Calleja, K., Steliaros, M., and Thomas, D. C., "A note on cost stickiness: Some international comparisons", Management Accounting Research, 2, 17, 2006, pp. 127-140.

[7] Chen, C. X., H. Lu, and T. Sougiannis., "The agency problem, corporate governance, and the asymmetrical behavior of Selling, General, and Administrative costs", Contemporary Accounting Research, 29, 2012, pp. 252-282.

[8] Dierynck, B., W. R. Landsman, and A. Renders. , "Do managerial incentives drive cost behavior? Evidence about the role of the zero earnings benchmark for labor cost behavior in Belgian private firms", The Accounting Review, 4, 87, 2012, pp.1219-1246.

[9] Kama, I. and D. Weiss, "Do earnings targets and managerial incentives affect sticky costs?", Journal of Accounting Research, 1, 51, 2013, pp. 201–224.

[10] Kong Yusheng, Zhu Naiping and Kong Qinggen, "Study on Cost Stickiness"(in Chinese), Accounting Research, 11, 2007, pp.58-65.

[12] Noreen. E, and Soderstrom, N., "The accuracy of proportional cost models: evidence from hospital service departments", Review of Accounting Studies, 1, 2, 1997, pp. 89-114.

[13] Sun Zheng and Liu Hao, "The Expense Stickiness Behavior of Chinese Listed Companies"(in Chinese), Economic Research Journal, 12, 2004, pp.26-34.

[14] Weiss,D. , "Cost behavior and analysts' earnings forecasts", The Accounting Review, 4, 85, 2010, pp.1441-1471.

TABLE III
ROBUSTNESS TEST

	TC		OC		SC		GC	
	Model B	Model C	Model B	Model C	Model B	Model C	Model B	Model C
β_0	0.021**	0.042**	0.012**	0.025**	0.024*	0.050*	0.090**	0.191*
β_1	0.791**	0.819**	0.852**	0.877**	0.718**	0.739*	0.381**	0.435*
β_2	-0.229**	-0.146*	-0.190**	-0.092*	-0.048	-0.003	-0.308**	-0.296*
β_3	-0.042**	-0.040*	-0.030**	-0.016*	0.017	0.023*	-0.025	-0.053*
β_4	0.066**	-	0.062**	-	0.019	-	0.093**	-
β_5	0.180**	-	0.188**	-	0.113*	-	0.180**	-
β_6	0.018*	-	0.019**	-	0.003	-	-0.007	-
R^2_{adj}	0.861	0.872	0.877	0.894	0.433	0.415	0.262	0.283
F	2478.523**	5445.918**	2873.179**	6736.990**	307.926**	569.862**	143.646**	318.510**
Simple Size				2409				

Model B: Single-period with one-year -lag Model; Model C: Two-period Model
**significant at 1 percent level; * significant at 5 percent level

Game Model for the Allocation of the Cooperative Surplus in the Public-Private Partnership

Hui An*, Han-lu Zheng
College of Hydraulic & Environmental Engineering, China Three Gorges University, Yichang Hubei, CHINA
(17647800@qq.com)

Abstract - **The incompleteness of PPP contracts provides opportunities for the public and private partners to cooperate again in the consultative phase, and then both sides in the contract can get the cooperative surplus from the cooperation on the unclear defined matters by the initial contract. For the interest allocation of this part, an incomplete information game model is formed on this basis, the analysis of the factors in the allocation, it comes out that the type of the player and the patience have the more significant impact on the final result.**

Keywords - **Cooperative surplus, Game Model, Incomplete contract, PPP projects**

I. INTRODUCTION

PPP means a long-term agreement which signed between the government and private organizations, which form partnerships with each other. They invest resources jointly in construction, operate or manage the public infrastructure or provide public services and then share responsibilities and benefits. The success of PPP projects not only depends on the amount of the cooperative surplus, but also depends on the proportional allocation between the two sides. Effective coordination of the distribution is essential to achieving a win-win between the partners. At present, the PPP studies tend to focus on risk-sharing and management mechanism [1, 2]. Although the research on benefit distribution in cooperation is gradually increasing, a lot of them are for the model study on the stage of signing the concession agreement. Because of the incompleteness of the PPP contract, both sides can get the cooperative surplus through joint decision-making in the cooperative negotiation phase after the initial contracts were signed, there are few researches on this part of benefits distribution has few researches [3, 4]. This paper analyzed the factors which affect the distribution of benefits of PPP projects, established an incomplete information dynamic game model to study on the distribution of the cooperative surplus to enquire into the factors that affect the allocation results and get a Nash equilibrium solution.

II. THE COOPERATIVE SURPLUS AND ANALYSIS OF ITS INFLUENCING FACTORS

A. GHM Theory and the Incompleteness of PPP Project Contract

GHM theory points out that the contract is not complete. In addition to the specific rights prescribed in the initial contract, the remaining rights which can't be prescribed in advance are called "residual control". From the perspective of contracts, the essence of PPP is the connections of a series of contracts among the public sectors, private parties and all other participants [5]. Compared with the traditional cooperation between enterprises, PPP has three main different characteristics such as partners' different properties, products' high degree of public and long terms of cooperation .They combined with the information asymmetry endogenous from the project itself and exteriority from participants mix together, and make the PPP always has some rights that are not appointed in the contract or some items that are not listed. And then the incompleteness of contract could have created the conditions of cooperation for the later stage.

B. Negotiation and Cooperation

According to Grossman's assumptions about the PPP model [6], the public (G) and a private sector (P) jointly work on a project, and it has various stages as follows in Fig.1.

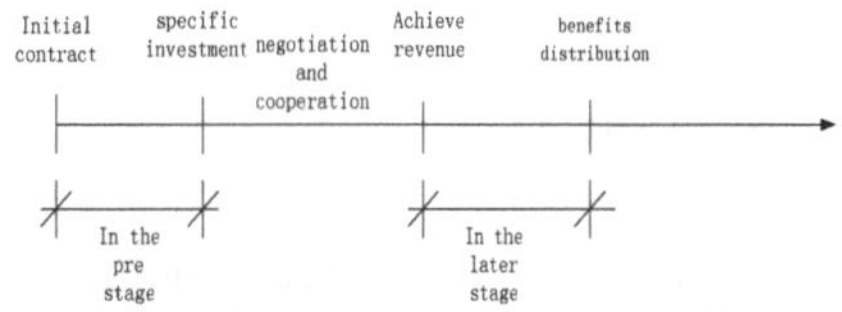

Fig.1. The stages of PPP project model

In the pre stage, after the cooperation relations are affirmed, the special investment decisions will be made on the basis of the respective control configuration in accordance with the initial contract. Once the assets form, its value will have a serious dependency on the cooperation between the two sides, and may thus result in opportunistic behavior problems, eventually leading to both sides' dread of being "ripped off" and making the initial specific investment insufficient, so important special assets owners should have its corresponding control, and the residual claim and the control rights should match [7]. In the initial allocation of the project control rights, it is usually distributed by a certain percentage.

After the investment occurs, because of the incompleteness of PPP project contract, if there are afterwards earnings which are observable but are unable to calculate or confirmed in advance, the two sides will need to make decisions on whether they should jointly maintain and operate the project .Benefit produced by cooperative decisions proceeds that only by the initial control configuration.

In benefits distribution phase, deterministic benefit is distributed according to the regulations of the initial contract, while the part surplus that comes from

E. Qi et al. (eds.), *Proceedings of the 5th International Asia Conference on Industrial Engineering and Management Innovation (IEMI2014)*, Proceedings of the International Asia Conference on Industrial Engineering and Management Innovation 1, DOI 10.2991/978-94-6239-100-0_67, © Atlantis Press and the authors 2015

cooperative decision will be done by negotiation [8, 9]. In fact, this part of the cooperation surplus is not necessarily evenly distributed, but by negotiation game to reach a final plan. For both sides, no matter from time or from the finial actual income, a good negotiation is a better choice than a collapse of negotiation which will have to resort to a third party to rule.

C. The Factors Influencing Cooperative Profit Distribution

While there are many factors influencing the PPP profit distribution in benefit negotiations [10], the main influence factors include information difference, risk sharing, capital investment, the degree of patience and consideration of other comprehensive factors, etc.

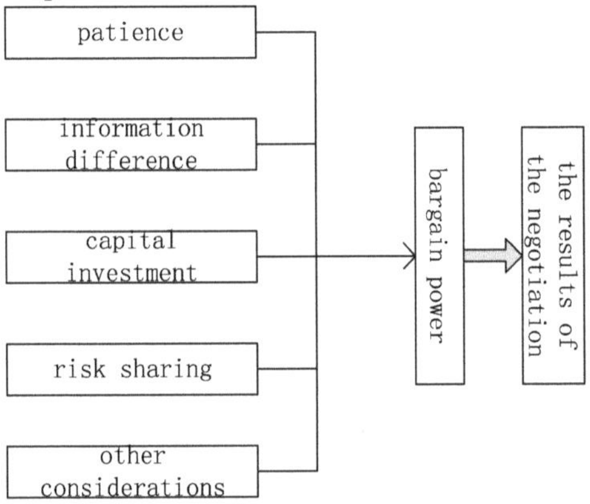

Fig.2. Factors affecting the distribution of income

(1) The information difference. The real cause of renegotiation phenomenon is often not the disagreement about the allocation or participants' greed, but one intentionally concealing information and resulting in information asymmetry. Participant who has information superiority, makes his own status higher and bargain power stronger than that of the other one. In PPP projects, different sources of information let both partners' innate advantages or acquired accumulation of favorable conditions get more information superiority, and then occupy a good position in the negotiations. In the allocation, this kind of difference in access to resources and information, typically causes the strong partner in the negotiation with its higher status to ask for the partial interest points obtained from the other one. In practice, the public sector doesn't always occupy the strong position, for some reason, such as time limits, or unknown of the private's status.

(2) Risk-sharing. "Risk-sharing, benefit-sharing" principle in PPP is actually a way by shifting some of the risks to the one who can operate manage it at a lower cost, thus reducing the risk cost borne by the traditional one, which greatly reduces overall project risk and operational cost, and even increase overall earnings. According to the risk characteristics and both sides' willingness, first they take the some part of risk by their own voluntary, and then the rest remaining risk will be distributed by their agreement. As an incentive, the risk that they take and rewards they get should match. In accordance with their respective quantity of risk and the synthesis important level determined by ISM method, risk sharing is another important factor in the distribution of benefits.

(3) Capital investment. PPP joint decision-making project investment includes not only capital but also non-capital investment incurred, such as patents, redeployment of resources and manpower and so on. The major aspects of capital impacts on surplus distribution are the amount, the degree of specificity and the structure. As capital incurred increases, investors expect to get paid more from the project, the mount of investment and the allocations of expected income is proportional. Normally when one's investment is specific, there is a possible that he will be threatened and exploited which will make his negotiations at a disadvantage position. However PPP project is based on partnership and trust mechanisms, both sides are willing to make more investments to achieve the best goals of the project. For PPP projects, financing capital structure includes debt capital and equity capital. Debt capital usually comes from private sectors, and it should be deducted first during cooperative surplus distribution, and the rest will be allocated in accordance with their respective equity capital ratio.

(4) Degree of patience. The patience level in negotiation between the two partners slightly differs from the general sense of meaning. It also contains a delay meant. For the PPP participants' negotiation costs, time delays, loss of interest and other factors may cause them miss the opportunity to invest in other projects. The opportunity costs are generally higher than in the public sector, thus they are significantly different in levels of patience. Weiying Zhang refers to the degree of participants' patience in game theory as the discount factor to represent the degree of participants patience. The bigger the number is, the higher the degree of the patience is. In the long negotiations, patience has a quite significant impact on the final result of the distribution.

(5) Other considerations. Other factors that participants consider in the assignments, such as intensity of effort, evaluations of their contribution to the collaboration, changes in the external negotiation environment, strategic partnering plan and so on, will affect those involved in policy and negotiating capacity.

III. ALLOCATION MODEL OF COOPERATIVE SURPLUS IN PPP

A. Description of Allocation of Cooperative Surplus by Game Theory

Defining participants in PPP from the stakeholder perspective, the core of stakeholders include the public and private partners. So assume that there are only two participants in the cooperative surplus distribution, and they are rational people, so they do not want negotiations broke down. Due to the surplus cannot be determined in advance, and it can only be allocated for negotiations afterwards, which is significantly different from the certainty benefit distribution. Negotiating distribution plan first put forward by the public sector, private sector choose to accept or reject. If the later accepts, the negotiations conclude; if rejects, the private sector propose another distribution plan and then the public sector chooses, and alternating negotiations proceed accordingly. During the course of game, both sides are not willing to limit the negotiation times and thus lose the advantage, nor appoint the compulsory provisions for ending the negotiation, so the game can be seen as infinite stages of bargaining game.

B. Assumptions in the Model

(1) The information between the two sides is symmetrical, and the judgment about the information type of each other is not entirely accurate. So it is an incomplete information game. The public sector coerces the private sector into transferring hi of its acquired shares with probability q; with 1-q probability does not take the strong position to obtain that additional benefits. As it is impossible for the public sector to claim more than the total share, so 0<ki+hi<1.

(2) Patience of the public sector is better than that of the private sector, so the discount factor δg is greater than δp, namely 0<δp<δg<1. As the negotiations advance, the single cost of negotiation will gradually increase for both of them, that is $0<\delta_{ip}<\delta_{2p}<\delta_{1p}<1$, similarly, for the public sector $0<\delta_{ig}<\delta_{2g}<\delta_{1g}<1$,and $0<\delta_{ip}<\delta_{ig}<1$.

(3) Assume that capital investment and risk sharing hardly changes in collaborative consultative phase. Capital ratio and risk-sharing ratio mainly act as determinants of certainty benefits distribution during the initial contract phase. However, their effects to the cooperative surplus distribution on the negotiation phase are less critical than its to the static certainty benefits distribution in the early phase, which mostly reflect on the impact on bargaining power, and the value of h obtaining extra benefits from public sector.

(4) Not taking the other factors into account, assume that both evaluations were satisfactory and no special plan on the strategy.

C. Building the Model

First round: public sector proposes his benefit proportion to be k1.Considering that the imperfect information led to its asymmetry status, public sector uses its own strong status to coerce or force private sector to transfer hi of the acquired shares to it with probability q, so that the actual portion of public sector increased h1, and the portion of private sector correspondingly reduced H1. Hence, public sector (G) and private sector (p) obtain respectively:

$$G_1 = q(k_1 + h_1)\ S \tag{1}$$

$$P_1 = q(1 - k_1 - h_1)\ S \tag{2}$$

When public sector doesn't take a strong position to obtain extra benefit from the private sector with probability 1-q, public sector (G) and the private sector (p) obtain as follows:

$$G_1 = (1-q)\ k_1 S \tag{3}$$

$$P_1 = (1-q)(1-k_1)\ S \tag{4}$$

The expected benefits of public sector (G) and the private sector (p) are:

$$G_1 = (k_1 + qh_1)S \tag{5}$$

$$P_1 = (1 - k_1 - qh_1)S \tag{6}$$

If the private sector declines the first round assignment which is made by the public sector, then the game goes into the second round; conversely, the game ends.

Second round: private sector proposes that public sector gets a proportion k2. Due to the increased negotiating rounds, both parties suffer a little loss; moreover, the asymmetry statue between the two parties led the public sector takes a strong position with probability q. So the expectations of both parties are:

$$G_2 = (k_2 + qh_2)\delta_{1g}S \tag{7}$$

$$P_2 = (1 - k_2 - qh_2)\delta_{1p}S \tag{8}$$

If public sector refuses the second-round assignment, then the game goes into the third round; conversely, the game ends.

Third round: public sector proposes the share is k3.similar to the second round, the expectations of both parties are:

$$G_3 = (k_3 + qh_3)\delta_{1g}\delta_{2g}S \tag{9}$$

$$P_3 = (1 - k_3 - qh_3)\delta_{1p}\delta_{2p}S \tag{10}$$

The game goes on until one accepts the other's allocation proportion, and then the negotiations are concluded.

IV. SOLVING THE MODEL

Using the Harsanyi transformation, introduce a player 0 named "nature" into the game. It decides types of each of the other players, and only let themselves know their own type .Then the games of incomplete information will be converted into games of complete but imperfect information. According to Shaked and Sutton's solution, for an unlimited period game, starting from the first round or the third round (if it reaches the third round), the result should be exactly the same. Assume the game has a backwards induction solution. How this solution comes out is not important, and the solution is that the public gets G shares, while the private gets P shares. It is the result that the public sector proposes the allocation ratio of k in the first round, and the private sector accepts. According to their conclusions, the assuming backwards induction solutions should also be results from the third round of the game. That is the allocation ratio of k proposed by the public sector in third round, and accepted by the private sector .

The game is reduced to a game of the three-stage dynamic game as shown in Fig.3. Because both sides know that each one more round means a little more loss, so dragging the negotiation too long is bad for the both sides. It is wise to let the other side get the necessary benefits earlier, so as to avoid its own deterioration of benefits.

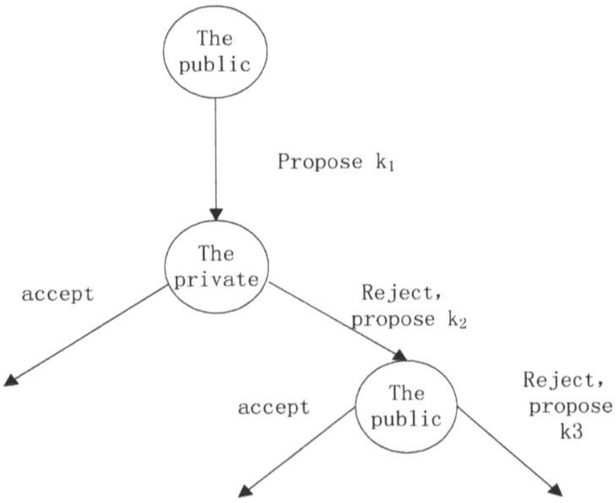

Fig.3. Game Model of the Allocation of the Cooperative Surplus

Along the above ideas, in the first round the public sector proposes the balanced ratio $k^* = (1-\delta_{1p} - qh_1 + qh_3\delta_{1p}\delta_{2g}) / (1-\delta_{1p}\delta_{2g})$, and the private sector receives. If hi=h(constants), then $k^* = (1-\delta_{1p})/\left(1-\delta_{1p}\delta_{2g}\right) - qh$. k is the share not relying on the public sector's strong position to obtain from the private sector, and qh is the extra share that the public sector obtain from the private sector.

When q=1, the public sector is fully aware of the private sector's type. It identifies itself as the strong one and uses its position to obtain benefit in the proportion of h , so its real benefit is $(1-\delta_{1p})/\left(1-\delta_{1p}\delta_{2g}\right)$, and when h is bigger , indicating that the more powerful the public sector is, the greater proportion of the benefits will be claimed from the private sector.

When q=0, the public sector is fully aware of the private sector's type as well, but this time the private sector is more initiative than the public sector .The public sector cannot coerce the private sector in strong position .

When 0<q<1, the information types are both not clear. The public sector cannot fully perform its strong position. Different from the complete information, the public sector ask for fewer additional shares.

Therefore, the public and private not only have the certainty benefits stipulated in the initial contract. They also cooperate and negotiate about the part not including in the contract after the specific investment occurs, and then negotiate with each other about the new cooperative surplus in the late phrase. During the negotiation phase, refined the Nash equilibrium solution relates not just to both sides of every phase of the relevant discount factor δi and also to the probability q that the public sector take a strong position with, and the value of share h which obtains from the private sector.

V. CONCLUSION

Contract of PPP is incomplete. After the initial allocation of control rights in the early phase of the project, there may be unlisted and observable benefits from some projects. If both sides jointly make decision and maintain the projects, its efficiency and eventual benefit are higher than that of only following the initial allocation of control rights. The uncertainty cooperative surplus resulting from the decision, is not simply equally distributed between the public and private in the late phrase. The distribution of benefits is closely related to a variety of factors. From the analysis of the game model above, the assignments associate with the degree of patience of the two partners and the types of their information. The degree of partner's patience depends on the opportunity cost and time cost during the negotiations, and it is relatively more critical to the result. In the cooperative surplus assignments between the public and the private, the acquisition of the other partner's information is quite important as well. Differences in information will affect the two collaborators to judge the position in the negotiations. So try to change that, it would be more useful to the result of the negotiations.

REFERENCES

[1] Ye, Xiaosu, and Chunmei Xu. "Review and Research on PPP Pattern in China." Proceedings of the 17th International Symposium on Advancement of Construction Management and Real Estate. Springer Berlin Heidelberg, 2014.

[2] Ng, A., and Martin Loosemore. "Risk allocation in the private provision of public infrastructure." International Journal of Project Management 25.1 (2007): 66-76.

[3] Iossa, Elisabetta, and David Martimort. "Risk allocation and the costs and benefits of public-private partnerships." The RAND Journal of Economics 43.3 (2012): 442-474.

[4] Xu Xia, Zheng Zhilin. "Benefit Assignment in PPP". Urban Studies, 2009, 16 (3): 104-106.

[5] Scharle, Peter. "Public-private partnership (PPP) as a social game." Innovation: The European Journal of Social Science Research 15.3 (2002): 227-252.

[6] Hart, Oliver. "Incomplete contracts and public ownership: Remarks, and an application to public-private partnerships." The Economic Journal 113.486 (2003): C69-C76.

[7] Reeves, Eoin. "The practice of contracting in public private partnerships: transaction costs and relational contracting in the Irish schools sector."Public Administration 86.4 (2008): 969-986.

[8] Zheng, Jurong, Jens K. Roehrich, and Michael A. Lewis. "The dynamics of contractual and relational governance: evidence from long-term public–private procurement arrangements." Journal of Purchasing and Supply Management 14.1 (2008): 43-54.

[9] Medda, Francesca. "A game theory approach for the allocation of risks in transport public private partnerships." International Journal of Project Management 25.3 (2007): 213-218.

[10] Zhang, Xueqing. "Critical success factors for public–private partnerships in infrastructure development." Journal of Construction Engineering and Management 131.1 (2005): 3-14.

Economic Design of a Nonparametric Control Chart for Shift in Location

Cheng-long Li[1,3,*], A. Mukherjee[2], Q. Su[1], M. Xie[3]
[1] School of Management, Xi'an Jiaotong University, Xi'an, China
[2] Indian Institute of Management Udaipur, Rajasthan, India
[3] Department of Systems Engineering and Engineering Management, City University of Hong Kong, Hong Kong
(liassy@stu.xjtu.edu.cn)

Abstract - **Most control charts economically designed rely on the assumption of normality or some specific process distribution. However, when identifying a specific distribution is not possible or unlikely as full knowledge is not available in practice, the economic effectiveness of conventional charts is very likely to be heavily discounted. In this paper, an economic model based on the Duncan-type cost function is researched on, which is applied to design a nonparametric sign chart to monitor the location parameter of a univariate process. Numerical results show that the proposed design performs fit well for various continuous distributions.**

Keywords – **Adaptive plug-in technique, Duncan-type cost function, Nonparametric, Sign statistic**

I. INTRODUCTION

The economic design of control chart aims at maximizing the profit or minimizing the cost associated with the implementation of the control procedure in a process [1]. Duncan [2] introduced the original economic model for Shewhart-type control charts, and after that the design approach of control charts has been widely used and extended in subsequent studies [3-5].

In spite of so many existing contributions related to economic design, almost all assume that process information is fully known. In fact, there are many situations where it is not reasonable to assume a specific form for the process distribution such as the normal, or accurately knowing the in-control (IC) population parameters. To avoid this problem, some economically designed control charts for non-normal data are proposed [6-7], but these charts are essentially parametric control charts. On the other hand, Linderman and Choo [8] and Vommi and Seetala [9] attempted robust economic design of control charts for not knowing the true values of input parameters, whereas a normal distribution is still assumed in each scenario. Thus, under the circumstances where the underlying distribution is not sufficiently known, the effectiveness of conventional economically designed control charts could be heavily discounted. Arnold [10] considered a typical loss function in designing sign chart but it differs from popular approach of economic design based on Duncan-type cost model. As a consequence, there is an urgent demand for a new economic design procedure that does not require a distributional assumption, that is, a nonparametric chart.

In recent years, research on nonparametric control charts has received a lot of attention. Chakraborti and Eryilmaz [11] studied the Shewhart-type control chart. The nonparametric CUSUM chart was put forward by Qiu and Hawkins [12]. Graham et al. [13] proposed a distribution-free EWMA control chart. A thorough review of the literature on nonparametric control charts was well presented by Chakraborti et al. [14]. These charts have the advantage that their IC performance such as the IC average run length is completely known and remains the same for all continuous distributions. However, good statistical performance does not necessarily acquire good economic performance. The existing nonparametric control charts are practically designed with respect to statistical criteria, whereas the design from an economic perspective has not been addressed yet. This is the main motivation of the present work leading to a design procedure of control charts from an economic perspective, without knowing the concrete form of the underlying distribution. Economic design usually involves a more complex model than the statistical design, as an initial study, we consider a nonparametric control chart based on the sign statistic to detect shift in location parameters.

The rest of this paper is organized as follows: The next section provides a brief description of the chart upon sign statistic. Section 3 develops an economic model for the sign chart based on Duncan-type cost function, followed by an introduction of adaptive plug-in technique. Numerical results are presented in Section 4 to demonstrate the performance of the proposed chart. Finally, Section 5 ends with a summary and conclusions.

II. CONTROL CHART BASED ON SIGN STATISTIC

Let X be the random variable denoting the process characteristic under consideration for statistical monitoring via control chart. We assume that the random variable X has an unknown but continuous distribution function F_X with a location parameter θ. Let $(X_{i1}, X_{i2}, \dots, X_{in})$ be the ith test sample from the random variable X for $i = 1,2,\dots,\dots$. Clearly, n is the fixed sample size for i^{th} test sample or the so called rational subgroup. Introduce the indicator function

$$u_{ik} = \begin{cases} 0 & if\ X_{ik} < \theta \\ 1 & if\ X_{ik} \geq \theta \end{cases}, k = 1,2,\dots,n\ and\ i = 1,2,\dots,\dots;$$

and consider the statistic $U_i = \sum_{k=1}^{n} u_{ik}$.

It is easy to see that $U_i's$ follow a Binomial distribution with parameters n and p.

In the present context, we consider that the standard is given and in-control set up, the location parameter $\theta = \theta_0$. By denoting $p_0 = Prob\,[X \geq \theta_0 | IC]$, it is easy to see that if a rightward shift in location occurs from θ_0 to $\theta_0 + \delta$ (for some $\delta > 0$) in the distribution of process variable X, we have $p_1 = Prob\,[X \geq \theta_0 | OOC] > p_0$. On the other hand, if a leftward shift occurs from θ_0 to

E. Qi et al. (eds.), *Proceedings of the 5th International Asia Conference on Industrial Engineering and Management Innovation (IEMI2014)*, Proceedings of the International Asia Conference on Industrial Engineering and Management Innovation 1, DOI 10.2991/978-94-6239-100-0_68, © Atlantis Press and the authors 2015

$\theta_0 - \delta$ in the distribution process of variable X, we have $p_2 = Prob\,[X \geq \theta_0 | OOC] < p_0$. Therefore, using U_i/n as the plotting statistic for i^{th} test sample, monitoring the unknown process distribution can be carried out via monitoring Binomial proportion p.

Note $E\left(\frac{U_i}{n}\middle|IC\right) = p_0$ and $Var\left(\frac{U_i}{n}\middle|IC\right) = \frac{p_0(1-p_0)}{n}$. Therefore, the control limits for the two-sided nonparametric chart based on sign statistic may be given by $p_0 \pm k\sqrt{\frac{p_0(1-p_0)}{n}}$, where k is the suitable chart parameter. A test sample of size n is drawn at the interval of h hours from the production line, U_i is computed at every stage and U_i/n is compared with specified control limits. These steps are repeated uninterruptedly as long as there is no OOC signal. If some points fall outside the control limits, the process is terminated and a search for assignable cause is begun.

III. ECONOMIC DESIGN OF SIGN CHART

A. Construction of Duncan-type Cost Function

In the present context, we assume that our sampling interval is h hours, that is, the test samples of size n are to be collected at intervals of h hours. Our main aim is to determine n, k and h via optimal design of the control chart based on sign statistic.

The process is assumed to start in an IC state and therefore, we can expect on an average np_0 observations per test sample will exceed θ_0 where θ_0 (and p_0) are known. However, in the current context, unlike economic design of a traditional p-chart for proportion of defective, $p = Prob\,[X \geq \theta | OOC]$ is not known. Note that, even though δ is known, exact value of p cannot be obtained since the process distribution F is unknown. In the next subsection we shall discuss a data dependent plug-in technique based on a pilot sample and use the estimate of p to design other parameters.

The assignable cause is assumed to occur according to a Poisson process with a rate of λ incidences per hour of operating time. Therefore, subject to the occurrence of the assignable cause between the jth and $(j+1)^{st}$ samples, the well-known expression for expected time of occurrence within this interval is:

$$\xi = \frac{\int_{jh}^{(j+1)h} \lambda e^{-\lambda t}(t-jh)dt}{\int_{jh}^{(j+1)h} \lambda e^{-\lambda t}dt} = \frac{1-(1+\lambda h)e^{-\lambda h}}{\lambda(1-e^{-\lambda h})} \quad (1)$$

When the assignable cause occurs, the probability that it will be spotted on any following test sample is:

$$\beta = \sum_{x=\lfloor np_0 - k\sqrt{n p_0(1-p_0)}\rfloor}^{[np_0 + k\sqrt{n p_0(1-p_0)}]} \binom{n}{x} p^x (1-p)^{n-x} \quad (2)$$

where $[q]$ denotes the largest integer contained in q, while $\lfloor q \rfloor$ denotes the next higher integer if q is a fraction or q if it is an integer. The quantity $(1-\beta)$ is the power of the test while β is the underlying Type-II error. Note that the expression for false alarm rate in the present context can be obtained by replacing p by p_0 in the expression of power. So, the expected length of the out-of-control period can be given by $\frac{h}{1-\beta} - \xi$. Following standard practice and as described in Montgomery [15], we consider that the time required to draw a sample and interpret the results is a constant g proportional to the sample size. Therefore gn is the length of this segment of the cycle. The time required to detect the assignable cause after an OOC signal is a constant D. Therefore, the expected duration of a cycle is

$$E(T) = \frac{1}{\lambda} + \frac{h}{1-\beta} - \xi + gn + D \quad (3)$$

The net income per hour of production in the IC state is V_0, and the net income per hour of production in the OOC state is V_1. The cost of taking a sample of size n is assumed to be of the form $a_1 + a_2 n$; that is, a_1 and a_2 represent, respectively, the fixed and variable components of sampling cost. The cost of finding an assignable cause and adjusting the process to in-control state is a_3, and the cost of investigating a false alarm is a_3'. The expected number of false alarms generated during a cycle is a times the expected number of samples taken before the shift, or

$$\alpha \sum_{j=0}^{\infty} \int_{jh}^{(j+1)h} j\lambda e^{-\lambda t} dt = \frac{\alpha e^{-\lambda h}}{1-e^{-\lambda h}} \quad (4)$$

Therefore, the expected net income per cycle is

$$E(C) = \frac{V_0}{\lambda} + V_1\left(\frac{h}{1-\beta} - \xi + gn + D\right) - a_3 - \frac{a_3' \alpha e^{-\lambda h}}{1-e^{-\lambda h}} - (a_1 + a_2 n)\frac{E(T)}{h} \quad (5)$$

Let $a_4 = V_0 - V_1$, the expected net income per hour $E(A)$ is found by dividing the expected net income per cycle by the expected cycle length, resulting in

$$E(A) = \frac{E(C)}{E(T)} = V_0 - \frac{(a_1 + a_2 n)}{h} - a_4 + \frac{\frac{a_4}{\lambda} - a_3 - \frac{a_3' \alpha e^{-\lambda h}}{1-e^{-\lambda h}}}{\frac{1}{\lambda} + \frac{h}{1-\beta} - \xi + gn + D}$$
$$= V_0 - E(L) \quad (6)$$

where the expression $E(L)$ represents the expected loss per hour incurred by the process. $E(L)$ is a function of the control chart parameters n, k and h. Clearly, maximizing $E(A)$ is equivalent to minimizing $E(L)$.

B. Adaptive Plug-In Technique

Assuming F is absolutely continuous and twice differentiable around θ_0 and under a standard location shift model $G(x) = F(x-\delta)$, for any $\delta \neq 0$, we can write,

$$p = Prob\,[X \geq \theta_0 | OOC] = 1 - F(\theta_0 - \delta) \quad (7)$$

This may be approximated by empirical distribution function $F_m(x)$ based on the pilot sample of size m. In general, it works as a very good choice if m is moderately large. An empirical study shows that, for small m, an ad-hoc plug-in estimate performs more effectively.

Applying Taylor's expansion, we get

$$1 - F(\theta_0 - \delta) \approx p_0 + \delta f(\theta_0^*) \quad (8)$$

Now, for most of the practical purposes, one can approximate p by $p_0 + \delta f(\theta_0^*)$, where $f(\theta_0^*)$ is the maximum value of $f(x)$ in the neighborhood of θ_0. This may be approximated by using suitable kernel density estimate available with most of the software, such as R or MATLAB. If estimated p exceeds 1, we must take $p = 1$.

Note that for large shift more often we may observe under alternative $p = 1$. Subsequently, we can estimate the Type II error β.

IV. RESULTS

A. Performance of the sign chart

Three loss situations, namely 'True case', 'Estimated case' and 'Actual case' are considered for numerical studies. These are explained in Fig. 1. From the economic perspective, the actual $E(L)$ is one of the most important indicators of performance (or economic effectiveness) of nonparametric charts, in addition to the true optimal $E(L)$. An effective nonparametric economic design requires that the design scheme is close to the true optimal design scheme and the actual $E(L)$ is close to the true optimal $E(L)$. In other words, the smaller the actual $E(L)$ is, the better the performance is.

In fact, the actual $E(L)$ is conditional on the pilot sample X_m. To evaluate the unconditional performance, we introduce the mean of actual $E(L)$ as the performance indicator. The actual $\overline{E(L)}$ can be obtained by calculating the mean for adequate replicates. In this section, we use the MATLAB software and 500 replicates which provide a pretty stable estimate. The standard deviation is also studied to indicate the variability.

For purpose of overall examination, we investigate both symmetric distributions and asymmetric distributions. Normal (0, 1) distribution, t (3) distribution and Gamma (3,2) distribution are studied as trial distributions. Gamma distribution (3, 2) is a positive-skewed distribution. Note that when we study a leftward shift in a positive-skewed distribution, it can be recognized identically as a rightward shift in a negative-skewed distribution. Thus, we focus on four cases, a rightward shift in Normal distribution denoting as $N_+(0,1)$, a rightward shift in t distribution denoting as $t_+(3)$, a rightward shift in Gamma distribution denoting as $\Gamma_+(3,2)$, a leftward shift in Gamma distribution denoting as $\Gamma_-(3,2)$. Note that we only consider a two-sided chart in this paper out of convenience, and the charts can easily be enhanced by adopting asymmetric control limits according to the actual situation.

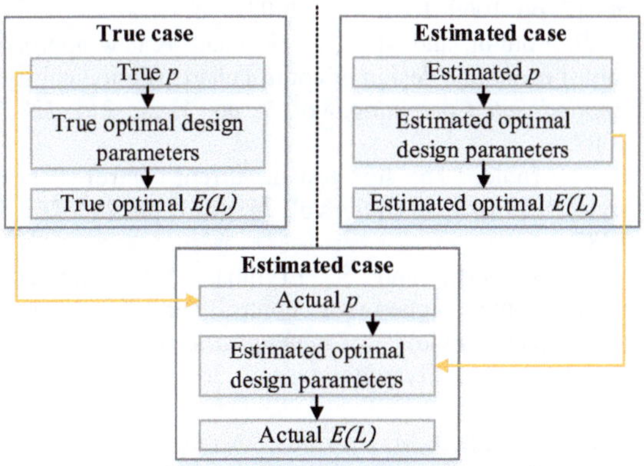

Fig.1. Three loss situations

Each numerical investigation entails 44 scenarios of interest as shown in Table I that covers a broad range of process parameters and cost parameters, which are adapt from Duncan [2]. In particular, we consider the target shift in location, $\delta = 0.5, 1.0, 2$ and 3 to cover small, moderate as well as large shifts. The results for the four cases are summarized in Table II and some remarks are made:

(1) The proposed nonparametric design performs effectively and expectedly under most of scenarios in terms of the proximity between actual $\overline{E(L)}$ and true optimal $E(L)$, and only few scenarios cause a relatively poor result (the scenarios leading to a high percentage (5%) of actual $\overline{E(L)}$ increase are boldfaced). According to Table II, we see that the design is especially effective for $\Gamma_+(3,2)$. In addition, the chart under $N_+(0,1)$ operates better than under $t_+(3)$.

(2) The distribution has an influence on the performance of nonparametric charts. The true optimums from symmetric distributions are obviously smaller than the ones from asymmetric distributions, which is attributed to a larger gap between p_0 and true p obtained from these distributions, given the same shift.

B. Comparison with a parametric $\bar{X}$ chart

To further examine the performance, we compare the nonparametric chart with a parametric $\bar{X}$ chart. The results of performance comparison are shown in Fig. 2. Standard normal distribution is studied here and 44 scenarios are examined. We see that, either the true optimums or the actual $\overline{E(L)}$ of the proposed chart stays close to the true optimums of the parametric $\bar{X}$ chart in most cases. That is, when we face a problem of parametric design, use the sign chart can obtain a close performance with the $\bar{X}$ chart. However, when the underlying distribution is unknown, the performance of the $\bar{X}$ chart with a normal distribution assumption will be heavily discounted, whereas the performance of the proposed chart still remains effective and stable. In general, the proposed chart could very well replace the $\bar{X}$ chart under the circumstances.

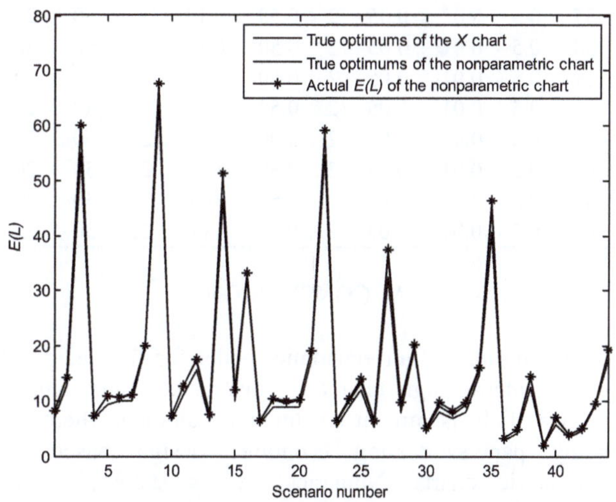

Fig.2. Performance comparison with a parametric $\bar{X}$ chart

TABLE I
44 SCENARIOS OF INTEREST

No.	Process parameters and cost parameters								
	δ	λ	g	D	a_1	a_2	a_3	a_3'	a_4
1	3.0	0.01	0.05	2	0.50	0.10	25	50	220
2	3.0	0.02	0.05	2	0.50	0.10	25	50	220
3	3.0	0.01	0.05	2	0.50	0.10	25	50	2200
4	3.0	0.01	0.05	2	0.50	0.10	2.5	5	220
5	3.0	0.01	0.05	2	0.50	0.10	250	500	220
6	3.0	0.01	0.05	2	5.00	0.10	25	50	220
7	3.0	0.01	0.05	2	0.50	1.00	25	50	220
8	3.0	0.01	0.05	2	0.50	10.00	25	50	220
9	3.0	0.01	0.05	2	0.50	1.00	25	50	2200
10	2.0	0.01	0.05	2	0.50	0.10	25	50	180
11	2.0	0.02	0.05	2	0.50	0.10	25	50	180
12	2.0	0.03	0.05	2	0.50	0.10	25	50	180
13	2.0	0.02	0.05	2	0.50	0.10	25	50	90
14	2.0	0.01	0.05	2	0.50	0.10	25	50	1800
15	2.0	0.01	0.50	2	0.50	0.10	25	50	180
16	2.0	0.01	0.05	20	0.50	0.10	25	50	180
17	2.0	0.01	0.05	2	0.50	0.10	2.5	5	180
18	2.0	0.01	0.05	2	0.50	0.10	250	500	180
19	2.0	0.01	0.05	2	5.00	0.10	25	50	180
20	2.0	0.01	0.05	2	0.50	1.00	25	50	180
21	2.0	0.01	0.05	2	0.50	10.00	25	50	180
22	2.0	0.01	0.05	2	0.50	1.00	25	50	1800
23	1.0	0.01	0.05	2	0.50	0.10	25	50	100
24	1.0	0.02	0.05	2	0.50	0.10	25	50	100
25	1.0	0.03	0.05	2	0.50	0.10	25	50	100
26	1.0	0.02	0.05	2	0.50	0.10	25	50	50
27	1.0	0.01	0.05	2	0.50	0.10	25	50	1000
28	1.0	0.01	0.50	2	0.50	0.10	25	50	100
29	1.0	0.01	0.05	20	0.50	0.10	25	50	100
30	1.0	0.01	0.05	2	0.50	0.10	2.5	5	100
31	1.0	0.01	0.05	2	0.50	0.10	250	500	100
32	1.0	0.01	0.05	2	5.00	0.10	25	50	100
33	1.0	0.01	0.05	2	0.50	1.00	25	50	100
34	1.0	0.01	0.05	2	0.50	10.00	25	50	100
35	1.0	0.01	0.05	2	0.50	1.00	25	50	1000
36	0.5	0.01	0.05	2	0.50	0.10	25	50	20
37	0.5	0.02	0.05	2	0.50	0.10	25	50	20
38	0.5	0.01	0.05	2	0.50	0.10	25	50	200
39	0.5	0.01	0.05	2	0.50	0.10	2.5	5	20
40	0.5	0.01	0.05	2	0.50	0.10	250	500	20
41	0.5	0.01	0.05	2	5.00	0.10	25	50	20
42	0.5	0.01	0.05	2	0.50	1.00	25	50	20
43	0.5	0.01	0.05	2	0.50	10.00	25	50	20
44	0.5	0.01	0.05	2	0.50	1.00	25	50	200

V. CONCLUSION

In this paper, an economic model for the design of nonparametric sign charts is constructed. Surprisingly, such a study is rare in the literature although there are many papers dealing economic design based on parametric charts. Numerical results show that the proposed nonparametric chart can provide a good overall performance for various continuous distributions. Even in a parametric situation, the obtained performance remains close to that of an $\bar{X}$ chart. Yet it does not require the practitioner to make the assumption of normality or any other distribution for the validity of the decision. Thus, the proposed chart can be very useful in practical applications.

ACKNOWLEDGMENT

This collaboration was made possible by a grant from City University of Hong Kong (Project No.9380058). The research by Prof. Xie and Prof. Su is also supported by grants from National Natural Science Foundation of China, (Project No. 71371163) and (Project No. 70872091), respectively.

REFERENCES

[1] H. Y. Zhang, M. Xie, T. N. Goh, and M. Shamsuzzaman, "Economic design of time-between-events control chart system (Periodical style)," *Computers & Industrial Engineering*, vol. 60, no. 4, pp. 485–492, May 2011.

[2] A. J. Duncan, "The economic design of $\bar{X}$ charts used to maintain current control of a process (Periodical style)," *Journal of the American Statistical Association*, vol. 51, no. 274, pp. 228–242, Jun. 1956.

[3] K. Linderman, and T. E. Love, "Economic and economic statistical designs for MEWMA control charts (Periodical style)," *Journal of Quality Technology*, vol. 32, no. 4, pp. 410–417, Oct. 2000.

[4] C. Park, J. Lee, and Y. Kim, "Economic design of a variable sampling rate EWMA chart (Periodical style)," *IIE Transactions*, vol. 36, no. 5, pp. 387–399, May 2004.

[5] G. Nenes, and G. Tagaras, "An economic comparison of CUSUM and Shewhart charts (Periodical style)," *IIE Transactions*, vol. 40, no. 2, pp. 133–146, 2008.

[6] C. Y. Chou, C. H. Chen, and H. R. Liu, "Economic-statistical design of $\bar{X}$ charts for non-normal data by considering quality loss (Periodical style)," *Journal of Applied Statistics*, vol. 27, no. 8, pp. 939–951, Nov. 2000.

[7] Y. K. Chen, "Economic design of $\bar{X}$ control charts for non-normal data using variable sampling policy (Periodical style)," *International Journal of Production Economics*, vol. 92, no. 1, pp. 61–74, Nov. 2004.

[8] K. Linderman, and A. S. Choo, "Robust economic control chart design (Periodical style)," *IIE Transactions*, vol. 34, no. 12, pp. 1069–1078, Dec. 2002.

[9] V. B. Vommi, and M. S. N. Seetala, "A new approach to robust economic design of control charts (Periodical style)," *Applied Soft Computing*, vol. 7, no. 1, pp. 211–228, Jan. 2007.

[10] B. F. Arnold, "The sign test in current control (Periodical style)," *Statistische Hefte*, vol. 26, no. 1, pp. 253–262, Dec. 1985.

[11] S. Chakraborti, and S. Eryilmaz, "A nonparametric Shewhart-type signed-rank control chart based on runs (Periodical style)," *Communications in Statistics: Simulation and Computation*, vol. 36, no. 2, pp. 335–356, 2007.

[12] P. H. Qiu, and D. M. Hawkins, "A rank based multivariate CUSUM procedure (Periodical style)," *Technometrics*, vol. 43, no. 2, pp. 120–132, May 2001.

[13] M. A. Graham, A. Mukherjee, and S. Chakraborti, "Distribution-free exponentially weighted moving average control charts for monitoring unknown location (Periodical style)," *Computational Statistics & Data Analysis*, vol. 56, no. 8, pp. 2539–2561, Aug. 2012.

[14] S. Chakraborti, P. Van der Laan, and S. T. Bakir, "Nonparametric control charts: An overview and some results (Periodical style)," *Journal of Quality Technology*, vol. 33, no. 3, pp. 304–315, Jul. 2001.

[15] D. C. Montgomery, *Introduction to Statistical Quality Control* (Book style), Sixth Edition, John Wiley & Sons, 2009, pp. 463-472.

TABLE II
NUMERICAL RESULTS FOR THE FOUR CASES

No.	$N_+(0,1)$			$t_+(3)$			$\Gamma_+(3,2)$			$\Gamma_-(3,2)$		
	True	Actual		True	Actual		True	Actual		True	Actual	
	$E(L)$	$\overline{E(L)}$	SD	$E(L)$	$\overline{E(L)}$	SD	$E(L)$	$\overline{E(L)}$	SD	$E(L)$	$\overline{E(L)}$	SD
1	8.07	8.07	0.00	8.67	8.74	0.00	10.03	10.38	0.46	**12.02**	**13.20**	**2.07**
2	14.23	14.23	0.00	15.04	15.14	0.00	17.14	17.58	0.59	**20.06**	**21.48**	**2.61**
3	59.91	59.91	0.00	61.69	61.88	0.00	68.17	68.99	0.79	76.44	79.02	4.09
4	7.22	7.22	0.00	7.54	7.57	0.00	8.39	8.51	0.16	9.42	9.71	0.57
5	10.87	10.87	0.00	11.72	11.92	0.00	13.50	14.07	0.84	**16.57**	**18.89**	**4.61**
6	10.52	10.52	0.00	**11.17**	**11.78**	**0.00**	12.44	12.86	0.77	**14.33**	**15.76**	**2.92**
7	11.00	11.00	0.00	11.88	12.04	0.00	14.24	14.66	0.50	**17.13**	**18.18**	**1.83**
8	19.98	19.98	0.00	21.82	22.11	0.00	23.44	24.00	1.14	26.28	27.36	2.82
9	67.70	67.70	0.00	70.34	70.53	0.00	79.51	80.28	0.88	89.29	92.44	4.51
10	7.29	7.33	0.00	**7.95**	**8.68**	**0.08**	10.89	11.12	0.29	12.49	12.80	0.67
11	12.61	12.67	0.00	**13.65**	**14.49**	**0.14**	17.94	18.25	0.34	20.24	20.59	0.70
12	17.43	17.49	0.00	18.76	19.65	0.14	24.01	24.37	0.37	26.79	27.12	0.66
13	7.34	7.38	0.00	**7.96**	**8.65**	**0.09**	10.72	10.92	0.25	12.20	12.47	0.58
14	51.10	51.20	0.00	54.04	55.09	0.17	66.73	67.39	0.69	73.10	73.71	1.13
15	11.98	12.00	0.00	12.87	13.10	0.03	17.11	17.24	0.19	19.08	19.21	0.22
16	33.07	33.11	0.00	33.59	34.23	0.08	35.91	36.09	0.23	37.18	37.40	0.54
17	6.34	6.35	0.00	6.85	7.00	0.02	8.29	8.38	0.06	8.68	8.80	0.19
18	10.22	10.35	0.01	**11.04**	**12.88**	**0.35**	15.60	16.06	0.56	19.25	19.77	0.95
19	9.59	9.95	0.02	**10.17**	**12.38**	**0.48**	12.91	13.16	0.31	14.50	14.82	0.73
20	10.18	10.26	0.00	**11.46**	**12.21**	**0.14**	15.73	16.08	0.46	17.55	17.81	0.45
21	19.05	19.12	0.00	**19.81**	**22.90**	**0.47**	23.70	23.74	0.18	24.81	24.88	0.40
22	58.92	59.04	0.00	63.21	64.31	0.16	78.80	79.94	1.23	84.30	85.14	1.48
23	**6.04**	**6.40**	**0.79**	6.51	6.82	0.83	11.13	11.46	0.45	11.89	12.06	0.23
24	9.97	10.40	0.90	10.67	10.96	0.64	16.92	17.36	0.46	17.88	18.03	0.24
25	13.38	13.91	1.07	14.23	14.69	1.11	21.60	22.06	0.34	22.51	22.63	0.20
26	**6.06**	**6.40**	**0.69**	6.50	6.76	0.56	10.66	10.96	0.35	11.31	11.43	0.14
27	36.44	37.38	1.71	38.50	39.27	1.65	55.41	55.52	0.29	55.83	56.33	1.13
28	9.57	9.71	0.25	10.34	10.48	0.24	14.50	14.53	0.14	14.62	14.65	0.18
29	19.98	20.26	0.58	20.35	20.69	0.81	24.02	24.30	0.32	24.60	24.72	0.16
30	4.84	4.96	0.27	5.08	5.17	0.20	5.98	5.98	0.00	6.03	6.03	0.00
31	**9.12**	**9.82**	**1.71**	**9.90**	**10.59**	**1.98**	20.96	21.20	0.28	23.03	23.20	0.28
32	**7.55**	**8.06**	**1.33**	7.98	8.35	1.10	12.53	12.68	0.11	12.86	13.01	0.27
33	9.26	9.65	0.74	9.99	10.28	0.56	13.12	13.12	0.11	13.25	13.25	0.00
34	15.58	16.02	1.68	16.49	16.66	0.91	18.96	18.96	0.00	19.13	19.13	0.00
35	45.15	46.37	1.84	47.88	48.74	1.78	57.87	57.88	0.01	58.33	58.33	0.01
36	3.22	3.26	0.07	3.45	3.49	0.08	5.21	5.22	0.03	5.22	5.23	0.02
37	4.79	4.84	0.06	5.09	5.14	0.09	6.83	6.83	0.02	6.83	6.83	0.01
38	14.23	14.37	0.22	15.02	15.23	0.34	19.60	19.60	0.00	19.63	19.64	0.20
39	2.00	2.01	0.03	2.02	2.03	0.03	2.17	2.17	0.00	2.17	2.17	0.00
40	6.86	6.93	0.09	7.32	7.42	0.12	12.22	12.26	0.04	12.41	12.43	0.01
41	3.83	3.89	0.08	4.08	4.12	0.07	5.66	5.66	0.02	5.67	5.68	0.03
42	4.99	5.02	0.05	5.15	5.17	0.05	5.88	5.88	0.00	5.89	5.89	0.00
43	9.52	9.52	0.00	9.65	9.65	0.00	10.40	10.40	0.00	10.42	10.42	0.00
44	19.16	19.30	0.32	19.42	19.49	0.27	20.78	20.78	0.00	20.81	20.81	0.00

Application Status and Development of Rorschach Ink-Bolt Test in Talent Assessment

Xiao-li QU [1,*], Jian-nan SHEN [1,2]
[1] Department of Business Administration, Suzhou Institute of Trade & Commerce, Suzhou, P. R. China
[2] ACEM, Shanghai Jiao Tong University, Shanghai, P. R. China
(quxl@szjm.edu.cn)

Abstract - **Background: Enterprises tend to adopt the current structured psychological scale when the psychological assessments on the employees are in process. However, due to various reasons, the assessed persons generally fail to give the reliable answers and personality performance in several situations, which leads to the inaccuracy of the final decisions of the enterprises in terms of human resource management. Research process: After reviewing the development and current status of the Rorschach Ink-Bolt Test (RIT), as well as taking into account the characteristics of talent assessment, this research demonstrated the principle, the scoring systems and the application features of RIT and analyzed the reliability and validity, in the talent assessment, of such a test. Furthermore, this research is dedicated to a vision and hypothesis about the application of RIT in the talent assessment area in respects of the research pattern, group tests and scoring codes. Conclusions: the application of RIT in talent assessment is an important topic worthy of being studied.**

Keywords - **Psychological Assessment, Rorschach Ink-Bolt Test, Reliability, Talent Assessment, Validity**

I. SKEPTICISM AGAINST THE PSYCHOLOGICAL ASSESSMENT APPROACH APPLIED IN TALENT ASSESSMENT

Currently, persons including astronauts, leaders in some enterprises and public institutions, senior managers pursuing jobs in the foreign-owned enterprises and graduates who just left the universities are going to, inevitably, be tested through various talent assessments. Indeed, the talent assessment technologies have already been one of the key tools for human resource management.

The three major technologies which are most common in the talent assessments contain the psychological assessments, interviews and the assessment center, among which the psychological assessment are the core instrument. The reason for this is that the modern corporation tends to adopt the psychological explanations in order to thoroughly understand talents [1]. Introducing the psychological assessments as one of the talent assessment approaches to the human resource management is conducive to the scientific evaluation of the managers, who are pursuing the effective personnel allotment, cultivation and management for talents, on employees' knowledge, capability and skills and the further understanding on their working motives, working attitudes, tempers, temperament, personality and other psychological features.

Nevertheless, due to the popularity of psychological assessment and the fact that a lot of evaluator do not have a firm professional basis, the assessed persons who have been familiar with the raised questions and assessing procedures may give the fake answers in an attempt to get through the assessment. Under these circumstances, managers have to be skeptical about the result and the reliability of talent assessments. The introduction of some classical projective tests has, to some degree, relived such problem but also left a though-provoking question to scholars and experts [2].

II. INTRODUCTION OF THE RORSCHACH INK-BOLT TEST INTO THE TALENT ASSESSMENTS

Rorschach Ink-Bolt Test (RIT) was compiled by Herman Rorschach, a Swiss psychologist, in 1921. Its materials are ten pieces of ink-Blot cardboard which about 24 X 17 cm sizes. There are five black and white pictures, tow black and red pictures, three colorful pictures of RIT. The whole testing procedure is unfolding by the one-to-one form of interviewers and subjects. The procedure divided into two stages of the free association and inquiry [3]. Indeed, the form and procedure of RIT have already been in the talent assessments special assessed the managers in higher-up of a company. Because of that is precondition and foundation of the Rorschach Ink-Bolt Test into the talent assessments.

A. Projection Principle of the Rorschach Ink-Bolt Test

RIT originated from ideas of the Gestalt and Psychoanalysis, putting its basis on the assumption that there is the "Verhältnis des Spiegelns" between personality and perception [4].

Freud, the founder of psychoanalysis, pointed out in his sub-consciousness theory that most of the human behavior and reactions were determined by their sub-consciousness. A man's authentic incentive, desire and any other mental activities will be necessarily reflected, either directly or indirectly, to his other mental processes or mental activity products. It is called the projective process. When demonstrating a group of pictures which do not embrace any specific meanings to various individuals, they will necessarily give the entirely different reactions and understandings on the meanings, as they believed, of these pictures. However, these reactions and understandings were not

E. Qi et al. (eds.), *Proceedings of the 5th International Asia Conference on Industrial Engineering and Management Innovation (IEMI2014)*, Proceedings of the International Asia Conference on Industrial Engineering and Management Innovation 1, DOI 10.2991/978-94-6239-100-0_69, © Atlantis Press and the authors 2015

generated randomly, but were being projected by their sub-consciousness. Despite how strong the consciousness people have, they are still under control and influence from their sub-consciousness [5].

Based on this principle, many renowned projective tests were developed. It is their advantages that compensate for the flaws of a lot of popular structured psychological tests.

The Rorschach Ink-Bolt Test is the more matured one among a lot of the projective tests. Therefore, the evaluators are convinced that the RIT is the comparatively good approach for talent assessment because it is not possible for the assessed to understand the purpose of the test in advanced if they have not been trained for a long time.

B. The Scoring System of the Rorschach Ink-Bolt Test

The scoring and interpretative system of the RIT is the key reason why this projective test can be so successful. In history, there were totally five scoring systems ever being adopted by the RIT, including the Beck system, Kopfer system, Hertz system, Piotrowski system and Rapaport system.

Currently, the comprehensive system developed by Exner J. is the most commonly used system for the RIT [6]. This scoring system is dedicated to standardizing the RIT with its implementation, scoring and interpretation strictly fitting the unified standards. Based on substantial statistical analysis, Exner J. preserved the reliability and validity of the previous Rorschach systems while added some unique scoring approaches which are based on the clinical experience to the comprehensive system. The 34 items were ultimately determined to be scored [7].

The comprehensive scoring system of RIT mainly involves the following aspects: Location, Cognitive Function, Form Quality, Determinant (includes Colour, the Colour, Shadow and Images in pairs), Content, Popular and so on. As shown in Table I [8].

TABLE I
THE SCORING AND CODED OF RIT

Aspect	Answer/ Reaction	Coded	Aspect		Answer/ Reaction	Coded
Location	Entirety	W	Determinant	Colour	Pure colour	C
	Common	D			Colour form	CF
	Uncommon	Dd			Form colour	FC
	Blank parts	S			Colour description	Cdes
Cognitive Function	+, V/+, o, V			The Colour	Impure colour	C'
Form Quality	Excellence	F+			The colour form	C'F
	Normal	Fo			Form the colour	FC'
	Rare	Fu		Shadow	Texture	c, Fc, cF
	Defect	F-			Atomization	K,KF,FK
Content	28 items				Dimensionality	V,VF,FV
Popular	Coded "P", no objective standard				Likeness dimensionality	kF, Fk, k
Blend	(.)				Image in pairs	Fr, rF,(2)

It is because of this standardized scoring system which is consistent with the principle of "being objective, scientific and fair" in the talent assessment that increasing number of the scholars introduced the RIT to the talent assessment area in recent years. And the Cognitive Function module of the scoring system is being studied most intensively at present [9].

C. Features of the Rorschach Ink-Bolt Test in the talent assessment

The purpose of RIT is to test the deeply-rooted quality and personality of the assessed like the thinking characteristics, judging approaches, life experience, emotion and personality tendency. The assessed will expose their authentic psychological status without being conscious about the test, which is entirely different from the structured psychological assessment in which the assessed can guess what answers the evaluators want to get and therefore cover their authentic ideas and characteristics. This is because when he or she is telling of a story on the pictures, he or she has already project himself or herself into that circumstance. It is also because the agencies of this test are ambiguous without a fixed model that people have to use their imagination for interpretation. Through different answers and reactions, the personality and mentality of the assessed will be exposed. The largest advantage of this projective test is that the true intention of the evaluators have been covered, which creates a comparatively subjective circumstances in a pursuit of the correct, reliable and deep understanding on the mental activities of the assessed [10] .

However, the disadvantages of RIT in talent assessment cannot be avoided. Regarding individuals' free performance and unrestricted expression, it is difficult to set the analytical framework and then analyze that whether the performance of the assessed is good or not. In order to do so, the judgments and comments from human resource specialists or professional consultancy enterprises are demanded. In addition, this approach is time-consuming and also efforts-consuming, which will possibly influence the assessment efficiency. This is why the projective test is seldom applied in the employee recruitment but only used when employing the managers in higher-up of a company.

III. RELIABILITY AND VALIDITY OF THE RORSCHACH INK-BOLT TEST IN TALENT ASSESSMENT

By now, RIT has become one of the most widely adopted projective tests. A survey on over 400 American psychologists demonstrated that 82% of them frequently used the RIT. The frequency of being used is only less to the one of the Wechsler Intelligence Scale (WIS), Minnesota Multiphasic Personality Inventory (MMPI) and Sentence Completion Test (SCT), ranking at the fourth position.

Meanwhile, the adaptability of RIT for various groups has been verified by psychologists from many countries through numerous race samples [11]. Solid evidence has been presented to prove that RIT is widely recognized by the world of psychology and embraces a high value for application. Exner J. called RIT as "the X-ray in psychology" [12]. Despite that, the reliability and validity of RIT still arouses plenty of controversies. "No projective tests can cause so many controversies as the RIT does" [13].

A. Reliability of the Rorschach Ink-Bolt Test

The earliest report about the research on the reliability of RIT was from Hertz M R. in 1934. With a group of 100 high-school students, the reliability coefficient by the split-half method (corrected for attenuation) for the total number of responses is .89 and for various test factors from .67 to .97. Consideration is also given to test standardization, procedure of administration, scoring, and previous investigations [14].

After that, skepticism against reliability of RIT is mainly focused on the test-retest reliability of each sub-item. In talent assessment, RIT is mainly adopted in order to test the intellectual capability, creativity and potential of the assessed. Therefore, its scoring codes are concentrated on items like forms, colors and contents [15].

In recent years, there are substantial reports regarding the test-retest reliability of these items, whose conclusions were various. In 2003, Gronnerod C adopted the meta-analysis approach in order to launch the comprehensive analysis on the reliability of RIT. The analysis indicated that under the integrated scoring and interpretative system established by Exners, the comprehensive effect of most of the content codes is Rw>.80 while the comprehensive effect of the form and color code is Rw>.70 [16].

Certainly, scholars conducted substantial research on the intercoder reliability of RIT. Initially, the rank order test and variance analysis were adopted as the statistical approach for the projective tests like RIT, which subsequently were marginalized while the generalized analysis became the mainstream approach, as a result of the development of statistics and generalization theory [17].

Recently, scholastic news about intercoder reliability of RIT in talent assessment included articles published in Personality Assessment in 2011 by Sahly J. and her colleagues. This study examines the intercoder reliability of Rorschach Comprehensive System protocol-level variables. A large international sample was combined to obtain intercoder agreement for 489 Rorschach protocols coded using the CS. Intercoder agreement was calculated using an Iota coefficient, a statistical coefficient similar to kappa that is corrected for chance. Iota values for the variables analyzed ranged from .31 to 1.00, with 2 in the poor range of agreement, 4 in the fair range, 25 in the good range, and 116 in the excellent range of agreement [18].

Form Driskell's (2012) research, the analysis was based on a total of 16 studies with 30 separate hypothesis tests, representing the behavior of 2847 trainees. Results indicated that the effect of deception detection training on detection accuracy was positive, significant, and of medium magnitude [19].

Such an empirical analysis provides solid evidence to the belief that RIT has the good test-retest reliability in talent assessment.

B. Validity of the Rorschach Ink-Bolt Test

Due to the fact that any kind of projective tests are, to certain degree, subjective when interpreting something, RIT is also believed to be subjective in some situations. The complicated implementation process and scoring system of RIT will also generate errors to the testing results [13].

Furthermore, psychoanalysis as the guiding philosophy of RIT is also controversial. First, RIT is a kind of non-structured measurement with purpose being covered. The assessed is less likely to guess the purpose of RIT and therefore is impossible to cover the thoughts during the test. Due to this feature, RIT has the possibility to measure the authentic personality.

Second, Exners conducted a research in 1991, which demonstrated that RIT has a high detection rate regarding to personality disorders. Scholars adopted the matured personality questionnaires to measure the validity of RIT. In the result, some sub-items like content, form and color of RIT, such as Ta, PER, MORb, AG, Al, AgC, MOAS-H, DEVAL, FC+CF+C and IDEAL, had been proved to be correlated (.05) with Pd of MMPI-2 [20][21].

Meanwhile, in an attempt to refute skepticism against validity of RIT, supporters of this test adopted the meta-analysis approach again, conducting two sets of comparisons in terms of the average comprehensive validity-- RIT and WAIS, MMPI and WAIS. Despite that the comprehensive validity of RIT is lower than both MMPI and WAIS, but it failed to reach the significant level [22][23][24][25].

The newest report is that Mihura, Joni L, et al (2013) systematically evaluated the peer-reviewed Rorschach validity literature for the 65 main variables in the popular Comprehensive System (CS). Across 53 meta-analyses examining variables against externally assessed criteria (e.g., observer ratings, psychiatric diagnosis), the mean validity was R = .27 (k = 770) as compared to R = .08 (k = 386) across 42 meta-analyses examining variables against introspectively assessed criteria (e.g., self-report) [26].

Some researchers verified the validity of RIT in talent assessment through comparing the consistency between RIT and competency quality [27][28][29] [30].

Although most of the reports regarding the validity of RIT are about construct validity, their predictive validity is satisfactory. People like Matsunaga N applied RIT to test the "potential of becoming success" of 125 persons. Subsequently, they followed the high-rating group and the low-rating group respectively for 11 years, figuring out that RIT has a good predictive validity [31]. This is what we want in the talent assessment.

IV. DISCUSSION AND PROSPECT OF THE RORSCHACH INK-BOLT TEST IN TALENT ASSESSMENT

Above is the overall review on the research about the application of RIT in talent assessment. In general, despite that clinical psychology has demonstrated the burning interests in RIT and has more achievements, RIT's application in recruitment also has a better future. RIT's future application in recruitment is also a hot topic for discuss. Followings are some key points for thinking:

A. Change of Research Pattern

The psychological research on RIT mainly concentrated on the clinical psychotherapy area, leading to some constraints by researchers on the application area of RIT. Although there is research on talent assessment recently, it is still limited to the psychiatry pattern or it did not have the empirical reports [32]. As stated above, the most important module of RIT in the talent assessment is the "Cognitive Function", which has been applied the most widely.

The reason for its intense popularity lies at its function as the typical indicator for the awareness of being tested and the quality of way of thinking. Some scholars commented that some of the assessment codes of RIT, such as color, form quality and content, were suitable to make a distinction between the normal and the abnormal. However, if substituting the central concepts of the modern human resource management to this "psychiatry" pattern, the assessing targets of RIT can be changed into the excellent and the common, as well as the right ones and the unfit ones. Simultaneously, RIT will necessarily be identified as a matured talent assessment tool only with support from various assessment investigation and experimental data.

Therefore, in pursuit of the application of RIT in talent assessment, it is necessary to integrate the pattern of talent assessment center with RIT and utilize advantages of RIT for talent assessment.

B. Development of the scoring system

The scoring codes, variant items, reliability and validity of RIT have been verified by the 50-year empirical experiments and research. The same as its research pattern was originally applied in the clinical psychiatry.

Therefore, some evaluating criteria which incorporated a few major code categories of RIT failed to be directly applied for the daily talent assessment. The new code categories should be developed on the basis of current comprehensive system while the evaluating and scoring criteria need to be remade in pursuit of the application of RIT to talent assessment. Meanwhile, it is necessary to further improve the efficiency of the scoring approach of RIT. With the high efficiency, convenience and accuracy of the scoring setting, the value of applying RIT to talent assessment will be intensively improved.

In addition, the assessment variants and scoring codes of RIT can only figure out what sort of personalities fits some specific coding items, and their degree of correlation. It fails to explain how the coding items interpret the correlation between themselves and the related personalities. The correlation still demands further research from experts in psychology and behavioristics. As long as the question has been solved, it can be widely applied to each field (certainly including the talent assessment).

C. Modification on group tests

The largest advantage of RIT is that it can cover the evaluator's intention so that the answers from the assessed are comparatively objective and reliable. However, it is also not possible to ignore its disadvantages, which are that it is difficult to analysis performances of the assessed in addition that the professional analysts engage in the test. Furthermore, this approach is time-consuming and demands substantial resources. Therefore, it is hardly adopted in the employee recruitment but only in the manager recruitment.

Nevertheless, it does not suggest that only the managers will pretend themselves in the tests. In fact, ordinary employees will tend to cover themselves in the tests. As the development of projective tests for group, the solution to this phenomenon emerged. Japanese scholars has suggested that Group Rorschach Ink-blot Test (GRIT) could be applied to large-scale talent selection and assessment. They made contributions to the acknowledgement of the good validity and reliability of GRIT [33].

However, the current revised version of GRIT only inherited 5 pictures out of 10 in traditional RIT while eliminated the relevant codes of texture factors and 28 content codes in comprehensive system, in pursuit of a simplistic evaluation mechanism for codes of both the cognitive function and form quality. The assessment efficiency of the current version of GRIT on ordinary employees is almost equivalent to the one of general self-report inventories, which can reduce the pictures with function of stimulating reactions, and scoring systems [34].

Indeed, the resources and time the group protective tests require during the implementation will not go up in the condition that pictures and scoring codes are hardly deleted. What will be increased is the research cost for scholars during the modification.

All in all, the application of RIT in talent assessment is an important topic worthy of being studied and is also the sphere that human resource management seldom involves with. It is also the developing sphere that may be supplementary to other talent assessment techniques.

REFERENCES

[1] Judge T A, Bono J E, Ilies R, et al. Personality and leadership: a qualitative and quantitative review [J]. Journal of applied psychology, 2002, 87(4): 765.

[2] Gregory Robert. Psychological measurement: history, summarize and applications (English edition) [M]. Beijing: Peking University, 2005:1-101.

[3] Rieber R W. A History of the Rorschach Test and an Analysis of Bookplates of Famous Psychologists [M]. Freud on Interpretation. Springer New York, 2012: 105-136.

[4] Lilienfeld SO, Wood M.J, Garb HN. The scientific status of projective techniques [J]. Psychological Science in the Public Interest, 2000, 20(1): 27-66.

[5] Schwartz A L. Book Review: Choca, The Rorschach Inkblot Test: An Interpretive Guide for Clinicians [J]. Journal of personality assessment, 2013.

[6] Garb H N, Wood J M, Lilienfeld S O, Scott O. Lilienfeldc, M. Teresa Nezworskid. Roots of the Rorschach controversy [J]. Clinical Psychology Review, 2005, 25(1): 97-118.

[7] Leichtman M. The Rorschach: A developmental perspective [M]. Routledge, 2013.

[8] Exner J. A Rorschach workbook for the comprehensive system (5th Edition) [M]. Wiley and Sons, 2008.

[9] Carter N T, Daniels M A, Zickar M J. Projective testing: Historical foundations and uses for human resources management [J]. Human Resource Management Review, 2013.

[10] Thomas NW. Psychosomatic medicine and the Rorschach Test[J]. Psychosomatics, 2005, 46: 377.

[11] Rafiee Cher. Applicability of the Rorschach performance assessment system norms to an Iranian American sample [D]. Alliant International University, Ph.D, 2013.

[12] Exner J E. The Rorschach: a comprehensive system: basic foundations and principles of interpretation (4th ed.) [M]. New York: Wiley, 2003, 11(1):142- 143.

[13] Roger E. Millsap, L. Andries van der Ark, Daniel M. Bolt, Carol M. Woods. New Developments in Quantitative Psychology [J]. Springer Proceedings in Mathematics & Statistics, 2013, (66).

[14] Qu Xiao-li. The research of the reliability & validity of the teens group Rorschach Inkblot Test [D]. Suzhou University, 2007. (in Chinese)

[15] Hertz M R. The reliability of the Rorschach ink-blot test [J]. Journal of applied psychology, 1934, 18(3): 461.

[16] Gronnerod C. Temporal stability in the Rorschach Method: a meta-analytic review [J]. Journal of Personality Assessment, 2003, 80(3): 272-293.

[17] Vacha-Haase T, Henson R K, Caruso J C. Reliability generalization: Moving toward improved understanding and use of score reliability [J]. Educational and Psychological Measurement, 2002, 62(4): 562-569.

[18] Sahly J, Shaffer T W, Erdberg P, Philip Erdberg & Siobhan O'Toole. Rorschach Intercoder Reliability for Protocol-Level Comprehensive System Variables in an International Sample [J]. Journal of personality assessment, 2011, 93(6): 592-596.

[19] Driskell J E. Effectiveness of deception detection training: A meta-analysis [J]. Psychology, Crime & Law, 2012, 18(8): 713-731.

[20] Garb HN, Florio CM, Grove WM, The validity of the Rorschach and Minnesota Multiphasic Personality inventory: results from meta-analyses [J]. Psychological Science, 1998, 9: 402-404.

[21] Hiller, Jordan B., Rosenthal, Robert, Bornstein, Robert F., Berry, David T. R., Brunell-Neuleib. Sherrie. A comparative meta-analysis of MMPI validity [J]. Psychological Assessment, 1999, 11: 278-296.

[22] Finn S E. Assessment feedback integrating MMPI-2 and Rorschach findings [J]. Journal of Personality Assessment, 1996, 67(3): 543-557.

[23] Blais MA, Hilsenroth MJ, Castlebury F. Predicting DSM–IV cluster B Personality Disorder criteria from MMPI–2 and Rorschach data: a test of incremental validity [J]. Journal of Personality Assessment, 2001, 76(1): 150-168.

[24] Pertchik K, Shaffer T W, Erdberg P, Philip Erdberg , David Ira Margolin. Rorschach Comprehensive System data for a sample of 52 older adult nonpatients from the United States [J]. Journal of personality assessment, 2007, 89(S1): 166-173.

[25] Latife Yazigi, Norma Lottenberg Semer, Terezinha de Carvalho Amaro, Maria Luiza de Mattos Fiore, Julieta Freitas Ramalho da Silva, Nara Lucia Poli Botelho. Rorschach and the WAIS-III after one and two years of psychotherapy [J]. Psicologia: Reflexão e Crítica, 2011, 24(1): 10-18.(in German)

[26] Mihura, Joni L., Meyer, Gregory J., Dumitrascu. Nicolae, Bombe. George. The validity of individual Rorschach variables: Systematic reviews and meta-analyses of the comprehensive system [J]. Psychological bulletin, 2013, 139(3): 548.

[27] Elisa de Villemor-Amaral, Anna. Executive performance on the Rorschach comprehensive system [J]. Rorschachiana, 2006, (01):119-133.

[28] Qu Xiao-li. The Research of applying the intelligence variable of RIT to the talent Assessmen of enterprises [J]. Corporation Research, 2010, (18): 47-49. (in Chinese)

[29] DeGroot T, Aime F, Johnson S G, Scott G. Johnson, Donald Kluemper. Does talking the talk help walking the walk? An examination of the effect of vocal attractiveness in leader effectiveness [J]. The Leadership Quarterly, 2011, 22(4): 680-689.

[30] Del Giudice M J, Brabender V M. Rorschach assessment of leadership traits in college student leaders [J]. Comprehensive Psychology, 2012, 1.

[31] Matsunaga N, Kabashima T, Asukata I, Shirahama K. Psychological characteristics of unsuccessful pilot trainees [J], Uchu Koku Kankyo Igaku, 2000, 37(1): 1-7.
[32] Edward Aronow, Marvin Reznikoff & Kevin L. Moreland. The Rorschach: projective technique or psychometric test? [J]. Journal of Personality Assessment, 1995, 64(2):213-228.
[33] Weiner IB. Using the Rorschach properly in practice and research [J]. Journal of Clinical Psychology, 2000, 56: 435-438.
[34] Laudel G. What do we measure by co-authorships?[J]. Research Evaluation, 2002, 11(1): 3-15.

Research on the Innovative Development of Higher Engineering Practice Education Based on the Engineering with a Big E

Zhuo-ping WU [1,*], Hong-zhe ZHANG [2], Lin JIANG [1]
[1]Faculty of Humanities and Social Sciences, Dalian University of Technology, Dalian, 116024
[2]School of Mechanical Engineering, Dalian University of Technology, Dalian, 116024
([*]wuzhuoping1987@126.com, zhanghongzhe@dlut.edu.cn, jianglin66@163.com)

Abstract - **Innovation is the soul of higher engineering practice education. It is a though-provoking problem that how to achieve the innovative development of higher engineering practice education so that China can turn into a powerful country in engineering education. The concept of engineering with a Big E education which emphasizes on integrity and systemic providing the innovative development of engineering education with important gist. The concept of engineering with a Big E education has the logical relationship of intrinsic unification with the innovation of higher engineering practice education in terms of similar background, consistent target, systematic knowledge, practical content and innovative result. To cultivate innovative engineering talents, set up systematic and practical courses, construct practical environment with engineering culture and cooperate with colleges and enterprises, we can find a way for the innovative development of higher engineering practice education.**

Keywords - Big E, higher engineering practice education, innovative development

I. INTRODUCTION

To construct of an innovative country and promote engineering technological progress and industrial development, China has carry out the higher engineering practice education. This program aims to find an appropriate way to cultivate the practical ability of engineering talents under the background of engineering with a Big E. CAE (Chinese Academy of Engineering) pointed out that “China is facing important strategic opportunities for the acceleration of cultivating innovative engineering qualified scientists and technicians in the next decades” in the project of “Innovative engineering qualified scientists and technicians training research”. Even though China has established a great system of engineering education, there still exist many problems such as industry-academy cooperation is not close, theory teaching and practice training disjoint and innovative ability cultivation is ignored. Practice is the core of engineering education, and the higher engineering practice education is an important part of engineering education. Innovation is the soul of higher engineering practice education. The innovative development of higher engineering practice education is the key point to turn China into a powerful country in engineering education. “Theory is the forerunner for practice, and thought is the guide for action.” Chinese state education commission delegation of engineering education to the USA firstly put forward the concept of engineering with a Big E education in <return of engineering ·diversification· macro-management-the investigation report of the USA> and <“return of engineering” and higher engineering education reform in the United States> in 1996 [1]. After that, Chinese academics have made a certain degree of development for the engineering with a Big E education in both connotation and denotation. On the basis of these studies, we think the engineering with a Big E education providing the research on innovative development of higher engineering practice education with important gist. This article will use the view of engineering with a Big E education to lead the innovative development of higher engineering practice education, to enrich our country's theory system, guide the practice of higher engineering education, and provide reference for reform and innovation of engineering education in our country.

II. CONNOTATION AND DENOTATION OF THE ENGINEERING WITH A BIG E EDUCATION AND HIGHER ENGINEERING PRACTICE EDUCATION

A. *The engineering with a Big E education*

As the powerful nation of higher education, the USA attaches great importance to higher education of engineering in the role of promoting economy development. It has been exploring a new idea of cultivating engineering qualified scientists and technicians that cater for the times demand. The Principal of Massachusetts Institute of Technology (MIT), put forward the concept of Engineering with a Big E in 1993. Joel Moses, Dean of college of engineering at the Massachusetts institute of technology, put forward “Engineering with a Big E: Integrative Education in Engineering: Long Range Plan for the School of Engineering, 1994-1998”. In his point of view, the term of Engineering with a Big E is a return of engineering education that services for the engineering practice, which is contrary to engineering science which is oriented on research [2]. National Science Foundation of United States has pointed out the importance of engineering with a Big E that emphasized on knowledge

E. Qi et al. (eds.), *Proceedings of the 5th International Asia Conference on Industrial Engineering and Management Innovation (IEMI2014)*, Proceedings of the International Asia Conference on Industrial Engineering and Management Innovation 1, DOI 10.2991/978-94-6239-100-0_70, © Atlantis Press and the authors 2015

integrity and systemic in <Refund engineering education: Focusing on reform-NSF report of engineering education workshop > in 1995. "In the Engineering with a Big E, modern engineering is a complex system and it requires engineers break the barrier of different subjects and restore engineering separated by subjects as a whole. Higher engineering education should not only let students learn knowledge and theory of engineering science, but also let them contact the large-scale complex system analysis and management. Should not only integrate to the related technical knowledge, but also enhance the understanding of economic, social, political and technical system" [3], as shown in Fig.1.

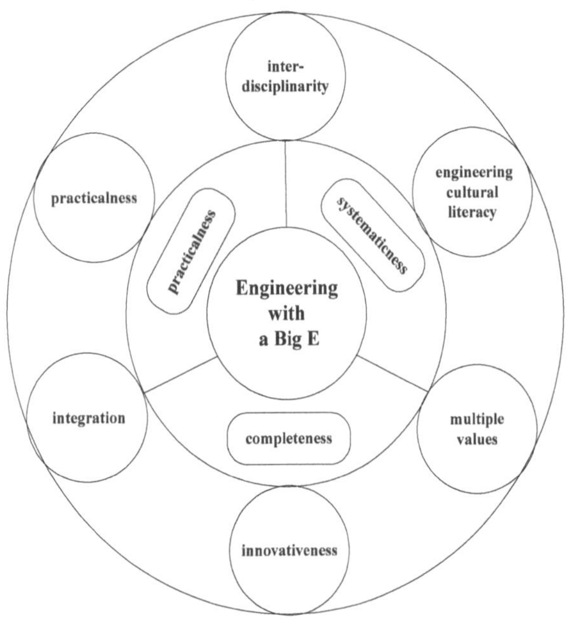

Fig.1. Evolution dynamic graph of the Engineering with a Big E

B. *Higher engineering practice education*

Higher engineering practice education is a program based on colleges and universities and in virtue of Engineering training center and Engineering practical education center. In order to solve the problem of the engineering practice, accumulate engineering practice experience and have influence on students through practical education activity, educators should have a target to guide students to participate in project design. The purpose of higher engineering education is to cultivate students to grasp more knowledge of engineering, possess the ability of practice, cooperation, innovation, adaption and engineering cultural literacy. Fu Shuigen, professor of THU (Tsinghua University) said "Higher engineering practice education includes ordinary labor practices, teaching experiment practice, engineering training practice, industrial training practice and social investigation practice. The engineering training practice has been one of the most representative and epochal way of practice education for modern higher engineering practice education." [4]

III. THE LOGICAL RELATIONSHIP BETWEEN THE CONCEPT OF ENGINEERING WITH A BIG E AND THE INNOVATIVE DEVELOPMENT OF HIGHER ENGINEERING PRACTICE EDUCATION

The innovative development of higher engineering practice education is a complex dynamic project. The concept of engineering with a Big E education has the logical relationship of intrinsic unification with the innovation of practical higher education in terms of the similar background, the consistent target, the systematic knowledge, the practical content and the innovative results, as shown in Fig.2.

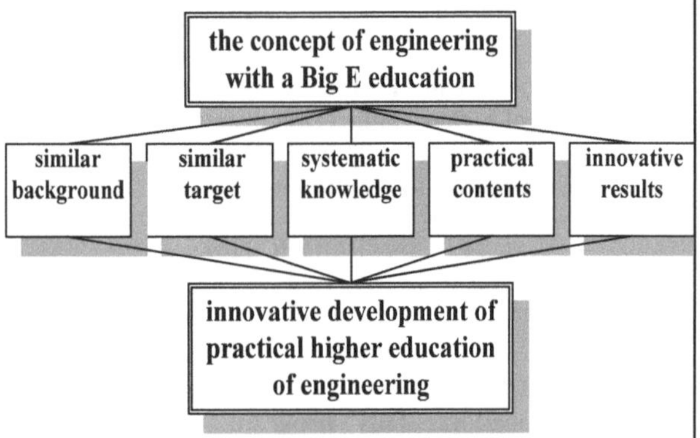

Fig.2. The graph of logical relationship between the concept of engineering with a Big E and the innovative development of higher engineering practice education

A. *The similar background*

Since the 20th century, the education model in the USA has experienced technical mode, scientific mode and engineering mode, as shown in Table I [5].

TABLE I
THE EDUCATION MODEL IN THE USA

Decades	Before 1940s	1940s-1980s	1990s to now
Mode	Technical mode	Scientific mode	Engineering mode
Educational concept	Traditional engineer-ing concept	"Engineer-ing science program" oriented education concept	Engineer-ing with a Big E

The higher education of engineering in China is now focusing on cultivating academic talents for universities and research institutions. Not enough attention is paid to practical innovation, which affected the process of building an innovative country. All sectors of society also call up for "attach great importance to practice". From the analysis of the American engineering education development, the backgrounds of Chinese Higher engineering practice education is similar with the background of American engineering education under the engineering with a Big E.

B. The similar target

Since the 1990s, American "return to engineering movement" contributed to a new wave of reform of higher engineering education. As the result of "return to engineering movement", the engineering with a Big E aims to reform the engineering education system which is over engineering-scientization, attaches great importance to the engineering practice and realize the innovative development of engineering technology [6]. Formal President Jiang Zemin said that: "Innovation is the soul of a nation's progress, is the inexhaustible driving force for the prosperity of a country." As an important part of the Chinese engineering science and technology talents cultivation, higher engineering practice education need to realize the innovative development.

C. The systematic knowledge

Because society is an integral whole, the concept of engineering with a Big E emphasizes on "Let students contact with the large-scale complex systematic analysis and management." This not only refers to the related technical discipline knowledge integration, but also refers to the increasingly increase of understanding to wider economic, social, political and technical system [7]. In order to realize the innovative development, the systematicness of knowledge is becoming more and more important for the higher engineering practice education. It will emphasize on the combination of natural science, technological science, engineering science and engineering disciplines. And pay attention to teach humanities and social science disciplines such as moral philosophy, sociology and management knowledge. Making an improvement of students' engineering culture accomplishment.

D. The practical contents

Practice is the soul of the project. The core content of American "return to engineering movement" is to reform the "over engineering-scientization" in engineering education system [5]. The essential of engineering with a Big E is "Set up an integrated, innovative and practical system of engineering education through the engineering practice."This is an innovative integration for engineering discipline based on the emphasis of engineering practice [8]. The target of higher education of engineering is to cultivate engineer rather than scientist. There is different teaching method and content for higher engineering practice education. As the shining points of Chinese higher engineering practice education, training is a unique way and has got a great achievement through decade's development. Fu Shuigen, a professor of THU (Tsinghua University) said: "In the system of cultivating engineering talents, it needs strong ability of engineering practice no matter for undergraduate or postgraduate. To solve this problem, we ought to improve the social engineering practice environment and strengthen the engineering practice education consistently through practice teaching and experiment teaching."

E. The innovative results

Professor von karman from the United States at the California institute of technology said: "Scientists study the existing world, engineers create the world in the future." Under the guidance of the engineering with a Big E, the Key ability of engineering talents is innovative ability. Engineering talents with innovative ability can obtain creative achievements and promote technical progress [9]. To promote force of China turning into a powerful state of engineering education, the higher engineering practice education can not limited to the number of talents any more. We need to pay attention to the quality of talents. Only by realizing the innovative development, can the higher engineering practice education achieve connotative ascension.

IV. THE DEVELOPING WAY OF THE HIGHER ENGINEERING PRACTICE EDUCATION UNDER ENGINEERING WITH A BIG E

To improving the quality of higher engineering practice education and cultivating innovative engineering talents, needs joint efforts of higher education of engineering field inside and outside. As for the higher engineering practice education , it means under the guidance of engineering with a Big E, the target of innovative development can be achieved through setting up the education objectives and curriculum, constructing engineering environment and selecting training mode, as shown in Fig.3.

A. Education target: cultivating innovative engineering talents

Xu Kuangdi, the Vice chairman of the CPPCC national committee, dean of the Chinese Academy of Engineering said in the third engineering education meeting of Chinese academy of engineering committee: "Our country is still in its medium-term of industrialization. In order to realize the goal of quadrupling GDP by 2020, we urgently need a large number of innovative engineering talents." Colleges and universities are important positions to cultivate engineering talents, and higher engineering practice education is the fertile soil for breeding innovative engineering talents. Pan Yunhe academician pointed out that: "To take a new road to industrialization and build an innovation-oriented country, we must give full play to the human resources of our country, build a creative engineering talents personnel." The innovative development of higher engineering practice education need to set a goal as cultivating innovative engineering talents for constructing a creative country and provide national engineering technology progress with talent support.

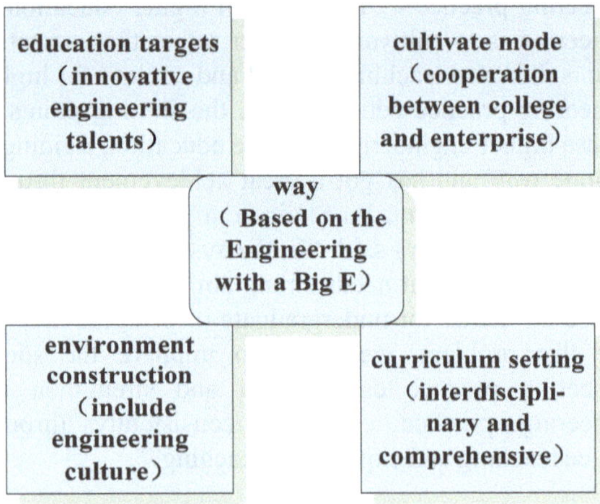

Fig.3. The developing way of the higher engineering practice education under engineering with a Big E

B. Curriculum setting: systematicness and practice

As modern engineering crosses and integrates with social, political, economic, ecological environment and multi-culture, higher engineering practice education needs to break the barrier of the disciplinary and professional boundaries, set interdisciplinary practical courses according to the characteristics of different profession. Take engineering technology as the main line, integrated practice teaching content systematically and practically. For example, MIT requires students study 19 subjects such as natural science, engineering, information technology and humanities and social sciences. Stanford University stipulated that: "Students should have broad understanding of engineering project as some kinds of social activities are very important. In order to increase students' intellectual and professional development ability in this respect, the university requires all engineering students must choose courses that have great help in solving problems occurred in the interaction of engineering, technology and society." [7]

C. Environment construction: including engineering culture

Engineering environment is the place where engineering talents do engineering practice and accept engineering education. It not only includes physical environment, but also includes culture environment. These will affect both engineering talents' body and mind. Engineering culture is a branch of cultural studies, with the general characteristic of organizational culture and the particularity of engineering practice. It is the summarization of behavior expressed during the conscious practice activity. These behavior include many aspects like values, mental state and emotional tendencies. In the construction of education environment, higher engineering practice education needs to pay attention to include engineering culture, create a cultural atmosphere and cultivate literacy of engineering talents. Only in this way can make them set off from the development of human being and people's revenue, break the limitation of pure technical perspective and correctly identify the engineering activities consciously. In addition, it can guide engineering activities, develop creative projects and promote the innovative development [10].

D. Cultivate mode: cooperation between college and enterprise

Cooperation between college and enterprise is a combination of the market demand and whole benefit. It is the docking and fusion of technical innovation of midstream, downstream and upstream. There still exists disconnection between practices and learning in recent higher engineering practice education Enterprise's enthusiasm for participation in engineering education in colleges and universities is not high due to lots of reasons. Universities and enterprises lack of effective cooperation in training students. In order to solve this problem, the Ministry of Education has set up a national engineering practical education center. It aims to build platform of talent training between college and enterprise. Under the background of new era, college and enterprise need to have close contact between each other, adapt diversified means of cooperation, provide diversification practice innovation platform for students, and promote the practical higher engineering education innovation and development.

V. CONCLUSION

Daniel Gross, Director of the center for European policy said that: "China is still in the stage of sweat type growth". With the improvement of Chinese economic development level, we must attach importance to the independent innovation ability [11]. The key to achieve economic transformation and accelerate the scientific and technological conversion efficiency is to promote the innovative development of higher engineering practice education, cultivate a batch of innovative engineering talents to adapt to the social and economic development. Only through exploring a new road for innovative development to adapt to the demand of eras, can we cultivate innovative engineering talents meeting the needs of national construction.

ACKNOWLEDGMENT

This work was supported in part by a grant from the Humanities and Social Science Foundation of Ministry of Education of China (13JDGC008).

REFERENCES

[1] The delegation of State Education Commission Engineering Education, "Back to Engineering Diversification Macro-administration" ,*Research in Higher Education of Engineering*, vol. 7, Dec. 1996, pp. 7-15. (In Chinese).

[2] Joel Moses, Engineering with a Big E: Integrative Education in Engineering. *Massachusetts: School of Engineering Massachusetts Institute of Technology Cambridge*, 1994.

[3] NAE, *The Engineer of 2020: Visions of Engineering in the New Century.* Washington DC: the National Academies, 2004.

[4] Fu Shuigen, "Historically looking back and developing trend on the higher engineering practice education", *Experimental Technology and Management,* vol. 2, Dec. 2011, pp. 1-4. (In Chinese).

[5] Xie Xiaozhen, "An Exploration of the Connotation and Essential Features of <Large-scale Engineering>", *Research in Higher Education of Engineering,* vol. 3, Dec. 2008, pp.35-38.(In Chinese).

[6] Wang Xuefeng and Cao Rong, "Engineering with a Big E and the Higher Engineering Educational Reform", *Research in Higher Education of Engineering,* vol. 4, Dec. 2006, pp. 19-23. (In Chinese).

[7] Zhao Tingting and Mai Nannan, "The characteristic analyses of curriculum setting in America higher engineering education based on the idea of large-scale engineering --A comparative research between MIT and STANFORD", *Journal of Higher Education,* vol. 6, Dec. 2004, pp.94-101. (In Chinese).

[8] Ju Likai and Xu Jiancheng, "Exploration and practice of Engineering Practice Teaching Reform Based on Engineering with A BIG E", *China University Teaching,* vol. 10, Dec. 2013, pp. 68-70. (In Chinese).

[9] Zeng Lijuan, Du Min and Ma Yunkuo, " Exploration of the reform of Higher Engineering Education under the concept of value engineering", *Education Exploration,* vol. 9, Dec. 2013,pp. 35-36. (In Chinese).

[10] Li Peigen, "Engineering Education Needs the View of Engineering with A Big E", *Research in Higher Education of Engineering*, vol. 3, Dec. 2011, pp. 1-3,59. (In Chinese).

[11] Daniel Gross. The National Development: from "sweat growth" to "innovation growth" [EB/OL]. [2013-11-1]. http://www.qstheory.cn/zl/bkjx/201311/t20131101_285318.htm.

Research on Real Estate Project Risk by Life Cycle Theory

Li-Jun GAO*, Fu-zhou LUO
School of Management, Xi' an University of Architecture & Technology, Xi'an, China
(Gaolj_0817@163.com)

Abstract - **There are a lot of risk factors in the various stages of real estate project such as investment decision-making, land ownership obtaining, project implementing and marketing & property management. Project risk management quality affects the economic benefits of real estate investment projects, and then determines the success or failure of the real estate investment projects. Therefore, real estate project risk management research has important realistic pertinence and necessity. On the basis of analysis of the current real estate project risk management problems, and then analyzes systematically the life cycle of real estate project risk management process, organization structure and basic procedure in the paper.**

Keywords – **Basic procedure, Life Cycle, process matrix, project risk, real estate**

I. INTRODUCTION

Real estate is the pillar industry of national economy, as well as a vast power in promoting social and economic development. The real estate industry is capital intensive, labor intensive industry, which has the characteristics of large-scale investment, long construction period and slow recovery of funds, so it can face more operating risks than other enterprises. The research on how to improve consciousness and ability in the project risk management of real estate enterprises, not only can ensure project proceed as planned, but also can promote the sustained and healthy development of real estate enterprises.

Project Risk Management constitutes an important part of project management. Based on the perspective of risk management process, it can be divided into five steps: risk management planning, identification, analysis, response plan and monitoring risks [1]. The fundamental means of real estate project risk management is to increase the probability and impact of positive events, and decrease the probability and impact of negative events in the project. However, the concepts or consciousness of real estate enterprises in the real estate project risk management is still lagging behind, the focus of risk management for real estate project mainly distributed in the phase of the project investment decision-making in china. The work scope of real estate project risk management is usually confined to the project feasibility study stage, is far from throughout the whole process of project management. There are two drawbacks of aspects at least in this situation. In essence, the current risk management in real estate project is divided into two independent phases: risk analysis and risk management by the project implementation timing. The enough and excessive concern is given to the former, however the latter is quite opposite. It is often in a semi ring management or single loop management state, which is not conducive to draw a lesson from failure and accumulate management experience through success [2]. On the whole, the current level of real estate project risk management is not adapted to the needs of work practice. Therefore, it is necessary to break the untimely idea and to seek the proper concept in the process of real estate risk management.

II. THE PROCESS MATRIX OF RISK MANAGEMENT IN THE REAL ESTATE PROJECT LIFE CYCLE

The basic characteristics of the real estate project are large investment, long construction period and fixed location. These features make the real estate project has a lot of uncertain factors in the various stages of investment decision-making, land ownership obtaining, project implementing, marketing and property management etc. The life cycle of risk management of real estate project refers to the identification, assessment, response, monitoring of risk throughout the life cycle of project by using systematic theories and methods in order to reduce the risk threshold of the project and improve risk management benefit [3].

The risk identification is the important prerequisite and foundation in the risk management of real estate project. According to the time and range of risk factors, the real estate project risk factors can be divided into comprehensive risk and phased risk. Comprehensive risk factors refer to risk factors throughout the whole life period of real estate project. Such as natural risk, political risk, economic risk, social risk, technology risk etc. Phased risk factors refer to risk factors unique to a certain stage of real estate project development. For example, there may be a location selection risk, regional development risk and property type selection risk etc.

The risk assessment is generally divided into qualitative and quantitative analysis method [4][5]. The most commonly used qualitative risk analysis methods including Fault Tree Analysis, Trend Extrapolation, Brain Storming, Delphi Method and Subjective Grading Method etc. The most commonly used quantitative analysis methods including Expected Value Optimization Method, Program Evaluation and Review Techniques, Analytic Hierarchy Process, List Sorting Method, matrix analysis method etc. Effective risk assessment works are good for us to deeper understand the meaning of the major risk factors of real estate project.

Risk response refers to formulate corresponding measures to deal with risks and reduce the negative impact of the project risk according to the results of the risk assessment. General risk countermeasures include

E. Qi et al. (eds.), *Proceedings of the 5th International Asia Conference on Industrial Engineering and Management Innovation (IEMI2014)*, Proceedings of the International Asia Conference on Industrial Engineering and Management Innovation 1, DOI 10.2991/978-94-6239-100-0_71, © Atlantis Press and the authors 2015

risk aversion, risk transfer, risk mitigation, risk retention etc.

Risk monitoring refers to tracking identified risks, identifying remaining risks and new risks, adjusting the risk management plan timely and assessing the risk reducing efficiency. On the basis of the foregoing analysis, we can establish the risk management process matrix in the life cycle of real estate project (as shown in TABLE I).

TABLE I
RISK MANAGEMENT PROCESS MATRIX IN THE LIFE CYCLE

Risk Management Process			
The Life Cycle of Real Estate Project	investment decision-making	identify	Risk(natural, political, economic, social, technical) > Risk(location selection, regional development, property type selection)
		assess	qualitative > quantitative
		response	design risk strategy
		monitor	tracking risks
	land ownership obtaining	identify	Risk(natural, political, economic, social, technical) = Risk(land ownership obtaining, resettlement, financing)
		assess	qualitative = quantitative
		response	formulation and implementation of risk response
		monitor	tracking risks, identifying new risks
	project implementing	identify	Risk(bidding, contract, duration, quality, cost) > Risk (natural, political, economic, social, technical)
		assess	quantitative=qualitative
		response	formulation and implementation of risk response
		monitor	tracking risks, identifying new risks, adjusting the risk management plan, assessing risk reducing efficiency
	marketing, property management	identify	Risk (market, price, marketing, property type selection) > Risk (natural, political, economic, social, technical)
		assess	quantitative > qualitative
		response	formulation and implementation of risk response
		monitor	tracking risks, identifying new risks, adjusting the risk management plan, assessing risk reducing efficiency

III. THE BASIC PROCEDURE OF RISK MANAGEMENT IN THE LIFE CYCLE OF REAL ESTATE PROJECT

Once the content of risk management in the life cycle of real estate project being analyzed and clarified, we can put forward an overall, dynamic risk management system framework of real estate project (as shown in Fig.1). Based on the framework, we can also understand deeply the basic procedure of risk management in the life cycle of real estate project [6][7][8]. Real estate project risk will be effectively reduced only if we implement dynamic and continuous multiple cycle management to real estate project comprehensive risks and phased risks in the various stages of real estate project life cycle [9][10].

Based on the theoretical framework of real estate project risk management, we implement project risk management has the following advantages: First of all, we can acquire the basic risk information of each stage for the purpose of to respond risk effectively in the project through the effective feedback mechanism. Second, we have accumulated a lot of risk information by dynamic cycle management for many times, it is valuable for establishing a valuable risk management database [11].

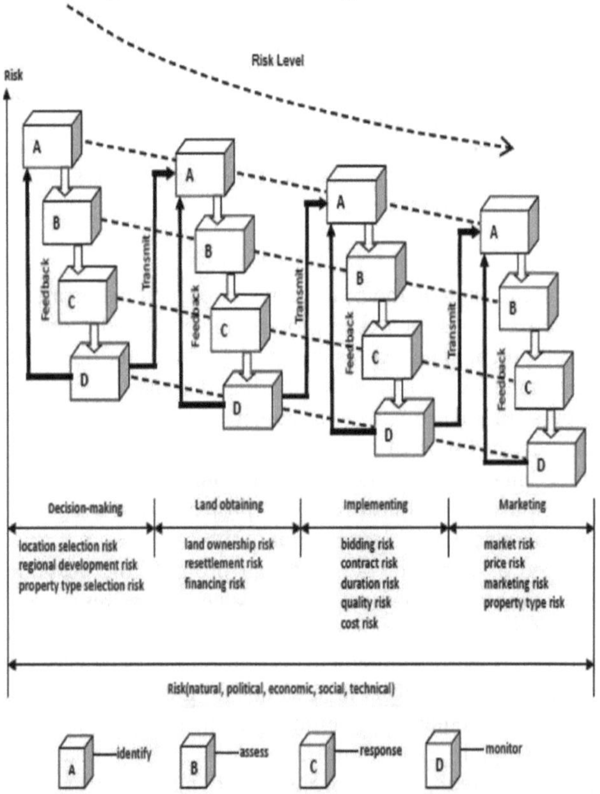

Fig.1. Procedure of risk management in the life cycle

IV. THE ORGANIZATIONAL STRUCTURE OF RISK MANAGEMENT IN THE LIFE CYCLE OF REAL ESTATE PROJECT

Classical organizational structure of project risk management mainly divides into three types: functional organizational structure, matrix organization structure and projectized organization structure [12][13]. Real estate project risk management organization structure is based on time and space scope of risk management that should be open, flexible and dynamic organization structure. Real estate project risk management organizational structure depends on various factors. One of the most important of these factors is a project risk in time and space distribution characteristics. The function of project risk management in the organization to perform showed the characteristics of the combining of centralization and decentralization.

A. Functional Organizational Structure

The functional organizational structure is suitable for project risk management in the stage of investment decision-making, land ownership obtaining and marketing or property management. The personnel derives mainly from engineering department, marketing department, finance department is responsible for duties and responsibilities of risk management under the leadership of project manager. Specifically, these people may include real estate appraiser, cost engineers, consulting engineers, lawyers, insurance consultants, security and environment evaluation engineers etc.(as shown in Fig.2) Generally speaking, with differences in the scale of the enterprise, its risk management functions have different centralization degree [14][15]. When the lower level of centralization of risk management functions in the real estate project, the management responsibility should be refined corresponding.

B. Matrix Organization Structure

In the project implementing stage, using the matrix organization structure is suitable for real estate project risk management (as shown in Fig.3). At this stage, the focus of real estate project risk management in the project management department. When at the beginning or the end of project construction, a weak matrix organization structure should be adopted by project risk management. Project risk management must have a strong matrix organization structure only if when the project is in the middle construction.

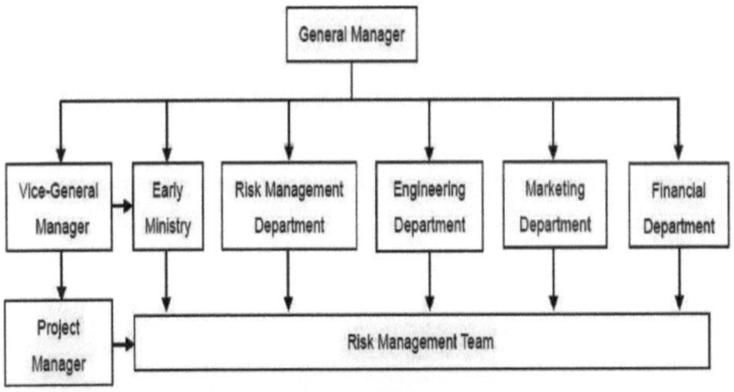

Fig.2 Functional organizational structure of risk management

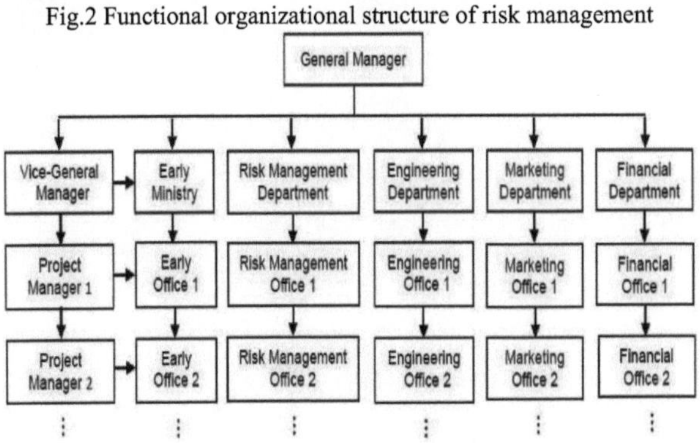

Fig.3 Matrix organization structure of risk management

REFERENCES

[1] Qi Ming Li, Lei Zheng, Xiao Peng Deng, Liang Feng Shen. *"Modern life cycle of real estate project management"*, Jiangsu, Nankin: Southeast University Press, 2006, pp.34-40.

[2] Yuan Hua Qiu. *"Modern project risk management theory and practice"*, Peking: Science Press, 2003, pp.85-94.

[3] Project Management Institute. *"A guide to the project management body of knowledge"*(5th Edition), Newtown Square, Pennsylvania: Project Management Institute Inc.2013, pp.47-60.

[4] Wei Hui Zhang, Zhi jun Yin, Liwen Chen. "Based on the whole life cycle of a project risk analysis method of comparative study," *Science and technology management research*. no.9, pp.104-106, 2007.

[5] Zhuo Fu Wang, Fei Tan. *"The theory, method and application of management in engineering project"*, Peking: China Water Conservancy and Hydropower Press, 2003, pp.203-208.

[6] Xin Guo Jiang. *"Commercial real estate development and management "*, Peking: China building industry press, 2007.pp.47-50.

[7] Jian Min Cai. *"Study on the Risk Management Methods for Real Estate Project"*, Beijing Jiao tong University, Beijing, China, Ph.D. dissertation, 2009.

[8] Jun Liu. *"The Risk Analysis and Evaluating of the Real Estate Project Based on Theory of Life Cycle"*, Northeastern University, Shenyang, China. Ph.D. dissertation, 2010.

[9] Jian Min Cai. "Research on real estate project risk management project based on the dynamics", *Modernization of Management*.no.4.pp. 55-57,2006.

[10] Fang He. "Research on Risk Recognition of Real Estate Projects Based on MIV-BP Neural Network Test", *Operations Research and Management Science*. no.2. pp.229-234, 2013

[11] Bin Jiang Li. "Research on the Intergrate Risk Evaluation of the Life Cycle High-tech Project Based on AHM", Value Engineering, no.14.pp.162-163,2012.

[12] Li Cai Lei, Jing Zhou. "Risk Control Model for Large-scale Project Based on Lifecycle Synthesis", *Soft Science*.vol. 25, no.10, pp. 27–31, Oct. 2011.

[13] Shu Kuan Zhao, Li Ma. "Research on the investment risk evaluation system of large-scale real estate projects", *Science & Technology Progress and Policy*, no.4. pp.118-119, April. 2002.

[14] Seung H H, Kim D Y, Kim H. "A Web-based Integrated System for International Project Risk Management", *Automation in Construction*, no.17, pp.342-356,2008.

[15] Chang Hong Jin. *"Modern real estate project dynamic integrated risk management"*, Peking: Guang Ming Daily Press, 2011, pp.74-83.

A Study of the Procurement Reformation of Government Technology Entity by Backcasting Approach to Transform to an NDPB

Ming-Te Hsu*, Dah-Chuan Gong, Cheng-Chang Chang
Department of Industrial and Systems Engineering, Chung Yuan Christian University, Chung Li, Taiwan
(*hsumingtetwn@gmail.com, gongdc@cycu.edu.tw, g9502404@cycu.edu.tw)

Abstract - **Under the global trend of adopting arm's-length body as an edged tool in organizational change of public entities, the Taiwan authorities also have realized the legislation of its Non-departmental Public Bodies (NDPBs) Act. Chung-Shan Institute of Science and Technology (CSIST) is one of the NDPBs transformation in Taiwan. This paper focuses on the practices of procurement and supply management (PSM) reform of CSIST. Through the exerting of backcasting method, this thesis derives the PSM reform towards the desirable goals, centralized purchasing and e-procurement. The flow charts and Unified Modeling Language (UML) sequence diagrams are used to present some details about the ideal PSM organizations, procedures and IT system in the future. This thesis not only can be used as the basis in the PSM reform of CSIST while transforming into an NDPB. Moreover, when other large entities, with similar PSM features to CSIST, need to transform into NDPBs and execute the PSM reform, they also can refer to this study.**

Keywords - **Backcasting, non-departmental public bodies, procurement and supply management**

I. INTRODUCTION

After decades of preparation, the Taiwan authorities eventually started to realize broad reform of its government organizations since 2013. This reform, i.e. modifications to the organization of the Executive Yuan, was mainly focused on building a streamlined, flexible and effective government [1]. Besides, a type of NDPBs, was also utilized by Taiwan. They created four NDPBs into the government to let its administration structure more flexible so as to have enough dexterity to deal with special public activities [2, 3]. The concept of Taiwan's NDPBs was resulted from "arm's-length bodies/agencies", a term called by the Organization for Economic Co-operation and Development (OECD) and scholars in new public management domain [4-6]. OECD once described the arm's-length bodies that: government distributes some responsibilities to bodies at arm's length from the control of politicians, with different organizational structures from traditionally ministries, and management autonomy from too much political influence. As to the rationale of creating these entities, OECD further argued that: to make the government system more efficient and effective or, to legitimize the right of decision making by providing some independence from direct political intervention. Such special public-private mixed arrangement gives considerable flexibility to these non-traditional government bodies in the aspects of organization, personnel, accounting, purchasing as well as the ways to do public activities. Because of such flexibilities and benefits, in the past few decades the most important organizational change that has taken place within the governments in the world has been the erections of "arm's-length bodies" [7-10].

Under such global public reform trend, the Taiwan authorities also evaluated this kind of non-traditional entity for a period of times and tried to introduce arm's-length bodies into the overall organizational change as a supporting policy. As to the approximate features of Taiwan's NDPBs, they consists of: a board as the top management with considerable decision-making rights, partly grant from the central government, deregulation in organization, personnel, accounting and procurement rules, and performance appraisal executed by internal and external auditors. From these features we found that the Taiwan government wants to widely deregulate the law limits of NDPBs and to delegate more operational rights to these special public entities. CSIST is the largest one and could be the most complex case in Taiwan's NDPBs policy [11, 12]. Myriads of institutions and regulations need to be altered. Organization structure, promotion system, report system, strategy reorientation, even culture adjustment are also in need of to be changed. In order to assist CSIST elevating its overall operation performance while transforming to an NDPB, this thesis conducted a study focused on the topic of PSM reform [13]. The premise of this study is that the procurement legislation of Taiwan had been partly deregulated for NDPBs. While CSIST becomes an NDPB, it is estimated that more than half of its purchase cases, generally indicating which value are below specific threshold or which fund are not mostly granted from government, don't need to follow the public procurement legislation again. Instead, CSIST can introduce the methods of business management to improve its procurement performance. This legislation alteration gives CSIST unprecedented opportunities to greatly change its procurement plight [14-16]. If CSIST can effectively emulate the expertise in business style PSM field and change its overall procurement model, then elevating overall procurement efficiency and effectiveness will come true and thereby successfully promoting the overall operational performance.

In consideration of above situations, we especially selected the backcasting as the research approach. Because the PSM reform of CSIST is a complex, large-scale, multi-discipline, having to leap dominant public procurement trend and long-period lasting issue. It is needed to thoroughly and systematically check all the facets about PSM institution of this large entity and followed by exploring the possible method to change. Therefore, a more comprehensive and problem-solving approach is needed. Under such context, we prefer the

E. Qi et al. (eds.), *Proceedings of the 5th International Asia Conference on Industrial Engineering and Management Innovation (IEMI2014)*, Proceedings of the International Asia Conference on Industrial Engineering and Management Innovation 1, DOI 10.2991/978-94-6239-100-0_72, © Atlantis Press and the authors 2015

backcasting which is especially suitable to handle large-scale and complicated project [17].

Meanwhile, the backcasting research framework of this study is also proposed is this study. Through deep exploration of private-style PSM literatures and practitioner experiences, critical strategies in contemporary PSM field are identified. Thereby, the desirable PSM future goals for CSIST are constructed. This helps us to understand the outline of the dominant public-style PSM activities of CSIST [18-20]. Thereby, the baseline of the reform can be gotten. About the PSM organizations and procedures in this reform are presented. Also, UML sequence diagrams are applied to present these procedures' dynamic behaviors and to be used as a guideline for successive programming in the PSM reform of CSIST. Thereby, a future depiction of a new private-style e-procurement system can be clearly presented.

II. METHODOLOGY

The aim of this thesis is to do a research on the PSM reform of a large government technology institute in Taiwan. According to the situations applicable to using backcasting, argued by Dreborg (1996), this study was just the case which was suitable to done via backcasting. Firstly, "when problem is complex". Reforming the PSM institution of CSIST, from public-style to private-style, is very complex. The complexity results from many facets: from tangible political, legal and institutional to intangible logical, conceptual and cultural aspects. Especially when this reform will be done in a developing country, the path to thoroughly reform is indeed an intricate problem. Secondly, "need for major change". The procuring activities of a public entity have to be compliant with the Government Procurement Act all the time. Now, the government of Taiwan has released the limits of the procurement regulation for NDPBs and permits them to introduce private style procurement skills. Therefore, if CSIST can't conduct a major change on its PSM model, it will soon face an awkward situation, mainly means: bad PSM performance and unsatisfied whole organization's performance. Thirdly, "dominant trends are part of the problem". This is right the situation for the PSM reform for CSIST. The dominant trend is the public-style PSM model: from ordinance, regulation and institution to value system, knowledge base and behavior pattern. All of them are bound to strict legal system. So, when a group of public style officials start to conduct a transformation towards the business style, they are prone to be tied to the dominant trend. Thus, the dominant trends, i.e. public PSM system, are doubtlessly the major part of the problem while reforming. Forth, "time horizon is long enough to allow considerable scope for deliberate choice". Transforming a large entity, like CSIST, to an NDPB is indeed a complicated project. It nearly has to totally change the overall operational mode. Therefore, enough time to execute such transformation is needed. In other words, the project team has to spend enough time to elaborately and carefully analyze, research, and construct the PSM reform details in all aspects. Under such context, backcasting method is suitable to do this study. In summary, backcasting is suitable for CSIST to conduct its PSM reform, from a public-style to a private-style. Since backcasting was selected, the next step was to devise the backcasting method of this study. As to the constructing of the backcasting method of this study, we observed that The Natural Step (TNS) four-phase (A-B-C-D) methodology was clear, simple and easier to utilize. Also, the goal of TNS was to do strategic planning in companies and other organizations which was just the case of this study. Further, such method had been applied in more than 60 major corporations in Europe and the United States, e.g. IKEA, Scandic Hotels and Electrolux, with substantial quantitative and qualitative results (Holmberg, 1998) [21]. Therefore, though TNS was often used on sustainability development relevant topics, We still considered that the TNS four-phase (A-B-C-D) methodology was applicable to do this study. In addition to TNS, the theses of MacDonald (2005) [22] and Gong and Chen (2012) [23] were also cited. We find that the steps proposed by these three literatures are very similar, just with subtle differences among them. But the logics of their four-step methodologies are identical. In this same methodology, we decided to construct a four-step backcasting method for this study.

III. PROBLEM ANALYSIS

We noticed that the gaps between As-Is PSM model of CSIST and the reform targets constructed are really huge. Lots of institutions, systems, organizations, activities, job assignment logic etc needed to be altered. Even though there are so many tasks need to be changed in PSM reform for CSIST, but through deliberation we know that these tasks are not the real questions. Basically, the most important problems of this reform are "the thoughts". Inevitably, the prepossession of public-style PSM knowledge has deeply embedded in the brains of the government members. If all the PSM members, including the top management level, lack the awareness of the importance of learning the private-style PSM knowledge, how can they conduct an effective private-style PSM reform? Therefore, the vital problems of the PSM reform for CSIST are that: the reform ideas are still be bound by prejudiced public-style PSM thoughts and the reform is still conducted by original public-style PSM units and members. It needs to be emphasized that these members are still untrained in private-style PSM expertise. Under such situation, the gaps of the PSM reform may never be filled. These situations can be depicted like Fig.1.

First of all, the management level still lacks attempt on largely introducing private-style PSM expertise into the entity. This will result in the insufficient know-how in the reform team. Without sufficient private-style PSM know-how as the foundation, successive design of private-style PSM organizations, procedures and skills as well are nearly impossible to be properly constructed. Consequently, it can be estimated that the PSM reform in

CSIST will still stay at the original status. Even though a private-style PSM model will be designed, it could probably looks like another public-style one. After analyzing above problems and gaps, the next step is to propose the possible solutions. These processes of finding solutions are elaborately done through literature review and brainstorming.

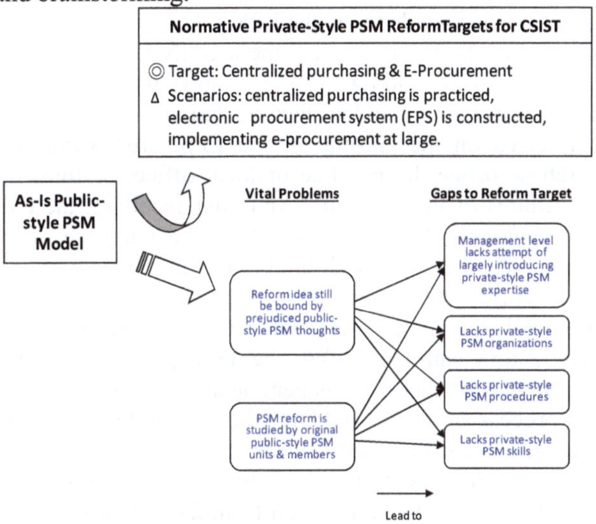

Fig.1. PSM reform targets for CSIST

IV. REALIZATION TOWARDS FUTURE PSM TARGETS OF CSIST

The outline of how to reform the PSM institution of CSIST has been primarily presented. It's focus on the succession of the redesign of PSM functions and procedures combined with IT tools. Only when these counterparts are thoroughly presented can let the government having an entire view of how to execute the PSM reform of CSIST. First of all, Figure 2 depicts the outline of the new arrangement of PSM organizations, functions and purchasing procedures. Also in the purchase life cycle, there adds a new step: integrate and assign. Besides, the purchasing procedures change a lot, dotted line means new procedures. In the beginning CSIST receives an order. The project units contract with customers and then assign works to the engineering units. After that, the engineering units deal with the R&D task in the design stage. After the specifications of products are designed, needed items are listed. Those changed procedures are depicted in blocks with dotted lines. Figure 2 shows that after the needed items are listed, the engineering units successively apply for needed items via the new built e-procurement system (EPS). After that, a new erected unit, purchase head office, executes the "integrate & assign" stage. It not only collect the PRs coming from all the needing units, mostly from the engineering units, but rather, it has the ability to classify and aggregate the needed items according to their categories, possible suppliers, and market conditions.

After gathering and classifying all the needed items, the purchase head office has to decide whether each purchase case should be purchased by itself or assign it to the local purchase office. One thing special is that the staffs in this head office are assigned according to their professional categories, e.g. electronic parts, machinery parts, chemicals etc. instead of original public-style purchasing stages. Therefore, they are all experts in respective commodity categories. Under such prerequisite, the advice of item reduction and aggregation can be made by these experts. This action will bring about considerable cost saving because of the scale of economy. If such advice can be accepted, the engineering units then modify the original purchase requisition (PR) and submit to the purchase head office again.

Afterwards, there is another revolutionary change in this new arrangement: the purchase office takes charge of the jobs from compiling purchase case to submitting it to the authority unit to be ratified. Therefore, jobs from issuing the RFx to possible providers, receiving the quotations, compiling the PRs, submitting to the PR review units, and submitting to the authority unit to ratify, are all done by the purchase office. Thereafter, from turning purchase case into tender documentation and bidding to delivering and payment. The only difference is that: transit original paper works to the novel EPS.

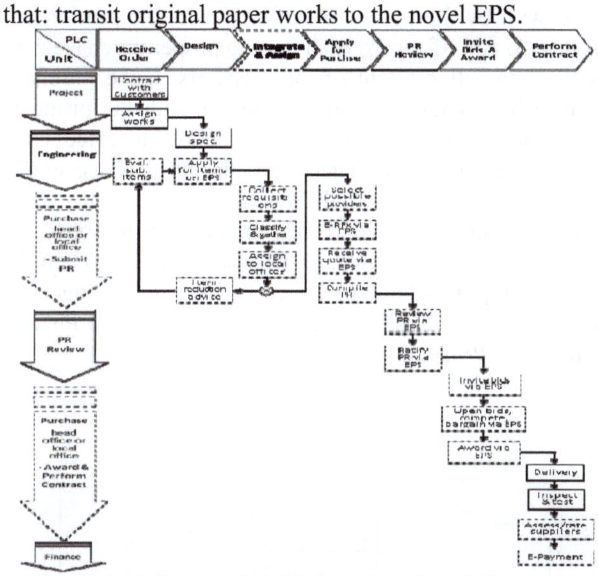

Fig.2. The modified PSM procedures for CSIST

Although Fig.2 has illustrated the outline of the life cycle of the purchase activities, but some details are not so clear. Hence, there is a strong need to utilizing the UML sequence diagram to present some more details. Thereby the interactions among departments as well as the activities in sequence can be presented clearly.

First of all in Fig.3, when the EPS receives the requisitions applied from the engineering units, it automatically classifies the needing items and puts them into proper purchase carts. This automatic action is based on two conditions: (1) all the items are recorded according to their part number which are consistent with the descriptions among the practitioners; (2) each part number has been identified with a proper category on the database of the EPS. After each item is put into proper purchase cart according to its category, the EPS then notify the purchase head office to deal with these carts. Then, the staffs of purchase head office check the details of each purchase cart and modify them and turn the purchase cart into purchase case as well. When more and more requisitions are applied, above process of item classified is repeated.

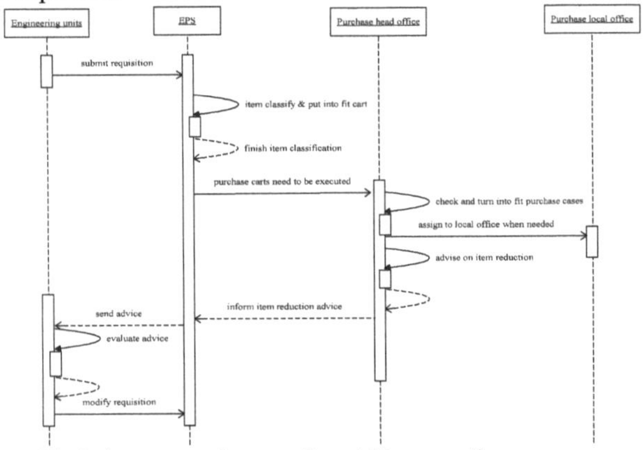

Fig.3. A sequence diagram of requisition execution process

It needs to be noticed that the EPS only can do the job of classification at the start. Proper modification also needs to be done by the purchase staffs. This can only be dealt with by the purchase staffs. Hence, the proficiency of how to aptly forward each item to proper purchase case is the imperative talent of the members in the purchase head office. They have to be familiar with the products categories, from specification and source to price level and market condition. Thereafter, the head office exerts its proficiency to advise the issue of item reduction/aggregation, and then via the EPS to notify the engineering units to evaluate these suggestions. When the engineering units have reviewed the advice coming from the purchase head office, they modify their requisition and submit to EPS again.

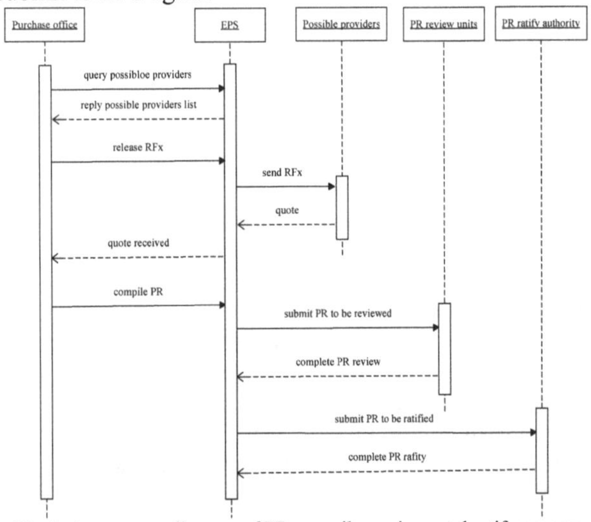

Fig.4. A sequence diagram of PR compile, review, and ratify process

No matter the needed items are dealt with by the purchase head office or local offices, the successive jobs of next stage "apply for purchase" are: select possible providers, e-RFx, receive quote, compile PR, review PR, and ratify PR. The action sequences and interactions among relative units can be depicted like Fig.4. We observe that the purchase firstly query the EPS to get the list of possible providers. Then, the purchase office sends the RFx (request for information/ proposal/ quotation) to these possible providers via EPS. After receiving the reply from the providers, the staffs of purchase office compile the purchase case and form a formal PR. Afterwards, the PR is reviewed both by the comptroller and finance units and then submit to the authority officer to ratify. All these activities also are done via the EPS.

After the PR is ratified, the activity stage moves to the phase of bid and award. Fig.5 shows that the PR will be turned into a tender documentation which will be publicized on the Gazette of the EPS. Afterwards, the purchase office, head office or local office, perform the procedures of inviting bids. And after receiving the bids and up to the due date of bidding, the purchase office opens bids and notifies all the bidders to compete online. This will be an imperative way for cost saving. Only when all the bidders acknowledge that there are many competitors can they offer the best price to contend among one another. This is beneficial for the buyer. After the process of online bidding, the best bidder emerges. The purchase office then rechecks the price and the terms and conditions with the best bidder. Before formally awarding, apply for ratification from the authority officer is necessary. We can see that above purchase processes are all done via the EPS. Therefore, not only the efficiency is far much better than ever, the cost saving resulting from the perfect competition is considerable. At the same time, the opportunities of fraudulence on procurement, both by internal purchase staffs and external bidders, are very small. At last, e-awarding is set.

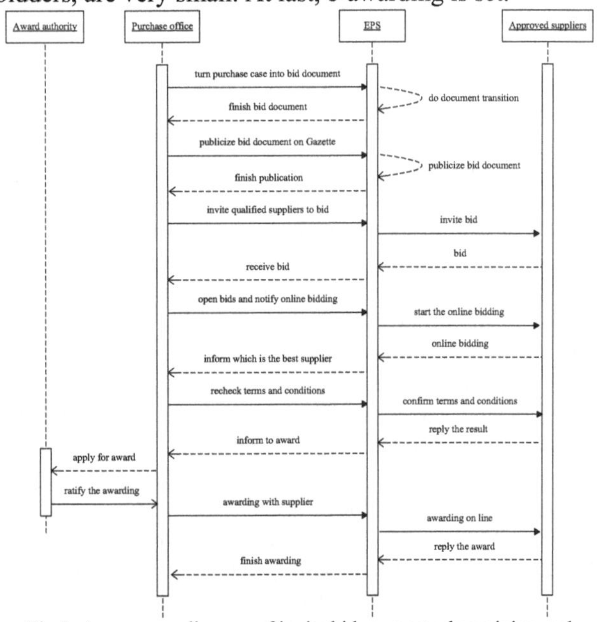

Fig.5. A sequence diagram of invite bid, compete, bargaining and awarding

In the long run, the activity stage comes to the delivery, inspection, and payment. Fig.6 presents that the EPS will automatically check the delivery cycle of every order. When the date of delivery is near, e.g. a week ago, the EPS will automatically send a message to check the actual delivery date with the supplier. After receiving the supplier's reply, EPS forwards it to the purchase office. Thereby, the purchase office can previously notify relative units to prepare the activities about acceptance. Before the date of shipping, the supplier sends a shipping notice to the EPS and immediately passing to the purchase office. Once the goods have arrived, relative units come to the warehouse to make sure the delivered items, quantities, and whether relative certifications are correct and ready. After that, the goods are passed to the inspection units. Following the finish of inspection, the notice of payment will sent to the financial unit which will exert the e-payment job via the EPS.

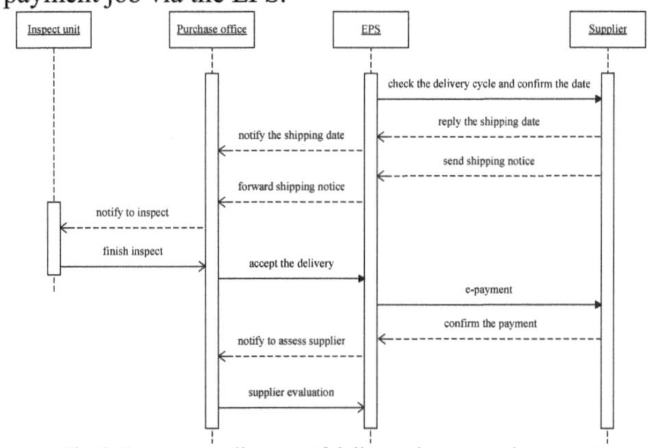

Fig.6. A sequence diagram of delivery, inspect and payment

Through the presentations form Fig.3 to Fig.6, the whole picture of how to erect the new private-style PSM organizations, procedures, and operations of the EPS are completely constructed. The steps to change the PSM model from original public-style to the desirable private-style have been clearly presented. The A-B-C-D steps of this backcasting study also have been accomplished. Till the end of this chapter, CSIST not only can hold the critical private-style PSM strategies, but also, they have clear referenced action items to do this PSM reform.

V. CONCLUSIONS

The global trend of exerting arm's-length bodies in organizational change to drive public reform is irresistible. At the start of this study, adopting the backcasting approach was an important decision. The proactive problem-solving attitude as well as effective research steps laid solid foundations in doing this research. Looking back to the course of this study, backcasting approach urged to bravely pursue a desirable target of PSM reform so as to struggle to break away from the constraint of the dominant public procurement institution. Thereby, critical PSM literatures could be deeply explored and the benchmark in PSM reform of private sectors could be emulated. Consequently, the desirable and normative targets of PSM reform for CSIST are reasonably and effectively constructed. Comprehensively reviewing this study, the results of this research not only can systematically and reasonably offer a valuable framework for CSIST to move the PSM reform, but also, the details of the proposed action items can instantly give the reform team a clear guidance to transform the PSM institution step by step. Furthermore, when other large public entities in Taiwan ready to transform to NDPBs and need to do the PSM reform, they also can use this thesis as a reference.

REFERENCES

[1] Allix, M., and Van, T. S., 2005. "Mapping the Field of Quasi-autonomous Organizations in France and Italy", *International Public Management Journal*, vol. 8, no. 1, pp. 39-55, 2005.

[2] Chang, S.C., "Reverse Electronic Auctions: Cases in Taiwan's High-tech Industry", *Technology in Society*, vol. 29, no. 4, pp. 490-496, 2007.

[3] Chen, Y.J., Chen, C.Y., Tsai, R., and Chen, W., "The Study of E-Marketplace: A Case Study of Formosa Technologies E-marketplace", *International Journal of Electronic Business Management*, vol. 3, no. 4, pp. 283-291, 2005.

[4] Croom, S., and Brandon-Jones, A., "Impact of E-procurement: Experiences from Implementation in the UK Public Sector", *Journal of Purchasing and Supply Management*, vol. 13, no. 4, pp. 294-303, 2007.

[5] Feisel, E., Hartmann, E., and Giunipero, L., "The Importance of the Human Aaspect in the Supply Function: Strategies for Developing PSM Proficiency", *Journal of Purchasing and Supply Management*, vol. 17, no. 1, pp. 54-67, 2011.

[6] Holmberg, J., and Robèrt, K.H., "Backcasting from Non-Overlapping Sustainability Principles - A Framework for Strategic Planning", *International Journal of Sustainable Development and World Ecology*, vol. 7, pp. 291-308, 2000.

[7] Karjalainen, K., "Estimating the Cost Effects of Purchasing Centralization: Empirical Evidence from Famework Agreements in the Public Sector", *Journal of Purchasing and Supply Management*, vol. 17, no. 2, pp. 87-97, 2011.

[8] Kickert, W.J., "Public Management of Hybrid Organizations: Governance of Quasi-autonomous Executive Agencies", *International Public Management Journal*, vol. 4, no. 2, pp. 135-150, 2001.

[9] Kumar, N., and Srinivasan, B., "Implementation and Performance Effect on Electronic Procurement and its Ship Management Companies", *I.J. Information Engineering and Electronic Business*, vol. 5, no. 5, pp. 10-16, 2013.

[10] Robinson, J.,"Future Subjunctive: Backcasting as Social Learning", *Futures*, vol. 35, no. 8, pp. 839-856, 2003.

[11] Ronchi, S., Brun, A., Golini, R., and Fan, X., "What Is the Value of An IT E-procurement System", *Journal of Purchasing & Supply Management*, vol. 16, no. 2, pp. 131-140, 2010.

[12] Soares-Aguiar, A., and Palma-dos-Reis, A., "Why Do Firms Adopt E-procurement Sstems: Using Logistic Regression to Empirically Test A Conceptual Model", *IEEE Transactions on Engineering Management*, vol. 55, no. 1, pp. 120-133, 2008.

[13] Tassabehji, R., and Moorhouse, A., "The Changing Role of Procurement: Developing Professional Effectiveness", *Journal of Purchasing and Supply Management*, vol. 14, no. 1, pp. 55-68, 2008.

[14] Teo, T.S., Lin, S., and Lai, K.H., "Adopters and Non-adopters of E-procurement in Singapore: An Empirical Study", *Omega*, vol. 37, no. 5, pp. 972-987, 2009.

[15] Trautmann, G., Bals, L., and Hartmann, E., "Global Sourcing in Integrated Network Structures: the Case of

Hybrid Purchasing Organizations", *Journal of International Management*, vol. 15, no. 2, pp. 194-208, 2009.
[16] Trkman, P., and McCormack, K., "Estimating the Benefits and Risks of Implementing E-procurement", *IEEE Transactions on Engineering Management,* vol. 57, no. 2, pp. 338-349, 2010.
[17] Van Thiel, S., "Trends in the Public Sector: Why Politicians Prefer Quasi-autonomous Organizations", *Journal of Theoretical Politics*, vol. 16, no. 2, pp. 175-201, 2004.
[18] Van Thiel, S., "Comparing Agencification in Central Eastern European and Western European Countries: Fundamentally Alike in Unimportant Respects", *Transylvanian Review of Administrative Sciences*, vol. 34, pp. 15-32, 2011.
[19] Wynstra, F., "What Did We Do, Who Did It and Did It Matter: A review of fifteen volumes of the (European) Journal of Purchasing and Supply Management", *Journal of Purchasing and Supply Management*, vol. 16, no. 4, pp. 279-292, 2010.
[20] Zheng, J., Knight, L., Harland, C., Humby, S., and James, K., "An Analysis of Research into the Future of Purchasing and Supply Management", *Journal of Purchasing and Supply Management*, vol. 13, no. 1, pp. 69-83, 2007.
[21] Holmberg, J., "Backcasting: A Natural Step in Operationalizing Sustainable Development", *Greener Management International*, vol. 23, pp. 30-51, 1998.
[22] MacDonald, J.P., "Strategic Sustainable Development Using the ISO 14001 Standard", *Journal of Cleaner Production*, vol. 13, no. 6, pp. 631-643, 2005.
[23] Gong, D.C., and Chen, J.L., "Critical Control Processes to Fulfil Environmental Requirements at the Product Development Stage", *International Journal of Computer Integrated Manufacturing*, vol. 25, no. 6, pp. 457-472, 2012.

Collaborative Production in Networked Manufacturing

Ruo-xi Lei[1,*], Sheng-liang Zong[2]
[1] School of Accounting, Lanzhou University of Finance & Economics, Lanzhou, China
[2] School of Management, Lanzhou University, Lanzhou, China
(*leiruoxi88@163.com, zongshl03@gmail.com)

Abstract - **In this paper, we considered the synergy effect and effort of networked manufacturing and constructed the collaborative production model for both measures without guarantee and measures with guarantee. Our analysis shows that the moral hazard exists in the enterprise collaborative production without guarantee measures, and the enterprise will take a tradeoff effort level in order to maximize their profits when guarantee measures exist. Meanwhile, the enterprise can obtain more profits from the increase of synergy effect, the expansion of the networks size and the enhancement of leaning ability. Finally, we provided a numerical example.**

Keywords – **Effort, guarantee measure, moral hazard, networked manufacturing**

I. INTRODUCTION

Networked manufacturing as the main manufacture model has played an increasingly important role in modern society. Many developed countries have put Networked Manufacturing as an important part of their countries' strategy system. Such as "21st century network plan of Korea". Both "Boeing 787 aircraft" in USA and "Toyota Motor Co" in Japan are all the typical cases of Networked Manufacturing.

So far, many scholars have launched a wide rang research about the global manufacturing. Abelson [1] first proposed the concept of global manufacturing. Later, Fawcett [2] studied the role of strategic logistic in coordinated global manufacturing's success. Monroy and Vilana [3] analyzed the main facts that determine the formation of GMVNs (global manufacturing virtual networks) as well as the strategy of the companies involved in these organizations and how GMVN will evolve in the future. A conceptual framework is proposed based on four network features: strategy, structure, communication systems and culture. Later, Wadhwa et al. [4], Huang et al. [5], Tso et al [6] also studied this problem.

Networked Manufacturing is the main manufacturing model in modern society. Ortiz et al [7] studied the information technologies for network manufacturing. Shin [8] analyzed the vulnerabilities of the biometric system used for access control and the authentication of access to confidential information in the networked manufacturing system. Foreman et al [9] described the development, implementation, and impact of the process and optimization-based control system now used by Dell to address this supply-routing challenge for its U.S.-bound monitors. Their divided three phase. Tursi et al [10] proposed a novel approach, postulating that the product, represented by its technical data, may be considered as interoperable per se with the many applications involved in manufacturing enterprises as far as it embeds knowledge about itself, as it stores all its technical data, provided that these are embedded on a common model.

Collaboration is a key issue to rapidly answer market demands in a manufacturing company, through sharing competencies and resources. And it is also a common strategy used to increase competitiveness. The collaborative networked organizations (CNO) area focused on this type of organizational models that use ICT. Camarinha-Matos et al [11] described the key concepts related to CNO, provides a high level classification of collaborative networks, and presents some application cases in the manufacturing industry. Verdecho [12] present an approach based on the analytic network process (ANP) to manage collaborative relationships under an integrated approach by considering the inter-enterprise performance elements and the factors that influence collaboration. Furthermore, Macedo et al [13] introduced an approach based on causal models and graph theory for the analysis of core value alignment in collaborative networks. The potential application of the approach was then discussed in the virtual organization's breeding environment context. Romero and Molina [14], Ulbrich et al [15] and [16] also researched this problem.

In this paper, we studied the collaborative production for

horizontal enterprises in networked manufacturing. The model assumptions and variable description are introduced in Section II. In Section III we analyzed the model that without guarantee measure. Furthermore, we introduced and analyzed the model with guarantee measure in Section IV. Finally, a numerical example is given in Section V.

II. ASSUMPTIONS

To construct the model and solve the problem, we first introduced the basic assumptions:

(1). There are many enterprises product the same products. It is a perfectly competitive market. The anti-linear demand function is $p = a - bQ(a > 0, b > 0)$. Q and p represents the demand quantity and the price of the products in market, respectively.

(2). Let n be the number of the manufacturing enterprises.

(3). Let w_i be the market share of enterprise i, and $\sum_{i=1}^{n} w_i = 1$.

(4). Let c_o be the average unit cost of production. e_i ($0 \le e_i \le 1$) is the effort level of enterprise i, which can reduce the production cost by r_i, and $r_i = ke_i$. If $e_i = 0$, which indicates that the enterprise i do not take any measures to reduce the production cost. On the contrary, $e_i = 1$ indicates that the enterprise try their best to reduce the production cost. Let $s\sum_{\substack{j=1\\j\neq i}}^{n} e_j (s > 0)$ be the reduced unit production cost from other enterprises' effort. Therefore, the unit production cost of enterprise i is:

$$c = c_0 - r_i - s\sum_{\substack{j=1\\j\neq i}}^{n} e_j \tag{1}$$

III. MODEL ANALYSES OF NONGUARANTEE MEASURE

According to the above assumptions, we can get the profit function of enterprise i is:

$$\pi_i = (p-c)w_i Q = (a - bQ - c_0 + r_i + s\sum_{\substack{j=1\\j\neq i}}^{n} e_j)w_i Q \tag{2}$$

π_i^* and $w_i^* Q$ represents the maximum profit and the equilibrium quantity of enterprise i, respectively, when Eq. (2) acquire the maximum value. We take the first derivative respect to $w_i Q$ for Eq. (2), and make $\frac{d\pi_i}{dw_i Q} = 0$. Then, we have

$$\frac{d\pi_i}{dw_i Q} = (a - c_0 + r_i + s\sum_{\substack{j=1\\j\neq i}}^{n} e_j) - b\sum_{\substack{j=1\\j\neq i}}^{n} w_j Q - 2bw_i Q = 0 \tag{3}$$

We make $i = 1, 2, \ldots, n$. Then, we can reach the following equations through simultaneous these equations.

$$w_i^* Q = \frac{(a - c_0 + nr_i - \sum_{\substack{j=1\\j\neq i}}^{n} r_j - (n-1)se_i + s\sum_{\substack{j=1\\j\neq i}}^{n} e_j)}{(n+1)b} \tag{4}$$

Therefore, we substitutes Eq. (4) into Eq. (2)

$$\pi_i^* = \frac{(a - c_0 + nr_i - \sum_{\substack{j=1\\j\neq i}}^{n} r_j - (n-1)se_i + s\sum_{\substack{j=1\\j\neq i}}^{n} e_j)^2}{(n+1)^2 b} \tag{5}$$

Lemma1. For Eq. (5), we have

(1) π_i^* is non-increasing in e_i.

(2) e_i is non-increasing in *n*.

(3) π_i^* is non-decreasing in *n*.

Proof. Because $w_i^* Q \ge 0$, and then we have $a - c_0 + nr_i - \sum_{\substack{j=1\\j\neq i}}^{n} r_j - (n-1)se_i + s\sum_{\substack{j=1\\j\neq i}}^{n} e_j \ge 0$. Moreover, in terms of any integer *n*, we always have $n - 1 \ge 0$. Then, we evaluate $d\pi_i^* / de_i$.

$$\frac{d\pi_i^*}{de_i} = -2(n-1)s\frac{(a - c_0 + nr_i - \sum_{\substack{j=1\\j\neq i}}^{n} r_j - (n-1)se_i + s\sum_{\substack{j=1\\j\neq i}}^{n} e_j)}{(n+1)^2 b}$$
$$\le 0$$

Therefore, π_i^* is non-increasing in e_i.

We know that π_i^* is the function with respect to e_i according to Eq. (5). Therefore, the first derivative with respect to e_i must equal to zero in order to reach the maximum value. Then

$$\frac{d\pi_i^*}{de_i} = -2(n-1)s\frac{(a-c_0+nr_i-\sum_{\substack{j=1\\j\neq i}}^{n}r_j-(n-1)se_i+s\sum_{\substack{j=1\\j\neq i}}^{n}e_j)}{(n+1)^2 b} = 0$$

Because $s>0$, and for any integer *n*, this equation is always established. Therefore, we have

$$a-c_0+nr_i-\sum_{\substack{j=1\\j\neq i}}^{n}r_j-(n-1)se_i+s\sum_{\substack{j=1\\j\neq i}}^{n}e_j = 0$$

If $s=k$, and we have $e_i = \frac{a-c_0}{s}$. Meanwhile, If $s \neq k$, and then we have $e_i = \frac{a-c_0+(s-k)\sum_{\substack{j=1\\j\neq i}}^{n}e_j}{n(s-k)+s}$. Because $e_i > 0$, and then we have the result that e_i is decreasing in *n*. Therefore, e_i is non-increasing in *n*.

Obviously, π_i^* is non-decreasing in *n* according to the Proof listed above.

Theorem1. The enterprise will not take any measures in order to pursue the maximum profits as the information is asymmetry and the environment is imperfectly. The effort level e_i and the maximum profit π_i^* of enterprise i will not increase while the number of networked manufacturing enterprises increasing.

IV. MODEL ANALYSES OF GUARANTEE MEASURE

We know that no enterprise will take any measures to reduce the unit production cost in order to obtain the profit according to the analysis in Section III. Therefore, the resource of the networked manufacturing can not be effective integrated and the collaborative effect of the network can not be fully played. Thus, it is necessary to take guarantee measures to ensure the collaborative effect is fully played. Assuming that the Coordination function is $h = l^{\sum_{i=1}^{n} e_i}$ $(l>1)$ [17], which can bring unit production cost reduce $e_i l^{\sum_{i=1}^{n} e_i}$ [18]. Other assumptions are the same as Section III. Therefore, the unit production cost of enterprise i is:

$$c_i^s = c_0 - r_i - s\sum_{\substack{j=1\\j\neq i}}^{n}e_j - e_i l^{\sum_{i=1}^{n} e_i} \tag{6}$$

Then, the profits function of enterprise i is the following:

$$\pi_i^s = (p-c_i^s)w_iQ = (a-b\sum_{i=1}^{n}w_i^sQ-c_0+r_i+s\sum_{\substack{j=1\\j\neq i}}^{n}e_j+e_i l^{\sum_{i=1}^{n} e_i})w_i^sQ \tag{7}$$

Eq. (7) is the same as Eq. (2), and can get the maximum value. π_i^{s*} and $w_i^{s*}Q$ represents the maximum profit and the equilibrium quantity of enterprise i while it gets the maximum value, respectively. We take the first derivative respect to w_i^sQ for Eq. (7), and let $\frac{d\pi_i^s}{dw_i^sQ}=0$. Then, we have

$$\frac{d\pi_i^s}{dw_i^sQ} = (a-c_0+r_i+s\sum_{\substack{j=1\\j\neq i}}^{n}e_j+e_i l^{\sum_{i=1}^{n} e_i}) - 2bw_iQ - b\sum_{\substack{j=1\\j\neq i}}^{n}w_jQ = 0 \tag{8}$$

We make $i = 1,2,\text{L}\ ,n$. Then, we can reach the following equations through simultaneous these equations.

$$w_i^{s*}Q = \frac{(a\text{-}c_0+nr_i+2s\sum_{\substack{j=1\\j\neq i}}^{n}e_j-\sum_{\substack{j=1\\j\neq i}}^{n}r_j-(n-1)se_i+l^{\sum_{i=1}^{n} e_i}(ne_i-\sum_{\substack{j=1\\j\neq i}}^{n}e_j))}{(n+1)b} \tag{9}$$

Then, we can obtain the following equation by substituting the result of Eq. (9) into Eq. (7)

$$\pi_i^{s*} = \frac{(a\text{-}c_0+nr_i+2s\sum_{\substack{j=1\\j\neq i}}^{n}e_j-\sum_{\substack{j=1\\j\neq i}}^{n}r_j-(n-1)se_i+l^{\sum_{i=1}^{n} e_i}(ne_i-\sum_{\substack{j=1\\j\neq i}}^{n}e_j))^2}{(n+1)^2 b} \tag{10}$$

Because $w_i^{s*}Q \geq 0$, thus $a\text{-}c_0+nr_i+2s\sum_{\substack{j=1\\j\neq i}}^{n}e_j-\sum_{\substack{j=1\\j\neq i}}^{n}r_j-(n-1)se_i+l^{\sum_{i=1}^{n} e_i}(ne_i-\sum_{\substack{j=1\\j\neq i}}^{n}e_j) \geq 0$. We know that $-(n-1)se_i$ is decreasing in e_i, which indicates that π_i^{s*} is decreasing in e_i. Meanwhile, $l^{\sum_{i=1}^{n} e_i}(ne_i-\sum_{\substack{j=1\\j\neq i}}^{n}e_j)$

and nr_i will increase while e_i is increasing, which means that π_i^{s*} is increasing in e_i. Therefore, the maximum value of π_i^{s*} is decided by an equilibrium effort level e_i of enterprise i.

From Eq. (10) we know that π_i^{s*} is a function respect to e_i. Its first derivative must equal to zero in order to obtain the maximum value. Thus, we have

$$\frac{\partial \pi_i^{s*}}{\partial e_i}=\frac{2(nk-ns+s+nl^{\sum_{i=1}^{n} e_i}+l^{\sum_{i=1}^{n} e_i}(ne_i-\sum_{\substack{j=1\\j\neq i}}^{n} e_j)\ln l)}{(n+1)^2 b}$$

$$(a\text{-}c_0+nr_i+2s\sum_{\substack{j=1\\j\neq i}}^{n} e_j-\sum_{\substack{j=1\\j\neq i}}^{n} r_j-(n-1)se_i+l^{\sum_{i=1}^{n} e_i}(ne_i-\sum_{\substack{j=1\\j\neq i}}^{n} e_j))$$

$$=0$$

Because

$$w_i^{s*}Q=\frac{(a\text{-}c_0+nr_i+2s\sum_{\substack{j=1\\j\neq i}}^{n} e_j-\sum_{\substack{j=1\\j\neq i}}^{n} r_j-(n-1)se_i+l^{\sum_{i=1}^{n} e_i}(ne_i-\sum_{\substack{j=1\\j\neq i}}^{n} e_j))}{(n+1)b},$$

which is the production quantity of enterprise i. Then, we have

$$nk-ns+s+nl^{\sum_{i=1}^{n} e_i}+l^{\sum_{i=1}^{n} e_i}(ne_i-\sum_{\substack{j=1\\j\neq i}}^{n} e_j)\ln l=0 \tag{11}$$

Lemma2. For Eq. (11), we have

(1). $k>s-\dfrac{s}{n}$

(2). $n>1+\dfrac{1}{\ln l-1}$

Proof (1). For Eq. (11) we have

$$l^{\sum_{i=1}^{n} e_i}=\frac{n(k-s)+s}{\sum_{\substack{j=1\\j\neq i}}^{n} e_j\ln l-n(1+e_i\ln l)}$$

We know that $l^{\sum_{i=1}^{n} e_i}$ is increasing in e_i. Furthermore, we have $\dfrac{\partial l^{\sum_{i=1}^{n} e_i}}{\partial e_i}=\dfrac{n(n(k-s)+s)\ln l}{(\sum_{\substack{j=1\\j\neq i}}^{n} e_j\ln l-n(1+e_i\ln l))^2}$. Thus, we have $n(n(k-s)+s)\ln l>0$. Consequently, we can get $k>s-\dfrac{s}{n}$.

(2). because $n(n(k-s)+s)\ln l>0$, and have $l^{\sum_{i=1}^{n} e_i}>0$. Therefore, we have $\sum_{\substack{j=1\\j\neq i}}^{n} e_j\ln l-n(1+e_i\ln l)>0$. We can get $n<\dfrac{\sum_{\substack{j=1\\j\neq i}}^{n} e_j\ln l}{1+e_i\ln l}\leq\sum_{\substack{j=1\\j\neq i}}^{n} e_j\ln l\leq(n-1)\ln l$. Consequently, we have $n>1+\dfrac{1}{\ln l-1}$.

Theorem1. The enterprise will take some measures in order to pursue the maximum profits under guarantee measures. The effort level is in an equilibrium status. The number of enterprises is correlation with the synergistic effect coefficient of the networked manufacturing.

V. NUMERICAL EXAMPLES

We assume that there are 10 enterprises produce the same products. The average unit production cost c_0=10. The effort level of all the enterprises except enterprise i is $\sum_{\substack{j=1\\j\neq i}}^{n} e_j=1$, and the reduced cost is $\sum_{\substack{j=1\\j\neq i}}^{n} r_j=2$. According to the assumption (3) we know that $r_i=ke_i$, which shows that the reduced cost is increasing with the effort extent of enterprise i. In this numerical example, we set $k=1$. Other parameters values are as follows: a=100, b=3, s=2, l=3. The result is showed in Fig.1.

Curve 1 shows the profit of enterprise i that the networked manufacturing without any guarantee measures, which indicates that the profit of enterprise i obtained is decreasing with itself effort level. The enterprise which is a rational individual will not take action to reduce cost in order to obtain the maximum profit, which causes the existence of "moral hazard". Curve 2 shows the profit of enterprise i that the networked manufacturing with guarantee measures, which indicates that the profit of enterprise i obtained is decreasing at begin and then increasing after it reach the minimum value. The result matches the fact. The enterprises have to pay for reduced cost, which is positively related with the effort level of them.

The decreasing of initial profits is the benefit, due to the synergistic effect, which cannot offset the cost paid for the effort. The benefit will increase with the increasing of synergistic effect when the effort level reaches a certain level.

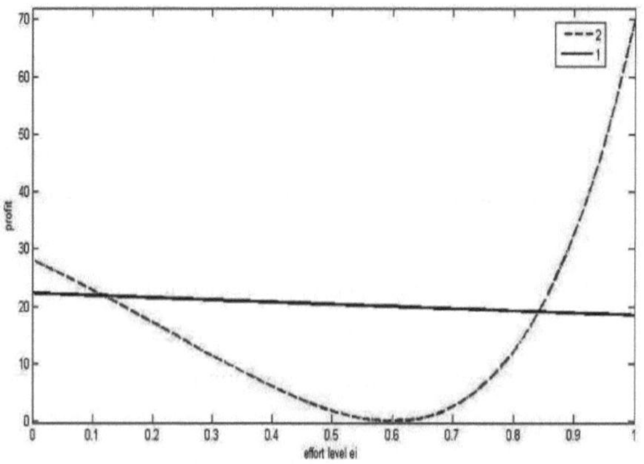

Fig.1. The relationship between π_i and e_i

From Curve 2 we also can have that when $e_i = 0$, and $\pi_i^{s*} = 28.10$. While $e_i = 1$, and $\pi_i^{s*} = 69.64$. Furthermore, (69.64-28.10)/28.10=247.83% shows that the profit obtained through effort is 2.5 times more than that of the situation when no action is taken. This indicates that the enterprise has motivation to take action to reduce the unit production cost. Moreover, the enterprises will improve the effort level while realizing that effort can obtain more profit.

Then, we evaluate the sensitivity of π_i^{s*} respect to *l*, *k*, *s* and *n*, respectively. The variation of the profit is $\Delta\pi$, and π_0 is the initial profit. The variation of the parameters is Δa, and a is the initial parameters. $SE = \dfrac{\Delta\pi / \pi_0}{\Delta a / a}$. The results are showed in Table I.

Table I shows that π_i^{s*} is positively related with *l*, *s*, *n*, and negatively related with *k*. The results indicate that the profit of enterprise *i* will increase with the increase of synergistic effect coefficient of the network, the expansion of the scale of the network, and the growth of the learning ability. Meanwhile, the benefits of the enterprise would be more stable or even lower when the synergistic effect and scale of the network increased to a certain level, otherwise, the enterprise will get more benefits with the improvement of the learning ability.

TABLE I

SENSITIVITY ANALYSIS π_i^{s*}

l	π_i^{s*}	SE	k	π_i^{s*}	SE	s	π_i^{s*}	SE	n	π_i^{s*}	SE
2.5	8.833	--	0.5	74.093	--	1.5	66.612	--	10	69.644	--
2.6	15.160	17.907	0.6	73.192	-0.061	1.6	67.213	0.134	11	80.946	1.623
2.7	23.901	14.976	0.7	72.297	-0.073	1.7	67.817	0.144	12	91.173	1.389
2.8	35.538	13.148	0.8	71.407	-0.086	1.8	68.423	0.151	13	100.423	1.218
2.9	50.579	11.865	0.9	70.523	-0.099	1.9	69.032	0.160	14	108.801	1.084
3.0	69.644	10.924	1.0	69.644	-0.112	2.0	69.644	0.169	15	116.407	0.979
3.1	93.387	10.237	1.1	68.771	-0.125	2.1	70.259	0.176	16	123.332	0.892
3.2	122.545	9.665	1.2	67.903	-0.138	2.2	70.876	0.184	17	129.655	0.820
3.3	157.928	9.223	1.3	67.041	-0.152	2.3	71.496	0.191	18	135.446	0.760
3.4	200.433	8.881	1.4	66.184	-0.166	2.4	72.119	0.200	19	140.767	0.706
3.5	251.042	8.588	1.5	65.333	-0.180	2.5	72.744	0.208	20	145.669	0.661

VI. CONCLUSIONS

In this paper, we studied the effect of the effort on profits without guarantee measure and with guarantee measure. There exists "moral hazard" due to the information asymmetry and the network environment imperfectly. Enterprises will not take any action to reduce production cost in order to obtain the maximum profit without guarantee measure. On the contrary, the enterprise will try their best to reduce production cost in order to obtain maximum profits with guarantee measure. The effort level will reach an equilibrium status at last. The guarantee measure including: (1) Creating a good network environment, which could make the enterprise be willing to form strategic alliances. (2) Building the platform of information sharing, which make the enterprise can share one another's information inside the network. (3) Constructing the credit reporting system of the enterprise. (4) Perfecting the mechanism of governance.

In this paper, we only consider the horizontal cooperation product enterprises in Networked Manufacturing. The contract among enterprises need further study. Moreover, the relation and the cooperation production among longitudinal enterprises also need further study.

ACKNOWLEDGMENTS

The authors are grateful to the editor and referees for their useful comments and constructions suggestions which have improved the presentation of the paper. This work is supported by the Fundamental Research Funds for the Central Universities (Nos. 14LZUJBWZY074).

REFERENCES

[1] P. H. Abelson, Global manufacturing competition, Science 232 (1986), no. 4755, 1181-1181.

[2] S. E. Fawcett, Strategic logistics in coordinated global manufacturing success, International Journal of Production Research 30 (1992), no. 5, 1081-1099.

[3] C. R. Monroy and J. R. Vilana Art, Strategic positioning of global manufacturing virtual networks in the aeronautical industry, Technology Analysis & Strategic Management 22 (2010), no. 5, 631-647.

[4] S. Wadhwa, M. Mishra and F. T. S. Chan, Organizing a virtual manufacturing enterprise: An analytic network process based approach for enterprise flexibility, International Journal of Production Research 47 (2009), no. 1, 163-186.

[5] S. Huang, Y. J. Hu and C. G. Li, A tcpn based approach to model the coordination in virtual manufacturing organizations, Computers & Industrial Engineering 47 (2004), no. 1, 61-76.

[6] S. K. Tso, H. Lau and J. K. L. Ho, Coordination and monitoring in an intelligent global manufacturing service system, Computers in Industry 43 (2000), no. 1, 83-95.

[7] A. Ortiz, R. D. Franco and P. G. Gasquet, Information technologies for manufacturing networks, Studies in Informatics and Control 20 (2011), no. 1, 5-6.

[8] Y. Shin and W. Shin, A telebiometric system mechanism model and biometric network protocol for the security of networked manufacturing, Journal of Intelligent Manufacturing 21 (2010), no. 5, 595-605.

[9] J. Foreman, J. Gallien, J. Alspaugh, F. Lopez, R. Bhatnagar, C. C. Teo and C. Dubois, Implementing supply-routing optimization in a make-to-order manufacturing network, M&Som-Manufacturing & Service Operations Management 12 (2010), no. 4, 547-568.

[10] A. Tursi, H. Panetto, G. Morel and M. Dassisti, Ontological approach for products-centric information system interoperability in networked manufacturing enterprises, Annual Reviews in Control 33 (2009), no. 2, 238-245.

[11] L. M. Camarinha-Matos, H. Afsarmanesh, N. Galeano and A. Molina, Collaborative networked organizations - concepts and practice in manufacturing enterprises, Computers & Industrial Engineering 57 (2009), no. 1, 46-60.

[12] M.-J. Verdecho, J.-J. Alfaro-Saiz, R. Rodriguez-Rodriguez and A. Ortiz-Bas, A multi-criteria approach for managing inter-enterprise collaborative relationships, Omega-International Journal of Management Science 40 (2012), no. 3.

[13] P. Macedo, A. Abreu and L. M. Camarinha-Matos, A method to analyse the alignment of core values in collaborative networked organisations, Production Planning & Control 21 (2010), no. 2, 145-159.

[14] D. Romero and A. Molina, Collaborative networked organisations and customer communities: Value co-creation and co-innovation in the networking era, Production Planning & Control 22 (2011), no. 5-6, 447-472.

[15] S. Ulbrich, H. Troitzsch, F. van den Anker, A. Pluss and C. Huber, How teams in networked organisations develop collaborative capability: Processes, critical incidents and success factors, Production Planning & Control 22 (2011), no. 5-6, 488-500.

[16] S. Saetta, L. Tiacci and L. Cagnazzo, The innovative model of the virtual development office for collaborative networked enterprises: The gpt network case study, International Journal of Computer Integrated Manufacturing 25 (2012), no. 11.

[17] A. N. Berger and D. B. Humphrey, Efficiency of financial institutions: International survey and directions for future research, European Journal of Operational Research 98 (1997), no. 2.

[18] L. R. Christensen, D. W. Jorgenson and L. J. Lau, Transcendental logarithmic production frontiers, Review of Economics Statistics 55 (1973), 28-45.

The Applications of Industrial Engineering in Agriculture Production Management

Mingjian Yuan*, Tzongyin Shieh, Guanyin Lu, Yenhung Lin
Department of Industrial Engineering and Management, Taiwan Yunlin University of Science & Technology, Douliu, Taiwan
(yuanmj@yuntech.edu.tw)

Abstract - **Industrial engineering has been successfully applied in many manufacturing, logistics, health-care, and service industries. This paper explores its applications in agriculture production. Agriculture relates closely to people's livelihood but usually suffers as a weak sector in economics. Attentions to agriculture are regained as food safety and shortage become world-wide concerns. This paper addresses an analogy between industrial manufacturing and agriculture production to demonstrate that industrial management can adapt to agriculture. In particular, enterprise resource planning system (ERP), which is the core management information system used by most manufacturing enterprises, can be implemented in farming. Our short-term goal is to transform traditional farms to "managed factories" to produce safe products effectively. The ultimate goals are to provide the public with safe products, to increase farmers' income and to elevate farmer's economic status to a new middle class. In particular, new job opportunities are expected to be created in farms management.**

Keywords - **Agriculture production, enterprise resource planning system, farms management**

I. INTRODUCTION

Agriculture relates closely to people's livelihood but usually suffers as a weak sector in economics. Take the agriculture development in Taiwan as an example; it evolves through a number of stages: agrarianism or the so-called agriculture reform was the starting point, in which farmers were granted land ownership so that they could do planting in their own land. The reform solved the economic imbalance of farmers who used to renting land and paying heavy duties. However, it is also the cause of "small farmers" who possess only small size of land that is hard to achieve economic scale. The early agriculture production in Taiwan is labor intensive. Yet, the agriculture population dwindled when manufacturing industries emerged and employed a great number of man power. The shrinkage has continued to this day. Agriculture mechanization was the second important milestone, in which various farming machines were developed to make up for the losing work forces. At the same time, scientific research was initiated to improve crop breeds and cultivation techniques to increase productivity. Remarkable achievements have been made in that period of time. Thereafter, farmers found chemical fertilizers and herbicides extremely effective to boost production. Consequently, more and more chemicals were used to the extent that they were abused. Crops production did increase; however, at the cost of safety, that has become a major concern in most nations. Nowadays, high techs continue to be applied in agriculture production. Genetically modified crops (GM crops or biotech crops) are well-known but controversial examples. Between 1996 and 2011, the total area of land cultivated with GM crops had increased by 94 times, from 17,000 to 1,600,000 km^2 [1]. Ten percent of the world's agriculture lands were planted with GM crops in 2010 [1]. In parallel, organic cultivation emerged against conventional farming. Organic crops are healthful but the quantities are very limited by nature. In the other direction, plant factory has also emerged as a recent development. A plant factory is a closed growing system that enables a farmer to keep constant production of vegetables all year around [2]. The plants artificially control LED light, temperature, irrigation, and carbon dioxide concentrations to produce non-toxic crops. At present, the plants produce only high-valued vegetables or medical herbs. After years of evolution, part of agriculture problems were solved, but newer ones have arisen as the results of previous solutions. The present agriculture challenges are outlined in the following:

1. Small farmers [3]

Most individual farms are of small size; this is particularly true in Taiwan. Since economic scale cannot be achieved, the farmers need to discover new niches to be competitive.

2. Low product prices [3]

The average price of agriculture products is much lower as compared to industrial products. For example, the price of an iPhone can pay for more than a thousand kilograms of oranges in Taiwan. Small farms can barely maintain breakeven.

3. Worldwide competition [3]

World Trade Organization (WTO) and Free Trade Agreements (FTA) enforce open markets. The already low prices face up to further challenges from world competitions. Naturally, local small farmers can hardly fight against the international competitions.

4. Insufficient and aging work forces

Faced with the challenges above, the youth generation cannot find a reason to stay on the farms. Work force is losing in most farms, particularly those far from cities. Aging agriculture population is already a trend. Table I shows the shrinkage and aging trend of agriculture work forces in Taiwan.

5. Safety

In order to produce more quantities of products with superior "appearance," excessive herbicides and chemical fertilizers were used. Not only the ecological environment was seriously affected, but the public health was also at

E. Qi et al. (eds.), *Proceedings of the 5th International Asia Conference on Industrial Engineering and Management Innovation (IEMI2014)*, Proceedings of the International Asia Conference on Industrial Engineering and Management Innovation 1, DOI 10.2991/978-94-6239-100-0_74, © Atlantis Press and the authors 2015

risk. Not to mention the potential hazard from the GM crops.

TABLE I
SHRINKAGE AND AGING TREND OF AGRICULTURE WORK FORCE IN TAIWAN [4]

Year	Agriculture employment (unit: k persons)	Male age (%)				Female age (%)			
		15~34 years old	35~54 years old	>64 years old	Sub Total	15~34 years old	35~54 years old	>64 years old	Sub Total
2004	642	8.9	52.5	9.8	71.2	2.6	22.8	3.4	28.8
2005	591	8.1	52.8	10.3	71.2	2.7	22.7	3.4	28.8
2006	555	7.4	51.5	12.3	71.2	2.3	22.5	4.0	28.8
2007	543	7.2	49.4	13.0	69.6	2.5	23.3	4.7	30.4
2008	535	7.2	50.3	12.4	69.8	2.6	22.9	4.7	30.2
2009	543	7.4	50.0	12.5	69,9	2.8	22.8	4.5	30.1
2010	550	7.7	49.6	12.5	69.8	2.8	22.9	4.4	30.2
2011	542	7.6	50.9	12.7	71.3	2.6	21.8	4.3	28.7
2012	544	7.8	51.8	13.0	72.6	2.7	20.7	4.1	27.5

The above challenges come from all aspects: economic, ecological, public health, societal, culture, and even political. It appears there is no one-time or one-for-all solution to all the challenges. Fortunately, similar dilemma was encountered in the 70's in Taiwan, when small manufacturing companies emerged. The companies faced similar challenges at that time: labor-intensive, small-scaled, short of capital and R&D capability. It was with the help of management professionals and information systems, the companies eventually overcame the challenges and made huge contributions to Taiwan economics. In addition to economic side, these companies became successful small businesses that comprised the economic middle class in society. The solid middle class created welfare that was shared by most of the general public and thus laid a foundation for a stable and prosperous society. With this experience, this paper explores the applications of industrial engineering in answer to the existing agriculture difficulties. In section 2, the analogy between industrial manufacturing and agriculture production is built to demonstrate that industrial management can be applied to agriculture. In section 3, a cloud farm example is presented to illustrate the applications. Conclusions are given in section 4.

II. APPLICATIONS OF INDUSTRIAL ENGINEERING

According to the IIE official definitions [5], industrial engineering (IE) is concerned with the design, improvement and installation of integrated systems of people, materials, information, equipment and energy. The initial IE applications were devoted to manufacturing industries. The applications extended to logistics, and then health-care and service industries. Numerous businesses have been running successfully with IE assistance. From these valuable experiences, we are motivated to explore IE applications in agriculture production.

The exact event that triggered the exploration was the Melamine incident in 2008 [6]. In the wake of the event, we developed a product tractability system (TRC) [7] to reveal materials, production, and quality investigation information to consumers while they are shopping for goods. The system was experimented for fruits in 2009. Since product traceability relates to every production stage from materials acquiring, IQC, materials retrieving, production, PQC, stocking until shipping, it has to be integrated with a comprehensive management information system to make the routines workable. ERP is the system to be linked with, since it is the core management system used by most manufacturing enterprises. Three tough questions have challenged us since then: The first one is "Will manufacturing-oriented system be applicable to agriculture production?" The second is "Are farmers capable of using the system?" And the last is "Will this solve agriculture dilemma?" Here the exploration started. Since all the management contents involved here are IE disciplines, the application is attributed to an industrial engineering extension. In section 2.1, we will build an analogy between industrial manufacturing and agriculture production to demonstrate the manufacturing-oriented ERP can adapt to agriculture production. In 2.2, we will explain that not farmers but rather management professionals will operate the system. In 2.3, we present a vision of duplicating the manufacturing experiences on agriculture.

A. Analogy between industrial manufacturing and agriculture production

At the first sight, industry and agriculture seem two different worlds. But if we think again, agriculture is in fact a kind of "factory" that manufactures natural foods. Since ERP used to be manufacturing-oriented, the analogy can be built if we show that ERP can also be implemented for agriculture production. The "proof" comes in two folds: (1) an ERP system can be set up using agriculture data, and (2) the standard ERP managerial cycle can work for farming. For the fold 1, Table II lists the comparisons: the first column are the ERP building elements; in the second column are the elements usages in manufacturing industries, while the third column is the usage interpretations in agriculture production.

TABLE II
ERP IMPLEMENTATION COMPARISONS

ERP building elements	Manufacturing usage	Agriculture usage interpretations
Company	basic enterprise unit	refer to a farm
Factory	plant where goods are produced	farmland where crops are cultivated
Operator	factory employees	farm workers
Items	finished products, semi products, raw materials, molds, tools	crops, vegetables, fruits, seeds, fertilizers, herbicides
Bill of materials (BOM)	components structure to produce goods	components structure (items and usage) to plant crops
Material Requirement Planning (MRP)	analysis to generate procurement plan and production plan	analysis to compute the items to be purchased and the area of land required to plant crops
Materials acquirement	purchase raw materials or components	purchase seeds, herbicides, fertilizers
Work orders	authorized orders to produce goods	authorization to grow seedling and crops
Assembly line/workstation	flow line of machines and equipment	refer to specific farmland where a crop is planted

Equipment	machines, fork lifters, wrappers	tractor, seeding machine, mowing machine
Manufacturing process	cutting, lathing, welding, milling, drilling, grinding, etching, ion-planting, sputtering, diffusing	ploughing and weeding, fertilization, thinning, irrigation, herbiciding, picking
Route	the sequence of operations to produce goods	the sequence of cultivations to plant crops
Outsourcing	part of manufacturing is done by outsourcers	part of the planting is done by outsourcers
Quality control	sampling to investigate quality of raw materials or finished goods	sampling to investigate herbicide residuals
Warehouse	buffers for temporarily storing semi products and finished goods	building for temporarily storing materials and crops
Shipping	delivering to distributers	delivering to markets or consumers
Product traceability	tracing the materials used and the production process	tracing the fertilizers, herbicides used and planting process

It can be observed that for each ERP building element in manufacturing, there exists a comparative usage in agriculture production. That is, an ERP system can be truly set up in an agriculture environment. Some elements may need modifications; for instance, there is no assembly line but farmland. A manufacturing ERP can be readily converted to an agriculture version. This completes the first-fold of the analogy. Next, we will investigate if the standard ERP management cycle can work for crops planting. Fig.1 is a typical order-to-delivery (OTD) process in ERP. We will go through it for crops planting. It is noted that there are actually eight major management cycles in ERP. Each one is the best practice in industries. By going through it for crops planting, we *implicitly* duplicate the best practices to agriculture production and treat farming as a *business*, this is an important goal of this paper.

OTD process is activated by external or internal (prediction) orders. Similarly, farming starts from a prediction order at the beginning of a growing season. It is followed by requirements planning, which computes the material types and quantities to be purchased and the area of land to be planted. The former goes through a procurement cycle, which directs the acquirement of seeds, herbicides, and fertilizers. The lot/batch number will be registered while materials are delivered to a farm. And the latter goes to a production cycle that directs the planting according to a route. The planting route usually need to reference a good standard; TGAP (Taiwan good agriculture practice) is one example. At the beginning and end of each process of the route, one has to check in and out the ERP system so that the execution *evidence* is prepared. The evidence is called planting "foodprint", a newly created term similar to "footprint" in spelling and meaning. When materials are retrieved through ERP, evidence is prepared to create "materials foodprint". After harvest, herbicide residuals are tested to create "QC foodprint". The completely documented crops are then delivered to the market. It can be observed that the management cycle used by manufacturing industries fit in well agriculture production. This will suffice the second-fold of the analogy. It is worth noting that ERP integrates and directs each process to make sure goods are manufactured on time, with good quality, and at reasonable cost. It is indispensable to manufacturing industries, and it will also work for the agriculture.

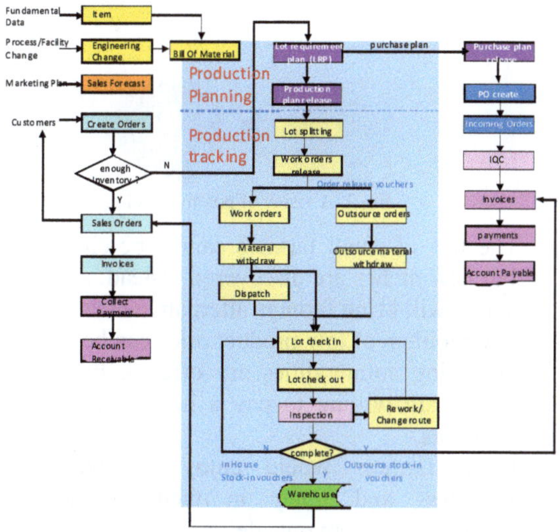

Fig.1. ERP OTD process

B. Farm managers

After demonstrating that agriculture can be treated as a special manufacturing industry, and that ERP applies to agriculture production, a natural question is that who will use the system to manage the farms. People asked this in that they do not think farmers can play the role. Indeed, the same myth occurred 40 years ago, when "black-handed" mechanics were not believed they can use computers to manage their factories. What happened was that mechanics may not know computers; they hired professionals to do the job. It was the professionals that helped manage the manufacturing industries and kept them prosperous, many of whom were well-trained industrial engineers. The same reasoning applies to farmers. A single small farmer may not be able to afford a dedicated manager; however, a group of farmers together should be able to hire a management *team* with production, logistics, and marketing specialties, as long as the farms can profit. Fig.2 shows such a vision. A positive cycle may realize the vision: It starts from management that brings effective and safe products to agriculture. The effectiveness controls the cost and quality. And evidence-based *safe* products come as the results. Although small farmers do not have the economic scale advantage, *safe* products are very appealing commodity in most markets. Prices are expected to create reasonable profits, which in turn can hire more professionals to run effective marketing by use of Internet of things or e-commerce. Eventually, a positive cycle can repeat itself.

C. Solutions to the agriculture dilemma

Will the positive cycle in Fig.2 solve the agriculture challenges as described at the beginning of the section? Let's conduct an investigation one by one:

Fig.2. A vision of positive cycle

1. The fact of small farmers cannot be changed easily, but new niches are discovered instead: safe products. These will bring farmers alternative niches.
2. The problem of low product price can be solved when appealing commodities are offered. In general, the public are willing to pay a higher price for health concerns.
3. Local farmers would enjoy the advantages of *freshness* and *safety* in their products against competitions from outside.
4. When farms start to profit and offer a better pay to their employees, work forces will flow back from other industries. More importantly, the farm would offer a "hope" and a promising "future" to the youth generation, as the manufacturing industries did before.
5. Safety problem can be solved now, in that farmers do not have to over-use chemicals to sale their products. The public health can be assured.

Not only the dilemma can be solved but more profound influences will take place in economy and society, which are briefly discussed in section 4.

III. A CLOUD FARM EXAMPLE

The vision described in section 2 is not imaginary. We have carried out years of capstone projects in NYUST to experiment with the vision. This section describes the project which is called "Cloud Farm."

A. The cloud farm

The project was initiated in 2009, when oranges were oversupplied in Taiwan and the prices almost hit the record low. The project was to assist farmers with marketing. A local farmer in Yunlin, which is a conventional agriculture county, participated in the project. The farm was called a *cloud farm* in that significant cloud techniques were applied. Eight senior IE students initially spent a whole academic year to manage the farm. The project team came up with six value-added services for the farm:

1. Add value by management: The team treated the farm as a business and implemented a commercial ERP system to execute the management cycle as described in section 2.1. A tracking system (SFT) was also set up to track the daily operations. In each process of the "planting route", a teammate checked in and out the system and took pictures.
2. Add value by packaging: The team cooperated with the Design College of NYUST to design elegant fruit boxes and labels to enhance the fruit's value. Green packaging was also adopted for group buyers.
3. Add value by marketing: The team explores various marketing resources to promote sales. First they gave the fruit a brand name "cloud orange," which sounds more valuable. The cloud oranges were then promoted in campus activities. In addition, a press conference was held to introduce the fruit to news reporters and the general public, with assistance from local government. Various sales channels were also extended to industrial park enterprises and hospitals, where healthy products are welcomed.
4. Add value by EC: A website was developed to narrate the farm story. The website was also empowered with EC capability.
5. Add value by safety: Safety is the project focus that can't be overemphasized. In ERP, the project team set up a TGAP-compliant route, which follows the good practices of cultivation and herbicides usage. The safety was built in from the very beginning. Each daily operation was executed according to the route. Before sales, the oranges were sampled and sent to a SGS agency for testing. More than 250 items were tested to generate official test reports. All the evidence-based information from materials, cultivation process, and test results are compiled to a card called FoodPrint card as is showed in Fig.3. A card is placed in each box to be shipped to customers. It is safety that distinguishes cloud oranges from the others in general markets.
6. Add value by convenient delivery: The project team signed a contract with a logistic company to make sure fresh fruits are delivered to customers on time. Fig.4 shows the shipping and the project team.

Fig.3. Shipping of cloud oranges and project team

Fig.4. FoodPrint Card

B. A FoodPrint system

Most commercial ERP systems do control major management cycles, but may not compile the database to trace the production history. Before the capstone project, we developed a TRC system to do the tracing in two directions [7]. Up-stream tracing traces back the materials structure, level by level upwards, until the item that caused a defect is found. Down-stream tracing traces the outgoings of the semi products or finished products that used a defect component so that they can be recalled in time. This project continued to develop a system called FoodPrint to further facilitate product traceability.

The FoodPrint system consists of two modules. One is to customize the item attributes and the location to be traced in that different crops may have different properties to be traced. The module offers an interface to maintain the attributes synchronously with the farming. Fig.5 shows the interface, both PC and hand-held versions are displayed. This module is designed for farm managers. The other module is to display traced information for consumers while they are shopping for goods. Four standard "foodprints" are provided: Item foodprint reveals the general product information such as lot numbers, nutritional ingredients, and valid dates. Material foodprint reveals the materials used and their *source*. Cultivation foodprint reveals the cultivation process. And QC foodprint reveals the test results and legal values. Fig.6 shows the four foodprints. A responsive design was adopted for the foodprints so that the contents can automatically change formats according to the size of device used by a consumer. The FoodPrint system supplements ERP traceability and makes information available to the public. The system uses a number of cloud techniques and is deployed as a cloud service. There is no effort to maintain the system. Users who can link with Internet can use the service anytime or anywhere.

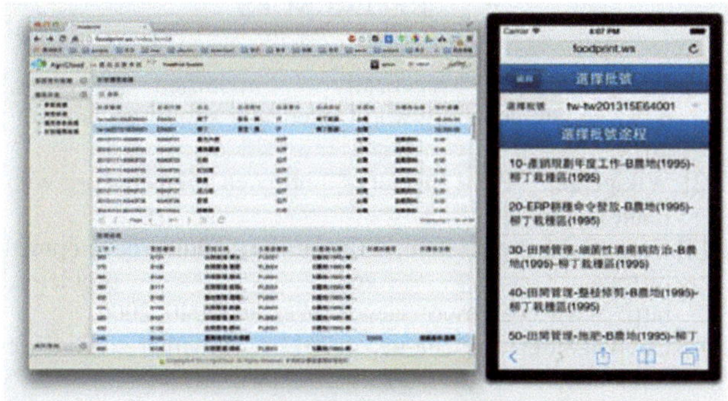

Fig. 5 FoodPrint for farm managers

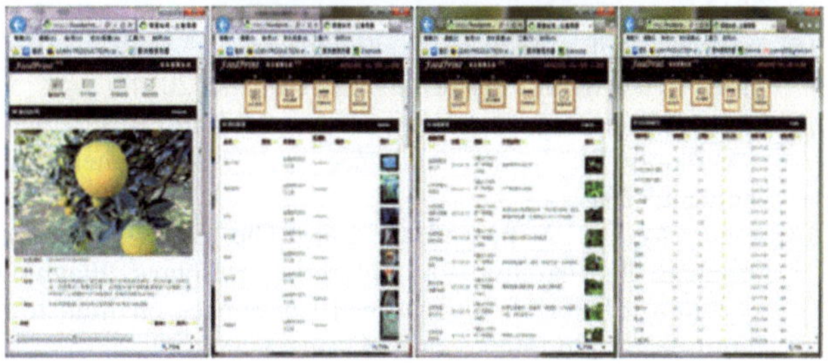

Fig. 6 four FoodPrints for consumers

IV. CONCLUSION AND FUTURE DEVELOPMENT

Agriculture is a nation's infrastructure. However, the public appears to have a misled impression that farm jobs are inferior. It prohibits talented personnel from joining the profession. It is high time for us to pay more attention to agriculture.

This paper explores the solutions for agriculture challenges. The core is to transform farms into small yet promising business. And management is the way to make farm business profitable in the right way. Safe foods are the output of well-managed farms. Simply the management thinking will not make it happen. It is learned that an information system is needed. Fortunately, ERP can be readily converted to an agriculture version. In addition, cloud techniques can be used to make the services easily deployed and implemented.

The final comment that is that information system alone will not suffice the solution. Education is the final piece required. Since there is significantly large quantity of farms, equally a great many management professionals will be in demand. Industrial engineering would be the appropriate source to cultivate the professionals. In fact, the manufacturing professionals have been over-supplied. Recent graduates have had difficulties to find an appropriate position. Since there are huge demand in farm management, work forces can split into agriculture to solve the insufficient and aging agriculture work forces, while relieving the oversupply elsewhere. A more profound influence is that the transitions of the work forces will help balance society: the equilibrium will be maintained among people in different sectors in terms of their social economic status; each individual will make equally important contributions to society. In the future, likely explorations will continue in the fishing and livestock industries.

REFERENCES

[1] Wikipedia, "Genetically modified crops," http://en.wikipedia.org/wiki/Genetically_modified_crops, 2014.

[2] Wikipedia, "Plant factory," http://en.wikipedia.org/wiki/ Plant factory, 2014.

[3] Council of Agriculture, "Taiwan agriculture development history and problems," http://www.coa.gov.tw/view.php?catid =3040, 2000.

[4] Council of Agriculture, "2012 Agriculture Statistics Summary," http://agrstat.coa.gov.tw/sdweb/public/book/Book. aspx, 2013.

[5] Institute of Industrial Engineers, "About IIE," https://www.iienet2.org/details.aspx?id=282, 2014.

[6] Lwo, Shih-hsiung, "A study of risk index and measurement uncertainty for food surveillance – A case of melamine incident," 2012.

[7] Yuan, M., H. Yeh, and G. Lu, "The Development of Products Traceability for Enterprise Resource Planning system", IEEE 18th International Conference on Industrial Engineering and Engineering Management, 2011.

Evaluation of Development Motivity in Chinese Cultural Creative Industry Cluster

Qiu-xia Qu
College of Economics & Management, Qingdao University of Science and Technology, Qingdao, 266061
(qqx001@126.com)

Abstract - **Development of cluster of cultural and creative industries needs enough motivity. The evaluation system of culture creative industry clusters mainly consist of the following key elements: natural factor, humanity factor, financial factor, enterprise interaction factor, intelligence factor, government service factor and facility factor. Results about Factor Analysis show that clusters of cultural and creative industries in Beijing and Tianjin, Shanghai, Guangzhou and Shandong province are different in development motivity.**

Keywords - **Cultural and creative, cluster of industries, development motivity, factor analysis method**

I. INTRODUCTION

In knowledge economy era, the economy development model is expanded to the culture field and the creativity plays more and more important role in the development of culture industry, which leads to the new industry conformation: the culture creative industry. It turns into a strong driving force for economies of clustering areas. How the culture creative industry cluster keeps a persistent progressive growth attracts attentions from scholars and governments. Currently, the research on the culture industry cluster focus on its evolution mechanism (G.Y. Fan) [1], motivation (H. P. Hou ect) [2]. There are many different techniques working on the problem. However, most of them are qualitative investigations and a few of them utilize the quantitative analysis. For instance, the results on the competitive evaluation in (J. He, 2011, Z. P Wu, 2007) [3] [4], where the systematic dynamics method is employed. The results mentioned above motivate the present paper.

We aim to investigate the driving force of the culture creative industry. We utilize the factor analysis technique, through introducing an index system, to evaluate the driving forces of China's main culture creative industry clusters.

II. INDICATORS ON DEVELOPMENT MOTIVITY FOR THE CULTURE CREATIVE INDUSTRY CLUSTERS

The key point determining the development of culture creative industry clusters is the strength of its driving force. We summarize and categorize these motivations as the natural factor, the humanity factor, the service factor, the financial factor, the technology factor, the of enterprises interaction factor, the intelligence factor, the government service factor, the facility factor.

A. The natural indicators (x_1)

(1) The cultural reserve on natural resources x_{11}. The cultural reserve of a specific area, among other natural resources, directly affects the growth of the culture creative industry.

(2) Habitat suitability x_{12}. The physioclimate and geographical conditions also impact if the talented persons and enterprises organize creative activities in a specific area.

(3) Number of natural landscapes x_{13}. The larger number of natural landscapes has a gathering effect and has a great attraction to the industry cluster.

(4) Natural landscape attraction x_{14}. Elegant scenery fits in well with the culture creative industry cluster provide further appeal.

(5) Level of urban greening x_{15}. Greening is also an important element for locating the culture creative industry cluster.

(6) Climate comfort degree x_{16}. Pleasant climate is an important factor for attracting creative enterprises and talented persons.

(7) Natural recognition of creative talented people x_{17}. The location is important elements for talented people since creative ideas might be motivated by elegant sceneries.

(8) Distance from landscapes to industry clusters x_{18}. The location should take account of both life and work and the shorter distance the better life and work quality.

B. Humanity indicators x_2

(1) Area total population x_{21}. This refers to the population of talented people.

(2) Population density x_{22}. Here, the population density refers to the population density of creative talented people.

E. Qi et al. (eds.), *Proceedings of the 5th International Asia Conference on Industrial Engineering and Management Innovation (IEMI2014)*, Proceedings of the International Asia Conference on Industrial Engineering and Management Innovation 1, DOI 10.2991/978-94-6239-100-0_75,

(3) Number of cultural center x_{23}. This element gives clue on the culture level, which is the foundation base to develop creative industry.

(4) Number of local folk brands x_{24}. Local folk brand is characterized by local culture. The large number of local brands has the cluster effect.

(5) Attraction of local folk activities x_{25}. Only the well recognized folk activities have the cluster effect.

(6) The number of heritages x_{26}. This element can be used to judge the historical accumulation related to local culture.

(7) The number of creative industry clusters x_{27}. The number of creative industry clusters gives hints on its attraction and which also has a demonstration effect to the other industry clusters.

(8) Historical accumulation x_{28}. The culture creative industry is based on culture resources, which implies that historical accumulation is one of the important elements.

(9) Cultural atmosphere x_{29}. A city's culture atmosphere is characterized by its architectures, history and so on, which is important to motivate the culture creative industry cluster.

C. *Financial indicators* x_3

(1) Science technology funding x_{31}. It refers to the funding planned to organize the creative activity.

(2) External investment x_{32}. The amount of the funding from outside of the industry cluster reflects the attraction of the local cluster.

(3) Risk investment x_{33}. It shows the level of how creative industry attracts the funding.

(4) The research development funding x_{34}. The funding used to support the researches of the enterprises in the industry cluster.

(5) Talented people funding x_{35}. The funding used to support the further study, training, exchange of talented people.

(6) Creative product purchasing funding x_{36}. This refers to the annual average expense on creative product for enterprises in the industry cluster.

(7) Profits of creative products x_{37}. This element reflects the return on investments.

D. *Technology indicators* x_4

(1) The number of researchers x_{41}. The researchers here refer to the people working full time on the development of science techniques, organizations and services to support researches.

(2) The number research units x_{42}. The research organizations provide intelligences to enterprises units on the industry cluster.

(3) Annual average patents granting number x_{43}. It is an index reflecting the ability of original innovation, knowledge properties usage and penetration.

(4) Annual amount of creative producing x_{44}. This element reflects the activeness of creative activity.

(5) Successful rate of the creative production x_{45}. This shows what the possibility of turning of a creative idea into a production.

(6) Market competitiveness of the creative production x_{46}. It reflects the market effect on the creative productions.

(7) Domestic leading level x_{47}. It gives clues on the competitiveness of the creative production.

(8) Information usage rate x_{48}. To some extent, the culture creative industry can be taken as the combination of the creative idea with advanced technologies, where employing information techniques inspire the development of industry clusters.

E. *Enterprises interactions indicators* x_5

(1) Division of labor and specialization x_{51}. The division of labor and specialization of the enterprises is helpful for the development of the industry cluster through division of cooperation.

(2) Cooperation of enterprises in the cluster x_{52}. Enterprises cooperation

(3) Trust degree among enterprises in the cluster x_{53}. The trust built in the cooperation of enterprises in the cluster is helpful for the stable growth of the cluster.

F. *Intelligence indicators* x_6

(1) Research quality of colleges x_{61}. The culture industry cluster needs to cooperate with colleges and research units to get supports on innovations and the quality of services they provide is important in this respect.

(2) Flexibility of the labor market x_{62}. Flexible labor markets are important guarantee for the enterprises, where the labor can freely flow in and out markets of different industries, careers, locations.

(3) Services provided by agents x_{63}. An agent can provide adequate supporting services and enhances the innovation skills of the cluster.

(4) Learning ability of the cluster x_{64}. The capability of getting knowledge of the cluster is useful to accumulate the knowledge resources.

G. Government service indicators x_7

(1) Policies x_{71}. It refers to the incentive policy, the guidance policy, the coordination policy and their validity in helping growth of the innovation activities.

(2) Training policy x_{72}. The training policy provides intelligence support to the labor of the cluster to improve the level in aspects of techniques, organizing, operating and learning.

(3) Planning and guidance x_{73}. The guidance of the government is helpful in persistent developing of the creative industry cluster.

H. Facility indicators x_8

(1) Infrastructure construction x_{81}. Convenient traffics and communications, water facilities, electrical and gas facilities are basic elements for the industry cluster.

(2) Soft investment facility construction x_{82}. This refers to the network construction of the cluster, in particular, the computer network and the information system.

(3) Public infrastructure facilities and services x_{83}. In general, excellent facilities on public infrastructure and services attract immigratory population and industry clusters.

III. METHOD

The factor analysis method is used in this paper.

The present section illustrate how to construct the indicator system given above with taking Beijing, Tianjin, Shanghai, Guangzhou and Shandong province as examples. We take convenience sampling to get the sample data and enterprises in the clusters are information providers. The survey uses Likert five scale and 209 pieces questionnaires are taken back, of which 203 pieces are valid and can be taken as samples of this investigation.

A. Data analysis

Results of KMO values and results obtained from the bartlett test of sphericity show that most KMO values are larger than 0.8, which shows that the correlation coefficients among variables are not distinct. The Sig values are less than 0.05 and this implies that variables are not independent and the sample data are applicable to do factor analysis.

B. Factoring

Principal components factoring approach, with the condition that eigen-values are larger than 1, is used to determine the number of components.

As a result, two factors are in x_1, x_2, x_3, x_4 respectively. One factor is in x_5, x_6, x_7, x_8 respectively.

C. Principal components

Expressions of the principal components are listed below by the SPSS analysis.

$$F_{X11} = 0.325X_{11} + 0.306X_{12} + 0.417X_{13} + 0.364X_{14} + 0.329X_{15} + 0.344X_{16} + 0.352X_{17} - 0.261X_{18}$$
$$F_{X12} = 0.435X_{11} + 0.342X_{12} + 0.235X_{13} + 0.064X_{14} - 0.22X_{15} - 0.376X_{16} - 0.335X_{17} - 0.426X_{18}$$

In particular, the function of x_1 is

$$F_{X1} = 0.48FX_{11} + 0.26FX_{12}$$
$$F_{X21} = 0.275X_{21} + 0.365X_{22} + 0.332X_{23} + 0.339X_{24} + 0.304X_{25} + 0.358X_{26} + 0.347X_{27} + 0.326X_{28} + 0.314X_{29}$$
$$F_{X22} = 0.365X_{21} + 0.425X_{22} + 0.371X_{23} + 0.354X_{24} + 0.301X_{25} + 0.308X_{26} + 0.323X_{27} + 0.366X_{28} + 0.345X_{29}$$
$$F_{X2} = 0.51FX_{11} + 0.14FX_{12}$$
$$F_{X31} = 0.426X_{31} + 0.482X_{32} + 0.45X_{33} + 0.41X_{34} + 0.42X_{35} + 0.51X_{36} + 0.42X_{37}$$
$$F_{X32} = 0.51X_{31} + 0.414X_{32} + 0.45X_{33} + 0.62X_{34} + 0.71X_{35} + 0.42X_{36} + 0.53X_{37}$$
$$F_{X3} = 0.47FX_{11} + 0.17FX_{12}$$
$$F_{X41} = 0.34X_{41} + 0.41X_{42} + 0.34X_{43} + 0.35X_{44} + 0.30X_{45} + 0.32X_{46} + 0.36X_{47} + 0.38X_{48}$$
$$F_{X42} = 0.52X_{41} + 0.41X_{42} + 0.44X_{43} + 0.55X_{44} + 0.36X_{45} + 0.42X_{46} + 0.41X_{47} + 0.28X_{48}$$
$$F_{X4} = 0.53FX_{11} + 0.3FX_{12}$$
$$F_{X5} = 0.474X_{51} + 0.535X_{52} + 0.520X_{53}$$
$$F_{X6} = 0.525X_{61} + 0.642X_{62} + 0.42X_{63} + 0.34X_{64}$$
$$F_{X7} = 0.421X_{71} + 0.51X_{72} + 0.41X_{73}$$
$$F_{X8} = 0.472X_{81} + 0.62X_{82} + 0.35X_{83}$$

D. Evaluations and factor scores

Table I provides the evaluation results for all areas, which are obtained by calculating synthesis scores using the equations given above.

TABLE I
SYNTHESIS SCORE AND EVALUATION RESULTS

	FX_1	rank	FX_2	rank	FX_3	rank	FX_4	rank	FX_5	rank	FX_6	rank	FX_7	rank	FX_8	rank
Beijing and Tianjin	6.32	1	13.72	1	8.11	3	8.59	3	9.85	3	11.26	2	9.32	1	10.25	2
Shanghai	7.34	2	12.05	2	9.36	2	9.46	2	9.94	2	15.39	1	9.29	2	12.3	1
Guangzhou	6.29	4	10.29	4	9.69	1	10.03	1	10.15	1	10.26	4	8.71	3	10.21	4
Shandong	6.58	3	11.51	3	7.51	4	7.06	4	9.00	4	11.25	3	8.33	4	11.6	3

IV. CONCLUSIONS

The culture creative industry cluster in Beijing-Tianjin area prevails in its natural environment, humanities and government services elements. Shanghai, as an economically developed city, its industry clusters has great attraction to intelligence and financial investments. Moreover, its facilities are well established which provide sufficient driving forces to the growth of the clusters. Guangzhou's culture creative industry has advantages in the financial support, technology support and interactions among enterprises. Shandong area, in contrast to other areas, lags behind. The development of its culture creative industry needs great improvements on the financial support, the technology support and government services.

ACKNOWLEDGMENT

This paper is supported by the following funds: the Planning topics of the Ministry of Education of China (Grant No. 11YJA630207); The Soft Science Research Program of ShanDong Province of China (Grant No.2011RKGB4014; 2013RKB01228); A Project of Shandong Province Higher Educational Humanities and Social Science Program (Grant No.J12WF84, J11WG10); The topics of Bureau of statistics of ShanDong Province of China (Grant No. KT13124)

REFERENCES

[1] Fan Guiyu, Culture Creativity Industrial Clusters-Evolution theory and Empirical research [D], Capital University of Economics and Business, 2007.
[2] Hou Hanpo, Song Yanjun, Xu Yanqing. Mechanism analysis and Empirical research on the Power of Culture Creativity Industrial Clusters [J], Develop and Research, 2010, 5: 55-57.
[3] He Juan, Evaluation and Research on the Competitive power of Culture Creativity Industry [D], Jilin University, 2011.4
[4] Wu Zhipeng, Mechanism analysis on the Creativity Industrial Clusters which is on the emergence [D], Beijing Jiaotong University, 2007.6.

Effect of Game Logic in the Private Enterprise Competition Strategy[1]

Ai-hua Gao
Continue Education College, Yanshan University, Qinhuangdao, Hebei, China
(gaoaihua@ysu.edu.cn)

Abstract - **This paper applies the method of game logic, to private enterprise's competition strategy for the analysis of several points, the object of competition and cooperation, first of all players that private enterprises; followed by the use of optimal strategy of private enterprises by the game logic virtual model and dynamic game process, enterprise advantage; and from the perspective of game theory how to enhance the credibility of enterprises, using matrix method is necessary to the national macroeconomic regulation and control of the private enterprises are discussed.**

Keywords - **Competitive strategy, enterprises, game logic, private**

I. INTRODUCTION

Private enterprises after more than 20 years of development, has become one of the important main body of the market in the allocation of social resources. Private enterprise distribution industry more and more broad; the proportion of its total output value in GDP has increased year by year; employment is rising year by year [1-3]. Because of their own factors, market environment, such as the role of policy factors, China's private enterprises from the very beginning, showing strong institutional advantages, as well as caused by the operation of market-oriented product competitive advantage. But in the process of development of private enterprises, there are still many problems, such as the problem bigger and stronger enterprises, enterprises of the problem of sustainable development [4, 5].

Game logic is the application of logic branch developed based on game theory, it is a research on "rational" actors or participants how to choose strategies or how to make the action logic in the process of interaction. Rational people want to maximize their own goals or benefit the people, in the economic activities of rational people is to maximize the economic goal, in this paper, refers to the private enterprise also refers to all other business enterprises. The enterprise in the competition the key your "thanks" maximization is reasonable "reasoning" [6]. In order to realize the analysis of corporate strategy, must use the analysis method of game logic. Cooperation and competition in the competition of the enterprises constitute the game relationship is very typical and complete [7, 8]. Only the analysis is clear, can clear enterprise decision is reasonable, can we understand the enterprise in the competition position, on this basis in order to guide the behavior of enterprises, both to ensure the maximization of enterprise profit, and maintenance of the best interests of consumers, and promote the sustainable and stable development of China's economy [9-12].

Game logic has opened the door for the enterprise managers and researchers, game logic, Nash equilibrium has been and is changing the status quo of enterprise management. Game logic is category, is a new interdisciplinary subject intersected with game theory and logic ", the game logic is the study of the 'rational' actor or actors in the process of interaction and how to choose the strategy or how to make the logic of action". [13] in the management of the enterprise competition strategy mainly deal with the issues of competition and cooperation game theory, is a research through the analysis of decision-making information interaction and balance. The game logic into the competitive strategy of private enterprises, not only effectively combines the characteristics of private enterprises, to fully consider the antagonism competition, dynamic characteristics make competitive strategy has stronger, more conducive to the development of private enterprises. By using of game logic analysis of large enterprises, analyses the dynamic competition and similar large enterprises, can private enterprises by means of analysis of enterprise choosing what confrontation or cooperation strategy, and focus resources building, through the use of game logic, enterprises can be more targeted analysis of their own resources, foster strengths and circumvent weaknesses in understanding objective market situation, to get more competitive advantage. On the other hand, the game logic can also be used to analysis between equal scale private enterprise's competition, no matter in which enterprises, game logic can be roughly play its positive role through the following aspects.

II. GAME LOGIC ABOUT THE COMPETITIVE STRATEGY OF PRIVATE ENTERPRISES

Comparative analysis of person in the game, its resources and core competence, private enterprises to truly become the core competitiveness with and create the value of the enterprise, must first take game logic based on the comparative analysis of the same industry competition bureau people, clear the advantages and disadvantages of features and core competition ability through the analysis, a comprehensive understanding of the condition of the enterprise internal, eventually combined with competitor conditions and its own characteristics to choose reasonable competition strategy,

Foundation: China Hebei province soft science research project,_Project name: Study on Hebei private enterprises from the perspective of game logic ,project number: 134576239

strategic plan more for selection for enterprises to gain more market opportunities.

Focus on the analysis of rationality, As in the process of the Strategic Management Bureau, limited rationality is the most fundamental characteristics, such as applied in the processing of information and imperfect memory constraints, participation ability analysis opponent behavior etc.. The Bureau will be restricted by their cognitive ability in forecasting has maximum often, namely players' bounded rationality. Rational will usually players by the game information, pre decision standard and logical reasoning ability limit. [14] in the complete information static game benefit information for players are the common knowledge, in this case, in reason by their game knowledge background limit, at the same time player rationality often will be affected by logical reasoning ability, it includes the player can draw the inference results and reasoning can be paid resources limit two. But in the Bureau of "limited rationality" instead of "no", should pay full attention to the rational development of players, after all, human beings are completely rational and completely irrational biological wisdom, private enterprise should pay attention to the improvement of the quality of decision-making layer, focus on the above several aspects as far as possible to improve the "limited rationality", in order to make the game situation in favor of its favorable direction.

III. THE GAME LOGIC METHOD TO DETERMINE THE OPTIMAL STRATEGY MANAGEMENT SCHEME

Game logic is advantageous to the virtual manufacturing ideas, implementation strategy. Virtual enterprises and manufacturing method is the enterprise strategy management model application in the past ten years more extensively, especially for private enterprises, is an important tool to participate in the market competition and win the advantage. Analysis of their own resources to get competitive advantage by private enterprises, the remaining resources in virtual building their core competitiveness of other enterprises under invest more time and material resources. According to the disadvantages of [15] technology enterprise or its hostile competition, private enterprises can through the virtual manufacturing methods combined with large enterprises or other small and medium-sized enterprises, to get as much market share in the strategy of implementation and cooperation win-win. Game logic for the decision makers and countermeasure and analysis of specific strategies for strategic management model of the virtual enterprise and virtual enterprise manufacturing method provides good theoretical guidance.

The strategic plan will use the game logic method considering the competition form more management strategy, more dynamic than the general strategy. Private enterprises is very limited in resources, which requires the management strategy should have the flexibility and comprehensiveness more, as far as possible the comprehensive consideration to competition and competition results, through the game logic analysis of the management strategy can make the private enterprises react quickly to face large enterprises or similar enterprises competitive counter when reduced to a certain degree, survival risk. Using the game model of game logic to establish enterprise competition, payment matrix can be used to analyze the game logic. Through the overview of the market and competition, can be learned in a market, have certain relations of cooperation between enterprises, but this relationship is not stable. The cooperation between enterprises is not empty talk, but has the strength to match, the cooperation of enterprises must be based on a certain foundation. Most of the private enterprises and state-owned enterprises, joint ventures and other large enterprises is not possible as an equal. So in reality, not only the non cooperative situation than cooperation in general; but, cooperation is often only temporary and the strategy, therefore, non cooperative game theory is the main competitive strategy of private enterprises. The enterprise strategy formulation should consider to maximize profits single competition, should also take into account the impact of this decision to the entire market and the effect on the subsequent game. Only a comprehensive consideration, can the private enterprises to obtain more development opportunities for its powerful in competition.

IV. DYNAMIC GAME LOGIC ANALYSIS IS CONDUCIVE TO PLAY THE ADVANTAGES OF PRIVATE ENTERPRISES

The private enterprise has distinct characteristics in strategic management. Compared with large enterprises, the strategic management of private enterprises have a greater flexibility, the magnitude of the strategic management mainly focus on the competitive strategy of one or several products of the enterprise production, management range and the relatively low degree of concern. According to the characteristics of private enterprises, the flexible use of game logic. The game is divided into static and dynamic game. In the static game, decision makers must also make their own strategy, the decision-making body actions are performed simultaneously. On the other hand, although not at the same time, but did not know the first action after actor actor's action, no one can know other people's decisions and actions, it is impossible in the decision to implement immediately observe other people choose action information, and then to respond; decision-making body only according to their own estimates of other decision makers and each strategy pattern of their pay to make a choice, this game is the static game. In dynamic game, policymakers do not act at the same time, such as in the classic strategy story "Tian Ji racing", Qi Wangxian elected to participate in the competition of the horse, let Tian Ji observed, Tian Ji again according to the intake to the information selected to participate in the competition of the horse, so "Tian Ji race" in the game is a kind of dynamic game. First, they will provide new information to the actors, and then they will adjust their strategy, it must consider the impact of their own choice of actors as well as the impact on the final results so the forerunner in choosing his action. As can be seen, the standard to distinguish the static game and dynamic game is the main decision-making body of the other party's information. While private enterprises in the overall strength than that of state-owned enterprises, but can play the local advantages. As competition in customer process to provide products or services to some one or two links or, rather than each link in the competition. Can also be although each link not better than competitors, but the comprehensive efficiency of the entire business process was obviously higher than that of rivals, the requirement is the enterprise has particular advantages in management skills.

V. TO PRIVATE ENTERPRISES BIGGER AND STRONGER, TO REALIZE THE SUSTAINABLE DEVELOPMENT OF ENTERPRISES

We should establish the credibility of private enterprises. The specific implementation steps are as follows: 1, repeated game, which is enterprise transaction must have high repetition rate maintained, more enterprises are to consider the future more credit. 2, through the implementation of threats and promises, can obtain a reputation. In repeated games with unified participants, this reputation is the most useful. In different game with various participants in, it is also useful. If failing to fulfill your threat or promise in a game, your reputation will be damaged, and you will be in the other game, get a low pay. Therefore, when you consider the game at any one time, you should adjust your and consider its impact on your payment in the game in which the payment. In an ongoing relationship, good reputation will bring benefits to you. 3, the game is divided into a plurality of parts. The whole game is divided into many continuous small part, so that the reputation mechanism into effect, as the building project, usually by the progress payment. 4, the information transmission speed fast enough, perfidious deception can be observed in time, and quickly spread, in order to reduce, avoid dishonest behavior. Fifth, must have the punishment mechanism of false. Game logic assumes that all decision makers are rational, they all act to benefit maximization as the objective. In order to standardize the decision-making body to take in the pursuit of interests means, government should intervene in the transaction, often on the people involved in the decision influence.

VI. EXTERNAL FACTORS CONSTRAINTS OF PRIVATE ENTERPRISES TO STRENGTHEN THE COMPETITIVENESS OF THE

External factors fifth, constraints of private enterprises to strengthen the competitiveness of the individual private enterprise cannot break through, which requires the government to play the main role of macro-control, the external obstacles to get rid of private enterprises to strengthen the competitiveness of the. Macro control of the government intervene the competition between the enterprise is very necessary. Can the government two aspects means of intervention on competitive strategy is: on the one hand is the formal rulemaking, such as strict legislation and judicial procedures, improve the legal system, the rule of law to the rule of law in the refining business, not trustworthy behavior of severe punishment, increase the cost of false. After the government intervention in the market intervention, the theme of the sanctions decision not trustworthy behavior will not only by the law, will suffer to social ethics strongly condemned morality; on the contrary, trustworthy enterprise by the social praise and encouragement, favored by consumers, the next opportunity will continue, more to promote access to enterprise profit. [16] if in every competition, have joined the government intervention, then the game both sides not trustworthy behavior will be found in time, and get the corresponding punishment, at this time, both sides of the payoff matrix, as shown in Table I

TABLE I
PAYMENT MATRIX A AND B OF GOVERNMENT INTERVENTION

		B	
		promise	lose promise
	promise	6, 6	5, -1
A	lose promise	-1, 5	4, 4

The decision of A according to the changes of B and changes of decision. Assume that B selection and trustworthy, if A is honest can get the payment of 6, if the A does not keep your words will lose 1; assume that B chose not to promise, if A and trustworthy, can get paid for the 5. If A is not trustworthy, can only get 4. That is to say no matter what is B's decision, A will not change the trustworthiness of the decision-making attitude. This analysis also applies to B, B will also adhere to the honest decision. In this way, the Nash equilibrium of this game is the obvious (and trustworthy, and trustworthy) portfolio strategy; and, on both sides of the faithless cost is high, (and trustworthy, and trustworthy) the equilibrium strategy stability is strong. In this case, if A and B are not trustworthy, B would lose 1 point return, and the best situation is obviously A, B both in good faith, the B can obtain 6 point return. Two kinds of decision making such comparison, can be said that if the B is not trustworthy, easily lead to not only caused a unit of losses, and also lost the should be 6 point profit opportunities, equivalent to a total loss of 7 units of interest. In real economic life, has a high reputation of well-known enterprises as soon as there is a non trustworthy records, is tantamount to the career plan future poured water, not only will be the spit, will provide consumers despised. This causes the loss will be very heavy, is completely contrary to the enterprise profit maximization desire. Therefore, the more reputable businesses, often more willing to adhere to faith belief. With this faith as a guarantee, whether a transaction or multiple transactions, (and trustworthy, and trustworthy) equilibrium situation will steadily maintained.

VII. CONCLUSION

The game of private enterprises in reality competition is complex, game logic analysis inevitably flawed, but game reasoning analysis to guide the role of competitive strategy of private enterprises cannot be ignored, but, with the development of the game logic, its defects are gradually improving. Therefore, the game logic analysis plays an important role in the growing of private enterprises

REFERENCES

[1] Hiirmann W. and Sak H. Copula Generation for Control Variates [J]. Mathematics and Computers in Simulation, 2010, Vol. 81, Issue 4: 782. 790.

[2] Hagiu, A. Two-Sided Platforms: Product Variety and Pricing Structures [J]. Journal of Economics & Management Strategy, 2009, Vol. 18, No. 4: 1011-1043.

[3] Gardner, Ct and A. Stone. Competition between Payment Systems [EB/OL]. RBA Research Discussion Paper No 2009-02, 2009.

[4] Chioveenu I. R. F. Oiler, J. Sandonis, and J. Santnmaria. ATM Surcharges: Effects on Deployment and Welfare [J]. The Journal of Industrial Economics, 2009, Vol.57, No.3: 613-635.

[5] Bolton, P. M. Dewatripont, Contract Theory [M]. MIT press: Cambridge and London, 2005: 47-48.

[6] Cabral, L. Dynamic Price Competition with Network Effects [EB/OL]. Working paper, New York University and CEPR. 2008.

[7] Caillaud, B. and B. Jullien. Chicken and Egg: Competition among Intermediation Service Providers [J]. Rand Journal ofEconomics, 2003, 34: 309-329.

[8] Calton, D. Wj and A. S. Frankel. The Antitrust Economics of Payment Card Networks [J]. Antitrust Law Journal, 1995, 63: 643-668.

[9] Calzada, J. and T. Valletti. Network Competition and Entry Deterrence [J]. Economic Journal, 2008, Vol. 118, Issue. 531: 1223. 1244. 124

[10] Zhang Ronggang. Liang Qi. Credit risk credit paradox, information transmission, repeated game - group affiliated enterprises logic analysis [J]. Social scientists, 2006 (1):70-73

[11] Han Zhiming. Structure and logic analysis, credit paradox of public policy [J]. Journal of Tianjin Institute of public administration, 2008 (5): 52-56

[12] Zhang Ronggang. Liang Qi. Credit risk credit paradox, information transmission, repeated game - group affiliated enterprises logic analysis [J]. Social scientists, 2006 (1):70-73

[13] Zhang Feng, game of logic using [J]. Journal of Beifang Univesity of Nationality (PHILOSOPHY AND SOCIAL SCIENCES EDITION), 2010, (4): 133-136

[14] Li Guangjiu. Game theory based tutorial, Beijing: Chemical Industry Press, 2004, 4-5

[15] Zhou Yan sheng. Huang Hao. Game theory and competitive strategy thinking of contemporary economy based on [J], 2012 (1):42-43

[16] Victoria Song. The analysis of the game logic of enterprise competition. Master's thesis, 2012-12, 39-40

Contracting on Project Time in Knowledge Process Outsourcing

Zhi Ouyang*, Qin Su
School of Management, Xi'an Jiaotong University, Xi'an, Shaanxi, 710049
(ouyangzhi1987@stu.xjtu.edu.cn)

Abstract - **Knowledge Process outsourcing enjoys the rapidest development in recent decades. Based on exogenous quality performance of KPO, we establish a theoretical model for optimal project time of outsourcing. we show that enough time should be allocated inside for follow-up knowledge assimilation and integration particularly when they can bring out significantly knowledge gain In comparing between risk-neutrality and risk- averseness of the client, we find that the risk- averseness of the client will decrease its profit, generally extend the optimal project time and shorten knowledge integration phase, and even sometimes it can transform the outsourcing decisions.**

Keywords - **Geometric Brownian Motion, Knowledge Process outsourcing, Project time**

I. INTRODUCTION

Great development has been witnessed in a new type of outsourcing form, Knowledge Process Outsourcing (KPO). According to statistics from the Department of Trade in services and Commercial Services of the Ministry of Commerce of PRC, the contract value of KPO undertaken in China reaches $21.45 billion in 2013. Although it remains less than that of Information Technology Outsourcing (ITO), which is $31.17 billion, it has risen 60.8% and thus been much higher than ITO's contract value increase ratio as 36.8%. Thus far, KPO has become the outsourcing type that's experiencing the fastest growth.

Typical KPO service involves role- or analysis-based tasks rather than rule-based ones, such as market analytics, financial data analysis, knowledge management, legal process supports, etc. Generally, these services are governed with contracts similarly as outsourcing projects. In order to obtain knowledge benefit, the firms must continue engage in the knowledge process after the contracting is determined and knowledge results have been reported. Due to the outsourcing characteristics of knowledge collaboration, the project time allocated to the service provider becomes a critical consideration to determine the ultimate knowledge performance. If the project time of the contract is constrained tightly, the service provider could rush in knowledge creation and engender median quality of the knowledge process. In the contrary, if the project time is allocated sufficiently, the firms might struggle for follow-up knowledge assimilation and integration within a short deadline. In order to amplify the KPO benefits, we will investigate how to decide the optimal project time for outsourcing contracting.

Opposite to the present status of KPO practice, researches on the KPO have been less focused. Most scholars have discussed the differences with conventional business process outsourcing (BPO), the opportunities and challenges of KPO development. For example, based on the Resource View of the firm, Laumer *et al.* [1] discuss the short-term revenue and long-term benefit from KPO with longitudinal case studies. Sen and Shie [2] examine some issues when the firms embark on KPO and start to transfer knowledge process. Mansingh *et al.* [3] describe the knowledge mapping of KPO with ontology-based framework. Mudambi and Tallman [4] adopt qualitative analysis to examine whether the firm should engage knowledge process production or knowledge ally, so as to suggest the possible alliance environments. These literatures enhance the understanding in the concept and development of KPO, but insufficient attention has been paid in KPO decision and operation.

Other studies in outsourcing seldom focus on the

E. Qi et al. (eds.), *Proceedings of the 5th International Asia Conference on Industrial Engineering and Management Innovation (IEMI2014)*, Proceedings of the International Asia Conference on Industrial Engineering and Management Innovation 1, DOI 10.2991/978-94-6239-100-0_77, © Atlantis Press and the authors 2015

outsourcing time, though the project time as a critical metric play important part in contracting. Most scholars mainly discuss the contract coefficients as staffing, contract payment, service time, *etc* in [5-7], but rarely project time. Besides, the collaborative working in knowledge creation distinguishes from the conventional service outsourcing like ITO and BPO (as in [4]). As to the duration of contract, Reference [8] examines how it affect investment, contract negotiation. And Reference [9] analyzes how it affects the possibility of collusion in a repeated price-setting game. Reference [10] discusses optimal expected length of a franchise contract and adopt Spanish cases to validate the outcome empirically. However, no research has been paid attention to project time in KPO. Therefore, examining project time of KPO is important in outsourcing contracting and knowledge results required.

The reminders present as follows: Section 2 describes the basic assumptions of KPO modeling. Following that, the optimal project time are calculated under both the assumptions of risk-neutrality and risk-averseness of the client. Section 4 examines how the project time affects the client's knowledge gains in different environment factors. Section 5 summarizes our findings and discusses possible future research.

II. MODEL SETUP

We consider a conventional outsourcing dyad, which consists of a client (he) and a service provider (she). Not a product or non-core business process but a knowledge-intensive process is involved, such as market analytics, equity research, etc. In general, the service is paid at fixed remuneration within a negotiated time, which is the main focus in this paper. The time can be as sufficient as it could be, to facilitate the quality accruing to the client. But it must be earlier than a expected deadline at least, in case the knowledge results are ineffectual and even counterproductive. In the light of peculiarity of knowledge results, the client must exert extra efforts to adopt, modify, assimilate, and integrate these results so as to promote his corporate performance. Therefore, firms should spare enough time for this integration in outsourcing decision-making rather than negotiate project time exactly as to the expected deadline. Here, we denote the deadline $T > 0$, and consider how to balance the merits and demerits in term of the project time.

We adopt the final knowledge performance to express the overall quality of the knowledge process service at the time T. At each unit of the time, the quality of the knowledge process accrues in fluctuation, so the uncertainty of quality performance at each time needs to be introduced. Therefore, no matter who implements the knowledge process, we assume the quality performance of the knowledge process evolves over time as Geometric Brownian Motion (GBM):

$$dp_1(t) = \mu_1 p_1(t)dt + \sigma_1 p_1(t)dW(t) \quad (1)$$

$$dp_2(t) = \mu_2 p_2(t)dt + \sigma_2 p_2(t)dW(t) \quad (2)$$

where the subscript denotes whether it is operated by the service provider (when it is 1) or the client himself (when it is 2); $p_i(t)$ is the quality performance at time t; $dW(t)$ is standard Wiener process; μ_i is the shift rates of expected future changes; and σ_i is the uncertainty rates of such process. We also assume the initial value $p_i(0) = 1$ .In case of the trivial solutions, we assume $\mu_1 \neq \mu_2$ and $\sigma_1 \neq \sigma_2$. Thus the final knowledge performance is the summarization of quality performance in each unit-time.

Different arrangements bring about distinct cost. For outsourcing, the client will pay contract-based payment. Here, we consider the payment directly depends on the project time, assumed as $a\ln(t+1)$, where a is the payment coefficient and the logarithmic function adopted to depict the marginal decreasing effect of project time. Note that when the client does not choose to outsource, the payment will be none. On the other hand, when the client implement the knowledge service by himself, we assume he has to take out part of total performance to offset the expenditures of assimilation and integration, here denoted the cost ratio as c.

With considerations of simplicity, we neglect the situation that the project time surpasses the expected deadline. Thus its definition domain is only $[0,T]$. When $t = 0$, it implies that all knowledge processes are operated inside the client firm; but when $t = T$, they're all outsourced to the service provider.

III. MODEL ANALYSIS

We will investigate the optimal project time from the perspective of the client. In this section, first we will calculate the expected performance gains from the knowledge process service under the assumption that the client is risk-neutral. Second, we will discuss the effect of risk-averseness on the optimal project time.

A. The client is risk-neutral

In knowledge process outsourcing, the client needs to determine the project time in the contract to obtain expected performance gains. Afterwards, he utilizes the remainder time to integrate with his corporate characteristics. Thus, if r is the discount rate ($\mu_i > r$), then the final knowledge performance of the client at time T is:

$$\pi(T) = \int_0^t e^{-r\tau} E[p_1(\tau)]d\tau + (1-c)\int_t^T e^{-r\tau} E[p_2(\tau)]d\tau - a\ln(t+1) \quad (3)$$

Under the assumption of risk-neutrality, the expected quality performances at each time respectively are:

$$E[p_i(t)] - e^{\mu_i t} \quad (4)$$

In order to solve the problem above, Proposition 1 describes the optimal project time under different circumstances.

Proposition 1. When the client is risk-neutral, the optimal project time is

(1) $t^* = T$, if $f(t) > 0$

(2) $t^* = 0$, if $f(t) < 0$

(3) $t^* = \hat{t}$, which $f(\hat{t}) = 0$ and satisfies

$f'(t)|_{t=\hat{t}^-} > 0$ and $f'(t)|_{t=\hat{t}^+} < 0$

Among them,

$f(t) = (t+1)(e^{t\mu_1} - (1-c)te^{t\mu_2}) - ae^{tr}$

From the Proposition 1, we show there are at least three possibilities for the optimal project time as it describes above. According to the formation of the first-order condition of (1), *i.e.* $f(t)$, it is clear that the larger the payment coefficient and the smaller the cost ratio, the higher likelihood that KPO will not be adopted, in which case the cost structures need to be consider for trade-off. Meantime, from the view of the quality performance, the more performance increment is obtained from the service provider and the less one from the client himself, the larger probability that he implement knowledge process outsourcing. The common cases are not-outsourcing-at-all and collaboration. Note that there might be more than one critical points where $f(t)$ equate to zero, so second-order conditions have to be satisfied to reach the maximum.

In fact, under some circumstances might only two possibilities take place. Here the Corollary 1 to describe one of these situations.

Corollary 1. When $\mu_1 > \mu_2$, either $t^* = T$ or $t^* = 0$ will happen.

This corollary implies that benefit from knowledge assimilation is inferior to one from outsourcing, in which case the client will never consider this integration. In KPO practice, knowledge integration is the indispensable step to obtain final outsourcing benefits and acquire competitive advantages. Therefore, we will only consider $\mu_1 < \mu_2$ then.

B. The client is risk-averse

When the client is a risk-neutral agent, the mean changes of the GBM are only considered. After the upheaval of worldwide economic climate, firms are becoming more and more risk-evaded. Hence, we assume the client here is risk-averse and has the mean-variance preference. Then the expected quality performances at each time respectively become:

$$E[p_i(t)] = e^{\mu_i t} - \eta e^{\mu_i t}\left(e^{\sigma_i^2 t} - 1\right) \quad (5)$$

where we assume more uncertainty the performance has for integration, i.e. $\sigma_1 < \sigma_2$. Analogously, we will have Proposition 2 to describe the optimal project time under the assumption of risk-averseness.

Proposition 2. When the client is risk-averse, the optimal project time is

(1) $t_a^* = T$, if $f_a(t) > 0$

(2) $t_a^* = 0$, if $f_a(t) < 0$

(3) which $t_a^* = \hat{t}_a$ $f_a(\hat{t}_a) = 0$ and satisfies

$f_a'(t)\big|_{t=\hat{t}_a^-} > 0$ but $f_a'(t)\big|_{t=\hat{t}_a^+} < 0$

Among them,

$$f_a(t) = \eta e^{tr}(t+1)\cdot \left[(1-c)e^{2t\mu_2}\left(e^{\sigma_2^2 t}-1\right) - e^{2t\mu_1}\left(e^{\sigma_1^2 t}-1\right)\right] + f(t)$$

The results are rather similar to the proposition 1, apart from a different first-order condition. The basic implications from proposition 1 are also applicable here. However, what we concern is how the optimal project time deviate due to the performance variance and risk-averse coefficient, which will be discussed in Proposition 3 as follows.

Proposition 3. (1) When $1 > c \geq 1 - R_V$, (i) if $f_a(t) > 0$, then $f(t) > 0$; but not vice versa necessarily. (ii) If $f(t) < 0$, then $f_a(t) < 0$; but not vice versa necessarily.

(iii) $\hat{t}_a > \hat{t}$, only if they exist.

(2) When $1 - R_V \geq c > 0$, (i) if $f(t) > 0$, then $f_a(t) > 0$; but not vice versa necessarily. (ii) If $f(t) < 0$, then $f_a(t) < 0$; but not vice versa necessarily. (iii) If $1 - R_V > 1 - R_M > c > 0$ and both critical points exist,

then $\hat{t}_a > \hat{t}$. (Where $R_V = \dfrac{\text{var}_1}{\text{var}_2} = \dfrac{e^{2t\mu_1}\left(e^{\sigma_1^2 t}-1\right)}{e^{2t\mu_2}\left(e^{\sigma_2^2 t}-1\right)}$,

$$R_m = \frac{e^{t\mu_1}(\mu_1 - r)}{e^{t\mu_2}(\mu_2 - r)})$$

According to the Proposition 3, we show under some circumstances, risk-averseness will transform client's outsourcing decision significantly. For example, it is possible that the client can adopt KPO when he neglect risk ($f(t) > 0$), but if he is risk-averse, he will switch to no-outsourcing ($f_a(t) < 0$). Besides, we also indicate some environment where the optimal project time when the client is risk-averse is a bit greater than that when he is risk-neutral. From multiple numerical tests, we have not found the opposite results, partly because the function (1) will reduce a simple convex one for which the maximum are obtained at the endpoints of its definition domain.

IV. NUMERICAL SIMULATION

In this section we will carry on numerical simulation to exhibit the effect of some coefficients and risk-averseness.

At first, we set $\mu_1 = 0.2$, $\sigma_1 = 0.01$, $\mu_2 = 0.37$, $\sigma_2 = 0.05$, $a = 1$, $c = 0.8$, $r = 0.1$, $\eta = 0.05$, $L = 10$, and we will get Fig.1(a). If we change μ_2 to 0.4, Fig.1 (b) is generated. In comparison, they show that raising μ_2 will reduce optimal project time from $t = 9.23$ to $t = 7.78$ when the client is risk-neutral, and even sometimes it will transform the outsourcing form from entire to part outsourcing (for example, when the client is risk-averse in Fig.1(a) and (b)). So does diminishing the client's cost ratio. It indicates that when the benefit from knowledge integration is remarkably significant, the client should spare sufficient time to implement integration on his own rather than letting the service provider continue to make quality improvement of the knowledge process.

Besides, for both settings, consonant with Proposition 3, the risk-averseness decrease the client profit, which is considered as risk premium, and also extend the optimal project time. It implies that due to the uncertainty of knowledge integration, the risk-averse client will shrink the working time inside the company and extend the project time of KPO.

If we change c to 0.5 based on Fig.1 (b), the Fig.1 (c) is drawn. In this situation, no matter in which attitude the client tends to risk, he will never adopt KPO. It implies that when the self-implementation inside the firm incurs modest cost, the outsourcing will not be considered although the service provider can bring higher quality performance. Also, it shows that the effect of risk-averseness on outsourcing decision is quite limited, and it highly depends on the variation of quality creation.

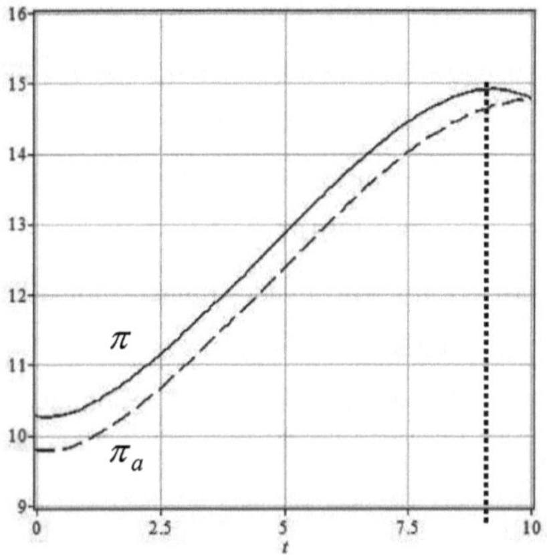

(a) When μ_1 =0.37 and c=0.8

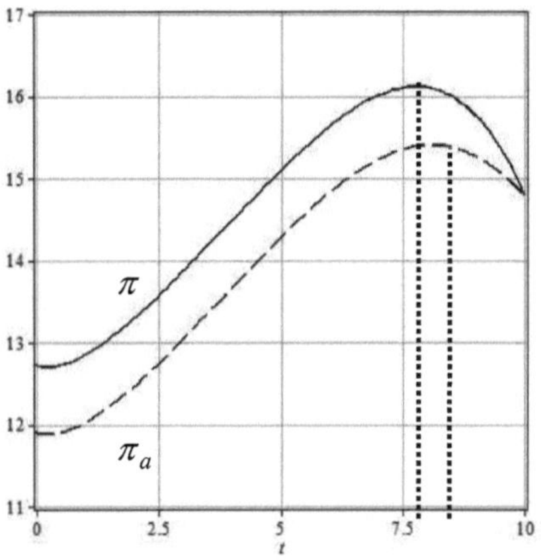

(b) When μ_1 =0.4 and c=0.8

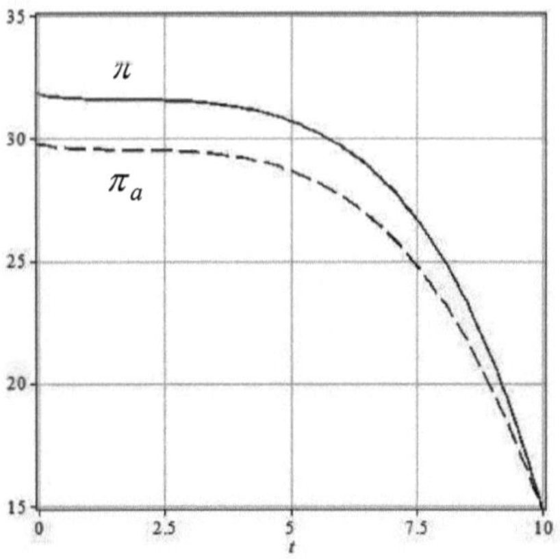

(c) When μ_1 =0.4 and c=0.5

Fig.1 Effect of project time on client's performance gains

V. CONCLUSIONS

Due to the collaborative characteristics of knowledge process, we consider the optimal project time in KPO and its effect on client's knowledge performance. We find out that KPO will be adopted in a very critical environment where outsourcing cost allows for the client and self-implementation encroaches on most knowledge revenue. Also, we show that enough time should be allocated inside for follow-up knowledge assimilation and integration particularly when they can bring out significantly knowledge gains. At last, we point out that the risk-averseness of the client will decrease its profit, generally extend the optimal project time and shorten knowledge integration phase, and even sometimes it can transform the outsourcing decisions.

This study can be extended in several ways. First, we adopt the knowledge creation in outsourcing exogenously, so a more detailed endogenous process of KPO can be considered. Second, we assume both parties are successively working for analysis convenience; hence, interactive knowledge creation could be further discussed. Finally, an empirical study on the determinants of project time could be conducted to validate the conclusions of the paper.

ACKNOWLEDGMENTS

This work was supported in part by a grant from the National Natural Science Funds (71371151).

REFERENCES

[1] Laumer S, Blinn N, Eckhardt A. Opening the black box of outsourcing knowledge intensive business processes--A longitudinal case study of outsourcing recruiting activities[C]. 45th Hawaii International Conference on System Sciences, IEEE, 2012: 3827-3836.

[2] Sen, F, Shiel, M. From Business Process Outsourcing (BPO) to Knowledge Process Outsourcing (KPO): some issues [J]. Human Systems Management, 2006, 25(2): 145-155.

[3] Mansingh, G, Osei-Bryson, K M, Hirata, T. Building ontology-based knowledge maps to assist knowledge process outsourcing decisions [J]. Knowledge Management Research and Practice, 2009, 7(1): 37-51.

[4] Mudambi S M, Tallman S. Make, buy or ally? Theoretical

perspectives on Knowledge Process Outsourcing through alliances [J]. Journal of Management Studies, 2010, 47(8): 1434- 1456.

[5] Ren, Z J, Zhou, Y P. Call center outsourcing: coordinating staffing level and service quality [J]. Management Science, 2008, 54(2): 369-383.

[6] Lee, H H, Pinker, E J, Shumsky, R A. Outsourcing a two-level service process [J]. Management Science, 2012, 58(8): 1569-1584.

[7] Kim S H, Cohen M A, Netessine S, Veeraraghavan S. Contracting for infrequent restoration and recovery of mission-critical systems [J]. Management Science, 2010, 56(9): 1551-1567.

[8] Guriev, Sergei, and Dmitriy Kvasov. Contracting on time .American Economic Review (2005): 1369-1385.

[9] Green, Richard, and Chloé Le Coq. The length of contracts and collusion. International Journal of Industrial Organization 28.1 (2010): 21-29.

[10] García-Herrera, Alicia, and Rafael Llorca-Vivero. How time influences franchise contracts: the Spanish case. European Journal of Law and Economics 30.1 (2010): 1-16

.

Analysis of Lead and Lag Relations between Tourism and National Economy

Jian-tong Zhang*, Chen-cheng Fang
Department of Management Science and Engineering, Tongji University, Shanghai, P. R. China
(zhangjiant@163.com*, fcc2007@foxmail.com)

Abstract - **Opinions on the acceleration of tourism development released by the State Council in December 2009 pointed out for the first time that tourism should be positioned as a strategic pillar industry in national economy and needs to be recognized as a modern service profession satisfied by the public. Over these years, scholars from home and abroad have widely agreed that tourism exerts multiplier effects on national economy by empirical data analysis. However, the lead and lag relationship between Chinese tourism and economy hasn't been studied extensively yet, which influences such crucial issues as the study and drafting of tourism policies. Hence this paper takes advantage of cross spectral analysis, a method to test relations between time series to find out that the development of tourism always lags behind that of economy and the coherency between them becomes greater when period is around 5 to 8 years. According to the results, this paper puts forward some effective ways in the conclusion section to balance the tourism demand and supply, thus helping to maintain the steady growth of tourism and the national economy as well.**

Keywords – **Cross spectral analysis, lead and lag relations, tourism economy**

I. INTRODUCTION

With the rapid development of Chinese economy and gradual release of tourism functions, the positive impact tourism has on national economy is becoming more prominent, attracting the intention from central and regional governments as well as the people of all circles. Opinions on the acceleration of tourism development released by the State Council in December 2009 pointed out for the first time that tourism should be positioned as a strategic pillar industry in national economy and need be recognized as a modern service profession satisfied by the public [1-3]. Also, the high level seminar on regional economy development of tourism held in Beijing, July 2012 reminds that tourism industry has been integrated into the national strategic system, being a mainstay of promoting the innovation of modern service industry and regulating national economic structures [4, 5].

Chinese scholars began to study tourism economy theories in the late 1970s, and during 1980s, their discussions were still mainly about introductions of foreign researches. Until early 1990s, home researchers started to investigate Chinese tourism economy. Yi-fang Chu [6] conducted relative studies on the economical effects of tourism, and Jiang-fan Lee [7] discussed the calculation methods of added value in tourism, which was the relatively early study on the economical effect of tourism. Furthermore, Shan-qing Sun et al helped to put tourism theory into practice and Bing Zuo [8] was the first researcher to test Chinese tourism multipliers. Moreover, Hui-lin Song [9] comprehensively evaluated the former researches on tourism economy. However, over these years, researches on the lead and lag relation between tourism and national economy have seen little progress but it really concerns about whether Chinese tourism is supply induced or demand driven, which relates to tourism policy making [10-12]. Hence, this paper is aimed to determine the lead and lag relations between them mainly through the cross spectral method. Section II is the introduction of cross spectral method and section III is about data processing. Empirical analysis is in section IV and discussions are included in section V.

II. METHODOLOGY

Cross spectral analysis is an effective method to test correlations between two time series at different frequencies [13]. To be specific, the phase spectrum it generates can explain the lead and lag relation which also helps solve the central problem of this paper. Furthermore, the coherency spectrum and gain spectrum can explain coherent relation and dependency relation between time series respectively [14, 15].

1) *Cross spectrum:* According to spectral theory, the frequency function of cross spectrum between two time series $\{x_k\}$ and $\{y_k\}$ is the Fourier transformation of cross covariance function, and the corresponding relationship between cross spectrum frequency function $f_{xy}(\mu)$ and cross covariance function $C_{xy}(k)$ can be shown as follows:

$$f_{xy}(\mu)=\sum_{k=-\infty}^{\infty} C_{xy}(k)e^{ik\mu} \quad (1)$$

$$C_{xy}(\mu)=\int_{-\pi}^{\pi} e^{ik\mu} f_{xy}(\mu)d\mu \quad (2)$$

The domain of positive function $f_{xy}(\mu)$ is $\mu=(-\pi,\pi)$ while the one of $C_{xy}(k)$ is $k=0,1,2,\cdots,n-1$. However, it is hard to analyze and explain cross spectrum due to its complex nature, so when it comes to practical use, cross spectrum is usually presented in the form of polar coordinates, thus deriving phase spectrum, coherency spectrum and gain spectrum to help explain the relations between time series.

2) *Phase spectrum*: When cross spectrum function is put in the polar coordinates, this paper gets:

$$f_{xy}(\mu)=P_{xy}(\mu)-iQ_{xy}(\mu)=A_{xy}(\mu)e^{-i\theta_{xy}(\mu)} \quad (3)$$

$P_{xy}(\mu)$, $Q_{xy}(\mu)$ and $A_{xy}(\mu)$ are co-spectrum, orthogonal and amplitude spectrum respectively.

E. Qi et al. (eds.), *Proceedings of the 5th International Asia Conference on Industrial Engineering and Management Innovation (IEMI2014)*, Proceedings of the International Asia Conference on Industrial Engineering and Management Innovation 1, DOI 10.2991/978-94-6239-100-0_78,

$\theta_{xy}(\mu)$ is the phase spectrum which can be transformed as:

$$\theta_{xy}(\mu) = \arctan(-\frac{Q_{xy}(\mu)}{P_{xy}(\mu)}) \quad (4)$$

Phase spectrum represents the mean of phase discrepancy values at corresponding frequencies between two time series, which is limited within $(-\pi,\pi)$ at usual. When $\theta_{xy}(\mu) > 0$, $\theta_y(\mu)$ which is the phase of $\{y_k\}$ at frequency μ is greater than that of $\{x_k\}$, indicating $\{y_k\}$ leads $\{x_k\}$ with $\theta_{xy}(\mu)$ phases and vice versa. Furthermore, time length between $\{x_k\}$ and $\{y_k\}$ can be drawn with the help of relationship formulation $T = \frac{2\pi}{\mu}$ between vibration period T and frequency μ :

$$L(T) = \frac{\theta_{xy}(\mu)T}{2\pi} \quad (5)$$

When $L(T) > 0$, $\{x_k\}$ lags behind $\{y_k\}$ at the frequency μ and vice versa. It has to be pointed out that the time difference given by phase spectrum can be only applied to the whole vibration period.

3) *Coherency spectrum*: Coherency spectrum is supposed to tell the coherent relation between time series at the frequency μ:

$$R_{xy}^2(\mu) = \frac{|f_{xy}(\mu)|^2}{f_x(\mu) f_y(\mu)} \quad (6)$$

$f_x(\mu)$ and $f_y(\mu)$ are the frequency functions of power spectrum for $\{x_k\}$ and $\{y_k\}$ respectively. Since auto-power spectrum is a real function, this paper gets:

$$f_x(\mu) = P_{xx}(\mu),\ f_y(\mu) = P_{yy}(\mu) \quad (7)$$

And because $|f_{xy}(\mu)|^2 = P_{xy}^2(\mu) + Q_{xy}^2(\mu)$, this paper gets:

$$R_{xy}^2(\mu) = \frac{P_{xy}^2(\mu) + Q_{xy}^2(\mu)}{P_{xx}(\mu)P_{yy}(\mu)} \quad (8)$$

Apparently, $R_{xy}^2(\mu) \le 1$. And the coherency degree between $f_x(\mu)$ and $f_y(\mu)$ at the frequency μ gets higher when $R_{xy}^2(\mu)$ becomes closer to 1.

4) *Gain spectrum*: Gain spectrum is a reflection of the dependency relation between $\{x_k\}$ and $\{y_k\}$ at the frequency μ, and the formulation is as follows:

$$G_{xy}(\mu) = \frac{\sqrt{f_y(\mu)R_{xy}^2}}{f_x(\mu)} = \frac{f_{xy}(\mu)}{f_x(\mu)} \quad (9)$$

Obviously, $\{y_k\}$ is more dependent on $\{x_k\}$ at the frequency μ when $G_{xy}(\mu)$ gets larger.

III. DATA COLLECTION AND PRE-PROCESSING

After the introduction of cross spectrum method in Section II, this paper will then come to empirical analysis. But before that, procedures of data collection and pre-processing should be illustrated firstly [16].

According to the accessibility of data and the requirements of cross spectral method, this paper decides on the time length of studied data to range from 1995 to 2012, 18 years in total which is obtained from National Bureau of Statistics. What has to be mentioned is that this paper uses interpolation method[1] to tackle with missing official data. And to make data more applicable, this paper has to pre-process it before actually putting them into use [17, 18].

A. *Sustainable domestic production*

Gross domestic production is the most commonly used index to measure national economy and it is of great guiding significance. Nevertheless, in view of this paper, the development of tourism can not only bring about the increase of tourism income but also the destruction to the local environmental resources and the consumption of fixed assets, so to make the economy development much more sustainable and greener, this paper takes off the impact of these negative elements by referring to the calculation of green GDP issued by Research Group of Macro Economy Institute in National Development and Reform Commission [19]. Also, owing to the big population in China, this paper considers average index should be much more reasonable.

$$SDP = \frac{1000 \times (GDP - FAD - EPI)}{P} \quad (10)$$

SDP represents sustainable domestic production with the unit of RMB per person. GDP is gross domestic production with the unit of 100 million RMB. FAD means fixed asset depreciation and EPI represents environmental protection investment, both of which share the same unit as GDP. And P is the population with the unit of ten thousand.

B. *Total tourism income*

This paper considers total tourism income as an appropriate measurement of tourism development level which is the expense of both home and abroad tourists when traveling within China. In fact, there are many indexes relative to the development of tourism, including tourism income of home tourists, tourism income of

[1] To be specific, this paper uses linear interpolation method to solve the problem of missing data because the length of studied data is long enough and the missing data only takes a small fraction of the data set.

foreign tourists, tourism investment and the alike. However it is hard and unnecessary to take all of them into account and meanwhile, the unsoundness of Chinese tourism statistical system makes some of these data unavailable [20]. By the contrary, both tourism incomes of home and abroad tourists travelling within China are collected and counted successively by governments, which make it ready and convenient to be obtained. Formulation of total tourism income is as follows:

$$TTI = HTI + \frac{FTI \times ER}{100} \tag{11}$$

TTI represents total tourism income with the unit of million RMB. HTI with the unit of million RMB represents the income generated by home tourists while FTI with the unit of million dollars means the income obtained from foreign travelers. ER is the RMB exchange rate to dollar (dollar=100).

IV. EMPIRICAL ANALYSIS

Section II and Section III show the method and processed data this paper requires and now it comes to the empirical analysis about tourism economy.

A. Phillips-Perron unit root test

As illustrated in Section II, cross spectral method needs time series to be stable during studied period, which implies that the mean and variance values should remained unchanged relatively [21]. Hence, this paper takes advantage of Phillips-Perron test to check the stationarity of both series. The results are listed below.

To SDP, the original data turns out to be unstable even with the significance level of 10% while first difference data presents stability (see Table I).

TABLE I

PP UNIT ROOT TEST STATISTIC OF ΔSDP

		Adj. t-Stat	Prob.*
		-5.504968	0.0024
critical values	1% level	-4.667883	
	5% level	-3.733200	
	10% level	-3.310349	

The null hypothesis of PP test supposes that time sequence is not stationary. From table I, t-statistic is smaller than any critical values at any significance levels, so it is safe to conclude that SDP=I(1). In the same way, this paper checks out that TTI=I(2).

B. Transformation of time sequence

Now, SDP and TTI are integrated at the first and second orders respectively, this paper has to make them transformed before doing cross spectral analysis [22]. The common way of transformation is the difference method. Hence, SDP is differenced at the first order while TTI is differenced at the second order (See Table II).

In Table II, SDP1 is the first difference data of SDP and TTI2 is the second difference data of TTI. To make the table look neat and clean, this paper keeps one digit after the decimal point. NA appears since difference method can lead to the missing of early data, and because of it, the studied time span is narrowed to 16 years totally ranging from 1997 to 2012.

TABLE II

DIFFERNCED DATA OF SDP AND TTI

Year	1995	1996	1997	1998
SDP1	NA	678.8	457.0	263.1
TTI2	NA	NA	-77.5	245.7
Year	1999	2000	2001	2002
SDP1	251.4	587.0	603.2	603.7
TTI2	-306.2	243.6	-45.0	-43.1
Year	2003	2004	2005	2006
SDP1	892.7	1540.3	1583.5	1959.6
TTI2	94.2	-1253.3	2640.7	-1113.4
Year	2007	2008	2009	2010
SDP1	3263.0	2976.3	1440.7	3793.6
TTI2	405.6	771.8	-1394.3	680.2
Year	2011	2012		
SDP1	4783.5	2747.7		
TTI2	1479.3	3967.1		

C. Cross spectral analysis

As discussed above, cross spectral analysis is an ideal method to test the relations between two time series in a comprehensive way [23]. The phase spectrum, coherency spectrum and gain spectrum that cross spectrum derives can tell an overall relations between time series. Hence, by using the processed data in Table II and SPSS18.0, this paper gets the analysis result (See Table III). What has to be noted is that the default result of cross spectral analysis in SPSS18.0 only includes graphs which could not tell specific values, so this paper takes advantage of the syntax function in SPSS18.0 to solve this problem. Syntax is showed below.

```
*CROSS SPECTRAL ANALYSIS.
TEST PRINT=DETAILED.
SPECTRA.
/VARIABLES=SDP1 TTI2.
/WINDOWS=HAMMING(5).
/CENTER.
/CROSS.
/PLOT=P S K CS QS PH A G BY PERIOD.
```

TABLE III

RESULTS OF CROSS SPECTRAL ANALYSIS

Period	Gain	Coherency	Phase
.	0.252	0.207	
16.000	0.327	0.323	0.421
8.000	0.589	0.711	0.902
5.333	0.655	0.685	0.949
4.000	1.007	0.721	0.905
3.200	0.956	0.383	0.810
2.667	0.940	0.248	1.013
2.286	1.526	0.175	0.726
2.000	3.890	0.348	0.000

In Table III, Estimated values of phase spectrum are all larger than zero except when the period is within 2 years, indicating in general that the development of tourism lags behind that of national economy during the studied 16 years. This result can be verified by development process of Chinese tourism shown below.

Before 1978, the year in which the policy of reform and opening-up released, tourism is mainly about receiving foreign governmental visitors and civil tourism activities were nearly zero. In 1980s, the industrialization of tourism were put into government agenda. In 1986, tourism was added into national plan of economy and society by the State Council and has been taking a part in the national economy ever since. Furthermore, in 1992, central government pointed out that tourism had become a key in third industry, and afterwards, Opinions on the ninth five-year plan of economy and society development proposed by central government rated tourism as number one in the list of newly emerging third industries. Moreover, China's accession to WTO in 2001 has also raised new opportunities and challenges to tourism. Hence, judging from all the issues above, it is also safe to infer that tourism development always lags behind that of national economy and may copy the development process of national economy in some way [24].

In Table III, the estimated values of coherency spectrum become larger during 5 and 8 years, which are about twice of those in other periods. The conclusion can be drawn that the coherency relation between tourism and national economy in short and long period is not obvious, but they becomes strongly related in midterm, especially 5 to 8 years, which can be explained by the hysteresis nature of tourism economy because the multiplier effect can make tourism income turns several times by the distribution and re-distribution of it which may take some time. Also, the evaluated values of gain spectrum in table 3 are larger when the period gets shorter, telling that the dependency relation between them are greater at shorter period. This paper has to denote that gain spectrum can only be used to test relative level of dependency between series, instead of absolute level.

V. DISCUSSION

Section IV gives out the SPSS implementation and empirical results of cross spectral method, so in section V, this paper mentions about the conclusions drawn from empirical results above [25].

A. Balance between demand and supply

The results of cross spectral analysis show tourism development lags behind that of economy, so tourism can be interpreted as demand driven in some degree. Only when people have extra time and money to travel, the tourism supply appears and progresses. Hence, it is of great significance to keep a balance between two of them. When tourism supply lacks, tourism demand gets surplus, making this part of demand ineffective. Followings are some potential ways to tackle with this problem.

B. Improvement of vacation system

After 1995 when the system of five working days per week established, the prosperity of tourism proves that centralized vacation time is crucial to the marketization of tourism. However, heavy congestions in tourist attractions recently raise another new challenges to tourism since many of potential tourists decide to stay home in vacations to avoid crowds. Hence, this paper considers that a much more flexible vacation system needs to be formulated, making them staggered to release the potential tourism demand.

C. Enhancement of tourists' ability to pay

This is an essential problem because the tourism income is a key element for tourism to realize its economical effect. This paper thinks the following methods may enhance tourists' ability to pay. For instance, special accounts of tourism can be established and non-cash payment system could be tried out.

D. Development of international tourism

The trend of globalization is irresistible and China's accession to WTO has put forward new opportunities to tourism. This paper considers the tourism income from foreign tourists traveling within China can be an original point of growth for tourism. Both simplifying the tourism visa system and implementing export rebates to tourism shopping can be effective ways to develop international tourism.

E. Diversification of tourist sites

In spite of the fact that some of tourist sites have monopolized, many of them can still be substituted very easily which makes the operations of those tourist destinations very unstable. Hence, this paper takes diversification as a good method to prevent this problem, that is making every destinations different and localized.

F. Mild stimulation to tourism

The result of cross spectral analysis shows that tourism development lags behind that of economy, so this paper considers tourism as a third industry needs the support of second industry which takes a large part in national economy. Only when the second industry has developed into a new stage can tourism get progress on the base of it. Hence, abrupt stimulation of tourism may lead to the imbalance of industries, which may ultimately influence the stable growth of national economy in the long run. This paper thinks mild stimulation to tourism can be a better way.

G. Proper time span for tourism plan

According to the coherency spectrum in Section IV, tourism and national economy are strongly related from 5 to 8 years because of the hysteresis effect of tourism economy. This paper then thinks the proper time span for tourism plan is 5 to 8 years when the impact of plan can finally come out, which also coincides with the time length of five-year plan of national economy implemented in China for decades.

ACKNOWLEDGMENT

Thanks a lot for the authors of pre-researches in related fields, without whose efforts, hardly could we reach the current results. Also, thanks for the careful work of staff in the Bureau of Chinese Statistics (www.stats.gov.cn) since most of our initial data is obtained from it. The meticulous performance of them ensures the accuracy of our analysis.

REFERENCES

[1] Alan T.K.Wan., Xin-yu Zhang., "On the use of model averaging in tourism research". Hong Kong, China

[2] Archer., *Tourism Multipliers: the state of the Bangor. Bangor*: University of Wales Press, 1980, pp32 – 47

[3] Archer B., Fletcher J., "The Economic Impact of Tourism in the Seychelles". *Annals of Tourism Research*, 23, 1, 1996, pp32 – 47

[4] Bao-jia Shen., Research Progress of foreign tourism, *Tourism Studies*, 1996, pp 1-5 (in Chinese)

[5] Bloch M., "The impact of electronic commerce on the travel industry – an analysis methodology and case study". In: *Proceedings of the thirtieth annual Hawaii international conference on system sciences*, IEEE, Maui, Hawaii, 1997, pp 48–58

[6] Yan Zhang., "The analysis of tourism economy in Qingdao city under the concept of multipliers", *Tourism Economics*, April 2013, pp115-116

[7] Isidoro Romero., Pilar Tejada., "A multi-level approach to the study of production chains in tourism sector". *Tourism Management*, 32, 2011, pp 297-306

[8] Bing Zuo, "The initial calculations of multipliers of Chinese tourism output and employment", *Academic Journal of Yunnan Financial College*, 18, 6, 2002, pp 30-34 (in Chinese)

[9] Da-wu Wong., Xiao-an Wei, *Tourism Economics*, Shanghai People Press, 2000 (in Chinese)

[10] Buhalis D., Law R., "Progress in information technology and tourism management: 20 years on and 10 years after the Internet", the state of e-Tourism research, *Tour Manage*,29,2008, pp 249-272

[11] Chang-sheng Liu., Yu-feng Jian., Relations between Tourism and National economy, *Tourism Science*, 22,5,2008 (in Chinese)

[12] Chun-zi Deng., *Researches on the accounting system of green GDP*, Yanshan University, Master Dissertation, December 2005 (in Chinese)

[13] Hui-lin Song.,"Literature review on economical effect of Chinese tourism", *Xuzhou Education Studies*, 4, 2007 (in Chinese)

[14] Jiang-fan Lee., Mei-yun Lee.,"The calculation of tourism added value and tourism industry", *Tourism Studies*, May 1999, pp 16 – 19 (in Chinese)

[15] Jie Zhang., Bjarne Madsen., and Chris Jensen., "Regional Economic Impacts of Tourism - The Case of Denmark", *Regional Studies*, 41, 6, 2007, pp839–853

[16] Keogh B., *Social Impacts of Outdoor Recreation in Canada.* Toronto John Wiley Press, 1985

[17] Li Tian., *Tourism Economics*, Beijng, Science Press, 2004 (in Chinese)

[18] Luis Robles Teigeiro., Barbara Diaz., "Estimation of multipliers for the activity of hotels and restaurants", *Tourism Management*, 40, 2014, pp 27-34

[19] Mathison A., Wall G., *Tourism: Economic, Physical and Social Impacts*. London, Inc, 1982

[20] Research Group of Macro Economy Institute in National Development and Reform Commission., "Statistical Method and Definition of Green GDP", *Macro economy studies*,1,2005, pp 30-33 (in Chinese)

[21] Shou-xiang Shi, "Dialectication of tourism multipliers", *Tourism Studies*, 22, October 2007, pp30-34 (in Chinese)

[22] Sinclair., "Tourism and economic development: a survey", Journal of Development Studies,34,5,1998, pp1-51

[23] Tian-yuan Lee., *Tourism (Second Edition)*, Beijing, High EducationPress, 2006 (in Chinese)

[24] Wood N., Liang K., "*The impact of tourism sector on the Vermont Economy*", the department of community development and applied economics, University of Vermont, Blington, Vermont, 2001

[25] Yi-fang Chu. *Space economical analysis of tourism*, Xi'an, Shanxi People Press, 1992 (in Chinese)

The Research on Household Financial Multi-objective Decision Based on Assets and Liabilities Management

Yan-wen Liu*, Lu-wa Pu, Chen Cao
Faculty of Management and Economics, Dalian University of technology, Dalian, china
(liuyanwen@126.com)

Abstract - **Household finance is an important emerging area of financial studies. According to asset and liability management, this article uses multi-objective optimization theory to build the model, the author also wants to use statistical and random programming method to deal with the stray parameter, and finally transfer the model to ascertained decision problem and determines the optimal decision strategy of household financial. Decision making analysis method of qualitative and quantitative composition is applied in selecting the model parameter. The results show that family can simultaneously achieve assets income maximization and predetermined solvency target configuration of assets and liabilities. And by comparing with the actual data, we found that household assets returns after the model optimization are higher than the actual value, and household debt paying ability is lower than the actual value, which shows that the improved model have played an important role in optimizing family assets and liabilities management decisions.**

Keywords - **Asset and liability management, household finance, multi-objective decision**

I. INTRODUCTION

With the continuous improvement of our country residents' income level, household financial assets have increased rapidly in quantity, together with the deepening of the marketization, asset allocation will become more and more diversified, only by allocating the assets properly and improving assets yield, can family guarantee the efficacy and safety of the household finances [1]. To ensure the effectiveness and security of the household finances, families should make their debt effective, and get financial leverage income, reduce the cost of the family fund [2]. So, how to allocate and manage the family assets and liabilities reasonably has become a major issue on household financial decisions

According to asset and liability management, this article uses multi-objective optimization theory to build the model [3], the author also wants to use statistical and random programming method to deal with the stray parameter, and finally transfer the model to ascertained decision problem and determines the optimal decision strategy of household financial [4]. Decision making analysis method of qualitative and quantitative composition is applied in selecting the model parameter to make the forecast more scientific [5].

II. MODELS

A. Multi-objective programming model

Due to the amount of assets and liabilities and deadlines are closely related, so the family assets and liabilities management goal presents the characteristics of the complexity and diversity [6]. Family assets should meet the requirements of profitability, safety, liquidity, and also meet the spending needs of family members and meet the requirements of household debt, such as pay family housing loans, consumer loans, etc. It means that family can simultaneously achieve assets income maximization and predetermined solvency target configuration of assets and liabilities [7]. Therefore, the family financial assets and liabilities management itself is a multi-objective decision problem [8]. Multi-objective programming theory can achieve the goal that optimizing family assets and liabilities management decisions [9].

B. Random programming model

As an individual investors, just like institutional investors, families will meet all kinds of uncertain factors in the process of participating in the market, combined with the limit of the fund, families may have greater risks in the aspect of asset allocation [10]. Therefore, in the family financial decisions, it is bound to consider these uncertain factors. We usually treat these uncertain factors as random variable. The model has become a random programming model whenever there are random variables in an ordinary mathematical programming model.

According to the general mathematical programming method [11], we obtained the optimal value under the expectations X[E(c_j)], but we hope to get the optimal solution of original problem E[X(c_j)], namely, the optimal solution of the stochastic variable expectations. In general both values should meet the following relations:

$$\max Z\left[E(c_j)\right] \le E\left[\max Z(c_j)\right] \tag{1}$$

$$\min Z\left[E(c_j)\right] \ge E\left[\min Z(c_j)\right] \tag{2}$$

C_j is a random variable coefficient of the objective function. For linear programming problem, type (1), (2) take the equal sign, so after alternative the solution is meaningful. The proofs of type (1), (2) are omitted in this paper.

For household financial, the asset-liability ratio, debt ratio and liquidity ratio all have general rules. Therefore, the constraint of household financial decision-making model is a certain feasible region. Because the yield for the family is uncertain, household financial decisions using statistics random linear programming model is consistent with the actual situation [12].

E. Qi et al. (eds.), *Proceedings of the 5th International Asia Conference on Industrial Engineering and Management Innovation (IEMI2014)*, Proceedings of the International Asia Conference on Industrial Engineering and Management Innovation 1, DOI 10.2991/978-94-6239-100-0_79, © Atlantis Press and the authors 2015

III. THE ESTABLISHMENT OF THE HOUSEHOLD FINANCIAL MULTI-OBJECTIVE DECISION-MAKING MODEL

A. Household balance sheets

The main purpose of this paper is to establish a random multi-objective programming model for family financial assets and liabilities management. So, first of all, we are going to explain and illustrate the household balance sheets involved in household financial activities. Household balance sheet is shown in Table I.

TABLE I
HOUSEHOLD BALANCE SHEETS

assets				Liabilities and net worth		
financial assets		cash	A_1	current liabilities		L
	bank deposit A_2	3months	A_{21}	Fees payableL_1	utilities	L_{11}
		half year	A_{22}		telephone bill	L_{12}
		1 year	A_{23}		fuel charge	L_{13}
		2 years	A_{24}	credit card		L_2
		5 years	A_{25}	insurance benefit		L_3
		due on demand	A_{26}	the current long-term liabilities		L_4
	national debt		A_3	long-term liabilities		
	stock、funds		A_4	housing loan balance		L_5
	financial products		A_5	Auto loan balance		L_6
	house		A_6	other loan balance		L_7
real assets	Investment in real estate		A_7	the final net worth		
	Furniture and home appliances		A_8	the current net income		V
	automobile		A_9	beginning net assets		W
	Jewelry and collectibles		A_{10}			
	others		A_{11}			
total assets				total		

B. Objective function and constraint condition

a. Assets revenue maximization goals

In order to achieve the household assets revenue maximum goal, we define family assets revenue function:

$$\Pr = \sum_{i=1}^{11} R_i A_i \tag{3}$$

A_i and R_i means that every content in the balance sheet We assume that R_i is a random number because R_i is influenced by the national economy, and it will change with the adjustment of national economic policy. So we use the statistics random goal programming method to seek the expectation $E(R_i)$ of random variable R_i. So the objective function changes into:

$$P = \sum_{i=1}^{11} E(R_i) A_i \tag{4}$$

A_i means that every content in the balance sheet items. $E(R_i)$ is the expected value of the return on assets. According to the above statistics random goal programming approach we know that type (3), (4) are equivalent.

b. Household debt paying ability goal

Household debt paying ability refers to a family's ability to repay debts; there can be many indicators to measure the ability. The solvency can be divided into short-term and long-term. Considering from long-term debt paying ability, household debt paying ability refers to the household net worth shown on the balance sheet. It is defined as:

$$C = W \tag{5}$$

W means:

$$W = \sum_{i=1}^{11} A_i - \sum_{i=1}^{7} L_i \tag{6}$$

We can also consider from the current period, and then household debt paying ability refers to the ability to repay and must repay the debt currently [12]. Because this paper is to build the current assets profit maximization goal, in order to match with it, here the family debt paying ability refers to the solvency of the current period. We define it as:

$$C = L_1 + L_2 + L_3 + L_4 + 0.8T + A_2 \tag{7}$$

T is the family net income of the current period, coefficient of 0.8. Generally, families will save 20% of their income in the bank. The family's bank deposits has strong solvency [13]. At the same time, during next year family's earnings and solvency P, C is not fixed values. According to the historical data we can predict them as P', C', and then we can get the solution of variable under the constraint condition.

c. Constraint condition

Family belongs to individual investors, and there are big differences between family structures, income level, education level, personal preferences, etc. [14], so the configuration standard of family assets and liabilities is vague. But there is a recognized standard in theory for household financial activities [15].

Firstly, it is family long-term solvency indicators. In theory, the ratio of total assets of the family and household debt must not exceed 0.5, also said the family debt ratio, namely the ratio of total family assets and household net worth should be between0.5 to 1. That is to say:

$$0.5 \le \frac{\sum_{i=1}^{11} A_i - \sum_{i=1}^{7} L_i}{\sum_{i=1}^{11} A_i} \le 1 \tag{8}$$

or:

$$\begin{cases} 0.5\sum_{i=1}^{11} A_i \ge \sum_{i=1}^{7} L_i \\ \sum_{i=1}^{7} L_i \ge 0 \end{cases} \tag{9}$$

Secondly, it is family short-term solvency indicators. They are the so-called liquidity indicators, namely the ratio of the family monthly spending and liquid assets. As the household balance sheet shown, liquidity liabilities are generally monthly deficit spending, and cash in household assets project is retained by the month, so they can be used to measure the short-term solvency. In addition, liquid liabilities is only part of their monthly spending, family daily life spending should also be taken into consideration, so that the quality of the family life can be ensured. Generally, household monthly maintenance costs are relatively fixed values, here is assumed to be S, S is a parameter that can be measured, not a variable. It is advisable when liquidity ratios are between3-6, namely:

$$3 \le \frac{A_1}{\sum_{i=1}^{4} L_i + S} \le 6 \quad (10)$$

or:

$$\begin{cases} A_1 \ge 3(L_1 + L_2 + L_3 + L_4 + S) \\ A_1 \le 6(L_1 + L_2 + L_3 + L_4 + S) \end{cases} \quad (11)$$

Thirdly, household debt paying ability also reflects on the relationship of debt and income. We define liabilities ratio as the ratio of debt spending and stable income. Stable income T refers to the monthly income, mainly the labor income of family members, such as salary, bonus, or business income. As T also can be measured, so T is not a variable. In general, we assume that household debt income ratio is no more than a third. Whenever the ratio is more than a third, it means that family income is not very normal, and it will influence the quality of family life, so:

$$\sum_{i=1}^{4} L_i \le \frac{1}{3} T \quad (12)$$

In addition, usually, deposits should not be higher than a quarter of family assets:

$$A_2 \le \frac{1}{4} \sum_{i=1}^{11} A_i \quad (13)$$

Family emergency reserve fund equals to a monthly income of all family members:

$$A_1 \ge T \quad (14)$$

Risk investment should be less than 4/5 of the financial assets, or the family will face a huge risk of assets loss:

$$A_4 \le 0.8(A_1 + A_2 + A_3 + A_4 + A_5) \quad (15)$$

According to the regulations of housing loans and car loans, loans shall not exceed the real value of a certain proportion δ_1, δ_2. The loans are also under the influence of interest rate and loan fixed number of year. Here, r, t are housing loans and auto loans interest rate, n, m are the numbers of the months to pay debts.

$$L_5 \le \delta_1 (A_6 + A_7) * (1 + r)^n \quad (16)$$

$$L_6 \le \delta_2 A_9 * (1 + t)^m \quad (17)$$

According to the current market, most financial products for individuals purchase amount of at least 50000, so we say:

$$A_5 \ge 5 \text{ or } A_5 = 0 \quad (18)$$

$$A_i \ge 0 (i = 1, 2, \cdots, 11), L_i \ge 0 (i = 1, 2, \cdots, 7).$$

C Establishment of the model

Based on the description of the objective function and constraint conditions, the problem change into goals P1 and P2 under the constraint condition.P1 is that the family investment returns in the next year are more than P', and P_2 is that family's solvency in the next year reach C'.We set up the multi-objective random programming model under the condition that household assets income reach the maximization and maximize solvency. According to the standard form of goal programming, the model is as follows:

$$M \text{in} Z = P_1 d_1^- + P_2 (d_2^- + d_2^+) \quad (19)$$

$$s.t. \begin{cases} \sum_{i=1}^{11} E(R_i) A_i + d_1^- - d_1^+ = P' \\ L_1 + L_2 + L_3 + L_4 + 0.8T + A_2 + d_2^- - d_2^+ = C' \\ 0.5 \sum_{i=1}^{11} A_i \ge \sum_{i=1}^{7} L_i \\ A_1 \ge 3(L_1 + L_2 + L_3 + L_4 + S) \\ A_1 \le 6(L_1 + L_2 + L_3 + L_4 + S) \\ \sum_{i=1}^{4} L_i \le \frac{1}{3} T \\ A_2 \le \frac{1}{4} \sum_{i=1}^{11} A_i \\ A_1 \ge T \\ A_4 \le 0.8(A_1 + A_2 + A_3 + A_4 + A_5) \\ L_5 \le \delta_1 (A_6 + A_7) * (1 + r)^n \\ L_6 \le \delta_2 A_9 * (1 + t)^m \\ A_5 \ge 5 \\ A_i \ge 0 (i = 1, 2, \cdots, 11) \\ L_i \ge 0 (i = 1, 2, \cdots, 7) \\ E(R_i) = \frac{1}{n} \sum E(R_{ij}), i = 1, 2, \cdots, 11; j = 1, \cdots, n \\ R_{ij} = H_k R_{ijk} \end{cases} \quad (20)$$

Here, E(Ri) is the expectation of the random variable R_i, R_{ij} is the adjusted value for No.i variable after k round adjust by specialist j.Hk is the probability of R_{ij}. S, T is measurable constant. δ1, δ2 are the maximum amount of housing loans and auto loans specified by the China banking regulatory commission. r, t are housing loans and auto loans interest rate, n, m are the numbers of the months to pay debts. These variables can be obtained according to the actual circumstance of the family. Besides, this model only based on the family assets to maximize return on investment and the target of solvency, not consider family "profits" , namely "Income - debt cost".

IV. A MODEL INSTANCE ANALYSIS

A. The determination of random variable parameters

We use Delphi method to qualitative predict the determination of random variable parameters. We get the returns by using Delphi method questionnaire of experts combined with expectancy method

We gain the family assets income and solvency according to the family balance sheet and income and spending in 2008-2013. We found that there is a linear relationship between investment income and year. We assume a function P=a+bt, and we get fitting coefficient a=-1.3079,b=1.3218. So the function is P=1.3218t-1.3079, R2=0.9562, it means that it fits well. It is a common sense that the higher the number of fitting, the approximation is the better. However, the data is limited, we only do the fitting twice. Assume a function $P=a+bt+ct^2$, the relationship between benefits and year is as shown in Fig.1.

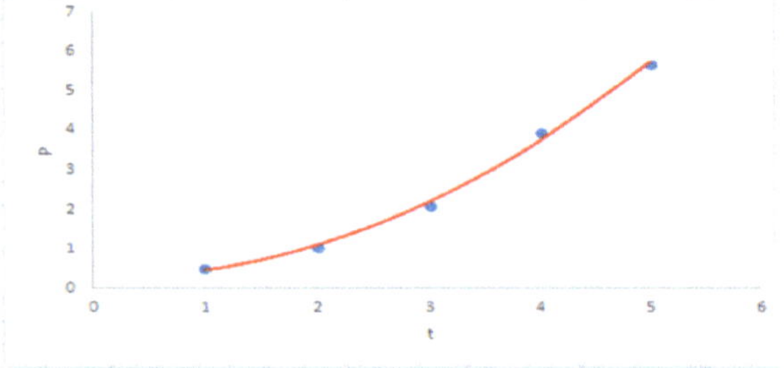

Fig.1. Capital Gains and Year Second Fitting Diagram

We get the coefficient a=0.2883, b=-0.0464, c=0.228, so the function is $P=0.228t^2-0.0464t+0.2883$, R2=0.996. It fits better. We put t=6 in the function, we get P=8.7747, Combined with experts' qualitative analysis we determine family asset returns of 8.4.

We can see the relationship between solvency and year in the same way. We can get the fitting coefficient a=5.281, b=10.828, and the fitting function C=5.281t+10828, R2=0.9695. We put t=6 in the function and get C=42.514.

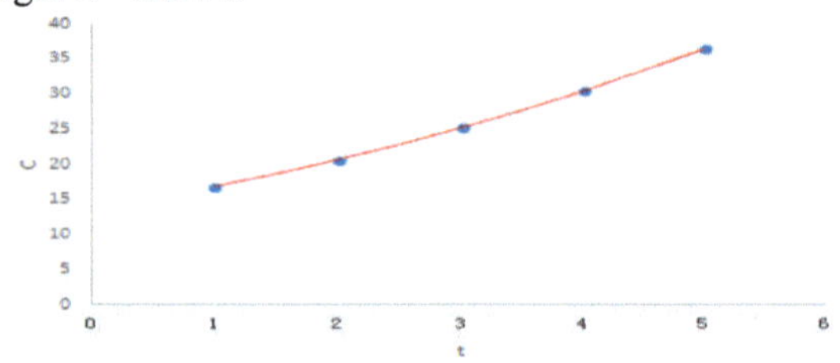

Fig.2. Solvency and Year Second Fitting Diagram

From Fig.2, we get the fitting coefficient a=0.2929, b=2.5237, c=22.379 and fitting function $C=0.2929t^2+2.5237t+22.379$, R2=0.9984. It fits better, too. We put t=6 in the function and get C=48.0656.

Considering expert analysis, the correct value of solvency is C=45.34.

B. Solution method for model

We obtained the parameters under the qualitative and quantitative comprehensive judgment in section 21, and the value can reflect the real situation well, the accuracy of the prediction is high. According to the household income spending, we determine that T=22000, S=7000, $\delta_1=\delta_2=0.8$, then we put them into the model. First, we should adjust the model by making $L=L_1+L_2+L_3+L_4$, and we get:

$$Min\,Z = P_1 d_1^- + P_2(d_2^- + d_2^+) \tag{21}$$

$$s.t.\begin{cases} 0.0327A_2+0.0370A_3+0.0610A_4+0.0507A_5+0.0598A_7+d_1^- -d_1^+=8.4 \\ L+0.8T+A_2+d_2^- -d_2^+ =45.34 \\ 0.5\sum_{i=1}^{11}A_i \geq \sum_{i=1}^{7}L_i \\ A_1 \geq 3(L+7000) \\ A_1 \leq 6(L+7000) \\ L \leq 7333 \\ A_2 \leq \frac{1}{4}\sum_{i=1}^{11}A_i \\ A_1 \geq T \\ A_4 \leq 0.8(A_1+A_2+A_3+A_4+A_5) \\ L_5 \leq 0.8(A_6+A_7)*(1+r)^n \\ L_6 \leq 0.8A_9*(1+t)^m \\ A_5 \geq 5 \\ A_i \geq 0(i=1,2,\cdots,11), L_i \geq 0(i=1,2,\cdots,7) \\ E(R_i)=\frac{1}{n}\sum E(R_{ij}), i=1,2,\cdots,11; j=1,\cdots,n \\ R_{ij}=H_k R_{ijk} \end{cases} \tag{22}$$

Here, $E(R_i)$ is the expectation of the random variable R_i, R_{ij} is the adjusted value for No.i variable after k round adjust by specialist j.Hk is the probability of R_{ij}. S, T is measurable constant. δ_1,δ_2 is the maximum amount of housing loans and auto loans specified by the China banking regulatory commission. r, t are housing loans and auto loans interest rate, n, m are the numbers of the months to pay debts, they are determined value for the current period. The assets for family use are also fixed value. This is the place where family financial difference from corporate finance, for the reason that families can't sell necessities in order to get greater returns in asset allocation.

According to the above equation, using linear programming software Lingo11, we get the efficient solution of equations, as is shown in table (unit: million￥)

TABLE II
Efficient Solution of X Household Assets and Liabilities

assets				Liabilities and net worth		
financial assets		cash	*2.42*	current liabilities		
	bank deposit A_2	3months		Fees payableL_1	utilities	
		half year			telephone bill	
		1 year			fuel charge	
		2 years		credit card		
		5 years		insurance benefit		
		due on demand		the current long-term liabilities		*0.6* *0*
	national debt		*4.0*	long-term liabilities		*108*
	stock、funds		*13*	housing loan balance		*108*
	financial products		*15*	Auto loan balance		*0*
real assets	house		*80*	other loan balance		*0*
	Investment in real estate		*83*	the final net worth		
	Furniture and home appliances		*2*	the current net income		

automobile	*15*	beginning net assets	*18*
Jewelry and collectibles	*3*		
others	*0*		
total assets	259	total	

Note: some items are not covered in the model, so there is no specific value in this table, but their relationship meet the basic accounting equation.

V. CONCLUSION

In this paper, based on the family assets and liabilities management perspective, we built a household financial multi-objective decision-making model based on assets and liabilities management. The conclusions are as following:

Family can simultaneously achieve assets income maximization and predetermined solvency target configuration of assets and liabilities. It solve the previous studies which only consider the limitations of a single goal, and it make family decisions more close to real life

By comparing with the historical data, family income targets in the multi-objective programming optimal solution and solvency target are increased. It shows that after introducing concept of random variables, the yield is more according with the practice of family life, and the model is feasible.

By comparing with the actual data, we found that household assets returns after the model optimization are higher than the actual value, and household debt paying ability is lower than the actual value, and target solvency can meet the needs of the family. It illustrate that the results conform to the set that make the asset return target as the first priority in the model.

References

[1] Campbell J. Y.. "The journal of finance", Household finance, 2006, 61(4): 1553-1604.

[2] Li Jianjun, Tian Guangning. "The residents financial assets structure and its change trend analysis in China". Financial Forum, 2011(11): 2-8.(language:chinese)

[3] DeVaney S. A, Lytton R. H. "Household insolvency: A review of household debt repayment, delinquency, and bankruptcy". Financial Services Review, 1995, 4(2):137-156.

[4] Campbell J Y, Chocó J F. "Household risk management and optimal mortgage choice". The Quarterly Journal of Economics, 2003, 118(4): 1449-1494.

[5] Bertaut C. "Stockholding behavior of US households: Evidence from the 1983–1989 survey of consumer finances". Review of Economics and Statistics, 1998,80(2): 263-275.

[6] Hess B, Bekker H, Berendsen H J C. "LINCS: a linear constraint solver for molecular simulations". Journal of computational chemistry, 1997, 18(12):1463-1472.

[7] DeVaney S.A, Lytton R.H. Household insolvency: "A review of household debt repayment, delinquency, and bankruptcy". Financial Services Review, 1995, 4(2):137-156.

[8] Casper L.M, Bryson K.R. "Co-resident grandparents and their grandchildren: Grandparent maintained families". Washington, DC: Population Division, US Bureau of the Census, 1998.

[9] Alessi S.M, Trollip S.R. "Multimedia for learning: Methods and development".Allyn and Bacon, Inc., 2000.

[10] Heaton J, Lucas D. Portfolio choice and asset prices: "The importance of entrepreneurial risk". The journal of finance, 2000, 55(3): 1163-1198.

[11] Guiso L, Paiella M. "The role of risk aversion in predicting individual behaviours". Centre for Economic Policy Research, 2004.

[12] Bertaut C.C. "Equity prices, household wealth, and consumption growth in foreign industrial countries: wealth effects in the 1990s". Board of Governors of the Federal Reserve System, 2002.

[13] Campbell J Y, Cocco J F. "Household risk management and optimal mortgage choice". The Quarterly Journal of Economics, 2003, 118(4): 1449-1494.

[14] Bilias Y, Haliassos M. "The distribution of gains from access to stocks". University of Cyprus, 2004.

[15] Campbell J.R, Hercowitz Z. "The role of collateralized household debt in macroeconomic stabilization". National Bureau of Economic Research, 2005.

Implementation Plan Design of Labor Cost Management System

Juan-juan XIAO
Department of Information Technology, Luoyang Normal University, Luoyang, Henan,471022, China
(jan_dan@126.com)

Abstract - **Labor cost management system implementation needs to be worked out a practical implementation plan. Proposed design labor cost management system of enterprises in implementing programs not only needs to protect the legitimate rights of employees, but also needs to ensure economic efficiency and competitiveness of products. It also requires appropriate labor cost control method to get the best artificial cost in order to achieve the purpose of controlling labor costs. This paper focuses on present situation of labor cost management in China, and the implementation plan design of the lab the labor cost management system was carried out from the ten areas.**

Keywords – **Enterprises, implementation plan, labour costs, management system**

I. INTRODUCTION [1]

The high efficiency of labor is an important aspect of competitive advantage and high efficiency of labor depends on high-quality and labor enthusiasm of the workers [1]. High-quality and labor enthusiasm of the workers with the input level of labor costs. If wages and benefits paid to workers, the lower level of employees by enterprises, especially smaller attraction of high-quality staff, and even discourage the labor enthusiasm of the staff. In the labor market situation will inevitably result in the loss of talent [2-5]. Therefore, the high returns to attract qualified workers to become an important means of competition for talent in the market economy conditions. In addition, enterprises effort to improve the income of workers, not only improving the workers' enthusiasm means, but also to promote workers to consciously improve the quality of skills a guide or "investment" [6, 7].

Labour cost management is a systematic project. With the deepening of economic reform, in the process of establishing a modern enterprise system must pay attention to the management of labor costs and to strengthen the statistics and analysis of labor costs, reduce the invalid people spending to increase labor productivity, labor costs, and input-output ratio optimization. Corporate profit should be implemented, the cost of human input and output are at higher or lower operating mechanism, and thus the formation of enterprises employing more than can less jobs at higher or lower, and income can be increased to a virtuous cycle can be cut to enhance the vitality ability to adapt with the market and increase market competitiveness [8-10]. Design labor cost management system implementation program, the status of the labor cost management for China's enterprises, in line with scientific and rational, user-friendly principles to the design [11].

[1] Project document: Soft Science Foundation of Henan province Science and Technology Department under No. 132400410562: Research on management and implement of enterprise's labor cost.

II. LABOUR COST MANAGEMENT SYSTEM OF DESIGN

A. Awareness-raising on strengthening the labor cost management

From the views of market, Competition between firms is none other than the price, quality, and service competition. Low cost is an important factor to expand market share, opening up space for profit. Enterprises must adapt to the transition from plan to market, as the configuration resource, cost adjustment optimization of products, market standards and giving them veto of decision-making. Labor costs are a top priority, because only the people, is the most dynamic and active factors of production [12].

Importance of strengthening statistical analysis of labor cost. Labor cost is an enterprise in a certain period, in the production, management and delivery of services activities in the work force and to pay the sum of all direct and indirect costs. Labor costs are important components of total cost. Labor costs are too high, will allow the employee to obtain higher than their labor theory of value of pay and benefits, reduced corporate profits will ultimately affect the power and potential of sustainable, stable, and healthy development of enterprises; Low labor costs, staff salaries and benefits will pay less than the value of their labor, thus affecting the work enthusiasm and creativity of employees, ultimately makes enterprise development lost vitality. How to scientifically and rationally determine the level of enterprise labor cost, maximize labor cost effective role, the important task of improving human resource management and development, and focus more and more enterprise managers. Statistical analysis of management system of establishing and perfecting the labor cost is the foundation of improving the labor cost management, as well as scientific basis for predicting, controlling Enterprise labor cost.

Labor cost management is a weak link. Awareness-raising on strengthening the labor cost management problems, first is strategic, we must recognize that it is related to the enterprise in diversified markets important strategic factors for survival in the competition; Followed from the distribution point, recognizing that it is the correct state, enterprises, workers' interests of the three important economic lever, it is the main body by regulating workers the benefits of economic behavior, thereby regulating labor resource allocation, evolve Enterprise incentive and motive mechanism of economic factors; On the third is from the management, recognizing that it is related to human resources development related to the increase of economic efficiency of enterprises on the important work of monitoring the consumption of living labor.

B. Serious statistical accounting of labor costs, enhanced artificial cost accounting analysis

Establish evaluation criteria. There is no specification for labor costs of statistical standards, it is hard to avoid human factors on the statistical, objective evaluation criteria could not be established [13]. Enterprises should seize the favorable opportunity of the implementation of the new accounting standards, labor costs of all project requirements and interpretation of science, unity, as far as possible to refine the scope of labor cost statistics; Scope and control provisions, self-inspection, Headquarters inspection to enterprise supplement range of labor cost statistics in the form of inventory, excluding non-standard statistics, standardization, specification of projects, all in labor cost statistics. At the same time, enterprises should focus on clearly artificial cost extraction, distribution of diameter, to reflect the annual labor cost control situation.

Establishing labor cost analysis system. Enterprises to establish a statistical system consisting of total labor costs and, the establishment of labor cost statistics account, provided statistical standards, at the same time, clearly artificial relationship between cost accounting and financial accounting of enterprises. In order to facilitate accurate statistics labor cost, can be set up in the financial accounting labor cost statistics, labor costs to summarize all the items, changing artificial fragmentation of cost accounting of projects in the past. First determine competitive enterprise in labor costs, home and province level comparison of objects, building frames; Secondly we should use the labor cost index for vertical and horizontal analysis, clear standard for dynamic control of labor costs; Finally to establish a labor cost analysis system on a regular basis, timely detection of problems, research strategies.

C. Improving efficiency to form the best combination of productive factors

Streamlining personnel, reasonable staffing levels is essential to strengthen personnel management, Foundation of living labor savings, reducing labor costs. If too many redundant staff in the enterprise bound to create artificial cost is not reasonable growth and labor costs without benefits, income levels to increase [14]. This is a long-term problem left over from the planned economic structure to State-owned enterprises Streamlining is not just for a few less people, with the aim of increasing efficiency by downsizing, overcome overstaffed, inefficient, application of the fittest, and selection for posts, talents, improve productivity, and truly realize the value of human capital.

The economic efficiency of enterprises is an important factor determining artificial inputs. Profit is the source of wages on the one hand, on the other hand wages to ensure effective implementation. Therefore, should be taken into account, finding business profit and loss of balance, or achieve the expected profit of enterprise and employment, labor cost is the highest limit must be controlled at this time.

Average Community salary is an important reference. The average community salary is the important reference standard should not be ignored; otherwise the enterprise wages low will cause a brain drain.

D. Determining the correct labor cost control objectives

Ideal labor cost analysis should be inverted u-shaped, that is: the second low-high, low staff cost rate or labor allocation rates. Labor cost management is not to reduce the absolute amount of enterprise labor cost, absolutely necessary as the amount of labor cost of social development continues to improve, this is a general trend.

From the perspective of economic benefit of input-output, output of certain artificial costs should bring some benefits. When enterprise labor cost per capita growth, per capita increased value and sales per capita income should also be increased, and the growth rate should be higher than the per capita rate of growth of labor costs. Bring economic benefits in order to improve and effect of increasing labor costs. Therefore, Enterprise labor cost control is to control the relative number. In other words, the labor cost control of enterprise's aim is to reduce the proportion of labor costs in total cost, enhance the competitiveness of the product; reduce the proportion of labor costs in sales revenue, enhance human cost of capacity to pay; reduce the proportion of labor costs in labor allocation rate, strengthen the capacity of human resources development.

Entrepreneurs need to convert, go the way of connotation of controlling labor costs. Content control labor cost concept, is relative to the extension of control labor costs, it is not relying on the artificial compression unit labor cost an absolute amount, but by improving labor efficiency, compression of output and employment. Reduced labor cost per unit of output costs, which contains the living labor savings, this connotation is controlling labor costs, that is, to cut labor costs by saving labor.

According to the conventional wisdom, to control labor costs got the idea to cut wages and reduced benefits. Reducing artificial cost in extension is the only way. In market economy, in order to achieve talent competition advantage, enterprise's wage offer should not be too low, rather moderate growth momentum should be maintained.

E. Enhancing the ratio control of labor cost

Present at low levels in terms of ratio control problems of income to individuals on the basis of inclined. For example, some enterprise labor distribution rates, the personnel expense ratio and proportion of labor costs in total cost is higher than the industry average, major companies to create added value in the vast majority of labor costs, while accumulating there for reproduction, apparent tilt of revenue over to the enterprise trade union issues [15]. Enhanced rate control measures are in the followings:

Refining Control standards. Control standard of refinement, refinement to industry within various different type Enterprise, established industry within various type enterprise of artificial cost analysis and control system, is to all type Enterprise average of labor distribution rate, and personnel costs rate, and artificial cost accounted for total cost share this three a ratio index for reference, to investigation its belongs enterprise and of corresponds to ratio index of deviation rate, from level State investigation Enterprise artificial cost of ratio control situation.

Strengthening macro-control. Strengthening macroeconomic regulation and control, contrast rate control policy to take appropriate measures. Ratio for better control of the enterprise, in keeping labor costs on the basis of reasonable proportion, allow modest increases in wages, in increasing income of staff and workers, mobilize the enthusiasm of workers at the same time, guaranteed profit goals to achieve, improve the economic efficiency of enterprises. In this way, the enterprise labor cost ratio index stays above the competitive level, is conducive to human resource development, utilization and forming a virtuous cycle of operation.

F. Strengthening the elasticity control of labor cost

Strengthen the flexible control, keeping labor costs per capita growth per capita increase of per capita sales value and revenue growth, labor costs and output efficiency to maintain reasonable proportions, this core problem is controlling labor costs and labor cost control of most of the key early warning line. Labor cost is a consumption factor, which of necessity must be consumed it enterprises output size, from the perspective of enterprise capital management, decision of Prime is based on the economic benefits of high and low labor costs, labor costs expenditure limit values must be income more than costs.

To strengthen the flexible control of labor costs, measures can be obtained from the following four key areas:

Establishment of enterprise labor cost analysis and control system. Business Enterprise labor cost analysis and control system should be established, sum up experiences and lessons, increasing labor costs management. Flexible control of effective enterprise to sum up experiences, while continuing to consolidate effective, on the basis of, through the control of labor costs to seek ways to further improve enterprise management. Labor cost flexibility out of control factors analysis of the enterprise, and find out the specific reason for improvement.

Achieving value added and sales revenue growth. Found on the way of production and management, increasing output that value added and sales revenue growth. For example to open up the market, expansion of sales, expanding the production of high value added products; Increasing scientific and technological input, adjusting products structure, adopt a favourable product mix strategy, increase the technological content of products, and so on.

Accelerating the transformation of economic growth mode. Accelerating the transformation of economic growth mode, strengthening intensive management, and constantly improve the production technology, improve labor productivity, reduce artificial consumption per unit of product, reduce material consumption costs, lower total costs, spend little or no additional investment in improving enterprise on the basis of volume of production and sales, increase profits.

Reducing invalid loss of human resources.Controlling limits, reducing consumption of invalid labor costs expenditure, reducing redundancy and plugging the loopholes to minimize the loss of human resource is not valid.

G. Playing Wage incentives to regulate labor cost structure

Total wages of State regulatory policy from management and gradual transition to the total labor cost control, formed a "wage salary guide Leader, labor market and macro-wage labor cost forecasting and early warning of Trinity monitoring guidance system." Particularly in 2008 after the implementation of the new tax law, abolish the wage tax system, unified by the enterprise and the actual payment of wages deducted is truly, actually had to cancel a total wage control [16]. From the perspective of these changes, the competitive advantage of Nations to keep labor costs, putting the State, enterprises, and distribution relationships between individuals, on the adjustment of income distribution policy has been changed from total wages of management to manage labor costs. Enterprises should adapt to the new requirements, must carry out comprehensive planning, labor costs overall, landscape analysis, the sustained improvement of functions, in-depth analysis of relevant issues, study the inherent law of labor cost management, explore practical management approach.

Labor cost structure, wages is the most motivating factor, also constitutes a main part of the labor costs. Visible, total level control and a variety of staff reasonable wage level opened the grade, fully reflect the principle of distribution according to work, giving priority to efficiency, is the key link of current control labor costs.

H. Improving the quality and skill level of workers, talent to play benefit

Labourers' quality content includes physical qualities, intellectual qualities, skills, and quality and taste qualities, such as character qualities. One of the more important is improving the quality of intellectual qualities and skills of workers. Market competition ultimately a talent competition, improve the overall quality of the workforce, and realize the full potential of enterprise talents is conducive to corporate streamlining, reduces labor costs. On the rapid development of science and technology and keen competition in comprehensive national strength, to seek business development, standing on competition in an invincible position, on the need to improve the quality of staff. Improve staff quality means to improve efficiency, reduce costs, and increase efficiency. Strengthen education and training, improving the quality and skill level of workers. Scratch staff education and training, means a lot. Training will require a fee, but a strategic vision in the long term; benefits arising after the training, the role of play are endless. Through training and education, improve the quality of managers and technical skill level, changing the concept of people's thinking, quickly adapt to the market economy and enterprise restructuring; Through the training, education, learn new skills, understand | new knowledge in the new age, increasing the sense of urgency and, enhance employee incentives.

I. Artificial cost budget and final accounts system

Budget is a method widely used in management control activities; it is in digital form to prepare plans of a certain period in the future. Budget will help managers in all units and departments to evaluate labor cost management and control. Departments in charge of labor cost budget management system should be developed so that labor cost budget management of institutionalization and standardization of procedures. Labor cost budget to finance annual budget year as labor cost budget, financial and other Professional based on budget data, carried out by the labor Department budget, which remained relatively independent.

Budget should be based on relevant principles and policies of the State in accordance with the business plan and task, actually occurred under the previous year, taking into account the special increment decrement factor this year, expenditure control, the principle of living within our means, careful, scientific and rational arrangement of the funds, preparation of the labor cost budget, should not make a deficit budget. In order to guarantee the seriousness of the system, budget, once approved, no adjustments. If under the influence of national policy in the annual performance, large changes in personnel agencies, do need to adjust, subordinate units, departments in charge of various departments to the labor cost to apply approved could not pay.

J. Constructing control models, sound evaluation system

Established labor cost control objectives. Labor cost input-output efficiency objective should be to improve the management of labor costs to ensure asset appreciation, realize economic benefits and employees ' income levels continue to improve the stability of enterprise and development. Labor cost management of enterprise's aim is to reduce the proportion of labor costs in total cost, enhance the competitiveness of the product; Reduce labor cost in the sale of your share, enhance human cost of capacity to pay; Reduce the proportion of labor costs in labor allocation rate, strengthen the capacity of human resources development. Therefore, enterprise labor cost control is to control the relative number.

III. CONCLUSIONS

Affecting the labor cost is the most important factor of productivity. When the enterprise controls labor costs, it needs to protect the legal rights of employees, and the need to ensure economic efficiency and the competitiveness of products. Therefore, enterprises need to use proper labor cost control method, seek the best labor cost inputs, to achieve the aim of controlling labor costs.

Implementation artificial cost management is enterprise of a systems engineering, to established and modern enterprise system phase adaptation of enterprise artificial cost self constraints mechanism for target, adhere to enterprise artificial cost and statistics phase adaptation of principles, to established artificial cost statistics for Foundation, to artificial cost analysis for pursuant to, to artificial cost control for means to established enterprise artificial cost management system, enhanced enterprise market competition ability, improve enterprise artificial cost output efficiency, optimization Enterprise artificial cost structure, Promoting the enhancement of economic efficiency and management level of enterprises will become the future exploration of management direction.

ACKNOWLEDGMENT

The research work was supported by Soft Science Foundation of Henan province under No. 132400410562: Research on management and implement of enterprise's labor cost.

REFERENCES

[1] Economics, [the US] Paul Samuelson, William Nordhaus, translation of Hu Daiguang (14th edition), the Beijing Economics College Press, June 1996. (In Chinese).

[2] On labor cost management in the harmonious enterprises. Mall modernization, 2007(04). (In Chinese).

[3] Important significance of strengthening the labor cost management, Liaoning economy, 2007(5). (In Chinese).

[4] Status, problems and countermeasures of labor costs in China. Economic research reference, 2006(44). (In Chinese).

[5] Enterprise labor cost management and control, Yu Huiping, Yang Yujie, Mall modernization, 2009(15). (In Chinese).

[6] Wang Hugang, management and control methods of enterprise artificial cost [J], Journal of petroleum chemical industry management cadre College, 2009 (04). (In Chinese).

[7] Yang Si Wei, labor cost management under the new accounting rules [J], China power enterprise management, 2009 (01). (In Chinese).
[8] Yang Yujie, Yu Huiping, important significance of strengthening the labor cost management [J], the Mall modern, 2009(14). (In Chinese).
[9] Lv Yuhai, On labor cost management in enterprises [J], Technology news 2010(20). (In Chinese).
[10] John Lvancevieh. Attitudes and performance [J]. Journal of Applied psychology, 1980, 65(2): 139-146.
[11] Gary Becker. Investment in Human Capital: A Theoretical Analysis [M]. The Journal of Political Economy, 1962, 70(2): 29-41
[12] Jaeob Mineer. Human Capital and Earnings. Studies in Human Capital: Collected ESSays of Jacob Mineer, Edward Elgar Publishing Limited [M], 1993: 74-79
[13] Maureen Weieher, William W.: Business Process Reengineering Analysis and Recommendations [J]. Baruch College, 1995, 32(l): 11-23.
[14] Jacob Mineer. On-the-Job The incentive factor on HR [J]. The Journal of political Economy, 2003, 70(2):50-79
[15] Jeff Wilkins, Bert Van Wagon. Understanding and Valuing Knowledge Assets [M]: Overview and Method. Expert Systems with application, 1997, 13(1): 55-72
[16] Chung Yi Ts. Assess factor of human value resource [J]. Journal of Development Economies, 2000, (61):137-174

Regional Differences in Medical Service Quality: Based on 2013 China Consumer Survey in 92 Provinces

Yun Liu
Institute of Quality Development Strategy, Wuhan University, Wuhan, China
(joanna0616@gmail.com)

Abstract - **There lie a great gap of medical service quality between eastern China and central and western China in 2013. It is revealed by a national consumer satisfaction survey in 29 provinces in China by Institute of Quality Development Strategy of Wuhan University. In 2013, consumer satisfaction score on local medial service is relatively higher, as high as 63.36, in eastern China; a relative low, 61.90, in central China; and lowest in western China, 59.28. In a multiple linear regression analysis, household income is correlated with Chinese consumers' satisfaction on medical service. It means that consumers with higher income grade higher on medical service. It is income factor that led regional differences in China medical service quality.**

Keywords - **Consumer survey, medical service quality, regional differences**

I. INTRODUCTION

In the past decades, China constantly increases investment of medical service to improve medical service quality. According to statistical data released by the Ministry of Health, the proportion of the total expenditure on health in GDP has been increased from 4.62% in 2000 to 5.36% in 2012 [1]. In the contrast, consumers' satisfaction of medical service quality keeps at a low level. Based on the consumer survey by Wuhan University Institute of Quality Development Strategy (WUIQDS), Chinese medical service quality has remained low level both in 2012 and 2013 [2]. Especially, in western China, the score of satisfaction on medical service is only 59.28, which is below the national average score 61.85. Therefore, it is necessary to evaluate the current situation of China medical service quality, to analyze regional differences of medical service quality, to propose feasible policy proposals.

II. METHODOLOGY

A. Evaluation Object

Previous studies have defined that medical service quality evaluation is based on the users of medical service, which is the consumer. In 1960s, the method of consumer satisfaction has been applied in medical consumers. From the perspective of patient satisfaction, doctor-patient relationship has been studied [3]. Then, influenced by market theory, study on medical consumer becomes more popular [4], and satisfaction survey is applied widely in hospital quality management system. In the 1990's, research on medical consumer satisfaction has peaked [5]. In addition, consumer-orientation is also the important way for quality medical service promotion in countries. European hospital performance evaluation by the World Health Organization contains six aspects, and patient satisfaction is the one of six evaluation indexes. The American Medical Association (AMA) regards patient-orientation service as one of the six goals of quality promotion. One of the three index systems of NHPF is medical institution achievement, in which appropriateness refers that whether medical care satisfies customer needs [6]. Since Chinese medical and health system reform in 2009, China government has proposed that medical service should shift from "provider orientation" to "consumer orientation" [7]. Both in China and other countries, consumers' evaluation is generally considered as a significant way to improving medical service quality [8]. In this paper, data is come from consumer survey and consumers' grade is used to evaluation the quality of medical service.

B. Calculation method

Data in this paper is from 2013 China quality development observation by WUIQDS. There are two dimensions to evaluate medical service by consumers. One is safety and the other is satisfaction. Consumer grades local medical service in these two indexes respectively and score ranges from 1 to 10 (1 stands for the lowest score, 10 is the highest score). The evaluation level division is as Table I:

TABLE I
EVALUATION LEVEL DIVISION

Score interval	Rank & Level
0~29	Extremely poor
30~59	Poor
60~69	Qualified
70~85	Good
85~100	Excellent

Safety of medical service quality could be regarded as consumers' evaluation on safety of medical service. While satisfaction of medical service quality refers to the degree medical service meets consumers' needs. Therefore, the difference between consumers' satisfaction degree on medical service quality and safety can be considered as the gap of medical service quality from safety to satisfaction [9, 10]. In this paper, the difference is defined as Medical Service Quality Index (MSQI). It results from subtracting safety of medical service quality and satisfaction of medical service quality. Evaluation medical service quality by MSQI has two advantages: First, it will present how much effort need to do for improving medical service quality; Secondly, it retains the benefits of consumer survey. The formula is as follow:

$$MSQI = \mathrm{SMS}_{\mathrm{safety}} - \mathrm{SMS}_{\mathrm{satisfaction}} \quad (1)$$

Quantitative range of qualitative analysis applied in safety and satisfaction of medical service quality is shown in the Table I.

SPSS.16 is applied in this paper.

E. Qi et al. (eds.), *Proceedings of the 5th International Asia Conference on Industrial Engineering and Management Innovation (IEMI2014)*, Proceedings of the International Asia Conference on Industrial Engineering and Management Innovation 1, DOI 10.2991/978-94-6239-100-0_81, © Atlantis Press and the authors 2015

C. Sample Range

Valid sample data with 4,893 in 29 provinces is applied in the paper. According to regional division in *Yearbook of 2013 China Health Statistics*, 29 provinces are divided into eastern, central and western regions. Eastern regions contains 11 provinces (cities): Beijing, Tianjing, Hebei, Shanghai, Jiangsu, Zhejiang, Fujian, Shandong, Guangdong, Hainan; central regions includes 8 provinces: Heilongjiang, Jilin, Shanxi, Anhui, Jiangxi, Henan, Hubei, Hunan; western regions involves 10 provinces: Inner Mongolia, Guangxi, Chongqing, Sichuan, Guizhou, Yunnan, Shannxi, Qinghai, Ningxia, Xinjiang [11].

Population characteristics distribution of the samples is as Table II:

TABLE II
POPULATION CHARACTERISTICS OF SAMPLES

Characteristic	Samples Distribution	Proportion
Gender	Male	48.3%
	Female	50.3%
Age	18-30	38.6%
	31-40	23.5%
	41-50	26.1%
	51-60	7.4%
	Above 60	3.2%
Household Registration	Urban	71.1%
	Rural	27.6%

III. RESULTS

A. Regional Results

TABLE III
REGIONAL DISTRIBUTION OF CHINA MEDICAL SERVICE QUALITY SAFETY

Province	Provincial score	Region	Regional score
Shandong	69.21	Eastern region	64.56
Zhejiang	69.07		
Tianjin	68.78		
Liaoning	68.67		
Fujian	67.25		
Shanghai	66.64		
Beijing	64.70		
Jiangsu	61.92		
Guangdong	60.97		
Hebei	59.82		
Hainan	53.18		
Hubei	67.95	central region	64.16
Jinin	67.71		
Henan	65. 50		
Jiangxi	63. 93		
Shanxi	63.06		
Hunan	62.75		
Anhui	61.33		
Heilongjiang	61.01		
Sichuan	67.40	western region	61.03
Yunnan	66.61		
Guangxi	64. 36		
Guizhou	63.04		
Chongqing	61.36		
Shanxi	60.85		
Ningxia	60.59		
Xinjiang	58.63		
Qinhai	57.30		
Neimenggu	50.19		
Total	63.57		

According to consumers' evaluation on medical service quality safety of each province (city), the scores of medical service quality safety in eastern, central and western regions are acquired.

China medical service quality safety in 2013 appears significant regional feature of "eastern-central-western". The score of medical service quality safety in eastern region is 64.56; the score of medical service quality safety in central region is 64.16; the score of medical service quality safety in western region is 61.03.

TABLE IV
REGIONAL DISTRIBUTION OF CHINA MEDICAL SERVICE QUALITY SATISFICATION

Province	Provincial score	Region	Regional score
Shanghai	69.31	eastern region	63.36
Tianjin	69.11		
Shandong	68.10		
Zhejiang	67.71		
Liaoning	67.71		
Fujian	64.13		
Jiangsu	60.66		
Guangdong	59.92		
Beijing	59.40		
Hebei	58.99		
Hainan	51.82		
Jilin	65.36	central region	61.90
Hubei	65.19		
Shanxi	63.25		
Henan	61.09		
Anhui	60.79		
Jiangxi	60.48		
Hunan	59.72		
Heilongjiang	59.31		
Sichuan	66.63	western region	59.28
Yunnan	64.86		
Guangxi	63.83		
Shanxi	61.02		
Ningxia	60.00		
Guizhou	58.82		
Xinjiang	58.21		
Chongqing	57.45		
Qinghai	56.40		
Neimenggu	45.56		
Total	61.85		

The score of China medical service quality satisfaction in 2013 is even lower than that of medical service quality safety, which is more significant in the regional feature of "eastern-central-western". The score of medical service quality satisfaction in eastern region is 63.36; the score of medical service quality satisfaction in central region is 61.9; the score of medical service quality satisfaction in western region is 59.28 (Table VI). Medical service quality satisfaction in western region is below the pass line 60, and it is also much lower than national average.

According to the difference value between consumer quality safety and quality satisfaction, medical service quality index is acquired. China medical service quality index in 2013 appears a significant tendency of "Forgotten Central China". The gap between satisfaction and safety of medical service quality is greatest in central China, lower in western China, and lowest in eastern China [12]. The MSQI are 2.26, 1.75, and 1.21 respectively (Table V). Eastern China is also the region with best medical service quality, while central and western China is far behind on the aspect of MSQI.

TABLE V
REGIONAL DISTRIBUTION OF CHINA MEDICAL SERVICE QUALITY INDEX

Province	Provincial score	Region	Regional score
Shanghai	-2.67	eastern region	1.21
Tianjin	-0.33		
Hebei	0.83		
Liaoning	0.96		
Guangdong	1.05		
Shandong	1.12		
Jiangsu	1.26		
Zhejiang	1.30		
Hainan	1.36		
Fujian	3.13		
Beijing	5.30		
Shanxi	-0.19	central region	
Anhui	0.54		
Heilongjiang	1.70		
Jilin	2.36		2.26
Hubei	2.76		
Hainan	3.03		
Jiangxi	3.45		
Henan	4.41		
Shanxi	-0.17	western region	
Xinjiang	0.42		1.75
Guangxi	0.53		
Ningxia	0.59		
Sichuan	0.77		
Qinghai	0.90		
Yunnan	1.74		
Chongqing	3.91		
Guizhou	4.22		
Neimenggu	4.63		
Total	1.71		

B. *Influencing Factor Analysis*

Why is there a significant regional difference of China medical service quality? Three aspects will be taken into consideration to analyze the influence factor of regional difference: medical service quality safety, medical service satisfaction, and MSQI. We introduced several possible influencing factor: gender, age, household registration, education degree, household monthly income, and household monthly expense [13].

With linear regression on the above factors and service quality safety, it is found that the sig. of household income is 0.002. It indicates that household income is a significant factor influencing safety of medical service quality [14].

TABLE VI
REGRESSION RESULTS OF INFLUENCING CHINA MEDICAL SRVICE QUALITY SAFETY

	t	Sig.
(constant)	35.641	0.000
Gender	1.074	0.283
Age	2.761	0.006
household registration	0.052	0.959
Marriage	2.626	0.009
education degree	-0.358	0.720
household income per month	-3.165	0.002
household expense per month	35.641	0.745

With linear regression on the above factors and service quality satisfaction, it is found that the sig. of dependent variable, household expense per month, is 0.003. Household expense per month exerts a significant influence both on medical service quality satisfaction.

TABLE VII
REGRESSION RESULTS OF INFLUENCING CHINA MEDICAL SRVICE QUALITY SATISFACTION

	t	Sig.
(constant)	34.576	0.000
Gender	1.108	0.268
Age	2.039	0.041
household registration	-1.663	0.096
Marriage	2.350	0.019
education degree	-.0703	0.482
household income per month	-2.969	0.003
household expense per month	0.173	0.862

Considering the influencing factors of MSQL are far more complicated than that of medical service quality safety and satisfaction, we introduce 2 more economic factors, that is , work and numbers of family members. With linear regression on influence factor of medical service index, it is found that the influence of numbers of family members and household income is the most significant influencing factor on medical service quality index, the values of sig. are 0.000 and 0.002 respectively.

TABLE VIII
REGRESSION RESULTS OF INFLUENCING CHINA MEDICAL SRVICE QUALITY INDEX

	t	Sig.
(constant)	34.445	0.000
Gender	0.765	0.444
Age	1.866	0.062
household register	-1.240	0.215
Marriage	1.599	0.110
education degree	-1.135	0.893
Work	-1.853	0.064
num of family members	3.883	0.000
household income per month	-3.122	0.002
household expense per month	0.525	0.600

IV. DISCUSSION

Based on data analysis, it is clear that regional difference of China medical service quality is significant. Medical service quality in eastern region is better than that in central region, and much better than that in western region. Especially for medical service satisfaction, western region is particularly backward [15]. As for medical service quality index, the difference value between medical service quality safety and satisfaction in central region is the highest. It is indicated that when medical service quality safety in central region has reached national average, medical service quality is not capable of satisfying local consumers' needs, which is because consumer needs increases rapidly in central region [16].

Additionally, income level has a significant influence on medical service quality. The reason is that people with higher income might spend more in medical service. Meanwhile, it also indicates that when people with higher income could not get satisfying medical service, it is easier for them to accept service in regions with better medical service [17].

V. CONCLUSION

Based on quality observation data from WUIQDS, this paper has found that China medical service quality appears difference of "eastern-central-western" in 2013. Especially, in the scores of China medical service quality index, there is a phenomenon of "Forgotten central". Therefore, in the future public investment, government will emphasize more in central and western regions. With further data verification, difference of household monthly income is the reason for regional characteristic of medical service quality. Consumers in cities with higher income will give a higher evaluation on medical service quality.

ACKNOWLEDGMENT

This paper is funded by Major Project of National Social Science Foundation of China (Project No. 11 & ZD158) and National Quality Inspection Non-Profit Industrial Scientific Research Special Project Subject (Project No. 201210117). All the data in this paper is from National Quality Observation survey conducted by Wuhan University Institute of Quality Development Strategy. Supported by the guidance and encouragement of my PhD supervisor, Cheng Hong, I have an idea and prove it in this article. But I am the one who is responsible for the consequence of this article.

REFERENCES

[1] Ministry of Health of the People's Republic of China, 2014: Yearbook of China Health Statistics in 2013, http://www.nhfpc.gov.cn/htmlfiles/zwgkzt/ptjnj/year2013/index2013.html.

[2] Y. Liu, "Factors Influencing the Quality of Chinese Medical Service: Technology or Function?", Journal of Macro-quality Research, Vol1, 2013.

[3] A. Cartwright, "Human Relations and Hospital Care", Routledge and Kegan Paul, London. 1964.

[4] S. Mclver, "Obtaining the Views of Outpatients", King's Fund, London, 1991.

[5] J Sitzia. and N Wood., "Patient Satisfaction: A Review of lssues and Concepts" , Pergamon, vol.45,pp. 1829-1843, Dec. 1997.

[6] F. Yang, Y.H. Qin, J. Dong, "Implementation and Thought of Overseas Medical Care Quality Evaluation", Hospital Management Forum, No. 3, 2005.

[7] S. S. Andaleeb, "Service quality perceptions and patient satisfaction: a study of hospitals in a developing country", Social Science & Medicine, Vol. 52, lssue 9, pp.1359-1370, 2001.

[8] J.A. Baker, and Lamb, C. W, "Measuring architectural design service quality", Journal of Professional Services Marketing, Vol,10, No,1, pp.89-106, 1993.

[9] D. Coddington, and K. Moore, "Quality of Care as a Business Strategy: How Customers Define Quality and How to Market it", Healthcare Forum, pp. 29-32, 1987.

[10] A. Donabedian, "The quality of medical care: a concept in search of a definition", The Journal of Family Practice, pp. 277-284, 1979.

[11] C. Gronroos, "A Service Quality Model and its Marketing Implications", European Journal of Marketing, Vol. 18, pp. 36-44, 1984.

[12] H. Hiidenhovi, P. Laippala, and K. Nojonen, "Development of a Patient-Orientated instrument to Measure Service Quality in Outpatient Departments", Journal of Advaned Nursing, pp.696-705, 2001.

[13] G. D. Kang, "The hierarchical structure of service quality: integration of technical and functional quality", Managing Service Quality, Vol.16 No.1, pp.37-50, 2006.

[14] G. W. Mangold, and E. Babakus, "Service quality: the front-stage perspective vs. the back-stage perspective", Journal of Services Marketing, Vol.5, No.4, pp.59-70,1991.

[15] A. Parasuraman, V.A. Zeithaml, and L. L. Berry, "Servqual: a Multiple-item Scale for Measuring Consumer Perceptions of Service Quality", Journal o f retailing, Vol.64, pp.12-40, 1988.

[16] V.Sower, J. Duffy, W. Kilbourne, "Dimensions on Service Quality for Hospitals; development and Use of the KQCAH Seale", HealthCare Management Review, pp.47-59, 2001.

[17] J.S.Weissma, R.Stern., L. F Stephen, and A.M. Epstein," Delayed Access to Health Care: Risk Factors, Reasons, and Consequences", Medicine and Public Policy,15 February, 1991.

MIX
Papier aus verantwortungsvollen Quellen
Paper from responsible sources
FSC® C105338

If you have any concerns about our products,
you can contact us on
ProductSafety@springernature.com

In case Publisher is established outside the EU,
the EU authorized representative is:
Springer Nature Customer Service Center GmbH
Europaplatz 3, 69115 Heidelberg, Germany

Printed by Libri Plureos GmbH
in Hamburg, Germany